Big Trees Not Big Stumps

25 years of campaigning to save wilderness
with the Wilderness Committee

by Paul George

*To Patrick Britten
Keep on protecting
the Wilderness!
Paul George*

Big Trees Not Big Stumps

25 years of campaigning to save wilderness with the Wilderness Committee

First Edition

Copyright © 2006 by Paul George and Western Canada Wilderness Committee

The views expressed are those of the author and do not necessarily represent those of Western Canada Wilderness Committee, its directors, officers, staff or funders.

All rights reserved. No part of this book may be reproduced or transmitted in any form by any means without permission in writing from the publisher, except by a reviewer, who may quote brief passages in a review.

CREDITS:
Editors: Tony Eberts and Adriane Carr
Final editors: Eve Adams, Richard Krieger, and Andrea Reimer
Contributors: scores of people mentioned in the text
Book design: Sue Fox
Cover design: Andrea Reimer
Layout and page design: Amanda Brown
Photo and document scanning: Amanda Brown
Video editing: Ben Graeme and Jeremy S. Williams
DVD production: Anton van Walraven
    conversion: Jeremy Lutter — *24 Frames Films and Video*
    menu: Marcela Noriega and Alejandra Zamorano
    mastering: Shane Korytko
    replication: *Effectuality Inc.*
Printing: Nik Cuff — *Navigator Communications Inc.*
Collating: dozens of WCWC volunteers

Front cover photo: *Randy Stoltmann standing by the Cheewhat cedar, Canada's largest known red cedar, Pacific Rim National Park. Photo: WCWC files.*
Back cover photo: *Joe Foy sitting beside red cedar stump near Vancouver. Photo: WCWC files.*
Spine photos: WCWC files.

Library and Archives Canada Cataloguing in Publication

George, Paul, 1941–
Big Trees Not Big Stumps:
25 years of campaigning to save wilderness with the Wilderness Committee / Paul George.

Includes chronology.
ISBN 1-895123-03-8

1. Western Canada Wilderness Committee—History.
2. Nature conservation—Canada—History.
I. Western Canada Wilderness Committee II. Title.
QH77.C3G46 2006   333.720971   C2006-901325-X

Published by Western Canada Wilderness Committee
277 Abbott Street, Vancouver, BC V6B 2K7 CANADA
1-800-661-WILD (9453) www.wildernesscommittee.org

Printed in Canada. **Suggested price $39.95 CDN.** Proceeds from sales go towards saving wilderness.

Contents

Foreword by Adriane Carr		vi
Introduction		vii
Chapter 1	WCWC's origins "Galapagos of the North" trip in '77 changes my life	1
Chapter 2	Saving South Moresby becomes an obsession	11
Chapter 3	Launching WCWC Our first Endangered Wilderness Calendar a success	19
Chapter 4	WCWC's campaigns extend into new territories and encompass controversial issues	25
Chapter 5	Forest industry and IWA fight against new parks Big WCWC fundraiser and big push for the Valhalla Park	31
Chapter 6	An unexpected wilderness win WCWC in first financial crisis	39
Chapter 7	Assuming full responsibility for a nearly bankrupt WCWC	43
Chapter 8	Winning a battle against government censorship	49
Chapter 9	Learning how to publish educational newspapers	51
Chapter 10	Meares Island – first big B.C. wilderness campaign to go confrontational Government subsidies end	55
Chapter 11	Meares Island takes centre stage in campaign efforts South Moresby heats up	59
Chapter 12	Taking on our third big campaign – the Stein	65
Chapter 13	Breakthrough in the campaign to save South Moresby B.C. thwarts new park creation with WAC process	71
Chapter 14	Haida defend Lyell Island A big push to save South Moresby	77
Chapter 15	Campaigning to protect "pocket wilderness" areas Battling to save Windy Bay	91
Chapter 16	Moving into a warehouse office South Moresby campaign reaches a peak	95
Chapter 17	Gwaii Haanas National Park Reserve becomes a reality WCWC sets big new goals	99
Chapter 18	Battles rage on many wilderness fronts in B.C.	107
Chapter 19	WCWC gets a full-time employee, wins its first big court case and launches a big campaign to save Carmanah Valley	117
Chapter 20	Meares trail work blocked Clayoquot conflict erupts	125
Chapter 21	Carmanah becomes our "flagship" campaign	129
Chapter 22	The Carmanah court case is a win but the war goes on to save the valley	135
Chapter 23	Expanding activities Going national, going retail, going WILD	143
Chapter 24	More staff, new campaigns, our first Branch and door-to-door canvass	147
Chapter 25	WCWC moves to Gastown and pulls off a colossal Carmanah Artists' project	151

Chapter 26	Vital projects, hectic growth and a Herculean work schedule	**157**
Chapter 27	WCWC builds the world's first temperate rainforest upper canopy research station	**163**
Chapter 28	B.C. government makes a half-park decision WCWC wins prestigious award	**169**
Chapter 29	Vandals leave Carmanah Research Station in ashes Project Phoenix Biodiversity bonanza	**173**
Chapter 30	WCWC branches out WILD holds an international conference and plans to help the Penan	**179**
Chapter 31	The WILD Conference – an international effort to map Earth's remaining wilderness	**189**
Chapter 32	Desperate financial shape WILD moves out Drinking water defended	**197**
Chapter 33	WCWC publishes Penan and Clayoquot books Fights dams and mines	**201**
Chapter 34	WCWC and WILD reunite WCWC gets top Canadian award	**207**
Chapter 35	New B.C. NDP government is pro-parks WCWC struggles to stay solvent	**211**
Chapter 36	WCWC's new "outside" board tackles our internal problems and imposes a solution	**219**
Chapter 37	Restoring our financial credibility; new personnel and campaigns revitalize WCWC	**225**
Chapter 38	We maintain our independence UNCED hopes dashed UN uses WILD's unique calendar	**233**
Chapter 39	Campout at the Legislature gets meeting with the Premier Rally turns into a riot	**241**
Chapter 40	Infamous April 13, 1993	**249**
Chapter 41	The Tatshenshini gets saved Other campaigns expand	**253**
Chapter 42	Clayoquot conflict heats up A *Conservation Vision* for Vancouver Island	**259**
Chapter 43	Building a Witness Trail Making a stand at Sutton Pass	**263**
Chapter 44	Stumpy's extraction Stumping to Ottawa seeking a Clayoquot Biosphere promise	**273**
Chapter 45	Stumpy goes to Europe Stumpy tours the States Stumpy lost in Montreal	**279**
Chapter 46	Parts of Paradise saved Clayoquot campaign continues Victoria watershed logging nixed	**289**
Chapter 47	WCWC fights for Boise WILD fights for Chilean wilds	**299**
Chapter 48	Walking the Ahousaht Wild Side	**303**

Chapter 49	Stoltmann Campaign launched Spotted owls doomed RPAC sell-out begins	**311**
Chapter 50	Stein is finally saved WCWC's prolonged protest at B.C. Legislature sets record	**319**
Chapter 51	Stoltmann Wilderness' blackest day "Old men's" boreal road show	**327**
Chapter 52	WCWC adopts Marr's BET'R Campaign	**333**
Chapter 53	A concerted effort to ban bear hunting	**337**
Chapter 54	Creative campaigning pays off Clayoquot hit with landslides and another blockade	**347**
Chapter 55	Environmentalists labeled "Enemies of BC" Violence erupts in Elaho	**351**
Chapter 56	A new head office and store The new fisheries and pro-Great Bear Rainforest campaigns	**367**
Chapter 57	Tackling global warming Lillooet LRMP table flounders	**377**
Chapter 58	Logging in Vancouver watersheds halted Canadian Senate challenged to fix a bad law	**381**
Chapter 59	Stoltmann campaign grinds on CMT research conducted MOU "saves" Clayoquot's pristine valleys	**385**
Chapter 60	No to Makah grey whaling No to land trade for parks No to foreign multinational takeover of MB	**391**
Chapter 61	Stoltmann wilderness campaign peaks in conflict	**395**
Chapter 62	WCWC-Victoria takes off WCWC-Manitoba gets started	**405**
Chapter 63	Cathedral Grove under attack Elaho campaign rolls toward resolution	**411**
Chapter 64	Defeating the NDP's "Working Forest" Legislation Protection prospects look good for the Great Bear Rainforest Continuing the fight to save endangered species	**425**
Chapter 65	Digging in to protect what we have gained	**435**
Chapter 66	Callous Canadian governments refuse to protect Canada's endangered species	**439**
Chapter 67	"Working Forest" monster held off Brand new campaign	**445**
Chapter 68	An organization that grows, keeps on creatively campaigning and never gives up its goals	**459**
Appendix I	Protected Area Accomplishments	**465**
Appendix II	A chronology of the key events pertaining to the environmental movement primarily in Canada and especially in B.C. with an emphasis on Western Canada Wilderness Committee's involvement	**467**
Appendix III	Campaigning Insights Summarized	**499**

Foreword

Two weeks before this book went to press, two pieces of bad news shocked me into realizing that the fight to protect wilderness is never over, even for the wilderness we think is protected.

On July 27, 2006, a joint board of the B.C. government and Nuu-chah-nulth Central Region First Nations approved plans to log up to 60 percent of the old-growth forests in eight wild valleys that everyone thought were saved in Clayoquot Sound. I couldn't believe it. Clayoquot has the largest extent of old-growth forest left on heavily-clearcut Vancouver Island. It has one of the largest roadless remnants of ancient temperate rainforest left in the world. The B.C. government's decision in 1993 to protect only one-third of Clayoquot Sound and open the rest to clearcut logging launched the biggest "war in the woods" in Canada's history. The campaign to save Clayoquot Sound consumed the Wilderness Committee and dozens of other groups for years, and inspired people planet-wide to get involved in the battle to save Earth's precious wild places.

The mass protests and international boycotts ended only because the environmental groups negotiated an agreement with the Clayoquot Sound First Nations, who hold a logging tenure in Clayoquot, not to log any of Clayoquot's unprotected pristine valleys. As the Wilderness Committee's lead Clayoquot campaigner in the '90s, I helped negotiate that agreement. There's no doubt in my mind that breaking this bottom line commitment in the signed agreement would put Clayoquot back at war.

Within a week of the announcement, environmental groups met with the First Nations. Together they put their agreement back on track and conflict was averted... for the time being. But it worries me that the B.C. government continues to push for logging in every watershed in Clayoquot that isn't a park.

On that same July day the B.C. government opened up twelve of B.C.'s big wilderness parks, including Cape Scott, Mount Robson, Golden Ears and Wells Grey to private-for-profit lodge development. Within days, the Wilderness Committee joined other environmental groups and individuals around the province to fight the building of commercial tourist accommodations in the wild hearts of B.C. parks.

Both of these incidents made me realize just how vigilant we have to be in our defence of wilderness. Our campaigns never really end.

In fact wilderness-saving campaigns aren't likely to end unless a bigger shift happens in our whole society. Until people truly value their quality of life more than consumer goods, until companies have to pay the full cost of their impacts on the environment, and until politicians start considering the welfare of future generations as more important than their party's success in the next election, every bit of Earth's wilderness is under threat.

Until we make the grand shift onto a saner path, we're going to have to count on groups like the Wilderness Committee to defend wilderness. Thank goodness our trust is well placed.

There's no group quite like the Wilderness Committee. In fact, there's no one quite like Paul George whose uncompromised integrity, humour, fearlessness and passion for the cause defined the Wilderness Committee's style and make this book such a good read.

The stories and the campaign hints contained in this book are priceless. Seasoned politicians and cabinet ministers have told me that no environmental group does grassroots education and mobilizes the masses better than the Wilderness Committee. That's why some of them, at least, listen to what the Wilderness Committee has to say. It's why the Wilderness Committee has the reputation of being the publishing and citizen mobilization arm of the movement.

But the Wilderness Committee doesn't do it alone. Through my 18 years of volunteering and working for the Wilderness Committee, what I loved best was the collaboration: the dynamic brainstorming inside the office as well as the experience of being part of a big extended family of activists. Visit the office and you will likely bump into an individual dedicated to a personal mission of finding big trees, or representatives from environmental groups or First Nations fighting to protect an important wilderness, or a fisheries or forestry expert, a scientist, or a citizen who wants to share some information or help on a campaign.

Paul was always generous with his time and with the Wilderness Committee's money (often to the detriment of the organization's bottom line). Although he'd talk one-on-one with anyone, his frustration with meetings that slowed down action and his zero tolerance for any tactic that undercut another group's wilderness-saving goals sometimes made him hard to work with. But it made the movement stronger and won a lot of campaigns.

If my admiration for the Wilderness Committee seems excessive it's probably because I'm biased. I joined the Wilderness Committee as member number 15 in 1980, fell in love with Paul in 1982 and gave up a good paying job and career to work there from 1988 to 2000. The Wilderness Committee was in our home for years, dominated our family life for decades, and will be in our hearts forever.

I really like this book, mostly because it brings back such fabulous memories and great feelings of accomplishment. I'm happy that it reveals how many different people and creative tactics have been engaged in the campaigns to save wilderness. No one person or group ever has all the answers, although the Wilderness Committee has some pretty good expertise. Its bottom line advice is sound: be sure your strategy leaves the door open for more wilderness protection down the road. Never sell out. Never give up.

- Adriane Carr, August 10, 2006

Introduction

The Wilderness Committee is a big extended family of nature-loving people self-taught in environmental activism. This book chronicles 25 years of our wilderness-saving campaigns. I have aimed to explain what worked and why; what did not work and why not. I've included adventure and success stories, heart-wrenching tales of our losses, and candid information about behind-the-scenes activities that few people, until now, have known about.

Not strictly chronological, you can read this book in bits and pieces, skip from here to there, look at the photos and read the captions. It makes this book a bit like the Wilderness Committee itself. At any one time we held the threads of dozens of campaigns together. Each day dealt us something new, while at the same time required us to continue with ongoing commitments.

In a pocket on the inside back cover is a DVD. It contains a two hour long collection of news clips, raw footage of our campaign activities, out-takes of our videos, etc., and a few ads that were never publicly aired. Where appropriate, the menu on the DVD refers to the corresponding page in the book that relates to that selection. The Wilderness Committee has a great website www.wildernesscommittee.org with other video clips and videos for sale. Posted on the website also are all of the educational newspapers, research reports and media releases that will enrich the information found in this book for those interested.

It was relatively easy for me to describe our various tactics and strategies. What was hard to adequately capture is how much fun we had. As a group we constantly brainstormed, over coffee in the morning, and beer and pizza at night. We always moved fast to put our ideas into action. We even made all-night newspaper production sessions and weeks of stuffing mail outs fun! Many people believe that environmentalists are drudges. It's simply not true.

It is my greatest hope that this book inspires a new generation of environmental activists.

If this book motivates even one person who has never experienced the mystic nature of wild places like Gwaii Haanas, Stein Valley, Carmanah Valley or Clayoquot Sound to begin trekking gently through some wilderness, I'll consider this book a success. If your heart is touched by a place where nature rules, even if that place is small or just beginning its long successional journey back to wildness, I know that you will be drawn into environmental activism. Once heart-touched, you'll no longer be able to sit on the sidelines. If your wilderness comes under threat, you will have to help protect it.

This book celebrates many people I came to know through their efforts to save wilderness. They include a forester who crossed the line from bureaucracy to activism, First Nations' leaders who defended their own precious piece of the Earth and many indomitable volunteers: men, women, children, elders, seasoned activists and young starry-eyed campaigners. I know there are people I should have named and written about but haven't. In fact there are boxes of Wilderness Committee files that still hold fascinating untold stories. Five hundred pages are too few to cover everything! An apology will not right these omissions, so I make none. I leave it to others to fill in the gaps. I hope that this book inspires other environmental groups to write their own histories.

Gripped by writer's block at the start, I began by constructing the chronology (see Appendix II). It contains the crucial framework of facts on which I hung my historical narrative. In a way, the chronology is the indisputable history. The other pages are my interpretation of all that happened in between the relevant concrete dates.

Like nearly all Wilderness Committee publications, this book has been a collaborative effort. Dozens of people submitted stories and provided interviews. Many who played a major role in the Wilderness Committee at one time or another reviewed the draft manuscript and provided their corrections and additions. I thank them all, for they improved the accuracy and enriched the contents of this book.

There are four people I want to especially thank for their invaluable and indispensable help. Richard Krieger, who co-founded the Wilderness Committee with me and hosted me at his home when I first began writing this book For four days and nights we relived the early days of WC-squared including our heated conflicts. He has since provided encouragement and helped edit this book.

As the book started to take shape, I met regularly with Tony Eberts. Every couple of weeks we would meet for an afternoon in a pub to review the manuscript and reminisce over a few beers. He offered many good ideas to improve the text. But, most importantly, when I got discouraged with my writing and felt daunted by the monumental task of finishing the book, he encouraged me not to give up.

Amanda Brown started volunteering to help me finish this book in November of 2005. A recent PhD graduate in biology, she spotted inconsistencies and questioned every sentence that didn't make sense. She edited my manuscript with fresh eyes, scanned the photos and documents used in the book and used her masterful computer skills to graphically design the pages and layout the text. Her intellectual curiosity, artistic ability, hard work and patience have made this book both beautiful and accurate.

Lastly and most importantly, I thank my wife and love of my life, Adriane Carr. When it became apparent that I could not single-handedly write this book to the standard needed for publication, she came to my rescue and rewrote everything! She untwisted my sentence structure, polished my thoughts and added details that I had forgotten—or never knew. I then edited what she had revised and submitted the text to others for their input. This process made this book, like all Wilderness Committee publications that we worked on together from 1983 to 2001, something that both Adriane and I are proud of.

Working at the Wilderness Committee has been inspiring and exciting. To have helped found it and helped it grow gives me deep-souled satisfaction. I revel in the wilderness we've saved and the people we've inspired. What better thing can you say about an organization than it makes the world a better place? That has been the Western Canada Wilderness Committee's mission and its success.

Chapter 1

WCWC's origins
"Galapagos of the North" trip in '77 changes my life

"The wilderness needs you," we often tell our members and donors. We kept working nonstop knowing that there was only one chance left to protect this planet's natural ecosystems that had taken millions of years to flower—and that chance happened to fall within our lifetime. We couldn't miss that chance for lack of effort.

Three short explanations as to how WCWC began

Western Canada Wilderness Committee is rooted in three different origin myths. The one I've most often told is that WCWC germinated in a brainstorming session around a campfire one magical summer night in 1977. Thom Henley, Richard Krieger and I were camped on a beach in Windy Bay on Lyell Island in the heart of the South Moresby wilderness area on Haida Gwaii (Queen Charlotte Islands). That campfire conversation was the genesis of not only WCWC, but also a remarkable nature-based outdoor education program called "Rediscovery," and lots of new tactics for the campaign to save South Moresby.

The second origin myth, which has been retold almost as often, is that WCWC was created because of the U.S. Sierra Club's refusal to publish a Canadian version of their annual wilderness wall calendar. Determined to see a Canadian product on people's walls, my friend Richard Krieger, a wilderness photographer, and I started our own non-profit society to make it happen.

The third story, never told before, is that after visiting the headquarters of Greenpeace in Vancouver in 1980 it dawned on Richard Krieger and me that we, too, could start and run our own creative organization focused on wilderness preservation. It looked like it would be a lot of fun.

All three stories have some elements of truth and on August 7, 1980 the Western Canada Wilderness Committee was born.

Richard Krieger backs my wildest dream

The whole thing began one fateful day in 1976 when my neighbour Richard Krieger unexpectedly stopped by my house to see me. We were working together as volunteers on a Parks Committee of the local Neighbourhood Improvement Programme (NIP), a Federal/Provincial/Municipal project that was infusing one million dollars into our working-class Victoria neighbourhood. We were plotting ways we could create more parks by strategically closing some streets in our over-roaded urban jungle. I was working in the kitchen in my usual messy fashion, with piles of paper spread out all around, sorting through information about the Queen Charlotte Islands. I'd started studying this archipelago, the most biologically unique in all of Canada, while writing a Grade 11 Biology correspondence course for the B.C. Ministry of Education, a task I had just completed. When Richard asked what I was doing, I divulged my dream of helping to save a spectacular wilderness—called South Moresby—in the southern part of the island chain. Local residents had been campaigning to protect South Moresby for the last two years. I wanted to explore it; moreover, I wanted to write a definitive coffee table book that would help save it from being logged. I even had the book's name picked out: *South Moresby – The Galapagos of the North*. I planned to pattern it after the famous U.S. Sierra Club/ Ballantine book series that was so successful both artistically and politically in saving wilderness areas in the U.S.

It was my <u>big</u> fantasy.

After hearing my story, Richard offhandedly said, "Let's do it!"

"Sure," I replied, carrying on in jest. "It'll be the adventure of a lifetime." Then, clicking back into reality, I said, "But you know I have no money." Richard, a few years younger than me, had a dry, impish sort of humour and I couldn't tell if he was pulling my leg. He was entertaining, smart and well-meaning. I got along really well with him but sometimes I just didn't recognize his boundaries for practical jokes and impulsive, off-beat behaviour...

"I'll finance it!" replied Richard.

He wasn't kidding! He even paid me a $300/month stipend to help support my wife at that time and my three teenaged kids while I was away.

This was the start of the journey that, inextricably and inevitably, led to the founding of the Western Canada Wilderness Committee.

But it wasn't as simple as that. Richard and I agreed that we couldn't do the book project by ourselves. We had to work with the local environmental group, the Islands Protection Committee. After all, this local citizen group was already campaigning to save South Moresby.

I immediately wrote to them and suggested we meet. It took a long time to get a reply, but finally a date was arranged. Richard and I strategized on how we should approach them and decided that, for the first meeting, I should go alone to broach the idea of the book. I

Richard Krieger (below) and I (left facing page) beside Windy Bay Poster Cedar. Photos: Left Richard Krieger, Right Paul George.

flew to Sandspit (paid for by Richard) and hitchhiked up to Masset to meet with the IPC directors at Dan Bowditch's house. Dan was the head of B.C. Hydro on the Charlottes and one of the key directors of IPC.

First meeting with the Islands Protection Committee

At this first meeting with the IPC directors I met Guujaaw (known then as Gary Edenshaw, before he was given his Haida name) who years later became the President of the Council of the Haida Nation. He was in his early 20s and one of the few really active Haida on IPC. Thom Henley (nicknamed Huck when a kid because of his love of huckleberries) was also at the meeting. Huck was an American draft adventurer who had done a lot of solo kayaking on the west coast, who had come to the Charlottes to kayak and ended up staying. Guujaaw and Huck had co-authored the "South Moresby Wilderness Proposal" that IPC championed. The proposal consisted of a map on which they'd drawn an infamous horizontal line just below the Tangil Peninsula with the statement, "No logging below here."

About ten or so other IPC directors were also in Dan's living room. Dan introduced me to everyone. Contrary to rumours, Guujaaw did not refuse to shake my hand, but he did ask me whether or not I was a police agent. Of course, I denied such a ridiculous notion and then gave them a short version of my life history. I heard later that I dispelled his and other IPC directors' suspicions of me with my genuine zeal for wilderness preservation and my knowledge about biology.

Guujaaw and I hit it off right from the start. I respected his ideas and blunt style of discourse. While everyone at IPC was interested in what I had to say, most were quite standoffish. IPC was a tight-knit island in-group. They had their own plans and I was an outsider.

The IPC directors didn't make a decision. They simply put me off, asking me to return with more details at a later date about our proposal for the South Moresby expeditions and book. I assured them that we would not embark on the project without their approval. A day later I flew back to Victoria to tell Richard the news.

IPC says 'No' to our book project

Richard and I worked hard on our proposal. For our second try, we decided to go up as a delegation of three. My friend Mark Horne—a smart, well-spoken young law student—came along for support. "This will impress them," I thought. In early spring of 1977 we returned to Haida Gwaii and got our second hearing at an IPC Board meeting at Dan's home in Masset. This time we had a detailed proposal.

Huck and Guujaaw demonstrate in front of the B.C. Legislature to protect South Moresby in Haida Gwaii, Queen Charlotte Islands. Photo: Richard Krieger.

We even had a name for ourselves: 'The Galapagos of the North' Book Collaborators. After hearing us out, the IPC directors reserved their decision saying they had to talk it over. Mark immediately went back to Victoria. Richard and I stuck around waiting to hear the decision. Late the next day Dan Bowditch broke the bad news to us. IPC would be doing their own book; therefore, they wouldn't be able to assist us, or be part of our efforts. Richard and I re-affirmed our decision that we couldn't do it alone, so that was the end of our dream. The proposed expeditions and book would not happen.

Richard was deeply disappointed and left for home the next day. I was depressed too, but tried not to show it. What would I do with my life? I had come to count on this project. Feeling sorry for me, two of the IPC directors, Lark Clark and Jack Latrell, offered to let me stay in their cabin on Kumdis Island in Masset Inlet for a few days. "It's a great place to get away from it all." They also suggested that I come the following weekend to the first All Island Symposium where South Moresby would be a big topic of discussion. As I paddled off in their small canoe with my backpack and outfitted with a crude map on how to get to their cabin, I told them I'd stay a whole week and would not attend the symposium: "I need the solitude," I said.

After staying three days, I had full-blown cabin fever. I had never been alone for so long. My day hikes into the incredible wilderness inspired only one small revelation: I discovered purposelessness was not for me. On day three the weather rapidly began to deteriorate and the winds picked up. I figured I might be stuck there for days if the storm brewing up was a big one. "It's time to go home to Victoria and my family," I said to myself. I quickly packed and closed up the cabin. In less than a half-hour's time, I was paddling back to Port Clements. I arrived in the late afternoon at the exact moment when Lark and Jack were leaving to go to the symposium at Skidegate Museum. There was room for one more in the van, so I hopped in. They said that I could probably stay that night on the floor in Skidegate in Joe Tulip's bachelor house. Huck and Guujaaw would be staying there too.

About 150 people attended the symposium. IPC was right: South Moresby was the big topic. Amongst the many who gave speeches that day was Guujaaw. He got sort of tongue-tied, stumbled over words and didn't say much. I learned later that day that it was Guujaaw's first big public speech. One of the Haida Elders told Guujaaw afterwards that it was a good one because, "It gave everyone there a chance to think about what you might have said."

Preliminary government report gives the South Moresby Wilderness Proposal a big thumbs down

Near the end of the symposium, Ric Careless, who then worked for the B.C. government's Environment and Land Use Committee Secretariat (ELUCS) and was attending the symposium as an "observer" for the government, unexpectedly gave a progress report on the Secretariat's unfinished study of the South Moresby Wilderness Proposal. During the previous two years Ric had visited the Charlottes and gone to several IPC directors' meetings like the one I had attended at Dan Bowditch's house. The IPC directors were sure that Ric supported their proposal: but circumstances had changed. The Dave Barrett NDP government that initiated the ELUCS study in 1974 had been voted out of government in late 1975. Now the Bill Bennett Socreds (members of the Social Credit Party) were in and their sympathies lay with the big logging companies.

Careless authoritatively informed the symposium audience that ELUCS had concluded that Rayonier, the big logging company that held Tree Farm Licence (TFL) 24 (accounting for 99 percent of the logging in South Moresby), was not a problem. It was the small-scale, independent "gypo" hand loggers who were causing problems and the B.C. Forest Service was going to handle that. It was obvious to me that the South Moresby Wilderness Proposal was dead.

The book dream is revived

Numbed, alone and discouraged, I left immediately at the end of the symposium. I didn't have the heart to wait around to speak with anyone from IPC. In the pouring rain, I walked slowly back from the museum along the highway to Joe Tulip's house in Skidegate Village. It was time to fly home.

Halfway there, Huck caught up to me. He said, "Paul, things have changed. We're going to need your help...all the help we can get!" I didn't ask him if they needed to have the other directors' approval to make that decision. I knew Huck wouldn't make such a statement unless the other leaders in IPC, especially Guujaaw, had already agreed. Huck promised to go on one of our expeditions. He—we—talked excitedly about what needed to be done. What an emotional roller coaster ride of changing fate! Now full of enthusiasm, I returned home and told Richard the good news.

A few months later, after a lot of preparation (mostly handled by Richard), we drove from Victoria to Prince Rupert in an old van that Richard had purchased for the trip. We flew across Hecate Strait to Masset and waited for our van, loaded with our gear, which arrived a day later on the Rivtow barge. We had a 10 ½-foot Avon inflatable boat (better than a Zodiac, Richard claimed) powered by a brand new 20 HP Mercury outboard motor. We had tried it out only once in False Creek in Vancouver before we left. We also had six empty 20-

Loading gas into the tiny Avon inflatable boat in Skidegate Channel. Note the clearcut and landslide in the background. Photo: Richard Krieger.

gallon jerry cans for gas and lots of fancy freeze-dried food including lots of freeze-dried ice cream. We were prepared.

To gather information for the book, I took a small portable tape recorder and dozens of cassette tapes and Richard, being a professional freelance photographer, brought along tons of camera gear and film. All his photo equipment was packed into a huge waterproof aluminum case that looked like one of those giant trunks you see porters carry in jungle movies. For both of us it was our first big wilderness expedition and our first involvement in wilderness preservation campaigning. We were greenhorns in every sense of the word; but we tried not to show it.

"Galapagos of the North" greenhorns face a steep learning curve on Haida Gwaii

People were friendly and helpful on Haida Gwaii. One of the first we met was Lynn Pinkerton. She was working on her Ph.D. studying the social dynamics of new and old immigrants to the Islands. She had a cabin that she'd built herself with the help of friends in Queen Charlotte City. It was on "Hippie Hill." She kindly offered to let Richard and I stay there while we prepared for our first expedition into South Moresby. It later became our regular place to stay in-between our trips into the wilderness. We built her an outhouse in lieu of rent. She provided insight into the social dynamics of the local "island culture," which had strong individualists who both supported and opposed the South Moresby Wilderness proposal.

To me, Haida Gwaii was a microcosm of the world at large and the struggle between two fundamentally opposing viewpoints regarding the essence of nature and the need to preserve wilderness. We quickly became immersed in the social network on Haida Gwaii and IPC's campaign to save South Moresby. I recorded conversations with everyone I met, from loggers and old-timers to back-to-the-land hippies and government officials in various resource ministries.

I was trying to understand everything about the proposed wilderness area, including what people thought about logging practices and old-growth forests. Maybe it was my training in sociology (a decade earlier I had earned a double major's degree in Zoology and Sociology

Sitka Spruce near the mouth of Jiinaga (Government Creek)—the last unlogged watershed in Skidegate. One of the protected areas. Photo: Ralph Nelson.

Taking the first load of supplies down the ramp to our awaiting ten-and-a-half-foot Avon inflatable on the float in Moresby Camp on July 11, 1977 in preparation for Richard and my first trip into South Moresby. Photo: Richard Krieger.

from the University of Minnesota), but I believed that saving a wilderness area like South Moresby depended upon people's attitudes towards wilderness and the industrial resource extraction that threatened it. Learning how to change and modify those perceptions was the key to wilderness protection.

I had the naive idea that if I collected sound bites from everyone I encountered, eventually it would all come together. I would gain some incredibly new and profound insight into the meaning of wilderness and discover an unfailing way to convince people to protect it. That single profound insight never came. But I did learn a whole lot. The information and little bits of insight I did gain were useful in the intense years of campaigning that followed. Now I know for sure. **There's no magic formula. It takes constant hard work, an innovative team of people and long term commitment to achieve social change—even to protect one small wilderness area.**

No one from IPC was able to join Richard and I on our first expedition. Perhaps they wanted to see us survive our maiden voyage before they risked going with us. After a few weeks of procrastinating in Queen Charlotte City waiting for the rainy weather to abate (which it never did), Richard and I zipped up our red floater jackets, pumped up our tiny inflatable boat, piled it high with gear and gas and launched out from Moresby Camp in the drizzling rain. Less than 10 minutes from the wharf, we had our first crisis. I freaked out at having Richard at the outboard helm. The waters were choppy and I was more experienced and cautious than he was. I had at least run an outboard motor several times before. I mutinied and said that I had to run the boat or I wasn't going. Richard simply swore at me and motored on. When he finally got tired, we changed places. As we both got more experienced, we shared "captaining" duties easily. When I look back I realize we took risks that no one takes today. We had no radio communication and much of the time we were a long way from anyone else. But we weren't foolhardy. We always wore our floater jackets and watched the weather.

First destination – Talunkwan Island, documenting clearcut devastation

On our first expedition we didn't get into the South Moresby wilderness. We didn't even reach the Tangil Peninsula. We pulled into Thurston Harbour just to the north of it and stayed at the site of the logging camp that had been abandoned only two years earlier. Talunkwan Island was an impressive monument to the destructiveness of clearcut logging. I felt from the first sight of it that **images of its ugliness were as important to the campaign to convince people to save South Moresby as images of the magnificent wild forests that remained.**

We spent ten days on this small eight square mile island, walking every abandoned logging road and documenting the logging. The unbelievable extent of the devastation was one of the reasons we spent so much time there. When the loggers had finished with Talunkwan in 1974, they proposed to move much further south in the archipelago, to Burnaby Island. Their plans are what triggered Guujaaw and Huck into creating the South Moresby Wilderness proposal. As a compromise, the loggers moved to Lyell, a bigger island, closer to Talunkwan. I couldn't help but think, "Won't Lyell, within the proposed South Moresby Wilderness area, look just like Talunkwan in ten years?" Richard photographed every clearcut, landslide and every piece of trash the loggers left behind—until he rebelled when I asked him to photograph a rusting old wrench in a puddle. Besides arguing a lot, inspired in part by the devastated landscape, we got the goods on clearcut logging. It wasn't right. Pictures speak stronger than words and Richard's photos revealed the ugly truth of nature being wrecked.

Garbage collected by the three of us on our clean up of Hot Springs Island in the summer of '77. From left to right, me, Guujaaw, Thom Henley. Photo: WCWC files.

South Moresby expeditions with IPC leaders

In the months that followed, we went on a series of two to three week expeditions into South Moresby. On most of them IPC members accompanied us. Huck came along on one. We talked about the countless threatened wilderness areas that were going down without anyone even knowing about them. Guujaaw guided us on our expedition circumnavigating Moresby Island. Although he had never been to some of the places we visited, he had a keen eye and good intuition. We saw many incredible things. Wilderness, I learned, is the most complex and complete expression of nature.

Near disaster on returning from our final expedition

The closest we came to disaster was on our final trip in the fall. In late afternoon, we stopped so Richard could take a quick photo of the Traynor Creek cabin on Louise Island. Returning to our Avon we'd left in the intertidal zone, Richard slipped on a rock, struck his

Richard Krieger posing beside an eroding road on clearcut Talunkwan Island. Photo: Paul George.

Wrestling for the fun of it with Guujaaw on Kunghit Island. Photo: Richard Krieger.

Looking north over giant clearcut on Talunkwan Island with Louise and Moresby Islands in the background. Photo: Richard Krieger.

head and knocked himself out cold. What followed was one of the scariest moments of my life as I held him up, waiting and hoping for him to recover. He finally came to and said he was OK, but I wasn't so sure. Less than an hour later a heavy rainstorm and high winds hit us as nightfall descended. Both of us were chilled into a state of hypothermia. We arrived at Moresby Camp to find the fish guardian shack we had planned to stay in locked. Sheer willpower and the drive to survive took over. We found the toolbox in the bottom of the Avon and brought it up to the cabin. I found the screwdriver and with numb hands managed to remove the screws that held the hinges of the lock and succeeded in breaking in.

We got the airtight stove going and slowly warmed up. We cooked up our last two freeze-dried dinners and ate our last package of freeze-dried strawberry ice cream that we had been saving all summer. Richard suggested we have one last chess game. Despite suffering from a raging headache (which he only told me about later) Richard won. The final tally for the whole trip (Richard kept score in his journal) was 52 to 48. Richard was the winner.

Lessons learned on Haida Gwaii in 1977

There were several lessons I learned that summer and fall of 1977 on Haida Gwaii. An old time retired logger took me out to see a small patch of old growth right in his backyard in Port Clements. As we wandered through the thick moss and climbed over big old rotting trunks he kept pointing out the rot and defects in the standing trees and the recent dead fallen giants on the forest floor. "This forest should have been logged 80 years ago. The timber was much more valuable then," he explained to me. "Most of the forests on the Queen Charlotte Islands are decadent and over-mature like this one. We've got to log them to let a young vibrant forest take their place." The logging industry had coined the term "silvicultural slum" to characterize the coastal forests as an unhealthy place inhabited by old trees full of fungus and rot which must be cut before they fall down. In the following years it took a great deal of effort to change this negative characterization into a positive one: ancient temperate rainforests as healthy ecosystems vibrant with inter-dependent life forms and radiant with majestic beauty.

An older forester with Rayonier Logging was kind enough to spend the whole day with me, taking me around to various logging sites in North Moresby explaining the company's logging practices. I asked him many questions and he patiently tried to answer them. Regarding one massive landslide we were looking at, he told me, "It was fortunate we got the timber off the slope before it slid: we saved a lot of valuable wood from being wasted." He had a rationalization for everything that looked bad. Finally, exasperated by my persistently probing questions, he said, "You'd have to take Forestry at UBC (the University of British Columbia) to understand." This incensed me and inspired me to question the logging decisions made by RPFs (Registered Professional Foresters) at every opportunity I could. I especially criticized their approval of clearcuts on steep unstable slopes. It galled me even more that forestry professionals, at the same time as defending ecologically damaging "silvicultural practices," always opposed wilderness preservation that "cost" their industry any amount of commercially valuable timber.

That summer (and in every campaign I've waged), I heard many people say, "You can never win protection for the area, even though it deserves it. It's a hopeless cause." It may have seemed like a hopeless cause to them, but it never seemed that way to me. Although it was

The triple cedars with me on Lyell Island. Photo: Richard Krieger.

hard work just to keep South Moresby an ongoing issue for people and not let it be overshadowed by everything else going on, I always saw no end of concrete things to do to advance this "hopeless cause." Whenever the campaign prospects seemed bleak, I just got more motivated. That's what needs to happen in order to win. If everyone gave up, well then the cause would be hopeless.

The more I explored South Moresby the more obsessed I became with saving the place. "The cause" became a full time volunteer job for me for the next three years leading up to the establishment of WCWC. I was motivated by the obvious need for public education on a much larger scale. I believed that if a critical mass of Canadians knew how wonderful and biologically rich the place was and spoke out about what would be lost in logging it, our elected representatives would have no choice. They'd have to do what it takes to protect the area. The big challenge and most difficult task was to get that critical mass of people aware and involved. In the years since then **my faith has been shaken in the power of the public will, as governments no longer feel obliged to protect an area or an endangered species even if a large majority of citizens wants it. Moneyed interests and industry have become much more powerful and more sophisticated in countering and discrediting environmentalists, and governments seem to have lost the belief that they have an overriding obligation to protect the public interest and think past their term in government.**

Spending many hours night after night around the campfire out in the South Moresby wilderness stokes the most primitive and basic thoughts and feelings in a person. Out of casual conversations come revelations. "There's lots of wilderness out there that in its own way is just as beautiful and spectacular as this place. These have no champions but they need saving, too," commented one of our expedition companions who had grown up in B.C.'s Interior.

Although I hadn't personally seen the areas he so vividly described for us (the Southern Chilcotin Mountains wilderness) I agreed with him. I vowed I would go and see it soon. Another person said, "If you succeed in saving South Moresby they will just cut some other wild place even more special and do it faster, like the tropical rainforests. There is so much demand for wood and paper that has to be met." I didn't have an answer for that one, except that possibly reducing the amount of good "waste" wood currently left behind would make up the difference. I needed more answers. And wilderness areas needed more campaigners.

Educating the loggers on Lyell Island

On one of our trips we took up an offer by Frank Beban, the owner of Beban Logging Ltd., to visit his logging camp on Lyell Island and tour his operations there. The deal was that in return, we had to come back and give a South Moresby slideshow for the loggers in the camp. We agreed.

"Stop here," cried Thom Henley as we climbed up the logging road in a crew cab pick-up with the camp supervisor at the wheel. We were right beside a landslide obviously caused by company road builders' side castings when they gouged the road across the steep mountainside. Richard got out and took pictures. During the two days we were guests at Lyell logging camp we documented other damage caused by the logging company, including the gray water pond outside the cook shack and the bulldozer tracks left in the mud of the estuarine inter-

Frank Beban and his Lyell Island operations manager looking out over a clearcut by his logging camp on Lyell Island. Photo: Richard Krieger.

tidal zone. We documented the eagles and ravens feeding in their camp dump along with a half dozen smashed up vehicles stashed beside the road near the fuel dock. How they could have so many accidents in two years, on only a couple of kilometres of road, was hard to fathom. All the while, our hosts were proudly showing us clearcuts they believed were examples of their good logging practices. Our investigations culminated in following up on a tip one of the loggers gave us. "Look around in the bush behind the camp."

Snooping there we discovered less than a minute's walk away from the cookhouse trailer the remains of a huge 600-year-old seven-foot-diameter Sitka spruce. It was all still there including the bucked up tree trunk and remnants of the eagle nest that once was cradled in its top limbs. In the sticks of the smashed nest we found pork chop was in the larger public and political arena that we had to convince people that this particular wilderness was more valuable left intact.

The next night I went alone to Sewell Inlet, the site of the other logging camp in TFL 24. It was just north of Huck and Guujaaw's "no-logging-below-here" line. The parent company Rayonier ran this camp and asked us to put on the same slideshow because their camp would be affected by the creation of a park, too. I had to give this slideshow alone (no one else from the IPC crew was willing to go) and slept in the camp bunkhouse after it was over that night. It was not as unruly as the Lyell Island show. In the morning as I waited on the wharf for the seaplane to come and pick me up, a wife of one of the loggers came down and talked to me. "When I first came to live here and saw the clearcuts I burst into tears. They got the company forester to tour me around and explain that it wasn't bad. It was just making way for a new forest to grow. But I still feel awful and feel it's wrong. Every time I see a new clearcut I still have to fight back tears." Just then the plane came and all I had time to say was "Don't believe the forester. Trust your heart. It is wrong!"

In the years that followed, many people accused WCWC of using emotional arguments to win people over to "our side." It's true we do appeal to emotion, especially through visual images. There is nothing wrong with it. **Most people are naturally emotional about nature and I'm glad they are.**

Thom Henley by the giant felled Sitka spruce Eagle nest tree behind Beban's logging camp on Lyell Island. Photo: Richard Krieger.

bones, indicating that the nest had been occupied recently—the eagles had fed off camp garbage. The tree had been felled, so the informant told us, because Workers' Compensation was worried about it blowing down and falling onto a nearby trailer. We measured the distance from the stump to the trailer and concluded that if they had moved the trailer a few yards it wouldn't have been necessary to cut this eagle nest tree down.

We presented our hard-hitting slideshow to the loggers in the Lyell Island camp's pub trailer later that summer. The rule established by the camp supervisor that evening was that there would be no drinking until our slideshow and talk was over. Huck brought a couple of women along from IPC to attend and their presence helped keep things under control. The first question came from the camp supervisor before the show even started: "Why do you guys have a hard-on for Rayonier logging?" he asked. I answered that it was nothing personal. It just so happened that this area they were logging was a very special one that should remain wild and not be logged. Everything went fine until the slides of the eagle nest tree.

Then all hell broke loose! The loggers objected vigourously. It was the Workers' Compensation Board that had forced them to cut the tree down. They didn't want to. They had other comments that showed they understood the issue. The slideshow was worth it because by listening to the loggers and understanding their side of the story we were able to strengthen our case. These guys wanted to log right. But they couldn't be convinced that, even if their logging practices were the best ever, they still shouldn't be logging in this place. It

Culturally Modified Tree (CMT), a cedar with test hole in it in Windy Bay. Photo: Richard Krieger.

Slideshows with the ugly vs. the beautiful work best

We also put on two other slideshows that summer, in the Haida village of Skidegate near Queen Charlotte City. The first show was in the middle of summer. We arranged to use the big old Skidegate Hall. We put up posters everywhere hoping to draw a large crowd. Only about a dozen people showed up. I was disappointed, but Guujaaw pointed out that Chief Percy Williams, who had taken Guujaaw many times into South Moresby to hunt and fish when he was young, was one of those who came. "The word will get out. The next time there will be lots of people," Guujaaw told me. The next show was much later that summer and we advertised it for a much longer time. Richard and I got back late on the day of the show from a three-week-

Richard photographing longhouse remains in old Haida village site on the Reserve on Tanu Island. Photo: Paul George.

long expedition in South Moresby. There was another meeting I had to go to that night, so I left Richard with the job of putting on the show. He only had two hours to pick and arrange the slides and get to the hall on time. Richard had never done a slideshow alone before. This time, the hall was packed with Haida.

They had all come to see the "beautiful pictures of South Moresby." They were deeply shocked. Richard had two trays of slides. In the first he just threw in slide after slide of recent clearcuts and landslides caused by the logging on Haida Gwaii. Most were of Talunkwan Island just to the north of the South Moresby wilderness; some were recent ones of Lyell Island. Richard showed them without any comments. People were in tears by the time he was done.

The next tray had the beauty shots. Although, as the raven flies, South Moresby was closer to Skidegate than was Masset, few visited the area because no roads led to it. Only fishermen and a few others—mainly hippie kayakers—were witnessing the wilderness' rapid destruction by logging.

I asked Richard the next day how the slideshow had gone. He didn't think it went very well. But I knew it was a great one. It woke people up. I heard for weeks afterwards about how moving it was. That show was one of the big turning points in the campaign.

On our expeditions into the wilderness—especially the trips with Guujaaw—I saw for the first time the extensive evidence of Haida use and traditional occupancy of the land. I saw my first culturally modified trees—cedars with test holes chopped into them—in Windy Bay on Lyell Island. As I learned more about the history of "land claims" in B.C., the treaty process that was designed to "extinguish" rights and settle the injustices of the past, I became completely convinced that the Haida would ultimately have to be the ones to "save" South Moresby. They had the power to protect the area. They just needed the will to do it. I felt strongly that the wilderness itself—the roots of their culture—would be that motivation. Cultures are inseparable from the natural ecosystems that support them. But until this dormant Haida power became engaged, we focused on every opportunity to protect the area through regular political processes.

Relaxing at Ninstints on our circumnavigation of Moresby Is. Photo: Richard Krieger.

GAAWA – HANÁS

Haida, pronounced Koo-ah-honass, meaning "a wonderful place"

Proposed Gaawa-hanás / Southern Moresby Wilderness:

- ■ Public Sustained Yield Unit
- ▨ Tree Farm License No. 24 (Blocks 3 & 4)

Falco peregrinus pealí

SOUTHERN MORESBY WILDERNESS PROPOSAL

AREA:
- 1436 sq. km (less than 15% of the entire Queen Charlotte Islands)

COASTLINE:
- 1696 km (more than now protected in all existing Parks, Park Reserves, and Ecological Reserves on the B.C. coast combined)

RELATIVE SIZE:
- less than 2/3 the size of Strathcona Provincial Park on Vancouver Island (smaller than 7 other established B.C. Provincial Parks)

OUTSTANDING FEATURES:
- provides nesting sites for over 25% of the total population of seabirds known to nest on the entire B.C. coast
- contains over 75% of the remaining largest and tallest virgin Western Red Cedars, Sitka Spruces, and Western Hemlocks in the southern half of the Queen Charlotte Islands
- shelters the largest rookeries of Sea Lions in the Canadian Pacific
- is the nesting site for over 25% of B.C.'s Peale Peregrine Falcons (40% of the Queen Charlotte Islands currently endangered population)
- encompasses a majority of the smaller islands in the Queen Charlotte Archipelago (over 138 islands, 37 of them larger than 16 ha)
- contains 42 lakes over 16 ha (some with biologically-unique fish)
- supports an exceptional variety of intact ecosystems ranging from alpine to intertidal, fresh-water to marine
- serves as a major feeding grounds on the Pacific Coastal Migratory Flyway
- is over 95% undisturbed by man (less than 60 sq. km logged or mined during historical times)
- has numerous Haida archaeological sites including several major villages abandoned less than one hundred years ago
- will provide a haven for endemic species and subspecies found only on the Queen Charlotte Islands (more endemics than any other place in Canada - eg. the world's largest black bear)

THREAT:
- clear-cut logging on Crown Land held by Rayonier Can. Ltd. under Tree Farm License 24 began within the proposed wilderness area in 1975 and is continuing at an accelerating rate (39% of the area is currently part of this T.F.L. as shown in map above)
- Rayonier is planning to completely log all merchantable timber on these lands over the next 40 years (indications are, projecting the current rate of cut, that they will be finished much sooner)
- much of the area to be logged is "ecologically sensitive" (steep slopes subject to landslides, productive salmon streams subject to siltation, islands with bird colonies, and shoreline with high recreational potential)
- only 280 ha of Height Class 7 trees (the tallest 59.8 m - 68.6 m) remain in the entire T.F.L., 89% of them are within the proposal area
- Rayonier claims that eventually 100-180 jobs in the forest will be lost if the area is preserved; however should a timber-trade be successfully concluded no net job loss would result; furthermore, Government research indicates that the mere elimination of waste in milling operations would result in substantial savings which could presumably be invested in labour-intensive forestry activity such as silviculture
- proposed multiple-resource development of the area will permanently destroy the remaining wilderness value of the beautiful sheltered East Coast Waterways
- despite recognition by Parks Canada and Provincial Parks officials of this area's "International Significance" no action has yet been taken to protect it

ACTION:
- write to the Honourable William Bennett, Premier of B.C., and the Honourable Sam Bawlf, Minister of Recreation and Conservation. Inform them of your concern and ask them to seriously consider this proposal (Parliament Buildings, Victoria, B.C. V8V 1X4)
- write to the Honourable Thomas Waterland, Minister of Forests, requesting full public renewal hearings for Tree Farm License 24 which expires in May, 1979 so that this wilderness proposal can get full consideration (Parliament Buildings, Victoria, B.C. V8V 1X4)
- join Islands Protection Society as an Associate Member in order to support this proposal and to keep informed of its progress. (I.P.S., Box 302, Masset, B.C. V0T 1M0)

Chapter 2

Saving South Moresby becomes an obsession

Returning to Victoria – failure as an author

I returned to my family in Victoria in the late fall of 1977. It was difficult for me to adjust to life there, and to the task of writing the *Galapagos of the North* book. In fact, it was one of the more frustrating times in my life. The chance of South Moresby being protected began looking bleak even to me. My head was swimming with ideas and I didn't know what I wanted to say or how to say it.

I worked for three months at the typewriter—there weren't any computers or word processors then—painfully producing a page or two every day. What I wrote was so bad that even I didn't like it...and everyone else I showed it to thought it was pretty weak as well. I was a failure. I was trying to write about the campaign to save South Moresby; but the campaign had just begun. I was far more interested in being an active player in the campaign to save the place than in being an author. I was worried that there was no master plan to the campaign—at least none that I could discern. We were all just "winging it"—making it up as we went along, taking advantage of every break and every opportunity to get the message out and motivate more people to help.

With a lot of help from others, I might have been able to put together a beautiful coffee table book with Richard's pictures; but I wasn't interested in doing that. It wasn't the goal when we embarked on this project. The goal was to save this magnificent wilderness.

Making a timely proposal to protect Windy Bay as an Ecological Reserve

One thing that I did manage to accomplish during this time was to put forward to the Ecological Reserves Unit of the B.C. government a proposal that Windy Bay be studied for ecological reserve status. I rushed to finish it during the last week of December and made sure that the envelope got a 1977 postmark. This was important, for if it was postmarked in 1978, the proposal would have to wait to be put on the Ecological Reserve Unit's 1979 work agenda. That would very likely be too late for Windy Bay. During our summer expeditions, I got the impression that Frank Beban was going to move fast to get the B.C. Forest Service to approve cutblocks in Windy Bay. He was smart enough to understand the significance of this place as an icon for the whole wilderness proposal.

Getting Windy Bay declared an Ecological Reserve study area was the only way that I could think of to get a moratorium to stop the place from being logged right away. Although he tried to maintain an air of bureaucratic aloofness, Bristol Foster, then the head of the government's Ecological Reserves Unit, was clearly on "our side." By luck, Dodge Point, adjacent to Windy Bay to the north, had a huge colony of nesting ancient murrelets. An estimated 60,000 nesting pairs used Dodge Point, making it the largest known colony of its species on the Canadian West Coast. This small seabird nests in burrows in the forest floor and obviously needs the intact old-growth forest—and no rats to eat its eggs—to survive. One of the major purposes of ecological reserves is to protect seabird-nesting colonies. The presences of this colony made it much easier for Bristol to investigate the merits of protecting Windy Bay as an eco-reserve. Nobody knew whether or not Windy Bay watershed being undisturbed was a reason why this bird colony was located there or if logging there would scare the murrelets away.

I was ecstatically happy when I heard that the Ecological Reserve Unit formally adopted my proposal. It was the first concrete, successful thing I had ever done in a wilderness campaign. It put logging on hold giving us more time to mount the campaign. However, there was another event on the horizon, which required an immediate response—and another excuse to postpone working on the book. It was pivotal to the campaign. The government was revising the Forest Act. Although Dr. Pearse's Royal Commission on Forestry had completed its work in 1976, there was a follow-up review that included public input and I made a submission. I'm sure that the review commission found my oral presentation entertaining. I talked about the current rate of cut being unsustainable, the need for more protected areas, and questioned the wisdom of automatically renewing Tree Farm Licence (TFL) tenures. I thought I presented compelling reasons why there should be legislated requirements to have public hearings prior to any renewal.

But the government and the big forest companies were bound and determined to strengthen corporate control over the publicly-owned provincial forests. I believe the forest companies knew the forests were overcut. Because they weren't practicing even-flow sustained yield (cutting equal or greater crops of trees each rotation with no "falldown"). They sought to absolve themselves of any obligations to do so by removing from the Forest Act all sections that mandated the forests be cut no faster than the rate at which a new forest with the same volume of wood grew back. Even-flow was a first principle in the original TFL agreements the companies signed; but they had never abided by it.

The new Forest Act introduced shortly thereafter by Minister Tom Waterland was a disaster for the South Moresby Wilderness proposal for it would automatically renew all TFLs for 25 years without any requirement for public hearings. It had an "evergreen" clause that provided for continued licence renewals—forever—despite the fact that the TFLs might be overcut and unsustainably managed. It also put social considerations—like the need to supply wood to existing mills, over biological considerations when it came to the Chief Forester's setting the rate of cut—an easy way to justify overcutting.

All the best-forested lands in South Moresby were sewn up in TFL

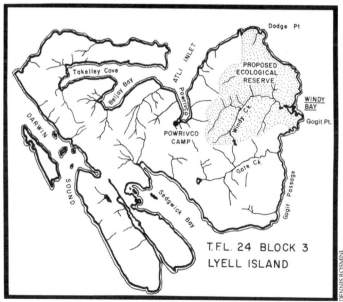

Left: First "action alert" information sheet that Richard and I produced for IPS.
Above: Map of Windy Bay. From page 66, All Alone Stone Spring 1980, Volume IV published by Island Protection Society. Cartography: Dennis Rosmini.

Forest Minister Waterland featured in a cartoon showing the Committee for Responsible Forest Legislation's brief opposing the proposed New Forest Act. Artwork: Bob Dalgleish, WCWC files.

24 and it just so happened this TFL would be the first one to come up for renewal under the new Act. We had hoped to quash this TFL renewal in a public renewal hearing. If the Act were passed, there wouldn't be a hearing.

Losing battle to get B.C. to enact ecologically-responsible forest legislation

We had only a few weeks to try to stop the passage of the new Forest Act. Richard and I and a few others formed an ad hoc group we called the Committee for Responsible Forest Legislation (CRFL). At first there was no political opposition from the opposition parties in the Legislature. CRFL created a short brief, complete with great cartoons penned by Bob Dalgleish—a Masset teacher and IPC director—and gave a copy to every MLA. The results: every opposition party spoke out against the legislation in the Legislature during its third reading using arguments contained in our brief. The new Act was obviously skewed in favour of the big companies, and against the interests of the public and the First Nations who theoretically, at least, owned these forests. But the Socreds had a large majority in the Legislature and, at the end of June 1978, six weeks from the date of its introduction, the new Forest Act was passed.

The fourth and final reading of the Act was held at a late night sitting of the Legislature. Richard and I were the only people still in the public gallery when the final vote was taken. Immediately after it passed, the Legislature recessed. It was past midnight. On the way out of the building I couldn't believe my eyes. Pouring out of the front doors of the Legislature were dozens of big forest company executives dressed in suits and half-cut on booze. They were laughing, joking and celebrating. They obviously had been drinking all evening, confidently starting their victory celebration early. We figured that they must have been lounging in the cabinet offices waiting for this moment to arrive. It was a dark day for wilderness preservation in B.C. and for the publicly owned forests.

Returning to Haida Gwaii to further study South Moresby and promote its preservation

Shortly thereafter I moved back up to Queen Charlotte City to live with Lynn Pinkerton. My relationship with my wife Carol had come to a dead end. I didn't want to go back to full-time teaching; I wanted to focus on the fight for South Moresby. I needed to find out more about TFL 24 and work on the campaign up in Haida Gwaii. I spent time with and learned a great deal from Guujaaw and Captain Gold, a Haida who was keenly interested in Haida Gwaii archaeology and worked with Guujaaw to protect aboriginal forestry utilization sites where there were Haida culturally modified trees and canoe building sites.

Guujaaw felt strongly that the renewal of TFL 24 must be opposed. He knew that renewal of this tenure for a minimum of 25 more years would impinge on Haida rights as well as be a huge blow to the chances of protecting the South Moresby wilderness area. In talking to lawyers he found out that the government had a duty to "act fairly" to those affected and if it didn't, there was remedy through the courts under the Judicial Review Procedures Act. He and I believed that "acting fairly" meant having public hearings. Government should listen to the reasons why the Haida and other people didn't want TFL 24 renewed. Given that the new Forest Act was silent on the requirement for public hearings, it seemed likely that we would have to go to court to get one. I decided to help build the case and

Above and Right: Demonstration in front of Rayonier's headquarters in Vancouver to try to get the government to hold a public hearing before renewing (replacing) TFL 24 which encumbered much of the South Moresby Wilderness Proposal Area. Photos: Richard Krieger.

urged IPC to become a registered society so it could take the case forward. I also made it a goal to have some sort of article or mention of South Moresby in every issue of the islands' weekly newspaper, the *Queen Charlotte Observer*. Of course the ultimate goal was protection of South Moresby.

During this period Richard and I produced our first poster: *Supernatural Windy Bay...let it be*. We printed 1,000 copies—which

Another remarkable cedar growing nearby the "Poster Cedar" in Windy Bay on Lyell Island in the proposed Windy Bay Ecological Reserve. Photo: Richard Krieger.

we thought was a lot at the time. We hand-lettered the type using Letraset. On the bottom of the poster in fine print we put, "*To obtain more information or express support write to the Ecological Reserve Unit*" with the government's address. I heard later that there were a few government noses out of joint because, although we did not do it deliberately, to some people it looked like the Ecological Reserves Branch had produced the poster. There was no publisher's credit printed on it because it was an independent project done by Richard and I. We just wanted to get it out, knowing it would be a great campaign hit.

The poster featured a high-resolution black and white photo taken by Richard of an unusually huge cedar growing in Windy Bay, with me standing at its base looking up in wonder. It took the Rayonier foresters a while to find the tree because it was quite short and "pitch-forked-topped" (having multiple tops) so it did not stand out in aerial photos. It was also quite remote. When they did locate it they cried foul, claiming that our Windy Bay poster tree was actually two trees that had grown together. Two trees or one, it was magnificently huge, truly a marvel of nature. Richard had taken the photograph with a standard lens and it was un-retouched. The "one or two tree" controversy didn't detract one bit from our poster's effectiveness in promoting the Windy Bay ecological reserve proposal. In fact, it helped.

Creating the first public education flyer for the South Moresby Wilderness proposal

With a possible public hearing and a court case looming, IPC needed to get the message out, appeal for greater public support and raise a lot of money for the lawyer they needed to fight the renewal of TFL 24. Richard said, "Why not use the Sierra Club's proven formula for 'action alerts'?" He offered to put together such an action alert for South Moresby. I helped gather the statistics for it.

My first task was to figure out the bio-geographical attributes that

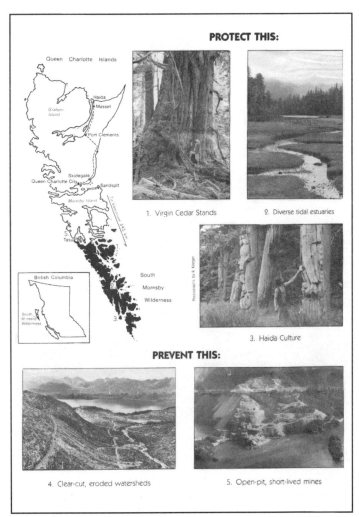

Excerpt from the Save the South Moresby wilderness flyer Richard and I produced for Islands Protection Society. Document: WCWC archives.

would astound people. I used a little tool that looked like a pencil with a small wheel and counter on the eraser end. For hours I rolled it along the coastline of the South Moresby proposal on a set of 1:10,000 maps, adding up the total length of the shoreline. I counted the number of islands and lakes within the proposed protected area. I also totaled the estimated number of nesting seabirds in all the various known bird nesting colonies and compared it to the estimated number that nested along the entire B.C. coast. South Moresby stacked up well. The facts were very impressive. We illustrated "what was at stake" and "the threat" with photos from our expeditions the year before. We included a map clearly outlining the wilderness area we wanted protected. There was even a locator map pinpointing the "Canadian Galapagos" spot on Earth. Richard did a beautiful job of laying it out and illustrating this black and white flyer. The most important part, however, was the appeal form—a perforated tear-off coupon at the bottom, asking for donations to be sent to IPC.

Richard paid for and printed up 10,000 copies on 8-1/2 inch by 17-inch glossy paper—a large print run for those days. We sent out packages of the flyers to all of the environmental groups we could find addresses for. The appeal worked. IPC directors reported that the money came flowing in.

An all-out effort to thwart the renewal of TFL 24

The push to get the Islands Protection Committee directors to formally register as a society finally paid off. In December of 1978, Islands Protection Committee received its registration papers and began to launch the TFL 24 court case.

My job was to prepare an affidavit with information proving that the TFL had been overcut. We aimed to prove that the licence-holder had contravened the first principle of even-flow sustained yield clearly stated in the original licence agreement. We figured it was the best argument to force government to pay attention to "public interest."

During the winter of 1978-79 I spent weeks sitting in the Queen Charlotte City Forest Service office poring over the original management and working plans, cutting permits, maps, inventory and research reports. The staff working there were kind enough to give me full access. At that time, nearly everyone in the Forest Service thought they were doing a good job and were sure that close scrutiny would prove it. No one from the public had ever really questioned their cut controls before. In less than two years' time, Forest Service greatly restricted public access to their files keeping "sensitive information" from public viewing.

The data I saw convinced me that TFL 24 was already seriously overcut well beyond an even-flow sustainable rate. I found no single smoking gun—just many smouldering embers that added up to a significant overestimation of the forest resource. I built the case as to how the overcut had happened gradually over many years. Every five years, when the TFL's Management and Working Plan came up for renewal, every rationalization conceivable was employed in the company's relentless drive to increase the Annual Allowable Cut (AAC). It was sort of like tax accountants exploiting every loophole to maximize short-term gains. Nowhere could I see evidence of long-term or holistic thinking. While each AAC increase might have made sense in isolation, no one bothered to do any global auditing to see if, added together, the cut increases really made sense in terms of the needs of future generations or sustaining the complex, multi-faceted forest ecosystem.

To calculate the TFL's overall sustained-yield cut level, I used the Forest Service's own data taken from their reports in the Queen Charlotte City office. I then applied the same formula used by the Forest Service. The TFL was at least 30 percent overcut. I proved that it was not being managed on an "even-flow" sustained yield basis as required by the original licence. I was ecstatic, sure that this failure to meet the fundamental terms of the original licence would render that licence null and void and thus force a renewal hearing before a new 25-year licence was issued. I was very naive.

For months we had been encouraging the public to send in letters

The Queen Charlotte Public Advisory Committee to B.C. Forest Service meet ever month in the basement of the Royal Canadian Legion to discuss forest issues including South Moresby. Membership included representatives of IPS the forest Service and the logging companies Photo: Richard Krieger.

Bruce Fraser conducts in 1977 one of the very first public involvement exercises sponsored by the B.C. Forest Service in Queen Charlotte City. Photo: Richard Krieger.

and petitions asking government for a public hearing on the renewal of TFL 24. We were getting pretty frustrated by the lack of government response and discouraged by Forest Minister Tom Waterland's statements in the media that indicated he was going to proceed with an automatic TFL renewal without any hearing. People's letters hadn't made a whit of difference.

It was my first reality check regarding B.C. politics. Government was there to facilitate "business as usual"—and that included automatic renewals of Tree Farm Licences. I don't think those in power even understood what it meant to act on behalf of the greater public good. Moreover, I think they saw First Nations and local environmental groups as simply getting in the way.

Our best chance—albeit slim—was the legal action we'd thought about for months, challenging Waterland's automatic renewal of TFL 24 based on the obligation he had to "act fairly" to those who would be potentially hurt by it.

Guujaaw lined up a young Vancouver lawyer, Garth Evans, to take the case forward. Island Protection Committee became a registered society (renamed Island Protection Society) so it could take this court case forward. Not all IPC members were happy with this change to become a formal organization. IPS became a petitioner on behalf of its members and committed to pay the bills. Guujaaw got Nathan Young, Chief of Tanu, the Trap Line Licence holder on Lyell Island (whose licence was overridden by TFL 24) to be a co-petitioner. He also included himself as a co-petitioner, as a Haida "Hunter and Gatherer." Some thought this move was a bit cheeky. But I thought it was brilliant. One of the biggest issues was how logging in South Moresby was ruining both the wildlife habitat and diminishing the seafood that the Haida relied on. Garth insisted at the last minute that a white trapper, Glen Taylor, who had a registered trap line on Burnaby Island in South Moresby, also be included as a co-petitioner to ensure court standing just in case the judge threw out native petitioners and IPS.

The petitioners sought a court order that would force the Minister of Forests to share information with them, as well as hold public TFL renewal hearings so their interests could be fairly addressed. It was a long shot legal argument that required lots of effort to assemble the documentation and build the case.

Garth worked hard for low pay on the case and his secretary worked long hours of overtime for no extra pay—all because they believed in the cause of saving this beautiful bio-rich place.

Garth said that my affidavit was good, but it would have little credibility in court because I was not a registered forester. The only way I could beat this was to have someone with forestry credentials verify my method—not the data (which would take too long to verify) or the conclusion. Ray Travers, then a government-employed forester, who had taken the time to teach me some of the fundamentals of forest cut control, referred me to a progressive professor on the UBC forestry faculty: Al Chambers, a sympathetic forester, he thought might be willing to help. I called him up and asked if he could take a look at my work to see if I was making any obvious mistakes or misapplying the mathematical formula used to calculate the cut for TFL. I did not mention that it was going to be a court document, or that I wanted him to testify that my study was legitimate scholarly work. I didn't want to scare him off before he even had a look at it.

Chambers agreed and I gave him a copy. A week later I went back to see what he thought of it. He said that I had used the equations properly and he could not find anything wrong with my reasoning, but that he could not vouch for my conclusion because he could not verify the data I was using to put into the equation. "That's great," I said. Then I sprang the question. "Would you swear that in an affidavit?" His answer was something I couldn't believe. He told me that he would like to, but he just couldn't. In short, it would ruin his career. He would never again get money for research projects for his graduate students, etc., etc. The forest companies were that powerful! So much for academic truth and freedom. All of a sudden the odds against our winning protection for South Moresby crushed in on me. I had naively thought that the information I garnered during six months of full time research in the Queen Charlotte's Forest Service District Office would deliver the knockout punch to Garth's bag of arguments.

The day before our court hearing was to commence, in February of 1979, Guujaaw insisted that all the documents of IPS pertaining to South Moresby be put before the court. That way, Guujaaw reasoned, no matter what the outcome, this information would forever be stored in the court's vault and be part of history so that "future generations could see what happened." Garth, however, said it wasn't necessary and wouldn't help the case. But finally he conceded, saying that if we gave him seven copies of this huge "foot-tall" file stack of IPS letters and documents by 8 a.m. the next morning, he would file it. I believe that Garth thought it was impossible to get that much copying done...it was already 4 p.m. in the afternoon.

We called the only place we knew that might let us use their photocopier: the West Coast Environmental Law Society. They were having a meeting until 8 p.m., but after that they said we could use their office. They had one of those ancient photocopiers where the top sluggishly slid back and forth for each copy. It was slow and laborious work. About 4 a.m. the machine, which was running very hot, caught on fire. I opened it up, yanked out the burning piece of paper and stamped it out, figuring that we had lost our battle. After letting the machine cool down for ten minutes we turned it back on and, to everyone's amazement, it still worked! From then on we paused every so often to let the machine cool down. At 7:30 a.m. the task was completed and we rushed the seven copies back to Garth's office at Pender and Granville. He filed this exhibit just before the court hearing began.

My overcut affidavit was never challenged in court. It was, in the eyes of the tunnel-visioned court, not germane to the case. Neither was the big stack of IPS documents. This case hinged on whether or not the petitioners had a legitimate interest in the land and whether or not they had been treated fairly by the forest minister. In an unusual exception to the rule that all the testimony in this kind of court proceeding be done entirely by affidavit, Nathan Young, chief of Tanu, was allowed to orally testify in his Haida language using a translator. The first thing the government lawyer did was to try to destroy Nathan's credibility, asking him about the sea otters that he

The last giant old-growth red cedar being cut down in the Cowichan Valley on Vancouver Island. Photo: Anonymously sent to WCWC in the 1980s.

trapped. Minnie, a Haida elder, translated the lawyer's question for Nathan. They talked back and forth in Haida and laughed. Then Minnie relayed Nathan answer correcting the lawyer, saying Nathan trapped river otters, not sea otters. Sea otters had been hunted to extinction on Haida Gwaii for the fur trade long before Nathan was born. Nathan clearly established he was a trapper and that he knew about the land and the harm clearcut logging would do to the animals he trapped.

On March 6, 1979, two weeks after the three-day-long court hearing was over, Justice Murray ruled that all except IPS had "standing." Guujaaw established that he had used and occupied the land for gathering seafood, fishing and trapping for a long time. Chief Nathan got standing in court as an owner of a traditional trapline but not for being chief of the area and Glen Nailor also got standing as a trapline owner. He also ruled that the Minister of Forests had a duty to treat these petitioners fairly. But in a twist no one had expected, the justice also said that he expected that the minister would treat them fairly, given there was almost two months for him to do so before the end of April when the licence was scheduled to be renewed. The petitioners had not proven beyond reasonable doubt that the Minister of Forests would not act fairly during this time. So he dismissed the petition.

Within a few days of this court decision, the government released the ELUCS study on South Moresby. It was obvious that is was rushed into print for it was full of typos and thin on content. At the same time, the government announced it would be setting up a South Moresby Planning Team where all interests could be heard regarding the land use options for South Moresby.

The government lawyers acting against the petitioners offered another option. They contended that the petitioners' interests could be accommodated as well in the five-year management and working plan for the TFL after the TFL was renewed. Perhaps the government thought that that was all it had to do to act fairly.

"You have to respond," I told Guujaaw. "The judge said that the Minister of Forests has to treat you fairly. All you have to do now is to send off double registered letters to Waterland and ask to be heard and they'll have to give you a hearing." He and Chief Nathan Young did that, and waited and waited. A week and a half before the deadline for renewing the TFL, with no answers to the letters, Garth went back to court with the evidence that Guujaaw and Nathan had tried to get fair treatment and had been rebuffed. A court hearing date was set for April 30, 1979, the last day before the licence was to be replaced.

All of a sudden the government scrambled to "act fairly." Guujaaw was summoned to Victoria to meet with the top brass in the Forest Service. He brought me along as an advisor. We spent about four hours in negotiations with the Chief Forester, the Deputy Minister and several other senior people in the B.C. Forest Service going over the proposed new TFL agreement clause by clause. In these negotiations the government agreed to make important changes in this TFL document to accommodate the Haida's interest in TFL 24, such as allowing the Chief Forester to ask that the company do studies when information was lacking to properly manage the area.

The Deputy Minister subsequently swore an affidavit attesting to the fact that they held this negotiation meeting, citing the changes they made as evidence of good faith consultation. The government's lawyers presented it in the courtroom the next day. The justice, accepting that affidavit as evidence the government had discharged its duty to act fairly to the petitioners, ruled against the petitioners.

A few days later the B.C. Forest Ministry scrapped all the agreed-upon modifications we had negotiated in the new TFL 24 licence agreement, leaving the document exactly as it had been before we had started to negotiate. It was the most cynical exercise of raw power that I had ever seen and it made a mockery of the whole court-ordered "fair treatment" process. We heard much later through the Forest Service grapevine that the other large forest companies had vigorously objected to the changes and refused to have these clauses in their agreements. TFL 24 was the template and it had to be the same as the others.

"So much for justice and winning environmental protection through the courts," I thought to myself. Garth launched an appeal. The appeals court hearing was postponed several times and finally the case was heard in 1981.

Last TFL 24 court case concludes in a whimper

On March 25, 1981 I got the phone call from Dan Bowditch urging me to attend the Appeals Court hearing on the TFL 24 renewal court case the next day—March 26, 1981 at 10 a.m. I got there later than I had intended, just before the proceeding started. There was no trouble finding a seat: I was the only person in the public gallery!

The three austere judges sat in front, with only Garth Evans (the lawyer for Islands Protection Society and the other petitioners) standing before them. Garth started out rambling. He spoke no more than a few minutes when one of the justices asked him if he could explain briefly and simply what the problem was. Garth said that he would like to have just a few minutes' recess to consult with the person in the audience—me. They refused. Garth rambled on for a few minutes longer and then folded the case.

Without taking a second to deliberate, the Appeals Court Justices dismissed IPS's appeal. Perhaps by now it was academic. The TFL had been renewed for 25 years (replaced by a new agreement) more than a year ago. But I couldn't believe that Garth didn't raise what happened: the government negotiating with the petitioners, winning the case with their affidavit attesting to changes made in the TFL 24 agreement to accommodate the petitioners' interests, and then a few days after the judgement cynically changing the TFL agreement back to the original wording. It was so fundamentally unjust.

Perhaps if I had arrived earlier and had a chance to consult with Garth beforehand, there would have been an entirely different result. But then again, it most likely wouldn't have made any difference at all.

Reflecting on this appeals case while writing this book, I came to a new understanding. We were in the wrong law court. Instead of appealing the second Supreme Court of B.C.'s decision to the Appeals Court of B.C., where they only look at errors made in law not at the merits of the case, we should have gone back for a third time to the Supreme Court of B.C. after the Minister of Forests actually renewed TFL 24 with its original wording in the new TFL agreement. Here we could have presented the new facts that proved the petitioners had not been treated fairly; but even this probably wouldn't have worked.

I truly believe that the government at the time had no intention of treating the Haida petitioners fairly by holding a public hearing about the renewal of TFL 24 or by protecting the South Moresby area from logging. It would have found some way of getting around any court order that headed in that direction.

After the fact, "what if?" mind games are fraught with frustration because you cannot change the past. I've learned that you can strive not to repeat your mistakes, always focus on the present and quickly exploit to the fullest every new opportunity and turn of events when they arise.

"They must make a lot of money with those calendars," I said to Richard. "I wonder how much of it they give to their Sierra Club here? Hey, Richard: you should be able to find that out." Within a day, Richard, a Victoria Sierra Club member, had the answer. It was nothing. They got no share of the profits their U.S. parent organization made by selling thousands of them through the bookstores in Canada. It had to be a tidy sum—tens of thousands of dollars—we figured, now that we knew a bit about printing costs. We felt it was wrong that the B.C. Sierra Club was struggling on its own and had to raise all of its own funds.

"Why don't they produce a Canadian Calendar featuring threatened Canada wilderness areas like South Moresby?" I asked. Richard, who was going to San Francisco to visit his relatives for Christmas, said he would go to the Sierra Club's head office there and make a proposal to them about publishing a Canadian calendar.

Meanwhile I was quite busy with my first long-term job since finishing the Biology 11 correspondence course two years earlier. The Nuu-chah-nulth Tribal Council had hired me on contract as a part-time forest researcher and environmental consultant. My first job was to represent the Clayoquot Band (Tla-o-quiaht First Nation) on the Meares Island Planning Team. The team had just been struck by the B.C. Forest Service to deflect opposition to the proposed clearcutting of this magnificent island (the source of Tofino's drinking water) by MacMillan Bloedel. I was getting to know a new environmental group—the Friends of Clayoquot Sound—that had initiated the fight to save Meares Island.

One of many of Krieger's photos of the native kids around a big cedar along the Big Cedar Trail on Meares Island for a poster for the Nuu-chah-nulth Tribal Council. All except one had one or two slapping at mosquitoes except the "perfect one" used for the poster. Photo: Richard Krieger.

U.S. Sierra Club refuses to produce a Canadian wilderness calendar

In the fall of 1979, after TFL 24 had been renewed, I moved from Haida Gwaii down to Vancouver with Lynn. Richard and I got together and brainstormed how we could further help the South Moresby cause. One thing that irked us was the fact that the U.S. Sierra Club sold thousands of their beautiful wilderness calendars in B.C. and across Canada every year, and this American publication almost exclusively featured U.S. wilderness—places that were already fully protected. They sometimes included one token area in Canada—usually one of our large national parks.

We both thought we had more and better wilderness areas in Canada than the U.S. has. Most were endangered and would soon be destroyed by roads and logging. And, because hardly anyone knew about them, most of them would go down without a peep of protest.

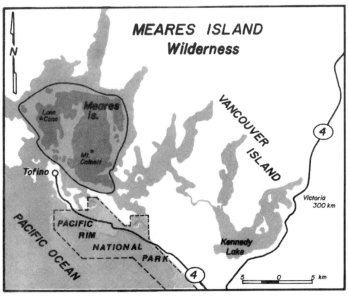

Map used in WCWC's 1982 calendar. Cartography: Gary Crocker.

Excited about the possibilities of a Canadian Wilderness Calendar, the first thing I said to Richard when he arrived back home in January 1980, was, "How did you make out with the Sierra Club?"

"They weren't interested," he replied.

I was disappointed. I hated accepting "no" for an answer to a good idea.

Years later, I found out from Richard that he had been turned away by the receptionist at the publication sales and marketing division in the Sierra Club's head office. No one in authority even deigned to talk to him about our Canadian calendar idea.

Chapter 3

Launching WCWC
Our first Endangered Wilderness Calendar a success

After visiting Greenpeace's office in Vancouver, Richard said, "We can do that!"

A few weeks after his return, Richard visited me in Vancouver and we went down together to visit the Greenpeace office on West Fourth Avenue only a few blocks from where I lived. We were seeking their help on the South Moresby Campaign. Patrick Moore, Greenpeace's executive director, was in an expansive mood. He toured us around the entire office, introducing us to everyone. He bragged about Greenpeace's campaigns and entertained us with stories about his past exploits. However, while expressing verbal support for the South Moresby cause, he made it clear that Greenpeace couldn't do much for us. They already had their campaign plate full and their campaign collection plate was nearly empty.

As we descended the steps from their second floor office, Richard turned to me and said, "We can do that!"

"What do you mean?" I asked.

"We could run an organization like that and have our own groupies," he replied referring to the volunteers we had met there. Richard was single then and looking for a girlfriend. "Let's start our own society. We could have an office and staff like that and publish our own wilderness calendar. It can't be that hard."

"Let's do it!" I replied. With that, we were off and running. It was the only way there would ever be a Canadian Wilderness Calendar. No one else was going to do it. Another motivating factor was that it had become increasingly difficult to do things through IPS now that neither of us lived on Haida Gwaii.

Registering and selecting a name for our own wilderness preservation society

We talked to Mark Horne, my lawyer friend, and got his help. Neither Richard nor I had ever started a society before. In consultation with us, Mark came up with a simple all-inclusive set of constitutional objectives. Public education was the key component. It was what we intended to do, and was in accordance with the criteria that would make us eligible for federal charitable tax status. We wrote our constitutional objectives as follows:

1. *To educate the public concerning Canada's wilderness heritage and to preserve representative areas for future generations.*
2. *To conduct research concerning wilderness values.*
3. *To obtain and distribute information on areas in Western Canada which have potential for protective status.*

These objectives have remained exactly the same ever since. Several times directors have suggested changes. But when they realize that this would entail going through the hassle of re-registering as a charitable society, they opt for the status quo. From day one until today, these constitutional objectives have never blocked anything we wanted to do.

When first registering, we chose the standard "Schedule B" bylaw package to make things simple. We also talked some great people into sitting on the first board. Two of them, Dan Bowditch and Bob Dalgleish, were involved in IPS but had since moved from Haida Gwaii. The other was Murray Rankin, a brilliant young University of Victoria environmental law professor who had graduated from Harvard Law School a few years earlier.

Everything went smoothly until it came to choosing a name. As usual, Richard and I disagreed. The name Wilderness Society was already taken by the Australians and also by a U.S. group. Wilderness Association was already in use by the Alberta Wilderness Association. We did not want to restrict our activities only to B.C.; so we didn't want a name with "B.C." in it.

I got the idea of going by the name "Wilderness Committee." Richard didn't particularly like it, but I went into the Corporate Affairs Branch of the B.C. government anyway to reserve this name. They immediately informed me that the name was too generic. I'd have to add some sort of qualifier. On the spot I thought up Western Canada as a prefix.

The person at the desk wasn't really happy with that name—Western Canada Wilderness Committee. He objected to the word "Committee." It sounded too government-like. I argued that we worked like a committee with other environmental organizations. It was a bit of a stretch, but after I calmly insisted that nothing else would really do, the keeper of names let me reserve Western Canada Wilderness Committee as our society-in-formation's tentative name. I thought Richard would like the name because of the WCWC initials. He did get a bit excited about calling ourselves WC-squared. I argued that, in mathematical terms, WC^2 translated into WCC. "It should be written with brackets around the WC—$(WC)^2$," I insisted. "That wouldn't be right either," Richard pointed out, "it would be

First calendar cover features South Moresby. Cover image: Richard Krieger.

like Western Canada Western Canada or Wilderness Committee Wilderness Committee. Both terms would have to be exactly the same to be written like that." With this playful bantering, I thought that there was no problem with this cumbersome name. A few days later I found out that this wasn't true. In a letter to me Richard wrote:

> Went to the registrar of Co. yesterday. I tried to get Western Canada Wilderness Council but they said it won't do because Council doesn't imply a society (the rest is OK and shouldn't be in conflict with anything else existing). Today I'll try for Western Canada Wilderness Confederation–if that doesn't work I'll try for Western Canada Wilderness Association (aside from meaning a shared partnership, it also is an ecological term)–and if that don't work, then I'll try Western Canada Wilderness Cult. At least I've tried. I hate the word Committee and it doesn't fit our aims at all....

The WCWC name was already reserved so Richard finally gave in. I told Richard the name really didn't matter—a rose by any other name is still a rose—what did matter was what we did with it as an organization. But the name did matter a lot more than I realized at the time. If we had thought up one as catchy as Greenpeace perhaps we would have been even more successful in a shorter amount of time.

Designing and producing our first (1981) *Western Canada Wilderness Calendar*

Our society's registered papers were accepted on August 7, 1980. I immediately sent off our application for charitable tax status to the federal government. Although Richard and I did most of the work, it was very good to have other directors on our board. They kept us on a business-like plane and got us to think through and plan out our projects and campaigns more clearly. They also mediated between Richard and me when we squabbled.

Our first project was the publication of the Endangered Wilderness wall calendar (for 1981) that Richard and I had contemplated producing for several years.

In talking to printers, we quickly discovered that we had to do a big print run (10,000 or more) to keep the costs down in order to retail them at $5 a calendar. This selling price was considerably less than the Sierra Club's calendar—our biggest and only competitor in the field of wilderness calendars at the time—and it seemed like a very affordable price. The only way to get rid of that many calendars was to recruit the help of other environmental groups to sell them.

We could not have a calendar that just featured South Moresby, although the thought crossed our minds. But of course a South Moresby image would grace the cover, since furthering this campaign was one of the main motivating factors for the whole project. We had the brilliant idea of using our calendar as a vehicle to strengthen the whole wilderness preservation movement, while at the same time solving our distribution problem. Each month would feature a different wilderness area sponsored by a different group that was campaigning to save that area. We both agreed on the calendar's basic format without any arguments. It would follow the formula used for the South Moresby information and appeal sheet that Richard had designed and put together the year before. It had been a huge success in raising money.

Each month would feature a gorgeous photo image that captured the essence of the threatened wilderness area. We'd include a map showing the area's location, and a brief statement of about 150 words lauding the area's outstanding features and highlighting the imminent threat(s). It had to be written in such a way so as to captivate people who knew nothing at all about the issue. At the bottom of the calendar pad we had a perforated tear-out donation form that hopefully the calendar purchaser would use to send in a donation and help out the environmental group fighting to save the area. Naively, we thought that we could convince other groups to buy into the press run at cost (half the retail price) and sell them to their members and the public as a fundraiser.

Getting other environmental and wilderness preservation groups to sponsor a month that featured the area they were campaigning to save turned out to be the hardest thing to do for our calendar project. We quickly found out that, although they liked the idea, not one group would take any risk—put any money up front or commit to selling a given number of calendars. All of them were smaller than the impression they presented to the public. All of them had tiny budgets and none of them had any money to spare. To gain their participation, we had to make a no-risk deal for them. We only asked them to do their best to sell some—as many as they could—with no financial obligations attached. We'd give them the calendars on consignment and they would only have to pay the $2.50 wholesale price for those they sold. They could return the unsold calendars when calendar-selling season was over. Not only was there no risk, they'd make $2.50 on every calendar they sold for their cause!

Both of us lined up the groups, picked the issues and selected the images. I did the write-ups. I knew nothing about typesetting, layout and the rest of the preprinting process: that was Richard's end of things. He had friends he hired to draw up the maps, typeset and paste it all up. Richard was also responsible for the finances; he used his house as collateral for the printing.

I was busy working for the Nuu-chah-nulth and let Richard design the date pads for each month. I figured nothing could go wrong there. A month later, Richard took me to the home of the woman who was doing the layout sheets to proudly show off the progress. Horrified, I found that instead of neat boxes for each date, there were hexagons (our logo was a hexagon with a stylized old-growth tree inside—designed by Richard). Because of their shape they did not stack solidly one directly underneath another. They were staggered. It looked really goofy and I told Richard so. Richard, who had recently taken up beekeeping, was attracted to the hexagon shape of the bees' honey cells.

He thought it was really cool and added a unique design flare. The work was too far along and not worth fighting over...so I gave in. I got my way about featuring a "mystery" wilderness area, a concept that Richard wasn't too keen about. In retrospect, that wasn't such a hot idea either.

During the next year we received several complaints about the calendar pad. People actually missed important appointments because they misread the day of the week associated with the date! In the 1982 calendar this innovation was abandoned, although Richard still persisted in a stylized box that I thought unnecessarily detracted from the wilderness image. We were learning. We often fought about details, but rarely squabbled about the big picture or the concepts.

Among the areas featured in our first calendar were the Valhalla Mountains, the Grand Canyon of the Stikine, Southern Chilcotin Mountains, Gitnadoix, Kakwa-Sir Alexander, Robson Bight, Tahsish-Kwois and the Akamina-Kishinena. All were unprotected then; all are protected in parks today. The "mystery wilderness" featured on the inside of the front cover of our premier calendar was the Yakoun Lake basin on Graham Island on Haida Gwaii—the headwaters of the largest river in the archipelago with exceedingly important salmon spawning habitat. It is still intact and now part of the system of Haida protected areas that are gaining formal recognition. The Haida earmarked it for preservation in the '80s and successfully kept logging out for nearly a quarter of a century.

After publication, then came distribution

Our calendars finally rolled off the press in mid-November, a couple of weeks later than we had originally planned. Both Richard and I worked hard to sell them before year-end. I thought that they would be easy to sell but they weren't. It was a hard sell—even amongst environmentalists. I quickly found out that there was little entrepreneurial spirit in the B.C. wilderness preservation movement. But the biggest problem was that the movement itself was tiny. Our market niche was a lot smaller than we had projected.

Everywhere I went I took a box of calendars. I went to every environmental club meeting I heard about. At one, the ROSS (Run Out Skagit Spoilers) executive meeting—this group was fighting the building of the high Ross Dam in Washington State that would flood the beautiful Skagit River valley in B.C.—I promised to include their issue in next year's edition. Here for the first time I met Adriane Carr, an instructor at Langara College who took 50 calendars—a really big sale at that time. I put down in my notebook "A really good calendar seller" to describe Adriane...having no clue as to the huge role she would be playing in my life and WCWC in just two years' time!

Some environmental groups were "good sellers" too. IPS did especially well on Haida Gwaii although they, like many of the others, were slow to pay. It was only because many people worked hard at selling our calendar, like Colleen McCrory of the Valhalla Wilderness Society who sold hundreds, that our fledgling effort was a success.

To get rid of the calendars that first year, we also developed a "moccasin network" of distribution that involved hundreds of individuals.

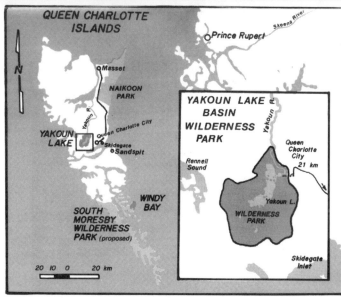

Top above: Mystery area featured on our first calendar. Photo: Mike Moon.
Above: Map of Yakoun Lake Basin Wilderness we published in our second calendar to reveal the location of the mystery wilderness. Cartography: Gary Crocker.

Many of them needed the "pin money" they made. Trustingly, we always gave them calendars on consignment—asking only that they send us the money (which was always half the retail selling price) after they'd sold them. Occasionally we got burned, but not often: that was business. There was no other way to sell them in large enough numbers to make it an economically viable venture and to change

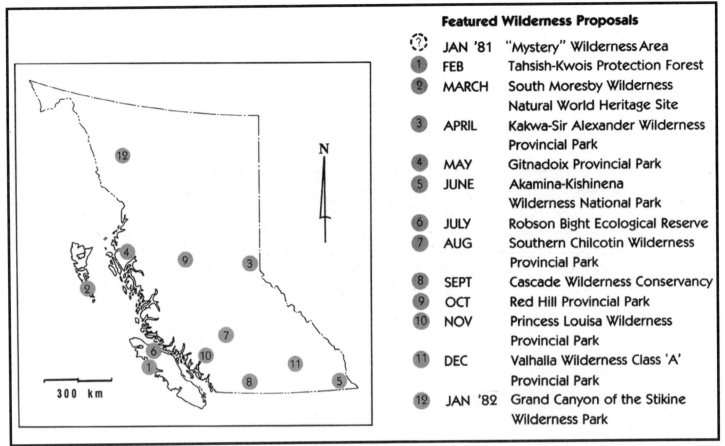

Areas featured in our 1981 Western Canada Endangered Wilderness Calendar. Graphic artwork: WCWC files.

public consciousness, which was our ultimate goal. Over the years we continued to rely heavily on this method of sales. To our great disappointment, all the normal outlets like bookstores would take only a few copies each year. We would have been out of business before we had even started if we had counted on them.

WCWC's charitable status placed in jeopardy

Just as we were beginning our calendar sales push we got a letter from Revenue Canada regarding our application for Charitable Status. It had several questions asking us to give more detail as to what our proposed activities would be. Richard said, "You handle this one, Paul." I gave it my best effort. I was aware that charities were not allowed to lobby government; so I stressed the educational aspect of our work, the publication of the wilderness calendar and posters, and the slideshows we put on. I thought I did a good job.

About a month later came the response: I hadn't done a good job. The bureaucrat's letter had many more questions for us to answer. I panicked. It was entrapment and I was probably already trapped by not having crafted my answers in the previous letter to fit the letter of the law regarding what charitable societies can and cannot do. I needed help. I showed the letter and the file of earlier correspondence to Mark Horne. He said, "I'll handle this," and took it away. He showed it to his father, Ian Horne, QC, who was the clerk of the provincial Legislature. Ian Horne did not bother trying to answer the questions. He merely wrote a very brief and polite letter, which essentially said that we were upstanding citizens and that our organization complied with the law.

A few weeks later on January 12, 1981, another letter arrived from Revenue Canada. With trepidation I opened it. It simply stated that we had been granted charitable status retroactive to August 7, 1980—the date we had registered as a society. What a relief! In the years to come we had to fight several times to retain our charitable status. Other groups, like Greenpeace, lost theirs.

A good way to get rid of unsold calendars

It became evident in late December that our first calendar was not going to sell out. Facing a surplus of quite a few hundred, we decided to give a complimentary copy to every B.C. MLA and to every Canadian MP. The postage to the House of Commons in Ottawa was free; and we hand-delivered them in Victoria, so it cost us practically nothing. The cover letter simply lauded the Canadian "wilderness resource"—which was the finest in the world—then urged the representative to use our calendar, and keep it as a reference for decisions they may be making regarding the fate of the areas featured. Thus began an annual WCWC tradition that has continued without interruption to this day. It's not lobbying; it's education—providing useful information to our elected representative lawmakers.

We originally set our overall preservationist goal post for B.C. very low. We publicly stated that we only sought an additional one percent of Crown Lands in B.C. to be turned into park-protected status by 1990. This would increase parklands from five percent to six percent of the total provincial land base. Even this modest goal was vigourously opposed by the logging industry. Over time, as our understanding grew of what a fantastic wilderness resource B.C. had and what it would take to protect big carnivore species, we developed a grander vision of a large, sustainable, interconnected system of wilderness areas. We kept shifting the goalpost outwards. This incensed our opponents who said we were never satisfied. It actually only reflected our naïveté about the extent of wilderness and the consequences of its loss.

Mailing our calendars to elected politicians made some think that a new supergroup had emerged on the conservation scene. It prompt-

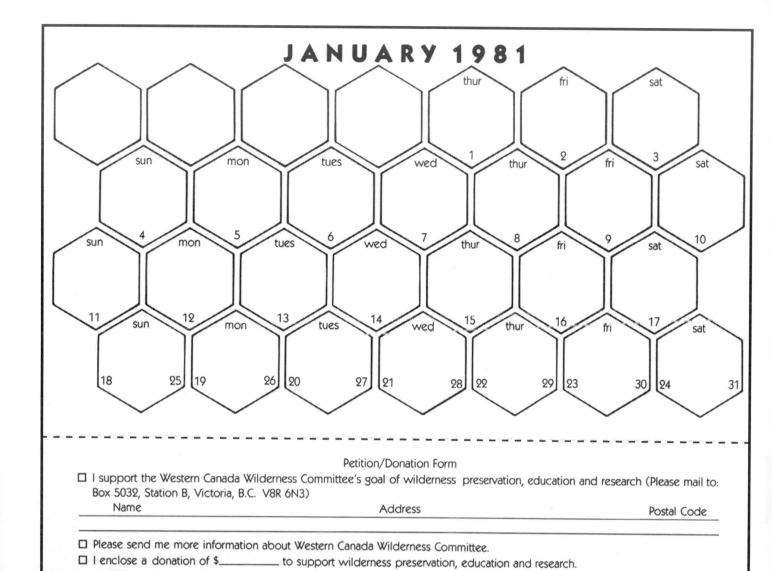

The 1981 calendar had perforated tear out coupons for people to send in to support the environmental groups fighting to save the threatened wilderness area featured that month. Graphic artwork: WCWC files.

ed James Chabot, B.C.'s Minister of Lands, Parks and Housing to write me this strange letter dated February 2, 1981. In it he asked a multitude of questions.

>...The name "Western Canada" suggests that you have active representation over a wide region. Could you please indicate the basic support groups included within the committee designation and identify their particular areas of interest and geographical locations?
>
>I would suppose that this group has an executive group which is linked with various associations. Could you please provide me with the names of these people and the basis by which they receive their appointment? There obviously is a budget which permits the operations of offices in Vancouver and Victoria and the employment of yourself as Director. Are there other employees which would allow for effective liaison with interest groups throughout all of Western Canada?
>
>Is there some active executive group which establishes the guidelines for action of this association and outlines work which you, as Director, should perform? [And it gets more bizarre.] As Minister, I cannot be expected to hold paid employees responsible for activities which are initiated and controlled by others. I do not have that privilege in my Ministry and your executive officers must expect to accept the same responsibility if representations are to be taken seriously.
>
>Information concerning your new committee will be received with genuine interest.

I didn't know how to answer this letter. I felt that Richard and I were, in some odd way, like the Wizard of Oz—appearing a lot more powerful than we really were. I never did answer it. Anyway, a couple of months later, Tony Brummet replaced Chabot as Minister.

In April 1981, after tallying up everything, our first calendar turned out to be a resounding financial success. Richard got repaid for his loan, and all the bills associated with the project were paid. The effort had made a profit of about $1,000 for WCWC after Richard and I each received a $1,000 honorarium for our efforts. Neither of us was good at accounting, but Richard was a lot better than I was, so he took care of the financial end of things. His method was to open up a new bank account for each project. Not the most sophisticated of systems! I took care of handling correspondence and memberships. At most it involved answering a few letters a month. We had fewer than 20 members in total. Adriane, my future wife, was member number 15, joining through one of our calendar tear-offs.

CANADA'S TALLEST TREES... Must not be logged

Write the Minister of Lands, Parks, and Housing, Parliament Buildings, Victoria, B.C. V8V 1X4, request that the Nimpkish Island Ecological Reserve on Vancouver Island be established.

Chapter 4

WCWC's campaigns extend into new territories and encompass controversial issues

Visiting Alberta to find areas for our calendar to make us truly a Western Canada organization

In April of 1981, I hitchhiked to Alberta to attend the Alberta Wilderness Association (AWA)'s annual general meeting being held that year at the college in Red Deer. I arrived in the evening, later than planned. I had wanted to get there for the afternoon's early registration so I could meet someone and find a place to billet. I didn't have enough money to get a motel room; nor had I a tent, foamy or groundsheet—but I did have a sleeping bag. I found a pine tree with low sweeping branches to sleep under in the middle of the lawn in front of the building where the next day's meeting was to take place. I spread out some newspapers to sleep on (they provided an amazing amount of insulation from the solidly frozen ground). Thank goodness it didn't rain that night! Actually, it couldn't have; it was far too cold. It was one of the longest nights I can remember. When I awoke the frost was an inch thick on the ground. I have never been so happy to see a meeting start.

I was very impressed with the AWA and with how many projects they had on the go. They were over 15 years old—founded in rural southwest Alberta by backcountry enthusiasts, ranchers and outfitters to protect their province's wilderness values. Like IPS, they were a well-knit friendship group. But I noted that the AWA was far more moderate than either the IPS or WCWC, working on Alberta government committees and on government-funded projects. What really amazed me was that over a hundred people attended this meeting. WCWC had grown but still had only 30 members in total.

I gave them samples of our 1981 calendar and got on their agenda under "new business" at the end of the meeting. After hearing about my no-risk offer, they agreed to participate in our 1982 calendar by providing images and write-ups for some key Alberta issues. It was great news. We ended up featuring three Alberta areas in the 1982 calendar: Milk River, Porcupine Mountains and Rumsey Aspen Parklands (our second and last "Mystery Area"). Including Alberta areas in our calendar was a great boost and cooperating with the AWA increased our distribution. I was most happy because we were no longer just B.C. focused: we were becoming truly Western Canadian.

Making Nimpkish Island an Ecological Reserve becomes WCWC's second major campaign

Inspired by the AWA's breadth of campaigns, Richard and I embarked on several new projects in 1981. This was hard because WCWC was only a part-time volunteer commitment for both of us. Drawn by the news that local people had recently discovered unique killer whale rubbing beaches beside the Tsitika River estuary, we decided to help gain protection for the Lower Tsitika Valley and get the Robson Bight Ecological Reserve established. This campaign was already a big priority for the Sierra Club of B.C., of which Richard was still a member.

We also decided to make the establishment of the Nimpkish Island Ecological Reserve a major campaign. This one was Richard's baby.

It seemed weird to me to have to put so much energy into trying to save 18 hectares of old-growth Douglas fir trees. Working on saving the 147,000-hectare South Moresby seemed much more significant and worthwhile. Nimpkish Island was like a tiny tree museum, but it was reputed at the time to have the tallest trees in Canada growing on it. Perhaps Bristol Foster urged Richard to take on this campaign to bring us down to reality and show us just how strong the resistance was in government and industry circles to any reduction in "harvesting" the old-growth coastal trees...no matter how few were involved. There were only about 40 big firs on Nimpkish Island. On the other hand, it was part of the big picture: increasing the awareness of the plight of B.C.'s old growth.

We made our first (and my only) expedition to Nimpkish Island shortly after WCWC's first AGM. It was memorable. Richard had purchased an ancient milkman's home delivery van, which had only two pedestal seats in it in the front. Richard drove; his girl friend sat on the other pedestal, while my girlfriend Lynn and I sat on milk crates in the back. Beside us was Richard's infamous camera gear trunk. It was a long five-hour drive to this "island of old growth" in the middle of the Nimpkish River on north central Vancouver Island. It was bumpy. The truck rattled incessantly and the back of the van captured exhaust fumes.

But it wasn't until we hit the gravel road that the real nightmare began. Clouds of dust poured in with no way to stop them. The van acted like a vacuum cleaner, sucking up every tiny particle of silt kicked up by the wheels. After several hours of this, we arrived coated in dust. Happy to be in fresh air, we crossed over the small branch of the river via a small logjam to reach the site. Within an hour I had explored the entire island and crossed over the main branch of the

"The most beautiful thing that people can experience is the mysterious. It is the basis of true art and science. They to whom this reaction is a stranger, who can no longer pause, wrapped in awe, are as good as dead, for their eyes are closed."
— Albert Einstein

Our second calendar cover again features South Moresby. Cover image: Richard Krieger.

river via another logjam. Here, the remnant of old-growth forest was even more spectacular than on Nimpkish Island and I felt it must be added to the proposed reserve.

Meanwhile, Richard found a perfect spot to set up the tripod with his Hasselblad camera: on a fallen trunk across an old river channel. It put him above the salmonberry bushes with a great view of one of the largest Douglas firs growing 50 metres away. Using his huge $2,000 fish eye lens, Richard took at least a dozen images of his petite girlfriend posing beside the giant tree. One of Richard's photos was a winner and it became WCWC's first poster: *Canada's Tallest Trees...must not be logged.* Not being able to afford colour printing, we printed it in black-and-white like an Ansel Adams classic.

We decided in the end not to propose an increase in the size of the proposed Nimpkish reserve because it was just not politically possible to save more according to Bristol Foster. The word "biodiversity," referring to the totality of species and genetic diversity of an area, had not yet been coined. (E.O. Wilson, the world-famous ecologist, first used this term in 1986.) Anyway, what is the difference between an 18-hectare and a 100-hectare protected area? Even then, I knew on a gut level that both would be ecologically unsustainable and fail to protect the natural ecosystem.

We passed out free copies of our black-and-white Nimpkish Island poster to every MLA and sold a few over the years. I'm sure it contributed to this area finally achieving Ecological Reserve status in 1988. However, by 1990, all the surrounding old growth had been cut. The Nimpkish Island Ecological Reserve was suffering on the edges from windthrow (trees blown over by the wind), while erosion was eating away at the island because of the increased spring runoff in this large nearly completely clearcut watershed. How much old growth is left standing today on tiny Nimpkish Island I don't know, but I fear the worst.

B.C. Chainsaw Massacre — the 8 mm film

In the summer of 1981, I got involved in making my first and only film: *The B.C. Chainsaw Massacre.* It was a joint project with Steve Lawson and Susan Hare, ardent environmentalists who lived a "back to the land" lifestyle on an island in Clayoquot Sound. Both were determined that not another old-growth tree should fall in Clayoquot Sound. The film featured a mammoth landslide in Hot Springs Cove, as well as other clearcutting-caused disasters. The concept of the 17-minute film was great; but the execution left a lot to be desired. Steve used a top-of-the-line super 8mm movie camera, a dying technology soon to be replaced by video camcorders.

Technological glitches plagued this movie. The worst was the faulty sound. I simply could not get the soundtrack recorded onto the copy of the film I was hurriedly preparing to show at a Sierra Club event at Robson Square in Vancouver. There was no time to get another copy made.

I talked my eldest daughter, Athena, into reading the script with me from a set of microphones onstage. My attempt to lip-sync the dialogue was a disaster. Most embarrassing were the scenes where I was on-camera explaining the massive landslide in Hot Springs Cove that had occurred several years after the whole mountainside had been clearcut and how the landslide had ruined the salmon-spawning stream there.

People rolled in the aisles with laughter at our attempt to be the soundtrack. But according to Adriane (whom I didn't know yet, but who was in the audience), they got the message that logging on steep slopes in Clayoquot Sound was wrecking the land, too.

Learning how not to run an organization while being on Greenpeace Canada's Board of Directors

In 1981, I also became a director of Greenpeace—ostensibly invited to join their board because they were going to increase their involvement in forestry campaigns. I thought that by being on their board I would learn how they operated, which would help me develop ideas on how to run our fledgling organization. It was indeed a learning experience. I saw how their autocratic executive director, Patrick Moore—who funneled all the media and power through himself—limited the organization. I learned that fundraising by lottery and bingo was hard to do and diverted energy and attention away from campaigning. I learned how not to run a board meeting. I much preferred the Nuu-chah-nulth council meeting style that maximized participation of all involved. The Nuu-chah-nulth Tribal Council (NTC) office had open files (hanging two-ring clipboards on the outer office wall) where all incoming and outgoing correspondence was posted.

It kept everyone informed about everything going on. This is the system we later instituted at WCWC's head office and still use to this day. The NTC also encouraged everyone's participation at council meetings. The chairman never cut any speaker off during debates—even if the person was off-topic or long-winded. It made for slower but much more democratic meetings.

By being on Greenpeace's board, I also learned lots about the downside of using non-violent civil disobedience as a tactic to gain media attention for a cause. It reaffirmed my belief that WCWC should eschew this tactic and work within the law. At nearly every board meeting we'd get a report from Greenpeace's lawyer regarding the Spatsizi affair. Greenpeace activists had camped in the Spatsizi Park and tooted horns and made noise to scare big game animals away from a party of trophy hunters. The angry hunters, who landed a helicopter near Greenpeace's remote camp, roughed up the activists and burned and destroyed all their equipment including their cameras. The hunters weren't charged; instead, Greenpeace was charged and found guilty of disturbing a legitimate hunt and eventually had to pay thousands of dollars in compensation to the hunters. It was a real eye-opener. Civil disobedience doesn't always work!

"Headhunting" — our first divisive conservation issue

I convinced Greenpeace to participate in our 1982 wilderness calendar. They decided to feature the issue of trophy hunting within Spatsizi Park; I added an expansion to the park to make it fit our format of featuring only proposed protected areas. Greenpeace then helped with sales, by including our calendar in their Christmas catalogue and sold several hundred copies for us.

After all the stress involved in the 1981 calendar, we were hoping our second calendar would be a lot easier. Richard and I arranged to publish the 1982 calendar on contract for WC^2. This removed any liability the other directors might have borne. We envisioned a better product, a bigger press run, higher sales and a lot more profit. The carrot for us was the huge theoretical profit that Richard and I would reap after the bills had been paid—a royalty of 17 cents for each calendar that sold. It was a good deal for the WC^2, too—because we (Richard, that is) took all the financial risk. It motivated us to get started working early. We succeeded in getting it out in August (not in November like the first one), giving us more time to sell them. We printed 15,000 copies—one-third more than the first year. The federally-funded B.C. Environmental Network (BCEN) was just getting started; we thought that the network groups would help sell the extra copies.

Above: Spatsizi Stone Sheep feature in the 1982 calendar. Photo: Bristol Foster.
Facing page: WCWC's first poster (black and white) featuring tiny Nimpkish Island's tall old-growth Douglas firs—a candidate for Ecological Reserve protection.
Poster image: Richard Krieger.

We sent boxes of them out to all the participating groups—and relaxed, thinking that they were all busy selling them. In November when we checked to see how things were going, we discovered that most of the recipients had put the box in their closet and forgotten about them! Again it was a mad scramble to sell them.

Then, the Spatsizi controversy broke. The Greenpeace-sponsored calendar month featured a photo of Stone Sheep in Spatsizi Park with the big headline *Spatsizi - Head Hunters Not Wanted*. The write-up, briefly mentioning the expansion of Spatsizi Provincial Park, focused mainly on opposing trophy hunting (Greenpeace's real campaign). This prompted quite a few strong letters of protest including a very articulate one from Dick Ferris, one of the founders of the Alberta Wilderness Association.

I wrote back explaining that each group's page was theirs and we did not censor them. I also explained that while WCWC was not against all hunting, for reasons of keeping the gene pool strong, we were against trophy hunting. It was a very emotional subject and I'm sure this limited the sale of our calendar in Alberta that year. Greenpeace came to our defence not only using biological arguments (which are very convincing), but also pointing out the fact that the Spatsizi Protected Area's mandate in law specifically said that natural processes were to go on there unaltered by man. Despite the objection to the Spatsizi headhunting issue, the AWA continued to help out with our calendars in the years to come.

Our 1982 calendar was a much bigger success than the first one, despite the controversy and late sales. In our report to WCWC's board, Richard and I were proud to announce that WCWC had made a profit of $2,228. This was after settling all the bills and paying ourselves $1,840, the 17-cents-per-calendar-sold royalty (10,888 calendars sold of the 15,000 produced), which Richard and I split between us. The norm was established: calendar season would always involve a lot of work; and no-one would get rich on calendar sales.

Gaining House of Commons support for creating a South Moresby National Park Reserve

Our main project in the early years of WCWC was pushing the South Moresby issue onto the national stage. The failed IPS court cases made us realize that the door was shut in Victoria to protecting South Moresby. The corporate logging lobby was simply too powerful in B.C. We needed to shift our efforts into making the protection of South Moresby a national issue. All of us active on the campaign reached this conclusion at the same time. The area certainly had the qualifications for such a designation, and the nation could afford it. Early in 1981 Jim Fulton, the Member of Parliament for Skeena and the Queen Charlotte Islands who was extremely supportive of protecting South Moresby, got his colleague Ian Waddell, MP for Vancouver-Kingsway, to introduce a Private Member's Bill in the House of Commons (Bill C-454) that would facilitate the establishment of a South Moresby Wilderness National Park. Fulton didn't forward it himself because it would have alienated his supporters who worked in the logging industry. Getting this Bill introduced was a major breakthrough in the campaign, and it presented a tremendous opportunity.

I decided that WCWC should test for "agreement in principle" for this Bill by individually polling every MP. I spent a month in the fall of 1981 doing this. I put together an information package with the IPS information sheet and a copy of the Bill. I sent each MP a handwritten letter asking them to fill out an enclosed questionnaire (a small strip of paper with their name on it) regarding Bill C-545. Did they support it in principle, not support it, or were they undecided? I provided a return envelope hand-addressed to WCWC. I put stamps on the envelopes even though I knew the MPs had free mailing privileges, hoping that this expression of naiveté and earnestness would increase the returns. It worked.

I learned years later about how important personal touch is. On my visits to Ottawa, I saw cartloads of cards and form letters being delivered to MPs' offices each day. In talking with their executive assistants, I learned how they screened the mail and how little actually got through to their MP.

While the presentation of this South Moresby polling package was perhaps amateurish, the information about the outstanding attributes of South Moresby was professional, accurate and compelling. The whole effort took over a couple of weeks of full-time work. I put what I had finished each day in the mail with a hope and a prayer. Three weeks later the first polling envelopes began to arrive, and slowly over the next couple of months they continued to trickle in. Many Liberal backbenchers expressed support, as did all the NDP caucus. Then one day an official-looking letter arrived. It was from Joe Clark, Leader of the Opposition, pledging the support of the Conservative Party. That was it. Adding them all up we had the support of a solid majority of MPs in the House of Commons for a South Moresby National Park Reserve! This fact I repeated many times over in the following years, thinking that actual protection was just around the corner. Only gradually did I realize just how far away it actually was and how much greater an effort it would take to get there.

David Suzuki's *Windy Bay Show*

Things really took off for WCWC in 1982. Campaigns were heating up. On January 27, David Suzuki's *Nature of Things* TV series aired a special "Windy Bay Show" focusing on the incredible biological attributes of the South Moresby area and the logging threats to Windy Bay on Lyell Island. Suzuki's show sparked much controversy. Most of the controversy centred on the film footage of some black bears taken in another part of the South Moresby Wilderness. If you were not paying attention you might think that the bears were found in Windy Bay. But, in reality, there were no bears living in Windy Bay or on Lyell Island at that time. The logging lobby protested vigorously saying that the video falsely represented the story.

To test out their contention, I obtained a copy of Suzuki's program and showed it to the students in the Quest outdoors program at Prince of Wales Senior Secondary School. After they watched the show, I asked them if there were bears in Windy Bay. All except two raised their hand to say "yes," even though the show clearly stated *"in another nearby watershed"* when the footage of the bears was shown. To me it made no difference. I always thought that the bear-less-ness of Lyell Island was a temporary phenomenon, anyway. Perhaps the loggers working on Lyell in the early 1900s who had cut all the island's major valley bottoms except Windy Bay had shot them. It was a very short distance from Moresby Island across Darwin Narrows to Lyell: no barrier for the "world's largest black bears" to swim and re-populate this island.

Suzuki's Windy Bay show pushed South Moresby into the national consciousness overnight. Guujaaw's interview featured in the show was most remarkable. When asked by David Suzuki what would happen to the Haida if South Moresby were logged he replied, *"Well, I guess we'll just become like everyone else."* Guujaaw opened people's eyes to the inextricable connectedness of traditional Haida culture with the natural ecosystem in which it developed. Their cultural survival depended upon the survival of wilderness.

Aiming for the big time – Richard becomes WCWC's "Interim Office Director"

Both Richard and I had big plans for 1982. He wanted to put on an art show and auction, raise money to buy a computer and open a WCWC office. I planned to produce and publish more posters. We had grown. Now we had 37 paid-up members!

Despite the small membership, running WCWC had grown into a full-time job. In June of 1982, we made a board decision to hire Richard as the "Interim Office Director" (executive director) with a salary of $400 per month. Where were we going to get that kind of money? I had no idea. We opened our first office in Victoria, a windowless closet on View and Vancouver, but moved shortly thereafter (June 1982) to a much better location at 620 View Street, a grand old heritage building where many other non-profit groups had their offices. Rent was only $200 per month. The office came in handy; I slept there many nights on the couch. In July Richard got a loan from B.C.'s truly alternative credit union—Community Congress for Economic Change (CCEC)—in Vancouver for $3,000 to print 10,000 Valhalla posters and $4,000 to purchase an Osborne computer, a 64K portable machine.

I was getting busier with my contract work; increasingly more of the WCWC workload fell onto Richard's shoulders. My attitude was that, since Richard got paid, I wouldn't have to do as much. But I still was deeply involved in campaigns and hands-on in projects like the calendar.

Richard and I decided to produce a bigger and better 1983 calendar. We enlarged it to the size of the Sierra Club's calendar and switched from a stapled to a coiled binding. We increased the price from $5 to $7, which covered the increased costs and also gave us a theoretically bigger profit margin. We dropped production back to 10,000 copies, more accurately reflecting our market capacity. After some argument we also changed to regular square boxes for the calendar pad, with large numbers to make the calendar more utilitarian. But our big new venture was to challenge another of the Sierra Club's products with our own version: a small engagement calendar date book that people could carry around with them. The datebook nearly proved to be our undoing.

Federal government begins to fund environmental groups – a mixed blessing

As we expanded, we began tapping into a new source of funding: Canadian tax dollars funneled to environmental groups through Environment Canada. Despite self-righteously warning other environmental groups that relying on government money to fund the environmental movement wasn't a good idea, we held out our hand for the easy money too. In 1981 the federal government provided the funds to set up the B.C. Environmental Network (BCEN) as a loose umbrella for environmental non-government organizations (ENGOs) to collaborate and share information with each other. It was also a way for the federal environment minister to more easily consult directly with the myriad of environmental groups. Later on, the B.C. government provided funding for the BCEN too.

WCWC did not play much of a part in the BCEN except to attend the annual province-wide meeting. We were too busy with our own campaigns. It almost appeared in the early '80s as if the federal government, by handing out goodies, was consciously trying to soften the stance of outspoken environmental groups. At the very least, such easy money diverted many ENGOs' efforts away from the grassroots public outreach needed to build an ever-larger base of individual members and donors. Federal government handouts included money for delegations of environmental groups to attend annual networking conferences in B.C., and money for representatives to go to Ottawa once a year to meet with the Environment Minister for an hour or two.

Being perpetually poverty-stricken, environmentalists always find any source of money alluring. So, too, is the "carrot" of a shortcut to political influence via easy access to high levels of government behind closed doors. I strongly believe that being on the government dole blunts the chances of success and the only political influence that will last over time lies in public support. But that didn't deter us from using the government photocopy machines in the offices of friendly elected officials or other offices when we got a chance to make many copies 'courtesy of the government.'

Environment Canada funds our first colour poster
Canadian Wilderness – Environments Worth Protecting

Word filtered down that Environment Canada was looking for some good environmental projects to fund. Richard and I decided that this rumour was worth investigating. We got an appointment to meet Paul Mitchell in Environment Canada's communications office in Vancouver and…by golly it was true! Of course, there were strings attached. Environment Canada wasn't going to fund anything that was critical of governments or could somehow turn out to be embarrassing in any way. Mitchell could not, for example, fund a poster promoting the creation of a South Moresby National Park. In the course of the conversation, it became obvious that he could make the decision himself, on the spot, if we could come up with something suitable. If the project were to be the publication of a poster, it would have to be a generic one. They were looking for something along

Wilderness Committee's first colour poster featuring Burnaby Narrows within the South Moresby Wilderness Area. Poster image: Richard Krieger.

the wavelength of "It's Respectful to Recycle" or "National Parks are National Treasures" or some other such trite, motherhood slogan.

Richard and I left and immediately began to brainstorm. We had already picked the image for the poster: a spectacular panoramic view of Burnaby Narrows at low tide, rich with sea life including colourful leather starfish in the foreground and the ancient rainforest of Burnaby Island in the background. We had used it in our first calendar to epitomize South Moresby. It was one of many winning images Richard had captured with his huge, expensive fisheye lens that we had lugged around South Moresby in the summer of '77. All we needed was a catchy statement: a poster slogan with a punchy message, not a sappy one. But at the same time, it had to be acceptable to government; it couldn't offend anyone.

Would "Canada Needs More National Parks" do? Probably not, we both agreed. Although true, it would undoubtedly be too controversial. After bantering around for quite a while with lots of jokes like

"Starfish unite to save Starship Earth" (there were dozens of colourful leather stars in the shallow waters of the photo's foreground) we hit upon one that we knew ought to work. It was the product of intellectual "jamming," a way of working out things together in a freewheeling humorous way. Over the years this method was especially effective in germinating ideas for writing projects as well as campaigns. In the end it never was clear exactly whose idea it was in the first place.

In the blue sky above the starfish-encrusted inter-tidal zone, emblazoned in big letters, we proposed to print *"Canadian Wilderness – Environments Worth Protecting."* Richard and I went back to see Paul Mitchell and proposed this title to him, showing him the image at the same time. He immediately gave his approval, telling us to go ahead with the layout but to be sure to bring in the proof for his final approval before going to press—or else Environment Canada wouldn't pay for it. "No problem," Richard and I replied simultaneously. He also okayed printing 10,000 copies—as long as the entire bill was under $4,000. It was our job to arrange the printing and afterwards submit the bill to him for payment. It was so easy spending taxpayers' money!

In a few days it was ready to go. Besides the slogan on top, on the bottom we put *"Burnaby Narrows, Queen Charlotte Islands, Photo: Richard Krieger"* along with *"published by Western Canada Wilderness Committee"* and our address. We took the blueline (the full-sized proof of the film to be used to make the press plates) to Paul Mitchell. He gave us final approval.

Richard and I immediately headed over to press with it. On the way, I said to Richard, "You know, the description of the location is incomplete. We should add the words '*in the Proposed South Moresby Wilderness Area*'."

"Yes, that makes sense," agreed Richard, who added, "Shouldn't we take it back to Environment Canada for approval?"

"No," I said, "I don't think it's a substantive change. We're just more accurately describing where the photo was taken. That's got to be OK."

In a couple of days the poster was printed. Richard and I took a couple of packages—100 in each—to the Environment Canada office and departed quickly without seeing Paul Mitchell. We heard indirectly a short time later that Environment Canada was not pleased with the last-minute change, but the addition was not so outrageous that they could justify not paying for the posters. No doubt, someone realized it was better to just ignore it than risk the bad publicity we might stir up if they formally complained.

Posters become a powerful tool in WCWC campaign kit

This poster was a big hit. It was another "weapon" we had on our side in the "fight" to save South Moresby, as we so politically incorrectly described our campaign literature and campaign efforts in those days. Our first priority was to roll up, put in mailing tubes and mail 280 posters with personal cover letters to every MP in Ottawa asking again for their support, in principle, for the cause. I was in the process of doing this when Adriane Carr came by to interview me about South Moresby for an academic paper she was writing. I kept rolling posters as I talked for several hours with her.

Just before the post office was about to close, I told her I had to go and mail them. She offered to help and, as we dropped off the last of the tubes, on impulse, I asked her out for supper that night. This opened up a whole new chapter in my life and the life of WCWC. Within a few weeks I was living with Adriane.

After this South Moresby poster blitz, I became convinced of the power of this tool to save wilderness. But like most tools, to be effective, a poster has to be a good one. For years it cost nothing to mail anything to an MP. Later the rules changed and only flat first-class envelopes qualified for free postage. By then we had figured out a way to get around that rule by taking flats of posters to Ottawa, rolling and addressing them there in a sympathetic MP's office, and taking them directly down into the parliamentary mail room. The money saved in postage and cardboard tubes more than covered the cost of a discount round-trip plane ticket.

When I first went to Ottawa in 1986, I noticed that many MPs had our wilderness posters beautifully framed and hanging on their office walls. Curious as to why they were so popular, I asked a secretary and found out that we were one of the very few groups that sent MPs posters. The main reason, however, was that all MPs had free framing services: all they had to do was to send the poster out to the government framing shop. Many years later this perk for MPs also disappeared. But, by then, our posters were already up on many office walls on the "Hill."

Chapter 5

Forest industry and IWA fight against new parks

Big WCWC fundraiser and big push for the Valhalla Park

IWA vs. WCWC? – Recognizing a fundamental stumbling block thwarting new park creation

In the early days of WCWC we spent many hours preparing briefs—before we realized how little effect such efforts had—to present to commissions and committees. There were two such briefs that I remember distinctly. One was to the NDP Resources Committee that held a public meeting on March 20, 1982 in Victoria on *The Future of the Forest Industry in B.C.* The NDP was the official opposition at the time. We were working hard to get the provincial NDP to commit to creating a number of new parks if they got elected in the next general election. The election had to be called, at the latest, in the spring of 1983. The other was a brief I presented to the Pearse Commission on Fisheries. Both briefs highlighted problems that existed then and still exist today. As far as I could tell, both presentations fell on deaf ears and made no difference at all to future events. Here is the text of the one I presented on behalf of WCWC to the NDP's forestry forum, in its entirety:

> *During the past few years the NDP has gradually been drawn into an unfavourable and untenable position. The growing environmental movement has sought and received support for the creation of Wilderness Areas such as South Moresby and the Valhallas, the expansion of the Ecological Reserve System and the adoption of environmental policies that mean the withdrawal of sensitive land, the curtailment of certain projects.*
>
> *The labour movement, the traditional backbone of the NDP, has by and large opposed the proposed Wilderness and Ecological Reserves because they reduce the land base which provides their bread and butter.*
>
> *The conflict between established labour and the upstart environmentalists' aspirations is eroding the credibility of the NDP and perhaps sapping its strength. Other political viewpoints capitalize on the covert conflict. The more rabid environmentalists, suspicious of the NDP's commitment to labour's policies, and noting that the IWA's policy is to log the existing parks, look towards a Green Party or Rhinoceros Party cop-out. The more redneck unionists, suspicious of the environmentalist leaning of the NDP, and noting that the NDP has passed resolutions to preserve Windy Bay and the Tahsish, see their fortunes in rapid resource development, and become voting-booth Socreds.*
>
> *The negative spin-off from this rift is the unfruitful focus of debate on a false issue: jobs vs. the environment. Meanwhile in the forestry sector, the cures, more-than-token intensive silvicultural projects and systems to eliminate wasteful practices, never get implemented.*
>
> *The proposed Wilderness Areas and Ecological Reserves presently on the drawing boards encompass less than 2 percent of B.C.'s productive forest land base. Few doubt that their establishment would diversify B.C.'s economy and protect B.C.'s ecological heritage. Many believe we can't afford them. However, there is no doubt we could afford them if:*
>
> *1) the "good site" forest land, which ranks with the best tree-growing land in the world's temperate zone, which has already been cut over, were put back into production. Now over half of these lands are growing "weed trees" or brush.*
>
> *2) The needless wasteful practice of free-falling trees, especially cedar on the coastal sidehills was replaced by cable-assisted uphill falling. This <u>alone</u> would produce more than enough wood to counterbalance all the environmental withdrawals. Yet for several years a labour-management wage dispute over the pay rate for the additional workers needed to save these trees (which are needlessly smashed to kindling at the rate of thousands per week) has prevented the implementation of this proven waste-saving falling method.*
>
> <u>Recommendation 1</u> - *that the NDP invite the environmentalists and unions to a bargaining table to discuss the wilderness and environmental issues,*
>
> <u>Recommendation 2</u> - *that the NDP seek to redirect the debate away from superficial jobs vs. wilderness and the environment towards the underlying issue of job-producing waste abatement and intensive silviculture vs. short-term corporate and government power and profit.*
>
> <u>Recommendation 3</u> - *that the NDP replace the worn-out concepts of "maximum wood fibre production" and "multiple use" (which are merely slogans meaning, "log everywhere") with the concept of "optimum sustained yield" of all values which flow from the forest. Optimum sustained yield means that wildlife, wood products, fish and wilderness recreation will exist and be maintained without one predominating at the expense of the other.*

Colleen McCrory fights to use an inferior image on our WCWC-sponsored Valhalla "Shangri-La" poster

My next priority project after publishing the South Moresby *Canadian Wilderness - Environments Worth Protecting* poster was a full-colour Valhalla Wilderness poster. Posters are relatively easy to do compared to putting out a calendar and datebook. Right from the beginning of WCWC, I had been hitchhiking or taking the bus regularly to New Denver to see Colleen McCrory and the other people active in the campaign to save the Valhalla Wilderness. Colleen and I had met under unusual circumstances two years earlier.

One day in 1980, a call came in from Mike Halleran, a person I only knew from seeing him on his regular show on CBC. He had just aired a piece about the Valhalla Wilderness proposal (I didn't see it). He asked me to get in touch with Colleen McCrory, "a woman with a cause" from New Denver, on the chance that I could help her out with her Valhalla Wilderness campaign. According to Halleran, Colleen was very emotional and didn't understand the reality of politics in B.C. She had broken down crying the last time he talked to her. I agreed to see her. I phoned Colleen and she told me she would come down to Vancouver to meet me the following week.

I vividly remember the day I first met her. There, standing at the door of my messy apartment, was a woman all dressed up in fancy clothes looking like she was going to a high-paying job interview. She didn't look at all like the outdoorswoman Halleran had described. Colleen relayed to me a completely different story about the situation that had caused her to burst out in tears. Halleran had done a number on her. His program on the Valhallas was supposed to focus on the tourism potential of the proposed park and the widespread local support it had, including town councils and mayors from surrounding towns. Instead, Halleran used his TV show to twist the Valhalla

Poster promoting park protection for the Valhalla Mountains Wilderness area. Poster image: Dewitt Jones.

issue into a typical B.C. battle between loggers and treehuggers. Although, like many of his fellow rod-and-gun-club types, Halleran had earned credibility as an environmentalist for opposing dams, he was not sympathetic towards establishing more parks. I told Colleen about what we were doing to try to save South Moresby, and suggested that maybe some of the same tactics would work for the Valhallas. At the end of our conversation she said, "You'll have to come up and see the Valhallas. They're beautiful."

I couldn't talk too long with her because I had to catch the bus to Port Alberni to attend a Nuu-chah-nulth Tribal Council meeting, but I generously offered to let her stay and use our apartment. My girlfriend at the time was away on a trip. "Help yourself to any of the food you find and make yourself at home," I said as I rushed off, leaving her a spare key.

Years later I found out that Colleen was as shocked about how I looked, as I was at how she looked. She had made a special trip to Vancouver—a 12-hour bus ride—just to see me and had taken special care to dress up to make a good impression, having been told by Halleran that I was a powerful person that could really help in her cause. And here I was, as usual, all scruffy and shabbily-dressed. Her immediate reaction was disappointment, thinking that she had wasted her time and wondering how I could possibly be of any help. She also had no money. When she looked in all our cupboards, they were absolutely bare. All she found was some popcorn and tea bags.

However, being oblivious to her opinion of me, in less than a month's time I took up her offer to host me on a visit to the Valhallas. I was curious as to how their group worked, and knew for sure that I could get some good ideas that would be applicable to the South Moresby campaign. From then on I regularly visited the Valhalla Wilderness Society. In certain ways their campaign was miles ahead of the South Moresby campaign. They had much greater local support, for example. One fundamental difference, however, was that—unlike Haida Gwaii—there were no First Nations groups living in the proposed park area or active in their campaign.

Although I always wanted to take a week and hike into the Valhallas, I must confess that I only got across the lake once for a day hike. Their Valhalla proposal seemed so reasonable. Compared to South Moresby, there was relatively little commercial timber involved; but still there was opposition from people like Corky Evans, a local NDP politician who eventually went on to become the area's MLA. During my first visit I attended a public meeting where Corky—in his homespun way—pompously said he preferred to see horse logging there rather than a park.

"What you need is a great poster," I kept insisting to Colleen. It seemed as if they'd done everything else possible to get the political will needed to save the place. Finally, in 1982, we were producing that poster together. Colleen was not easy to work with (and from her and some other people's points of view, I'm not either). We are both strong-willed. We quickly agreed on the poster slogan coined by Valhalla Wilderness Society director Richard Caniell—*Valhalla Wilderness – Canada's Shangri-La*. But choosing the right image for the poster was another matter! She wanted a monochromatic dull bluish photo taken from a local vantage point—Idaho Ridge—featuring the town of New Denver in the distant foreground with Slocan Lake and the proposed Valhalla Mountains Park in the background. I wanted another shot taken there at a different angle with gorgeous, colourful alpine wildflowers in the foreground and the Valhalla wilderness in the distant background. Colleen and I had a royal row. The town of New Denver wasn't visible in the image I wanted; it didn't show enough of the proposed Valhalla Park; it gave the impression that those flowers were in the proposed park. I argued back that the image I wanted was a lot more colourful. "Aren't there flowers just like that in the alpine meadows of the Valhallas?" I queried. "Yes, of course there are," she replied. Enough said! I insisted and prevailed. After all, we were the group paying for it.

Valhalla Wilderness – Canada's Shangri-La was the Committee's second major colour poster publication. On seeing the final product, Colleen had to admit that I was right. It was stunning—one of my all-time favourite posters. It became one of WCWC's all-time best sellers, too. The Valhalla Wilderness Society took half of the first press run of five thousand and promised to sell them quickly. Even retailing them at $1 each we'd still make a big profit and still have a lot left over to give away. In the future whenever I got into an argument with Colleen, I reminded her that I was right about the Valhalla poster and she should concede that I was no doubt right again.

Me in kayak with a recent clearcut on Lyell Island in the background. Photo: Adriane Carr.

Return to South Moresby with Adriane Carr in 1982

It had been five years since Richard and I had gone on our exploratory trips into South Moresby with IPC. WCWC was two years old and it still had fewer than 50 members. Membership was just $15 per year; but then it didn't include a calendar being automatically sent like the $35 annual membership fee does today. I was spending less time with Richard since I had fallen in love with Adriane Carr. My

Adriane in kayak in Juan Perez Sound. Photo: Paul George.

work as a consultant to several First Nations had also become more and more time consuming.

I rationalized my diminished involvement by telling myself that Richard, as the paid executive director, was managing fine. I felt no guilt when Adriane and I decided to spend a month kayaking in South Moresby during her college teaching break.

Meanwhile, Richard slaved away at WC^2, as he liked to call it, working especially hard to organize our big art auction planned for the fall. It was only years later when WCWC mounted the Stein and the Carmanah art auctions in '88 and '89 that I realized just how much work was involved in successfully putting on such a big event.

The big blowout with Frank Beban, owner of the contract logging company clearcutting Lyell Island

Adriane and I arrived on Haida Gwaii at a pivotal moment in the South Moresby campaign. The South Moresby Planning Team (SMPT), established over three years earlier, was just wrapping up its work. It had developed four options, ranging from "business as usual" to near complete preservation. All that was left was the SMPT's last public input meetings, final write-up and delivery to government. We decided to delay our kayak trip to South Moresby to attend that final public wrap-up meeting in Sandspit.

Sandspit was the only permanent town on Moresby Island and had Haida Gwaii's only commercial airport. Frank Beban, the contractor logging on Lyell Island and one of the leaders of the anti-South Moresby Park lobby, owned Sandspit's new hotel and pub. Adriane and I arrived early to the meeting and tried to sit inconspicuously in the back of the community hall. I was surprised at how civil the people were while the SMPT members made the presentation. I guess everyone figured that at this late date it would be useless to try and change anyone's mind. The issue was completely and fully polarized. Neither Adriane nor I said a word. Someone in the audience tried to get me to speak, but I refused, saying we were just observers at this meeting meant to gather input from people living in Sandspit. I knew the final decision would eventually be decided in the bigger provincial and national arenas.

When it was over Adriane and I walked over to the hotel to have a beer in the pub. As we entered, Frank, who was a really big man (over six and a half feet tall and well over 300 pounds), saw me and called us over to his table. We sat down and I said, "Let me buy you a beer," purchasing a round for the table. From the talk going down, Frank and all the industry people sitting there were pretty confident that the South Moresby Park proposal was going nowhere. The B.C. government must have given them backroom assurances.

As the evening wore on, we kidded back and forth and had more to drink. Frank kept putting me down and tried to hit on Adriane, saying to her several times (referring to me), "What's a nice girl like you doing with a four-flusher like that?" (I only learned years later that a four-flusher meant a bowel movement so big that it took four flushes of the toilet to make it go down.) Even without knowing what it meant, it got pretty tiresome listening to him. Frank kept challenging me to arm wrestle. I knew it was a losing proposition so I kept refusing. Guujaaw was there. While Frank kept putting me down he kept putting Guujaaw up. Near the end of the evening Guujaaw took up Frank's challenged to an arm wrestle. Guujaaw put him down and won. Shortly after that it was last call and no one at our table ordered anything. We had all had enough. Adriane and I were just about to leave when something snapped inside of me. I said in a loud voice, "Hey, Frank, I always thought you were a generous guy. I bought you a drink tonight and you never even bought me one."

A healthy Sitka spruce with a leader that grew more than a metre in one year in the big Talunkwan "greening-up" clearcut. Photo: Adriane Carr.

It was a royal insult to a guy like Frank. He jumped up from the table, livid with anger. "A round for the house," he shouted. Even though it was well past closing time, the waitresses brought around drinks for everyone in the still-packed pub. I quickly downed my beer and Adriane gave hers away. We left before anything ugly occurred.

I never talked to Frank again. He died of a massive heart attack as he was taking down his Lyell Island camp in 1987 after the area was made a park. The fact was, he was very generous with his workers and the community and this was one of the reasons it was so difficult to campaign for park protection in Sandspit where he had his headquarters. Some of the rednecks said he died of a broken heart because of losing the South Moresby battle. But other people said he had a weak heart and died from drinking too much whiskey, eating too much ice cream and blowing up too many times in anger.

Adriane and I slept that night in our sleeping bags on the beach by the airport runway with a million sand fleas for company. Several days later we departed from Moresby Camp in two old borrowed Frontiersman kayaks, fulfilling a vow I made when Richard and I were speeding around South Moresby in our outboard inflatable: to go back someday under paddle power and explore South Moresby again. It was a magical trip, 27 days of paddling and hiking, cementing both our relationship with each other and our mutual resolve to work for as long as it took to protect this magnificent place.

Confronting "brainwashed" students at the UBC School of Forestry

Shortly after coming back from our kayak trip, I got a call from Al Chambers, the professor at the UBC School of Forestry who had looked at my TFL 24 overcut study. I hadn't spoken to him since then. Hearing from him brought back a vivid flood of memories of the TFL 24 court cases and the frustration I had with not being able to get his assistance.

He asked if I would speak to his fourth year class (called *Issues in Resource Management*) about WCWC's views on forestry in B.C. I asked him what it entailed. In short, his class had already heard from forest company representatives, and he needed me to provide balance. I'd have the undivided attention of his class for a whole hour. I jumped at the chance.

Despite not coming through for me in the TFL 24 court case, Chambers was, in my mind, one of the few "good" professors at the UBC Forestry School. I considered most others to be nothing more than corporate pawns and apologists. I know it's a harsh judgment; but B.C.'s big forest companies had, from 1951, the year this forestry school was founded, set the agenda for the "professional training" (I call it "indoctrination") that goes on there.

As the maxim goes, "those that pay get the say." The big forestry companies pay the shot at the UBC Forestry School, providing much funding, including graduate student research grants. They also are the "employers in waiting," hiring most of the students who graduate. Even those hired by the government's Forest Service can't escape the corporate influence. Then and now, the B.C. Forest Service functioned with paltry little independence from the corporate world of forestry. Mostly, the role of the Forest Service was to permit and legitimize the liquidation of B.C.'s old-growth forests and make it as easy and as profitable as possible for the "majors" to operate. In short, the school taught how to industrially clearcut in the cheapest possible way. Forest engineering was its engine and silviculture the caboose—a 1970s add-on course.

Because of growing criticism from environmentalists, the UBC Forestry School had recently added a new course in public relations. This new course that I was invited to attend as a guest speaker mostly focused on how to argue effectively with detractors of B.C.'s forest practices, to explain forest practices to a more skeptical public and to use "public involvement" processes to defuse opposition. To be fair, in the milieu that existed then and continues today, it was exceedingly difficult—make that impossible—for any professor teaching at this school of forestry to buck the system. One could keep quiet and unquestioningly uphold the industrial forestry paradigm, meekly and mildly criticizing it on occasion, or hit the road. Chambers took the middle approach.

With my bad logging slideshow and lots of South Moresby information sheets and Windy Bay petitions in hand, I arrived only a minute late. As I walked to the podium, I handed small piles of South Moresby information sheets to the students on the aisle seats and asked them to pass them down the rows. Noticing that quite a few of the students handed the pile along without bothering to take a copy, my temper rose.

I stood before the class and waited until the students got absolutely quiet. It took almost a minute. Having taught for four years in a senior secondary school in B.C., I saw a familiar "we have no respect for you" look in many of the students' faces. I wasn't getting paid for this. I had nothing to lose. When they finally settled down and one could hear a pin drop, I blurted out, "I must be honest with you before I start. I've never met a forester that I've liked." Half the class appeared to be stunned, the other half groaned. I continued, "They're all a bunch of corporate sell-outs—and that's likely how you'll end up, too." Then I proceeded to tell them how their schooling was just brainwashing, and that they would inevitably become shills for the big forestry companies that put profit before what was good for the forest.

I was on a rant that quickly antagonized and polarized the students. After getting everyone riled up, I presented my slideshow that documented the destruction caused by clearcut logging, especially focusing on the landslides caused by logging and logging roads. I punched holes in the rationalizations that foresters gave to justify bad logging practices. I showed slides of slash burning and explained why it, like clearcut logging, was not an environmentally sound silvicultural practice in the temperate rainforest along the B.C. coast. I especially expounded upon the reasons why a place like South Moresby—including its heart, tiny Windy Bay—should be set aside and protected.

Near the end of my lecture I passed out our Windy Bay petition and asked students to sign it. Noticing that most students quickly passed it along as if it were a blanket encrusted with smallpox scabs, I said, "I bet you guys won't sign it because you think this type of logging is OK. You're trained now." It was obvious to me that most of them already knew the score. If they signed the petition, they might be blackballed by the industry that they were hoping would hire them. "Foresters aren't just part of the problem, they are the problem," I concluded. I know I made few friends that day. But at least I pushed everyone into recognizing where they stood on the issue. After the class quite a few students came up to talk to me. They were so eager and innocent, and so sure that they would never be like the foresters I talked about. I hoped they were right. But the odds were hopelessly stacked against them.

"Deceitful" wording in our Windy Bay petition

During this time I was constantly pushing WCWC's Windy Bay Petition. As petitions go, it was very controversial and resulted in my first experience with a defamation lawsuit. The *Vancouver Province* and the *Queen Charlotte Observer* published a letter to the editor by Doug McLeod, a Rayonier forester, saying that I was deceitful and dishonest in the wording of the petition, which stated:

> *Whereas the Ecological Reserve Unit of the B.C. government has recommended that the Windy Bay Watershed–Dodge Point Murrelet Colony comprising approximately 12 square miles of Lyell Island in the South Moresby Region of the Queen Charlotte Islands be preserved;*
>
> *Whereas this area contains one of the last remaining significant stands of virgin cedar spruce and hemlock of magnificent size, an outstanding salmon stream which is more productive than any other for its size on the B.C. coast, and the largest Ancient Murrelet (forest-floor-nesting seabird) Colony in B.C.;*
>
> *Be It resolved that we, the undersigned, support the B.C. government in the establishment of the Windy Bay-Dodge Point Ecological Reserve to conserve this special area for future generations to observe and study.*

McLeod contended that in order to be a "government" proposal it had to come from the elected government, specifically the cabinet. He believed that this ecological reserve was my private initiative, in spite of the fact that the Ecological Reserves Unit had adopted it and it was now under study by their department. I claimed that the Ecological Reserves Unit was part of government; therefore, it was a "government" proposal.

I convinced lawyer Jack Woodward to act for me *pro bono*, charging me only costs. We did an examination for discovery, which cost me a lot of money for a court reporter to record it and then type out a transcript. McLeod had a high-priced lawyer, John McAlpine. It became more expensive because McLeod launched a countersuit against me for calling him "unprofessional" in a letter to the *Queen Charlotte Observer* weekly. The case dragged on for a year. Jack no longer had time for it, so I took it over. I'd seen the TV lawyers in action! Foolishly, I dreamt it would bring big media attention to South Moresby.

After a well-respected justice told me, "All you will end up doing is spinning your wheels in court," I came to my senses and settled out-of-court in December 1983, the morning the trial was to begin. Both of us apologized publicly. I drafted the letter McLeod published in the two papers where he originally slammed the Windy Bay proposal and me. (Of course, the letter contained a subtle plug for the merits of protecting Windy Bay as an ecological reserve.) I wrote my own apology to McLeod for calling him unethical and it was published as a letter to the editor in the QCI Observer. **I learned a big lesson by this confrontation: that you have to keep your eye on the ball, focus on your ultimate goal and avoid being diverted into activities that waste time and do little to help further your cause, in this case saving the South Moresby wilderness.**

Our first big art auction is a success

When the evening of WCWC's big art auction rolled around in October '82, I arrived to find almost everything completely planned. We had to rush around and handle only a few last-minute details. Several hundred people were there at the Royal Museum's Newcombe Theatre in Victoria. I noticed that most were intently focused on reading our fancy printed program of the items up for auction—a good sign. Most were dressed in suits and formal dresses—an even better sign. It was the crowd we wanted. There was even a Japanese film crew present taping the whole thing for a show back home. Weeks earlier Richard had mailed a note card announcing the show to lists of art buyers supplied to him by several galleries. On the card was an image of a beautiful original harlequin duck painting that Robert Bateman generously donated as the silent auction centrepiece for our event.

Just a few moments before the live auction was about to begin, Haida artist Bill Reid, who was a passionate supporter of the South

Harlequin duck oil painting by Robert Bateman that we sold in our first auction and used as an image on a card to invite people to attend. Card: WCWC files.

Moresby cause, came over to us and donated a unique piece he had just completed. It was a small ivory pendant with a green scrimshawed frog on it. He told us it was his first scrimshaw (an engraving in ivory with pigment applied into the incised lines). Reid's work was stunningly beautiful, but it was not advertised in the auction program, and no one had seen it except us.

The auctioneer saved this frog until the end. Unbeknownst to me, Bill Reid had put a reserve bid of $900 on it. For some crazy reason the auctioneer started the bidding for the pendant at $100 dollars and raised it by only $25 at a time. Richard, who had been on stage all night holding up items for display, held the pendant up for everyone to see. But it was so small that, of course, no one could really see it. It took forever to creep up to $300. (Richard told me later he thought that the auctioneer's judgment may have been impaired by too many nips from a hidden bottle.) Finally the auctioneer started raising the bids by $100s. Bidding finally stalled out at $850: "Going once, going twice..." And then Richard raised his hand on the stage and said, "I'll bid $900." There was no other bidder and Richard got it for this ridiculously low price.

It was actually embarrassing how little this unique piece fetched. But at least, I thought, Richard had saved it so we could raffle it off later and realize its full value and make a lot more money for WCWC. That evening at the after-event party I complimented Richard for his quick thinking saving the Bill Reid piece for WCWC. "Oh no, I didn't get it for WC^2, I got it for myself!" he said. I argued as hard as I could that keeping it for himself wasn't the right thing for him to do. He argued back that he had done a lot for WCWC, putting in thousands of dollars of his own money and that if he hadn't bid on it, it would have gone back to Bill Reid and WC^2 would have gotten nothing out of it anyway.

I did not give up on this issue and I made a big thing of it at the next director's meeting. It seems petty now, but at the time I was really angry about it and it began a rift between Richard and I that continued to widen.

Despite the frog fight, this first big WCWC fundraiser was a resounding success. It raised $29,000 and netted over $15,000 with $11,500 of it coming from the sale of Robert Bateman's harlequin duck painting. All the artwork sold except one beauty, a smartly framed large limited edition print titled "On the Move—Bull Moose" by Robert A. Wyatt. It had an upset price of $500 that no one had met. We nicknamed it "Bruce the Moose." It hung around our office for several years until Colleen McCrory became a director. She said, "Let's present it to Tony Brummet (who was then B.C.'s Minister of Lands and Parks) at our meeting with him next week." Everyone agreed it was a good idea.

The downside to our auction was that it slowed our 1983 wall calendar and engagement date book production. The Osborne computer Richard bought for WCWC to manage our small database didn't speed things up either. It was useless. Dehn, a young computer whiz kid just out of senior secondary school, spent days writing a custom program to track our few members and calendar distributors. All of a sudden one evening when Richard was trying to input names—poof!—the program and database disappeared into the ether. There was no backup. We had the right idea but computer technology and Microsoft was not there yet.

Passing out the Valhalla posters to the delegates attending the Social Credit Party annual convention

One day that fall after our art auction, I happened to notice a small article in the newspaper about the Socred Annual Convention being held the next weekend in the Hyatt Regency Hotel. What luck! Our Valhalla Shangri-La poster had just come off the press a few days earlier. It was a great opportunity to hand out free posters and push the Valhalla issue. I tried to get other volunteers to go with me; but no one wanted to spend their Saturday doing this except Bruce Kahn, then a law student at UBC and a passionate Valhalla campaigner. He and his wife had lived in the Slocan Valley and were good friends with Colleen. But he could'help only for a short time in the morning. He had to get back home to study for a test.

I was determined to take advantage of this convention. This was B.C.'s governing party and they could create the Valhalla Park "with the stroke of a pen." An election call was expected soon, and I knew that the Socreds would be giving out the vote-buying "goodies" like all governments do during those times. The Valhalla Park could be one of them. The Socreds hadn't created any large parks in their term in government: here was their chance.

This was one of the rare times I put on a suit jacket and dress pants and even combed my hair. I wanted to look like a fellow Socred as best I could. Bruce, too, was in a suit. He looked more at home in it than I did in mine. The two of us got there early with several heavy boxes of posters. We placed them on one of the many coffee tables in the conference centre lobby and began rolling them up and putting elastics around them.

This was a quiet convention for a change. For some reason there were no placard-waving protesters outside and no sign of heavy-duty security. The delegates were just beginning to arrive. We had no letter accompanying the posters explaining why we were giving them away; all we had was the poster. We figured it said it all.

At first I stood at the bottom of the escalator leading up to the convention hall upstairs and asked, "Would you like a beautiful poster of the Valhallas?" Only about a quarter of the people would take one. Then I decided to go up the escalator to see what was happening upstairs, thinking that maybe it would be easier to give them away up there. Here was a big lobby with tables set up for registration. I positioned myself near the top of the escalator. One of the first people to come up to me was a little old lady.

"What do you have there?" she asked.

"Here, I'll show you," I replied as I unrolled one of the posters.

"Oh, how beautiful! I love flowers," she exclaimed.

"We have the finest alpine flower meadows in the world in B.C.," I asserted, explaining that we were trying to get this area made into a park. I told her I was here giving the posters away as a thank-you to the Social Credit government for putting a moratorium on logging in this area, and for initiating a study of it to see if it would make a good park.

"Do you know who I am?" she asked.

"No, I don't."

"May Richards Bennett." She was the Premier's mother! We exchanged small talk and I told her how much I admired her late husband, W.A.C. Bennett, who had launched the Socred Party and ruled B.C. during its glory days in the '50s and '60s. As she left me, I thanked her for her interest in the Valhalla Park proposal.

I handed out the last of my posters and went down to get some more. Midway through my handing out the second batch, two burly men dressed in plain blue suits came up to me and told me I could not protest inside the lobby of the hotel.

"I'm not protesting," I protested to them. "I'm handing out these educational posters as a thank-you to the Social Credit Party for placing a moratorium on development in this area and considering it for park status."

"Here, give me one. You stay right here and don't pass out any more," said one of them. "We'll have to check this out," said the other as they both left and went up the escalator. They were gone for what seemed like ages. Finally they came back and, to my surprise, they said that it was OK to pass them out on the ground floor. But if I caused any problem, they would have me removed. I thanked them and continued to pass out the posters for hours until the late afternoon, when the four hundred I'd brought were all gone. To everyone who stopped to talk, I used the "thank you for the moratorium" line—thinking all the while what a very good omen for the park it was that I wasn't kicked out. On the other hand, the poster itself was so beautiful, who could dislike it or object?

Passing out Valhalla Shangri-La posters and forwarding an emergency resolution on wilderness at the New Democratic Party annual convention

A few weeks later we planned to do the same thing at the New Democratic Party (NDP) annual convention: pass out the Valhalla poster to everyone attending. Their convention was being held in the Vancouver Hotel on the November 27-28 weekend. WCWC Director Murray Rankin wrote a letter on behalf of WCWC and obtained permission from the NDP convention committee for us to have an information table in the lobby there. If our wilderness calendars had been ready, it would have been a great place to sell them, but they weren't. A few days before the convention Adriane came up with the idea of forwarding a resolution at the convention to ensure that, if the NDP got elected, they would "park" the big wilderness areas we were campaigning to protect.

We were afraid that, even before the next election was called, the Socreds would reject the park proposals and decide to log these areas as a way to garner election support from the forest industry. Since it is customary for new governments to accept decisions of previous governments on land-use issues, we reasoned that we needed a resolution from the NDP convention to counter this possibility.

The resolution stated:

> WHEREAS the NDP is on record supporting the preservation of the natural heritage of the Valhallas, South Moresby, the Tahsish, and Meares Island; and,
>
> WHEREAS the present provincial government is rushing decisions, rejecting all recommendations to afford special protection for these areas and making a mockery of the "public involvement" process;
>
> THEREFORE BE IT RESOLVED that an NDP government will undo any decisions to industrialize these areas, institute comprehensive province-wide land-use planning with legitimate public involvement and pass legislation required to fully protect and preserve the Valhallas, South Moresby, the Tahsish, and Meares Island and other areas deemed to be of outstanding Provincial and Canadian significance in their natural state.

We knew it was too late to put this resolution through the normal process: the deadline had long since passed. The only way to proceed was to put it forward as an emergency resolution. There was a formal convention process to do this. The resolution would have to successfully pass through several committees to reach the floor, be debated, and be voted upon.

Adriane and I got to the hotel early that Saturday morning. I brought up lots of flats of Valhalla posters. We set up our table with information sheets (this was two years prior to our first tabloid newspaper publication) about the various wilderness proposals. I began rolling up individual Valhalla posters and putting an elastic band around each one of them. We planned to hand out a free poster to everyone, just as I had done at the Socred convention a couple of weeks earlier. This time, however, around each poster we rolled a letter from the Valhalla Wilderness Society to the NDP delegates. The letter read:

> *Dear N.D.P. Delegates and Supporters:*
>
> *This complimentary poster is in appreciation of the efforts that the N.D.P. has made, including the resolution of support at your last convention and the private member's bill by Lorne Nicholson, to establish the Valhalla Park in the Slocan Valley of B.C.*
>
> *Because the park proposal has overwhelming local support, the backing of the local villages and Regional District and a study that concludes that the park would produce more than ten times the permanent jobs than would logging in the area, it seems logical that the Social Credit government will create the Valhalla Park.*
>
> *In case they don't, we trust that the next government, an N.D.P. government, will act wisely and undo any harm that the Socreds might have done to the Valhallas. Thank you.*
>
> *Enjoy the poster as we enjoy the Valhallas.*

As usual, not many people stopped at the table to look at our materials. Our purpose in being there was to make good connection with the few who did stop to talk. This time I didn't have time to let the lack of interest discourage me. I had to continually roll the posters to build up a huge stack to give away.

Colleen McCrory was an NDP delegate from Nelson-Creston and had a seat on the floor. She introduced Adriane to one of the persons involved in the emergency resolution committee. He explained to Adriane how the process worked. The committee vetted and ranked the emergency resolutions in the order that they would be presented on the floor during the time slot devoted to emergency resolutions near the end of the convention late Sunday afternoon.

Adriane had to make a presentation to the committee in the morning and again answer questions at an afternoon meeting. There was only a little time to debate emergency resolutions; so it was important to get a position near the top of the list. She would know by the end of the day what position our resolution was assigned. Meanwhile, it was important to gain the support of all the delegates—because if the resolution got to the floor, they had to vote for it for it to pass.

Somehow, Colleen got Adriane—even though she was not a delegate—permission to go right onto the floor of the assembly before the formal meeting started and during breaks, and pass out the posters to the delegates sitting at constituency tables. Colleen and Adriane worked hard at it, coming back to the table often for more rolled-up posters. She handed out hundreds of posters, while tirelessly talking to delegates to gain their support. I rolled and gave out posters to delegates who wanted an extra one to give to a friend or relative when they got home. Just as at the Socred convention, the Valhalla posters were a big hit with the NDP.

Adriane found most of the delegates on the floor supported the resolution. Now, if she could just get it considered and ranked high by the emergency resolution committee, we would have a chance of seeing it put to the floor and passed. Adriane reported to me after the first meeting with the emergency resolutions committee that everything had gone well. Most everyone seemed very sympathetic. The afternoon meeting went well, too. At the end of the day the committee chair informed Adriane that our resolution was number seven on the list and had a good chance of making it onto the floor. At past conventions they had often made it that far down the list, he told her: "it all depended on how long people talked and how controversial the resolutions were." We went home that night feeling confident that we had a good chance of succeeding.

1983 calendar cover featuring Haida Tribal Park (Duu Gaust). Image: Thom Henley.

The NDP backroom betrayal

The next morning, Adriane and I got there early and talked excitedly at our table before the delegates had begun to arrive. Looking around, I noticed a large stack of papers on the main registration table. Adriane went over to take a look. It was the emergency resolution package for the delegates. She took one and brought it over and we looked for our resolution. The first one was against U.S. cruise missile testing over Canada. We kept going down the list looking for ours. It wasn't No. 7. Finally, we found it. It was near the bottom—No. 14—after a resolution supporting gay marriage rights. There was no chance that our resolution was going to reach the floor!

"How did that happen?" I asked Adriane, who was in shock. "I'll find out," she said. She stalked off to find the emergency resolution committee chairperson. She came back about an hour later, very angry and upset. No one would tell her what had happened. Everyone had just shrugged his or her shoulders and said that things like that sometimes happened—and we should not take it personally.

"So much for the D in NDP," I quipped, totally unaware of the structure of the party that gave organized labour block votes and extra privileges. We were also unaware of just how powerful the IWA (International Woodworkers of America) union was within the inner power structure of the NDP. It was just hard for us to understand how such a change could be made.

Someone obviously didn't like our resolution and was able to overturn due democratic process to make sure it wasn't debated. Was this the attitude of the leadership? Could we count on them to create parks to protect these places if they got elected? The rest of the day unfolded as predicted. The emergency resolutions came up. They took an inordinately long time on the anti-Cruise Missile testing resolution even though it was obvious everyone was going to vote for it. They got through resolution No. 7. So if ours had held this position on the list it most likely would have been debated. But they never got close to No. 14. At the end of the session, someone came out and said to Adriane. "Oh, don't worry. We took up the environment at our previous convention in Kelowna, and I'm sure the environment will come up again at another convention soon."

We stayed around to the bitter end of the day. Just as we were packing up, Graham Lee, MLA for the Queen Charlotte Islands and North Coast, dropped by and stopped to chat. We filled his ear. He replied there were several other important issues we should address, such as opposing the proposed northeast coal development. The parks issue was just another one amongst many to be dealt with.

The birth of the B.C. Green Party

"Come on, let's go down and have a beer," Graham said. That sounded like a good idea. We took our stuff to the car and, on the way back to the "Lumberman's Lounge," Adriane picked up a newspaper in the lobby of the hotel. In it was a story about the Green Party in Germany winning seats for the first time in the West German parliament. We were all pretty curious as to what the Green Party was all about. As we all downed several beers, Adriane and I bantered back and forth with Graham. It became obvious that the sort of breach of democratic procedure we had experienced that day was commonplace in the NDP. You just had to accept it and get over it. I remember saying, "Well that's it for me and the NDP. We'll just have to start a Green Party here." And, as they say...the rest is history. On February 4, 1983, Adriane, Richard and I—along with 14 other people—signed the papers to found the B.C. Green Party, the first Green Party organized in North America.

Chapter 6

An unexpected wilderness win WCWC in first financial crisis

Having two 1983 calendars creates problems

On the home front WCWC was not faring too well. Our late submission of camera-ready copy to the press—and the new coil binding process—considerably delayed the delivery of our 1983 calendars. In fact, the hand-done coil binding was so slow that the finished calendars only came trickling in, with several hundred not delivered until after Christmas. Our new product, the small engagement calendar, caused even greater problems. We decided to do it on the cheap: so we designed it to have small photos of threatened wilderness areas so it could all be printed on one large thin sheet of paper.

When we finally got the mock-up proof of the engagement calendar we realized it was way too thin. "No one will pay $6 for that!" I exclaimed. Jim Astell, our salesperson at Mitchell Press, suggested that we could "bulk it up" with blank paper. Then we got the brilliant idea of putting in art paper that one could sketch or paint on and calling our product a "date-sketch book." Christmas almost arrived before we got our first date-sketch books to sell. Most of them arrived after Christmas. We moved most of the calendars in a very short time; however, the date-sketch book was another story. They were not just a hard sell: it was hard to even give them away!

I was yelling more than ever at Richard—you had to speak loudly in low tones normally so he could hear due to his deafness—blaming him for not getting our date-sketch book out on time and for causing the confusion with Mitchell Press. He blamed Mitchell Press and wanted to go to court against them. It took a few months to see exactly how well we did or didn't do, because of the lag time it took to get the consignment sales money in. When all the bills came in we found ourselves in serious debt.

Valhallas are saved
– the first big protected area WCWC helped win

It was late February 1983. Although I was still in the doldrums over disappointing calendar and date-sketch book sales, there was something to celebrate: the Socreds had decided to create the Valhalla Provincial Park. It was the first big wilderness park that WCWC played a role in winning.

The government media conference was held at the Georgia Hotel in downtown Vancouver. I arrived a few minutes late. A number of bureaucrats and Tony Brummet, Minister of Lands, Parks and Housing, were milling around nervously. There were charts and maps on easels and a large stack of press kits. A slide of the beautiful Valhalla Mountains was projected onto a wall screen at the front of the room. Bruce Kahn, who was there to represent the Valhalla Society, arrived a few moments after I did. We were the first guests to arrive...and the last. Not one single member of the media showed up! This was not as disastrous as it might seem, however, because the government had already leaked the story to the press the day before. The story of the new 60,000-hectare Valhalla Provincial Park, along with photos and a map, were plastered on the front pages of both major provincial newspapers. There couldn't have been better coverage.

Thankfully, the Socreds had decided to go for park supporters' votes on this issue, rather than forest industry votes. We had a the big hint that they were leaning our way a week before. The government's new tourism brochure gave it away. It featured our Valhalla Shangri-La poster image on its cover, the "poster child" photo that Colleen and I had fought so intensely over just a year before.

The empty media conference provided an opportunity to chat off-the-cuff with Minister Brummet. He was happy that no one except us showed up. He told us that if the mining or forest industry strongly opposed the creation of this park, they would have been there to make a stink. He took pains to impress upon us how hard it was for a government to decide to create a park because the ongoing costs of maintaining the park go on forever.

A couple of days after the announcement, I wrote this letter to the Valhalla Wilderness Society:

> What a victory! Congratulations for all the work and effort over the years to make the Valhallas truly the Shangri-La of the Slocan!.... In a few months the wounds in the community should be healed, the process should speed up as everyone scrambles for one of those 200+ jobs being created in the tourism sector.... You know I couldn't let this letter go without a thank-you to Colleen for telling me that the Valhalla Society will pay back their debt to us with continued poster sales and the Robert Bateman Print lottery. No kidding about the Wilderness Committee's financial dilemma. And now there are Chilko Lake Posters (sell the enclosed tube's worth if you can). Our Committee must move on and not gloat over wins.

Richard Krieger went to New Denver as the WCWC representative to the special May Day event to celebrate the creation of the park.

Cover of WC²'s ill-fated date-sketchbook. Cover image: Richard Krieger.

Although I was invited too, I was too busy working on the provincial election campaign trying to get Adriane elected as a Green Party MLA. It is one of the few decisions I regret. I would have loved to see B.C. Parks present a massive picnic table identical to those used in B.C.'s parks to Colleen McCrory. Indestructible, it still dominates her back yard in Silverton (sister city to New Denver) today. Mike Apsey also attended. He was the deputy Minister of Forests during the crucial period when the Forest Act of 1979 was pushed through the Legislature, was now the head of the large forest companies' powerful Council of Forest Industries (COFI), a cartel that collectively looks after the companies' well being. I heard through the grapevine that Apsey liked the gutsy McCrory, and that he played a major behind-the-scenes role in making this park happen.

At the time I had no inkling of how long it would be before the next wilderness "win." It would have been a great opportunity to gossip with the bureaucrats and find out what secrets lay behind achieving this one.

Richard Krieger quits as WCWC's executive director

In the third week of April, right in the middle of the 1983 election in which Adriane was running for a seat in the Legislature in Vancouver-Point Grey under the Green Party banner, I received a heavy-duty letter from Richard. I knew that we had lost money on calendar and date-sketch book sales; but I didn't realize how bad the situation was, or how out of touch I was with how burdened Richard felt until I read his letter:

> *It saddens me that after years of a good friendship between us, feelings have degenerated to a degree that makes it even difficult to talk to each other. You have decided to embark on a single-minded devotion to the Green Party at the exclusion of everything else. You are probably accusing me of being lazy or irresponsible. I'm simply burnt out and you are free to apply any distortionary epithets you wish. I do wish somebody would pick up the calendar–it just can't be me...*

I knew the "somebody" Richard referred to had to be me—or WCWC's 1983 calendar would be its last. There simply was no one else who would continue WCWC's work.

Accompanying Richard's letter were the minutes of the directors' meeting, held at the end of March, that I hadn't attended. At that meeting they had decided to terminate Richard's job as executive director as of the end of April. At the next directors' meeting they proposed to discuss two options: one to *"close office; sell computer; pay off debts; continue meetings on a shoestring budget,"* the other to *"close down completely."* They scheduled the next directors' meeting to be held in Tofino at the end of April. The election writ was dropped on April 7, making it impossible for me to attend that meeting. In Richard's letter, he said the meeting had been rescheduled for May 7-8 (still in Tofino) two days after the election, and he hoped I could attend.

I knew things were bad—but I didn't think they were so bad that Richard would quit. I called Richard and found out that he'd already given notice on the office. "No, I don't want any of the furniture, desks or file cabinets," I told him. "Just ship the files over to me. And there's no way I can make a meeting in Tofino so soon after the election," I explained. He told me that they would reschedule again for a later date and that he would get back to me. The tone of the conversation was frosty.

Richard tries to disband WCWC

It was only through sheer luck or serendipity that I found out about WCWC's next directors' meeting. In fact, it was scheduled to be WCWC's third AGM. I didn't expect an AGM until the end of June, a year from the date of the last one. At the end of May I happened to call Tom Schneider, another Islands Protection Society activist who had recently moved to Victoria from Queen Charlotte City, to talk about the South Moresby campaign.

Just as I was hanging up, he said, "We can talk more about it at the meeting tomorrow."

"What meeting?" I asked.

"WC²'s AGM at Murray's (director Murray Rankin's) house."

"Oh, of course. What time does it start again?" I asked.

He told me the time

The next morning I took the first ferry over to Victoria and got to the meeting an hour before anyone else did. I chatted with Linda and Murray as if nothing special was happening. When Richard walked in, I saw a surprised look flash across his face. He obviously hadn't expected me to be there. I was sure that he was planning to pass a resolution to dissolve WCWC because he was burned out, and use my non-attendance as evidence that I was no longer interested in

Cover of brochure produced for IPS—Richard Krieger's last contribution as a director of WCWC. Photo: unknown.

Valhalla Wilderness Society directors at the Valhalla Park celebration ceremonies in New Denver in May 1983—from left to right Valhalla Wilderness Society director Colleen McCrory, Honourable Tony Brummet, Minister of Environment and Parks, and Wayne McCrory and Craig Pettitt. Photo: Valhalla Wilderness Society files.

it. There was no reading of the last AGM minutes. The first piece of business was Richard's verbal report on the state of the finances. The good news was that we had nearly $8,000 in the bank and about $2,000 more in collectable calendar debts. We had received nearly $32,000 from the sales of calendars and our date-sketch books. The bad news was that, despite negotiating a several-thousand-dollar reduction on our calendar printing bill with Howard Mitchell, the owner of Mitchell Press, we still owed $18,000 on our bill.

Mr. Mitchell was willing to wait for payment through sales of the 1984 calendar if we came up with the money up front to print the calendar with his company. In his report, Richard made no other mention of any other debts.

I pleaded with the other directors to be given a chance to rescue WCWC and proceed with the 1984 calendar. I would find "new blood" for WCWC and I would assume all risks, take the calendar project on under contract and come to a deal with Mitchell Press. I had high hopes that the Outdoor Recreation Council (ORC) would help more than it had in previous years. To that end Murray agreed to go with me and talk to Ken Farquharson, one of the more conservative and well-respected leaders in the environment movement, then the head of ORC, to seek help.

Directors give me the chance to rescue WCWC

Without really any argument, they gave me the chance to save WCWC. There was a proviso: I would have to assume all of WCWC's outstanding debts and liabilities. Dan Bowditch agreed to purchase the Osborne computer by taking over the payments, thus relieving WCWC of an ongoing expense. Richard stayed on as one of four directors: Richard and I, Linda Hanna as secretary, and Dan Bowditch.

Richard finished up one last major project—a South Moresby brochure titled *South Moresby Wilderness - What Price Paradise*—that we promised to mass-produce and distribute for IPS. But after he completed that, he had little more to do with WCWC. In a letter dated April 12, 1983, he informed all the directors that he was going to step down as of the next AGM. WCWC had entered into a new phase; the full burden was transferred onto my shoulders; and I, in turn, unloaded a lot of it onto Adriane's.

In retrospect, there was another thing besides the problems with the 1983 calendars that split Richard and I apart: politics. At first, Richard was keen about the idea of starting a Green Party in B.C.—he even wanted to be on the party's council—but I said he couldn't be-

WCWC's third colour poster—*Chilko Lake...a park proposal*. Poster image: Peter Jordon.

cause he was only a landed immigrant and not a Canadian citizen. (Both of us are immigrants from the States. I became a Canadian in 1974; Richard didn't become one until 1989.) Although I kept the Green Party activities completely separate from working on WCWC campaigns, Richard began to complain about my involvement. It was evident that his many NDP friends had 'got' to him. From then on all he could talk about was the Green Party "splitting the vote," how important it was to get rid of the Socreds, and how bad an idea it was to start a new political party that could ensure the Socreds would win another term.

Even before the start of the Green Party, we had been arguing about nearly everything. One of the bigger fights was over the printing bill of several thousand dollars for a Chilko Lake poster that I had gone ahead with, even though I didn't have board approval to do so. Bolstered by the success of the Valhalla poster, I wanted a poster to help save the spectacular Chilko Lake. The poster featured the wild Chilcotin horses in front of Chilko Lake, with the beautiful Coast Mountain Range behind. I had the posters printed, hoping that the Federation of Mountain Clubs would buy half (they eventually only bought $100 worth).

I also counted on Environment Canada to buy them. Paul Mitchell, Regional Information Director for Environment Canada, promised to buy 250 when the April 1983 budget money became available. Later, though, he decided not to because I had used the phrase "Super Natural British Columbia"—the world-famous slogan used by the provincial government in its tourism advertising campaigns—without the B.C. government's permission. "There could be possible copyright infringements that we wouldn't want to become entangled in," he told me. Richard, making a big point of my reckless spending and disregard for process, had urged the other directors to censure me.

Mostly, I think the fights between myself and Richard were over just how much work it took him to run WCWC; also Richard's frustration by my spending more and more time on political activities with Adriane. Richard wasn't an office type of person. He did a tremendous amount of WCWC work for little thanks and little money. He also got tired of my thinking that he could always bankroll everything. He wanted to move on with his life, find a wife and start a family. When I got the letter from Richard, my life was in turmoil. I had just found out that my contract as a part-time researcher for the Nuu-chah-nulth Tribal Council that lapsed at the end of April was not going to be renewed. A couple of months earlier, Adriane had become pregnant. And Richard was right: since I had met and fallen in love with her and helped her launch the Green Party, I hadn't paid much attention to him or WCWC.

Thinking it all over, I realized I didn't want all the work we had put into WC^2 to go down the drain. I just couldn't be a failure in this endeavour.

Chapter 7

Assuming full responsibility for a nearly bankrupt WCWC

Adriane gives WCWC her full support…by supporting me while I volunteered full-time for WCWC

Right after the AGM where WCWC's future was put in my hands, Adriane and I had a heart to heart talk. She wanted me to stay home and not travel as much. "I make enough money teaching to support both of us," she said. "When the baby is born, you can take care of it. I don't want to leave our baby with anyone but you."

"You won't mind if I don't earn money and pay my fair share of the living expenses? You'll support me to do the Wilderness Committee work?" Adriane answered with an unqualified yes.

Her complete support gave me a tremendous feeling of relief. It meant that I could devote myself full-time to WCWC and saving wilderness (except, of course, for taking care of the baby). We had bonded in the wilderness and I knew she wanted to see South Moresby saved as much as I did.

As an added bonus, I knew she would help by editing WCWC's publications. Her grasp of English was far superior to mine. I vowed to make a success of the calendar this year. It would be a calendar of endangered wilderness areas—not an endangered calendar.

Where are the books?

Our Kitsilano home was in total disarray when the three boxes of WCWC files finally arrived from Richard. We were in the midst of completely renovating our house—lath and plaster removed but insulation and drywall not yet installed. It was also "station central" for the budding Green Party, with people coming and going all the time. There were files and papers everywhere. As I took Richard's boxes into our bedroom—the only room in the house free of clutter—I thought to myself, "Is this all there is to WCWC?" One of the boxes contained financial vouchers. One contained old mock-ups, paste-ups and research relating to the calendars. Another had memberships information and files on wilderness issues. There were no financial books, only the big blue ledger books that we kept the calendar consignment and payment information in. I immediately called up Richard and asked him to send me the financial books. He insisted that he had sent me everything.

Both WCWC and the Green Party offices shared our home for only a few weeks. A month after the provincial election in which the Green Party fielded its first few candidates (including Adriane), it moved to an office on West Broadway and our Trafalgar Street house became just the Wilderness Committee's new home, shared only with our family. In the next couple of months, I tried to split my time between the two organizations evenly, but in reality I spent more time trying to get the Green Party growing.

Outside help to finance WCWC's '84 Endangered Wilderness calendar fails to materialize

A few weeks after WCWC's files arrived, Murray Rankin came over to Vancouver. He and I met with Ken Farquharson, president of the Outdoor Recreation Council. Murray was hoping that ORC would take the endangered wilderness calendar off our shoulders. I was skeptical, knowing that ORC was dependent on provincial government grants for nearly all of its funding. Our calendar was fundamentally critical of government for not creating enough new parks to adequately protect wilderness. Ken strongly supported our calendar and offered some help. We could use ORC as a mailing address and put an article in ORC's newsletter encouraging all ORC members to help sell the calendars. But that was it. He could do no more.

Murray and I then went to see the owner of Mitchell Press. He told us exactly what he had told Richard. We had to repay the $18,000 we still owed; but, because they wanted to keep our business and continue printing for us, they would extend us credit. I promised, optimistically, that we would begin paying the outstanding balance off at $500 a month. Mr. Mitchell must have believed in our cause, although he never said so in words. It was the only way to explain why he was so lenient with us. Over the months to come, it worked out that as long as I made progress towards paying our debt down, or at least didn't go deeper in debt, Mitchell Press would extend us credit at no interest and would keep printing for us.

I knew that Murray was skeptical about the whole thing. Now WC²'s survival was entirely in my hands. Linda Hannah, Murray's wife whom he had first met at a WC² meeting, offered to help with the Victoria sales if I managed to get out a calendar that year. Besides that, there would be no help from any other director.

On September 7, 1983, I signed a contract with the rest of the directors that put me fully in charge and made me responsible for the outstanding debt to Mitchell Press. I had to save 60 cents from the sale of each calendar, use it to pay back the debt to Mitchell Press for the 1983 calendars, and provide free educational calendars for all the MPs and MLAs. On the insistence of Richard, the contract had a special clause saying that I could not sell calendars at a wholesale price to any political party (to prevent me from possibly using the Green Party as a sales outlet). I agreed. WCWC needed to be non-partisan to maintain its credibility and its charitable tax status.

We didn't have another directors' meeting until the end of December, after the calendar had been out for a month and was already on its way to becoming a success—or at least not a disaster. I was not able to follow the letter of the contract regarding paying back Mitchell Press or paying myself 15 cents from every calendar sold. But I was able to pay off some of the outstanding debt, at least enough to keep the directors and the printer happy.

WCWC's bank account emptied

There was one last bone to pick. After taking over the responsibility of running WCWC, I began to get monthly statements from our credit union. Upon opening the first one I got a rude shock. There were only a few dollars in WCWC's account. Just before Richard turned over the records and the chequebook to me, he had drained the account dry. The $7,000 that Richard had reported in his financial report at the AGM was gone. Since both of us had to sign every cheque, I should have known about it.

I had been counting on that money to reduce the printing bill from $17,000 to $10,000. Richard had not reported to the directors any debts other than the amount owing to Mitchell Press. Actually, there was another outstanding debt. It was for $7,000 and owed to Richard himself. He paid it off before handing the accounts over to me. There was no question that WC² owed Richard the money. He had all his expenses documented including his back wages, which he hadn't taken for many months. It took me years to realize that it's not wrong to take the money that's owed to you when you quit your job. I had counted on Richard to bankroll and bail out WCWC for so long that I had completely lost perspective. Richard had helped launch

WC², co-captained it on its three-year maiden cruise and generously supported it. I should have felt gratitude, not anger.

I never did get any financial books from Richard, not because there was malevolence on his part—an opinion I held for years—but because there simply weren't any! He and I had both used the "shoebox" method of accounting: throw every receipt into a box to organize later. The only WCWC "books" were the big blue ledger books with all the transactions of the calendar sales carefully and accurately written down.

It took more than a decade for WCWC to master its finances. Volunteers were a lot better at helping save wilderness areas than at keeping the books.

Calendar concept redesigned

Being fully in charge of WCWC's annual endangered wilderness calendar for the first time, I decided to use a beautiful Meares Island image by Adrian Dorst on the cover. I also featured Meares inside as a "Tribal Park" candidate. I had a premonition that Meares would soon become the hottest wilderness preservation issue in B.C. For stylistic emphasis, I used bold futuristic shadowed numbers for the dates so people could see them from across a room, thus correcting one of the technical faults of earlier calendars. It gave the calendar a futuristic look to match Orwell's infamous book—*1984*.

The biggest change I made was to drop the perforated coupon at the bottom of each month that people were supposed to clip and send in with a donation. It was too time-consuming to find groups to sponsor each month. Besides, the novelty had worn off. The sponsoring groups discovered that only about 20 to 30 people bothered to use the coupons, a response rate of about 0.2 percent.

It was hard for a group struggling to save a wilderness area to stay optimistic, and too easy for them to conclude, "If so few people care, we don't have a chance." Some of the "sponsor" groups, too, failed to respond to the donors. Adriane found this out when she diligently sent in the tear-off every month along with a donation to the sponsoring group in our first calendar (1981). Every group cashed her cheque, but only a few (including myself at WCWC), acknowledged the donation with a thank you letter or receipt.

To make up for the lack of monthly coupons, I included on the inside cover of the '84 calendar an update on all the campaigns that we had featured in the past three calendars. I also provided contact information for the organizations currently working on them, and urged people to support them. The only tear-off I kept was the one for us on the back cover of the calendar. It gave people the opportunity to order next year's calendar for $6.95 and join WCWC for $15.

This "makeover" also freed us to campaign on all the issues without stepping on other groups' toes. Not one group was upset about the change. Many were relieved. Almost all of them helped sell the newly-formatted calendars, although some took a lot fewer copies. The Sierra Club of B.C., for example, took only 20 that year.

B.C. government's postal windfall saves us

A surprise turn of events made our '84 calendar more profitable than any of the previous three calendars even though it involved no increase in sales. In retrospect, without this fortuitous windfall, which would never have come our way if Richard hadn't quit and I hadn't moved our operations to Vancouver, WCWC undoubtedly would have disappeared into the quagmire of bankruptcy.

In the early fall, I set up the new mailing address that had been offered to us by Ken Farquharson of the Outdoor Recreation Council. It was: WCWC c/o ORC - 1200 Hornby Street, Vancouver. One week later when I went in for the first time to check for mail, Robin Draper, ORC's Executive Director, called me into his office.

"There is a new policy regarding mail service. It applies to everyone including your society—an organization affiliated with ORC that currently uses this address but does not actually have an office here," he began. Before I could even start to get worried, he continued, "From now on you can bring your outgoing mail into the mailroom and the staff there will put it through the postage machine and you won't have to pay postage."

"You're kidding!" I replied incredulously. "How can that be?" I couldn't believe the good news.

He explained that it was a B.C. government initiative to save money. The government bureaucrats had decided that it would cost the government less if all the groups in the Sport B.C. Building left out postage as a line item on their grant applications for government funding, and instead processed all their mail through the B.C. government postage machine in the building's mailroom. Because of its huge volume of mail, the B.C. government had a deeply reduced contract price on postage from Canada Post. WC² was one of a very few groups using 1200 Hornby as a mailing address that did not get B.C. government core funding and rely on such funds for its existence and survival. We were, however, a charitable society with a Sports B.C. Building address and that made us eligible. The free postal service began immediately.

"Does this cover packages too?" I asked.

"Yes, everything," replied Robin.

I was ecstatically happy. Postage, after printing, was WCWC's number one expense. It was so large that it severely limited our public educational outreach efforts. My imagination soared, thinking of ways to take advantage of this fabulous windfall.

I started out using the free postal service modestly, thinking that as soon as someone in government noticed, it would be pulled out from under us. Gradually, I threw caution to the wind. We quickly became the largest user of the mail service. No other group had ambitions like ours! We asked everyone we knew for their mailing list so we could send out educational information about South Moresby and order forms for our 1984 calendar. We got five different lists. Several times per week I would bring in stacks of envelopes. Each one was stuffed with a South Moresby brochure, a calendar order form and a request that they send a letter to the Prime Minister telling him how they felt about Canada establishing a national park reserve to protect this national eco-treasure. (We used this obtuse wording in order to conform to the rules governing charities that forbid us from directly telling people to lobby government.) The order form listed the areas featured in the 1984 calendar and had special deals for buying calendars in bulk—5, 10 or 25—to use as gifts.

Our incoming mail was increasing in direct proportion to the increase in our mail-outs. Thus I learned the secret of marketing: it's all to do with output volume. Pre-publication orders—with cheques enclosed—came in like never before. We used the cash flow to pay for envelopes, promotional printing and a down payment on the calendar printing. With no wages and no postage, these were our only expenses.

After a few months of intensive use of the free mailing service, Robin called me into his office and gently tried to persuade me to cool it a bit. I asked him outright if he was telling me that we had to use it less. He said no, he just thought it was a good idea, or we might blow it for everyone. By that time, we had sent out over 12,000 pieces of mail. While it may not seem like a lot today, it was huge number for that era. Fortunately, we had completed the mailings to every environment, conservation and citizen group's mailing list that we could get our hands on; so I complied and gave it a few weeks' rest while I worked on getting the calendar to press.

Producing the calendar harder than I thought

I found out that it was hard to produce a great calendar on deadline. There are so many details to handle. I hit one small snag after another. Even finding a profound Canadian quote about wilderness to put on the front cover was frustratingly difficult. I could find no Canadian equivalent to John Muir or Aldo Leopold. I decided to feature a quotation from the late Roderick Haig-Brown (1908-1976), one of the forefathers of the conservation movement in B.C. I read many of his essays searching for the right quote. He was far ahead of his time in his thinking regarding salmon and steelhead conservation, but he was wordy and obtuse. After hours of research, the best I could come up with was *"Conservation must stake its claims, aggressively and authoritatively, ecologies are the key..."* from his book *Some Approaches to Conservation* (1966). 1984 was the last year we featured a quotation on our calendar cover.

Calendar production dragged on and on. Photographers were slow in submitting their images. Maps had to be redrawn and corrected. There was much fussing around to find the errors and get copy re-typeset. Meanwhile, I was busy pre-selling them, getting the mail-in orders and building up some money to pay the one-third down to get Mitchell Press to do the pre-press work. Then there were the weekly prenatal classes that Adriane insisted I attend. During the relaxation exercises, exhausted, I always fell asleep!

Full page ad donated by the Outdoor Recreation Council of British Columbia and published in its *Outdoor Reports* Vol. 6 Fall 1983 No. 3. Clipping: WCWC files.

1984 calendar featuring Meares Island on the cover. Cover image: Adrian Dorst.

A teaching job I can't refuse delays the calendar

To further complicate things, in the midst of calendar production and promo mailouts, Adriane was hit with a case of acute appendicitis. It happened near the beginning of her fall teaching term, 11 weeks before she was to give birth to our first child. The doctor immediately sent her to the hospital for an emergency appendectomy. The dire consequences of her appendix rupturing while she was pregnant vastly outweighed the risks of the operation (although they, too, were considerable). Major surgery meant Adriane could not continue teaching that semester. Her department found substitutes for all her courses except her "B.C. Environmental Issues" geography class. I was asked if I would do the job. I couldn't say no. Although I'd done graduate work in genetics, I'd never taken even one university-level geography course. I prepared lessons (using Adriane's great lecture notes) and worked hard at it. Most students liked me, and I got an excellent recommendation. Good teachers are essential to a decent society; but I couldn't help thinking, "Why does Adriane do this? I know she loves it and the money's great, but we only have this one chance in the entire lifetime of Planet Earth to save wilderness from being lost forever. We're losing the fight because so much work that needs doing is going undone. The Earth needs her talents."

Bringing the Wilderness Calendar blueline instead of flowers to the maternity hospital

It was already mid-November; the baby was due any day; all I had to pre-sell the new calendar was the colour proof of the cover and the featured images for each month. On the evening of November 18, while I was trying to get calendar orders at the North Shore Hikers Annual General Meeting being held in a church near our Kitsilano home, a friend of ours rushed in and said, "Adriane's waters have broken." I hurried home, took some last minute pictures of Adriane (we had been so busy I had not taken one photo of her pregnant) and drove her to the hospital. This birth wasn't the best-timed event. The next morning I was scheduled to pick up the final calendar proof—the blueline—give it one last check for mistakes, approve it and get it back so the printer could go to press. Because of all the delays, there'd be only a month until Christmas to sell them. I was beginning to panic.

After 27 hours of labour, our daughter Kallie emerged, a healthy bundle of joy. The next morning while other new fathers were bringing their wives flowers and chocolates, I brought Adriane the calendar blueline to proofread. I held our baby girl, while Adriane gave the blueline a thorough look-over and found a few more typos (where do they keep coming from?). That night the calendar started rolling off the press. Two days later I held the finished calendar in my hand—the first one I was solely in charge of producing—as Adriane reveled in holding our first wonderful child in her arms. There was no turning back now.

A message we repeated over and over again in the years to come

"The Story behind this Calendar," our lead article inside the cover of our '84 calendar, has the basic message we have been repeating with slight modifications time and time again ever since.

Four years ago, a few environmentalists decided to try free enterprise to market the idea that the wilderness of Western Canada is worth protection. They believed that while Canadians might not be able to make their own cars and TVs, they certainly could make their own wilderness calendar. By promoting preservation of Canada's natural beauty in this way, profits could be kept in Canada and used to extend Canada's Parklands and Ecological Reserves.

Western Canada had and still has the best raw material for the job—an unrivaled variety of beautiful but threatened wilderness areas. Canadian Wilderness, like any naturally good product, has a tendency to sell itself. However, it needs greater advertising to compete in the crowded marketplace of ideas.

The Western Canada Wilderness Committee's calendar could not have been published these last four years without the generous help of many individuals and environment and recreation groups that contribute to its contents and believe in its causes. Their help with distribution and sales has been particularly important in making this effort a success.

This year's calendar is dedicated to all our friends, especially the Valhalla Wilderness Society, the ROSS Committee and the Alberta Wilderness Association, examples that go a long way to prove that good causes with well-informed dedicated people behind them can't fail.

Mark Haddock—employed on pittance piecework pay—launches a Widgeon Lake campaign

There was just too much work that had to be done in getting the calendars out in short order. I needed an assistant. One day in late November 1983, when I was getting the mail at 1200 Hornby Street, a young man named Mark Haddock, who'd been in the UBC forestry class for my tirade the year before, approached me. He told me he was working on saving Widgeon Lake, his favourite hiking spot, from Greater Vancouver Regional District's plan to dam, divert its water and cut off public access. He asked my advice.

I gave him this simple formula:

1. Make up a name that sounds important. Don't call yourself a society unless you become legally registered.
2. Go to a print shop and get some fancy, expensive-looking letterhead made up.
3. Write letters to the government, but find some doctors or someone important to sign them because otherwise the government won't take your letters seriously.
4. Go to the media and get them to write an in-depth piece about the issue.

Desperate for help, I then asked, "By the way, WCWC needs some help with calendar distribution. Are you interested?" I offered him a low-paying piecework job at 50 cents for each package of calendars he prepared for mailing. Hours were flexible. He could do it anytime as long as it was right away.

He took the job. Years later, this is what Mark has to say:

"Accepting it, I crossed the Great Divide from being an ex-Forest Service worker to a paid environmentalist. The first thing I learned on the job is that this was a committee of one—now two with me on board—and three if you count Adriane who was teaching at Langara and financing the whole operation. Like Arlo Guthrie said, "If there's at least three of you they'll think you're a movement." And that's exactly what we were—the Western Canada anti-wilderness-massacre movement. We looked much bigger when you added the directors. And if you included everyone we sold a calendar to, hell, we were the largest wilderness organization in Western Canada!"

When I brought Adriane and the baby home, our house was a beehive of activity. There were stacked boxes of calendars and boxes of envelopes everywhere. Because the calendar was so late, people were constantly stopping in to pick them up. We had to work around the clock packaging them up and sending them out. Adriane vividly remembers those nights. The loud ripping noise of the tape gun kept her awake as Mark and I worked all night long. But, thank goodness, it didn't disturb baby Kallie who slept soundly next to Adriane.

Postage freebee includes Calendars!

In the morning I went down to ORC with just a few single-calendar envelopes to mail. No one complained. That afternoon, I returned with a lot more. The next day, all day long, I kept bringing in packages of calendars to send through the government postage machine. I gave everyone working in the mailroom a free copy as well as complimentary copies to all the groups in the building, with an order form, of course, in the hopes that they would order more as gifts. Every day I brought down hundreds to mail, including packages of 25 and 50 to send to other environmental groups. No one complained. I didn't keep track of exactly how much money we saved; but it was well over $6,000. WCWC survived, thanks to ORC's B.C. government mail machine.

Quest students help out

The word was out about the plight of WCWC. Many of the groups put in an extra effort to sell our calendars, because they knew that we were on the ropes and this was the make-it-or-break-it year. They included the ROSS Committee, the Valhalla Wilderness Society and Alberta Wilderness Association. The Quest outdoor education students at Prince of Wales Secondary School, and its team of teachers including Tom Ellison, were exceptionally helpful. Twice a year a new group of grade ten students, selected from applicants throughout Vancouver, took the full-time Quest programme. English, Social Studies and Phys. Ed. were all part of the curriculum, along with lots of expeditions into the outdoors.

Instead of selling the usual chocolate bars or cookies, Quest students sold Christmas trees and WCWC calendars to help fund their activities. Tom Ellison sailed in South Moresby every summer, taking a group of select students along and turning them into avid campaigners for making the area a park. Students spent hundreds of hours collecting thousands of signatures on our Windy Bay petition.

Ellison's Quest program was unabashedly pro-wilderness preservation and encouraged the eager students to be environmental activists. According to one student, it was more like a philosophy course with a heavy reading list of books like *Silent Spring* by Rachel Carson, *Walden* by Henry David Thoreau, and *Northern Frontier–Northern Homeland* by Justice Thomas Berger.

In our 1984 calendar I featured a photo of Tom's sailboat in stormy seas near the southern tip of the South Moresby. That fall of '83, the 40 Quest students sold a record number of calendars—nearly 3,000. Over the years Questers played a crucially important role in saving South Moresby. One Quester, Jeff Gibbs, even organized a "Tree Club" which exclusively focused on saving South Moresby.

There's no outdoor education program that I know of like Quest today—one that teaches the principles behind environmentalism. Quest ran from 1973 through the spring semester of 1988, one year after South Moresby became a national park reserve. Starting in September of 1988, Prince of Wales Secondary School replaced it with a new program called Trek. It carried on the tradition of outdoor education, but with less emphasis on environmental issues. Tom Ellison, who left the program when it changed its focus, continued to promote the preservation of the B.C. coastal wilderness by donating the services of his sailboat every summer to take dignitaries and media to the Khutzeymateen, the "Valley of the Grizzlies" located 45 kilometres northeast of Prince Rupert.

After the B.C. government finally protected this 44,300-hectare wilderness as Canada's first grizzly bear sanctuary on June 2, 1992, Tom, along with bear biologist Wayne McCrory (Colleen McCrory's older brother), then proposed the creation of a large "Spirit Bear" sanctuary on B.C.'s mid-coast around Princess Royal Island. Designed to protect Kermode bears, a unique white-coated subspecies of the black bear, it still hasn't been fully established. Tom continues to take eco-tourists on *Ocean Light II* (his spacious 71' ketch rigged sailboat) to experience the west coast wilderness and encourages them to become involved in efforts to protect it.

Mr. Mom carries on

Despite our new baby, my temporary teaching job and the late publication of the calendar, the '84 calendar was a success. It wouldn't have been without the help of Mark, the Quest students, our great "moccasin" distribution network and the free mail service. By the time the last money dribbled in, I had paid off the printing bill and $5,000 on our outstanding debt owing from the 1983 disaster. We still owed $12,000, but Mitchell Press was not pressing us for it. Mark Haddock went on to become a lawyer. He claims that in following my advice, he only had to use 20 of the 500 sheets of the Widgeon Lake Preservation Society's fancy stationary he'd printed up to save this wilderness gem. He photocopied his wife's PhD thesis on the backside of the rest, long after Widgeon was saved.

Adriane went back to college teaching in January. I didn't let

Tom Ellison's sailboat the *Compass Rose* in South Moresby waters. Image featured in our 1984 calendar. Photo: Tom Ellison.

being "Mr. Mom" for seven-week-old Kallie slow me down. While Adriane was at work, I multi-tasked, playing with Kallie to keep her happy while doing WCWC or Green Party work. When Kallie cried, she came first. When she got a little older I put her in a backpack and she went with me everywhere, peeking over my shoulders at our active, vibrant world.

Below: Khutzeymateen bumper sticker. Artwork: WCWC files.

khutzeymateen
Save the Valley of the Grizzly

PARDON ME THOU BLEEDING PIECE OF EARTH THAT I AM MEEK AND GENTLE WITH THESE BUTCHERS.

Shakespeare

Photo by Husband/Pothier

Clear-Cut Logging, British Columbia 1985.

Chapter 8

Winning a battle against government censorship

Our Hot Spot map project gets federal funding

Early in 1984, I asked Paul Mitchell, Director of Information Services of Environment Canada Pacific Yukon Region, if a project I had in mind—an Environmental Hot Spot map for B.C. to highlight and explain the major environmental issues in their geographical setting—would qualify for Environment Canada funding. I showed him a copy of the first one that the Society Promoting Environmental Conservation (SPEC), one of Vancouver's oldest environmental organizations, had published years earlier. This map was now out-of-date and out-of-print. Without a moment's hesitation, Paul said yes, Environment Canada would fund it. They really had a 'big bucket' of taxpayers' money to spend then! There were conditions, however. Our map had to cover both B.C. and the Yukon. Environment Canada would hire a graphic artist to design the map, which Paul envisioned to be in brochure format that folded up for easy storage and future reference.

Yuppie-designed and subject to censorship

It wasn't exactly what I had in mind, but hey, who was I to argue when the government was paying the full shot. Paul Mitchell went on to explain diplomatically that I could not "wave a red flag." The material WCWC wrote had to be educational and not full of radical rhetoric that slandered governments. He assured me this did not mean he or anyone else at Environment Canada would be censoring the material or deciding which hot spot issues and places we could include and which ones we couldn't. It was how we said it that was of concern. To reassure him, I explained that college students under the able guidance and supervision of a college professor (Adriane) would do the research, which would be accurate and professionally written like the write-ups in our calendars. He made clear, too, that last-minute changes after approval, like we did when we added "in the South Moresby Wilderness Proposal" on the *Canadian Wilderness–Environments Worth Protecting* poster, couldn't happen again!

This time, he was going to arrange for the printing and the graphic artist he hired would take the blueline in. I agreed to all the conditions. Our goal always has always been to reach the general public, not just the "radical fringe" or the "already converted." I believed that under the conditions he imposed we could still get out a strong message that the environment needed a lot more protection.

I had wanted to do this project for a while. Adriane (and my brief experience teaching her course last semester) convinced me that her environmental studies students were capable of professional-calibre research. Our first step was to work up a list of the key issues and hot spots. Her students, as part of their course assignments, did the background research and writing. Together, Adriane and I edited and polished the text. I checked with many other groups including SPEC to make sure our new hot spot map covered all the major issues, from forestry to fisheries, pollution to parks, energy conservation to land claims. It provided a thorough thumbnail sketch of the problems (most of which are still with us and worse today) and solutions (most are still appropriate and more desperately needed than ever). The only part that is really out-of-date today, 20 years later, is the list of park candidates. Many of the areas under threat in 1984 have been protected— only a few were lost. I put special emphasis on the hottest topics like the proposed South Moresby Park proposal, proposed energy mega-projects including the Site C dam on the Peace River, offshore oil drilling in Hecate Strait and the Hat Creek coal development in the Cariboo.

Paul Mitchell, or someone on his staff, proofed the final draft. There were a few typos to correct and minor wording changes to fit the politically correct wording conventions of the day; then off it went to the printer. I wasn't too keen on the sophisticated (yuppie) design with glossy grey background everywhere and the large type "HOTSPOTS" headline on the cover in hot pink. Nonetheless, it was very educational, showing the boundaries of the big watershed, the B.C.'s Agricultural Land Reserve areas and the major population centres in the region. "It's really eye catching!" I enthused.

No doubt about it, it was very trendy. No one casually glancing at this brochure would ever guess that it was an environmentalist's tract, particularly with Environment Canada's logo and address on the back.

Hot Spot Row — Environment Canada put a "hold" on the distribution of our environmental brochure

About a week later, I got a call from Paul. "They're printed. Come on down and see." I hurried downtown to Environment Canada's office to get there before they closed. It's always a 'rush' to hold the first copies of your publication "hot off the press" in your hand.

"I only have one box here, the rest are at the printer," Paul explained to me as we both admiringly opened up one of the map-brochures and looked it over. I thanked him profusely and asked, "Can I have some of these?" "Take as many as you want. You can pick up the rest tomorrow when we get them from the printer." I took a big armful, rushed home and waited for Adriane. Finally, she arrived. Together we unfolded the map and looked it over carefully. "I can't wait to take copies to my students," she said. It was something for them to be proud of, for sure. She counted out a class set to take to them the next day.

Next morning, I dropped Adriane off at Langara College and headed downtown, arriving at Environment Canada's office at about 11 a.m. to pick up more hot spot maps. As I entered, the secretary at the front desk said Mr. Mitchell wanted to see me right away. She ushered me into his office where he informed me that there was a "hold" on the maps. He could not release them. I couldn't believe my ears. He explained that it had been read by "others" who found the language used in some write-ups "inflammatory." I asked him who the "others" were and what specifically was "inflammatory." He refused to be specific. Despite my protestations, the bottom line was Environment Canada was not going to release our hot spots map.

I stalked out of the office empty handed and fuming. "Censorship!" I couldn't believe it. "So much for federal funding," I muttered to myself. I was glad that I had at least a few copies, although not very many now that Adriane had taken most of them for her class. If I hadn't picked them up the day before, I would have been completely stymied and not able to show anyone what the conflict was all about. I went home, picked up a couple of the remaining copies and stormed down to the Pacific Press Building. The first person I got a hold of was a *Province* newspaper reporter. Just a few weeks earlier this paper had changed its format to a tabloid and started featuring more short and sensational stories. This story fit the bill. He would write it up. I just had to leave him a few copies. That night, at about 10 p.m.,

Facing page: Sierra Club's Shakespeare poster reprinted by WCWC.
Poster image: Husband/Pothier.

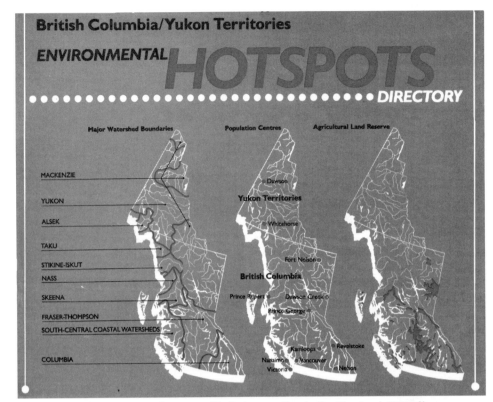

Cover of HotSpot brochure temporarily suppressed by Environment Canada. Graphic artwork: WCWC files.

I walked down to the Press Club to have a beer or two and wait to see an early copy of the *Province*. As usual the first copies rolled off the press a little after midnight and pressmen, taking a refreshment break, brought a few papers "hot off the press" over to the Club.

I didn't have to look hard to find the story. Boldly emblazoned on the front page, in the fattest font and biggest type I'd ever seen was "**HOT SPOT ROW.**" The brief article that accompanied this headline sensationalized the story. Boy was I happy!

Environment Canada relents

One day later, without making any big deal about it, Environment Canada released our *Hot Spot Directory*. No big row—just quiet acquiescence. I was given blank labels to obscure Environment Canada's name, address and the government logos on the back of the brochure; I promised we would put them on every one. Since then, I've never underestimated the power of the media or the importance of making issues public.

I found out later who had complained about the "controversial language": the B.C. government! Most of the *Hot Spot* write-ups were critical of B.C. government policy. A few weeks after "the row" we received a two-page letter from B.C. Environment Minister Tony Brummet. He complained about the tone of our publication and asked all kinds of questions about our organization. They boiled down to probing as to why we weren't like the other environmental organizations that came to meet with him to respectfully present their complaints directly.

Months later when we finally got a meeting with Brummet I explained to him that WCWC concentrated on building public support for our issues, instead of relying on the strength of our arguments to sway governments behind closed doors. In response to his complaint that the publication focused only on the negative, I explained to him that from our point of view there really wasn't much positive to say because his government had done so little to protect the environment, conserve energy or create more parks. At the end of this meeting Colleen McCrory and I gave Mr. Brummet our "Bruce the Moose" oil painting. He really liked it, commenting to us that even though he was a hunter he would never shoot such a majestic animal as that. It was a great way to end the meeting!

Brummet was not the only one unhappy with our *Hot Spot Directory*. In a biting letter to us, the Yukon Conservation Society pointed out that we had not consulted with them as to which topics were hotspots in their territory. We had completely left out a really major issue: placer mining and the siltation of rivers that it caused. A couple of other B.C. environmental groups complained that we had not consulted with them, either.

These letters acted as a sobering prod to do a better job of working with other groups when our activities or publications concerned their issues. We tried much harder to work collaboratively with other like-minded groups in the years to come.

One thing was for sure: this "hotspot row" dried up the government funding for WCWC. We never got another grant from Environment Canada. In fact, from then on, every environmental group seeking Environment Canada funding had to vet its project proposal through the B.C. Environment Network (BCEN). Soon the federal pot of money shrank drastically. There was only enough for core funding for the BCEN, and for environmental group leaders to attend annual meetings.

The last conservation group-initiated publication funded by Environment Canada was the "Shakespeare Poster," put together by Vicky Husband, Patrick Poitier and John Broadhead of the Sierra Club. It featured a slash-burned clearcut beside the road to Clayoquot Sound along with the quotation, "...*Thou Bleeding Piece of Earth...*" from the play *Julius Caesar*. Several years later when it ran out of posters, the Sierra Club granted WCWC the right to reprint it without charging any royalties. We've kept it in print ever since. It is one of the most popular Canadian environmental posters of all time.

Chapter 9

Learning how to publish educational newspapers

Success – Skagit spoilers are run out

In April of 1984, Canada and the United States signed the Skagit Valley Treaty. It scrapped the proposed High Ross Dam project that would have flooded the Skagit Valley in B.C. in exchange for B.C.'s agreeing to provide Seattle with equivalent power. WCWC had very little to do with the successful campaign to save the Skagit. We only featured the area and issue in our calendar two times. The concerted efforts of the ROSS (Run Out Skagit Spoilers) committee over the years, and people like Dr. Tom Perry who chaired it, proved that persistent campaigning for a good cause that most people thought was a lost one could bring victory. This win was inspirational. It gave us hope that we would eventually win our big campaigns too.

Above and below: Joe David, Tla-o-quiaht artist carves a Welcome Figure that represents the natives' efforts to save their lands from exploitation. About the figure Joe David said, "I wish to portray the beauty between man and nature." Photos: WCWC files.

Tom Perry and Curley Chittenden (right side) by the sign where the high water mark would have been. Photo: WCWC files.

Meares Island Tribal Park celebration in Tofino

On the April 20-22, 1984, Easter weekend, Adriane and I went to Tofino with our baby Kallie to attend the big Meares Island celebration. At it, the Clayoquot Band Council and Hereditary Chiefs declared Meares Island a Tribal Park and erected a "Meares Island Tribal Park" sign on Meares near Mosquito Harbour. Events included the raising of the *Weeping Woman of Meares* totem pole, carved by a non-native Tofino resident. Because a permanent place for it wasn't yet secured, it was raised on an old barge moored in Browning Channel where it had been carved. It was big and heavy and took many men to pull it upright. The weather was storming fiercely and raining hard. Just as the pole was teetering on the edge of going upright, the swift-moving clouds parted and the sun shone briefly upon us. Instantly a beautiful, richly-coloured rainbow arched over Meares Island. Moments later, when the pole came to rest upright, the sun and rainbow disappeared and the rains took over again. It was the briefest of magical moments. It was definitely an omen—a powerful and affirming one. Within six months, there were blockades to prevent logging on Meares. The weeping woman of Meares was eventually erected on a small island between Tofino and Meares Island.

Learning how to be our own "free press"

WCWC underwent a real growth spurt in 1984. It was the year that I learned how to produce newspapers and put into practice the maxim "the only free press is your own press." Although there are some good reporters and an occasional great article about the environment in the mainstream media, there is not nearly enough coverage to change consciousness. The only press that will tell a story accurately and fully is your own.

One afternoon when I was down at the new B.C. Green Party office in Kitsilano, a volunteer asked me why we didn't publish our

Meares Welcome Figure erected in front of Legislature at a Nuu-chah-nulth sponsored rally in 1984. The figure was eventually placed in front of the UBC Museum of Anthropology where it stands today. Photo: WCWC files.

own newspapers to get our message out. Before I could say, "I don't know anything about how to do it," he explained that he had worked for a small newspaper and that anyone could put one together. You simply get some free layout sheets from a web press company that prints on rolls of newspaper (they're free), and then paste the headlines and "body copy" in columns on them following the blue-coloured layout guidelines.

"It's the cheapest way to get the message out, if you're printing lots of copies," he asserted.

"How much?" I asked.

"About $600—everything included—for 20,000 copies of a four-page tabloid-sized black and white newspaper. It's equivalent to two 11" x 17" sheets of paper printed on both sides. I've done it before. I'll show you how. You need a hand waxer to make the back of the copy sticky. They cost about $40—but I've got one."

The next day, working together, we designed and began producing the B.C. Green Party's first newspaper. I typed all the stories in columns using a very old manual typewriter with a very small typeface. He showed me how to get headlines professionally typeset and specify the size and type font so they would look like those found in a regular newspaper. Without the aid of a light table (which would have helped), we waxed the backs of the pieces of copy and headlines, and stuck them on the layout sheets as straight as we could. He had some rolls of layout tapes with different kinds and sizes of lines to separate the articles. The final "trick" was to get the photos "PMTed" (Photo-Mechanically Transferred) into the right-sized dot screen and paste them in place. It was easy and fun.

I was amazed at the results. It looked great! It was a real newspaper. I realized that it was the way to go from now on to build public awareness. It was by far the cheapest way to get the message out to "the masses."

South Moresby — WCWC's first tabloid newspaper

In was a good thing I had a new tactic in my pocket because the fight to save South Moresby was coming to a head. In January 1984, after four years of deliberation (11 years after the Islands Protection Society first proposed that the South Moresby Wilderness area be protected), the South Moresby Planning Team finally handed its report to the B.C. government's Environment and Land Use Committee. The report had four land use options. Throughout the time this planning team had talked, logging continued on Lyell Island without any slowdown. Much damage had been done; but Windy Bay was still intact. The Haida, running out of patience with the "talk and log" situation, began to hold meetings with Tony Brummet, B.C. Minister of Parks. Knowing we had to get the issue into the minds of the public and the media, I put together WCWC's first newspaper publication: *South Moresby–A Special Issue—Summer 1984*.

On the front page of this paper I featured the full text of federal Environment Minister Charles Caccia's recent letter to B.C. Environment Minister Tony Brummet expressing Canada's interest in establishing a national park in South Moresby. Next to it was the federal government press release of May 29, 1984 that accompanied the release of Caccia's letter. This letter was a huge breakthrough in the campaign that, until then, had only received small mention in the regular media. In WCWC's paper I had the chance to give this letter the profile it deserved.

The story behind Caccia's letter

In late 1983 Colleen McCrory received the Governor General's award, presented by the Tourism Industry Association of Canada, for her work to protect Valhalla Wilderness Park. When Colleen stopped in Ottawa to meet Governor General Ed Schrier, she decided that she would try to meet the federal Environment Minister, Charles Caccia, to see if he would support South Moresby becoming a National Park.

Right after meeting with the Governor General, Colleen made her way to the office of Jim Fulton, the NDP MP from Skeena, the riding in which South Moresby lies. Colleen was running out of time and couldn't get in to see Environment Minister Charles Caccia. She knew that Jim was sympathetic and asked him if he would help her get her an appointment with Caccia. Jim invited her to attend question period and Colleen watched from the visitor's gallery as Jim wrote the note to the Honourable Charles Caccia asking him to meet with her. She saw a young page deliver the note across the floor.

Caccia agreed to meet that very night at 8:00 p.m. At that meeting Colleen presented all the information about South Moresby, telling him the importance of this ecological treasure to the nation. Caccia then asked Colleen how much it would cost to resolve the issue with the various interests. Colleen, having no clue of how much it would cost, bluffed. "It'll cost about $20 million," she said with an air of confidence. Caccia then told her that he thought that this was very reasonable and said he would support protecting this area

SOUTH MORESBY – A SPECIAL ISSUE
Summer 1984

Federal Government Wants National Parks in South Moresby
— BUT PROVINCIAL APPROVAL NEEDED FIRST

Caccia writes Brummet

[Text of letter]
Dear Mr. Brummet:

I was pleased to learn that the South Moresby Land Use Alternatives report was recently completed and released for public comment before decisions are made on the future of this magnificent natural area.

As the heritage values of the South Moresby area are of both national and international significance, and should therefore be protected for all time, I recommend that the Government of British Columbia seriously consider selecting Option 4 in the report.

In order to implement this option, I propose that we examine the possibility of creating a national park/national marine park in the South Moresby area. This proposal would include some of the islands and sea-bed to the 50 fathom depth in the eastern and southern portions of the area, as well as a coastal component in the San Christoval Range to protect some of the most outstanding fiord and mountain landscape in Canada. Ideally, this could be complemented by the creation of a Class A provincial park, which would contribute to the British Columbia park system while helping to ensure the long-term ecological integrity of the national marine park. The enclosed map illustrates this concept. [See map in this special issue.]

If you wish to explore this proposal further, please inform me so that a meeting can be arranged between Parks Canada and your officials. Because the area is of such national and international conservation interest and significance, I have received many letters and representations, as I am sure you have also.

Accordingly I would like to make my proposal known within a couple of weeks if you have no objection or other advice.

Yours sincerely,
Charles Caccia

Sitka Spruce Snag PHOTO – A. CARR
Lyell Island Old Growth – Important Part of Ecosystem Missing in Second Growth

Press release - May 29, 1984
PROTECTION PROPOSED FOR SOUTH MORESBY

OTTAWA -- Federal Environment Minister Charles Caccia today made public a letter to British Columbia's Minister of Lands, Parks and Housing, Anthony Brummet, in which Mr. Caccia proposes exploring the possibility of creating a national park and a national marine park in the south Moresby area, possibly complemented by a provincial park. Mr. Caccia's proposal was made in response to the release by the province of the South Moresby Land Use Alternatives report and the strong expression of public interest in preserving the South Moresby area.

Mr. Caccia added "I am open not only to discussion of this proposal but also to other possible options for federal involvement in national park or national marine park establishment".

Describing the heritage values of the south Moresby area as both nationally and internationally significant, Mr. Caccia recommends that the province consider selecting Option Four, "Natural Emphasis" as described in the South Moresby Land Use Alternatives report released in February. Option Four suggests that most of the south Moresby area remain in its wilderness state. These islands, situated west of Prince Rupert, are often referred to as the Canadian Galapagos for their unique plants and animals.

Western Forest Products Lobbies Government to End Logging Moratorium

Sources in the B.C. Forest Service have revealed that in a behind-the-scene move, Western Forest Products, a Tree Farm License holder in part of the South Moresby Wilderness Area, is in Victoria urging the Ministry of Forests to let them begin clear-cutting blocks on the south end of Lyell Island. The local Forest Service representatives had rejected their request.

The area in question has been under a logging moratorium during the South Moresby planning process so as not to prejudice the final land use decision. Conservationists contend that such logging would leave ugly scars visible from Hot Springs Island and the entire Juan Perez Sound, the heart of South Moresby.

Western Canada Wilderness Committee spokesperson Paul George claims that there is plenty of Western Forest Products' timber left in the northern part of Lyell Island near Beljay Inlet, enough for several years of logging at the present high rate. However he thinks that the company wants to log the southern part of Lyell simply to wreck the view, a ploy the company has already used successfully to ruin the water corridor south of Windy Bay on the east side of Lyell.

George said, "It's only a few months until the government decides the ultimate land use in South Moresby and it is inconceivable that the Forest Service will crumble to the company's pressure at this time.

But the fact that the company is pushing makes it all the more important for people to let the government know how they feel about South Moresby. With this world reknown biological treasure-house, an irreplaceable monument to nature's splendor, the chain saw lobby just can't win."

CLEAR CUT SO. LYELL HURTS VIEW PHOTO P.C.G.

Front page of WCWC's first newspaper publication. Document: WCWC archives.

as a national park. Colleen could not believe her ears. After visiting the Social Credit government ministers in B.C. for so many years to lobby for the Valhalla Park, she was thrilled to find this federal minister enthused about protecting wilderness. This was a huge boost to the campaign for South Moresby.

Paper promotes *Islands at the Edge*

Thom Henley and John Broadhead were nearing completion on Island Protection Society's book called *Islands at the Edge*, to be published by Douglas & McIntyre. "At least I can help sell the book," I thought to myself, still feeling guilty that I had not been able to write the book that initiated my involvement in South Moresby. Although I hadn't even seen a draft copy of IPS's book, I knew from the list of contributing writers and the fact that John Broadhead (who was a talented graphic artist active in IPS from the start) helped design it—that it would be great. In fact, the book featured a large number of Richard Krieger's photos that were originally destined to be in the book that I was supposed to write but never did. Anyway, I believed that *Islands at the Edge* could take the South Moresby campaign over the top if enough people—and the decision-makers—read it.

The key feature of our *South Moresby –A Special Issue* paper was a whole page devoted to promoting the *Islands at the Edge* book. I was worried that the publisher would print only a couple of thousand copies. So to show them that there was a big demand and prod them to print more copies we asked readers to make a pre-publication purchase of the sight-unseen book, by making a cheque directly payable to Douglas & McIntyre for $29.95 (the full retail price) and sending it to us. We printed only 20,000 copies of our first paper and they were soon gone. As the orders flooded in, we forwarded them to Douglas & McIntyre. I think this pre-publication interest led to the printing of an extra thousand books. It dawned on us that, in the future, we should buy the books at wholesale and sell them ourselves.

Impressed with the success of our first South Moresby newspaper, we decided to publish a second one right away and double the print run. This paper, simply titled *South Moresby–Fall 1984*, was a cooperative effort. The National and Provincial Parks Association of Canada (NPPAC)—now called Canadian Parks and Wilderness Society (CPAWS)—helped pay part of the printing costs, while IPS freely provided photos from the collection used to produce *Islands at the Edge*. The whole middle two-page spread of our paper featured the book. This time the cheques to purchase the books came to us; we in turn purchased 250 copies of the book from the publisher at wholesale. We were developing some business savvy! One of the first things I did with our "profits" from the book sales was to buy 20 copies and send them out with personal cover letters, as gifts from WCWC, to all the key political decision-makers. It was not lobbying–it was educating. Unlike lobbying, new information is always appreciated by politicians as they can use it as a basis for their decisions. We got sincere thank you letters from most of them for our thoughtfulness in sending them this educational book.

We agreed to share all donations from the appeal in the newspaper with IPS and CPAWS. There weren't as many donations as we'd hoped for and, after costs were paid, there was nothing left to distribute. This was pretty typical in the years to come for all our papers. They rarely made money but they served their prime purpose well: to get the message out to a huge number of people.

Our '85 wilderness calendar comes out late—as in previous years

Procrastinating again, and busy on a number of campaign fronts, I got a late start in producing the 1985 calendar. This year Mark Haddock did the layout and design. Because 1985 was the 100th anniversary of the establishment of Banff—Canada's first National Park—we decided to include in our calendar several northern areas that

Cover of 1985 calendar features Castle Mountain in the proposed South Chilcotin Mountains Provincial Park. Note Parks Canada's Centennial Logo on bottom right. Cover image: John Philip.

Parks Canada was in the process of protecting, and got permission to add Parks Canada's centennial logo on the calendar cover. Although we had planned to use an image of one of the northern National Park candidates on the front cover, none of the 35mm slides Parks Canada sent to us made the grade. Instead, we used one of a hiker walking towards the black crumbling volcanic cone called Castle Mountain, in the heart of the proposed South Chilcotin Mountains Provincial Park. With deep blue sky and rich red rock it was spectacular. I knew it would sell calendars—a very important consideration!

Chapter 10

Meares Island – first big B.C. wilderness campaign to go confrontational
Government subsidies end

Meares Island conflict comes to a head

During the final days of the 1985 calendar production, the simmering Meares Island conflict boiled over. For years the B.C. government's Meares Island Planning Team had been meeting to develop options for this scenic island backdrop to Tofino on the west coast of Vancouver Island. For a few of those years I had been the Nuu-chah-nulth Tribal Council's representative on this planning team. I had quit this Team when it started working on options on how to log the island. The position of the local Band, the Tla-o-quiaht First Nation, was that the land and trees on Meares were theirs. Their main village, Opitsaht, had existed on Meares for millennia and they didn't want any logging there.

The Team worked up several options; but even the most pro-logging option wasn't enough for MacMillan Bloedel (MB), whose Tree Farm Licence tenure contained most of the good timber on the island. Several months before the Team completed its work, MB quit the Team and submitted its own aggressive logging plan directly to the B.C. government. It was an arrogant move that demonstrated the company's complete disregard for what local citizens wanted.

Knowing this, it still came as a shock when, on November 10, 1984, the B.C. government's Environment and Land Use Committee (ELUC) announced that it had accepted nearly all of MB's independently-submitted plan. The government's decision allowed logging on 90 percent of Meares. It meant that all but 800 hectares of Meares Island's 8,000 hectares would be clearcut, leaving untouched only the watershed around the lake in the centre of the island that provided Tofino's drinking water.

Tofino residents and Tla-o-quiaht First Nation citizens were incensed. A week later the B.C. Forest Service issued cutting permits including one on the face of Meares fronting Tofino. Protesters gathered and camped at Heelboom Bay (C'is-a-quis) on Meares where the logging was to begin.

On November 21, 1984, when MB's loggers arrived at Heelboom Bay ready to start cutting, they were stopped by the first logging blockade in Canadian history. Tla-o-quiaht Chief Moses Martin spoke eloquently at the blockade line, telling the MB loggers that Meares was his people's "garden," the cedar on Meares had been used for countless generations, and the forest there bore evidence of that use. Culturally modified trees (CMTs) were everywhere. He told the loggers that they were welcome to visit, but they had to leave their chainsaws in the boat. Shortly thereafter, the loggers left. A series of court cases began, as the local citizens maintained the blockade and vigilantly made sure the company didn't begin to log somewhere else on the island.

While this was going on, I felt powerless in not being able to help the blockade in any direct way because of WCWC's commitment to work within the law and not participate in civil disobedience. But there were other things we could do besides standing on the blockade. I asked Ron Martin, a Nuu-chah-nulth artist, to draw something

Tla-o-quiaht Chief Moses Martin stands firm at the blockade line on Meares Island at C'is-a-quis (Heel Boom Bay). Photo: WCWC files.

that could be used as a button to symbolize the resistance. What he created was startling: it was a native mask starkly drawn with stringy hair and big lips.

I was skeptical about it at first but went ahead making up the buttons anyway. On the top, above the mask, I put in a semicircle PROTECT MEARES ISLAND; and below, in smaller red type, *Clayoquot Sound*. I rushed to get the buttons pressed and sent some to Tofino within a couple of days. People liked them. The button conveyed a powerful expression of determination and defiance. During the next few months we produced and distributed several thousand. At $1 each they raised some money for the cause, too.

Meares button art with mask drawn by Nuu-chah-nulth artist Ron Martin. Artwork: WCWC files.

For the Children poster. Poster image: Adrian Dorst.

Meares "For the Children" poster

The Meares campaign also needed a new poster. I had a great image in mind—the Adrian Dorst photo we were using in our 1985 calendar of a young person standing beside the "Hanging Garden Tree" on Meares. At 61 feet in circumference, this tree was the second-largest known red cedar in Canada. The heat of the campaign made this Meares poster a real rush job. I couldn't find any graphic artist to design it; so I did it myself. Unhappy with the typeface that I had typeset and desperate to get the poster to press, I crudely tore out large individual letters from a sheet of white typing paper spelling out "*For the Children.*" I pasted them on the poster artwork imprecisely at a slight angle above the typeset statement "*Make Meares Island a Park.*" Given that I have little artistic talent, it looked fittingly child-like and appealing. The 5,000 posters cost $2,000 to print. I gave half of them to the Friends of Clayoquot Sound on consignment at 50 cents each.

For the first time, we mass-mailed campaign posters to the media. Someone gave me a copy of the Vancouver PNE fair's media list, which had the address of every B.C. newspaper on it. I sent each one a copy of the poster in a mailing tube along with my personal "letter to the editor." Making a splash in the media was more important than making money. Here is the text of that letter:

Dear editor:

Here is a "For the Children – Make Meares Island a Park" poster. It features a full-colour photo of the largest uncut cedar in Canada. More than 19 feet in diameter, this tree grows within a designated clear-cut zone on Meares.

We have published this poster in co-operation with the Friends of Clayoquot Sound, the Tofino environmental society that is fighting to save Meares Island, which lies in their "backyard." The poster attempts to show those who have not yet visited the island what will be lost if it is logged.

Currently, the Friends are helping maintain the forest protectors camp which has been established by the Clayoquot Indian Band at Heelboom Bay, the site where MacMillan Bloedel Ltd. plans to begin logging in the next few weeks.

Our Committee will send a copy of this poster free-of-charge to any of your readers who want one. Of course, donations to help win this conservation cause are much needed and graciously accepted.

This is a big environmental battle. We must save Meares Island's magnificent old-growth forest for our children's children to experience, study and enjoy.

(originals signed)

Paul George – Director, Western Canada Wilderness Committee

I don't know how many newspapers actually printed my letter; but I heard from people that quite a few did. I knew from past experience that we wouldn't be overwhelmed with requests for free posters. As expected, in exchange for the priceless publicity, fewer than a hundred people took advantage of the offer. The donations that concerned citizens sent in to save Meares far outweighed the cost of sending out the posters. **I learned that every new stick that one added to a campaign fire helps it grow.** The Meares Island campaign fire was beginning to roar.

A year later a letter arrived in the mail from a Vancouver Island lawyer. In it was a bequest cheque for $2,000 specifically for our "Meares Island Fund." In a postscript to his letter the lawyer said: "This good lady was inspired by the '*For the Children*' poster put out by WCWC on the wall in my office."

Postering MB's office building

Although WCWC didn't participate in unlawful acts of civil disobedience, nothing stopped us from organizing lawful protests and rallies. Over the years we held lots of them. Shortly after our poster came off the press, we organized a demonstration in front of MB's building on West Georgia Street in Vancouver. Although only a few dozen people showed up, they held up Save Meares Island placards and handed out hundreds of free posters that volunteers had rolled up the night before to passersby.

On impulse, I took a big garbage bag full of posters in my hand, and with Kallie (she was so cute!) on my back, entered the building, boarded an open elevator and pushed the button to the floor where the MB's head office was. Just as the door was closing, a security guard shouted in an angry voice, "Stop, you can't go up there!" He rushed over and tried to put his hand in the door to stop it, but he was too late. I thought to myself, "Hey, they are going to be watching to see where this elevator stops, then they'll take the next one up to apprehend me." I pushed several more buttons on higher floors, and got out a couple of stops above MB's main floor and started to hand out the posters. The secretaries were delighted to get the poster and coo at Kallie. After finishing that floor, I took some evasive action and went down several floors below to hand out posters there. It took security about 15 minutes to find me. By then, I had almost finished passing out all the posters I had with me. The two guards who located me were very angry and threatened to call the police. I apologized, said that I didn't know it was private property and promised not to do it again. They let me go.

This was as close as I ever came to practicing civil disobedience.

As I was doing it, I kept thinking about the people who were camped out on the Meares blockade lines, and the court hearings underway in which MB was seeking injunctions against the protestors. I had personally experienced some of Meares' ancient forests and I was as passionate about protecting Meares as I was about South Moresby.

Fed's free postage gives '85 calendar a big boost

The 1985 calendars came out at the very end of November. Again it was a frantic scramble to sell them. Thankfully, free mailing continued at ORC through most of 1984. We were able to send out all of our pre-Christmas appeal letters and calendar order forms. In November, however, the 1200 Hornby mailroom began to clamp down. Anyone mailing more than ten of the same form letter had to have the contents inspected to make sure the mail service was not being used for personal or commercial purposes. Packages were no longer acceptable. This definitely was bad news for us. We hadn't budgeted for postage and there was no way to pay for it on credit.

Despite the Hot Spot row, WCWC's relationship with Paul Mitchell at Environment Canada had continued to be friendly. Casually mentioning to him that we were suffering a cash crunch because of the unexpected mailing costs, he offered to let us and all other B.C. environmental groups use their mailing machine. But we had to operate it ourselves. He put no restrictions on how much we could use it. I don't know if he truly understood what this meant. I got the sense that he was thinking in terms of dozens of envelopes and I was thinking in terms of thousands of envelopes as well as hundreds of large packages. We were at the height of our busiest season with nearly a hundred envelopes of one and two calendars and dozens of boxes of five, 10 and 25 calendars to send out every day.

The first day I casually walked into Environment Canada's mailing room with two boxes of addressed, stuffed single calendar envelopes to run through their metered mail machine. This wasn't easy to do because the Environment Canada office was located right in the middle of downtown Vancouver in a tall tower so it involved parking in an underground paid parking lot and taking the heavy boxes of calendars up and down the elevator.

A secretary gave me instructions on how to run the single calendar packages through the big Pitney Bowes mailing machine and then left the room. I had not brought down any of our packages of calendars. I thought I'd wait to do that until the next day so as not to freak them out about how much we were going to use the free mailing service. While I was putting the envelopes through the machine I noticed that they had left the manual on how to run the machine out on the table. I took a look and soon figured out how to print off postage labels with the automatic licker disengaged. Voila! I had dry "stamps" to take away. Knowing the weight and exact amount of postage the packages of five, 10 and 25 calendars needed, I set the dials and punched the print button in rapid succession. Zip. Zip. Zip. Out popped the postage labels which accumulated in a pile beside the machine. I repeated the process for each quantity, taking home only what I knew we could use that same day since the date was printed on each one. It probably also had some sort of code identifying it as federal government postage, but it wasn't obvious to me. I took the postage labels home, put them on the packages and took them to the local post office. No questions asked.

We repeated this every day taking a token number of envelopes with us to directly apply the postage there and printing off a stack of dry labels to take away for the rest. I wondered what the limit to our use might be, but never asked. One day I discovered that there was a button on the side of the meter to push to check on the remaining postage in the meter. I pushed it and couldn't believe my eyes. There was just over $67,000 to use up before it ran out! Whenever I visited this small regional office it was always seemed to be a sleepy hollow compared to the bustling WCWC office in our home. I figured it would take years for them to use up that much postage. Our use, at most five thousand taxpayers' dollars, hardly made a dent. This free mailing lasted the duration of our calendar season. Early in the New Year, Paul Mitchell informed us that we could no longer use the machine. The only explanation he gave was that his budget year was up. From the start I knew this privilege wouldn't last long. But while it lasted, it gave WCWC a big boost during the critical time when we were still paying off our '83 calendar debt.

All good things come to an end

The free mailing privileges at ORC were coming to an end, as well. On January 10th another directive came down from on high. If more than 50 pieces of one type were being mailed, a sample had to be left with them on file. Then, starting April 1, we were allowed to mail only small #9 envelopes weighing 30 grams or less. All others (like the standard #10 envelope that our tabloid newspapers fit in) no longer qualified. The writing was on the mailroom wall! On March 19, 1985, a few weeks after we moved our home office to Alberta Street near Vancouver's City Hall, I found a curt note in our mail box stating: "Please be advised that as of the above date there will be no more free mailing service. If you have any questions contact Robin Draper."

It applied to everyone. I'm sure our strongly worded wilderness preservation advocacy mass mailings contributed to its demise. Politicians don't like to fund their critics. I was philosophical about this loss. As they say, "all good things come to an end." The two years of unexpected government mailing machine handouts helped pull us through a make-or-break era in the life of WCWC. But more importantly, it taught us the power of mass mailings as a way to educate people. We may never have discovered that on our own. **Successful campaigners always take full advantage of every break they get!**

Chapter 11

Meares Island takes centre stage in campaign efforts
South Moresby heats up

Third WCWC newspaper a "masterpiece"

In January 1985, we produced and published our third tabloid newspaper—*MEARES ISLAND - PEACEFUL PROTEST HALTS LOGGING*. It was a masterpiece.

The front-page action shots of protesters and confrontation really grabbed peoples' attention. The photos of massive clearcuts shocked people, and the full-page photo of a huge ancient spruce on the back page awed them. It was the paper that took us to a new level of campaigning and public education.

The lead article began:

> On November 21, 1984, MacMillan Bloedel's crew boat, the Kennedy Queen, headed towards Meares Island. On board were a few chain saws, some loggers and a handful of company brass....
>
> In less than an hour, screaming saws in the hands of the 'disconnect technicians', as fallers are known in logger lingo, would be separating the stumps from the trunks, creating a clearcut landscape on Meares. The forests would be converted to two by fours, newsprint and slash.
>
> But not to be....
>
> On shore they were met by nearly one hundred people with songs and drumming.... Moses Martin, elected Chief Councillor of the Clayoquot Band, read from the original 1905 Timber Licence, the legal cornerstone of MB's claim to rightful ownership of the trees. This agreement exempts all Indian plots, gardens and grounds from logging. Chief Martin told them the Island was a Tribal Park, his peoples' garden; that they were welcome as visitors but that no logging would be allowed.

The images, supplied by the Friends of Clayoquot Sound, covered more than half of the newspaper. It was a winning 50/50 ratio of images to words that we've aimed for in every WCWC publication since then. The tendency is to fall short on the photo side, succumbing to our intellectual bias that words seem more important than pictures. They're not. **If there's one thing I've learned through the years of producing WCWC educational materials, it's that most people grasp ideas faster and better when they're presented visually as well as verbally. If you are aiming to get through to peoples' hearts: pictures speak way louder than words.** This Meares paper told the whole story in 15 photos

WCWC's anti-tree-spiking stance angers environmental movement's radical fringe

The odd photo in that paper, the one that shook people up the most, featured a person, back to the camera, "modelling" as a tree spiker in the act of spiking a tree. Our caption read "Tree spiking, a practice not advocated by the Clayoquot Band or the Friends of Clayoquot Sound." We didn't condone it either, but it was being talked about amongst many of the grassroots activists as an easy and sure way to save the forest. Our paper didn't stop some misguided activists from spiking some trees on Meares Island, but we remained adamant that tree spiking did our cause more harm than good. It

Above: Photo featured on page three of WCWC's *Meares Island - Peaceful Protest Halts Logging* January 1985 newspaper with this statement underneath it: *Tree spiking, a practice not advocated by the Clayoquot Band or the Friends of Clayoquot Sound* [and WCWC]. Photographer unknown.
Facing page: Small boat flotilla encounters MB's Kennedy Queen and explain their resolve to protect Meares Island from logging. Photo: WCWC files.

shifted public discussion to the issue of law and order and the concern that a faller or a sawmill worker might get hurt or killed if a saw hit one of the spikes. Concern for peoples' lives and safety then became the prime issue, not the ecological importance of saving the wild forest.

A couple of years later when the issue of tree spiking came up again, WCWC offered a $5,000 reward for information leading to the arrest and conviction of any tree spiker in B.C. This offer still stands today. At that time some person or group reacted violently to our reward offer, gluing shut the outside locks on our Gastown storefront office and defacing our big store window by etching into the glass the word "traitors."

We learned another lesson in publishing this paper. To be a good campaign tool, a publication has to present a clear vision of what you want as well as decry what you don't want. This Meares paper presented a great vision. Set into the big tree photo on the back page was the Clayoquot Band Council's recently proclaimed *Declaration to Preserve Meares Island as a Tribal Park*. It even spelled out the rules for public use.

In any campaign you have also got to ask people to personally do something concrete and useful to help. Every newspaper we've published since this Meares Island tabloid includes two fundamental requests from our readers: write letters to government and send us money. Our conviction that we would ultimately be successful was rooted in our belief that our democratic system worked. If we mobilized enough people to support the cause and write letters to government, we'd eventually persuade politicians to make the right public policy decision and win. From WCWC's cash-strapped point of view, it seemed like getting a donation was the most important part of every paper. If people put their money behind the cause, they really believed in it.

The majority of WCWC newspaper publications (collectively the total copies exceed 12 million over the last 25 years) have been jointly produced with another environmental group, like this Meares paper, which was co-published with the Friends of Clayoquot Sound.

Meares protests locked in court cases

Our Meares paper only briefly hinted at the legal quagmire that was deepening month by month around the Meares issue. On November 23, 1984, within days of MacMillan Bloedel being blocked from logging by protestors at Heelboom Bay, MB sought an injunction against Chief Moses Martin, Friends of Clayoquot Sound spokesperson Michael Mullin and "anyone else" found obstructing its logging operations on Meares. Shortly thereafter, the Clayoquot and Ahousaht Bands filed for their own injunction to prevent MB from logging Meares. For five days in January of 1985, the Supreme Court of B.C. heard MB's arguments why they needed to log Meares to sustain jobs and First Nations' testimony as to their traditional ownership and use of Meares. Justice Gibbs granted MB a temporary injunction against the blockaders and rejected the First Nations' request to delay logging until a full court trial determined the ownership of Meares. The First Nations immediately appealed the decision, which kept MB loggers temporarily at bay.

Lone Meares Island picketer gets big publicity

I just happened to be in Victoria on the opening day of the Legislature on February 11, 1985. On the spur of the moment I decided to use two Meares Island posters to make up a Meares Island picket sign and be there during the opening ceremonies to protest the government's decision to log the island. By chance I turned out to be the only protester on the Legislature lawn that day. For some unknown reason, no other protest group showed up. My one-man protest made a big splash in the news.

It's vitally important to take advantage of every possible opportunity to get your issue in the media. Where the media is sure to be present is a good place to be. A protest doesn't have to be big to be effective. It's worth it to constantly "be out there" to build support and chip away at government defences.

From then on I made sure that WCWC was present at almost every Legislature opening day ceremonies to pressure for particular wilderness issues that were hot at the time. Some of these protests were large. Some were small. But the February '85 protest still holds the record as the smallest protest of all time.

Developing a sound formula for success

In the 1984-85 escalation of the Meares conflict, WC²'s role started to become defined as "the publishing arm of the movement." **There are basic secrets to building a movement through public education. Volume is key. You've got to reach out to the unconverted and appeal to their values. Educational materials are most effective when they are focused on one campaign issue, not a bunch of them 'cobbled' together. Each publication has to stand on its own—in other words, be understandable and interesting to someone who knows absolutely nothing about the issue.** The cover must be inviting and stimulate curiosity. Anyone casually picking it up must be able to read the headline, titles and the cutlines under the pictures and get the basic message in a matter of seconds. A campaign newspaper has to leave the reader with a good grasp of the issue, even if they are opposed to wilderness preservation. It has to include the pros and cons: images both of the natural splendour and what it would look like if the area were "industrially developed." These evoke compassionate responses and create new preservationists. It's also got to be well researched and factually correct. It cannot just be full of rhetoric and opinion. In later publications we often included a "Q&A" piece: the main arguments made by our opposition and our answers to them, thus better "arming" our supporters to debate the issue and convince their friends.

I take much pride in the fact that over the years seldom has anyone disputed the facts in WCWC's newspapers even though they may not have liked or agreed with our position. We've also heard that some of our papers have been used as briefing documents for cabinet ministers. That's a sure sign they're effective.

Our Meares publication had some new graphic twists. A part time worker at Pacific Press had given me a shoe box full of roll ends of different kinds of graphic layout tapes that he'd scavenged from the trash cans when paste-up workers threw them away. We could never have afforded the new rolls of graphic tape—they cost between five and ten dollars each. In this gift box were over 100 different kinds of tapes—thick lines, thin lines, dotted lines. They were like magic. Instantly they made our papers look more professional and exciting. I especially like the thick black tape with the white stars reversed out of the middle. I used it all over the Meares paper.

I also learned the magic of using "spot colour" for emphasis. It only cost another $50 for the whole press run and well worth it. We picked a dark green spot colour for this Meares paper. Spot colour also meant that paste-up had to be done on a light table to ensure proper registration of the second colour. With spot colour included, it cost only $1200 to print 50,000 copies—less than three cents a paper—cheaper than a single letter-sized photocopy.

This Meares publication was extremely successful, one of the few that brought in more donations than it cost to produce. However the Friends of Clayoquot Sound, with whom we were to share the "profits," owed us more than a thousand dollars for posters and calendars we'd given them on consignment, so we mutually agreed to apply all their share of the net donations towards retiring that debt. FOCS were financially stretched, too, keeping the protesters' camp going and patrolling around Meares Island to make sure that MB didn't try to sneak in and log there. They spent thousands of dollars on boat fuel alone during this intense period of confrontation.

Big decisions

Colleen and I had made a pact when we first met that whoever won their campaign first would devote their time and energy to the other's. After the Valhalla Park became a reality, I invited Colleen to join WCWC's board of directors and use her skills and energy to campaign for South Moresby. She did work on South Moresby, but on her own, staying with Elizabeth May and other friends when she went to Ottawa to promote South Moresby. Short of money for expenses, she was going into debt paying for her telephone bills, travel, meals, and taxis. I finally enticed her to join our board by telling her that we would cover all these expenses. We elected her to the board, by acclamation, at our AGM in October 1984.

During the next few years she visit Toronto and Ottawa on a regular basis. In Toronto she convinced environmental reporters at the

Globe and Mail to do a series of stories on South Moresby. She got key reporters at the *Toronto Star, Ottawa Citizen* and *Montreal Gazette* to start writing stories too. Colleen worked hard to raise South Moresby to a national issue. She met with national environmental groups, distributed WCWC's volumes of educational material to MPs, key media in Toronto, Ottawa and Vancouver and spearheaded the National Committee to Save South Moresby.

In late winter of 1985, Adriane and I had another heart to heart talk about our future. We both concluded that we couldn't continue to work full out on both the Green Party and WCWC and be effective. We'd have to concentrate our efforts on one or the other. The choice was not hard to make. The Green Party, despite our best efforts, had declined. The membership was a fraction of the 1,500 it had at election time in May 1983. There was also much friction in the party and focus on talking rather than action. There were too many single-issue people and quite a few who believed that the party shouldn't even run in elections. ("It must be a 'movement'," they said.) The party had adopted strict consensus decision-making and it was bogged down in process with stubborn people "blocking" decisions at every meeting. Many resented having any kind of leader or spokesperson and felt that everyone should have an equal say. The Green Party was a good idea whose time had not yet come. The B.C. public especially wasn't ready for it.

On the other hand, during the same time period, the wilderness preservation movement had taken off. Meares Island was the flash point. WCWC had grown. With a concerted effort there was a good chance that a great deal of wilderness could be saved in the next few years. There were also many good, hardworking positive people in this movement. Their conflicts were over strategies, not fundamental goals. One issue—creating more parks—is much easier to move forward than the whole gambit of issues on a political party's plate. Thus Adriane and I decided to continue to be Green Party members and supporters but not to be actively involved—at least for a time. We'd put our full efforts into building WCWC and educating people about the basic issues of wilderness conservation and ecological sustainability. It was a good decision.

On a mini-holiday in the spring of 1985, Adriane and I stumbled upon a piece of property near Gibsons on the Sunshine Coast. It was wooded, isolated and adjacent to a regional park with large second-growth trees in it. Almost as a lark we put in an offer at half of the asking price. The next day our offer was accepted! To get the money to build, we scrambled to finish and sell our Kitsilano home. Until our new home on the Sunshine Coast was finished, we rented the main floor of a duplex on Alberta Street near Vancouver's City Hall. It was small but it had a spare room that made a perfect WCWC office. Our new place became "WC^2 Action Central." We flourished and continued to grow.

An officewarming present

On moving day to Alberta Street, as we were putting the furniture into the new study, I casually said to Adriane, "This room looks like it's missing something—like a light table to paste-up newspapers." It was my gentle way of introducing the idea that she should lend WCWC the money to buy one. While getting layout materials from a graphics supply store a couple of months earlier, I'd spotted a "slightly damaged" light table on sale. It had a small crack in the top corner in the glass top and some paint scratches. Otherwise it was a beautiful big, tilting, sturdily built workstation. It cost only $395. After very little arm-twisting, Adriane agreed to buy it "for the cause," if it was still there. I raced down to the store. It was. I purchased it and got it

Covers of first two Meares newspapers. Documents: WCWC files.

delivered that same day.

During our first week at Alberta Street, our first really large donation arrived in the mail. It was a neatly handwritten cheque for $2,000 from a private individual. Prior to that letter the most we had ever received at one time was $100—except for the constant donations from Richard and now from Adriane that I simply took for granted. The $2,000 came from someone living in Ontario that I didn't know! At a directors' meeting we decided to use this wonderful windfall to produce our third South Moresby paper. The campaign desperately needed another big push.

The transformation of the efforts to save South Moresby to create a National Park Reserve

Since January, Colleen had been organizing a National Save South Moresby Committee. By spring of 1985, 20 famous Canadians had joined, including MP Jim Fulton, Hon. Charles Caccia (then the Federal Liberals' Environment Critic), artist Robert Bateman, zoologist Bristol Foster and writer Farley Mowat. Colleen traveled to Ottawa to build support for South Moresby on a regular basis. Of course, she was always educating, not lobbying, MPs to see the wisdom in supporting national park protection for this area.

Every time she went, we loaded her down with bundles of educational newspapers and flats of posters. More often than not Colleen was able to talk her way onto the plane without any extra baggage charges. This also included the time we piled her baggage cart so high with heavy newspaper bundles that a wheel bearing screeched ear-splittingly as it rolled along. Everyone in the airport stared at her as she approached the ticket counter. Many people, including airline workers, supported the campaign to save this special place. Colleen claims that her permanent back problems stem from toting tons of WCWC material in and out of cabs and around Ottawa on her frequent trips there during this campaign.

Colleen's national efforts obviously had an impact on B.C. politicians, too. On March 13, 1985, B.C.'s new Minister of Environment, Austin Pelton, met with four of our directors including Colleen and me. It gave us our first real glimmer of hope. Minister Pelton promised us that he would discuss possible park options for South Moresby with his federal counterpart, the Honourable Suzanne Blais-Grenier. Pelton told us he personally favoured the park but he "needed demonstration of widespread public support to win over more skeptical cabinet members."

Building that public support was our mission. We published our third South Moresby newspaper in June of 1985, a couple of months before the Parks Canada Centennial celebration in Banff. We titled it *South Moresby, Queen Charlotte Islands – Misty Wilderness Gem of the Canadian Pacific*. The lead story headline read "Famous Canadians Team-up to Save South Moresby." It featured quotes from some members of the august group that Colleen had pulled together, including this blunt statement by Farley Mowat: "*South Moresby is the last remaining region in Canada that has not been desecrated by the blind avarice of modern man. It must be saved not just for our time, but for all time. It will be saved if a shred of honest morality still exists in Canadian society.*"

Baking the mail to get our "Act now to save South Moresby" appeal out

We had to get this new South Moresby newspaper into our supporters' hands quickly. There were slightly under a thousand people on our mailing list at the time but they were a dedicated group. We could count on them to both write letters and send in donations!

As usual, money was tight. We had only enough in WCWC's bank account to pay the postage for a standard letter of 30 grams (34 cents in 1986). After I picked up the newspaper from the press, I emptied WCWC's bank account to buy the mailout stamps. I figured that we could squeeze in the new South Moresby tabloid newspaper and a single page appeal letter and come in just under the maximum weight allowed.

It took all afternoon to print the mailing labels. I had runoff the "perfect" letter, edited by Adriane of course, the day before. A volunteer had already rubber-stamped the envelopes with our return address. The volunteers were all scheduled to come to our Alberta Street house to apply the stamps and stuff the envelopes that night. As always, as it still is now with mailouts, timing was critical. We had to get the letter in the mail the next day both to generate letters to government to sway them away from the brink of a bad decision and to generate some money for our now empty bank account.

Everything was on track...but I was worried that the package might weigh too much. The newspaper was slightly larger than the last one and the recycled brown envelopes seemed a bit heavier that the white envelopes we had been using. It was 4:50 p.m. and I rushed to the local post office six blocks away arriving just in time to get a sample package weighed before closing time.

"I'm sorry," said the postal clerk, "It's three grams over the limit. You'll have to put on 17 cents more postage."

"Are you sure?" I asked.

"Yes. Our new electronic scales are extremely accurate."

Disheartened, I started back home. I just couldn't ask Adriane for another loan to pay for the extra postage. Moreover, the delay would cost us at least a couple of days and the message had to get out now.

Then I got an idea; a flashback to something I learned during my days as a high school science teacher. Newsprint is very hygroscopic. The freshly printed newspapers were slightly damp. If somehow I could dry them out, our envelopes were sure to come in just under the 30-gram limit. I picked up some large plastic garbage bags on the way home, determined to get the mailing done that night.

After supper, the volunteers came and began the labour-intensive job of folding, stuffing, stamping and sealing the envelopes. Meanwhile I preheated the oven to 225 degrees F and got out our cookie sheets. The first batch was a failure. The big pile of stuffed envelopes did not dry out—only the outside ones did. I figured out that success lay in baking a small number of envelopes at a time for 30 minutes each batch. Fresh from the oven you could really feel the difference...they almost crackled they were so dry. Immediately upon baking them I put them in a garbage bag and twist-tied the top tight so the crispy-dry envelopes would not reabsorb moisture from the air.

It was a slow, laborious process. Long after the volunteers had finished and gone home, I was still baking the mail. Finally, at 4 a.m. in the morning I took the last batch out, turned the oven off and went to join Adriane, who had gone to bed hours earlier.

Five hours later (at nine the next morning), in a heavy downpour, I loaded the tightly sealed garbage bags into the car, confident that the envelopes would pass the weight test. As bad luck would have it, as I entered the post office with one of the bags, the clerk who had weighed the test envelope for me the night before spotted me. "Let me see one," he asked. He took one randomly out of a bag and put it on the scale. A couple of seconds later he said, "That's odd, the weight isn't registering on the scale.... Oh, there it is—28 grams." He was about to take the letter off the scale and hand it back when he stopped and looked puzzled again...the weight on his display screen had disappeared and a few moments later suddenly flashed back on again, this time reading 29 grams.

"It must be right on the border between 28 and 29," I commented, thinking to myself that he better get it off the scale fast. The envelope and its contents, rapidly absorbing moisture from the humid air, soon would be overweight again. "It's great you have such an accurate scale," I began to banter. "We had to work hard last night to get the weight down to meet the limit."

He took the envelope off the scale, handing it back to me and I proceeded, with an inaudible sigh of relief, to unload the rest of them into the mail bin. Another volley was on its way to help save South Moresby.

Celebrity endorsements fail to help financially

This "Famous Canadians" South Moresby paper was not as successful as the previous newspapers, at least as measured by donation coupon returns. I couldn't accept that it was because people no longer cared about the issue. I came to the conclusion that it was because readers must have thought that: "If all these important people backed the cause, South Moresby will surely be saved without the help of a 'nobody' like me."

That was the last time we featured testimonials from well-known artists, politicians, writers and naturalists on the front page of one of our publications.

Campaign to make record-sized native trees "Canadian Landmarks"

During all of 1985 we tried to make a big thing of the 100th anniversary of Parks Canada. We could not understand why the federal government seemed to be downplaying this centennial. In our minds it was a perfect opportunity to push the South Moresby park campaign over the top. All year, we feverishly promoted the idea of creating a "South Moresby National Park Reserve" as the best possible way to celebrate.

We also promoted another "centennial" idea: to start recognizing and protecting B.C.'s record-sized trees as "National Landmarks." Parks Canada had taken the first steps to protect our country's rare and exceptional natural features as "Canadian landmarks" in 1983. In April of that year the federal Environment Minister at the time, John Roberts, wrote a four-page letter to us explaining Parks Canada's objective:

> *To foster protection for all time of exceptional natural sites of Canadian significance in a co-operative system of Canadian landmarks, and to encourage public understanding and appreciation of this natural heritage so as to leave it unimpaired for future generations.*

It was a lofty goal but government had moved at a snail's pace in getting the program going. The enabling legislation, the Historic Sites and Monuments Act, was finally enacted in 1985.

Meanwhile, in 1984 Bristol Foster had introduced me to Randy Stoltmann, a bright 22-year-old who was constantly exploring our coastal old-growth forests and already knew more about B.C.'s record-sized trees than anyone else. Randy was the first person, as far as I know, to advance the idea that Canada's most massive and ancient living things were perfect "Landmark" candidates. I encouraged him to put together a tabloid newspaper to promote the idea. Randy wrote the text, supplied the photos as well as a hand-drawn illustration of an ancient "candelabra" (multiple top) red cedar snag, and compiled a table listing some of B.C.'s record trees. In February of 1985 we scraped together $700 and published 20,000 copies of Randy's four-page paper entitled *"a proposal to create CANADIAN LANDMARKS – Protection for our largest, tallest and oldest trees."* We featured four Canadian Landmarks candidates: The Nimpkish Valley Douglas Fir, The Cheewhat Western Red Cedar, The Red Creek Fir and the Carmanah Creek Sitka Spruce.

Later that year, a commission toured Canada to hear proposals for Canadian Landmarks. I presented our paper to it. Kim Campbell, who later became Canada's first woman Prime Minister, sat on that commission as B.C.'s representative. On the afternoon of my presentation I couldn't find a baby sitter to take care of Kallie, so I put her in the baby carrier on my back and took her along. It worked out really well for the first couple of minutes of my speech. Then she began to act up. With her tiny hands, barely reaching around my head, she kept pulling on my beard and trying to cover up my eyes in a vain attempt to get my attention.

I know the commissioners were more amused by her antics than they were interested in our worthwhile proposal to give recognition to the outstanding trees in B.C.—trees that rank amongst the top ten oldest and largest species in the world. One commissioner commented that the idea of calling living trees "landmarks" was a rather novel one for they were thinking primarily of geological oddities (like the Pingos of Tuktoyaktuk, which became the first Canadian Landmark) and man-made structures. I still don't understand why our governments are not more interested in protecting these marvels of nature. Most people, fascinated by "Guinness Records," loved Randy's search for these huge marvels of nature. Randy had a way of comparing the tree heights to well-known heritage buildings to bring home the point of how incredibly massive they really were.

Our worthy proposal fell on deaf ears. But Randy continued this campaign and it helped him get an "official" registry of record-sized

Candelabra cedar tree drawn by Randy Stoltmann. Artwork: WCWC files.

trees in B.C. Unlike several other provinces and the U.S., B.C., with its spectacular forests, had neither a record of the largest native trees nor a mechanism to protect them. Ultimately, in 1986, the B.C. Forestry Association took on this registry task and the Association of Registered Professional Foresters took on the task of verifying measurements. Our directors' minutes say we "forced" these undertakings but I always felt it was more like the professional foresters were embarrassed into taking some action by Randy's efforts.

WCWC publishes its first book
Meares Island – Protecting a Natural Paradise

On March 27, 1985, the Appeals Court of B.C. made a pivotal decision that changed the fate of Meares Island. By a three to two decision, the court overturned Justice Gibbs' decision to allow logging on Meares Island and ruled in favour of the Clayoquot and Ahousaht natives. They stated that MacMillan Bloedel Ltd. (MB) could not log Meares Island before the issue of native land claims was fully settled in court. It seemed like a win, but at WCWC we were worried that MB and the Province of BC might take the case to the Supreme Court of Canada. We were skeptical about legal "wins" and knew that greater public understanding and support were the sure keys to lasting campaign success.

Shortly after this decision, Ron Aspinall, a Tofino doctor who was very active on the Meares campaign, phoned me with a great new project he wanted us to undertake. Ron, too, understood the importance of building public support. "We have to have a book that tells the story of Meares and spells out the logging threat. It must have lots of photos. And it has to come out in time for the summer tourist season," he said.

Daughter Kallie riding on my back at rallies often hung onto my beard. Photo: WCWC files.

"That's only two months away," I replied politely, thinking it was impossible. My comment didn't register with Ron, a determined, action-oriented person. "Besides we don't have the money or credit to do it," I continued. Money turned out not to be a barrier. Ron would lend us the estimated $20,000 to print the book. We'd co-publish it with the Friends of Clayoquot Sound who would also help with production and sales of the book. Adrian Dorst, the outstanding local nature photographer whose picture we'd used on the cover of our 1984 calendar, would provide most of the photos. Ron would find people to help with the writing.

There was no way our directors could turn down this book proposal. Everyone knew how successful *Islands on the Edge*, published six months earlier, had been in raising the profile and chances of success of the South Moresby campaign. I contacted Jim Astell, our helpful sales representative at Mitchell Press, and told him of our short time frame and budget. These dictated the size (64 pages), the press run (4,000) and the binding (soft cover). The cheapest way to go was to concentrate the colour pages on two press sheets and use only black and white for the rest of the book. We wanted to keep it inexpensive so that the widest possible audience could afford to buy a copy. I wanted it to be under $10 but economics dictated that the retail price be $12.95.

I contacted my daughter, Athena, who had helped with WCWC's 1985 calendar by handwriting nearly 2,000 thank you letters. (I paid her 25 cents per letter to write one to everyone who ordered a calendar that year to give our effort an added personal touch!) She had taken a course on layout and design as part of her schooling at David Thompson University in Nelson. I talked her into pasting up the Meares book in her "spare time." She did it for $5 an hour and she promised to meet our July 1 deadline if we got the material to her in reasonable time.

What Athena remembered most about the book production was having to simultaneously cope with our toddler Kallie. According to her recollection, taking care of Kallie and changing her diapers was an unwritten clause in her contract. Kallie certainly wasn't deprived. Most afternoons and evenings, young volunteers were at our house, stuffing envelopes and mailouts to help save South Moresby. They loved to play with her and she loved the energy and people...and the office "art" supplies. She had unlimited use of the pens, paper, tape and glue. She once stamped her entire body, except her diaper, with the WCWC rubber return address stamp.

Everyone knew how important it was to get the Meares book published and many people worked full out to get it done on time. Our house just buzzed with activity. Our new Kaypro portable computer, also purchased for WCWC by Adriane, helped in editing. Amazingly our Smith-Corona electric "daisy wheel" typewriter worked as a printer the first time we connected the cable. We felt so advanced in being able to switch wheels to get different fonts. There was no "computer to plate" printing. Even after word processing the text, it all still had to be retyped by a typesetter, which always introduced errors that had to be found and corrected through meticulous proofreading before it was "pasted up."

Ron recruited C. J. Hinke, nicknamed "Spike," to be one of the Meares book's authors. Hinke, who got his nickname by advocated tree spiking as the sure way to save Meares (a tactic he knew WCWC condemned), was a colourful local character and a published author whose claim to fame was that he translated *The Wonderful Wizard of Oz* into Latin. His history of the island was the last section of the Meares book to come in for final editing. Adriane, after spending a few minutes typing his manuscript into our Kaypro computer, called out, "Paul, you have to read this! It says here that Meares Island was the site of the first major settlement by early explorers on the entire West Coast of North America and they worked co-operatively with the natives building sailing ships. I don't think that's right."

"It sounds weird to me, too," I replied. "You better check it out." Adriane began checking. She soon found out that Hinke had written a historical fantasy. It was a great piece of creative writing, but it was entirely inappropriate. It was lucky that Adriane caught it. If we had published Hinke's "history" as written, it would have been a huge embarrassment to us and it would have tainted the credibility of the rest of the book. Adriane worked for several days researching the history of Meares and completely rewrote this portion of the text, delaying the publication by a week. She should have received credit in the book, but didn't because it was easier to just let the matter blow over rather than make a big thing out of removing Hinke's name as an author.

On the last day of July the first boxes of Meares books arrived at our home. We loaded as many in our car as it could carry and headed for Tofino, arriving very late at night. No one knew we were coming or that the book was finally off the press.

Informal book launch in the Common Loaf bakery

Early the next morning we went for a coffee to Tofino's famous Common Loaf bakery. By chance or serendipity, every single person who had helped on the book wandered in, one after the other. For some of them, a morning cup of coffee at the Common Loaf was a ritual. For some, it was just happenstance that they came in that particular morning. Everyone oo-ed and ah-ed over the book, and autographed each other's copies. C. J. Hinke didn't say a word about his heavily-edited history.

It was one of those magical moments that made all the long days and nights of work worth it. The 4,000 copies—everything included—cost $23,039. They were all gone by the middle of the next summer and we reprinted another 4,000 copies, this time hardcover binding 1,000 of them primarily for library orders. Several years later we had to reprint again.

Chapter 12

Taking on our third big campaign – the Stein

Ken Lay brings us a new campaign tool and involves us in a new campaign

Campaigning to protect Meares Island and South Moresby and tending toddler Kallie was all that I had the time to do that spring. While immersed in the midst of the Meares book, an energetic young volunteer named Ken Lay stopped by to see me. He, too, had a proposal I couldn't refuse. I really didn't know much about him, except I'd heard from Adriane that he had a reputation for being able to "bend spoons" with his mind. I had only met him briefly during the calendar blitz the previous fall.

To avoid the commotion inside our house, I took him out to the back porch to talk. Ken explained, "The cable car is gone and no one can get across the Stein River to the trail and hike into Stein Valley." The cable itself that crossed the river was in good shape and, according to Ken, needed no work. It just needed a cage to ride on it. The government had originally built this access system in the late 1970s, but now refused to maintain it. Ken explained how important he felt it was for people to see the spectacular Stein wilderness area located just north of Lytton in the Fraser Canyon. The Stein was threatened by logging, he explained, and public access was vital for the campaign to save it.

Ken proposed a novel low-budget solution. He proposed to build a new cage out of 2-inch steel plumbing pipe and make the floor out of thick-gauge chicken wire. All he'd have to get welded were the pulleys onto the top pipe. In fact, he'd come armed with sketches and plans, which he unrolled on the porch railing. He figured it would only cost about $300 for material and welding. He'd do the major construction work himself and other volunteers would help him install it.

Cable car design simple, but too heavy

It sounded like a bit of a hair-brained scheme to me, but the only comment I made was "Don't you think your cage will be too heavy?" Ken assured me it wouldn't be. His cage would be small and impossible for more than one person at a time to use it. "Why don't you get the Stein Alliance, the Stein Action Committee or SPEC to sponsor it?" I asked. Ken's face fell. He told me that he'd already gone to all of them and they'd turned him down.

It wasn't really the money. It was a question of liability. They didn't want to risk it. I knew it was a very honest answer. He wanted to get started right away and I wanted to say, "Yes!" right on the spot. But I knew I couldn't do that. I'd have to talk to the other WCWC directors and get their consent. Anyway, we didn't have even that small amount of money in WCWC's bank account. I said instead, "I'll see what we can do. Come back tomorrow afternoon."

During the following hours waiting for Adriane to return from teaching, I thought about Ken's proposal. As soon as she entered

Looking up to the pristine Stein Valley to its headwater mountains. Photo: Bob Seminuck.

the door I started talking to her about Ken's project, emphasizing the point that it wasn't that much money and it wouldn't be good to crush this young energetic activist's initiative and enthusiasm. We also discussed our liability and I dismissed it. "Putting up a sign 'Use at Your Own Risk' should handle that," I asserted. "Anyway it's the builder not the funder who'll be at fault if something goes wrong. If you always worry about liability you'll never do anything."

It would take at least a few days to get everyone's approval on WCWC's board and some of the directors were quite conservative and might not have gone along with it. Ken couldn't wait. Adriane thought for a moment and then said, "Why don't I just give you the money to give to him." It was the easiest solution. Adriane and I lived cheaply enough to make her college teaching salary stretch a long way.

I had a feeling that Ken had the drive and smarts to succeed and that he might help WCWC in the future. I had no idea that we were opening up a whole new era for WCWC and embarking on a major new strategy—trailbuilding—that would help create the support needed to preserve many wilderness areas. When Ken came back the next day I told him the good news that Adriane and I had personally decided to support the project. I drove Ken around while he got the pieces of pipe. But that was it. After that day, Ken was on his own. It was his pipe dream and it was up to him to make it a reality. Ken succeeded in building his "cable car" and, with the aid of friends, got it installed. It was well used that summer. Hundreds of hikers visited the lower canyon. The Stein Wilderness campaign was heating up.

Ken's uniquely constructed "cable car" only lasted that summer, however. Someone left it in the middle of the cable hanging over the river during the winter. The cable sagged and the river caught a hold of his handiwork, dragged, smashed and destroyed it.

Undaunted, the next year Ken had a "new improved model" made: a cage of aluminum struts bolted together. He also ensured the cable was tightened to avoid sagging. Ken, we quickly learned,

First Stein cable car made out of plumbing pipes. Photo: WCWC files.

One of the new improved aluminum framed cable cars for Stein river crossing. Photo: WCWC files.

was a resourceful fellow who never gave up on a project. By this time WCWC was fully engaged in outfitting and supporting volunteer work parties to restore and re-open the traditional heritage trail to the headwaters of the Stein. It involved building several more cable crossings, lots of trail clearing and raising of funds.

Ron Adams, volunteer trailbuilder, takes Scudmore Creek cable car. Photo: Leo DeGroot.

Big promotion for the first Stein Festival

In the middle of publishing our Meares book, we also produced our first Stein Valley newspaper. We were latecomers to this preservation battle, which had already been raging for fifteen years. Our four-page paper, *Wild Watershed: The Stein–Summer 1985*, was jointly sponsored by the Stein Action Committee, SPEC and the Institute for New Economics run by Michael M'Gonigle who, along with his wife Wendy Wickwire, were known as the "Stein Scholars."

Besides presenting the strongest possible case for protecting the Stein, which was the largest unlogged watershed remaining in southwest B.C., our paper urged people to attend the first "Voices for the Wilderness Stein Festival." The festival was sponsored by the Lillooet Tribal Council and scheduled for Labour Day weekend (August 31 – September 3, 1985) in an alpine meadow in the Upper Stein Valley. Not only was it free, the Tribal Council also planned to supply food. The promotional flyer said:

> *Easy access to Festival site – All Welcome. Although the Lillooet Tribe will see to it that no one goes hungry, whatever food you can bring will be appreciated. Try to choose items which lend themselves readily to combination with other fare.*

This 1985 Stein Festival was an unprecedented, historically-significant event: a First Nation reaching out to educate and seek the support of non-natives in their fight to save a sacred wild place. We decided that their campaign and their festival merited and needed a lot of publicity, so we ordered a big print run—50,000 copies of our paper. We really had to scramble to get all of them distributed to all the hiking groups and outdoor enthusiasts in time. Once the festival was over, our paper would be stale. Real deadlines are great motivators.

Adriane, Kallie and I take a break and paddle to Megin Lake in the heart of Clayoquot Sound

From May through July, Adriane and I worked on one WCWC publication after another, barely getting outdoors. We were ready for a break. We decided to take up Adrian Dorst's offer to let us borrow his beautiful cedar wood strip-fiberglass canoe for a weeklong trip into the Megin River in Clayoquot Sound. Kallie was 18 months old and it was time for her first big wilderness adventure. "Make sure you don't scrape it on the rocks," Adrian said as we departed.

Although Meares Island had all the headlines, the Megin River Valley was another big wild area in Clayoquot Sound that local people were fighting hard to save. I loved to get into the wilderness. I knew that **once you personally visit a wild place you naturally work harder to protect it. It's sort of like you own a piece of it...or it owns a piece of you.**

Steve Lawson, who worked on the *West Coast Chainsaw Massacre* movie with me a couple of years earlier, took us, the canoe and all our gear in his big aluminum herring skiff up through Sulphur Passage to the Megin. Along the way he pointed out all the places and outstanding features of Clayoquot Sound. It was an exhilarating trip. We arrived at high tide—the only time when you can get a boat in and easily disembark on a nice gravel bar. During low tide the Megin River empties into the ocean via a waterfall. There are steep cliffs and no beaches on either side of the falls. So it's high tide or no entry.

Once on the river we were told it would be easy going most of the way to Megin Lake. It would take us a half day at most to get there. But the water in the Megin was extremely low and we had to line (get out and pull) the canoe through the shallow water much of the way up the river. We even had to portage in several places. It took a full day. Kallie took to the trip like a duck to water. I lined the canoe with her on my back. She grabbed at huckleberry bushes, stripping the berries and leaves with her small hands and stuffing it all into her mouth whenever she got a chance. We had the whole place to ourselves for a week. Only one floatplane dropped into the lake for a brief visit and left. It was great to reconnect with nature, fall in love with one more area in Clayoquot Sound and get revitalized for the work ahead.

Voices for the Wilderness – high in the Stein

When we got back to Vancouver our home was abuzz with Stein Festival fever. Our volunteers, led by Ken Lay, were helping John McCandless, a dedicated Stein preservationist who worked for the Lytton Band Council, with the preparations for the big event. To pull off this event was a huge undertaking. Ken urged us to come and, since Kallie had been so good on our trip in the Megin, we decided to go. The festival was only a week away.

Getting to the festival site trailhead meant driving (bus transportation was provided for those who needed it) up the Fraser canyon and then a long way up logging roads to the top of the heavily clearcut Texas Creek, a watershed adjacent to the Stein. From there it took a hike over a pass into the Stein alpine. In our paper we had advertised

WCWC volunteer Michelle Evelyn, a Quest program graduate, building a cairn trail marker across a rocky alpine section of the Stein Trail above Stein Lake. Photo: WCWC files.

that the trail to the festival site was good, the grades relatively easy and the trip suitable for children and non-mountaineers. The hike, we said, would only take three hours.

The newly made "five-kilometre-long" trail mostly lived up to our description except for one short steep section where volunteers were stationed to help the "non-mountaineers" get to the top. The volunteers even carried peoples' backpacks and bags over this tough spot,

Black and white large format "Ansel Adams-like" image of the Stein Valley used in our first Stein Valley poster. Poster image: Ian Mackenzie.

Voices for the Wilderness...

Our Stein poster showing an aerial view of the festival. Poster image: WCWC files.

if they needed the help. At the top a young child gave each person a homemade cookie. It took us five hours to hike the trail because Kallie (who was very strong-willed) insisted on walking all by herself most of the way.

After that difficult stretch, which freaked out only a few novice hikers, we went up a ridge, along a scree slope and down a short traverse into the festival site, a beautiful meadow by Brimful Lake. We were almost two kilometres above sea level. It was a spectacular, awe-inspiring setting, with mountain peaks and meadows all around.

Adriane Carr and me with Kallie on my back descend to the Stein *Voices for the Wilderness* Site. Photo: unknown.

A huge ancient limber pine hanging on to a thread of life grew less than a hundred metres from our tent site. I'm sure this tree had to be over 2,000 years old.

Shortly after we arrived at Brimful Lake, a helicopter landed with some Stein First Nation elders who were too old to negotiate the trail. After the last hiker straggled in near dusk, the count exceeded 500. Among them were several First Nations representatives from Clayoquot who came to express their support. That evening our First Nation hosts provided a fantastic feast of barbecued salmon, potatoes, corn and bannock bread for everyone.

This, the first of many Stein festivals, was a transcendent experience. It empowered people for years, if not for their entire lives. It proved to me that you had to get people into the wilderness to get them fully motivated to help save that wilderness. It felt good knowing that our newspaper helped greatly in getting such a large number of people into the Stein.

That first night Adriane went to bed early, cuddled up to Kallie as the air turned from chilly to frosty. I stayed up late talking to people I had never met before around a campfire. All my life I've never needed more than a few hours of sleep a night so I was one of the last to leave the dying fire's embers. With the temperature sinking well below freezing, I walked back to our tiny tent and crawled into the sleeping bag warmed not only by Adriane but also by the fact that now three huge wilderness preservation campaigns—Moresby, Meares and the Stein—raged in B.C. and First Nations who passionately wanted to preserve their culture and natural heritage led all of them. I knew they had the land claim clout to succeed. While my passion for protecting wilderness sprang from a biological understanding of its importance, theirs came from life experience and the knowledge that they had to maintain the natural heritage of their ancestral lands to keep their traditional cultural heritage alive.

Over the three-day period I had the opportunity to talk to many different people. One of the elders told me that when he was young and hunting in the Lower Stein he happened upon a vision quest cave with hundreds of pictographs. He explained that the Stein Canyon was extremely important in their sacred traditions and he was thankful that so many non-natives were helping protect it. Much of the 107,000-hectare Stein watershed is extremely rugged wilderness. It still hides some important secrets. That elder's words redoubled my desire to do everything we could to stop the road from being built up the canyon into the mid-valley where the commercially valuable timber grew. Like others, I mistakenly thought it would be easy to save the Stein because the cost of building the 20-kilometre-long access road far exceeded the value of the logs that could be extracted from the Stein. But political decisions often defy economic sense. The political leaders of the day sided with the logging companies. Soon it became obvious that the government would rather spend taxpayers' money to subsidize their logger friends than to save money and see the Stein protected.

Ancient pictographs in a vision quest cave high on a cliff wall in the Stein Valley. Photo: WCWC files.

Salmon cooking for the feast hosted by the Lytton Band for all attending the first *Stein Voices for the Wilderness Festival* Labour Day weekend 1985. Photo: Pat Morrow.

Elder Louis Napolean of the Lytton Band speaks to those gathered in the Stein Alpine for the first *Voices for the Wilderness* festival. John McCandless, Stein Coordinator for the Lytton Band, stands behind in support. Photo: WCWC files.

Chapter 13

Breakthrough in the campaign to save South Moresby
B.C. thwarts new park creation with WAC process

Making Parks Canada's centennial celebration in Banff meaningful

On the hike out of the Stein, Adriane got cold feet about continuing on from the Stein to Banff to attend the Parks Canada 100th anniversary celebrations. Colleen had insisted that we and representatives from the other key groups working to protect South Moresby attend because the new environment minister would be there and we would have a real opportunity to get his attention. But Adriane was worried about how much preparatory work she had left to do for her upcoming semester of teaching at Langara College. Finding an easy solution I said, "I'll take Kallie and drive up there by myself and you can catch a ride back to Vancouver, finish your work there and fly up to meet me in a couple of days. Then we can all ride back together in the car. It'll cost us a little more, but we need you there."

It was the first time 19-month-old Kallie had been away from her mother for more than a day. But Kallie was a good camper and, secure with me, loved the trip. On our second day on the road, high in the mountains in Glacier National Park near Rogers Pass, I encountered the most remarkable meteorological phenomenon I've ever witnessed. It was about noon and we were driving through a thick cloud blanketing the mountains. There was fresh snow on the steep mountainsides around us.

All of a sudden it became very light, as if the sun was about to break through. At the same time, it began to both rain and drizzle. Suddenly we were inside multiple rainbows. Intense brightness and vibrant rainbow colours surrounded the car. "Kallie, look!" I said, but she was fast asleep in her car seat. I slowed down and, by luck, there was a pull off and I parked. I quickly woke her up. She looked around with her eyes wide open in what could only have been astonishment. I regretted that she was too young to remember it. After a few moments, the multiple rainbows surrounding us disappeared as suddenly as they'd appeared. Nature is full of events people seldom see, let alone can comprehend.

We arrived in Banff the day before the conference and set up our tent in the National Park campground. I cased the conference centre site looking for a good place to hang our giant "Save South Moresby" banner. That night it snowed. In the morning about 20 centimetres covered the campground. Kallie and I were the only ones sleeping in a tent. The few other brave campers there were in nice, cozy camper vans. I shook the snow off the tent and got Kallie up. It was freezing cold, way colder than it ever got in Vancouver. "Thank goodness Kallie is too young to complain," I said to myself. The short drive down to Banff village for breakfast was treacherous because they hadn't ploughed the road. Camping season was definitely over! Banff village, about 100 metres lower in elevation than the campground, had little snow. After breakfast, we went to the Banff Centre, arriving a couple of hours before registration for the conference began.

Someone had parked a large voyageur canoe advertising a guiding company on the lawn in front of the main conference building. "It's a perfect place to display our giant 'Save South Moresby' banner," I thought to myself. "Everyone will see it when they arrive to register." I draped our banner over the canoe and tied it down, knowing it would cause controversy. The banner lasted in this prominent position all day. That night we were told we had to move it. We then strung it up between two trees right in front of the main entrance. It stayed there for the duration of the conference.

I set up a WCWC display booth featuring all our newspapers and posters and, of course, some 1985 calendars that hadn't sold. Our display included free copies of a big black and white poster that we had just produced with a loaded logging truck rumbling through a huge clearcut with the statement *Don't let this happen to British Columbia's Treasures*. We paid only $400 to print 2,000 of them and were selling them for $2 each—and giving lots of them away—effectively shocking and motivating people at the same time.

It astounded me that our small, volunteer-run, five-year-old organization took up two large tables to display our educational publications. Most of the other groups, some founded decades earlier, shared tables because they had hardly anything besides their in-house newsletters, membership brochures and copies of briefs presented to government commissions. "No wonder park creation crept along at a snail's pace in Canada," I thought to myself. We were the only group, as far as I could see, primarily concentrating on building widespread public support and aggressively promoting amongst the "unconverted" the creation of new parks.

I didn't go to any of the meetings that day because I had to look after Kallie and I really didn't mind. The meetings were primarily geared to talking about park management, not park creation. Most of the activists there were from B.C.—Vicky Husband of the Sierra Club, Colleen McCrory and Grant Copeland of the Valhalla Society, Thom Henley of Islands Protection Society, Adriane and I from WCWC. I talked to people who came by our table during the breaks or who had skipped out of meetings and wanted to talk about campaigns. Mostly I made sure Kallie, by far the youngest "delegate" there, didn't get into any trouble as she ran around and explored.

It seemed weird to me that this was it. Our federal government had obviously decided that it wasn't going to make a big thing of our national parks' centennial. Around the world people recognize that we, in Canada, have one of the best national park systems, if not the best in the world. The way to celebrate it would be to hold a huge event with announcements of several large new parks including the one that millions of Canadians wanted—South Moresby.

I had little tolerance for the bureaucratic ways of Parks Canada's planning department. I'd visited their offices (cubicles) several times in Hull, Quebec, within walking distance of Parliament Hill and I had a hard time believing that so many people could work for so long and create so few parks.

One thing we had going for us at this conference was that Prime Minister Mulroney had just appointed Tom McMillan as our new Environment Minister. McMillan replaced Suzanne Blais-Grenier whose "brilliant" idea was to get environmental groups to raise $2 million to show good will towards protecting South Moresby in order to get the federal government to act to protect the area. It was a totally flaky idea presented in a totally bizarre way. Blais-Grenier arranged for Colleen McCrory and Diane Pahael, from Alberta Wilderness Association, to attend a meeting in Toronto with an Ad agency. The ad agency people sprang the idea on her and told her they'd help

Facing page: Poles standing in 1977 in the Haida village site of Skung Gwaii (Ninstints) in Haida Gwaii located at the remote southern end of the South Moresby Wilderness proposal area. Skung Gwaii became a UNESCO World Heritage Site in 1981. Photo: Richard Krieger.

raise the money.

Colleen telephoned Jim Fulton, and people working on South Moresby, who agreed with Colleen, that this would be a terrible direction for public land. She immediately went to the press and blew the whistle on it. It was difficult for WCWC to raise even $2,000 and we were one of the most active groups. This did not mean people cared little about creating a South Moresby National Park. They simply believed it was the Canadian government's job to do it, with every taxpayer sharing the cost.

Abandoning our tent in the snow to share tight indoor quarters with other campaigners

By the time Adriane arrived by bus from the Calgary airport in the late afternoon, Banff had already chilled down to below freezing. I told her about our "cozy little tent" set up in the frozen snow-covered mountainside campground. She asked me if I would please talk to Colleen to find out if we could stay in her room at the convention centre. Colleen took pity on us, especially toddler Kallie, and offered us a place in her hotel room on the floor. We had to come in late, leave early and be very quiet because she was sharing the room with Grant Copeland—she had promised him if he drove her down to the conference she would not invite others to stay in the room—and it was on Grant's credit card. I rescued our tent and frozen sleeping bags.

By the end of the evening, several other cold and miserable environmentalists had also moved in. Soft hearted Colleen just couldn't say no. By the last night of the conference seventeen of us slept in that room! It was very crowded on the floor, but toasty warm. We were all united in our determination to get some action on South Moresby out of this federal 'feel good' event. After the conference Grant received a bill for over $800, which really upset him. Obviously some staff person kept track of how many were sleeping there! Grant objected and refused to pay the extra charges.

Over the next two days I spent most of my time around our display table taking care of Kallie while Adriane attended the meetings on behalf of WCWC. I'm not very good at meetings except adversarial ones where it is OK to speak out strongly and occasionally yell at opponents in frustration. These Parks Canada meetings called for diplomacy. The conference organizers were determined to make sure nothing of real significance happened, like the passing of a resolution to establish a new park in South Moresby. We were just as determined to force the issue.

Putting South Moresby at the top of the new national park wish list

At the end of the conference's last plenary session, the "activists" were finally granted some time to talk to the conference attendees about South Moresby. We were told we had five minutes to present our case and it had to be done before the new Minister, Tom McMillan, arrived to give his speech. We all agreed that Thom Henley was our best spokesperson.

Thom gave a brilliant, emotional speech. In a lucky break for us, McMillan walked in unnoticed and sat down in the back of the auditorium shortly after Thom had started and heard most of it. Thom wrapped up his speech with a passionate plea that, as a way of celebrating the National Park Centennial, we must do something truly significant. Make South Moresby a National Park Reserve!

With the organizers politely trying to stop him, Grant, who was a very forceful kind of person, took the microphone from Thom and continued our "five minute presentation." He read out the resolution "...*to make South Moresby a National Park Reserve and the surrounding waters a National Marine Park Reserve.*" Before one of the conference organizers could say that resolutions were not in order, Grant called for a "vote." People spontaneously cheered and gave Thom and Grant (or more accurately, South Moresby) a standing ovation, unanimously endorsing the resolution.

Tom McMillan then presented his prepared speech. It was his first major one as Canada's new Environment Minister. The big news was supposed to be that negotiations were almost concluded to establish a new national park in the high Arctic on Ellesmere Island. But after making this relatively unexciting announcement, in an obviously impulsive move, McMillan deviated from the script and stated that he would "make working with the British Columbia government to protect the treasures of South Moresby a top priority." This commitment got a standing ovation.

Putting icing on the Environment Minister's promise

Afterwards we all gathered in Colleen and Grant's room, excitedly talking about our success. I was probably the only cynic thinking, "Talk is cheap, especially from politicians." Out loud I said, "We've got to give the Minister something to reinforce his commitment. How about getting a huge cake that says, 'Thank You Tom McMillan - South Moresby National Park' and presenting it to him during the closing reception tomorrow?" Everyone thought it was a great idea. Timing was tight. It was near the end of the day. Adriane immediately went out and found a bakery that was still open and would bake the cake, instructing them to write our message on it and decorate it with bright green frosting covered with dark green evergreen trees.

The next day Colleen managed to get the big cake box smuggled into the reception hall and then, in front of everyone, presented our cake to the unsuspecting Minister. She gave him a cake knife to cut the first piece.

McMillan, shrewdly understanding the implications, refused to do it. He proclaimed that it wouldn't be right to eat cake and celebrate the park before it actually was a park. "Put this cake in the freezer and we'll share it when South Moresby is saved," he declared.

On the way home in the car Adriane and I talked on and on about all the things we had to do, now that we had both the federal and provincial environment ministers saying they wanted to talk turkey about creating a South Moresby Park. Things were looking up in this campaign.

Stein festival success builds Stein support and spurs on the opposition to fight aggressively to get a quick logging decision

After returning home from Banff, I immediately poured on the heat to finish producing our 1986 Wilderness Calendar. We also rushed to print a poster of the Stein Festival, featuring all of the colourful tents that dotted the alpine meadows. To top it off, we also produced a new Stein newspaper to capture the energy from the Stein Festival and alert the public to the new big push by the logging industry to build a road and start logging the Stein in the spring. This tabloid was our hardest hitting newspaper to date. Instead of using green as the spot colour we used red! It was titled *The Stein Wilderness is in Danger of Immediate Destruction!* A bold headline on the front page stated: *Forest Minister's Stein Stand Costly to Public and Industry – Attempt To 'Buy' Votes In Own Riding May Backfire.* It seemed obvious to us that Tom Waterland, Minister of Forests and a former mining engineer, considered logging a form of mining. As the sitting MLA for the riding that encompassed the Stein Valley, not only did he favour logging the Stein, he was bent on seeing all the wilderness

Chief Ruby Dunstan speaks at a Stein rally. Note the Stein Valley Pictograph, which had become the logo for the campaign, painted on the banner in the background. Photo: WCWC files.

areas we were fighting to protect put to the chainsaw as soon as possible. Ken Lay, who was still a WCWC volunteer, wrote the hard-hitting lead article in this second Stein paper. It read in part:

> The Stein issue is accelerating to confrontation. Waterland is pushing it that way. His recent statement 'I'm not concerned with any Native claim on the Stein. The B.C. government, regardless of what party has been in power, has never recognized land title claims,' is motivating Natives—one third of his constituents—to get on the voters list. His closed minded refusal to meet with locals who want to find a way to save the Stein and use the logging subsidy more wisely may be a mistake. With an election just around the corner, The Stein could be Waterland's and the Social Credit's Waterloo.

On the back page of the newspaper, in the 'credits' box, I put "*Opinions expressed by the authors are not necessarily those of WCWC,*" to appease some of our board members who were not as radical as the rest of us. But the articles sure expressed my opinion.

It looked grim for the Stein. The only thing stopping the logging from starting immediately was road access. One piece of private property—Earls Court Farm—blocked the entrance to the valley. The owner, a Stein wilderness supporter, refused to grant a right of way through his property for the proposed logging road and the government was about to expropriate this access corridor. If we could preserve this "bottleneck" we'd save the Stein, because it was the only feasible way to get to the commercially valuable trees in the mid-valley. It was too costly to build a road and truck the logs up and over a mountain pass into an adjacent watershed.

Every Stein preservation group focused on raising funds for the Earls Court court case. WCWC's angle was unique. We launched our first *"Adopt-a-Tree"* project—a way to build public support as well as raise funds by getting people to symbolically "adopt" every tree along the proposed logging road right-of-way.

The appeal on our *Adopt-a-Tree* flyer read as follows:

> *Your adopted tree lives in a beautiful old-growth forest in Southern British Columbia's last remaining major wilderness watershed. It grows on the route of a proposed logging road into the heart of the rugged Stein.*
>
> *The B.C. government plans to allow it to be cut down in the very near future. Logging the beautiful Stein Valley is irresponsible. It will cost B.C. taxpayers millions of dollars in direct subsidies—the trees are not otherwise worth logging—while destroying an irreplaceable wilderness treasure.* BUT YOU CAN HELP SAVE THIS WILDERNESS!

We featured a stunning image of the beautiful forest blocking the proposed Stein mainline haul road in our 1986 calendar. We then donated 200 calendars to the Stein Action Committee who obtained a door-to-door canvassing licence from the city of Vancouver and subsequently sold them to raise money for the lawyer. Pursuing the "legal route" to save a wilderness area always proved to be expensive. But sometimes it is the only way to go. And it often delayed logging, giving us more time to build the public campaign for protection. **I believe in the tactic of delay. The longer you can delay the destruction of a wilderness area the better are your chances of saving it.**

In the spring of 1988 we also sent our dendrochronologist, Marion Parker, to look for culturally modified trees along the proposed logging road route up the Stein Canyon. He found some and wrote a detailed report. In many cases it was impossible to tell whether natives or early settlers and explorers had modified the trees. We submitted this report to the government and called for more studies to be done before it issued any road building approvals.

Anthropomorphizing the trees to save the Stein

The Stein *"Adopt a Tree"* operation was hugely labour-intensive. For a donation of $20, $50 or $100 (the vast majority donated $20) a person could select a cedar, fir or Ponderosa pine tree and give it a name. With every "adoption" the donor received a photo of "their" tree. This photo showed the tree with a person holding out a red ribbon tied around it that clearly showed the donor's chosen name written on it. In the adoption package was a map showing their tree's "exact" location. Every donor also got a tax receipt and *"unlimited visiting rights."* These tree adoptions were very popular. In the first three weeks we raised over $2,000. By May 31, 1986 we'd raised over $9,000 and had adopted out over 300 trees.

Our own "red tape" adoption system, however, proved to be challenging. We saved the Adopt-a-Tree "application forms" that came in the mail and on weekends a couple of our volunteers would head up to the Stein and complete the adoption on-the-ground. They'd select the trees, write the names on the ribbons and take the Polaroid photos. Back home we'd put the right photo along with a photocopied map with an X on it where the tree grew and the official donation receipt in an envelope with a thank you letter, then send them out. It was a great deal of work for a $20 donation. But eager volunteers did it all.

With so many steps, and such high expectations on the part of the adopters, things sometimes went wrong. The complaints from proud new parents of a Stein tree were few but often emotional.

"That's not my tree. My tree's name is Frits." We'd obviously put the wrong photo in the envelope. Someone else must have got Frits. "We'll undoubtedly be getting a complaint from them soon, too," I thought.

Appeal on the back page of our *Stein Wilderness is in Danger of Immediate Destruction!* newspaper. Artwork: Annett Shaw.

Dendrochronologist Marion Parker beside a bark stripped CMT red cedar. Photo: WCWC files.

"I asked for a cedar and you gave me a pine!" There were more pines than red cedars growing near the mouth of the Stein where it meets the Fraser River. Most people didn't mind, or even notice the difference. A few did.

"You spelled the name of my tree wrong. Our family name is 'Ferlinskchy' not 'Ferlenskchy'. I wanted to give the tree to our son as a birthday present at the end of the month. Could you please correct it and take another photo and send it back right away?"

"How did that happen?" I asked Ken about the misspelled name.

"Look at the handwriting on the adoption form. It's just a scrawl. We're lucky to have spelled it as well as we did!" he replied.

"Can you find the tree again and correct it?" I asked, "You know how the saying goes 'The donor is always right'."

"Not a chance. We have several hundred ribbons spread all over the place out there. I'll just go and adopt another one for them and explain to them how lucky they are to get two trees for the price of one!" replied an upbeat Ken.

Many people took their tree guardianship seriously. One day, several months after our program began, a frustrated adoptive "parent" called, "I went hiking in the Stein this weekend looking for my tree. Your map is inaccurate. I couldn't find it. Please give me better instructions!" I called him back and told him that we couldn't. His map made by one of our volunteers was the only record we had. I explained that an adoption was merely a symbolic form of protest and way of providing us with financial support for our campaign. He

Leo DeGroot beside a CMT along the Stein Valley trail through the lower canyon. Photo: WCWC files.

understood, but was not happy.

There were even a few people who were unhappy with the size of the tree we'd chosen to be their adoptee. "My tree is very small. I thought I was adopting an old-growth tree." I explained to this woman that in the proposed Stein roadway there were trees of all sizes and ages, just like in most typical old-growth forests. "We're finding adoptive 'parents' for all of them. Thanks for helping out."

We also sent quite a few photos of trees to Premier Bill Bennett on behalf of people who had adopted them as gifts for him. The volunteers picked extra big, old and beautiful trees for these. To give the campaign an additional boost, Ken, who had taken charge of it, named an adopted tree after each MLA, and sent each one the standard package with a picture of their tree and information about the Stein. It was a fun fundraiser. We used "adopt-a-tree" programs very effectively in later campaigns, especially Carmanah. But never again did we make it so complicated. In future adoptions we sent out beautiful "adopt-a-tree" certificates with photos of trees but never ones that had ribbons with names on them or maps indicating the locations. We'd learned from experience.

Haidas discover that the provincial government's South Moresby promises are worthless

In the fall of 1985, the South Moresby campaign was in worse shape than the Stein. Frank Beban's crew on Lyell Island was logging like there was no tomorrow. Expectations were that they'd be finished with their existing cutblock permits by mid-October and the logging would stop. Guujaaw told me that several cabinet ministers, including Austin Pelton, had promised the Haida in early fall that no more cutting permits would be issued for Lyell Island until the government met and came to an agreement with the Haida. Pelton promised even more: that the logging moratorium would continue until the government made a formal decision about the fate of the entire South Moresby area.

During that fall, federal Environment Minister Tom McMillan, following up on his commitment at the Banff Centennial parks celebration, visited South Moresby. The weather was great—no rain—a lucky omen for that time of year. McMillan also met with his provincial counterpart Austin Pelton. A provincial election was due soon.

After that visit I expected, at the very least, a B.C. government announcement of a moratorium and the commencement of formal talks between the feds and the province. But I underestimated the political power of the forest industry. Making sure that logging in South Moresby continued was a matter of pride for the industry. Despite our gains in public support, the loggers and logging companies still had a lot more influence in B.C. government circles than environmentalists or the Haida.

CMT found near the lower cable car crossing in the Stein Valley. Note pictograph graffiti that someone had recently painted on the scar. Photo: WCWC files.

Government establishes Wilderness Advisory (diversionary) Committee

I was bewildered on October 18, 1985, when I heard via the newspapers that the B.C. government had created a *Wilderness Advisory Committee* (WAC) to look at the issue of development vs. preservation of 16 key wilderness areas and provide recommendations to government. This announcement came as a complete surprise to me. The WAC study areas included the Stein Valley, the Tatshenshini and, of course, South Moresby. A week after WAC was struck, someone close to government told us that we should be proud—WCWC had provided the list of areas to be studied by WAC. In the rush to set the mandate, a bureaucrat took out a folder with all of our past calendars and selected the issues from them. It's true that WCWC had featured in our first five years of endangered wilderness calendars all but one of the areas under consideration.

Premier Bennett convinced a well-respected lawyer, Bryan Williams, to chair WAC. We could find no fault with him, but we sure found fault with the other members appointed and with WAC's ridiculously short mandate. The group had four months to study, deliberate and present its recommendations. All the members appointed were industry representatives, academics or government bureaucrats. There was not one person representing the environmental movement. After a massive public outcry, Ken Farquharson, a conservative but well-respected environmentalist, was added on. He had to be coaxed into returning from living in Scotland to serve on WAC.

Shortly after the announcement of WAC, we held a directors' meeting. Back then WCWC held directors' meetings only a couple of times per year. At that meeting there was a wide difference of opinion as to whether or not we should participate in the WAC process. Colleen was adamant that we should have nothing to do with it. In fact, she wanted to boycott their public input meetings. She thought the whole thing stunk and was destined right from the start to make pro-industry decisions regardless of any submissions we, or anyone else, made. She was worried that Ken Farquharson would not stand up firmly for South Moresby against such an industry-biased group. I, too, felt right from the start that the real goal of the government was to use WAC to distract and diffuse support for the South Moresby campaign and others. The WAC tactic did appease the powerful forest industry and put off dealing with the federal invitation to start real South Moresby park negotiations.

Director Murray Rankin quits our board

Murray Rankin saw it differently. He thought our participation in the WAC process would be useful. He argued that Bryan Williams, WAC's Chair, was sympathetic to wilderness preservation. We should take full advantage of this opportunity to be a strong voice for the areas we were striving to save.

I had to side with Colleen. I'd been on the Meares Island Planning Team and sat for a while on the South Moresby Planning team. Over the years I'd presented seemingly countless briefs to commissions and none of them had done one whit of good as far as I could see. It was public education and campaigning that made the difference. However, I disagreed with Colleen on one point. She thought that we shouldn't even table our publications with WAC. "Send them a letter telling them that this information is part of the public record and available to them if they want it," she said. I thought it would be better to get a set of all our newspapers and posters relating to the wilderness areas under study into WAC members' hands without submitting any formal brief or letter of explanation. I did that. That was the extent of our participation.

The week after this directors' meeting Murray Rankin quit WCWC's board. It came as a complete surprise. He had served on our board for five and a half years. Although he did not give any reasons in his letter of resignation except to say that it provided an opportunity to *"attract new blood"* to WCWC, I think that we, especially Colleen and I, were becoming too radical for him. Our WAC decision was the last straw. We missed Murray's input, his calm, rational way of handling internal disputes and especially his accurate minute taking. The minutes of directors' meetings held in the following few years after his departure were not nearly as thorough or professionally written as they were when he was in charge of them.

Colleen's and my gut instincts about the WAC were right. We soon learned that while the B.C. government was establishing the WAC, ostensibly to find a solution to the conflicts over wilderness preservation—especially South Moresby, the B.C. Forest Service was quietly issuing more cutting permits on South Moresby's Lyell Island ahead of schedule.

WCWC gets a "transfusion" of excellent "new blood" to reinvigorate its board of directors

Two weeks later, at our November 1985 AGM, "new blood" came forward. We elected by acclamation three new directors—Ken Lay, who was campaigning for the Stein, Ron Aspinall campaigning for Clayoquot Sound and Bob Broughton, WCWC's volunteer computer whiz. Dan Bowditch and Linda Hanna decided not to run again. That left only Adriane, Colleen and I as holdovers from the previous board. I was the only original director left.

Cover of *Haada Laas* newspaper. Document: WCWC archives.

Chapter 14

Haida defend Lyell Island
A big push to save South Moresby

Haida take action to save Gwaii Haanas

Letting logging continue on Lyell was the government's big mistake. In fact it was the big turning point in the campaign to save Gwaii Haanas. The new cutting permits made a lie out of the promises made to the Haida by Cabinet Ministers Pelton and Brummet that the government wouldn't "talk and log."

The Haida, tired of the duplicity, decided to act directly to stop the logging. In late October of 1985, Haida elders blockaded the main logging road on Lyell Island. Many were jailed for defending their land—or, in legal terminology, for being in contempt of court in defying a justice's orders get out of the way and allow the logging company to carry on its "legitimate" business as usual. In all, 71 Haida were arrested.

The Haida blockade boxed the governments of Canada and B.C. into a corner. I knew the strength and pride of the Haida from living for two years on Haida Gwaii and working with Guujaaw on the campaign. The Haida, now fully engaged, wouldn't give up or go away. The B.C. and federal governments had to protect South Moresby or face the consequences, specifically an all-out conflict with the Haida which would bring international embarrassment and censure.

My only fear was that this conflict might culminate in a typical Canadian "compromise" that would sacrifice the wilderness's ecological integrity for an illusion of "balance" and fairness to both the logging interests and the Haida. If that happened, what we had been fighting for would be lost.

By now, all sides agreed that it was no problem to protect 80 percent of the total area of the South Moresby Wilderness—the trees in this 80 percent of the "pie" were of little commercial value. I worried that people not familiar with the area would consider saving this 80 percent a win. I knew from firsthand experience exploring South Moresby that it was crucial that the other 20 percent with the "good trees" be protected, too. This one-fifth of the pie was the biologically-rich core of the area—and we couldn't allow it somehow to be negotiated away. Windy Bay was its heart. That's what we at WCWC kept campaigning as hard as we could to get protected.

During the blockades, there really wasn't too much we could do to help directly. Except for MP Svend Robinson, everyone taking part in the Lyell blockades was Haida. The Haida did not want participation from others, including local or off-island non-native environmentalists. There was one thing we could do, however, and that was to help mobilize massive public support. One of our small contributions towards this end was the production and distribution of a button that thousands of people wore to express support for the Haida blockade on Lyell.

Lyell Island Eagle

I had in mind a design like the Nuu-chah-nulth facemask we'd used for the *Save Meares Island* button we produced a year earlier. "Why not use the Eagle design that Haida artist Jim Edenshaw carved as a pendant for me," suggested Adriane. It had been my present to her for her last birthday. It was a powerful design.

It also was a good luck piece. A year earlier, in a "Gladstone Gander" lucky moment, I happened to glance down from the pouring rain into a gutter on Howe Street in Vancouver. Something unusual had caught my eye. I reached down and picked up a mangled money clip shaped like a dollar sign. The piece was quite heavy, with two tiny sparkling stones. I showed it to Jim Edenshaw, a young Haida carver, who confirmed it was gold with diamonds. We cut a deal. He traded in the crumpled piece of gold to a jeweler for a piece of new gold to carve, kept one of the diamonds as partial payment for his work and used the other as the eye in the proud fierce eagle pendent he created for Adriane.

I called Jim and asked if WCWC could use his eagle design for a *Protect Lyell Island – Haida Country* button. He said yes, reproducing the same design in the form of a two-colour black and red original print. I reduced the artwork to button size, produced 2,000 buttons and sent lots up to Haida Gwaii. Everyone loved it. It was a powerful icon of both the wilderness and the Haida culture. Volunteers sold them on street corners in Vancouver and later we published a limited edition of 500 *Lyell Island Eagle* silkscreen prints, selling them for $25 each to raise money for the campaign.

Poster for rally to support the Haida. Eagle artwork: Jim Edenshaw.

Record-sized stump near Sandspit—no longer possible to age. Left: Sign beside stump. Right: Stump. Photos: WCWC files.

No thousand-year-old trees in South Moresby?

That same October, I received an intriguing phone call from a fellow named Marion Parker. He had listened to Jack Webster, then the most popular radio talk show host in B.C., assert on radio that morning that there weren't any thousand-year-trees growing on Lyell Island. "I know there are!" he told me, "I'm a dendrochronologist."

"When did you age trees there?" I asked.

"I haven't," he replied. "But I have found them elsewhere on the B.C. coast. They're sure to be on the Queen Charlotte Islands. The conditions and species are the same."

I knew Parker's claim alone would not convince our opponents. We needed hard evidence from trees that actually grew there to do that. For weeks Jack Webster ranted on and on about the foolishness of creating a park in South Moresby. I can still recall his loud gruff voice saying "Log Lyell Island!" Webster, who more often than not was on the good side of an issue, was wholeheartedly on the wrong side of this one. He was championing the cause of the poor loggers who would lose their jobs if a park were created there. According to Webster, the claim of millennial trees was merely a propaganda ploy by environmentalists seeking to sway a gullible public.

Live on radio defending the South Moresby Wilderness Proposal. From left to right Jack Miller, resident of Port Clements, Tom Schneider, resident of Queen Charlotte City and IPS representative with host Jack Webster. Photo: WCWC files.

Curious as to how Parker would get the proof, I asked him how he would go about coring the giant trees in order to count the growth rings. "Oh, I don't need to do that," he replied, "there are enough stumps in clearcuts adjacent to the old growth. The trees are of similar age and it's much easier to cut samples off stumps with a big chainsaw and count the rings on them." I liked Parker's practical approach to finding living thousand-year-old trees.

Parker fervently believed that not another tree over 1,000 years of age should be cut down in B.C. or, for that matter, anywhere else on Earth. "They are too rare and valuable as living scientific data banks. Their growth rings have chronicled the changes in weather and climate over the centuries and continue to do so," he told me. His idea didn't seem very radical to me, except perhaps if there were more millennial-aged trees in B.C.'s old-growth coastal temperate forests than most people suspected.

The other thing I liked about Parker was that he immediately understood how little money we had. He agreed to work for a pittance. He'd go up to the Queen Charlotte Islands, find and document thousand-year-old-trees for us.

All we had to do was pay his expenses, provide him with a guide with a truck and a chainsaw to take him around and help take samples, and pay him an honorarium of $200 after he finished the study and delivered his written research report to us.

Research was one of WCWC's constitutional mandates. Prior to Parker's quest for thousand-year-old trees on Haida Gwaii, we'd sponsored no original scientific research at WCWC. This was our first chance. We issued a press release at the end of October announcing that we had hired a dendrochronologist to search for thousand-year-old trees on Lyell Island. We explained the scientific importance of such extremely rare *"living assets,"* and asked for the B.C. Forest Service and Western Forest Products, who now owned TFL 24, for full co-operation in finding big stumps for Parker to study. In the press release we asserted that if we found thousand-year-old trees on Lyell, *"absolutely no cutting"* should take place there. However, we had to wait until spring to undertake this research. It was impractical to go do it in the winter—too much rain and too little daylight.

Struggling to make our 6th annual Endangered Wilderness Calendar a financial success

Every fall producing and selling our wilderness calendars consumed most of my time and energy. We picked up our 1986 calendars from the bindery on the last day of October 1985. It was a real accomplishment to get it out so early! We'd set the price one dollar higher hoping to make enough money to finally pay off the $9,000 printing debt still owing from 1983. I hired Ken Lay full time at $5 an hour to help distribute them. The previous two years I had only had part-time help. It made quite a difference.

After only two weeks, there were less than 2,000 calendars left of the 11,000 press run. We'd given away 1,000 of them including, as usual, copies to every MLA and MP. We probably should have featured a South Moresby photo on this calendar cover considering the prominence of this campaign, but several months earlier I'd promised the Alberta Wilderness Association we'd feature, for the first time, an Alberta issue on the cover—the White Goat Wildland Recreation Area.

Despite our activity and hot campaigns, WCWC still had only 75 members. Ken Lay asked me one day, "Why aren't we asking people to join WCWC on all our appeals?" How could have I been so dumb? Thanks to Ken, in every publication from then on we asked people to join. We also set the goal of having at least a thousand members by 1987.

Our first big grant

That Christmas we got by far our biggest donation to date—ten times more than any other single donation—a cheque for $20,000 from a foundation that wished to remain anonymous. We were ecstatically shocked. We hadn't applied for a grant. In fact, we'd never even heard of this foundation before the cheque arrived in the mail! The letter accompanying the grant said that we could spend it as we saw fit, but the foundation was particularly interested in seeing the South Moresby area protected. I immediately called Guujaaw and asked him what we could do to help. He said they really needed to publish another edition of their newspaper Haada Laas ('good people' in Haida) to explain what was happening in South Moresby. I told him to come to Vancouver and he could use our light table and all of our resources to make it happen. We dedicated $5,000 of the Eden donation to this publication.

I, too, felt it was critical for the Haida to get out their side of the story as to why they resorted to blockades to stop logging on Lyell Island. Unlike the Meares blockades and trials, which were covered on an almost daily basis in the press, the major media really downplayed the Haida conflict on Lyell. In fact, there was hardly any news coverage at all of the subsequent court cases in which the Haida acted as their own counsel. Few people understood why the Haida had resorted to blockading, or knew that since logging began on Lyell Island ten years earlier the Haida Nation had continuously spoken out against it.

A couple of weeks later Guujaaw came down to Vancouver to work on this newspaper along with John Broadhead of Islands Protection Society who had graphic design skills and who earlier had helped put together the *Islands at the Edge* book on South Moresby. They took over our light table and our study "production room." In February of 1986, the Haida Nation's special eight-page edition of *Haada Lass* (good people) *Journal of the Haida Nation* (VOL II NO 2) rolled off the press. The lead article explained the Haida's efforts to protect Haada Gwaii and the South Moresby Wilderness Area along with statements of elders and photos of the blockade. It's the only place I know of where this full story is told. This edition of *Haada Lass* also had an article about the spectacular 149,550 hectare Duu Guusd wilderness on the northwest corner of Graham Island that the Haida declared a Tribal Park in 1981. At that time hardly anyone outside of Haida Gwaii knew about Duu Guusd even though it's slightly larger than South Moresby. This wilderness has remained to this day almost completely intact because of the Haida's firm commitment to protect it.

Colleen McCrory quits WCWC

At the beginning of January 1986, Colleen McCrory and I got into a big fight over—guess what? Money! As a director she continually wanted to see our budget (we didn't have one), our financial records (they were in disarray) and the cheque book balance (I didn't keep one. I only knew in my head approximately what we had in the bank and how much money was already committed). Colleen's phone bills were soaring and I felt that we needed to have a budget at least for her that limited them to $600 per month. I maintained that we just couldn't afford to spend any more. "We just don't have the money!" I shouted at her over the phone. Colleen insisted that any curb on her use of the phone (her major campaign tool) was unacceptable. I agreed that contacting people to gain more support was vitally important, but argued that it could be done at night and on the weekends when the rates were lower. Since the results of her "networking" were not readily apparent to me, it was not my highest priority. Even with the rush of money that had just come in from calendar sales, we still had empty pockets after all the calendar bills were paid. Colleen quit and hung up. I thought she was bluffing. I immediately wrote her a conciliatory letter. She replied a few days later.

Though your letter of January 9th has a far better tone than your telephone call a few days ago, beneath your improved approach is the same objectionable sentiment: that I have been doing unnecessary telephone work....

Your implication of me doing unnecessary work is also made in your statement that telephone work must be confined hereafter to 'set goals'. What kind of double talk is this? Is my work on preserving South Moresby or battling the Pelton Committee [WAC], or talking to high profile politicians in connection with these subjects, or the Haida, or the coming election, not connected to 'set goals'? Is developing strategies with regards to the Stein, Stikine, and other parts not part of our goals? Aren't the federal ministries part of this too?

What is it that I'm doing, working myself to death, that is both unnecessary and not connected to 'set goals'? Or is it a fact that the goals you wish to aim at are the ones that you, alone, choose or prefer, as in the calendars, posters and books. While these are important, you cite them as the sole source of contributions. What about the political work, the letters to the press, the radio and TV material, meetings with high profile people that associate Western Canada Wilderness [Committee] into the battles and targets I cite and which develop public appreciation and support for our environmental positions? You omit all of this in your language and tone repeatedly. In the face of the night and day labour I've expended, particularly in the unrelenting demands on me since last August, the tone of your letter and its evident disregard of all that I have put in leave me no choice but to resign....

Obviously my letter was not conciliatory enough! She asserted that WCWC <u>did</u> have the money to pay her phone bill. She pointed out that just a few days earlier I had asked for, and got, her approval for the spending of $6,000 for new posters. There was no use in trying to dissuade her.

Adriane wrote Colleen this letter accepting Colleen's resignation on behalf of the directors saying in part:

As I explained to you in my phone call of January 6, the post calendar season always brings to WCWC the necessity of tight budget-

ing and a focus on new fundraising endeavours.... I want to reassure you that this budgeting process was not aimed at you but at all the WCWC directors.... I must stress that we consider your work not just valuable but necessary to the preservation of Canada's wilderness and your resignation is a distinct loss to our Committee.... We all hope that in your continued work you will keep in touch with WCWC. Our goals are the same.

Colleen returned to work full-time for the Valhalla Wilderness Society and built that organization. Her vital work to save South Moresby continued too. Although it was traumatic at the time, it was the right move. It strengthened the movement. She was one of the key leaders of the Valhalla Wilderness Society and her friends there had missed her during her "stay" at the Wilderness Committee. My friendship with Colleen was broken only for a week or two.

Some believe that if all the environmental groups got together and spoke with one powerful voice for conservation we would save much more wilderness. However, I believe that one big voice could more easily be silenced by one big 'no' from governments. There is greater strength in diversity. A many-pronged, loosely coordinated effort is much harder to stop. I also believe there is no one right strategy that always works best. It's been a combination of tactics—each one better than another at times—that has worked to achieve the wilderness protection we've managed to win to date in western Canada. We have always encouraged new grassroots groups to form around local issues and existing environmental groups to get stronger by building their membership.

A train caravan whistle stops across Canada to save South Moresby

The "Save South Moresby National Caravan," a publicity-boosting brainchild of Island Protection Society's Thom Henley, was getting ready to roll that spring. The aim was to get people devoted to South Moresby to ride the rails together across Canada to draw media attention and boost public support for the national park proposal. It was a huge undertaking and all of us, including Colleen, pitched in to help.

The caravan started small on March 5, 1986, in St. John's Newfoundland and gathered steam along the 7,500 kilometres of rail tracks across Canada. WCWC had no representative on the train, but we took the lead in getting people out to the grand finale rally. Ken Lay was in charge and he worked full time at it. Our efforts to get people out to the rally were boosted by the fact that two representatives of the "Moresby Island Concerned Citizens," a group that opposed the creation of the South Moresby park, boarded the train

Braving cold weather, a couple of courageous South Moresby supporters await the caravan train at a remote rail crossing somewhere in the prairies. Photo: unknown.

near the beginning of the journey. Their presence made the caravan controversial and attracted media attention that it would otherwise never have gotten. At every whistle stop, even the ones in the wee hours of the morning, people came to the stations to greet the train. This campaign was fast gaining "critical mass." Just a dozen or so Haida and a few non-Haida went the whole caravan distance but more people joined in as the train crossed Canada. By the time it reached Vancouver on the afternoon of March 15th there were 200 on board.

Over 5,000 people waiting at Thornton Park in front of the train station at Terminal and Main greeted them with the rousing cheers when they finally arrived (late, as trains often do). This South Moresby gathering still holds the B.C. record for the largest number of people rallying in support of a wilderness area. After Haida drumming and some speeches at the park, we all paraded (led by a person in a giant fluttering eagle costume) to Canada Place at the foot of Burrard Street to hear more speeches. The streets were closed for this parade just as they are for Vancouver's major peace walks.

In April to keep the campaign momentum going and build on the media coverage of the caravan, we published an eight-page stapled booklet titled SPECIAL ISSUE - Save South Moresby National Caravan. Thom Henley wrote the lead story. Despite my vow never to use testimonials in our campaign publications again, we couldn't resist. This featured quotes from prominent people like former Prime Minister Pierre Trudeau who said, *"I wish all possible success to the effort in support of the South Moresby Wilderness"* and John Turner, federal Liberal leader, who said, *"I just cannot believe that this country will throw South Moresby away."* It also included statements of support from the U.S. National Parks Conservation Society and the International Coalition to Save South Moresby, which collectively represent over three million people. Everyone was climbing aboard the Moresby bandwagon. It seemed unstoppable. The 'big-name' quotes again had a dampening effect on donations. But I believe they helped persuade many Canadians to support this cause celebre.

By then WCWC had grown to 100 members! We were so tiny compared to the big U.S. environmental groups. That spring we published and distributed our first WCWC membership brochure. Why it took us that long to produce such an essential membership-building tool I still can't explain. The brochure did its work: by the end of May 1986, we had 250 members. We were one-quarter of the way towards achieving our goal of 1,000 members.

The Wilderness Advisory Committee makes disastrous recommendations

The South Moresby Caravan completely overshadowed the March 6, 1986 release of the Wilderness Advisory Committee's report (called *The Wilderness Mosaic*) recommending what the government should do about BC's highest profile wilderness issues. After four short months of superficial study, WAC advocated resource development in nearly all of the 16 study areas under consideration, including the Stein Valley and the Tatshenshini. From B.C. conservationists' point of view, it was a complete disaster. WAC recommended that all of Lyell Island be logged except for 10 percent of the Windy Bay watershed: a leave-strip along the ocean and along the lower banks of Windy Bay Creek. This was the logging company's preferred "option." Windy Bay had the largest salmon stream in all of South Moresby and the biggest trees. The fact that WAC couldn't recognize its value as a centrepiece for a South Moresby park or that such a leave strip would have little ecological value, showed just how much of a green-washing exercise it was. We continued to concentrate our

Five thousand march to Canada Place where the federal government had its offices. Photo: WCWC files.

South Moresby efforts that spring on saving the entire Windy Bay.

There were, however, a few silver linings to the black clouds in the WAC report. One was the recommendation that no road be constructed up the Stein Canyon without a formal agreement between the provincial government and the Lytton Indian Band. This helped hold off development there. Also, as a result of the WAC report, the Forest Act was amended in August 1987 to recognize wilderness as a distinct resource and a legitimate land use. Several areas that the forest industry and mining industry really didn't care about (because they had minimal exploitable timber and minerals) became

Haida elders from the train caravan. Photo: WCWC files.

South Moresby Park supporters walk from the train station on Main Street to Canada Place behind the large eagle wings. Photo: WCWC files.

"Recreation Areas" where no logging could occur. This was a major first step toward full provincial park protection for these areas with very significant ecological values. They included the 10,915-hectare Akamina-Kishinena Recreation Area, the 28,780-hectare Brooks Peninsula Recreation Area, and the 58,000-hectare Gitnadoix River Recreation Area, all of which were ultimately designed provincial parks. We had featured all of these fabulous wilderness areas in past WCWC calendars.

There are thousand-year-old trees on Haida Gwaii!

At the same time that we were getting out our *Save South Moresby National Caravan* booklet, Marion Parker embarked on his dendrochronological fieldwork in Haida Gwaii. Dennis Rosmini, a good friend of mine with superb bush skills, guided Parker in his search for stumps of thousand-year-old trees. For eight days Parker checked out stumps in the clearcuts around Queen Charlotte City, especially at higher elevations. The weather and politics made it too difficult to get to the clearcuts on Lyell Island. Certainly the "good will" that the logging company had extended to Richard and I nine years earlier during the fieldwork for our proposed "Galapagos of the North" book had evaporated. Anyway, Parker reasoned rationally that if he found 1,000-year-old trees on Graham Island, they had to exist on Lyell. Besides taking sample slices of stumps, Parker also took some core samples from Culturally Modified Trees (CMTs) that he was guided to by the Haida.

What we eventually got from Parker was not one report, but two—*Preliminary Investigations of the age and dendrochronological Quality of four Coniferous Species on Graham Island* and *Notes on Culturally Modified Trees on the Queen Charlotte Islands*. He found the trees he studied very old and slow growing, the oldest being a 1,294-year-old yellow cedar on a steep mountainside clearcut near Queen Charlotte City. Regarding the CMTs he studied, he stated in his report, "*The CMTs contain much information about cultural practices and occupation dates and need to be extensively studied and preserved.*"

Parker brought back several unique samples. One spruce "cookie" (a round cut from a stump) was only 30 centimetres in diameter and packed tight with growth rings.

"How old do you think this tree was?" Parker asked me.

"350 years?" I guessed.

"No, 700!" Even for Parker this was remarkable. A tree doesn't have to be huge to be ancient. He also collected several samples of yellow cedars that were just over one metre in diameter but over 1,000 years old. None of the samples had any rot in them—evidence that the trees were healthy when they were cut.

The loggers who felled these trees undoubtedly had no clue as to their age. It gave a whole new meaning to the phrase "mindless destruction." Every day fallers along B.C.'s coast were cutting down thousand-year-old trees. Knowing the antiquity of these trees made a mockery of the forest companies' PR line that claimed they were "improving" the forest by "harvesting" the old, decadent forest and planting three nice young seedlings to replace every old-growth tree cut. In fact, they weren't replacing decadent forests with healthy young forests, they were destroying healthy and still-evolving natural forests of different-aged trees of diverse genetic heritage, including some very ancient trees, and replacing them with vulnerable, man-made plantations of young, same-aged trees of similar genetic make-up.

Selling slices of antiquity to raise campaign money

As we were passing Parker's samples around for the volunteers to see, someone got the great idea that we should make souvenirs out of them: mini versions of the giant rounds of record-sized old-growth trees on display in many natural history museums. One of the slices from a yellow cedar stump was just the right size—60 centimetres from the first growth ring in the centre to the outside edge. We knew that the date of the last growth ring was 1982, the year the trees in that stump's clearcut were logged. There were over 1,100 rings in all. Parker agreed to do the painstaking job of marking the growth rings on each wedge and placing significant historic dates by the right ring and mounting the wedge on a plaque for only $75 each. We dreamed of making a lot of money selling them and using them as gifts to inspire politicians to protect South Moresby.

It took Dennis Rosmini a few months to get back into the cutblock and find the yellow cedar stump we wanted. He sliced off some slabs and sent them down from Haida Gwaii. After the wood dried (which took another six months), Parker sawed up dozens of thinly sliced pie-shaped wedges and sanded them with ultra-fine sandpaper to bring out the ultra-thin growth rings.

It took him hours to count the rings on each wedge, using a magnifying "eye loop" and marking every 100 years with a dot. We had a lot of fun selecting which historic dates to feature—like 1000 AD when Leif Erickson first "discovered" North America. Parker pasted our one-line descriptions of historic events next to the growth rings of the same date. He sealed it all with clear shellac. We figured it

Jim Fulton, M.P. for Skeena and the Queen Charlotte Islands, a long time supporter of the South Moresby Park, accepts from me a plaque with a 1,000-year-old pie shaped wedge of yellow cedar wood from a stump of a tree recently cut on Haida Gwaii. It is dated with historical events beside the corresponding growth rings. Photo: Valhalla Wilderness Society files.

made a perfect conversation piece to mount on a den wall. Parker presented his first finished plaque to Jack Webster on air. Webster immediately retracted his statement that no thousand-year-old trees grew in South Moresby. Over the following months we gave 1,100-year-old tree wedge plaques to a number of political decision-makers. Some were genuinely interested and impressed. Others didn't seem to think it was that special. We offered the plaques for sale at $125 dollars each. We occasionally sold the odd one but there was never the big rush of customers I'd anticipated.

Trailbuilding grows into a WCWC tradition

During the summer of 1986 while Adriane and I spent most of our time on the Sunshine Coast building our house, WCWC kept on campaigning. Under the auspices of the Lytton and Mt. Currie Bands, WCWC volunteers worked hard upgrading the Stein Heritage Trail from Cottonwood Creek to Stein Lake. Ken Lay, the volunteer head of WCWC's trailbuilding efforts, enlisted the help of dozens of volunteers including many ex-Questers (Prince of Wales Quest outdoor education program graduates) and avid outdoors people like Leo DeGroot, who became a stalwart WCWC trailbuilder.

Isha, Ron, Michelle Evelyn (our youngest WCWC director) and Wendy. Part of a large trailbuilding crew on the Cottonwood trail in the Stein Valley. Photo: Leo DeGroot.

We raised just enough money for our 1986 Stein trailbuilding activities through a "Sponsor a Stroll in the Stein" flyer. It informed our supporters that every dollar they donated cleared a metre of trail. The only really expensive part of this trailbuilding project was the helicopter. The mid-Stein section was 20 kilometres from the nearest road. It was just too much to ask volunteers to hike in a 30-kilo pack of supplies, hike back out with the empty pack and then do it again. We made the decision to helicopter in a major drop of supplies and equipment. It cost a little more than $2,000 but it paid off in keeping our volunteers happy and productive.

That summer a new WCWC tradition began. Every year from then on, WCWC focused its summer campaigning on trailbuilding, including restoring historic trails like the one into the Stein Valley and scouting out new hiking routes into endangered wilderness areas like Carmanah Valley. Our volunteers and staff love it. Being able to spend some time each summer in B.C.'s spectacular wilderness counter-balances a winter of putting out publications, stuffing mailouts and attending meetings. Virtually every one of the many hundreds

Stein trailbuilding stalwarts, right Leo DeGroot, middle Ken Lay. Photo: WCWC files.

Appeal for donations to support Stein trailbuilding. Artwork: WCWC files.

of volunteers who helped build WCWC's wilderness trails became a passionate advocate for protection of the area they worked in.

Grizzly bear cover photo helps sells our '87 Western Canada Endangered Wilderness Calendar

After a summer of hands-on outdoors work, we all returned to the intense effort to get out our 1987 wilderness calendar. Production went smoothly. We changed the format slightly, making it wider and shorter so that it would fit into a standard envelope that qualified for first class mail. That tiny change saved us several thousand dollars in postage. The changed dimensions also gave us room to print four postcards alongside the calendar cover on the standard cover stock paper that otherwise would have been cut off and recycled as "post consumer" waste. These postcards cost us only a couple of cents each. We used some to promote key campaigns; others just to sell. The only problem was that we got as many postcards as calendars. Ten to fifteen thousand postcards are a lot to sell!

That year, for the first time, our endangered wilderness calendar cover didn't feature a wilderness scene. It featured a grizzly bear. We made this decision because 1987 was the centenary of Canada's first national wildlife sanctuary (Lost Mountain Waterfowl Refuge in Saskatchewan). We wanted B.C. to do something concrete to celebrate this year, like making the 39,000-hectre Khutzeymateen Valley Canada's first grizzly bear sanctuary. The B.C. government thought differently. Instead, the Ministry of Environment and Parks launched a three-month competition to select a provincial bird that year. The winner was the Stellar's Jay, a noisy, colourful bird that befriends fallers, steals sandwiches from loggers' open lunch buckets and survives well in clearcuts.

The day our 1987 calendar was being printed, Jim Astell, our sales representative at Mitchell Press, arrived in a panic at our door. He just stood there, holding up the first sheet off the press and asked, "What's wrong with this?" I looked at it for a few seconds and gasped. Our calendar cover said 1984 not 1987. I'd given our graphic artist an old 1984 calendar cover with the instruction that I wanted similar lettering. Somehow he had used the old date in his artwork and not one of us had caught it at any of the subsequent proofing checks. Thank goodness a vigilant pressman stopped the press. That big goof up could have cost WCWC many thousands of dollars!

WCWC's last ditch efforts to save Windy Bay

In early November, just before Ken Lay began the big job of selling our '87 calendars, we sent him up to Lyell Island to check out a horrible rumour that logging had already encroached on Windy Bay. If so, we needed photographic evidence. We also needed images for a poster and another newspaper to intensify our campaign. We'd heard that road building and cutblock permits were about to be approved right in the Windy Bay watershed, with logging scheduled to start early in the new year.

Accompanied by his girlfriend Carleen (who later became his wife and worked for years managing WCWC's database), Dave Weir and Jeff Gibbs (both graduates of the Quest program). Ken drove up to Prince Rupert and took the ferry across to Skidegate. From there they hitched a ride on a Skidegate fishing boat down to Windy Bay and hiked across to Gate Creek to the south. They verified that the forest up to the edge of the Windy Bay was gone. A huge new clearcut extended right up to, and perhaps even slightly into the watershed. The area was quite level and the watershed divide indistinct. Road and cutblock flagging tape were all over the forest inside Windy Bay. It was obvious that Beban's loggers were about to move in.

Haida defend Windy Bay

The good news was that Ken also found the Haida working on a small, beautiful traditionally-designed longhouse in Windy Bay located by the shallow intertidal bay near the creek's mouth. A little more than a hundred years earlier a Haida village had thrived in exactly the same spot. Village sites like the one in Windy Bay should all have been designated as "Indian Reserves" during the reserve designation

Photo taken along Windy Bay Creek during the fact finding expedition to Lyell Island used on the cover of our last South Moresby newspaper Battle for Windy Bay. Jeff Gibbs sits by Windy Bay Creek and contemplates this watershed's future. Photo: WCWC files.

process in the late 1800s. But many, like the one there, were not. Therefore, the Haida's new Windy Bay longhouse was considered to be a "trespass cabin" by the B.C. government.

Members of WCWC's last ditch effort to save Windy Bay pose for a portrait by an ancient Sitka in Windy Bay. From left to right Dave Weir, Carleen Lay, Jeff Gibbs and Ken Lay. Photo: WCWC files.

But, according to Guujaaw, one of the longhouse builders whom our team members met at the site, the RCMP boat had visited and, after one tense night when it seemed like the RCMP might come ashore and order the Haida to stop and remove the structure, they left without doing anything.

This Windy Bay longhouse was obviously going to be the headquarters for the next wave of Haida "direct actions" if the government allowed more logging on Lyell Island, especially in Windy Bay. This time, it was obvious that the Haida blockades would be massive. It seemed insane to me that the B.C. government would provoke such a confrontation. This commitment made a sham of the Wilderness Advisory Committee's recommendation that 90 percent of the Windy Bay watershed be logged and only a 10 percent of it be left uncut as a leave strip along the ocean and creek in a meaningless "ecological reserve."

Our four-member WCWC volunteer crew helped the Haida bring in pebbles from a nearby beach for the longhouse's floor. They also took many good photos of the recent nearby clearcuts and the large ancient trees in Windy Bay watershed for a new WCWC newspaper and poster.

Ken returned from Lyell Island all fired up on the campaign. He was full of energy and marketed our calendars successfully. Having a grizzly on the front cover greatly helped sales. People love bears.

South Moresby Delegation "crashes" Socred annual convention to present a case for preservation

We constantly thought up new creative ways to get our message out. Nothing, as long as it was legal, was too outrageous or brazen. We learned through a *Vancouver Sun* newspaper article that the Social Credit Party would be holding its 1986 annual Convention in the Vancouver Hotel at the end of November. "Why not rent a hospitality suite on the same floor as the convention is being held to promote creating the South Moresby Park? It can't cost that much," I said, as we all relaxed around the kitchen table drinking a few beers after a long day working on a mail out. "Unless every room is already taken," I added, just a bit skeptically.

"Let's try it," said Ken. "What have we got to lose?"

We decided the best way to approach the hotel booking was to be assertive. Adriane was "elected" to be the one to call the hotel because of her air of positive confidence and good negotiating skills. We decided we'd call ourselves the "South Moresby Delegation" (even though there was a group of citizens from Sandspit with a similar name who were vocally opposed to the park proposal.)

Adriane called the next morning and, yes, there was one last room available. She booked it. I couldn't believe our good luck. It cost only $400 for the two-day convention. Only after we arrived at the hotel early on the morning of the first day of the convention and I saw in big letters "South Moresby Delegation - Room 322" underneath the heading "Social Credit Party Convention" on the hotel's bulletin board in the lobby, did I realize that the hotel thought we were Socred Convention delegates.

We hurried upstairs and quickly set up our room with our maps and photo displays. We filled several tables with our newspapers and copies of Parker's dendrochronological report on thousand-year-old trees in Haida Gwaii. We also set up a projector with a provocative slideshow. We even ordered some juice and coffee to serve to people. I was in a sports jacket but Ken Lay was decked out in a fancy pinstriped suit. He blended in perfectly, looking exactly like an up-and-coming young Socred. Ken circulated in the crowds of conference delegates outside our room, encouraging people to see our slideshow. During the day he managed to get into a big argument with Tony Brummet and to talk directly with Premier Vander Zalm.

Meanwhile Adriane and I stayed in our suite and talked to the delegates that began to drift in. After looking around for a few moments, many had a bewildered look on their face. Some left without saying anything, some stayed and talked to us. Most assumed that we were from Sandspit and were seeking support in our fight against environmentalists who were trying to shut down logging and destroy jobs. "Oh no, we support the idea that this special area become a park, and you should, too," Adriane explained.

Some were friendly and supportive. Others were not so supportive, but willing to civilly exchange ideas and opinions with us. Most went away with a rolled up copy of our beautiful South Moresby *Canadian Wilderness – Environments worth Protecting* poster tucked under their arm.

A few delegates to the convention, some of whom belonged to the Moresby Island Concerned Citizens championing the fight to log the area, angrily confronted us. We never lost our cool. When you stay cool, you always win in a situation like this. We had a perfect right to be there and to present our side.

In the afternoon, a representative of the hotel staff came in and said there might be some confusion and asked us if we were an "official delegation" at the conference. "No, we're not," I answered and he left. The next time I passed through the lobby I noticed that the listing "South Moresby Delegation – Room 322" had been moved from under the list of Socred Convention rooms to a position all by itself near the bottom of the board. But they didn't try to move us to a different room. I was elated. We were right in the midst of the Socred conference and having a mostly positive impact.

That afternoon Adriane and I took a break to join a "hospitality" event down the hall. There Adriane ran into Kim Campbell, who was elected in the next provincial election as a Social Credit MLA and subsequently elected as a federal Conservative culminating her political career, briefly, as Canada's first woman Prime Minister. Adriane knew Kim as a fellow instructor at Langara College. They often had lunch and dinner together in the faculty lounge.

Kim launched into a vitriolic tirade against Adriane, advising her to stop wasting her time with such a foolish campaign as South Moresby. She had expected more of Adriane, she said, than to be involved in something as "unintelligent" as that. Kim Campbell obviously had no sufferance for wilderness preservation. Adriane countered by saying she expected Kim, with her intelligence, to learn more about the issue

My favourite WCWC memories are of gutsy moves like this "infiltration" of the Socred convention.

We decide to rent office space and move WCWC out of our home

At WCWC's 1986 AGM, held that fall, the directors unanimously decided it was time for WCWC to find office space of its own. Adriane was tired of bundles of newspapers along every wall and volunteers in every room. We never had the house to ourselves and she was now pregnant with our second child. Besides, she and I would soon be moving into our new house on the Sunshine Coast, which would leave WCWC homeless. We budgeted $6,000 towards the yearly rental of a new WCWC office and directed Ken to start the search.

We also decided that, if Adriane purchased for $3,500 an IBM clone computer (something we all wanted her to do), WCWC would "lease to purchase" it from her. WCWC's computerization was spurred on by the fact that Bob Broughton, a systems analyst who became a director at our last AGM, was between jobs and had the time to custom-modify a database program to keep track of our members, donations and sales. During three weeks worth of full-time work he produced a program that was, as they say, "robust." With many updates and improvements over the years, this customized software is still being used by WCWC today to handle our database of over 150,000 members and supporters.

As usual WCWC was still in debt. But our bookkeeper, Ed McDonnough, reported to the directors that we were making good progress in getting our financial ship in order. When our inventory of unsold posters worth about $6,000 was taken into account, our debt seemed manageable. From then on it was the potential worth of our inventory that kept WCWC's books almost continuously in the black.

At that AGM, for the first time, WCWC directors discussed the classic "catch 22" idea to get rid of debt: cut back on our spending. McDonnough thought it was the only solution. I disagreed, contending that the only way we generated as much donation money as we did was because we were active on many projects. It was a treadmill we couldn't escape. Moreover, I feared that if we stopped working on a campaign no one else would take up the paddle and continue to build the massive public support needed to win. To get out of debt we'd have to be more effective and do more, not less, I maintained.

Our first federal audit

Shortly after our AGM, our bookkeeper finally filed our very late charitable tax reports for 1983 through 1985 with the federal government. Shortly after he filed, an official looking letter arrived in the mail. We were being audited by Revenue Canada. I called the phone number in the letter, talked to the auditor and arranged for him to come and pick all our financial documents, meeting minutes—everything—at our home the following week. I called McDonnough, who had all our financial records at his house, and went to pick them up. There were two apple boxes full of file folders, documents, vouchers, cancelled checks, bank statements—the works. Ed had unstapled all of the receipts from Richard Krieger's early expense reports and re-sorted them in a different way for some reason that I couldn't fathom. It looked like a complete unintelligible mess to me... but it was all there.

The government's auditor arrived as planned the next week. He was of English heritage and very proper in demeanour. He was most pleasant but unable to answer any of my questions like, "How long is it going to take?" and, "Is there any particular reason why we were selected for an audit?" Just as he was about to get up and leave with the boxes of financial records, three-year-old Kallie ran into the room. She was used to lots of different WCWC volunteers in our home and was at a particularly inquisitive and talkative age. She went right over to him, looked him over and said, "You have a penis and I have a vagina!" The embarrassed, now very agitated auditor quickly picked up both boxes full of files and hurried out the door. Embarrassed, too, I asked Adriane about it. Apparently, that morning, Adriane had straightforwardly answered Kallie's question about how girls were different from boys. Obviously Kallie had absorbed the information and wanted to try out her new vocabulary.

About three months later, the auditor returned everything neatly sorted. I asked him how it went and he said he would be sending us a written report. He told me he had spent a lot of time trying to figure out our calendar finances and, regarding that, he had one question for me, "Why did you keep on publishing the Wilderness Calendar when it obviously loses money every year?" I tried to explain that we did it because of its educational value. It brought to public attention threatened wilderness areas that needed park protection like no other avenue of communication could. "It doesn't lose that much money, does it?" I asked. I had no clue as to how much it lost every year and was genuinely curious as to what he had calculated. With a shrug, he left without answering the question.

Threatened with loss of our charitable tax status

I'd almost forgotten about our audit when, several months later, the auditor's promised letter arrived. I read the four-page report slowly, in shock. The list of things we were doing wrong was staggering. We had one month to show reason why our charitable status should not be taken away from us. This was serious. He noted that our official donation receipts lacked some of the vital information they were supposed to have (like the date we received the money). WCWC had funneled money to the Council of the Haida Nation ($5,000 for the Haada Laas newspaper) without having a contract with them as required by law. Some of our contract employees, like Ken Lay, should be regular employees requiring us to pay UIC and CPP. These were three of the more major "infractions." There were many more minor ones.

Remembering the magic performed by a well-respected lawyer when we were first attempting to get charitable tax status, I decided that we had to get a big name lawyer to handle this situation. I called Murray Rankin and Jack Woodward and several other lawyers. None of them felt qualified enough to take it on. Societies and charitable tax status was a specialty in the practice of law. They referred me to a lawyer who was an expert in this field; he worked for one of the big firms that had acted for logging companies on blockade injunctions. Of course, the company had not acted against us, for we had never taken part in any civil disobedience. I contacted the recommended lawyer and nearly choked at the $200 an hour fee.

He reassured me that it shouldn't take more than an hour or two to draft the letter. I delivered the auditor's letter to his office. And, while I sat in his plush office with the meter ticking at over $3 a minute, I watched him read it. I couldn't read any emotion on his face as he did it. When he finished he didn't give me any reassurance. All he said to me was that he happened to be going to Ottawa the next week on other business and he would stop in and see someone about this matter when he was there. He made another appointment in ten days, very close to the deadline for our reply. I spent an anxious week and a half waiting.

Finally, the day of our next meeting arrived and I again sat in his swanky office. He started out by saying that our breaches of the charitable tax law were very serious. "I can write the letter for you, but I think it would be far better if you wrote the letter yourself," he said, explaining that a letter from me, the founding director, would have a better impact than if the letter came from a hired lawyer like him.

He advised me that I should admit that we had unwittingly done all the numerous "illegal" things listed in the audit report but had not intended to flout the law. We simply were ignorant of the law. (Which was true, although everyone knows that ignorance of the law is really no excuse!) He further advised that I should apologize and state specifically how our organization was going to correct each error so we would be in complete compliance in the future. "You can only write this kind of letter once," he said. The lawyer's parting words were, "There's a good chance they will give you a second chance. I'll read the letter and check it over before you send it, if you like."

It took me a whole day to write it (with Adriane's help, of course). I took the finished letter to the lawyer and he made virtually no changes—except to improve its legal language. I sent it off with a prayer. With trepidation, I opened the response that came three weeks later. It worked! We'd been given a second chance! We immediately scrapped our old tax receipts, redesigned new ones to comply and got them printed. We made sure we wrote contracts whenever we did projects with other groups that did not have charitable tax status, and we started passing spending resolutions at directors' meetings. We made every effort to fully comply. Our charitable tax status was precious and precarious.

In the years to come we were audited several more times. I'd heard a rumour that some of our audits were instigated by disgruntled MPs who disagreed with the stands we took on issues. Each time, the auditor found a few small things we weren't doing right. Each time, we corrected the problems. I told one auditor who spent over two months in our office poring over our books, "I hope you spend the same proportion of time, relative to the amount of tax money involved, looking at the books of big forest companies." He didn't comment.

Last of the Giants poster

Meanwhile, Randy Stoltmann kept pushing on his campaign to have Canada's record-sized trees declared National Landmarks and protected under Canada's Historic Sites and Monuments Act. During the fall of 1986 he focused the campaign on Canada's largest known tree, the Red Creek fir that grows near Port Renfrew on Vancouver Island. This tree is Canada's largest living thing. He produced a four-page WCWC newspaper that included detailed instructions on how to visit this tree. We also published 2,000 full coloured *Last of the Giants* posters of the Red Creek fir and sold them for $5 each through a coupon in the paper. Our paper also included a photo of the Koksilah fir, a tree nearly as big as the Red Creek fir, which was located only a few kilometres east of it.

It was a "fallen giant." (It blew down in 1979 shortly after the surrounding forest was clearcut). People could walk along the trunk for several hundred feet. It was a poignant picture. We urged that the remaining old growth around the Red Creek fir tree be left as a buffer zone to protect it from the same fate. We were justifiably worried because the Red Creek fir was already flanked by a clearcut on one side and storm-driven winds often reach hurricane force on this part of the west coast.

We were really disappointed that our Red Creek fir campaign failed to bring National Landmark protection. I figured that it might be because Parks Canada feared that the tree might soon blow down, instantly destroying a National Landmark and generating negative publicity. In 2006 the tree still stands tall and still has no special status. In December of 1998, Dr. Robert Van Pelt from Washington State University re-measured this monster tree using the latest electronic measuring tools including a laser clinometer. He confirmed that it is the world's largest living Douglas fir, with a height of 74 metres, diameter at breast height of 402 centimetres, crown spread of 22.9 metres and a total wood volume of 349 cubic metres (enough wood to fill ten logging trucks, for those needing such a measuring stick to comprehend its enormity). This tree is also the largest known individual plant in its taxonomic family, Pinaceae.

Volunteers from the Port Renfrew Chamber of Commerce maintain the 15-kilometre-long Red Creek Fir Trail and road from Port Renfrew out to this tree. It's navigable only by four-wheel drive vehicles with high road clearance. It's incredible that such a remarkable tree is not recognized as a Canadian Landmark, granted full Parks Canada protection and made more accessible for people to visit.

As part of our efforts to get more protection for B.C.'s big trees (which rank among the top ten species in the world), I wrote a letter to the Postmaster General proposing that a series of commemorative stamps be issued featuring Canada's record-sized native tree species. I made the point that there were stamp series on Canadian wild flowers and wild animals but none featuring trees. I illustrated the point with dozens of flower and animal stamps on the envelope.

I got back a polite bureaucratic reply saying *"Please be assured that this suggestion will be brought to the attention of the Stamp Advisory Committee, a group of citizens knowledgeable in a variety of disciplines, who advise the Minister on the selection of subjects and designs for Canadian postage stamps...."* Nothing ever came of it. Not one record-sized native Canadian tree has a stamp commemorating it. For that matter, to this day there are no stamps commemorating any of the softwood tree species of Canada... it must be due to politics! But we do have stamps paying tribute to nearly everything else, including Superman in 1995.

We publish *Hiking Guide to the Big Trees* — Randy Stoltmann's first book

Randy kept at it. His next project with us was the publication of an amazing book: *Hiking Guide to the Big Trees of Southwestern British Columbia*. It was a project in which he'd been immersed for years, collecting information since he was a young teenager. It was our second book production, so we knew some of the ropes. At our November 1986 directors' meeting we allocated $16,000 to publish 6,000 copies of Randy's book. Luckily my daughter Athena, who did the layout and paste up on our Meares book, needed a job, so I convinced her to take on this project for starvation wages. She worked long hours to get the manuscript "camera ready" for printing. Randy, who was meticulous about details, provided her with numerous small scraps of paper filled with extremely tiny writing detailing his corrections and additions. Athena made sure every one was made. We published Randy's 145-page soft cover book featuring hikes in over 20 different

First edition of Randy Stoltmann's *Hiking Guide to the Big Trees of Southwestern British Columbia* published by WCWC in 1987. Cover image: Randy Stoltmann.

areas in June 1987. It had 40 black and white photos and 20 illustrated maps hand drawn by Randy.

The book sold for $9.95 was very popular, especially with outdoors enthusiasts. Randy didn't receive any money for his work: he got books to sell instead. We ran out and had to reprint it in 1992. He did not get any royalty payments on the second printing, either—just more books to sell. It was the only way we could afford to do it. Thank goodness Randy was interested in seeing these places saved, not in making money. The book worked miracles in this regard. Almost all of the places he featured in this hiking guidebook eventually got protected.

Facing page: Randy Stoltmann beside the Cheewhat cedar, the world's largest known red cedar tree. Photo: WCWC files.

Chapter 15

Campaigning to protect "pocket wilderness" areas
Battling to save Windy Bay

Meeting Joe Foy for the first time

One morning early in December of 1986, while he was packaging up calendars in our living room to mail, Ken Lay said to me, "I bet you don't know who has sold more of our calendars than anyone else this year." I hadn't a clue. "You should meet him," Ken continued, "and I'll bet we'll get a free hamburger for lunch, too." At noon Ken took me down to meet Joe Foy who was then the owner-manager of the now long-gone Dairy Queen on Water Street. It was located near the famous steam clock in Vancouver's historic Gastown, just a couple of doors away from where WCWC's third floor skylighted headquarters are today. We learned about Joe's promotional method that made him our champion calendar salesman of the year. And, sure enough, Joe gave us free hamburgers.

In his Dairy Queen he had a big floor display about the Stein Valley with our calendar turned to the month that featured the Stein. With the purchase of any meal in his restaurant you could get a copy of the calendar for just a couple of bucks. No wonder he was so successful. He seemed quite young to be owning and running his own business, but he obviously had good business savvy. It was noon hour and he was too busy to talk at length. He had just enough time to tell me that the 10-day trek he'd taken through the Stein Valley in 1981 had transformed him.

Joe had heard about WCWC and our calendars through Eugene Rogers, a long time conservationist who, like a number of other conservationists, was working within government processes to try to protect the Stein. I invited Joe to come over and work together on increasing our efforts to save the area.

Joe launches his "Pocket Wilderness" campaign

A few days after that meeting in the Dairy Queen, Joe showed up at our house after work. Ken introduced him to another WCWC volunteer who happened to be there that night—a forest technician working for the B.C. Forest Service named Clinton Webb. Both Joe and Clinton were knowledgeable and passionate about protecting the last remaining wild areas in the Lower Mainland around Vancouver. Over beers, they brainstormed which of these local areas needed protecting the most. A few weeks later Joe came to our home armed with maps and information about "pocket" wilderness areas that he and Clinton believed ought to be made into parks.

A new campaign—including a detailed newspaper to launch it—began to take shape on our light table. Since Adriane had purchased it 18 months earlier, the table had been in constant use.

Trust grant gives WCWC's campaigns a huge boost at a critical time

In December of 1986 we got our second $20,000 grant from the Canadian foundation that wished to remain anonymous. Again it was unexpected and sorely needed. We decided to use part of it to boost the Khutzeymateen grizzly bear sanctuary campaign led by the Friends of the Ecological Reserves and the Sierra Club of Western Canada. We also used some of the grant money to hire Arne Hansen, a former volunteer and newspaper reporter, on a part time contract to produce a paper about mining in Strathcona Provincial Park. A local group was actively resisting new mining exploration and development in the park. The public needed to know what was at stake there, for it affected all parks. We spent most of the money, however, on our South Moresby and Windy Bay campaigns.

Last South Moresby newspaper — *Battle for Windy Bay*

At the same time that Joe was working away on the "Lower Mainland Pocket Wilderness" paper, we began working on a special Windy Bay newspaper, using the images and information Ken brought back from his November trip to Lyell Island. As a reality check, we ordered a Landsat satellite photo of Lyell Island taken just a few months earlier. It cost us over a thousand dollars but it was well worth it. It showed the shocking extent of clearcutting that had occurred since Beban began logging there ten years earlier.

We finished our *Battle for Windy Bay – Last Chance to save WINDY BAY* newspaper in the first week of January 1987 and printed 100,000 copies. We sent two to every federal MP and B.C. MLA asking them to share one with their staff. We were trying desperately to fend off a decision to log there.

Guujaaw drums and chants Haida war song at our Save South Moresby public events

Knowing that the Haida would be resorting to a blockade if the government issued any cutting permits in Windy Bay, we scheduled two large public meetings—in Victoria and Vancouver—in March of 1987

Above: Guujaaw announces to a overflowing crowd that the Haida were at war to protect their lands at a Save South Moresby event we held in the Robson Square Theatre. Photo: Greg McIntyre.
Facing page: Eagle Mountain—a proposed Pocket Wilderness Park. Joe Foy is in the foreground on the left. Photo: WCWC files.

South Moresby protest with one person holding up a picket sign with our *Windy Bay – On the Edge of Destruction* poster on it. Photo: WCWC files.

and flew Guujaaw down to speak at them. A sell-out crowd packed the large theatre in Vancouver's Robson Square Media Centre. Guujaaw began by drumming and singing a powerful song I had never heard before. After he finished he said, "That was a Haida war song. We are at war to protect our lands." It gripped the audience. He gave a short speech about the Haida's resolve to protect South Moresby and the B.C. government's dishonesty in dealing with the Haida. Then Ken Lay gave his slideshow. He mostly used the images he had taken on his trip to Windy Bay seven weeks earlier. People were aghast at the sea of stumps in the fresh clearcut adjacent to Windy Bay. They were awed by the lush natural forest and huge old-growth trees that would soon be cut in Windy Bay unless the B.C. government changed its mind. The destruction of South Moresby was not an abstract fear; it was an ongoing tragic reality.

We had a long intermission in the middle of this event to give people time to do something concrete to help save South Moresby. We had everything prepared and set up in the theatre foyer—paper of different sizes, pens of various colours, envelopes of diverse shapes and several kinds of commemorative stamps—all so it wouldn't look like a form letter protest. I watched people struggle to write letters to the Premier and Prime Minister as they got stuck on exactly what to say. Some were new Canadians and felt that their English wasn't good enough. Our staff and volunteers went around and encouraged them by saying, "It doesn't have to be perfect or profound, it just has to let the government know how you feel about protecting this place. Please finish it tonight so we can mail it right away. The decision will be made soon." We explained that their letter would get read (at least by a ministerial assistant) and tallied. What counted was the number of pro preservation letters versus those pro logging. Several hundred people wrote letters that night. The minutes of our directors meeting held a few weeks after this event said: *"We believe that our efforts helped turn the decision around regarding Lyell Island."*

Windy Bay – On the Edge of Destruction poster blitz

A few weeks after publishing the newspaper, we printed 5,000 copies of a new colour poster: *Windy Bay – On the Edge of Destruction*. It featured an exposed big old-growth spruce tree next to the clearcut on the southern border of Windy Bay. We didn't produce this poster for sale. We just gave them away as fast as possible to get people to write letters of support. We got rid of all of them. There's not even one copy in WCWC's archive today.

We also gave one to every B.C. MLA. During this non-paranoiac pre-terrorism time, I was able to walk into the Legislature buildings in Victoria with a huge garbage bag full of rolled up posters and go to every MLAs office, including the Cabinet Ministers' offices, and personally hand them out. I was never stopped or questioned by any security guard. I always carried extra posters to hand out to all the secretaries who wanted one.

NDP Opposition MLA Mike Harcourt backs South Moresby National Park

Most of the MLAs and MPs never even sent an acknowledgement that they had received our posters or educational newspapers. The

Fresh clearcuts extend further along the east coast of Lyell Island south of Windy Bay as Beban Logging accelerates its rate of cut. Photo: WCWC files.

rest, for the most part, sent us a form letter that applied to all the correspondence they received. But there were enough thoughtful replies to make it worthwhile—like this one from a politician destined to become Premier in a few years.

> Dear Mr. George:
> Thank you very much for your note of February 26th with the enclosed complimentary education poster 'Windy Bay on the Edge of Destruction.'
> I have the poster on my legislative office wall.
> Dan Miller, our MLA from Prince Rupert, has been very aggressive in our Caucus on the whole issue of South Moresby and Windy Bay. I'm sure you'll appreciate that our Party and Caucus position is very supportive to the aims and goals of the Western Canada Wilderness Committee.
> We will be continuing to fight for a national park on South Moresby.
> Mike Harcourt, MLA – Vancouver Centre

South Moresby Rally at Fantasy Gardens

To increase the pressure on Premier Bill Vander Zalm, we mailed a copy of our Windy Bay paper to every household in his electoral district. Vander Zalm had become B.C.'s Premier in August of 1986. We thought that maybe, just maybe, because he was a professional gardener, he'd be more sympathetic towards protecting South Moresby and other wilderness areas than his predecessor, Bill Bennett, a real estate and investment businessman. This householder mailing used up over 10 percent of our 100,000-copy press run.

We also held a protest rally at Fantasy Gardens, then Vander Zalm's privately owned tourism operation in Richmond. I generally do not believe that people should move the political debate to the homes and businesses of elected officials and protest in front of them, but there were extenuating circumstances in this case that I believe warranted this action. The Premier was holding a cabinet meeting there—a wholly inappropriate place for government business to be conducted—and at that meeting they'd no doubt be discussing South Moresby.

With placards and a large Canadian flag waving, Ken Lay, Joe Foy and about 25 WCWC volunteers brought the message "home" that South Moresby must be protected. We urged Vander Zalm to accept the federal government's offer to work towards creating a National Park Reserve. Fantasy Gardens wasn't really set up to handle a protest and it's amazing that we did not get kicked off this private property. One motorist slowly driving into the parking lot got so distracted by Ken Lay waving his sign that he lost control of his big Cadillac and smashed into the rear of the car ahead of him, causing minor damage. This protest had a significant impact for it was reported province wide in the major papers and TV news.

"Pallbearers" parade TFL 24 RIP coffin at yet another South Moresby rally

That same March of 1987 Jeff Gibbs' "Tree Club" sponsored a large South Moresby rally in downtown Vancouver. Everyone marched from the steps in front of the Art Museum on Robson Street to the headquarters of Western Forest Products (WFP), the

Symbolic end to the logging rights in the South Moresby Wilderness Area. Pallbearer on far left, Ken Lay. Leading the funeral procession, Paul George. Photo: Martin Roland.

new owners of TFL 24. About 600 people attended. We wanted TFL 24 out of South Moresby. To make the point, we borrowed a grim theatre prop—a life-sized big black plywood coffin. Six of us dressed up in black suits with dark sunglasses carried this somber prop with a big sign on either side that read "TFL 24 R.I.P." to WFP's headquarters.

It caused quite a stir along the Vancouver streets. I'm sure most onlookers had no clue as to what the coffin was all about, but it certainly made good theatre! A video clip of our coffin was featured prominently in the brief TV coverage. We arrived at WFP's office building to find it completely shut down and all the doors locked. We heard later that the Tree Club's poster advertising a "special event" at WFP's headquarters made the company fear the worst. According to Jeff they just put the words "special event" on the poster because they didn't have anything specific planned yet, not because they were going to do some acts of civil disobedience.

Stopping the "legal" falcon poaching on Haida Gwaii

That spring WCWC took on its first court case. Someone upset in government tipped us off that the B.C. Fish and Wildlife Branch had issued a permit (Wildlife Sundry Permit CO 17993) to allow the capture of 10 Peales Peregrine Falcon chicks on the Queen Charlotte Islands. Some of them would come from nests within the proposed South Moresby National Park.

The provincial government granted this permit in an extraordinarily unorthodox way, circumventing input from local B.C. Fish and Wildlife officers on the Queen Charlotte Islands who knew best that the falcons on the islands were in a most precarious situation. If they had been consulted, they undoubtedly would not have approved the permit.

Previously, permits issued on B.C.'s coast had been for one or two chicks per year. Even this rate of chick stealing was wrong as far as we were concerned. We had biology and the public on our side. Only the falconers, who needed wild genes to keep their domestically-bred birds fierce, supported the big "harvest."

WCWC, jointly with the Islands Protection Society, petitioned the Supreme Court of B.C. to judicially review the decision to grant this permit and strike it down. Our case never went to court. The government pulled the permit, citing the "threat of violence" (not the extremely low numbers of birds left in the wild) as its reason.

The next year the B.C. government established a one-man commission to decide whether or not to allow the capture of falcon chicks from wild nests on the Haida Gwaii. On March 15, 1988, Cyril Shelford's hearing on Peales' Peregrine Falcons got under way under fairly bizarre conditions. A man with a live falcon perched on his shoulder—thankfully, with a leather hood over its head—sat right behind me. (I think the presence of this live falcon in the audience helped our side more than it did the falconers.) I presented to Shelford WCWC's brief with strong biological arguments for why the government should cease issuing permits to "capture" chicks from wild falcon nests. Our brief, along with others, led to Shelford recommending that no permits be issued.

Chapter 16

Moving into a warehouse office
South Moresby campaign reaches a peak

WCWC rents office space on West 6th Avenue in Vancouver that, by luck, happens to be directly across the alley from the Vancouver Press Club

At the beginning of 1987 our search for an independent WCWC office took on a new urgency. The new baby that Adriane and I were expecting was due in April. There simply wasn't enough room in our small Alberta Street home for the baby, rambunctious Kallie and the WCWC office. WCWC had to go.

After a long search we finally found a place—an office space on the second floor of a building on West 6th Avenue, less than a block away from the Pacific Press building, headquarters of the *Vancouver Sun* and *The Province*, the two large daily newspapers in Vancouver. It was a bit funky, with the entrance in an alley and the washroom located in a tiny closet halfway up the entryway stairs. But it was cheap and large. It was also quite a mess and strewn with broken glass. A huge chunk of thick safety glass that had shattered cracks but was still in one piece was too hard to remove so we stood it up on end and made it a feature wall in front of our reception desk. We had occupancy as of April 1, but it took several weeks of cleaning and fixing before we could move in.

Ken Lay answers the phone at his desk at West 6th. Note shattered glass feature wall to his left and the Toni Onley and Australian posters on the wall above behind. Photo: WCWC files.

South Moresby Summit Meeting in Ottawa

While fixing up the new office, I got an urgent call from Colleen McCrory. "You have to come to Ottawa right away for a meeting with Environment Minister McMillan. Don't worry about the cost, the government's paying for everything." When I got there environmentalist Elizabeth May, who was still McMillan's executive assistant, got everyone together to strategize before the big meeting: Colleen, Thom Henley, John Broadhead, David Suzuki, Vicky Husband, several others including Reverend Peter Hammel of the Anglican Church and Gregg Sheehy of the Canadian Nature Federation and me. We were the key environmental leaders who—not counting the Haida—had made the environmental movement to save South Moresby as successful as it had been... so far. She broke the good news to us that the government was going to create a South Moresby Park.

Then she broke the bad news that the deal for the park included letting Frank Beban log the rest of Lyell Island except for 10 percent of Windy Bay to be left as a pseudo-ecological reserve. This was unacceptable to all of us. We all agreed we would not give in. We wouldn't accept this sell-out.

The next day at the meeting with the Minister we all sat in chairs in a big open circle. We first heard from Tom McMillan about the "deal" that would bridge the economy on the Queen Charlotte Islands from a logging-based one to a tourism-based one. When my turn came up I made a passionate plea to protect Windy Bay. All the years of research paid off. I told him how little wood really was in Windy Bay—just five day's worth of the province's 80 million cubic metres of wood cut annually. I also showed a two-minute video Ken had taken of very recent, very ugly, clearcuts made on Lyell Island. Everyone else gave great moving speeches too. Together we had quite an impact on the Minister.

At the end of the meeting, McMillan asked us one last question: if he insisted on having all of the South Moresby Wilderness Area in the National Park, we might get nothing. Were we willing to risk that? Everyone said yes. A South Moresby National Park without Windy Bay and Lyell Island would not be much of a park. Canada could afford to save the whole thing. Now it was in the federal government's hands. I knew it would be a rocky road ahead, but I was privy to none of the bumpy details. I headed home the next day to help complete the renovations at our West 6th Avenue office space.

Moving WCWC to its new office hastens the Stork!

Moving day—April 21, 1987—arrived. The volunteers and truck were scheduled to come in the afternoon. In the morning Adriane decided to go shopping for an antique porcelain bathtub for our new home. Nine months pregnant, Adriane climbed in and out of quite a few different tubs to "test them out." In the evening, as the volunteers helped move the last of WCWC's office stuff into the truck and she helped lift some of the lighter loads, Adriane lay down on the couch complaining of a backache. After about an hour, one of the women said to Adriane, "You are complaining at regular intervals." All the volunteers dropped their boxes, sat on the floor in front of Adriane, and looked at their watches, timing Adriane's complaints. "Ow, my back," said Adriane. "Three minutes!" they yelped out in unison. Adriane was in advanced labour. I immediately took her to the hospital and Terren John (soon nicknamed TJ) was born a few hours later in the early morning of April 22—Earth Day.

The volunteers completed the final load and finished the move on their own. Adriane and I personally, and WCWC collectively, had entered into a new, exciting phase.

A special trip into South Moresby before the decision comes down as to its fate

Several months before TJ was born, I had accepted an invitation from WCWC's largest donor to be his guest on a week-long sailing cruise in South Moresby. I'd be sailing on Al Whitney's 71-foot ketch the *Darwin Sound*—all expenses paid—with five other guests. Adriane very generously said, "Go!" Her parents had agreed to help take care of her, Kallie and the new baby at their home in Burnaby while I was gone.

Guests and crew on South Moresby cruise to film ancient murrelet chick fledging Al Whitney far right, Bristol Foster to left of mast (wearing toque), Paul George far right. Photo: unknown.

There I was in the first week in May, less than two weeks after TJ's birth, back in South Moresby. We ended up exploring places that I'd never been to, either with Richard Krieger in the summer of '77 or with Adriane in the summer of '81. The trip re-affirmed for me the source of my energy and my devotion to the conservation cause—the wilderness itself.

The highlight of the trip was filming the emergence of ancient murrelet chicks from their burrows in the forest floor on the Copper Island Ecological Reserve. Bristol Foster, recently retired from the Ecological Reserves branch of government, was one of the guests on the *Darwin Sound*. He was making a film about this small seabird for a TV nature program. To the north of where we were, right next to Windy Bay, was another ancient murrelet nesting site at Dodge Point, Canada's largest known one.

The emergence of the murrelet chicks is a miracle of nature few have been privileged to witness. Near midnight we watched hundreds of tiny fluffy chicks emerge from their burrows and scurry to the sea, guided solely by the moonlight and urging cries of their parents who were waiting in the waters offshore. However, when we turned on the battery-powered light to film the cute little scurrying balls of feathers, they instantly became disoriented and swerved towards the light. It was hard not to step on them. Finally we decided to aim our lights and camera with our backs to the ocean and shoot the film in brief bursts, so that when the chicks swarmed toward the light, at least they were moving in the right direction. We didn't stay very long, for this was a place where nature had to be left undisturbed. Three-quarters of the world's population of ancient murrelets live on Haida Gwaii (an estimated 250,000 breeding pairs). They're the islands' most successful sea bird. Perfectly attuned with both land and sea habitats, they're a powerful symbol of this Galapagos of the North.

Every evening on board our luxurious yacht we listened to the short-wave radio, expecting to hear the announcement of the creation of the South Moresby National Park. But it never happened. Negotiations between the provincial and federal government had broken down... yet again. Every night in my tiny bunk bed, I couldn't sleep thinking about Adriane and our kids. I also thought a lot about our new office back in Vancouver and wondered whether people were settling in. Although we did not have the assured income to pay the $600 per month rent, we'd risked the move, figuring that with a much larger and totally independent space, WC^2 would quickly grow to afford it. I knew some of the volunteers had not come as often or stayed as late as they would have liked to in our home because they were worried about disturbing our family. The new office was also closer to Prince of Wales school where many of our young volunteers attended the Quest outdoor education program. I figured that, if we waited until WCWC was financially secure to move to a large enough space, it would never happen. It was a risk we had to take in order to expand and increase our chances of success.

The new office at West 6th teeming with activity

When I got back to Vancouver from South Moresby, I found our new office in full swing. Ken Lay and others had taken charge and were busy working on publications and campaigns. WC^2 had taken off. Our membership campaign was working. Membership had increased more than tenfold to 1,000. Most had joined in the last eight months, since we had added a membership appeal to every publication. While at West 6th office our byword became "never say no to a good idea or a worthwhile wilderness-saving cause" even if we could only help out a little. It wasn't a very well thought out strategic plan, but it worked.

At our first board meeting in our new office, one of our directors brought some homemade wine to celebrate our new home. The meeting rambled on and on for nearly eight hours until 3 a.m. as we laughed, joked and progressively got more inebriated.

We almost couldn't get it together to call the meeting to a close. From then on there was an ironclad rule we always kept: no drinking at any WCWC meeting.

Joe Foy completes the Pocket Wilderness paper

A few months after we moved WCWC to West 6th, Joe Foy's Pocket Wilderness paper was ready to print. It represented six months of work, mostly Joe's and Clinton Webb's volunteer labour at night after working their daytime jobs. It was our largest newspaper production to date, a 12-page, three-colour tabloid newspaper titled *Lower Mainland Pocket Wilderness Coalition – VANCOUVER AREA WILDERNESS – It's time to save the few pockets left*. Its centrefold map showed 20 areas around Vancouver that merited park protection. Six of them were prominently featured with a whole page devoted to each one. We printed 50,000 copies.

To raise money through the newspaper Joe thought of a novel idea: offering numbered limited edition *"Fair Share Certificates"* to acknowledge the donations. There were *"Common Shares"* at $20 each, *"Preferred Shares"* at $50 each and *"Blue Chip Shares"* at $100 each. They were elaborately designed using a real 1880s share certificate as base artwork, making them look like old-time mining stocks. Joe described the "stock offer" on the newspaper's cutout donation form this way: *"Forest companies have been issuing shares for decades, a way of raising capital and power. The results on our forested wilderness heritage have been devastating. The Pocket Wilderness Coalition of the Lower Mainland (PWC) is working hard to see that a few pockets of local wilderness are handed down to our children. Can you help?"* To keep the campaign rolling

Joe Foy at Flora Falls in the Chilliwack region, a "pocket wilderness" candidate. Photo: Leo DeGroot.

Fair shares certificate issued to donors for contributions to our pocket wilderness campaign. Document: WCWC archives.

he informally created a new environmental group called the Pocket Wilderness Coalition, which, as far as I could see, existed more on paper than in reality. But the few people involved were extremely devoted and the group gave legitimacy to the local campaign.

I strived to bring people with talent into the WCWC's fold. Shortly after we'd moved into our new office, a young man named Ian Mackenzie walked into our office. He brought with him a portfolio of stunning black-and-white Ansel Adams-like photos that he had taken in the Stein Valley with his large-format camera. Everyone in the office ooh-ed and ah-ed over them. I particularly liked one taken from a vantage point high up on the canyon wall looking upstream into the virgin valley. I knew it would make a great Stein Valley poster. Ian donated the image and within a few weeks a new Stein poster rolled off the press. It was the start of a long-term relationship in which Ian provided WCWC with priceless images and invaluable support. He also joined WCWC's board of directors, offering sound advice on our campaigns and business affairs too.

I've often thought about why I grabbed every chance I could to produce a new poster or campaign newspaper. **I've always believed in this publish-or-perish rule. The more you publish about a threatened wilderness area, the less likely it is that the wilderness area will perish.**

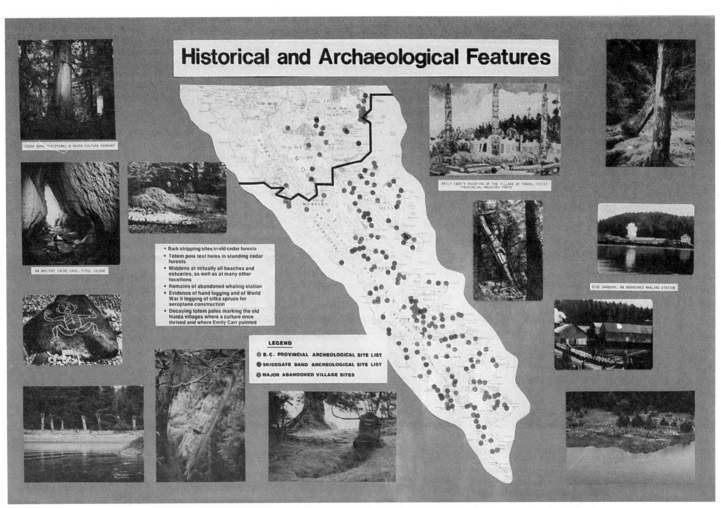

A display board IPS made in the late '70s to try to convince B.C. Parks Planners to seriously consider protecting South Moresby. Photo: Richard Krieger.

Chapter 17

Gwaii Haanas National Park Reserve becomes a reality
WCWC sets big new goals

Conservationists win "the big one" – South Moresby

For months we had been waiting for "the good news." Finally it came. In July 1987, the governments of B.C. and Canada finally cut a deal to create the South Moresby National Park Reserve. It involved a lot of money and compensation for the logging company. Some people felt that the price tag was too high, setting a bad precedent that would make the protection of other wilderness areas too expensive in the future. I didn't buy that.

After all, it was only a little over $100 million to save a very special place. That's just about the cost of one modern fighter jet plane and less than the cost of one of three "fastcat" ferries that the B.C. government built about a decade later that never really worked. It's important to put things into perspective. The precedent South Moresby set was that a large wilderness area could be protected. It didn't have to be a small, idiotic, ecologically-compromised one. The South Moresby success bolstered environmentalists' efforts to protect other big areas. We could win them too!

I was WCWC's only representative at the Government House signing ceremonies in Victoria on the afternoon of July 11, 1987— which coincidentally happened to be exactly ten years from the very day Richard Krieger and I took off on our first expedition into South Moresby in our little inflatable boat. It was surreal watching Prime Minister Brian Mulroney and Premier Vander Zalm joke around with each other before the actual ceremonies began; then, with a few strokes of a pen, sign the memorandum of agreement to establish the 147,000-hectare South Moresby National Park Reserve (also named the Gwaii Haanas National Park Reserve/Haida Heritage Site). It was all over in a flash. Afterwards I shook both their hands and simply said "congratulations." I couldn't think of anything else to say.

I knew full well it was the Haida who had forced the issue and it was not out of the goodness of these two leaders' hearts that this event took place. For about a half-hour after the signing everyone sipped champagne, ate hors d'oeuvres and chatted. I small-talked with all my environmentalist friends who had been a vital part of the colossal effort that helped push it over the top. As everyone was leaving, John Broadhead of Islands Protection Society came up to me and asked, "Do you want to fly up to the Charlottes with us? There's going to be a celebration in Skidegate. Someone dropped out and there's room for one more on the plane. Everything's paid for." Perhaps foolishly, I said, "No, I can't go. I have to get back to Adriane tonight."

I didn't have any money with me and I was not really part of the "in group" assembled by Elizabeth May, federal Environment Minister Tom McMillan's executive assistant, to help during the final government-to-government negotiations. I also rationalized that I wouldn't feel right taking advantage of this government freebie nor did I want to be part of a celebratory "government group" when confronted by the unhappy loggers who surely would be protesting at the Sandspit airport when they arrived. Some of the loggers I knew personally and liked, in spite of the fact they had fought so long and hard against us. I know I missed a really good time. I would have liked to have shared the cake that we had given Tom McMillan two years earlier at the Banff conference that he had put in the deep freeze to serve when the park became a reality. It was a lonely trip home.

President of the Haida Nation Miles Richardson and Federal Environment Minister Tom McMillan cut the South Moresby cake (not the same one we presented to the Honourable Minister in Banff two years earlier—it had been lost). Vicky Husband of the Sierra Club, far left and Thom Henley of Islands Protection Society, far right, look on. Photo: WCWC files.

On the ferry I re-read the press release accompanying this signing, noting that the Canadian government also committed to establishing *in the near future* a 305,000-hectare Marine National Park Reserve to protect the surrounding waters in the new Gwaii Haanas National Park Reserve. I wondered to myself, "How long is that going to take?" I knew there were offshore oil and gas leases in the area. Eighteen years later, this marine park reserve has yet to be established.

It's funny the letdown one can feel when one gets something one has worked so long and hard to get. I think in my case it was because the final days of the "battle" had been strung out for so long.

It was only when I got back and told Adriane about the signing that I cut loose and celebrated. It was easy to decide the theme for our '88 calendar: celebration of the creation of Gwaii Haanas National Park Reserve. We got on top of production right away and by the end of July we had all of the photos in and the maps under way. We printed 13,000 calendars and got them out at the beginning of October (almost a record for us). By the end of November, for the first time ever, we were running out of calendars and decided to order 3,000 more at a cost of $2.39 a copy. They went fast, too, and we sold out of them. With the money we made we finally retired the old Mitchell Press debt for the 1983 calendar.

Prime Minister Brian Mulroney shakes hand with Premier Bill Vander Zalm after signing the Gwaii Haanas National Park Reserve agreement. On the far left is Federal Environment Minister Tom McMillan. On the far right B.C. Environment Minister Bruce Strachan. Photo: WCWC files.

WWF's "payback"

In the final year of the campaign to save South Moresby, when success was definitely in sight, someone showed me an appeal letter by the World Wildlife Fund asking for money for its South Moresby Campaign. For years while so many others and I worked full out on this campaign, WWF was nowhere to be seen. Everything we did was on a shoestring. When I saw this fundraising letter I thought it was the ultimate expression of crass opportunism. I fired off a letter asking Monte Hummel, President of WWF, exactly what WWF had done for the cause. I don't remember ever getting a reply, but no doubt I must have. Perhaps they were lobbying federally at the highest levels. But where was WWF when we were desperately in need of money at the grassroots level to fund posters, newspapers and information sheets that had to get out to build the public pressure that ultimately compelled the governments to act?

In August 1987, less than a month after the successful conclusion to the South Moresby Campaign, WWF, the master of sophisticated fundraising, put out another appeal for South Moresby. This time Monty asked for money to help payback the courageous campaigners who went all out and into debt in the last year of the campaign to make the park happen. His letter said in part:

> WWF was only one player in the South Moresby effort and by no means the most important. The critical factor in this conservation achievement was the dedication and stick-to-it-ness of a small group of local individuals who sacrificed their personal lives and financial situations to fight for something they believed in. People like John Broadhead, Thom Henley, Vicky Husband, Colleen McCrory and Paul George....
>
> The way I see it, <u>we owe them</u> a debt. They managed to hang tough on behalf of all of us for many years. Now it's time to say thanks and repay the debt. They should come out of this back at square one....
>
> So far, WWF has invested over $50,000 in South Moresby. But the well has run dry and we need that much again. We need your help through a donation which will add your name to the long list of citizens who have personally gone on the line for South Moresby – people like Robert Bateman, David Suzuki and Pierre Berton....
>
> Many, many people called me in the hectic hours as we came down to the wire on South Moresby, asking me what they could do. <u>Well, this is what you can do.</u> Send in a donation to wipe the slate clean and to make Moresby the accomplishment it should be. Please lend a hand to make this conservation achievement complete.

What a powerful and effective appeal letter! And it really worked. In early September I got a call from WWF saying that its appeal was so successful that we should get together all of our receipts showing what we had spent on our campaign in the last couple of years and send them in. We found vouchers that totaled more than $14,000 in direct non-wage expenses including the bills for the last South Moresby newspaper we published. This certainly didn't include everything we spent for we did not have project accounting that kept track of these expenses then. It was wonderful to get WWF's cheque just in time to help launch our next big campaign—to save the ancient forest in Carmanah Valley.

A big picture decision – entering onto the world stage

A few days after the South Moresby victory, while we were still basking in the glow of this major win, a brochure arrived in the mail inviting WCWC to attend the Fourth World Wilderness Congress scheduled for September 1987 in Denver, Colorado. The other WCWC directors decided that if we could get a table there to display our publications, Adriane and I should go.

One phone call later we had a booth in the conference's convention hall, and confirmed registration at this congress that was held only once every four years. It was something to look forward to. We had no idea how dramatically it would shift our thinking and WCWC's campaign focus.

The Press Club becomes WCWC's hangout

One of the great advantages of our new office on West 6th was its location by the Vancouver Press Club where reporters, editors, sales staff and pressmen gathered to relax. The Club's back door was less than 10 metres away from our front door (directly across the alley separating us). At that time the Press Club opened every day at 11 a.m. and closed at 2 a.m. It was never very busy until the late afternoon and only really busy after midnight and on Friday nights.

During the day, we often used it as our off-premise lounge and meeting spot. Getting to know media reporters and constantly meeting new people with different ideas made many positive contributions to our various campaigns.

I had become a Press Club member years earlier. Down from "the Charlottes" for a brief visit in the fall of '78, I went to the *Vancouver Sun* on a Friday afternoon to meet Eli Sopow, then the paper's forestry reporter. I had heard that he was sympathetic and might do a story on South Moresby. We met in the Pacific Press building cafeteria. After a short while he suggested, "Since it's Friday, perhaps you'd like to accompany me to the Press Club across the street and continue talking there over a beer." Never one to turn down a 'cold one', of course I said yes. He signed me in and after a few beers and some interesting conversation, he asked me if I wanted, for just $20, to get one chance in a couple hundred to win return flight tickets for two to anywhere in the world. "What do you mean?" I said, not believing that I had heard right. He explained that a couple of times each year, the Press Club held an "airline's night" where Air Canada donated a pair of return tickets to anywhere their airlines went, to be raffled off! The catch was that the winner had to be a member of the Club and had to be present at the time of the draw to win.

I looked around. The place was small. It would be impossible to fit more than 150 people into the club. Eli continued to explain that the airlines had been donating the tickets for years as a way to help ensure positive coverage in the press. The next "airline night" party was going to be held on Saturday, the upcoming weekend.

"But don't I have to be a reporter or work for Pacific Press to join?" "No," he answered, explaining that practically anyone could join the Press Club. "You're working on a book about South Moresby, the Galapagos of the North, aren't you?" I nodded. "Then you're a writer. You qualify." Eli said that if I was interested, he would go up to the bar and get an application form that I could fill out. Then he'd endorse my application and find another member in the club to also endorse it and right then and there I'd be a member.

"I'll join," I said, even though I had only $50 in my pocket and it had to last until I flew back to Haida Gwaii at the end of the week. "Maybe I could change my flight, stay for 'airline night' and win!" I also thought to myself "Being a Press Club member might get us better coverage of the South Moresby issue." I never won the lottery. From then on, however, every time I came to town or if I thought I had a hot story—or even if I didn't—I would stop in at the Press Club. (I remained a Vancouver Press Club member until, sadly, it closed its doors in 1999, a few years after Pacific Press moved its head offices downtown and its presses to Surrey.)

Now holding a membership in the club where many sympathetic (and not so sympathetic) reporters hung out and having a friendly place to take guests visiting WC² turned out to be a wonderful thing. It was also a great place to take the "of age" volunteers to relax and brainstorm up campaign ideas after a long day's work. Some in our organization thought that having a watering hole so close had drawbacks, but they were definitely in the minority.

Baby Kallie attends "office playschool"

There was an ancillary advantage to our new office and my Press Club membership. After Adriane's maternity leave ended in January 1988, our kids were again regularly in the office under my care. It was one thing to have WCWC in our home with kids tagged onto the deal. It was another thing to have kids part of an independent WCWC office setting. No one complained too much. Volunteers, thankfully, helped keep them occupied. Once a day while TJ napped, Kallie would get to watch a Disney movie (her favourite was *Alice in Wonderland*) on the office VCR. It was my hour-and-a-half to concentrate on work. To believe some, like Ken Lay who was by now a full-

TJ gets babysat in the backpack while I work in our new office on West 6th. Photo: Steve Short, Photo/Graphics published in the June/July 1988 Canadian Geographic as part of an article titled *Saving our biggest and oldest trees.*

time volunteer and an honorary uncle to Kallie and TJ, volunteering for WCWC also meant volunteering to help raise Kallie and TJ.

One afternoon, as I was leaving the Press Club, Jay, the club's manager, asked me what my hurry was (I was the only person in the place at the time). "I have to get back to take care of my kids... especially Kallie who, by now, will be driving everyone nuts in the office."

"Why don't you bring her over here?" she said.

"I can't do that. It's illegal," I responded.

"No, it's not!" countered Jay who proceeded to explain that the Vancouver Press Club was a private members club and, as such, could allow children in, but of course, not to drink. In fact, the Club even had an annual Christmas party for members and their children.

From then on, Kallie and TJ were regular guests at the Press Club. Kallie learned to play the pinball machine and the Pacman video game at a very early age. When TJ's arms got long enough to reach the flipper buttons, he learned how to stand on a chair, put a quarter in and play pinball all by himself. A quarter didn't last very long! I spent a lot of quarters while Ken, I and others brainstormed and planned wilderness-saving campaigns.

Hot Spot II publication

The second big publication out of our new West 6th Avenue office was a massive tabloid-sized newspaper titled *B.C. Environmental Hotspots*. It had a full-colour centrefold map of B.C. showing the specific locations of as many hot spots as possible. Our 1984 silver and pink *Hot Spot* brochure sponsored by Environment Canada was out of print and out of date. This time we did the project without any help—or censorship—from anyone. It was a much expanded and more ambitious publication: 16 pages of in-depth writing about B.C.'s hottest environmental issues. We worked in conjunction with the *New Catalyst* newspaper, published by Chris and Judith Plant. They wanted to include our Hotspots newspaper as a special supplement in their June 1987 *New Catalyst Quarterly* newspaper. We also paid to print 20,000 stand-alone copies so we could continue to distribute them over the next few years

This 1987 *B.C. Hotspots* newspaper was a huge undertaking even with the New Catalyst doing the layout and paste up. Again, Adriane involved her environmental studies students at Langara College to do all the background research as a special class project. They completed the draft at the end of the spring semester, just before TJ was born and we finished the editing and polishing after I got back from my South Moresby sailing trip.

In the end, we found that such a publication was useful to already committed environmentalists, journalists and researchers, but it had limited use in reaching the not-so-committed average citizen. The list of problems was too overwhelming for people who tended to think that the government was adequately taking care of almost everything.

Our Khutzeymateen Bear poster "fraud"

We had several other papers on the "go" that summer besides the *Hotspot* paper. As part of our push to make the *Wildlife '87* year meaningful, bear biologists Wayne McCrory and Erica Mallam of the Valhalla Wilderness Society (VWS) researched and wrote an eight-page, two-colour, tabloid-sized newspaper for us titled *Special wildlife centenary issue – GRIZZLY BEARS Western Canada – A last stronghold of the monarch of North American wildlife*. We gave the VWS a lot of calendars to sell as a partial payment for this work. The VWS was always good at selling our publications. Our bear paper made a strong case for creating the Khutzeymateen Grizzly Bear Sanctuary. It gave people solid biological information about grizzly bear habits and habitat needs and the differences between grizzly and black bears. People love to learn about bears. It was a very popular paper.

To further boost this campaign, we produced a wonderful full-colour poster of a grizzly bear mother with two cute cubs by her side titled HER CUB DESERVES A FUTURE – KHUTZEYMATEEN – SAVE THIS GRIZZLY SANCTUARY. Debbie Duncan, who worked with my daughter Athena at the Lynn Valley Park Ecology Centre, designed it. We sold lots of them through the newspaper's coupon for the low price of $5 each including taxes and shipping. This poster was extremely popular and eventually went through several printings.

There was one glitch; the grizzly mom and cub pictured in the

Jennifer and Jesse, Joe's kids on a Pocket Wilderness jaunt. Jennifer is wearing our Khutzeymateen Grizzly bear questionnaire T-shirt, one of our most popular products. The answers to the questions are on the back of the shirt. Photo: Joe Foy.

A typical WCWC booth featuring WCWC products for sale with Adriane, myself and TJ our son "manning" it. Note Khutzeymateen poster with the U.S. grizzly bears on the display board behind. Photo: WCWC files.

poster were actually living in Alaska, not "the Khutz." We had hunted for good Khutz images, but not one of them was nearly as good as the one we chose. I knew at the time that it wasn't right. But others persuaded me that since the animals were of the same species and occupied similar habitats, it really wouldn't matter.

Several years later (just a few months before the B.C. government established the Khutzeymateen Grizzly Bear Sanctuary in June 1992) someone informed the media that our poster was a fraud. He knew that our "Khutzeymateen bears" actually lived in a sanctuary in Alaska—because he knew the photographer who took the image. Reporters swarmed us with questions.

It was big news. I took full blame and made no excuses. It was a bad decision! Industry people were always trying to discredit us and this was one of the few times they were able to. I learned my lesson from this mistake and never did anything like that again. I vowed to be even more diligent in preserving WCWC's reputation for honest and accurate information. It was, in my mind, absolutely essential to our effectiveness and achieving our goals.

Stein *Voices for the Wilderness* Festival III

Summertime meant another season of trailbuilding in the Stein and preparing for yet another calendar. Summer was also the key season to scout out wilderness. Ken used the Zodiac we bought for the Windy Bay expeditions a year earlier to explore Clayoquot Sound with Adrian Dorst, concentrating on Flores Island. We had several other publications in the works as well—many balls to juggle.

On the B.C. Day weekend (beginning August 1st) the Lytton and Mt. Currie Indian bands held their third annual Voices for Wilderness Festival. This time it took place back in the Stein alpine by Brimful Lake where the first Stein Festival was held. Again our volunteers helped John McCandless, the Lytton Band's Stein Coordinator, with preparations. This time the event was much big-

Khutzeymateen Bear poster called a "fraud" because the bears featured in it lived in Alaska. Poster image: Laurie Aumiller.

Big stage entertains and enlightens big crowd at the second Stein Festival. Photo: WCWC files.

ger. It took fourteen helicopter trips to bring in the stage and sound gear. Four times as many people—2,000—came to hear David Suzuki, Farley Mowat, Thomas Berger, Haida Nation's Miles Richardson and Lytton and Mt. Currie chiefs and elders speak.

The most senior politician attending was MP Charles Caccia, federal Liberal environment critic. John Denver, flying himself into the small Lytton airstrip, donated his time and talent. He sang "Rocky Mountain High" and many of his other hit songs. Long John Baldry, Connie Kaldor and other performers entertained, all donating their talents to save the Stein. The sound in the gigantic mountain amphitheatre was spectacular. The sun shone every day. The biggest downside was the overflowing porta-potties. Sadly, Adriane and I didn't go. We were immersed in finishing our house-building. This was the last year the festival was held in the Stein alpine. The movement to save the Stein had grown too big. The ecological footprint of 4,000 feet tromping around for three days on fragile alpine flora took its toll and took many years to heal. But the festival achieved its goal of inspiring the campaign troops. The problem was, the opposition was mobilizing, too.

John Denver sings at the Stein Voices for the Wilderness Festival III in August 1987. Photo: WCWC files.

Helicopter flying in porta-potties to the Stein Festival site. Photo: WCWC files.

Citizens say "No" to mining in parks

Meanwhile, during this busy summer, Arne Hansen completed our paper about industrial encroachment into Strathcona, B.C.'s oldest Provincial Park. The Socred government had given a small mining company permission to build a road into the park to do exploratory drilling near Cream Lake. Local conservationists were incensed. We co-published with the Friends of Strathcona Park an eight-page, two-colour, tabloid-sized newspaper titled SOLD OUT TO MINING INTERESTS 1987 – Strathcona Provincial Park printing 50,000 copies.

Our new second floor warehouse office had no elevator. When the papers arrived we had to "bomb the bales" up a flight of stairs by hand. Sometimes we had enough volunteers to use the "bundle brigade" method of passing each bundle from one person to another stationed on the stairs. It was easy and didn't take long to get all of them up to our office and temporarily stored until we could get them mailed out and distributed.

The Strathcona Park issue heated up fast. Local citizens started blockading the new mining road. The mining company got a court injunction against the blockaders. Over the following weeks 64 people were arrested for disobeying that court order. Those getting arrested were mostly local middle-aged and elderly citizens who were passionate about protecting their provincial park. There was widespread public support for Friends of Strathcona and our paper helped people understand the issue. The provincial government finally relented and revoked this permit. The new B.C. minister of lands and parks declared that: "A park is a park" and forbade mining exploration and the development of new mines within existing provincial parks. It was a clear victory for conservation.

September 1987 4th World Wilderness Congress in Denver, Colorado, shocks us into action

I was like a kid waiting for Christmas in anticipation of our trip to Denver to participate in the 4th World Wilderness Congress. I'd never been to Denver and I had never attended a wilderness conference. Adriane and I arrived knowing virtually nothing about the organization behind the congress. In 1987 there was no such thing as a Google internet search engine to check it out. The brochure explained that it was the first time that this gathering was being held in North America. By air, Denver, Colorado, is quite close to Vancouver, B.C. By booking ahead, we got cheap tickets and found an inexpensive hotel near the conference centre, which meant that we did not have to rent a car.

Our luggage was very light on clothes and very heavy on WCWC educational materials. We packed to the max, assuming that there'd be thousands of people there. Somehow we talked our way into checking numerous boxes and tubes of educational materials onto the plane without paying anything extra in baggage fees.

We arrived in Denver late in the evening, the day before the conference. It was quite a job lifting and packing everything into the taxi. Then it was a major chore finding an open entrance at the Denver Convention Center. I really didn't want to pack all the stuff up to our hotel room. Finally, after circling the building twice, we noticed a loading dock with a light on, knocked on the door and someone answered. Thankfully, we were able to leave our stuff there overnight.

The next morning we walked over early to set up our booth. As I glanced around in the exhibition centre I got a funny uneasy feeling seeing the other displays being set up. We were obviously different from almost all the other exhibitors. I took a quick stroll around. One booth in particular shocked me. On the floor were huge ivory elephant tusks. There were several elephant feet chairs to sit in, but I didn't try one. I couldn't believe it.

I rushed back to our booth to tell Adriane. "Someone's marketing elephant hunting safaris in Africa! There are lots of big game outfitters booths. I thought we were attending a conservation conference!" I exclaimed. The exhibition centre was huge and there were hundreds of exhibitors. Only one group was really similar to us—the Australian Wilderness Society (AWS). A couple of representatives from that organization had set up a table with lots of campaign materials, too. We got to know them well. They had lots of beautiful posters and strong messages urging citizens to help save specific threatened wilderness areas in Australia and Tasmania.

Over 2,000 people attended the conference. The exhibition hall was also open to the public every day and evening and many Denverites came to take a look. I got to attend only a few of the lectures because one of us had to stay with our booth. Adriane went to most of the sessions and took good notes while I 'booth sat'. I didn't mind because I constantly met interesting people and figured I was learning as much from them as I would from the lectures.

Shortly after Adriane left to attend the first plenary session, a well-dressed elderly woman came by and struck up a conversation. It quickly became obvious that she was hostile. "Why did you start up a new organization in B.C.? Couldn't you work with the Sierra Club of B.C. and strengthen that organization?" With restraint, I answered all her questions, politely explaining that we had formed because the Sierra Club wouldn't publish an annual Canadian wilderness calendar. I told her that we partnered with local grassroots environmental

Newspapers staked in the ally (front entrance of our West 6th Street office about to be moved up to our office via the "bundle brigade." Photo: WCWC files.

Friends of Strathcona Park resist mining exploration inside Strathcona Park. Cream Silver's geologist and security guard discuss what to do about two women lying down in their drilling area. They are eventually arrested and after a prolonged "siege" with more arrests of people trying to protect the park the government ultimately declares that B.C. parks are off limits to new mining activities. Photo: Anne Cubitt.

groups over specific issues and that this worked well in B.C. She eventually calmed down and informed me that she was a director of the Sierra Club of the U.S. She never told me how she got her negative impression of us.

It wasn't until the next day when talking to a conservationist from Europe that we learned more about the World Wilderness Congress that sponsored this conference. He told us that some South African big game hunters who wanted wilderness game preserves so they would have a continuing supply of trophy animals to shoot had started it. This is putting it a bit crudely, but the truth of this disclosure was supported by the preponderance of big game outfitters' booths in the exhibition hall. The Australian Wilderness Society, WCWC, the Sierra Club and a couple of other environmental organizations were the odd ducks here.

How much wilderness is left on Earth?

We looked forward with anticipation to one particular presentation scheduled for the last day of the conference. Michael McCloskey, chairman of the Sierra Club of the U.S., was reporting on the Sierra Club's big project to map the world's remaining wilderness. The results of this study were shocking. Industrial developments had already penetrated nearly all the wild places on Earth. All the really big wilderness areas left were a few obvious ones like Antarctica, the Canadian and the Russian Arctic, the big deserts and a bit of the Amazon Basin. None of the areas we were currently fighting to protect in B.C. were large enough to make it onto the Sierra Club's map. In fact, the map McCloskey presented didn't even have Vancouver Island on it! They had used a 1-million-acre "screening net" which meant that an area had to be at least 404,700 hectares in size and completely roadless to qualify as a wilderness and be put on the map. The Sierra Club's study was brilliant from a big picture "shocking truth" point of view but less than useless from a practical campaigning-to-save-a-wilderness-area viewpoint.

Adriane and I were aghast that the wilderness areas we were working to save were deemed "too small" to merit mention in the Sierra Club's global inventory of wilderness. Surely there were many other wild areas around the world that were on the brink of being destroyed; areas that were left out of this study but desperately in need of identification and campaigns to preserve them. This became the goal of WCWC's international WILD campaign that Adriane, with the help of some friends, launched a year later.

Australian Wilderness Society an inspiration

As we were packing up our display and leftover materials the woman leading the Australia Wilderness Society delegation came over and said, "Why don't we exchange the posters we have left instead of carting our own back home?"

"What a great idea," I replied. I was delighted. Their posters were truly an inspiration—most of them far better designed than ours. I was excited to learn that the Australian Wilderness Society actually made money on their poster sales. This became a new goal for me.

On our trip to Denver we'd learned a great deal, including the fact that the state of wilderness preservation on Earth was indeed grim. It motivated everyone at WCWC to redouble our efforts.

Chapter 18

Battles rage on many wilderness fronts in B.C.

Pocket Wilderness crisis

Right after we got home from Denver a crisis arose in our Lower Mainland "pocket wilderness" campaign. A key area was about to be logged. It wouldn't take much to move the cutblock and save the trail to beautiful Greendrop Lake. Time was tight and the only way to stop the logging was to go to court. We got an inexpensive young lawyer just out of law school who was sympathetic to the cause to act for us in seeking an injunction. On September 30, 1987, we appeared in the Supreme Court of B.C. One hour later we had an *ex parte* injunction to stop the logging above Greendrop Lake. I was in the courtroom observing and it was obvious to me that the judge was less sympathetic to our case than he was miffed that Whonnock Industries Ltd., the company doing the logging, did not even send a lawyer to present its position despite our lawyer giving them adequate notice. It was WC²'s first injunction and destined to be one of the shortest-lived on record. At 4 p.m. the same day, Whonnock got a hearing before another Supreme Court Justice without us even being informed of the proceedings and had WCWC's injunction quashed. The Greendrop Lake cutblock was quickly felled in a few days.

The loss of the forest backdrop to Greendrop Lake did not dampen Joe's enthusiasm for his pocket wilderness campaign. Half a year later he produced a great paper to help save a spectacular grove of big old-growth cedars near the world-famous resort town of Whistler. We published this paper, titled *Lower Mainland Pocket Wilderness – SAVE WHISTLER'S BIG TREES*, in April of 1988 and inserted this four-page, full-colour, tabloid-sized newspaper into the local paper—*The Whistler Question*. It, along with our full-colour poster (photo by Uve Meyer) of a cross-country skier (wearing skis) bracing himself about four metres off the ground between two giant cedars, was instrumental in saving the area.

This pocket wilderness campaign was an all-round success. The 50,000 papers brought in more than $4,000 in donations; this ancient cedar grove still stands and is a big tourist attraction today.

Countering the "share" movement with our own "spare" movement

At the same time we found out that Greendrop Lake was in dire peril, we also learned that a Stein pro-logging coalition had formed and was mobilizing like no other anti-wilderness lobby had ever done before in B.C. They made no secret of the fact that they were spurred on by their side's "defeat" in South Moresby. They wanted a no-compromise government decision to log the Stein and they wanted it fast, ignoring the recommendation by the Wilderness Advisory Committee (WAC) that government must first negotiate with First Nations. It was made up of 17 organizations, including the Council of

Facing page: Giant red cedar on Cougar Mountain near Whistler, B.C. From left to right Clinton Webb and Greg Stoltmann, Randy Stoltmann's brother. Photo: WCWC files.
Below: Old-growth forest beside Greendrop Lake quickly falls after WCWC's injunction is quashed. Photo: WCWC files.

Large logger rally on the lawns of the B.C. Legislature urging the government to approve logging in the Stein Valley. Partially visible sign on the far left "BOUCHARD BUTT OUT" refers to federal Minister of the Environment Lucien Bouchard considering national park status and not solving this provincial land use conflict. Photo: WCWC files.

Forest Industries (COFI), the powerful cartel that represented all the logging companies working on the coast, the Registered Professional Foresters (RPFs) of B.C., the International Woodworkers of America (IWA) of Canada and the Truck Loggers Association of B.C. It also included the towns of Hope, Lillooet and the village of Lytton that neighboured on the Stein Valley. In the brief they presented to government, they claimed that none of the proposed logging in the Stein would be visible from the *Voices for the Wilderness Festival* site, thus the Stein could be logged and still be enjoyed by wilderness enthusiasts. It urged the government to opt for *"balanced development"*–to "Share the Stein."

The brief failed to mention the First Nation's interest in the valley and the scientific studies regarding the negative impacts that logging would have on wildlife there. Instead, it echoed the ridiculously over-inflated claims made about logging job losses in South Moresby, claiming that not logging the Stein would mean the loss of three-quarters of a billion dollars worth of timber along with 237 direct jobs. Soon this pro-logging coalition had hats and T-shirts sporting their "Share the Stein" slogan.

Underneath this pro-logging slogan was a cute little logo with a

"Spare the Stein" slogan adapted from the pro-logging logo of the Stein Share Group. Artwork: WCWC files.

couple of hikers, a native wearing a headband with a feather in it, a bear, a deer, and a logger leaning on his chainsaw all standing in a row. I thought what they really meant was Shear the Stein.

Ken Lay immediately volleyed back, re-naming WCWC's campaign "Spare the Stein." He picked up one of their "Share the Stein" baseball caps and doctored it up, changing the "h" to a "p." He got lots of double takes and many nasty comments when he wore it to the next "Share the Stein" rally. We then made up our own "Spare the Stein" T-shirts and caps, blotting out the logger's chainsaw while leaving the rest of the logo the same. Within a couple of days I got

Ken Lay wearing a sweatshirt with WC²'s original logo and a baseball cap with the "Spare the Stein" slogan on it. Photo: Valhalla Wilderness Society files.

a call from a man who said he was the graphic artist who owned the logo. He told me that we had to stop using it in our modified way and destroy all the shirts and hats we had. "I didn't know it was trademarked or copyrighted." I responded, politely offering to pay him something for its use. "How much did the coalition pay you for it?" I asked. He refused to tell me and told me that he wouldn't give us permission to use it. I countered saying that we would be happy to pay whatever royalties we owed him for the unauthorized use. I told him if he did not want to negotiate such a price with me now, he could take us to court. He was not very happy when he hung up. The lawsuit never materialized.

Stein Declaration a turning point

On October 5, 1987, the Lytton and Mt. Currie Indian bands issued their "Stein Declaration." It proclaimed in part: "Our position, which will never waver, is to maintain the forests of the Stein Valley in their natural state forever; to share our Valley with other life forms equally; but also to share the Valley with those people who can bring to the Stein a respect for the natural life there similar to that taught to us by our ancestors."

This declaration was a turning point in the campaign. This assurance was exactly what was needed to counter the cynics who claimed, "The natives just want to get hold of the Stein so they can log it themselves." Everyone at WCWC was cheering. We knew that the B.C. government was still committed to negotiating with First Nations before allowing any logging in the Stein. It gave us real hope that the Stein would be saved.

To publicize this First Nations' declaration we rushed into production our third Stein newspaper. Within a couple of weeks we had 50,000 copies of a new eight-page full-colour tabloid titled STEIN VALLEY – The choice is ours. Where do you stand?

The cover was particularly powerful. It featured side-by-side aerial photos taken in the wintertime: one of the untouched Stein, the other of the heavily logged Kwoiek Valley immediately to the south of the Stein. The snow highlighted the clearcuts. Beside each photo was a list of groups supporting that vision. At the head of the pro-logging list we placed the Registered Professional Foresters who, we believed, had no business backing the logging of the last intact large valley in southwestern B.C. The centrefold featured the full text of the Stein Declaration signed by Lytton Chief Ruby Dunstan and Mt. Currie Chief Leonard Andrew.

That month, while we were distributing this paper, the B.C. government created two Recreation Areas—the Upper Stein and the Lower Stein—where no logging could take place. It was an empty gesture because neither of these two areas had timber of commercial value growing on them. Besides, these Recreational Areas could still be mined and have roads built through them. Nonetheless, we felt like we'd moved one step closer to a win.

The 1987 Okanagan-Kootenay's "uranium rush"

Other environmental issues were heating up that fall. Premier Vander Zalm lifted a seven-year moratorium on uranium mining in B.C. This moratorium was originally imposed in 1980 when Premier Bennett called off the Bates Uranium Commission after only four months of highly charged public hearings. The lifting of the moratorium prompted a "gold rush" of staking radioactive deposits in southern B.C. Some prospectors even staked under towns, including under the town hall and a schoolyard in Oliver. People were up in arms. When some concerned citizens asked if WCWC could help by putting out a newspaper about the issue, of course I said 'yes'.

This paper had a really short deadline. It had to come out in time to be householder-mailed to all the citizens in the Kootenay-Boundary Regional District (the "hot spot" of deposits and staking) in time to influence a crucial vote. Local citizens had managed to get a non-binding referendum question on the ballot in the November municipal elections. We pulled a few "all-nighters" to get this eight-page, two-colour, tabloid-sized newspaper published in time. On the cover, taking up the entire top half of the page, we featured the well-known bright orange triangle logo that is used to warn people about the presence of dangerous radiation. The inside of the paper was packed with information. The two-tiered title said it all: NO SAFE WAY TO MINE URANIUM! – Government in 'hot' water for lifting moratorium. We printed 10,000 copies and they were gone in an instant. They made it into the mailbox of every household in the "hot" area just a few days before the vote. On November 21, a larger then normal turnout of citizens at the polls overwhelmingly ticked "yes" in support of continuing the ban on uranium mining in B.C. Due to popular demand, we updated this paper and printed another 20,000

In response to the "grassroots" Share groups that sprouted up everywhere where we were advocating park protection, we contracted a local cartoonist to draw this *Share BC – Never had it so good* cartoon. Artwork: Geoff Olson.

copies and quickly distributed them. Some local unions helped pay for the re-printing. Shortly thereafter, the B.C. government re-established its ban on uranium mining. We count it as one of our wins, although, given the strength of the opposition, it's hard to imagine uranium mining ever being tolerated in B.C.

Help with data inputting

At WCWC we always had way more work to do than we could possibly handle. That's why we were always open to anyone volunteering who agreed with our strategies, had talent and wanted to put their energy into saving wilderness. One of the most onerous tasks was computer inputting the membership, donation and sales information into WCWC's customized database, preparing the bank deposits and depositing the money. I'd been doing most of this work from 1983 to 1987, but in the last few months it became too much for me. Many nights I'd work well past midnight.

One night I started to see double—one eye's image slightly off centre from the other. I was pretty frightened, especially since I was driving at the time it happened. After the doctor and optometrist checked me out, it turned out to be eyestrain. Too much computer work! Luckily we were able to talk Carleen Roth, Ken Lay's girlfriend and future wife, into volunteering to do the data inputting. She was so good that, shortly thereafter, we hired her. She was more than three times as fast as I was at the task and much, much more accurate. It was one of WCWC's best hiring decisions.

A stranger with new ideas comes to visit and we shoot the breeze over a few beers

Early one afternoon, a couple of months before Christmas, a man named Nikolas Cuff walked into our office. I found out later that the same Eugene Rogers who pointed Joe Foy in WCWC's direction also suggested that Nik see us too. Nik wanted to talk to someone about getting a campaign going to halt the destruction of an island in the Fraser River Estuary. To get away from the office distractions, I suggested we go across the alley to the Press Club to talk. He took me up on the suggestion and away we went. I returned three hours later. Some of the volunteers I'd left taking care of my children thought that spending the afternoon in the Press Club chatting with a stranger was excessive, especially when there was so much work still to be done on our 1988 wilderness calendar.

"What happened? Why did you stay for so long?" asked Adriane, who had just arrived at the office from teaching all day. I told her we had another campaign brewing; that I had promised to tour the 'wild' western half of the 445-hectare Annacis Island near New Westminster with a new guy named Nik. According to him, the last bit of wildness left there was being annihilated as sand dredged from the Fraser River was sprayed onto the island to create industrial park development lots. He had watched the progressive destruction of Annacis Island since the mid-1950s. This was the last chance to protect a small part of this privately owned island in a natural state.

I explained to him how difficult it was to protect someone else's private land. At WCWC we concentrated nearly all our efforts on protecting publicly owned lands.

I enjoyed my first meeting with Nik. He was colourful, entertaining and good-humoured. And he insisted on paying for all the beer! He was also full of new ideas and interesting information. A true bonus was that Nik owned a printing shop in New Westminster and had some great ideas on how we could print more effective campaign materials. The two of us had a lot in common: lots of energy, a love of nature, unlimited campaign ideas and a sincere love of beer. It was a great way to spend an afternoon.

During that first meeting Nik asked me what I thought was the most threatened wilderness area in B.C. I told him it was the Stein Valley, which the logging industry was pushing hard to log. It had become their "line in the sand" in the "battle" to protect the industry's

Lytton & Mt. Currie Indian Bands
"Stein Declaration"

As the direct descendants of those aboriginal peoples who have inhabited, shared, sustained, and been sustained by the Stein Valley for tens-of-thousands of years down to the present, our authority in this watershed is inescapable. The responsibility we bear for protection of the Stein has been passed us by our ancestors from our earliest memories, and should not be lightly dismissed. We, ourselves, have never dismissed this obligation: we have never entered an agreement with any nation or government which would abrogate our authority and responsibility in the Stein watershed.

It is our forebearers who developed the sustainable patterns of resource management in the Stein which leave the valley in its unmarred state today. Our tread has been deliberately light, but the spiritual and physical "footprints" of our peoples are evident for all to see throughout this watershed. To us, the valley is like the pages of a book upon which thousands of years of our history are written. There is no other record in the Stein Valley except our own, and we can never willingly abandon our committment there.

In sharp contrast to the relative silence of millennia of uninterrupted native habitation of the Stein are the shrill new claims which have arisen in the past few decades. With seemingly insatiable appetite, newcomers now clamour for our valley's legacy. Forests which for centuries have grown strong alongside our cultures will feel the hot bite of chainsaws if these people have their say.

We have waited patiently for those who now make these claims to consult with us about our homeland, and finally when it seemed they would never come to us, we felt we had to go to them. It was with misgiving that we entered into the hearings of the provincial Wilderness Advisory Committee in January 1986, but we made our concerns plain from the start. We said to the committee that if they made recommendations at the Stein which were fair to the native people, the provincial government would ignore them.

Sadly, time has proved us right. When the Wilderness Committee eventually acknowledged the contribution the Stein Valley continues to make to the spiritual and cultural integrity of our people, and recommended that no road be built into the area without the blessing of our people, the provincial government turned the proverbial blind eye and deaf ear. In contrast, the federal government accepted the overall land claims package of the Nlaka'pxm Nation, which includes the Stein Valley, on November 28, 1986.

We can wait no longer for other governments to come to their senses. For us to exist as a people and a culture we need to preserve certain of our lands, the only rightful place we have on this earth, in their natural state. We must continue to exercise our responsibility to protect these lands as we have since time began.

Our position, which will never waver, is to maintain the forests of the Stein Valley in their natural state forever; to share our valley with other life forms equally; but also to share the valley with those people who can bring to the Stein a respect for the natural life there similar to that taught us by our ancestors.

With the help of our elders we pledge to strengthen "Stein Rediscovery," the youth program located deep in the Stein's heart and in our own hearts as well. This creative wilderness experience for young people re-acquaints native youth with their own roots while cultivating essential understanding in non-native youth, and is a primary ingredient in the continuance of our cultures. Stein Rediscovery builds on the long tradition among our peoples of following Stein Valley pathways to spiritual maturity.

We will seek and form alliances with other native nations in the defence of the Stein watershed and in opposition to the common thread of aboriginal injustice which we suffer along with indigenous peoples everywhere.

We will further strengthen the alliances we maintain with those non-native peoples who can respect and share our values and perspectives, and with these people we will continue to enhance the ages-old system of trails which extends the length and breadth of the Stein Valley and first felt the feet of our forefathers. In this way we can extend the hospitality of our valley to all peoples, and re-create in others the awareness of and respect for the natural world which is our birthright.

Finally, and importantly, under the cooperative authority of our two bands we will maintain the Stein Valley as a wilderness in perpetuity for the enjoyment and enlightenment of all peoples and the enhancement of the slender life-thread on this planet.

In so doing we are but honoring those ancestors whose legacy to us is the Stein Wilderness, and in our turn we will extend this same opportunity, legacy, and responsibility to generations yet unborn.

At Lytton and Mt. Currie: October 5th, 1987

Mt. Currie Chief Leonard Andrew

Lytton Chief Ruby Dunstan

Councillors

Councillors

The Stein Declaration. Document: WCWC archives.

land base from the preservationists. They were still mad about losing South Moresby and reaching an injunction stalemate on Meares Island.

I explained that the government was ignoring the environmentalists and First Nations and still considering subsidizing a 23-kilometre-long road up the narrow canyon into the centre of the valley so loggers could access the good wood. The only thing holding back logging was the First Nations' *Stein Declaration* stating their resolve to save it. It was shaping up as a huge confrontation. "I don't know what else to do," I said.

Two days later, a volunteer opening the mail let out a loud whoop. "Two thousand dollars" she yelled out for all to hear. The only thing in the envelope was a $2,000 cheque from Mr. Cuff. It knocked everyone's socks off. And it totally vindicated my afternoon with him in the Press Club. No longer did anyone object to me using the club as a place to talk with guests. Nik called me early the following week. I thanked him profusely for his donation. Nik replied, "That's just a start! Can I come down later today? I have an idea for a project."

"Sure," I answered, curious as heck. When he came, we immediately went over to the Club to talk. This time there was no static from the staff.

How Nik Cuff's 20 Santas delivered joy to the Stein Valley preservationists' camp

Nik got right to the point. "The Stein Valley needs a 'big hit' right now to save it. I want to put out a spectacular poster and give them away as Christmas gifts to as many people as I can, especially to the businessmen and decision-makers in downtown Vancouver," he explained. "I'll pay for everything."

"It's a great idea," I said. "What do we have to do?" I asked, thinking that there must be a hitch. There was.

"I need the right image and I need it right away," he said. "You must have one in your files," he added.

"I'm not sure," I responded, racking my brain trying to think of the highest quality shots in our slide collection that might be of the emotionally-grabbing calibre Nik wanted.

One of the significant differences between WCWC and other environmental groups was that, right from the start, we featured the use of photographic images in all our campaigns and publications. We were constantly building our slide library. Each wilderness area we campaigned to protect had its own binder. We not only collected the beautiful images of wilderness, but also the ugly pictures of clearcutting and devastation. We had at least a thousand, perhaps even two thousand, 35mm slides taken in the Stein Valley. I looked through hundreds of shots of trailbuilding crews, camp fires and tents, rushing river, trickling streams, bubbling brooks, dry pine forests, wet cedar forests, alpine meadows, single ancient Douglas fir trees, devils club, canyons and pictographs. The area was vast and varied. You name it; we had documented it.

I still wasn't sure, however, that we had that one stunning shot for the spectacular poster Nik envisioned. I put together two plastic sleeved sheets of the best slides we had for Nik to review. He came back the next day to have a look at them. He wasn't impressed. I told him that the Stein Valley just wasn't photogenic that way... it was the sum total of its varied parts that made it so special. "I know they're not great, but I don't think you'll find anything better," I said. He took another look and picked out one showing the forested valley with some of the surrounding mountains in the background.

"Now," he said, "if only we had a panoramic shot like this one." I knew there were other similar shots in the file and pulled them out.

Nik pored over these slides on the light table with a magnifier eyepiece, carefully inspecting each one for sharpness and compositional appeal.

"Look at this," exclaimed Nik. "This shot overlaps this one. The photographer took them in horizontal sequence. Together they give an entire sweep of the mid-valley."

I couldn't understand what he was so excited about, until he explained how he could join the two shots together to make a panoramic poster image. They could be digitally scanned, knit together by a computer and recorded onto a giant magnetic tape for output onto film.

"You'll have to contact the photographer to get permission to use the slides first. His name is Jennings. Here, I'll give you his phone number," I said to Nik as he carefully put the two slides into protective plastic pockets to take them away.

"No problem," he replied as he left. It was the beginning of an intense five weeks.

Nik was confident that it could be done. He had just taken a night school course about the latest developments in printing technology from the "best in the business," a man named Wolfe who worked for Cleland Kent, then one of the biggest pre-press outfits in Vancouver. He got Wolfe to take on the job. It wasn't easy. To knit them together seamlessly required pushing the limits of the technology of the day. Using the crude software programs and the tiny computer memory available then, it took a very skilled operator hours and hours of painstaking manual work to match the colours accurately, millimetre by millimetre, along the adjoining edge. At first, the Stein image didn't work out perfectly because the photographer had slightly tilted the camera on one of the shots. This left a triangular piece in the lower right-hand corner missing when the two were matched up. Wolfe had a solution. He cleverly cloned an adjacent piece of the brushy foreground and neatly blended it in. Only if someone told you about it and you looked closely could you see it in the final product.

Gradually Nik revealed to us the scope, scale and detail of his—no, our project—for it all was being done under WCWC's name. Money was no barrier for Nik. He arranged everything himself. Perfectionism was his second nature. Making this project a huge success became his all-consuming obsession.

After much thought he decided that the poster's title should be *Stein Valley – Joy to the World*. It was an inspiring, uplifting Christmas message that matched the image perfectly. He talked his busy, talented brother, one of the best graphic artists around, into designing the poster—of course for free. He picked Hemlock Press as the best printer to do a perfect printing job. He chose the best heavy-weight, coated paper. Nothing was going to be cheap looking about this poster. In his shop Nik printed a small slip of paper that said in bold red letters "Caution – Frameable Print" to wrap around the outside of the rolled up poster under the ribbon.

"Are you sure you want to print that many?" I asked. A print run of 25,000 gave us two-and-a-half times more posters than we'd ever printed before. Nik insisted that we needed that many to do it right. It was less than three weeks to Christmas. Nik had almost all the pieces ready: rolls of red ribbon, the "frameable print" slips, and the posters. All that was left was the appeal letter. I wrote and rewrote it to make that just right too, with Nik providing useful criticism after each draft. The letter began with "*Joy to the World and let Heaven and Nature sing forever in the Stein Valley.*" I ended it with "*Don't let it be a clearcut decision.*" (We included a picture of a road and recent clearcut made in an adjacent valley, knowing that people don't act to save wilderness unless they see the threat.) The letter was short. It took up only one side of a regular sheet of paper. It was the best appeal

Stein Valley – Joy to the World poster was also published as a postcard. Photo image: David Jennings.

I could write to win peoples' intellectual support. The poster itself won their hearts.

Ahead was a Herculean assembly task. I had my doubts as to whether or not all of the seven-metre-high stack of flat posters could be rolled and ribboned in time. Nik delivered 7,000 of them along with the other printed materials to our West 6th office. This was WCWC's share of the job. Rolling each poster (with my letter inside) wasn't easy. We used round wooden dowels and rolled them smoothly to avoid making crinkles in the poster Adriane, Ken, Joe, I and many volunteers worked late, night-after-night, trying to get our share done.

Nik delivered the other 18,000 posters to an unused theatre that he rented. He convinced every one of his friends and relatives to help. More than 100 people participated in his gigantic poster rolling party that went on for eight days and nights. It took over a minute to roll up each poster, add the "frameable print" note, wrap around the ribbon, tie the bow and put the finished product into storage boxes specially ordered by Nik. Each box held 50 posters. The stack of filled boxes was enormous. They represented over 4,000 person hours of assembly work.

At WCWC's office we only managed to complete 5,000 of our 7,000 posters. Nik didn't complain. He took the remaining posters down to the theatre and had his volunteers finish the job. Seven hours before the appointed delivery time—three days before Christmas—all the posters were ready to deliver.

Nik planned the big day pretty much like a military operation. He had recruited 12 volunteer "Santas" and 25 "Santa's Helpers." He rented Santa suits and had silk-screened beautiful bright red T-shirts with "Joy to the World" on them (in the same large typeface as on the poster) for each of Santa's helper to wear. He must have had a direct pipeline to the powers above, for the poster give-away day turned out to be perfectly sunny. If the day had turned out to be rainy and windy, as days more often than not are at Christmas time, I hate to think of what would have happened.

Nik counted on the fact that a Santa Claus, right before Christmas, would be able to go anywhere without being questioned. Here's how he planned it and how the volunteers pulled it off. A Santa walked into the lobby of a big high-rise office building saying "Ho! Ho! Ho! Joy to the World!" and handed each security guard a rolled-up poster. Santa then rang for an elevator, got in and held the hold button while Santa's helpers completely filled the elevator with boxes full of posters. Santa then went to the top floor and put out a box or two there and punched every button for every floor on the way down, stopping at each to drop off the boxes of posters in every hallway. He then returned to the top floor, took the posters from the box he previously left behind and walked around the entire floor from desk to desk personally handing one out to everyone with a big "Ho! Ho! Ho! Joy to the World!" Our dozen Santas, going as fast as they could go, caught everyone by surprise. Not even Grinches could object! The poster, its message and its delivery were completely uplifting.

This process was repeated in building after building for seven hours with little incident. In one of the Bentall towers, however, our Santa was too successful. He waltzed his way past three sets of security guards right into an enormous office with only a single giant desk at one end. It was the ultra-secure top floor office of one of the richest, and perhaps most insecure, entrepreneurs in all Vancouver. The man freaked out. "How the - - - - did you get in here?" he yelled. "Get out!"

Back in our West 6th office I got an angry phone call from someone in one of the Bentall towers. "Get your - - - -ing Santas out of our G- - D- - - building!" he demanded.

"Yes, sir. Right away," I replied, not asking him which of the four towers he was referring to. I knew full well that it was impossible for me to do anything about it. At that time no one had cell phones. I didn't get upset... I knew it would blow over. It was Christmas time and there were very few Grinches and Scrooges out there.

Barry Jones, then a NDP MLA and one Nik's friends, was one of the Stein Santas. He reported later that he was caught and questioned by a security guard in one of the buildings

"What's your name?" the guard asked.

"Santa Claus"

"How do you spell your <u>real</u> name?"

"S. A. N. T. A. C. L. A....."

The guard gave up in disgust and told him to go.

Late in the day of this Stein Christmas caper I got another phone call, this time from someone informing me about several boxes of posters left in the lobby of a building.

"I'll get someone down there right away to pick them up," I assured the woman.

"No," she said, "That's not what I want. Can I pass them out?" She apologized profusely for their building's security staff not letting our Santa do his job.

"Go ahead! And have a merry Christmas!" I told her.

By the end of the day, there were only a few thousand posters left. The Santas stood by bus stops giving out these last few to commuters lined up for their buses home. People were ecstatic. It was such a great gift. Everywhere you looked in downtown Vancouver you could see people with a *Joy to the World* poster tucked under their arms. That same day in Lytton, Lillooet and other B.C. communities, Stein campaign volunteers gave away the posters too. In Victoria, volunteers gave away 4,000 posters in government offices. What more perfect way to celebrate the spirit of Christmas is there than to share the beauty of the Stein Valley wilderness with the people of B.C.?

Afterwards, framed posters began appearing everywhere. I was delighted one day to see one prominently displayed in the B.C. Supreme Court Registrar's office in plain view of everyone filing court documents.

Nik invents the three-part opinion poll mailer

Having a printer as a patron of the Wilderness Committee was a godsend. A month after the Stein Santa caper, Nik came up with an idea for another project—a Stein "three-part-mailer." He would print them on cardstock at his New Westminster print shop on his single-colour press, running it through four times in order to get full colour. It was labour intensive, but he assured me that it would work.

Nik designed what would become a standard WCWC campaign tool. One part of the mailer featured an ugly postcard of a recent clearcut near the Stein Valley. The backside of this part was addressed to the Premier with a statement about the uniqueness of the Stein and with several boxes to check where a person could express his or her opinion as to whether the Stein should be saved or logged like the picture on the front. The second postcard, easily separated from the first card by a perforation, featured a miniature *Joy to the World Stein Poster* on the front. On its backside, printed just like a regular postcard, was a short description of the Stein and the efforts that conservationists were making to save it. This "beauty shot" could be kept or sent to a friend. The point of the third part, separated by another perforation, was to bring in donations to our campaign. The front of this part featured a photo of the Stein with a rainbow over the valley. On the back were boxes to tick to donate to our Stein campaign and to join WCWC. Nik printed up 60,000 of these three-part mailers, for free.

We distributed them widely and they worked! From then on nearly every WCWC campaign had its own three-part-mailer. We even sent some of them out as unaddressed household mailers with varying degrees of success. Nik knew a lot about direct mail while printing fliers for experts that used it. It was a numbers game. You had to print up thousands in order to get dozens to respond. "There are a lot of busy people out there. A one percent return rate is great," he told me.

Fourth Stein newspaper features avant garde cover

In January of 1988, we published our fourth Stein paper, an eight-page, full-colour, tabloid-sized newspaper titled *Save the Stein – S.W. BRITISH COLUMBIA'S LAST MAJOR WILDERNESS VALLEY*. With a press run of 50,000, it featured Nik's *Stein Valley – Joy to the World* poster as the centre spread. Of course, we included a tear-off

Letter to our members and supporters that accompanied the Stein three-part mailer we sent out. Document: WCWC archives.

coupon on the back for our members and friends to order a *Joy to the World* poster for only $5. Joe Foy, Ken Lay, John McCandless and others wrote different parts of this paper. The cover, designed by Emily Carr College students Rod Rodenberg and Dave Coates, was spectacular. It was a collage of three photos, centering on Napoleon Kruger, an Okanagan Nation elder and spiritual advisor, holding high above his head his drum and drumstick, with a photo of the pristine Stein behind him in the blue sky—like a vision. At the foot of the collage, his feet were placed in another photo of a slash-burned clearcut. This cover was like nothing we'd ever produced before—the result of talented young artists using new computer-age graphic tools.

The lead article, titled *"New hope for the Stein,"* put into print the recent promise by the new Forest Minister Dave Parker that, as the Wilderness Advisory Committee recommended almost one year earlier, no logging would occur in the Stein without a formal agreement between the B.C. government and the Lytton Band. If we could hold the government to the Minister's word, I was sure the Stein would be protected.

It was a lot of fun writing the story about the Stein Santas caper. Here's what we said:

"*Even though we fell far short of reaching every household in British Columbia, the entire Joy to the World Poster Project brightened up Christmas for a lot of people. The reaction was overwhelming, with only one exception. (One person who said that the forest in* the poster looked like a wheat field ready to be harvested. We said get a pair of glasses!) Everyone else was delighted. One thing for sure—all the Santas claimed that Rudolf and the other deer are for a roadless Stein."

Not counting his time, Nik had spent more than $20,000 of his own money to make the "big Stein hit." Did it work? Yes. The Stein Valley never got logged and today it is protected as a provincial park! His contribution was a "big log" on that campaign fire when it really needed it. Every log counts. It was not the last of Nik's logs—his schemes and projects fueled many of our campaigns.

Putting the rate of wilderness destruction in perspective

After the Stein projects, Nik began doing most of our regular printing in his print shop—everything except our newspapers and our "high-end" calendars and posters. I spent thousands of hours keeping him company, helping where I could, and making sure we got materials in time for our volunteers to assemble, stuff and mail on our regular volunteer nights. I gradually understood the immense scale of the effort that was needed to win protection for even one single wilderness area. Educational advertising to sell wilderness preservation was an art that Nik helped WCWC perfect. It became the backbone of all of our successful campaigns.

I remember vividly an off-the-cuff remark Nik made late one night. Both of his AB Dick printing machines were clacking along, each printing for WCWC at full speed (about 9,000 sheets per hour). "Listen to the machines go. Imagine the total number of sheets coming out of them since I began printing 20 years ago. It's a huge number, but it's not even close to the number of humans added to this planet during the same time," said Nik.

It was a truly sobering thought. From that time on, every time I hear the machines going full out, I think about the sheer magnitude of the growing human population, the pressure that it's putting on every hectare of wild Earth and the ultimate demise of all wilderness if it is not curbed. I also think about the vast volume of wood being ground up daily to feed the "paper needs" in our culture of consumerism. Everyone switching to recycled paper won't solve this problem. And Malthusian checks undoubtedly will eventually curb the human population explosion.

Image used for a Stein newspaper cover collage. Photo: WCWC files.

Sign put up by the Lytton and Mt. Currie Indian Bands at the entrance to their Stein Valley Tribal Park across the Fraser River near Lytton. Photo: Leo DeGroot.

Chapter 19

WCWC gets a full-time employee, wins its first big court case and launches a big campaign to save Carmanah Valley

Joe Foy becomes our first regular employee and develops into an "extreme slideshow" expert

For nearly three years Joe had been spending most of his spare time volunteering for WCWC. But all of a sudden, in February 1988, things changed. The lease came up on Joe's Gastown Dairy Queen and the landlord refused to renew it. His only option was to open up another Dairy Queen in the Fraser Valley where other members of his family operated them. If that happened, he would be too far away to regularly volunteer for us. Joe also had a part-time job sexing chickens, (sorting female chicks destined to become laying hens from male chicks), but needed more money than that to sustain his family.

No one wanted to lose Joe. We called an emergency board meeting and talked long and hard about hiring Joe to work full time. Joe had grown to be indispensable. We had faith that if he worked full time, he would generate the tiny $1,000 a month salary he said he needed to live on. It was one of the best "risky" moves we ever made.

One of Joe's first assignments was to tour a WCWC slideshow about the Stein Valley through the communities in the Fraser Canyon. Ken Lay and Jessica Lightfoot, a Lytton resident who was active locally on the Stein campaign, accompanied him. Their slideshow schedule included Lillooet and Boston Bar, strongholds of opposition to the proposed Stein Park. Loggers packed into the little hall in Boston Bar to see Joe's show. *"It really impressed upon me the importance of slides,"* explained Joe later. "If you show up and just talk about an issue, many people won't believe you. If you have a slideshow, it's a whole different thing. At one point in this meeting, Chief Campbell of the Boothroyd Band got up on the podium and started taking our side. Half the building jumped to their feet and got real angry. A few guys came up on the stage and put their arms around the chief in a patronizing way saying, 'He doesn't know what he is doing. He's our friend,' trying to blunt his support."

This experience shaped Joe's and our efforts from that moment on. At the time it's hard to see how these slideshows helped the campaigns—all the polarization and people calling us names. But in retrospect they were all hugely important. They helped release pressure and often dramatically changed how people thought. Joe, who did the lion's share of these slideshows, has some philosophical insight regarding them:

"I believe that there is a thing that goes on in human beings. It's kind of like running the gauntlet. They want to know how strongly you believe in something. They want to know how determined you are. They want to know if you've thought it through. They want to know if you'll stand your ground. Then everyone goes away and thinks about it and it settles in.

"You've got to do it right. You can't lose your cool. You can't insult anyone. Hold your ground on your ideals and you'll come out ahead.

Facing page: Randy Stoltmann in an old-growth grove. Photo: WCWC files.

You have to be badly outnumbered. If there are more of your supporters than there are of theirs, you'll lose much of what you want to accomplish.

"If you are really outnumbered and they are calling you names and threatening and you continue on, what you are really saying is, 'I trust you that you are not going to kill me or vandalize my truck.' They understand that and do not harm you.

"We were scared, but never scared enough to stop doing them. We'd appear in front of any audience, no matter how hostile."

The "shame the foresters" campaign

During one of our late night Press Club meetings—it's impossible to remember which one—a totally new and different kind of campaign was launched. Joining us for the beer and brainstorm was Mark Wareing, a Registered Professional Forester who then worked for the B.C. Forest Service and who Clinton Webb had introduced to WCWC a few months earlier and who was now coming in regularly as a volunteer. Mark's Forest Service duties included the silvicultural beat in the three drinking watersheds of Greater Vancouver. He knew firsthand that clearcut logging and the road building associated with it were the main reasons for the increased siltation in the north shore reservoirs and the muddy drinking water that Lower Mainlanders were experiencing after big rains.

That night Mark made a startling statement. He said that the people ultimately to blame for the bad logging in B.C. were the members of his profession, the Registered Professional Foresters (RPFs).

"What do you mean?" I asked.

"Have you ever read our code of ethics?" he countered. "Well, I'll get you a copy and you'll see what I mean." He explained that the RPF's code contained many lofty commitments to good stewardship, but in reality they were meaningless because they weren't followed. Every cutblock in B.C. has to be approved—its paperwork stamped and sealed—by an RPF before the fallers can go in. The RPFs could stop bad logging practices. "All RPFs have to do is refuse to approve the cutblocks that break our code of ethics," Mark boldly asserted.

The next day he brought me a copy of the RPF's code. He was right. It specifically said that the foresters' first responsibility was to, *maintain the integrity of the forest resource* and *guard against conditions or practices that endangered the productive capacity of the land and that reduce its potential utility or value to society.* Not only that, but they also had a duty to bring such bad logging practices to the *attention of those responsible.*

Our next conversation was a lively one. Mark pointed out that in another section of their code, one that was rigorously enforced, was the mandate to *"refrain from criticizing other members [RPFs] in public"* and *"not engage in undignified controversial discussions in public or press."* According to Mark, those who ran his professional organization were more interested in punishing foresters who publicly criticized bad logging practices than in stopping bad logging practices. Out of our discussion that night came the idea to produce and publish an educational newspaper featuring pictures of bad logging and quotes from the RPF code of ethics in an effort to shame the RPFs into acting more responsibly as forest stewards. Mark would help, but he couldn't have his name on it anywhere or he would be subject to a disciplinary hearing. Thus was launched our newspaper aimed directly at B.C.'s professional foresters titled *FOR SHAME*.

It wasn't hard to find the photos of bad logging. We had whole slide binders full of them. We broke "bad logging" down into four main areas: overcutting and excessive wood waste, failure to regenerate a new forest "crop," erosion of the land base, and degradation of non-timber forest resources. Our use of the term "erosion of the land base" was a bit ironic for we were referring to soil erosion and massive landslides caused by clearcutting on steep, unstable slopes that should never have been logged in the first place. We knew that many foresters used this term to oppose the withdrawal of productive forestlands for the creation of new parks. Under each heading was a large photo, clearly labeled as to where it was taken, with a detailed description of the damage caused. On the back page in the *Acknowledgements and Credits* box we wrote: *Thanks to the underground network of concerned Foresters who helped provide information for this newspaper and who want to see their Professional Association advocate better logging practices.*

There wasn't any such network. It was Mark Wareing with the assistance of Clinton Web, who, although only a forestry technician and not part of the "profession," knew a lot about forestry. Because Clinton didn't belong to the "club" he could speak out and not be silenced through censure by the forester's professional organization.

Portrait photo of Mark Wareing, Professional Forester. Photo: WCWC files

Published in February 1988, our four-page, two-colour, tabloid-sized *British Columbia Professional Foresters FOR SHAME – Unethical Logging Practices Permitted* paper was very popular. When we ran out of the first press run of 25,000, we put it back on the press for a second run of 50,000. These went like hotcakes too.

Foresters visit our "hostility" suite

As part of our campaign of trying to get the professional foresters to clean up their act, we decided to rent a hospitality suite at the next RPF convention and AGM being held in the Empress Hotel in Victoria. Clinton Webb put together a "killer" 10-minute slideshow with all the "best" slides we had of bad logging practices. We got fancy cards printed up, inviting the foresters to see our show. Mark Wareing helped us pass them out on the convention floor. For a whole day and evening Clinton showed the show over and over again to groups of five to ten foresters. Joe Foy and I participated in the lively discussions and arguments during and following the shows.

Our room, nicknamed "the hostility suite" the by some foresters, was very popular. We estimated that more than a quarter of all the foresters attending the conference visited us.

The most amazing thing I learned through this effort was the powerful role that "denial" plays in perception. Far more than half of the foresters who saw our show couldn't see anything wrong with the forestry practices illustrated in our slides. They had an excuse for everything. They believed that it was WCWC's perception that was off base. What we saw as clearcut-caused problems they saw as natural processes. The forest would regenerate. The damage, at most, was trivial. They especially refused to admit that clearcut logging and building logging roads on steep slopes accelerated erosion and caused landslides. "The slopes were unstable. That slide started in the forest. Lucky they got the timber off before it slid or the damage would have been worse," they'd say. I knew after these encounters that it would be a long uphill fight to get better logging practices in B.C. The brainwashing at the UBC School of Forestry was complete. Most B.C. foresters believed their own mythology with the fervor, tenacity and blind faith of religious fanatics.

The book that WCWC tried to produce, but never did

In our *For Shame* newspaper we advertised a book titled *Halt the Destruction of the Forests of British Columbia* for a special low price of $3.98. The problem was there was no such book! This book had started out as a proposal by our director Ron Aspinall in 1985. It began as a small booklet that would simply spell out what was wrong with current forest practices in B.C. and how forestry could be done better. We contracted a well-known ecologically-minded professional forester, Herb Hammond, to write the text for us. But two things put this publication off the rails. Herb, as one of only a handful of B.C. eco-foresters, was very busy with other projects. And, a perfectionist by nature, Herb also found it impossible to treat the subject matter in such a brief way. Finally in the fall of 1987 he promised it would be done for sure at the beginning of the new year.

We even mocked up a cover for *Halt the Destruction of the Forests of British Columbia*, the title we gave the book, and put it as a tear-off coupon in our 1988 calendar. Our promo aroused peoples' interest:

Find out why the B.C. government's current old-growth forest liquidation policy, maximizing production of wood fibre to the detriment of everything else, is a dead end road. Specific problems in current forest 'management' are discussed, case by case. Learn about the common sense, practical, and feasible alternatives. Order your copy now of this well-illustrated booklet. Pre-publication price $3.95. Price after May 1 is $4.95.

Our hype sure made it sound like a great book at a great bargain price. Many people ordered it.

When this deadline came and went, Hammond informed us that he just couldn't condense the material down that much. There was just too much to explain to adequately cover the topic of "eco-forestry," the alternative to the industrial forestry practices of today, in a 48-page booklet. He convincingly explained that the book would have to be much larger and cost a lot more. But we just couldn't afford to publish such a book. We'd already delayed publication three times, writing to everyone who had ordered the booklet each time. It was getting very embarrassing. We couldn't maintain good faith with these people and change our minds again. In December 1989 we pulled the plug on the project, informing the many hundreds of people who had pre-ordered the booklet that we were not going to be publishing it and offering to refund their money or give them another book.

In this debacle we learned a major lesson. Don't sell something that exists only as a good idea. Eventually, in 1991, long after we had abandoned the project, Herb Hammond published his book *Seeing the Forest Among the Trees – The Case for Wholistic Forest Use*. With more than 300 photographs, 40 graphs and charts, and 15 illustrations, it was the definitive blueprint for ecologically responsible forest use and provided the theoretical basis for getting better forest practices. It was a great book—the book that Herb had all along wanted to do.

WCWC's court case establishes "standing" for all environmental organizations and halts the wolf slaughter in the Muskwa Valley in Northeastern B.C.

Unique opportunities knock only once and rarely come without risk. At WCWC, up until this unique opportunity arose in February 1988, we had focused nearly exclusively on protecting endangered wilderness areas through public education, on-the-ground trailbuilding and research. We left wildlife management and animal rights issues for others such as Greenpeace, the Sea Shepherd Society and Project Wolf to champion.

I always sympathized with those whose primary interest was in protecting wild animals. But I was careful never to get drawn into committing too much of WCWC's resources, personnel or time to those worthy causes. We had to stay focused. Our perpetual poverty made it easy to say that we couldn't help financially. If campaigns involved any sort of civil disobedience, that ruled us out too. But then, early one day in late February 1988 came a "unique opportunity" phone call. "Paul, some lawyer on the line wants to talk to you," one of the volunteers shouted out (we did not have a phone intercom system). I picked up the extension phone, expecting it to be Jack Woodward or Murray Rankin or some other lawyer I knew.

"My name is Don Rosenbloom of Rosenbloom & Aldridge," said the caller, "and I want to talk to you about Western Canada Wilderness Committee being party to a legal case seeking to stop the Fish and Wildlife Branch's helicopter predator control program in the Muskwa Valley in northeastern B.C. Are you interested?"

"We can't afford it," I immediately responded, thinking that he was some young inexperienced lawyer who was following this hot issue in the newspaper, who had some idea for a legal challenge and was looking for a client.

"No, it won't cost you anything," he said. "I already have a client who is paying all the costs. What I need to get the case before the Supreme Court of B.C. is an organization like Western Canada Wilderness Committee as a client that would have standing in court on the issue." After explaining to me the situation in greater detail, I wanted to say to Mr. Rosenbloom, "Sounds great, let's do it!" But instead I said, "I'm almost sure the answer is yes, but I have to talk to the other directors and get their approval. I'll get back to you early tomorrow. I understand that you have to act right away or it will be too late."

I talked to Joe Foy and the other volunteers in the office. Everyone knew about the conflict. No one was afraid of taking on this issue because it was too controversial. For weeks, the wolf hunt dominated the news. Activists, spurred on by Paul Watson, had set up a tent camp along the Alaska Highway in bitter 40-degrees-below temperatures and were planning to trek into the area where the proposed helicopter "cull" was scheduled to take place. Their strategy was to scare the wolves away to save them from Dr. John Elliot, the government wildlife control officer whom they characterized as being "wolf-hating and gun-happy." At WCWC we believed that this program was not only costly, but it was also unwarranted from an ecological or a game management perspective. Our information was that the helicopter hunts in previous years had already reduced the wolf population in the area to near extinction levels. Actually we felt the whole justification was wrong; to help guide outfitters entice more foreign big game hunters with the lure of a better chance to kill an elk or deer.

I called Adriane at her office in Langara College and enthusiastically explained the opportunity we had to fight the wolf hunt in court. She was skeptical, especially about the "costing no money" aspect of it. I also called a couple of my lawyer friends to get the scoop on Don Rosenbloom. They all knew him. He was one of the "best lawyers in town" and was known for winning labour cases. "You couldn't find a better lawyer," one of them said to me, not knowing that Rosenbloom had found us!

Everyone wanted to know who was paying the shot. The next morning I called Don Rosenbloom and told him that, pending his answers to a few more questions, the answer was "yes." "Could we get together to discuss the case?" He arranged to see me later that same day. At that meeting he explained that he had taken on the case by referral from "the largest law firm in the world"—Baker & MacKenzie. A lawyer in that firm's Toronto branch office had called him on behalf of a client in California who wanted to proceed legally to stop the wolf kill. Rosenbloom explained that the client was an extremely wealthy father of a young woman who wanted to come up to B.C. from California to join the campaign to save the wolves.

She and two women friends were planning to fly in by helicopter and parachute into the midst of a wolf pack to save it. They had already parachuted in once. Concerned for her safety, her dad desperately wanted to stop her from participating in this direct action jump in the frigid northern wilderness of B.C. He argued with her that there were better ways to stop the wolves from being killed—like a court action. She agreed, but her group did not have the money for such an action. He made a deal with her. If she would abandon her plan to make the parachute jump and not go to the winter protest camp, he would underwrite the costs of a legal case that would accomplish the same thing she wanted—saving the wolves. She agreed.

WCWC's participation was vital to the case's success

Rosenbloom's problem was that the young woman from California had "no standing" in a Canadian court on the issue of protecting wolves. In fact, in the narrowest sense that the law could be interpreted, only those directly affected financially could go to court to protect their interests. It made it difficult to launch the case because no environmentalist would suffer a financial loss if the wolf hunt went ahead. However, it had been established for some time in the U.S. that environmental groups could take the government to court to protect wildlife and challenge laws and bureaucratic rulings because they could "speak on behalf of wild animals." Courts in the U.S. recognized that conservation groups had a legitimate interest in these kinds of matters because they represented people who wanted to conserve and protect those animals and wilderness. But we had no such precedent in Canada.

WCWC, Rosenbloom thought, had the best chance to secure "standing" and the right to ask the court to judicially review the way the decision was made to conduct the Muskwa wolf control program. We were one of the largest membership-based charitable societies with a high profile on conservation issues. Also, on the plus side, we always worked within the law and did not participate in non-violent civil disobedient acts nor advocate any kind of illegal activities. The fact that we respected the legal system and believed in using the courts to settle disputes would positively influence a judge and increase our chances of a positive outcome, said Rosenbloom.

It came as a surprise to me that a key component of his case was going to be arguing that WCWC be granted standing in the matter. In my affidavit I swore that we had been developing a proposal for the establishment of a wolf sanctuary wilderness area in the Muskwa region for the past two years. While this was true, it was at the same time, a bit of a stretch. We had nothing formally written up on our proposed reserve. It was just one of many good campaign ideas that we had wanted to pursue... if only we had the money.

Not being a lawyer, I wanted to argue the moral issues. I knew that the Fish and Wildlife Act in B.C. charged the Minister with *"protecting and conserving wildlife."* How could the Minister's authorizing the killing of wolves so that hunters could have more deer and elk to kill qualify as protecting wildlife? I could find witnesses who would testify that the wolf population in the Muskwa area was already seriously depleted due to the two previous years' wolf kill programs.

For "our" lawyer Rosenbloom, the important thing was to win the case. To do that, he had to use the best legal arguments, not a moral one or a biological one. His aim was to get the decision that permitted the 'wolf control' hunt quashed by proving that the way the government decision was made was illegal, contrary to the way good government decisions had to be made. His case was based on a technicality. And his legal argument was fully backed by precedence—past court decisions. In short, he argued that Environment and Parks Minister Bruce Strachan had delegated his entire powers to a lowly local wildlife officer in the north without any restrictions. Common law requires that a Minister "fetters" (adds restrictions to fit the circumstance) the powers given to him by legislation when passing them down to a lower government employee, such as to the regional manager of the Fish and Wildlife Branch.

However, permit No. C012312 authorizing Dr. John Elliot and an unnamed employee to hunt and kill wolves from an aircraft had no restrictive conditions specified in it. Conducting such a hunt, argued Rosenbloom, was a very serious matter. Normally it is an extremely serious criminal offence to shoot a gun from an airplane or helicopter. It could be dangerous as well. In short, granting unfettered discretion in the form of this wolf hunt permit was *ultra vires* (outside the law i.e. illegal).

On leap year day, February 29, 1988, the B.C. Supreme Court hearing began on WCWC's petition on this matter. I watched Don Rosenbloom present all the evidence—the documents and affidavits—and argue his case. I was extremely impressed with his authoritative style and his obviously extensive knowledge of the case law that set precedence. He was articulate and confident. No wonder people say "get the best and most expensive lawyer and your chances of winning greatly improve." I thought to myself, "If we ever need a lawyer in the future, this is the one I want to hire. I hope we can afford him!"

All the while that the Honourable Madame Justice Huddard heard our case, she remained stony faced. It was impossible to tell if she was being swayed by Rosenbloom's arguments. She reserved judgment on the matter, including whether or not WCWC had "standing." Although there was no injunction, the hunt was stalled while waiting for the judge's decision. It would have been disrespectful to go ahead while the matter was before the court.

Precedent-setting court ruling granting us "standing" is a major accomplishment

On the morning of March 7, 1988, I put on my best suit coat and tie, and got down to the courthouse early. I wanted to be there at 10 a.m. sharp to hear Justice Huddard pronounce her decision on our case: "*Vancouver Registry A880554 - In the Supreme Court of British Columbia Re: the Judicial Review Procedure Act - Western Canada Wilderness Committee and Paul George petitioners vs Bruce Strachan, Minister of Environment and Parks, Steven Willett, Dr. John Elliot and John Doe respondents.*"

The attorney general's lawyers had vigorously argued that, although we were not "mere busybodies," simply having a strong view that the government program was wrong, this did not entitle us to standing in court. According to him, we did not have enough genuine interest in the matter to acquire it. He suggested that only the permit holders, MLAs, hunters or those who trapped wolves for pelts had a real interest and could bring such an application forward.

Madam Justice Huddard disagreed with this government lawyers' argument. She said:

"The 1,500 members of the Western Canada Wilderness Committee who have formed a British Columbia society to conduct research concerning wilderness values and to educate the public in Western Canada about our wilderness heritage and the reasons for preserving it are not mere busybodies. They must be regarded as concerned citizens interested in ensuring that lawful process is followed when decisions affecting the protection and management of wildlife in British Columbia are being made.... For these reasons I am granting the petitioners standing to bring this application to have Regulation 1(n) of the Wildlife Permit Regulations declared ultra vires, as an improper sub-delegation of authority, and hence null and void."

After about five more minutes of legal case citations and arguments about the fettering of ministerial powers that, frankly, I could not fully follow, she concluded: "*In summary, section 1(n) of Regulation 337/82 is ultra vires. Permit C012312 issued pursuant to it is ultra vires and thus null and void.*"

Rosenbloom immediately asked for costs to be awarded and the stunned attorney general's lawyer didn't object. Justice Huddard awarded us costs.

We'd won! Our petition was granted. The government permit for the wolf kill was ruled illegal. I asked Rosenbloom how we could get a written copy of Huddard's oral judgment. He told me that we could get it transcribed for a small fee, which we did. We soon had a typewritten court document granting WCWC standing in the court on wilderness issues, giving the concise reasons why based on our mandate and activities.

Our case proved to be a major precedent. From that day on, other environmental groups also had the right to use the Judicial Review Procedures Act to challenge government decisions that they believed to be unfairly arrived at and wrong, even if they were not directly, adversely affected financially by the decision.

On the way out of the courtroom I asked Don, "Because it's simply something technically wrong with the paperwork, can't the government just pass an Order in Council to make the changes needed to make it legal?"

"Yes, of course they could do that, but it would take time," he explained. Because of the weather conditions—it was approaching spring—there wasn't enough time and it would be too late to carry out the wolf hunt this year. Furthermore, Don conjectured, because the hunt was both unpopular and biologically unsound, it gave the government a way to drop the matter while at the same time saving face. He was right on both counts.

This was my first real taste of a courtroom victory and not the last time we would work with Don Rosenbloom. That night the news reached the brave wolf kill protest campers who had been living for more than a month in a tent beside the Alaska Highway. In that camp was the late Greg McIntyre who 13 years later became WCWC's mail and printing press operator. He vividly remembered the day they got the news that the hunt was cancelled. That night they watched the most spectacular display of aurora borealis he had ever witnessed.

Regarding the cost award money, which totaled slightly over $2,000, it remained in Rosenbloom's trust account gathering no interest until November 2005 when he donated it to WCWC at our 25-year anniversary *Tall Trees Gala* celebration.

Money starts flowing in as work load increases

Our seventh year finally brought WCWC the spectacular growth we'd been seeking. When the figures were tallied on April 30, 1988 (our fiscal year end) our revenues turned out to be $376,000—more than double the $177,000 that we budgeted. Of course, the money wasn't sitting in the bank. We'd spent all of it to boost our campaigns, mostly for printing new educational publications. At our May 1988 AGM we decided to open a wilderness store as soon as possible to sell all our products, just like the Australian Wilderness Society had done. We were on a roll.

The increased activity came with another price tag: long working hours. An 80-hour week was not unusual for Adriane and me. We always had at least a couple of newspapers plus a poster and mailout on the go, lots of correspondence and many phone calls to answer, as well as research and campaigning to do. We kept this pace up for weeks, months and years, with only a few breaks. After we moved into our new solar home on the Sunshine Coast in the summer of 1988 we found it impossible to get all the work done and commute back to Vancouver by ferry every day. So we bought a futon couch for the office. Adriane and I spent more nights putting our kids to sleep on that futon and pulling all-nighters than we would like to admit. I guess we were a bit possessed and fanatical. Someone told me you get that way when you're a "cause person." I always found fighting for "the cause" fun and exciting.

Randy Stoltmann discovers a "sneak attack" on Carmanah Valley's Sitkas

It was the first week of April 1988 and, as usual, I got to the office early on Monday morning. I had no sooner unlocked the door when Randy Stoltmann walked in. I was surprised. He had a full time drafting job and usually visited our office in the late afternoon after work.

Randy was fuming. He'd been out on the west side of Vancouver Island on the weekend with his friend Clinton Webb searching for record-sized trees in the old-growth forests. South of Nitinat Lake, they happened upon a new logging road that wasn't on MB's TFL 44 plans or maps. The road kept going and going. When it got too rough to drive, Randy and Clinton started walking. They hiked down several kilometres of road that was under construction and found themselves at the very edge of Carmanah Valley. They followed the road-right-of-way flagging tape down into the heart of Carmanah to the edge of a grove of the biggest and most beautiful Sitka spruce trees that Randy had ever seen. The whole grove was flagged as a cut block ready for clearcutting.

New logging road snakes towards Carmanah Valley through many kilometres of recent clearcuts. Photo: WCWC files.

Forester Mark Wareing beside a strange heavily burled Sitka spruce in Carmanah Valley mid-May 1988. Photo: Clinton Webb.

"MB is about to log the finest old-growth Sitka spruce in B.C. and no one even knows it's happening," Randy ranted. He insisted that the road he'd walked along was not part of the company's publicly-reviewed TFL 44 Five-Year Plan that the B.C. Forest Service had approved three years earlier.

"That road can't be legal. We've got to stop it!" said Randy. I totally agreed with him. We'd actually talked about this valley at the Wilderness Committee's first Annual General Meeting in October of 1981. The trees in Carmanah were of mythical status. I had been charged with trying to arrange a helicopter trip there through Parks Canada to check the area out as a possible addition to Pacific Rim National Park Reserve. It was one of those great ideas I hadn't followed up on. Partly this was because, MB's plans didn't show any logging in Carmanah until 2003 so, making this, in my mind, a low priority. Besides we had our hands full at the time with Meares and the Stein. I also knew that even though MB had clearcut one large patch in the upper valley a few years earlier, the steep mountainsides made road-building from the upper valley to the lush mid-valley impossible. Obviously something had changed in MB's plans.

I told Randy to come back that evening so we could get together with Joe and Ken and figure out what to do. MB's change in plans would require a major change in our plans, too.

First unsuccessful search for legendary mammoth spruce in Carmanah

Randy had first visited the mid Carmanah Valley in the summer of 1982 with Bristol Foster, who was then head of the B.C. government's Ecological Reserves division. They'd been flown in by helicopter as guests of Parks Canada, landing on a sandbar in the river next to this place where Randy now, six years later, found the spectacular grove that was flagged for cutting. In 1982 Randy and Parks Canada personnel had explored the surrounding old growth looking for a legendary Sitka spruce measuring 7.9 metres in diameter and 94.2 metres tall that a timber cruiser reported finding in the valley in 1956.

This tree was so much bigger than any other known tree of its species that it had achieved mythical status—sort of like a sylvan "Holy Grail." Alas, it was not to be found on that trip or on many later ones. But what Randy did find in 1982 was a "heritage forest" in Carmanah Valley that he described in our 1985 newspaper: "*A proposal to create... CANADA LANDMARKS – PROTECTION FOR OUR LARGEST, TALLEST AND OLDEST TREES.*" In this paper he called Carmanah Valley an *irreplaceable national treasure*.

Establishing a united front amongst conservationist groups to prevent a "living tree museum" compromise

That evening we talked about strategy. Randy wanted two other groups that Randy belonged to besides WCWC to be involved in the campaign, specifically the Heritage Forest Society (a volunteer group established in the late 1970s by conservationist Will Paulik) and the Sierra Club of B.C. He believed that the provincial government would be more willing to listen to those groups than to WCWC because they weren't so outspokenly critical of government and industry.

I told Randy that before WCWC got involved in a big way in this campaign, I wanted to make sure that all the groups involved worked together and stayed on the same page regarding our ultimate objective. Most importantly, we were interested in saving the whole valley. I was especially afraid that the other groups might be content with a small park that ostensibly protected the grove with the biggest trees while the rest of the valley got logged. If WCWC was going to put many thousands of dollars and many hundreds of hours into a campaign to protect Carmanah Valley, then everyone had to be on board for its complete protection. It would be awful if one group, willing to settle for less, helped legitimize a bad government decision.

I also argued that anything short of protecting the whole valley was biologically indefensible. The valley was just too steep-sided, the rainfall too heavy and the windstorms too strong for a small reserve ringed by clearcuts to last. Anyway, Carmanah was a relatively small watershed. At 6,730 hectares it was just over 16 times the size of Vancouver's famous Stanley Park. Every ten days, in total, an old-growth-forested area the size of the entire Carmanah was clearcut logged in B.C. Thus, full protection for Carmanah was reasonable and affordable.

After more discussion, Randy agreed that he would go to the Heritage Forest Society and the Sierra Club and get their commitment for a campaign to save the whole watershed. It turned out this was not a problem. Everyone readily agreed they'd settle for nothing less than the entire Carmanah Valley.

Discovering why Carmanah is under early attack

A couple of days later Randy took a day off work and went with Joe Foy to the B.C. Forest Service's District Office in Port Alberni. Their mission was to check out MB's paperwork to figure out when and why the accelerated pace of logging had been authorized. There they learned that a few months after the public viewing and approval of TFL 44's five-year-plan, MB applied for and got Forest Service approval for substantial amendments to the plan—without any public input! The revised plan advanced the date of logging throughout the middle and Lower Carmanah Valley, including the best spruce groves, to 1989. The allowable cut in this part of the TFL was also increased substantially. I figured this change was meant to make up for MB not being allowed to cut as many trees as it wanted to in Clayoquot Sound, especially on Meares Island.

It turned out that to facilitate this "transfer" of cut, in 1984 the B.C. government allowed MB to amalgamate its Tofino TFL 20 and its Alberni TFL 21 into one giant TFL 44 encompassing a whopping 452,826 hectares (three-quarters the size of Prince Edward Island). There were no rules as to how the annual cut had to be apportioned through whole of this larger TFL. Thus the accelerated cut in the Carmanah region was legal.

To make logging in this region even easier, MB applied for and got in 1986 an adjustment to the boundary of Pacific Rim National Park near Cheewhat Lake to facilitate construction of an access road to Carmanah. We had been informed about this small change to the national park boundary a few months before it took place—presented to the public as a minor "get a little, lose a little" land swap—and didn't oppose it because at the time it didn't seem like a big deal to us. Now, understanding the implications, we felt duped.

On May 2, 1988, the *Vancouver Sun* featured the first major article on Carmanah and Randy, titled *Tree hunter's claim of forest giants sparks preservation plea*. Since his discovery of the imminent threat, Randy had worked night after night writing up a formal proposal on behalf of the Heritage Forest Society and Sierra Club titled *A Proposal to Add the Carmanah Creek Drainage with its Exceptional Sitka Spruce forests to Pacific Rim National Park*. On May 13th he presented it to the B.C. government, MB and the media.

Hiking in to meet the MB chopper

Right after being given Randy's brief, MB offered to helicopter Joe, Clinton Webb and Randy into Carmanah for a tour and "to talk things over." I argued strongly that we should not accept this offer. Being guests of MB would put us at a distinct disadvantage in any negotiations because the gift of the ride would make us indebted to MB. Also, MB could make a big thing about it in the media. Joe agreed. Getting there on our own would show MB and the media that this place wasn't as remote or inaccessible as the company claimed. So we decided that Joe and Clinton would bushwhack into Carmanah one day early, in order to be on the sandbar to greet MB's helicopter party when it arrived the following morning. Randy chose to ride in on the helicopter.

It was a miserable weekend with buckets of rain pouring down. Regardless, on Friday afternoon as planned, Joe and Clinton drove to the edge of the valley, hiked down into the valley and set up camp. The following morning they tried to build a roaring warm welcoming fire on the sandbar; but the wood was so wet that their fire just smoldered, emitting tons of smoke. Randy told us later when they all got back together in Vancouver after the trip that it certainly got the talk going up in the chopper. The smoke plume was visible many kilometres away. Some thought it might be a forest fire; no one guessed it was a WCWC campfire!

Notwithstanding the smoke, Joe and Clinton's effort was good enough to bring the kettle to a boil and they greeted the surprised MB entourage with hot cups of tea. After tea and talk everyone toured the big grove of towering Sitka spruce that MB had flagged for logging. MB had to admit that not enough research had been done on Carmanah. No one knew how tall these trees were. Perhaps some of them in the grove were of record height, suggested Joe and Randy. MB's forester asserted that this grove was not unique—there were others just as grand—but when pressed he could not give any specific locations.

MB makes a move to avoid confrontation
WCWC establishes "Hell" and "Heaven" camps

On May 19, a few days after the helicopter tour, MB announced that it was voluntarily halting road construction into Carmanah Valley for one month to allow for a study of the area. It was a mini victory. By now we were completely pumped about Carmanah. It was an incredibly inspiring place; a living cathedral. None of us had seen groves of trees that tall in B.C.

The campaign took off like wildfire. All May, Randy, Joe and a ragtag assortment of volunteers made regular weekend expeditions to Carmanah to explore the valley. With the ferry ride and nearly two hundred kilometres of logging road to travel it took about six hours to get there from Vancouver. The trips in and out put a lot of wear and tear on everyone's vehicles, especially the tires, but no one cared. Carmanah was our passion.

One of the first things we did was flag a hiking route down into the mid-valley. It was too tough for most folks to bushwhack from the road to the valley bottom. If we built a trail to make it easy for people to hike in and see the awe-inspiring trees, we knew they'd come. Once they saw the majestic groves, they couldn't help but become passionate Carmanah preservationists and join the fight to save the valley.

On May 30th, in the pouring rain, four WCWC stalwarts left Vancouver headed for Carmanah to set up camp and build a good trail to the valley bottom. The crew consisted of Joe Foy, Clinton Webb and his brother and a new volunteer, George Yearsley. As usual rain poured down incessantly. Soaking wet, they pitched a camp right beside the new logging road. It was muddy and ugly: befittingly they nicknamed it "Hell Camp."

Despite the foul weather, our crew immediately started working on the trail. They'd hardly begun when some MB officials came by and asked them to stop. Joe politely refused. MB did not own the land, the company only owned logging rights to the trees growing there and we were not hurting the trees. By the end of the third day Joe Foy and our crew had completed a rough trail to the edge of the majestic spruce grove in the mid-valley and had set up a second camp. It was nicknamed "Heaven Camp"—a stark contrast to "Hell

Trailhead to the mid Carmanah Valley. Centre front Randy Stoltmann, behind Greg Stoltmann, on the left George Yearsley, on right unknown. Photo: Gary Fiegehen.

Camp"—with the grove of spectacular Sitkas beside it nicknamed "Heaven Grove."

George was a tough young kid who stayed in Carmanah for the rest of the spring and summer that year. He set a volunteer record working 90 days straight building trails and boardwalks. George worked rain or shine. No task was too tough. By the end of the summer he had become so attached to the place that it was hard to get him to leave. He told us his dreams had morphed from being about city stuff to being full of forests, birds and streams. Those who have spent several uninterrupted, unadulterated weeks in wilderness, with no TV, radio, newspapers or books, know this transcendent change that happens.

Chapter 20

Meares trail work blocked
Clayoquot conflict erupts

WCWC volunteers work hard to build a Meares Island Circuit Trail

Carmanah wasn't the only place where we were building trails that spring. At the same time Joe and his crew were snipping away in Carmanah, Ken Lay headed up another crew who were clipping away the brush to extend a trail around Meares Island. On this island there already was a well-used "Big Tree Trail" to the famous "Hanging Garden Tree" and other huge cedars. The map in our *Meares Island - Protecting a Natural Paradise* book, however, showed that trail continuing around the island (labeled *primitive trail*). In reality this trail wasn't "primitive." It was imaginary. All we'd done was flag-out a proposed hiking route. But we aimed to turn it into a well-built hiking trail soon.

In our trailbuilding the safety of the hiker was of prime importance. We also build our trails without cutting down any trees, although we'd have to cut through the occasional windfall. It actually made the trail route more interesting as it wove in and out around the huge trees along the way.

On Meares the task wasn't easy. Its terrain is rugged. We also had to completely avoid the water catchment area. The lake and its watershed in the centre of the island provides Tofino's fantastic-tasting, un-chlorinated drinking water. The toughest region to transverse, because of its steep and gullied nature, lay on the east side of the island from Mosquito Harbour (not the greatest name to engender wilderness preservation enthusiasm) to Heelboom Bay (not such a great name either). Heelboom called Tsis a Quis in Nuu-chah-nulth, was the place where the big standoff over logging had occurred four-and-a-half-years earlier. The section of the trail from Heelboom Bay back to the Big Tree Trail followed for a short way the route of MB's proposed logging road then continued along the mudflats bordering Browning Passage. This, too, was rough going, but in a different way from the steep mountainside. It was a lowland swampy traverse through gnarled cedars, blown-down ancient giants and nearly impenetrable salmonberry thickets.

Large grant from Taiga funds our Meares trail project

We were quite public about our plan to complete a circuit trail on Meares. In mid April, Taiga Works, a Vancouver-based company that manufactures and sells outdoor sports equipment and gear, gave us an $8,000 donation. It came with no strings attached. I don't know if using their donation for our work on Meares Island (this issue, which was hugely contentious in logging industry circles) worked from Taiga's point of view; but it sure worked for us. It was the first and only large donation this company ever gave us.

Mark Hume, then a *Vancouver Sun* reporter, covered the cheque presentation event. It was first time that WCWC created a two-metre-long cheque—a blowup of the real thing pasted on cardboard—like the ones at all the major charity events. Unfortunately, no TV cameras showed up. In fact, Mark was the only member of the media to come. A few days later, on April 22, 1988, (Earth Day) his article appeared in the paper. It was ace. There was a map of Meares showing our proposed trail route. He explained how WCWC would be using

Accepting donation cheque from the owner of Taiga Works. Tla-o-quiaht Chief Moses Martin on left is holding a limited edition Meares Island print—a gift for the owner. Photo: WCWC files.

the $8,000 to complete a trail that *...opens up for the first time the densely overgrown heart of Meares Island...It's always been difficult to get close to a rainforest like that because it's such a jungle.*

I'd told Mark that WCWC's goal was to make Meares so popular that it would never be logged. We had the full support of Moses Martin, the former Tla-o-quiaht Band Chief who spearheaded the blockade in the fall and winter of 1985, which had successfully thwarted MB's proposed logging on Meares.

Temperate rainforest bushwhacking the original extreme sport

Bushwhacking is the original "extreme sport" on the west coast of B.C. I'd struggled through thick coastal undergrowth along Juan de Fuca Strait from Jordan River to Port Renfrew almost every weekend in the early '70s, explored extensively the old-growth in South Moresby in '77 and trekked one cold, wet winter week in the early '80s from Steamer Cove to Ahousaht on Flores Island.

You had to be one determined bear to get through some places. Near the ocean the salal grows so thick in places you have to throw yourself onto the two-metres-high bushes and slither over top of them. Much worse are the thick patches of devils' club on steep slopes: it's impossible to avoid all their nasty thorns.

I discovered that thickets of evergreen huckleberries are the west coast's most impenetrable "bush" and that rare, but among the most hazardous of places, are ones where huge thousand-year-old trees blew down hundreds of years earlier. These fallen giants crisscrossed over each other, now rotten and overgrown with moss, create treacherous holes to fall through. Some places in the old-growth rainforest, however, are as easy to walk through as a manicured park. One of the best examples is Heaven Grove in Carmanah Valley, with clumps of two-metre-tall sword ferns carpeting the mossy forest floor between the well-spaced-out, majestic Sitkas.

Bushwhacking in the ancient west coast rainforest is challenging and addictive. Nothing in the world can match the thrill of slogging through a wild, un-trailed ancient rainforest. For me, it's the best of extreme sports, combining physical challenge with intellectual stimulation. Every hectare of wild primary forest is utterly unique and astonishingly beautiful beyond words.

Meares trail article probably provoked MB into action against WCWC

Mark Hume's article about our Meares Island trail had huge ramifications. I think MB may have started getting worried when I optimistically suggested that the Meares trail could become as popular as Pacific Rim National Park's West Coast Trail. But it was probably my statement, *"winning the hearts and minds is what the trail—which is being built without any government support—is all about,"* that enraged this corporate "bull of the woods." They treated their TFL lands like their own lands, and believed they could determine what activities were allowed and not allowed on them.

MB legally "blockades" Meares Island trail work

About two weeks after Hume's *Vancouver Sun* article about the Meares trail, Bill Ohs, an MB "Employee Relations Manager" (nicknamed Bill Ooze by Ken Lay) came poking around on Meares. Environmentalists who got to know him during the Meares confrontation told us that Ohs was "a smooth operator." He spent several days checking out our trailbuilding activity, without ever contacting or talking to our crew. We soon found out that he was gathering evidence to build a case that our trailbuilding was harming MB, especially the timber that MB had the "right" to harvest there. MB had also hired one of the best legal minds in B.C., John J. L. Hunter of Davis & Company, to try and stop the trailbuilding on Meares.

On June 29, 1988, MB launched a legal action. A courier served WCWC (delivered to my desk at our West 6th Street office) with MB's Notice of Motion. It informed us that MB would be in court at 9:45 a.m. on July 7th seeking a court order to force us *...to halt all trailbuilding activities and remove all traces of the trail including signs, ribbons, trail markings cabins, houses, bridges, floats and other structures* [on Meares Island]. Accompanying this notice were copies of affidavits by Bill Ohs and Norm Godfrey. MB had also recruited the Government of B.C. to support its effort to stop our trailbuilding, with the Attorney General of British Columbia entered as an intervenor. The letter concluded with a curt suggestion that we might wish to speak to counsel. Boy, did we ever!

I knew we needed a high-powered lawyer and Don Rosenbloom, who had won our Muskwa wolf case, was the one we wanted. I hoped he could defend us on such short notice...and that he would be affordable. Luckily, he was available; but he informed me that he would have to charge us, although at a charitably reduced rate.

I rushed Rosenbloom copies of the paperwork that had been filed that same day in the Supreme Court of B.C.: *Case No. C845874 - MacMillan Bloedel Limited Plaintiff and Mike Mullin, Linda Mullin, Ron Aspinall,* [a director of WCWC at the time] *Moses Martin, C. J. Hinke, Steve Lawson, Patrick McLorie, David LeBland, Rob Paxton, Adrian Dorst and Persons Unknown as respondents.* I guess that our WCWC trailbuilders were the "Persons Unknown." Ohs' affidavit clearly indicated that this court action was really directed at WCWC. In his affidavit Ohs swore that on June 20th he had a conversation with WCWC Director Joe Foy who told him we were responsible for building the trails on Meares. Ohs also swore that our cutting through old windfalls on the trail to make it easier to hike "reduced the value of the timber," and he attached 12 coloured photos to reinforce his case. It was quite a stretch to claim that our trailbuilding was diminishing the value of the trees MB "owned" on Meares Island; but his assertion that WCWC was responsible for the trailbuilding was fact. It was no secret.

Of greater interest to Rosenbloom was the other affidavit and exhibits; the 36-page document put before the court by Norman Godfrey, MB's Alberni Regional Forester. Norm was normally thought of as a "good guy" in MB, but his affidavit was intended to prove that MB had not only the rights to the timber, but also the authority and rights to manage all recreational activities on their TFL lands. According to Godfrey, our trail constituted unauthorized use of their TFL lands. His testimony amounted to saying that TFLs on Crown Lands were really like private "fee simple" lands. If this legal interpretation were to prevail, it would deal a fatal blow to one of our most effective tactics to save wilderness-trailbuilding.

To accomplish its objective, all that MB sought was a court order varying the original injunction (granted by honourable Mr. Justice Gibbs on January 25, 1985) that stopped the protesters who at that time were blockading logging activities to additionally forbid our trailbuilding activities. After looking at the legal documents, Rosenbloom told me bluntly that WCWC didn't have a chance. There were already two injunctions in force: MB's injunction that stopped blockaders and a countering injunction granted to the Nuu-chah-nulth that stopped MB from logging until the First Nations' claim of ownership to the forests of Meares Island was decided in court.

Together they essentially forbade all activities that altered the state of this disputed island and that, of course, included our trailbuilding. The big legal question—whether or not MB had exclusive rights to determine the recreational use of their TFL lands—was disputable, but it would never come up in this Meares Island court case because the overriding injunctions made that determination moot.

"But we only have a few more weeks of work left and then at least the roughed-in trail will be completed," I told him. Rosenbloom negotiated with John Hunter, pleading that it was difficult to get in touch with our trailbuilding crew to get them to immediately cease and desist and he got an extension of a few days time—to July 10, 1988—before we had to quit trailbuilding.

Meanwhile, Joe Foy rushed to Meares to let Ken Lay know what was happening and to spur on our volunteers to work as hard as possible to finish up the trail as best they could by this deadline. We also let Ken and his helpers know that if they wanted to continue building trail, they could go directly to Carmanah Valley from Tofino, and take all their trailbuilding equipment with them. The need for a trail there was now more pressing than on Meares.

Almost finishing the Meares Island Circuit Trail

July 10th came and there was still one section of trail that was really problematic, a kilometre-long, extremely steep and dangerous stretch in the Woods Inlets Creek area. Despite polite letters from Rosenbloom urging MB to give us temporary access to fix up this stretch for safety reasons, MB refused.

When Ken arrived at the government dock in Tofino with his exhausted crew and all the trailbuilding equipment, Bill Ohs, who "just happened to be hanging around," greeted them. He struck up a conversation with Ken and casually asked him what he would be doing next. "I'm exhausted," complained Ken. "I'll be taking the rest of the summer off, lounging on the beaches in Vancouver." What a whopper! The next morning Ken and his entire well-equipped trailbuilding crew took off for Carmanah.

Friends of Clayoquot's Sulphur Passage blockade saves the Megin but costs WCWC a director

Although we shifted our focus to Carmanah, all summer long conflict over logging continued in Clayoquot Sound. It focused on B.C. Forest Products' (now Interfor's) building of a logging road along the pristine shoreline of Sulphur Passage. Sulphur Passage is the rocky-

WCWC's Meares Island trailbuilding crews. Left: Ken Lay in the centre with flagging tape tied to the tassel of his toque. Photo: Leo DeGroot. Right: Group portrait of WCWC's second trailbuilding crew by the Meares Island Tribal park sign. Photos: WCWC files.

faced ocean gateway to the northern region of Clayoquot Sound including the Megin River Valley, the largest unlogged coastal watershed remaining on Vancouver Island. The Friends of Clayoquot Sound set up a blockade to stop the construction of this road. It had been four years since the Meares Island blockade.

The Sulphur Passage blockade lasted several months and resulted in the arrest of 36 people including Ahousaht Chief Earl Maquinna George, and well respected Nuu-chah-nulths Joe and Carl Martin. There was little play in the major news media regarding this confrontation, except for a few bizarre incidences like one scene where a guy paraded down the logging road nude, and another scene involving Ron Aspinall "riding" a drilling rig.

In his "direct action" effort to stop a hydraulic drilling rig from puncturing holes in the rock face for the sticks of dynamite that were needed to blast through the roadway, Aspinall climbed onto the rig as it was working and tried to cut the hydraulic hoses with his penknife. Instead of stopping the machine for safety reasons, the operator kept on drilling and Ron rode the rig's boom as it impacted away like a cowboy riding a bucking Brahma bull. It made dramatic TV footage and it cost WCWC a director. Knowing that we had a firm policy not to engage in civil disobedience, Ron stepped down from WCWC's board the next day. But Ron continued to financially support WCWC and played a major role in getting us working on the Ahousaht Wild Side Trail project on Flores Island in Clayoquot Sound in the mid '90s.

There is no doubt that the determination of Ron Aspinall and many other courageous members of the Friends of Clayoquot Sound and Nuu-chah-nulth First Nations saved the Megin and other areas in this remote part of Clayoquot from being clearcut. The blockaders slowed road building to a snail's pace, thwarting the company until it finally backed off. The Sulphur Passage road was never completed.

Songwriter Bob Bossin immortalized this piece of Clayoquot history in his song *Sulphur Passage*. It became an anthem for the huge 1993 Clayoquot protests. Nettie Wild, an award-winning Canadian film-maker, produced a super-emotional music video featuring the in-studio recording of the song by famous singers including Raffi, Valdy, Stephen Fearing, Roy Forbes, Ann Mortifee and others, interspersed with footage of protestors being arrested, clearcuts and loggers chainsawing down trees. The video and song brought many people, including Adriane, to tears.

Although civil disobedience wasn't WCWC's tactic, we were miffed by the media coverage of the conflict at Sulphur Passage. It was so obviously slanted against the protesters. To help pull the general public back onto conservationists' side, we quickly produced a newspaper titled *A time to sustain rather than destroy*. The front page featured an aerial photo of the ugly logging road being constructed along the steeply-sloped Sulphur Passage. The lead article explained from the point of view of the Friends of Clayoquot Sound why their blockade was necessary. The Atleo River, a once prolific salmon-spawning river just to the south of Sulphur Passage provided proof that logging caused severe damage, the fate that the protesters sought to prevent from happening further north in the Sound. The Atleo had been pretty much clearcut from stem to stern right down to the edge of the river a decade earlier. We got this paper out in August of 1988 and distributed it to our members, people living in Clayoquot Sound and to the tourists visiting Tofino.

CARMANAH
Big Trees not Big Stumps

Chapter 21

Carmanah becomes our "flagship" campaign

The perfect poster – *Big tree not Big Stumps*

I knew we needed a spectacular poster of Carmanah's big spruces to boost our campaign. I contacted Adrian Dorst, an outstanding nature photographer and friend of ours in Tofino, and asked him if he would do a shoot in Carmanah. He was keen. All we had to do was rent a medium format camera for him to use there and pay for his travel expenses and film. It was a deal.

It's hard to capture the perfect poster image of a coastal temperate rainforest. It's up to the photographer to select the right composition, but there are other factors beyond the photographer's control. The weather has to be perfect—high overcast cloud cover with a slight misting of rain. If it's sunny there's too much contrast between dark and light areas in the forest. It also must be calm or the vegetation will move in the wind during the long time exposure needed to get sharply focused depth of field. The best time of year is spring when the mosses and new growth are vibrantly green and lush.

Having the right "model" to pose by the big trees—so people can judge their enormous size and height—also helps. As luck would have it, on the same day Adrian arrived to pick up the camera and film, a petite biology student from eastern Canada serendipitously walked into our office. It was her first visit to the west coast.

She had never seen an old-growth rainforest, knew nothing about WCWC and had never heard of Carmanah Valley. But she had a backpack with camping equipment, liked the outdoors and had some free time before she started her summer job. When she heard about Adrian's expedition to Carmanah, she was keen to go. Everything worked out perfectly. Adrian came back with many good images, but we all agreed—one was truly a winner.

We brainstormed the catch phrase CARMANAH – *Big Trees not Big Stumps* as the slogan to accompany this image on the poster. Everyone who traveled to Carmanah reviled the clearcut destruction of the old-growth forest along the way as much as they raved about the old-growth forest in the valley. To put it bluntly, there were way more big stumps along the road to Carmanah than there were big trees in the valley. **The devastation covered thousands and thousands of hectares. The clearcuts were so enormous and so terrible that people used terms like 'nuked' and 'scorched earth' to describe them. Everyone who saw these clearcuts got motivated to save Carmanah, to stop it from becoming another addition to that moonscape.**

Two young Emily Carr art students helped design the *Big Trees not Big Stumps* poster. I wanted a classically clean look, which they delivered. We sent the original transparency to Wolfe at Cleland Kent who had done the outstanding job on prepping the *Stein Joy to the World* image for printing. He used his skills to digitally enhance the shadows and refine the contrast, which magically gave the end results a subtle look of super reality. We printed the largest poster that a colour press in B.C. could handle (26 inch x 39 inch). It became WCWC's most popular poster of all time. Since it first came out in 1988 we've reprinted it four times and sold over 50,000 copies.

The first thing we did when CARMANAH – *Big Trees not Big Stumps* came off the press was send complimentary "educational copies" to every B.C. MLA and all the MPs in Ottawa with a cover letter offering to take them into Carmanah. Conservative MP Bob Wenman

Bob Wenman, Conservative MP and elected Chief Peter Knighton cut the ribbon at the official opening of our trail to Heaven Grove in Carmanah Valley. Photo: WCWC files. Facing page: CARMANAH – *Big Trees not Big Stumps* poster featuring a biology student for scale—WCWC's most popular poster. Poster image: Adrian Dorst.

was the first one to take us up on this offer. Others followed. Long before Carmanah became a park, the Honourable John Fraser, Conservative MP hung a framed copy of this poster in the entryway to his Speaker's Chamber parliamentary office in the House of Commons in Ottawa. Later one even went up on NDP B.C. Forest Minister Andrew Petter's wall in his Victoria legislature office; but it was missing the "Big Trees not Big Stumps" title on the bottom. It had been cut off before the image was framed to render the poster politically correct for his office.

Fantastic Carmanah becomes focal point of conflict

In the summer of 1988, Carmanah Valley became WCWC's "flagship" campaign. Randy had the brilliant idea of making Carmanah look just like a U.S. National Park. "Build it and they'll designate it," he figured. He envisioned well-built cedar boardwalks and handrails and he made trail signs out of thick cedar planks with deeply routered-in lettering that he torch-singed to a nice dark brown.

Our multi-pronged public campaign strategy consisted of taking thousands of 35mm slides to use in different slideshow presentations and newspaper publications. We constantly sought out media to do feature stories and scientists to study the area. We also produced our first campaign video, *Carmanah Forever*.

To build public support, we mustered up many logical arguments as to why Carmanah ought to be saved. Because the valley was less than two percent of the productive forestland in MB's TFL 44, we reasoned that the company could easily afford to give it up. We calculated that MB could make up for the wood volume "lost" in Carmanah by recovering even a small portion of the huge volume of usable, "wasted" wood that it left behind to rot in the clearcuts.

We weren't stupid, however. Better utilization of already-felled trees could replace the "fibre lost" by forgoing Carmanah, support many additional jobs; but it wouldn't make the company nearly as much money. Carmanah's giant trees were worth a fortune and growing more valuable all the time.

Sitka spruce had always been a relatively rare coastal species compared to cedar and hemlock. Its clear, tight-grained wood is especially prized because of its outstanding qualities. It's the strongest wood for its weight in the world, claiming its fame first as wood to build airplanes, then as acoustical wood used to make guitars and piano sounding boards. Now, with less than two percent of the original big spruce trees that once stood on the rich flood planes of B.C.'s coastal rivers remaining, Carmanah was a timber company's gold mine.

Carmanah, we verified through personal experience, is an exceptionally rainy rainforest. On average, the annual rainfall totaled over three metres. But this statistic doesn't tell the whole story. When it comes to the amount of rainfall that dumps in sudden downpours over short periods of time, Carmanah Valley is one of the rainiest spots on earth. When temperature and wind conditions are just right, the thick moisture-saturated clouds are funneled into Juan de Fuca Strait and pinched between Washington State's Olympic Mountains and the mountains along the spine of Vancouver Island creating a whopper of a storm.

Meteorologists predict that, every 50 years or so, over 50 centimetres of rain pour down in one 24-hour period here! A huge sudden downpour in Carmanah once caught Randy Stoltmann, one of the most seasoned of west coast wilderness bushwhackers, by surprise.

One gray afternoon in that spring of '88 Randy pitched his tent high on a sandbar, at least a metre above the creek level. During the night it began to rain, heavily and steadily. By the middle of the night he was floating on his air mattress inside his tent. It was raining harder and harder and the water was rising fast. In the pitch black he moved his soaking wet tent and gear off the sand bar and up onto the alluvial flood plain two metres higher. In the morning, after the rain stopped, Carmanah Creek crested about a half a metre below the flood plain where the big spruces grew. During the next day, after the rain subsided, just as rapidly as it rose, the water level fell.

Debris deposited on the floodplain by the big flood when Carmanah creek overflowed its banks following heavy rains in 1989. Note sign-in kiosk beside Heaven Grove in the background. Photo: WCWC files

Scientists believe that a catastrophic flood about four centuries earlier wiped-clean the mid-valley flood plain enabling Heaven Grove's Sitka spruces to germinate and take root. Periodic deluges thereafter flooded and fertilized the young grove. The rich silts that settled each time the creek breached its banks provided the nutrients that sustained its exuberant growth. A major flood occurred in 1989—about a year after Randy's high-water encounter—when, thank goodness, no one was in the valley. Waters inundated the flood plain and left behind chunks of wood stuck in high in the bushes and around trees, including around our Heaven Grove visitors' kiosk.

As testimony to the richness of the soil, the salmonberry bushes in Carmanah's mid-valley floodplain grow to heights of over four metres with stems over six centimetres in diameter. Randy once fooled me with a round of wood asking, "What kind of 'tree' do you think this came from? It grows in Carmanah. I'll bet any amount of money you can't guess," he chided. I didn't bet, wisely, for I failed to guess it was a salmonberry "tree."

During the first weeks of that summer of 1988, it poured much of the time. This didn't deter our dedicated volunteers' determination to transform the hiking routes that Joe and his crew roughed out into well-built hiking trails. The *Sun's* Mark Hume was the first major reporter to hike down our trail to Heaven Grove. Joe showed him around and then left him to explore on his own. On returning to camp that evening Joe and the trail crew found a couple of candy bars that someone—it had to be Mark—had left behind. Joe said it was the best candy he'd ever eaten!

MB discovers Canada's tallest living tree

A few weeks after MB announced its self-imposed one month moratorium on road building and logging in Carmanah Valley, with fanfare, it announced that it had discovered (on June 10th) a 95-metre Sitka spruce, the tallest known living tree in Canada, growing in Lower Carmanah Valley just outside Pacific Rim National Park's West Coast Trail boundary. The company nicknamed it *The Carmanah Giant*. MB's foresters discovered this tree during a helicopter survey and dropped a chain down beside it to accurately measure its height. This news took us by surprise. We had no idea that MB was out looking for record height trees!

It was incredibly good news. "Who's side are they working for?" said one of our volunteers, knowing that this discovery would immensely help our campaign. The Carmanah Giant made big headlines and put Carmanah more firmly on the map and into public consciousness. It also strengthened our claim that the entire valley should be protected as a park. We immediately decided to push a trail down to the "Giant" abandoning, for the time being, extending our trail into the upper end of the valley. The steep terrain between the mid valley and The Giant made it the toughest trail project we'd undertaken to date.

MB did not resume road building when its self-imposed moratorium ran out. At the end of June, three weeks after they announced the discovery of the Carmanah Giant, MB proposed to create two reserves to protect Carmanah's big trees: a 90-hectare park around Heaven Grove and a nine-hectare park around The Carmanah Giant. MB also offered to build access roads to them and parking lots beside them as they logged. It envisioned people driving down to these little parks, parking their vehicles and strolling for a few minutes to see the marvelous trees. Few people bought into MB's vision. Parks Canada did not want a road that took people to within a couple hundred metres of the middle of their world famous West Coast Trail because people might sneak onto this trail from there. We argued strongly that these parks would only be temporary tree museums. After the surrounding forest was clearcut the trees in them would soon blow down. The proposed parks were far too tiny to protect the trees left standing in them, let alone sustain a dynamic Sitka spruce ecosystem over time.

The making of *Carmanah Forever* – our first video

Our response to MB's token parks proposal was to redouble our efforts to build public support for a whole-valley park. We needed a big show of support and figured the Canada Day long weekend was a prime opportunity. We started hyping a "Caravan to Carmanah," putting out flyers, letters to members and a media release inviting people to come and celebrate July 1st with us in Carmanah. Our volunteers worked long hours putting the finishing touches on the trail, including stairs in the steep spots, to make the hike to Heaven Grove

**Left top and bottom: Giant staircase using a fortuitously situated fallen cedar under construction on the trail down to the Carmanah Giant.
Above: Finished staircase, August 1988. Photos: WCWC files.**

as safe as possible.

Most of the WCWC full-timers headed into Carmanah the day before everyone was due to arrive. We traveled in with Susan Underwood, a talented videographer who had volunteered to produce for us our first-ever campaign video we had already titled *Carmanah Forever*. Several kilometres before Hell Camp we met our first big challenge: a new locked gate across the road. On the other side of it MB loggers had left a log loader and dumped a pile of logs in the middle of the road. Several days before, MB representatives had handed our trail crew a letter asking them to "cease and desist" trailbuilding in Carmanah. It wasn't an injunction or court order, so we didn't comply. The MB loggers were now obviously angry. There was no doubt that they wanted to thwart our Canada Day event. When we reached the gate, our volunteer trail crew stationed in the valley was standing by it waiting to greet us.

Our videographer was delighted. It was ideal footage for the video! We used the scene to "stage" Clinton Web handing me MB's "cease and desist" letter which I then read out loud, pretending that this was the first I heard of it. We taped the scene several times. The acting, especially mine, got worse with each successive take. I'm definitely a spontaneous kind of guy, not an actor. But the message I conveyed was clear. We were not going to voluntarily stop building trails at MB's request!

Carmanah Giant, September 1, 1988. Photo: Leo DeGroot.

Adriane Carr being interviewed by the poster spruces in Heaven Grove for *Carmanah Forever* video Photo: Gary Fiegehen.

Thank goodness, the weather that weekend was perfect. Over 150 people came, hiked in past the locked gate, camped beside the creek and celebrated Canada Day in Carmanah. They were in awe. A major part of the weekend was touring people around the spectacular spruce groves and valley bottom. We wanted to make sure they had experience enough of this magical valley to become fully committed to helping us win this campaign.

We also spent a lot of time shooting footage for our *Carmanah Forever* video. Kallie and TJ, along with some other kids, played on the sandbar as we made up the scenes and videoed them. In one, Randy Stoltmann waded across Carmanah Creek unrolling a tape

Susan Underwood sets up to shoot Clinton Webb in clearcut near Carmanah. Photo: Gary Fiegehen.

The crew in Heaven Camp. From left to right – unknown, George Yearsley, Adriane Carr with TJ, me with Kallie, Sylvia Algire and Joe Foy behind. Photo: Gary Fiegehen.

to measure the horizontal distance to a large, tall tree on the other side as I tried to use his clinometer to measure the angle from the ground to the tree's top to calculate its height. Then, all of a sudden, Joe popped out of the bush shouting out that he had just discovered another "really big" tree. Decidedly, none of us were actors. But the video, thanks to Susan and to David Suzuki who narrated it for free, turned out to be great.

Later that year, MB made its own slick half-hour video to counter ours. Together our two videos were shown in B.C. high schools for years to illustrate a "typical B.C. environmental conflict." We heard that many teachers let their students debate the merits of each position and we invariably came out on top.

Parkifying Carmanah

As the summer progressed, Carmanah began to look more and more like a real park. We built side trails to remarkable trees and other natural "points of interest." We put up more and more of Randy's U.S. National Park-like signs along the trails. We also built a visitor's kiosk at the foot of our main trail down to Heaven Grove, with a sign-in guest book, a map of the existing trails and locations of outhouses and a sign asking people *to be careful, tread the wild forest gently and leave no garbage behind.*

One of the outstanding features we discovered as we explored for trail routes was a "fallen giant." You could walk on top of this massive dead tree's trunk for more than 70 metres. At the rootwad end you stood more than three metres above ground. We built another side trail to a huge hollow cedar where bears had obviously denned during winters past. Children loved to climb inside and imagine what it would be like to be a hibernating bear. Another side-trail led to the "three sisters;" three giant Sitka spruces that had grown so close together that you could lie on the ground amongst them, press your feet against one, spread your arms wide and touch the other two with your hands. On one very windy day I spent several hours lying there looking up over 20 stories high at the waving branches and feeling the energy of the wind in the swaying massive tree trunks. The three trees worked together contrapuntally. The branches randomly

Kallie George inspecting a clump of moss in the Carmanah Valley. Photo: Joe Foy.

"Official" signs by Randy Stoltmann made the Carmanah park a virtual reality. Photo: WCWC files.

bobbed and weaved; the trunks danced and swayed, perfectly countering each other's harmonic wave energy. They obviously had been doing this for centuries, diffusing the fierce winds' energies preventing storms from toppling them over.

Randy found one giant Sitka that had a breast high circumference greater than any of the other trees. He thought that it might be the oldest living spruce in the valley until he looked up and discovered it was a snag (a standing dead tree). He named it the "Dave Parker Tree" after the Minister of Forests (placing beside it an "official" park sign), because it was "dead on the stump" to commemorate Parker's statement that the trees in Carmanah were "over-mature" and overdue for logging.

In stark contrast to Minister Dave Parker, was Conservative MP Bob Wenman. He was so inspired by his Carmanah visit that summer that he presented a special motion in the House of Commons to have the entire Carmanah Valley added to Pacific Rim National Park. It went down to defeat, but it did give our cause some more publicity and added credibility.

Chapter 22

The Carmanah court case is a win but the war goes on to save the valley

MB takes WCWC to court to try and stop our trailbuilding in Carmanah Valley

At WCWC we lived by an unwritten maxim: if a tactic is legal, affordable and had a chance of saving wilderness, we'll use it. Nothing was too audacious for us to try. And nothing was illegal unless a law or court decision made it so. There was no law saying we couldn't build trails on Crown Land. We reasoned that as long as we didn't damage the forest or wildlife, we had the legal right to build our trails on public lands.

In the third week of July 1988, MB served us with a Notice of Motion that it was seeking a court injunction to stop our trailbuilding activities in Carmanah Valley. Luckily, Don Rosenbloom was available and willing to represent us—again at a charitably reduced rate. This time the legal issue was not clouded by already-in-place injunctions as it was on Meares Island and he believed we had a good chance of winning. We had to move fast once we got MB's statement of facts and affidavits to gather and present our counter evidence. There were just a few days before the hearing. Don had already done a great deal of background research for the similar Meares case.

The Vancouver Sun, Friday, July 22, 1988 ★★★

MB seeks to halt building of trails

By MARK HUME

MacMillan Bloedel Ltd. was to appear in B.C. Supreme Court today seeking a court order to halt trailbuilding in Carmanah Creek Valley.

In documents supporting a notice of motion, the logging company accuses environmentalists of building unsafe trails, of damaging valuable timber and of spiking trees in a "life-threatening" manner.

Paul George, a director of the Western Canada Wilderness Committee, said the charges are groundless.

"MacMillan Bloedel really doesn't want people to see those trees," said George.

In recent months committee members have been constructing an extensive trail network in the Vancouver Island valley, site of some of the tallest Sitka spruce in the world.

Environmentalists want the entire 7,000-hectare valley preserved while MacMillan Bloedel has proposed saving two reserves of 90 and nine hectares.

While MacMillan Bloedel is moving against the trails in Carmanah, the committee is pleading with the company on another front for permission to improve trails recently built on Meares Island.

In addition to asking for an order restraining the defendants from building any trails in Carmanah, MacMillan Bloedel wants the court to order the committee to remove trails already built.

In its notice the logging company asks that it be at liberty to close trails "for the protection of the public."

Trail system extended

In an affidavit, MacMillan Bloedel employee-relations manager William Ohs states that a trail from Camp Hell to Camp Heaven "is very steep in places and is in my view dangerous to hike."

The two camps were named by the trail builders. Camp Hell was located at the trailhead, on a MacMillan Bloedel logging road. Camp Heaven was a few kilometres away, down a steep hillside, in a grove of giant trees next to Carmanah Creek.

Press clipping of *Vancouver Sun* article July 22, 1988. Clipping: WCWC files.

Contingency plan in place in case of court loss

Not being quite as confident as Don of our chances in court, we developed a "Plan B." It was Ken Lay's idea. He would stake the area around the Carmanah Giant as a mineral claim. As a "Free Miner" the law allowed him to build an access route to "work" his claim. So he and his helpers could build a trail, even if the courts said that MB could stop WCWC's activities. Mineral rights are king in B.C. and TFL rights cannot trump them.

This whole "mining project" was kept under wraps. The secret expedition to stake the mineral claim and the court proceedings both started on a Friday morning, July 22, 1988. It was summer break time at the Supreme Court of B.C. and many of the justices were on holidays. The few who remained were stretched thin. There were several shorter cases that came up before ours. I waited inside the courtroom and observed these proceedings. I didn't like the justice's demeanor. He was old and quite curt to the young lawyers arguing these relatively minor cases. It didn't look good for us.

The first Carmanah courtroom skirmish

Rosenbloom didn't say anything to me about the justice assigned to us by luck of the draw. Lawyers never discuss with clients the merits of the justice who is hearing their case—it's a form of contempt of court; but the truth is, as in any profession, some justices are better than others. More to the point, there are liberal and conservative justices. I could tell after an hour of listening to this justice's handling of other cases, that he was a super conservative one. I figured his sympathies would definitely lie with corporate interests. Finally, in mid afternoon that Friday our case got started. From Justice Scarth's questioning, it was obvious that he sought to finish our case that day. Rosenbloom outlined the case's complexity and explained that it was impossible to get through everything in that short a time. It would take at least two days, Rosenbloom politely insisted.

If Justice Scarth started hearing the case, he would have become "seized of the matter," which meant that he had to complete the case—and that meant losing part of his holiday, which was scheduled to begin on Monday. Wisely, he ordered that WCWC be restrained from any trailbuilding activities or *otherwise altering the area of Carmanah Creek watershed falling within Tree Farm Licence No. 44...until the matter was settled by the courts.* He put the matter over until 10 a.m. the following Monday when another justice would be assigned to hear the case. Although Rosenbloom never said a thing to me, I could tell by his change in mood that he was happy with this turn of events. Our luck had improved.

We then discussed the difficulty of informing our crew to immediately stop all activities because of the distance involved and the lack of direct communication. It wasn't a problem: MB did it for us. The ink was barely dry on the justice's order when, late that afternoon, MB helicoptered it into Carmanah Valley and served photocopies of the temporary injunction on our crew.

Joe was with Ken that day helping Ken stake a mineral claim around the Carmanah Giant. It was the first time that Joe had hiked all the way down to the Giant. Half way there they had to divert up and around a deep canyon that Carmanah Creek had carved over the ages through the limestone. From the edge, Joe peered down into it and watched an eagle soar by below him. He noted a logjam—"a brillo pad of woody debris" wedged high in the canyon's walls about 30 metres above the current water level. It was mute testimony that, in the not-so-distant past, an incredibly huge flood of water had raged through there.

Near the end of the day, after they had completed the staking and were on their way back towards Heaven Grove camp, they heard MB's helicopter. When it came close he and Ken scrambled to hide under logs. It circled several times and each time they successfully hid. They never were served, but other members of the crew were and everyone, of course, including Ken and Joe stopped trailbuilding.

Years later, one MB employee described the situation like this: "After a while it started to feel like Vietnam. We had all these helicopters and money and you guys were running around down there in your running shoes; and you were beating us."

A grand courtroom battle

On Monday morning I sat anxiously in the court's public gallery watching Don Rosenbloom present our case. Assigned to hear the case was a younger man, Justice Wood, who definitely seemed more interested and engaged. MB's affidavits stated that clauses in its TFL and Five-Year Management and Working Plan required MB, as the licensee, to include a recreation component and manage recreation on its TFL. MB argued that this requirement thereby gave them exclusive management rights.

Another one of MB's affidavits asserted that we were damaging their forest and implied that we might even be spiking trees. Bill Ohs included in the exhibits attached to his affidavit pictures of a handful of spikes lying beside the trail and a close-up photo of a two-centimetre-in-diameter stump of a tiny hemlock tree that a volunteer had loppered off. He also swore, that he had also heard a chainsaw running deep in the forest that must have been ours (inferring, ominously, that we were up to no good).

We countered all MB's affidavits with our own. We explained that the spikes were for a stairway down a steep portion of the trail. In fact, we were on public record for opposing tree spiking and had even offered a $5,000 reward for information leading to an arrest and conviction of anyone doing this illegal and dangerous act. We showed that our trail had minimal impact on the land and that it certainly did not harm or compromise the commercial value of "MB's forest." Joe swore our affidavit countering the MB's claim that we were damaging the forest. He included some large full-colour photos of huge MB clearcuts near Carmanah. Some of them had recently been slash burned. They were black and ugly, as if the areas had been hit by atomic bombs. I'll never forget the look on the justice's face as he looked at these photos. MB's giant clearcuts and piles of "waste" wood were a stark contrast to the insignificant impact our tiny trail through the old-growth forest made.

Rosenbloom argued persuasively that Tree Farm Licence agreements did not convey all rights over Crown Lands, but only the exclusive right to harvest the trees on those lands. MB did not have the exclusive right to manage everything else, including recreational use of the land, and did not have the right to decide who could or couldn't enter onto its TFL lands. If MB didn't have the right to limit the use of Crown Lands for other purposes, it didn't have the right to curtail our trailbuilding. Near the end of the hearing MB's lawyer tried a new ploy. He made a strong pitch that our trail was causing "irreparable harm" to MB because, by providing access for people to enter into the area, it was helping us build our case for Carmanah becoming a park. We might succeed and persuade the government to delete the area from MB's TFL. If that occurred, MB would lose the right to cut the trees and all the money it would make doing it and WCWC would not have enough money to compensate the company for its loss.

The Honourable Justice Wood did not buy this argument one bit. From the Bench he countered by stating that, if indeed the area were protected (which would be a political decision), the landlord (the B.C. government, not WCWC) would be obligated to pay adequate compensation for the loss of MB's cutting rights.

On July 26, 1988, in a written judgement in this case (No. C883767) Justice Wood dismissed MB's injunction application. He emphatically concluded that MB did not have exclusive possession of the Tree Farm Licence area, the foundation on which MB's case rested. In his eight-page written decision, Wood eloquently covered all the objections MB had made to our presence in Carmanah, including the one that our trail was dangerous and increased MB's liability because of the increased chances that someone could get hurt, and that we were cutting up valuable windfalls in clearing the way for our trail. He even stated that he believed the spikes that MB found were indeed being used to secure railings and anchor stairs and did not present a grave risk to MB's employees who might someday have to cut the trees or mill the wood. He said (WCWC) ...*appear to have adopted a responsible and careful approach to building the trail in question.*

The judgement ended on a sweet note. Since we were not interfering with MB enjoying its right (we were not blockading or trying to prevent logging directly), under the circumstances we were entitled to costs. Of course, these costs were only a fraction of the total legal costs; but it helped financially and signalled that we were totally vindicated. Even with MB paying its court ordered share, this court case cost us more than the entire trailbuilding work in Carmanah had cost us to date.

Rosenbloom had done a brilliant job against the big corporate hotshot lawyer. Best of all, we had set a precedent. Our case established the public's right to freely access TFL lands that encompassed nearly 30 percent of B.C. People could enjoy these lands in all ways except those that interfered directly with the licensee's logging operations. It was a victory for all B.C. citizens.

I was extremely happy. I'd never believed in the granting of TFL licences that B.C.'s citizens had got a fair deal. MB had obtained its TFL in Carmanah from the B.C. government 33 years earlier for a pittance and the company's promise to build a mill and provide steady local manufacturing employment. It was a lucrative deal for MB. Most of the security and wealth flowed to the company and little flowed to the people of B.C. as they closed mills and gradually exported more and more raw logs.

On the same day this court decision came down, the B.C. Forest Service asked MB to prepare a revised logging plan for Carmanah Valley and submit it by the end of September. This gave us some breathing room. Ken Lay never did register his mining claim. We redoubled our efforts to build a good safe trail to the Carmanah Giant. The route was steep and difficult and much further than it was from the logging road to the mid-valley. It would never be a trail for mom, dad and the kids to hike down and back up in a day. It was just too rugged. Although Parks Canada was not too keen on having a possible access point so close to the middle of the West Coast Trail, we figured they should be happy because our trail was a far better alternative to MB's proposed road.

The secret war over a locked gate

MB lost their court bid to stop our trailbuilding, but their employees continued to harass us. It was more than irksome to have them lock the access gate at night and on the weekends when MB employees weren't working. It was hazardous. We worried about our vehicles being locked inside and what we would do if someone got hurt and we had no way to get them out. Our complaints to MB got

Joe Foy holds large lock (now a paperweight memento on his desk at the Wilderness Committee). We cut it with our giant bolt cutters to open an illegally locked MB gate on a public logging road into Carmanah Valley during our Carmanah campaign in 1988. Photo: JP LeFrank.

us nowhere. We finally figured out that its workers' gate-locking ploy must be illegal. Even though MB built the logging road it actually was a public road paid for through a reduction in stumpage (the tax the company had to pay to the province for the trees the company cut) to reimburse the company for what it spent to build the road. So we got a bolt cutter and cut the lock off. The next week there was a bigger lock. Our cutter was too small to sever it.

"Somewhere, there must be a bigger bolt cutter," I surmised. Sure enough, one of our volunteers found a huge one that cost several hundred dollars. We bought it and the bigger lock was toast. So was next week's huge new lock. By the fourth week MB had redesigned the latch on the gate so that the lock was recessed inside a short thick steel pipe so bolt cutters couldn't get at it. We couldn't lose this fight. There was only one thing left to do.

"If there's a giant bolt cutter...there's got to be a giant pipe cutter, too," I told Joe. Sure enough, there was. We bought one and the next weekend our incoming crew cut and cut around the base of the upright pipes supporting the gate until the whole assembly fell over. It took the strength of our entire trail crew to pick the entire works and toss it over the embankment. This was the end of the lock wars. That gate was not replaced. As far as I know, no complaints were ever made to the RCMP regarding the removal of the locks or gate. But neither had we complained to the RCMP about MB's dangerous tactic of locking us in.

Our Carmanah Giant poster fiasco

Not everything in the Carmanah campaign went as smoothly as the court case and our trailbuilding work. The day that MB announced its discovery of Canada's tallest tree Nik Cuff called me. "You have to make a poster of that tree and get it out right away," he urged. "It'll become the icon that saves the whole valley."

"Great idea," I said, "but it's a really hard thing to photograph." This didn't deter Nik at all.

"We'll rent a helicopter," he replied, explaining that since the tree was growing next to the creek, we could land on a sandbar and a photographer could shoot a series of overlapping pictures as the helicopter slowly ascended beside the tree. We'd use a medium format camera for image clarity and pose Randy Stoltmann, who had blown the whistle on the logging in Carmanah, beside the tree. Wolfe at Cleland Kent, who knit together the Stein photos for the "Joy to the World" poster, would splice all the overlapping images together and we would have a spectacular, realistic full-length portrait of Canada's tallest tree. "It'll be the most phenomenal portrait of a tree in the world!" asserted Nik.

I told Nik that, despite how great a project it was, we just couldn't afford it. All our money was going into trailbuilding. Nik had a solution for that, too. "I'll pay for the helicopter and hire the photographer and we'll split the cost of the scanning. Then you pay for the poster printing. It really won't cost any more than a regular poster."

"We'll do it," I said, knowing everyone would love the project.

Nik arranged everything for the Carmanah Giant photo shoot. Everything worked as planned except, according to Nik, our photographer (Adrian Dorst) who was extremely afraid that he was going to fall out of the open door of the helicopter even though he was strapped in with a belt. A week later we got the medium format transparencies back. They were sharply in focus. However the tree didn't look anything like what I had imagined Canada's tallest tree to look like. It wasn't majestic. Instead of a tall towering trunk, there were big bushy limbs sprouting all the way up from the base of the tree to its top. Growing with one side facing the open creek, the lower limbs were never shaded and never dropped off. "What a goofy-looking tree," I thought. "But undoubtedly it will look better after Wolfe knits them together and works his magic."

Wolfe started to work on this project right away. There were a dozen photos that had to be digitally pieced together to make the tree whole. About a week later I got a call from Nik. We had a problem. Not all the photos overlapped. There was a segment of the tree missing. But, not to worry, Wolf would handle it like he had with the Stein Valley picture—do some cloning—and make it fit together. "You'll hardly be able to notice the difference," was the way Nik put it. And it wouldn't cost that much...only a few hundred dollars more. I was becoming increasingly sceptical but said, "Sure." We were too far into the project to quit now.

It's a huge understatement to say our full-sized proof of the Carmanah Giant image was a bust. When I unrolled the half-sized colour digital proof in our office everyone looked at it for a very long time sort of stunned. We all agreed: it looked like a giant shrub. A strange indentation in the trunk about a third of the way up marked the place where the missing photo had been digitally replaced. But there was something else wrong, too, something that I couldn't immediately put my finger on. I stood far back and looked at it for a long time. Then it came to me, "You know, the tree doesn't really look that tall, either." I mused to myself. I knew that Randy was exactly six feet tall. The tree was 312 feet tall. There should be exactly 52 Randy heights to its top. I took a ruler and measured Randy's height on the photo, then multiplied that height up the tree. For some unknown reason, our Carmanah Giant was only 42 Randy-heights-tall–256 feet–nowhere near the record height it was supposed to be. If I noticed this, other people would, too.

I called Nik and gave him the bad news. He talked to Wolfe and got back to me right away. It would be no problem! All Wolfe would have to do is to digitally cut Randy out of the picture, reduce him to the appropriate proportional size and clone him back in again. "The technology can do it perfectly," he said, adding, "It will only cost a few hundred dollars more." My scepticism was palpable.

A few days later the new revised proof came. The Carmanah Giant still looked like a shrub...just a bigger one. But now Carmanah Creek no longer looked like a creek...it looked like a mighty river. That was it. We scrubbed the poster. I moaned about WCWC spending over $4,000 on it. But Nik had spent a whole lot more—exactly how much more, he would never tell me. For a while we had the Carmanah

Giant poster proof taped up on the office wall, hoping we'd come to find it funny. But soon we took it down. None of us liked to be constantly reminded of such a costly failure.

More modest poster captures Carmanah's essence

In place of publishing the Carmanah Giant poster, we published a smaller poster using a 35mm slide taken by Gary Fiegehen, a professional photographer who had accompanied us on the weekend we shot the Carmanah video. It featured Joe Foy showing our five-year-old daughter Kallie a tiny fern growing on the forest floor with the towering Sitkas in Heaven Grove in the background. On the bottom we put simply *Ancient Forests Forever*. It is one of my all time favourite posters—for obvious reasons—although it was not nearly as popular as our *Big Trees not Big Stumps* poster.

Trail to Carmanah Giant completed

By the end of August '88 our large crew of volunteers finished the trail to the Carmanah Giant. On September 10, MP Bob Wenman and Peter Knighton, the elected chief of Oitadat, the First Nation whose main village is located nearby Nitinat Lake, simultaneously cut the ribbon officially opening the trail from Heaven Grove to the Carmanah Giant. About a hundred people came to witness the ceremony and celebrate this milestone. Shortly after this official trail opening, Arne Hansen arranged to have Joe Foy take Peter Knighton's mother on our trail into Carmanah. In her 80s, Susan Knighton had been born in Carmanah and she wanted to see it again before she died. As Joe began helping her walk down the steep trail to Heaven Grove, he realized this was a big, big mistake. She was frail and could never hike out of there.

It was a mistake in another way, too. Mrs. Knighton thought she was going to the beach at the mouth of Carmanah Creek where she was born and lived as a small child. When she arrived in the mid-valley she was a bit confused and disappointed. It was a place she had never seen before. Joe worried the whole time about how he would get her back out. Fortunately, Ken Lay who was camped in Heaven Grove came up with the idea of constructing a "sultan's chair" out of alder saplings with a seat suspended between two poles to carry her up the trail. The problem was that only two people could lift the chair—one at each end—because the trail was so narrow. Joe and an extremely strong native youth who accompanied Susan Knighton did most of the lifting. It took over three hours to carry her out. "It was the hardest thing I ever did," Joe told me years later. "The only way I got over the tough spots was to curse to myself under my breath. I swear that I lost a half an inch in height that day. She wasn't the only one really happy to get back!"

By now many people were using our trail to Heaven Grove. That same fall an elderly man died on it. He had a heart condition and his exertion triggered a massive heart attack. We were afraid that we would get some blame for it. But nothing of the sort happened. His widow was extremely kind. "That's the way he wanted to go, out in the nature he loved so much."

Fearlessly taking Carmanah slideshow into the heart of "enemy" territory

As the 1988 trailbuilding season came to an end, we shifted into our favourite fall activity: public events and slideshows. We had thousands of slides from the summer's trail work. Joe and Clinton Webb

Left: Randy Stoltmann standing beside the "Carmanah Giant," Canada's tallest tree. WCWC decided not to publish a poster featuring this image. Colour key: WCWC files.

> **LANGLEY TIMES, SATURDAY, SEPTEMBER 17, 1988 •19A**
>
> ## Wenman opens new wilderness trail
>
> VANCOUVER ISLAND – The Carmanah Creek Giant Spruce now are accessible from the Pacific Rim National Park's West Coast Trail.
>
> Robert Wenman, M.P. for Fraser Valley West, opened the recently completed Carmanah Creek Trail on September 10. The trail, constructed by members of the Western Canada Wilderness Committee throughout the spring and summer, runs 12 kilometres up the Carmanah Valley where it terminates at the base of Canada's tallest trees, the 312 foot high 'Carmanah Giants.'
>
> Wenman, accompanied by representatives from Langley and Burnaby, hiked the 12 kilometre trail and in a spirited ceremony cut the ribbon spanning the trail entrance.
>
> "This is my second visit to the Carmanah Giants and I'm just as impressed the second time," said Wenman.

Press clipping—report on opening of the new trail to the Carmanah Giant. Clipping: WCWC files.

put together a spectacular slideshow focused on why the whole of Carmanah Valley had to be protected and took it on the road. People loved it, especially the photos of Carmanah's big trees.

Our shows that fall were also motivated by the need to generate overwhelming public opposition to MB's newly revised Carmanah logging plans. Presented for input in October, MB's plans included its proposal to increase the areas to be protected to 175 hectares from the 99 hectares it initially proposed. This still only amounted to a mere two percent of the whole valley. Our slideshows fired up the public. The people who sent in letters about MB's logging plans overwhelmingly rejected MB's proposal as pure tokenism.

There was one show that likely didn't generate many anti-MB letters—the one in Port Alberni, the mill town where the wood from Carmanah (if it were to be logged) would go. Port Alberni was the home of most of the forest workers whose jobs were linked to logging TFL 44. Many environmentalists thought Joe and Clinton were pretty brazen to take our slideshow into the heart of "enemy" territory.

Over 200 people showed up at the Port Alberni Recreation Centre to see Joe and Clinton's show. A couple members of the Friends of Clayoquot Sound also came and they sat right in the front. The rest of the room was packed with angry loggers. There were some catcalls and heckling but most of the guys, to their credit, wanted to hear what our two campaigners had to say and they kept the rowdiest guys under control. I believe that just by having the courage to go to Port Alberni, we gained a lot of respect. It's also a lot better to talk face to face, than to communicate through the media, where myths too often drive opinion. I prided myself on WCWC being grounded in solid facts and accurate knowledge of forestry.

One of the biggest myths was that a colossal number of jobs would be lost if the whole of Carmanah was made a park. Many more B.C. jobs were being lost in technological upgrades to the mills and under-utilization of wood. I always felt that companies used the "job loss myth" as a way to shift attention away from their own excessive profit-making and labour-unfriendly policies. My hope was that if the loggers would find out that most "tree-huggers" were not against all logging and were actually pretty reasonable in wanting to see a shift to intensive eco-forestry practices that would ensure there will be an industry in the future, this would soften their opposition to us.

WCWC's first artists' project – the Stein – gets underway during the height of "Carmanah fever"

At the end of August 1988, in the midst of Carmanah trailbuilding and calendar production, we also launched our first collaborative project with artists to save a wilderness area. Arne Hansen got us into this through knowing Toni Onley, a well-known west coast artist who flew his own floatplane. Their idea was to have WCWC host artists on expeditions into the Stein Valley and then sell the art produced on the expeditions through a series of art show auctions. It would raise awareness of the Stein issue in new circles and money at the same time for our Stein campaign and for the recently started Stein Rediscovery summer camp that brought youth and native elders together in this wilderness.

Toni and Arne lined up artists to participate. Toni had often landed on Stein Lake in the remote Upper Stein Valley and said it was a perfect place to bring in the artists for a day's inspiration and painting. While a few of the artists went up to the Stein by car and van, hiked in the lower valley to find inspiration for art pieces, most flew up by chartered floatplanes into Stein Lake.

An aborted take-off creates a big scare

Everything went according to plans except for one incident. One evening when Adriane and I were working late at our West 6th office an upset woman called in. She was extremely worried about her husband. He was supposed to have returned home several hours earlier from the artists' floatplane trip to Stein Lake. It was already dark and she hadn't heard from him. A few minutes later we got another call from a worried partner of another artist on the expedition. I called Whistler Air and they informed me that their floatplane had not returned and that they had not been able to make radio contact with it. There was nothing their small airline company, or we, could do but wait until the morning. I knew that landing and takeoff from Stein Lake, situated in a small bowl in the upper valley alpine surrounded by mountain peaks, was a tight squeeze. Although the weather had been good all day in Vancouver, the weather conditions in the Upper Stein could have been entirely different.

It was already past 10 p.m. Adriane and I feared the worst. We were deep in conversation about who we should call next when the phone rang. "Hello, I'm the Captain on Air Canada Flight 689. I just talked to your people in the Stein and they are fine. They asked me to relay this message to you, knowing you'd be worrying about

Three sister Sitka spruces looking upward used on a T-shirt. Artwork: WCWC files.

Artist Victor Doray and his work produced for our art auction on location near Stein Lake. Photo: Arne Hansen.

them." He went on to explain that he was calling from the cockpit of a commercial jet descending to land at Vancouver Airport. He had just flown over Stein Lake and our pilot had made radio contact with him. Our pilot had told him they had taken off from the lake in the early evening but quickly circled back and landed because clouds had rolled in and completely obscured his visibility. He would fly out in the morning when the clouds lifted. I immediately called everyone who had called us and told them the good news. Boy, were they happy to hear it...and so were we!

However, a rough night lay ahead for those stranded at Stein Lake. We had a base camp already set up there. Our crew, including Ken Lay and Arne Hansen, gave up their sleeping bags and tent to the artists and a TV crew. Then the Whistler Air pilot, Ken and Arne moved the plane to a safe place along the lake. During the long night they had to scrape off the heavy snow that kept accumulating on the airplane wings to keep them from collapsing. Wrapped only in an old oiled canvas tarp, they managed not to freeze as they tried to sleep on the floor of the plane.

Ken Lay swears that it was the coldest night he'd ever spent out. The plane's aluminum body conducted the heat out of him a lot faster than his body could generate it. He soon gave up and waited for dawn with several others around the fire.

Here is an account of this hair-graying experience written by Cecile Helton, a nurse and one of WCWC's most dedicated volunteers.

I accompanied the CBC English and French TV crews to Stein Lake. They were going there to interview the artists who had been there already for one week. Connie Monk was the reporter. We met at the Vancouver South Terminal early in the morning. We were nine people in all—the two crews with cameras, tripods and other paraphernalia, the pilot with his BIG German shepherd dog, and myself.

We boarded a Whistler Air Beaver floatplane. The weather was somewhat cloudy. We left and gained altitude. Clouds covered the mountains; fog blanketed the valleys. After a while the pilot announced that we would land on another low elevation lake to wait for the clouds to lift. After some time, we saw blue sky and took off again, only to land on another lake to wait again. That is when the pilot announced that we had to fly to Whistler to refuel, which we did and waited there some more. Finally, the clouds lifted enough and we took off, headed over the mountains and descended onto Stein Lake.

The reporters and crew got to work. By then we had only about two hours before leaving again for a safe daylight return flight. What a beautiful sight, this alpine lake in the heart of the Stein Watershed. It was decided that the pilot would take back the CBC crews and some artists and we would await the return flight.

They boarded the floatplane. We watched as they took off towards the far end of the lake where a heavy fog was rolling in. The sky was gray. We watched as they lifted, turned, circled and came down and landed back on the lake. Then we could not see the top of the mountains. The fog rolled in. It got colder and started to snow. We huddled all evening, standing around the campfire. Our food was a pot of pea soup and tofu type stew in mugs and some bread. We passed around a flask of brandy that someone had thoughtfully brought along.

It was decided who should sleep where. Tents were scarce as the plan had been to fly in, do the report, and leave again. Some of the TV crew had come in with only jean jackets, jeans and runners—the style at the time. We went to bed early, trying to keep warm. In the meantime, the pilot tried unsuccessfully to make contact with the outside world. The pilot, Ken Lay and Arne Hansen slept poorly in the cold metal plane rocking on the waves of the lake. Next morning we heard that the pilot had managed to make contact but with a commercial pilot flying high overhead.

We awoke to blue skies, bright sunshine, crisp air, and snow on the ground. Hurriedly, we packed to send the first planeload on its way. That one-day in-and-out flight turned into an unforgettable adventure. Only later did we hear of the commotion we caused back in Vancouver.

It took four months to take our artists' project from the planning stage to the public art shows held in Vancouver's Robson Square Media Centre on November 15-18 and later in Kelowna and Lillooet. The silent auction of the Stein-inspired donated art grossed nearly $18,000. The expenses ended up being about $13,000, with most of it spent on air transportation and framing the art. We gave Stein Rediscovery a cheque for $3,000 and counted our first artists' project as both an artistic and a financial success.

Burns Bog another important issue we partnered with a local environmental group to help protect

WCWC was always busy on multiple campaign fronts. I liked the fact that we had become big enough to not only lead our own campaigns, but to help other groups on theirs too. During the summer of '88 we teamed up with a newly formed group, the Burns Bog Conservation Society, to co-produce a newspaper about saving *the lungs of the Lower Mainland*—a big bog of international ecological and recreational value located in Delta. I was impressed by the determination of the group's leader, Eliza Olson. **A wilderness preservation issue always has a good chance when a group forms around a clear vision of the protection it wants, has a strong dedicated leader and never stops campaigning until they've won.**

The Social Credit government's idea at that time to build a "super port" in the Fraser River next to Burns Bog and industrially develop the bog lands was a crazy one and we wanted to help stop it. Our four-page, coloured newspaper titled *Burns Bog mega-project would put farmland and Fraser fisheries at risk* was not our last publication in support of protecting this uniquely domed bog in the Fraser River delta. The Burns Bog campaign went on and on for years and years, even though poll after poll showed that the vast majority of people in the Lower Mainland wanted this bog protected. The big problem was that it was privately owned and the owners wanted a king's ransom for it. To Eliza's credit, she never gave up. Sixteen years later, in 2004, her group's sustained efforts got one of the most anti-environmental governments in B.C.'s history (the Gordon Campbell Liberals) to contribute the money to purchase most of Burns Bog. Today, the Burns Bog Conservation Society campaigns to protect the rest.

Sandhill cranes nesting in Burns Bog. Photo: WCWC files.

Chapter 23

Expanding activities
Going national; going retail; going WILD

Launch of WCWC's Canada Calendar

During this same summer of '88 we decided to entice people to join and renew their membership in WCWC by giving, as a membership "perk," a complimentary copy of our Western Canada Endangered Wilderness Calendar. This gave us a much larger print run and a lower cost per calendar. Eyeing the big Canadian market "back east," we also decided to branch out nationally and publish a 1989 Canadian Endangered Wilderness Calendar. We were dreaming big again with calendar sales. It was justified. We'd been gaining more and more members and donors in Ontario, especially through our Carmanah campaign. Several "easterners" had told me, however, that our wilderness calendars would never sell well in Ontario as long as they had "Western Canada" on the cover.

I wasn't about to change the name of our original calendar (it matched the name of our group!) but I liked the idea of putting out two calendars. It would allow us to showcase our flagship campaigns twice, with two different photos (I always had problems choosing between stunning images). I was sure that many of our members would buy our new Canadian calendar.

In October, we decided to send Ken and Carleen Lay to Ontario to build our support base and the market for our new calendar. They spent a month there, putting on 16 slideshows about Carmanah and the Stein Valley. Their slides of horrific clearcuts in contrast to the giant old-growth trees got the response we'd hoped for. Every show generated new members and generous donations. Ken and Carleen's success made it much easier for WCWC's board of directors to decide to embark on the Canada calendar project.

Keen to create a new look for our new pair of calendars, we hired the two young graphic designers just out of Emily Carr School of Art who had designed our hugely successful *Carmanah–Big Trees not Big Stumps* poster. I'd liked their clean and simple poster design, but with the 1989 calendars they cooked up a fancy graphic design with black shadows and fades. I hoped it would appeal to young buyers. The problem was, their design demanded a lot more pre-press work, on top of the fact that producing two calendars entailed almost twice the amount of research, writing, proofing and correcting. Both calendars came out late, giving us less than a month to sell them before Christmas. We knew that it was too late to sell a lot of them, so we cut back on our press runs and printed only 10,000 copies of each.

The inside covers of both of these calendars featured a photo of Joe Foy and his family having a picnic in a really big and ugly clearcut. Why the B.C. Forest Service decided to leave a picnic table in this spot after the surrounding forest had been logged, I'll never know. It made a perfect "photo op" for us. Our pitch was clear-cut too: *Why join the Wilderness Committee? Because Multiple Use doesn't always work.*

Notwithstanding the pre-press problems, I thought our 1989 calendars were great, especially our new Canadian one. Then I got a letter from the head of the Manitoba Naturalists Society's (MNS) Tall-Grass Prairies Conservation Project that stated in part: *In reviewing your material pertaining to the tall grass prairie I was most struck by the photo itself. The quality was certainly there–I thought the photography was great. The problem is that it's not a photo of tall-grass prairie!*

What was supposed to be a photo of Canada's most threatened ecosystem, was actually a stunning photo of one of the most common marsh plants in all of North America–*Phragmites comunis*–a giant reed grass! What was ironic about our photo, the letter informed me, was that it was taken less than a mile away from a 60-acre Tall Grass Prairie Reserve, one of only two small areas of that ecosystem-type protected in all of Manitoba. This two-page well-written letter really rubbed it in. *Errors in your material can seriously undermine your credibility and that of your cause.*" (As if I didn't know!) *Please know that my comments are offered only for the purpose of helping you improve on an already good product...*The letter concluded by stating that, because of this misrepresentation, their society of 2,500 members decided not to order any of our Canadian calendars that year.

I was extremely angry. I was counting on this group selling several hundred calendars. I wanted to write back sarcastically saying "Thanks a lot for turning us on to that photographer!" complaining to the MNS that it was the photographer that they had recommended who sent us the 35mm swamp grass slide with *Tall Grass Prairie* written on the side's mounting frame. But I didn't.

One thing she predicted correctly in her letter ...*To be honest, I very much doubt that anyone other than an avid naturalist would recognize the subject of that photo as being different than what the copy inferred.*" She was, indeed, the only person to point out our error. It illustrates how few people recognize this once-extensive plains ecosystem and how little tall grass prairie remains. In the end, there was nothing I could do except thank her for the correction. Once an error is in print, you just have to ride it through and commit to being more diligent and thorough in checking photos and facts to avoid similar mistakes in the future.

Grant enables WCWC to hire its first sales representative

Since our trip to the 4th World Wilderness Congress in Denver the year before, Adriane kept talking to people about it and the importance of figuring out how much wilderness we had left on Earth. Her conversations caught the imagination of Sue Fox, whom we'd contracted to do lots of WCWC's graphic design work. (We eventually hired Sue full-time as WCWC's in-house graphic artist.) Sue introduced Adriane to her friend Sunny Lewis, a remarkable journalist and global thinker, who in 1990 started up the ENS (Environmental News Service), an international wire service strictly focused on environmental news.

Visiting our office for the first time in the spring of 1988, Sunny was shocked by the huge piles of posters, postcards and other WCWC publications. "You've got to market these and get them into peoples' hands. This is money sitting here," she said. Then she gave us a substantial donation to hire our first marketing and sales staff person, Lisa de Marni, to focus on selling WCWC's products that summer.

Our trial Wilderness Committee store

Lisa convinced us that the best way to market our posters and postcards would be to get a store going, especially in the months before Christmas. In the early fall of 1988, our board of directors decided to open up a store at Christmastime to see if it would work. We found a temporary space with fairly good visibility–on West 3rd Avenue at the entryway to Granville Island. It was the best premises we could get for the low rent we had to offer. Our goal was modest: to at least break even.

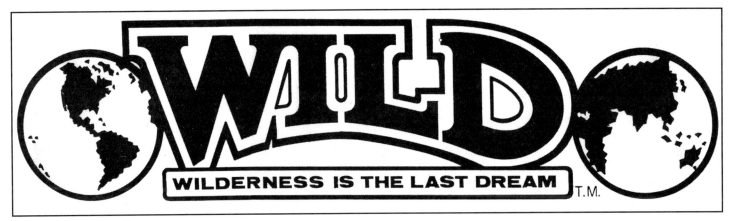

WILD logo. Artwork: WCWC files

Lisa hired a great store manager, Tim Howard, who was then a student at UBC earning part-time money in sales. Tim later went on to get a law degree and become a senior council for the Sierra Legal Defence Fund. We also hired Tim's girlfriend at the time, Lani. They both worked hard and, to our delight, made a small amount of money over and above all the expenses. Our first store's success spurred us on to find a storefront office the following spring.

Going WILD Internationally

Over the course of 1988, Sunny, Sue and Adriane became good friends. They met regularly and brainstormed about how best to follow-up on the mapping work that Michael McCloskey of the U.S. Sierra Club had presented at the 4th World Wilderness Congress.

They also invited Starlet Lum, former wife of Paul Watson (one of the original Greenpeacers and founder of the Sea Shepherd Society), to join their "wild women" team, as they referred to themselves. They came to the conclusion that involving environmental groups, First Nations and progressive scientists in a project to map in detail the world's remaining wilderness was key to successfully campaigning to save what was left.

At WCWC's October 1988 directors' meeting, the "wild women" presented a resolution, which we passed, that WCWC organize and hold a world mapping conference in Hawaii in June of 1990. Adriane was assigned the responsibility by the board to organize the whole thing. By December 1988, the "wild women" had chosen a name for their campaign: *WILD – Wilderness Is the Last Dream*. At the time we decided that WC² should strive to tithe (use 10 percent of its income) to further wilderness preservation campaigns elsewhere in the world.

Biggest oil spill to ever hit B.C. coast

That same December an event occurred that confirmed our belief that the world is small and vulnerable and that the ecologically threats to it are very great. Near the end of the month, the barge Nestucca spilled oil off the coast of Washington State. The ocean currents carried it up the coast of B.C. Nestucca's oil began fouling beaches in early January, especially hitting the shores of Clayoquot Sound. People rallied to Tofino and other coastal communities to help clean up the mess. We sent two directors and several volunteers to help. This spill eventually impacted every ecological reserve established to protect sea bird colonies on the west coast of Vancouver Island. An estimated 30,000 to 60,000 birds were killed. The Nestucca spill still holds the record as being the worst oil spill to occur on the B.C. coast although, in comparison to other spills around the world, it was only a tiny one—230,000 gallons of Bunker C fuel oil.

After the intensive weeks of clean-up ended, we put together a four-page, two-colour, tabloid newspaper with a huge red bannered headline *OIL SPILL*. Besides reporting on the Nestucca oil spill clean-up efforts in Clayoquot Sound, it made a strong plea to maintain the moratorium on offshore oil exploration and development in B.C.

Volunteers picking up clumps of "tar and birds" from the outside beach of Wickaninnish Island in Clayoquot Sound in January 1989. Photo: WCWC files.

We printed 50,000 copies and soon distributed all of them. We also got some small snap-top pill bottles and put a glob of the tarry, gooey pollutant that was washing up on the beaches into them, labeled them and sent one to every MLA in Victoria to physically show them what was coating the birds.

Big donor and dedicated daughter made *Rediscovery* book publication possible

One day that fall of '88, Thom Henley (Huck), who had been so active in saving South Moresby, came by and asked if WCWC would

Front cover page of our Oil Spill paper with volunteer holding two dead oil-covered birds. Artwork: WCWC files.

publish the "how to" guidebook he'd written about the outdoor educational Rediscovery camps he had been conducting for many years. He had tried to find a mainstream publisher with no success. I was with Huck around a campfire in Windy Bay when his idea of Rediscovery camps had germinated in the summer of '77. I like its goal: *to attune native and non-native youth to their natural and cultural surroundings, build connections between cultures and between people and their environment.* In the summer of '79 I spent two weeks at one of Huck's Rediscovery camps on the second year of the program in Lepas Bay on the north west coast of Graham Island on Haida Gwaii. There I saw first hand that the Rediscovery camp concept worked. In putting together native and non-native youth, most of whom had little experience in the outdoors, with native elders in a wilderness setting and then putting them through some new age sensitivity activities, the youth changed. They began to tune in to the natural world. I'm convinced that the key ingredient was being completely shut off from the buzz of our electronic world.

Of course I said yes. WCWC had some publishing know-how and

Athena George, senior guide in the Haida Gwaii Rediscovery program, studying a map of South Moresby with a Haida Gwaii Rediscovery participant planning their hundred-kilometre kayak expedition into the proposed protected wilderness area. Photo: Thom Henley.

the courage to undertake projects costing little money to initiate on the faith that somehow the rest of the money would materialise when it was needed.

The other ingredient needed to make this book happen was someone to edit and put it together. There was only one person I knew who was capable of doing it—my daughter Athena. Not only had she already proven her technical abilities in laying out and pasting up WCWC's first two books, but she also had hands-on Rediscovery experience. For several summers Athena had worked at a Sunship Earth camp in Oregon. Subsequently, she had spent several summers as a senior guide at Huck's Rediscovery camps on Haida Gwaii beginning in the summer of 1979-80. Her Sunship Earth experiences influenced many Rediscovery activities. She understood the philosophy behind the concepts. Still, I had to use every bit of my persuasive power to get her to work on the project, short of offering her a lot of money, which we didn't have. This included reminding her that a piece of her heart was still there. How else could she have written such a letter to the editor published in the *Sun* on May 22, 1981, entitled *Wilderness love letter*.

> *Each summer, guides and teenagers, both Haida and white, travel by kayak along the east coast of the Islands in the Southern Moresby area. Each day moving on the ocean one feels close to the water and the land. It is the highlight in my life; it heals and inspires me. I watch with amazement the positive effect that this wilderness has on all the participants.*
>
> *It was when I discovered that each beautiful place I visit has a logging obituary already written and pre-dated that my anger began to grow.... If all these islands are logged and left scarred and empty, I will not only lose a job, I will lose a piece of my heart.*

When Athena took on the Rediscovery book project, South Moresby had already been saved but the Duu Guusd where the main Rediscovery campsite was located was not. Working with Thom Henley wasn't easy. He was always off travelling somewhere, leaving most of the details in Athena's capable hands. It was also hard, I imagine, to work for her father.

Athena, who had a daytime job at the Lynn Canyon Ecology Centre at the time, convinced her graphic artist friend and co-worker, Debbie Duncan, to illustrate the book for next-to-no-pay. It was not an easy job, either! Athena took the rough manuscript of activities, games and sensitivity training exercises that Huck had developed over nine years at various Rediscovery camps and transformed it into an inspiring, easy-to-use guidebook.

Many of the 80 outdoor activities described in the Rediscovery book had never been written down before. Huck had always delivered his instructions orally. A few of the group activities were really complicated and difficult to describe on paper. Athena got volunteers in our office to try out her instructions to see if they'd work. In a few cases it took major editing to get the activity's instructions right. Athena devoted months to this project, painstakingly shaping the manuscript and a mountain of excellent 35mm images by Thom into book-ready form.

When we were getting close to pre-press and printing, with perfect timing, one of WCWC's big patrons came to visit me and asked what I wanted most to fund. Without hesitation, I said the Rediscovery book. "How much?" he asked. "$25,000," I replied. He wrote out the donation cheque for that amount without any comment. That single donation was far more than he had ever given at one time before. It was also his last big donation to WCWC.

Over a period of six years, he had given WCWC more than $100,000. His contributions had made most of our publications to save South Moresby possible. I suspect that he might not have been as convinced as I was that spending all that money on the Rediscovery book was the wisest choice. But it wasn't just the looming printing bill that influenced my request. **I believed then, and still do even more strongly now, that everything possible must be done to counter our culture's drift away from nature. Unless we change, our indifference to nature will spell doom for wilderness and our planet.**

We need more full-time staff members

By the end of 1988 the Wilderness Committee had grown significantly. We'd reached the point where I knew we had to start hiring more people full-time. Joe Foy was the only one on full-time salary. Ken Lay and I, along with a handful of others like Arne Hansen, Clinton Webb and Lisa de Marni, were on short-term work-specific contracts. Carleen Lay was on hourly pay doing data input. We'd hired to work part-time an engaging and dedicated young woman, Maria Hunter, to manage our growing number of volunteers. But the workload was greater than all of us could handle. I continually tried to convince Randy Stoltmann to quit his job and work full-time to save the big trees and wild forests on the coast. "The wilderness needs you full-time," I'd say to him every time we talked campaign strategy. **Everyone at WCWC worked full out knowing that there was only one chance left in our planet's lifetime to protect the magnificent natural ecosystems that had taken millions of years to evolve. That one chance fell within our lifetime. If you shared this belief, working only weekends and evenings to save wilderness wasn't enough.** Randy wasn't ready to make the leap right then but eventually, about a year later, he did.

Chapter 24

More staff, new campaigns, our first Branch and door-to-door canvass

Carmanah campaign heats up in January 1989 and WCWC forms its first branch in Victoria

After the busy Christmas season, we again picked up the pace on our Carmanah campaign. This campaign had become so big that WCWC itself was under pressure to expand. A group of Victoria-based volunteers, under the leadership of Derek Young, proposed forming a WCWC branch office. "A base in Victoria would help organize campaign volunteers," he argued. Derek, a middle-aged consultant in Victoria whose first big trip to Carmanah had motivated him to help save the valley, was dedicating every minute of his spare time to the cause. At a Special General Meeting in February

Derek Young speaks at a WCWC-Victoria Branch event. Photo: WCWC files.

'89, WCWC passed bylaw changes to allow for the formation of Branches. Soon our Victoria Branch was a vital hub of Carmanah campaign activists.

MB picked up its pace, too. In January '89, probably because they were getting worried about the rapidly increasing public support for Carmanah preservation, the company came up with yet another plan for partial protection. This time it included a narrow strip of land on both sides of the creek, connecting their proposed Heaven Grove Park with their proposed Carmanah Giant Park. Interestingly, it included the entire trail that we had completed to date.

We stood firm for a whole-valley park and called MB's new proposal a *blowdown and washout zone*, projecting that the first big wind storm would blowdown the narrow "leave-strip" of trees, and that the increased runoff from the clearcuts would wash out the flood plain that nurtured the biggest trees. We called for an independent scientific study of MB's proposal and collected signatures on a petition asking the federal government to add the entire Carmanah watershed to Pacific Rim National Park. MP Bob Wenman commissioned a study of MB's new plan. The forest ecologist who undertook it cautioned against allowing any logging in Carmanah for fear of destroying the dynamic spruce ecosystem.

Building our effort to get better B.C. forest practices

I always felt that continuous efforts to expose the problems with B.C. forestry practices were necessary to achieve our overall mission of saving more wilderness areas. It was great to have the help of an expert in this: Mark Wareing, the Registered Professional Forester and friend of Clinton Webb who as ghosteditors had helped write our *Registered Professional Foresters: FOR SHAME* newspaper a year earlier. Mark was, by now, putting in a great deal of volunteer time with us. In the spring of 1989 he left his job with the B.C. Forest Service and came to work full time for WCWC. He immediately launched into writing our second expose: *FORESTRY MALPRACTICE ON RISE IN BRITISH COLUMBIA*. It was an eight-page, full-colour, tabloid-sized newspaper. We printed 50,000 copies that spring and they went like hotcakes.

Ken Lay looking over our Forestry Malpractice layout on our light table. Photo: WCWC files.

As in our *FOR SHAME* paper, Mark focused on the professional foresters' duplicitous role in allowing bad logging practices to continue in B.C. Volunteers distributed this hard-hitting paper in the B.C. Forest Service's head offices in Victoria, pointing out that the paper called for the B.C. government to enact a "Forest Practices Act" to put in place eco-forestry standards. For nine years we'd been campaigning to get eco-forestry practices instituted in B.C. It was our most unsuccessful campaign to date, not because our campaign publications weren't strong enough, but because old-school corporate-controlled industrial forestry was so firmly entrenched in B.C.'s politics and economy.

Convincing Adriane Carr to quit her teaching job and work full time for WCWC

In the spring of '89, Adriane came home with the good news that her college had initiated an innovative program for full-time teachers at the top end of the salary scale to take a year's sabbatical on half pay. The time off had to be spent in relevant professional development. Given the fact that she taught geography, her international WILD campaign mapping work at WCWC should qualify. If her application to pursue this project were accepted, she'd be off from September 1, 1989 to August 1, 1990. In actual fact, she'd be able to start working full-time at WCWC when courses ended and her holidays began in July! "Could we live on that little money?" she asked me.

"Can we afford not to go all-out to save wilderness?" I replied.

During the six years that had elapsed since Richard Krieger quit and the running of WC² had fallen on my shoulders, Adriane had spent enough time in evenings and weekends editing material for WCWC and doing other volunteer work to constitute a second full-time job. Vancouver Community College accepted her sabbatical proposal. I could hardly wait for July and her two-fold full-time work for WCWC to begin.

Our risky move of starting up a door-to-door canvass brings exponential growth

In March '89, just as it was hitting home that we'd better start increasing our income fast to support our burgeoning campaigns and increased spending, we received a letter in the mail from a fellow named Frank Sloan. He said in it that he had been a door-to-door canvass director for several non-profit organizations including Pollution Probe in Toronto. He went on to claim that he could build up our membership, which was about 3,500 at the time, to 30,000 in a year's time through a door-to-door canvass in Vancouver. He also claimed he would raise hundreds of thousands of dollars for WCWC in the process.

His hype seemed so outlandish that I almost threw the letter away. There were always fundraisers looking for employment telling us that their schemes would bring in piles of cash to WCWC. But instead of dismissing this one, I called Joe, Ken and Adriane together and asked them whether or not I should follow up. We agreed to invest in a return flight for Frank to come out for a week, talk to us and give us a chance to see if he and his claims were for real.

On his visit Frank proved to us he was legitimate. He told us that Pollution Probe began door-to-door canvassing in Ontario in 1985 and went from 1,000 to 30,000 members. He was their star canvasser. After that he went to Halifax and set up a canvass for the Ecology Action Centre and increased the membership and profile of that organization "faster than their two paid staff could handle it." He told us they said to him "Hey Frank, slow down!" Then, after bragging about these past results, he actually went out and canvassed a couple of nights in Vancouver to prove canvassing would work for WCWC. His impressive results convinced us to take the plunge.

Frank began working for us full time in mid-April. He immediately started to hire and train canvassers. He vowed to raise our membership to 10,000 by August. His timing was perfect: it was the height of "Carmanah Fever" that was sweeping B.C.

Portrait of Frank Sloan who launched WCWC's fist door-to-door canvass. Photo: WCWC files.

Soon Frank had a crew of 20 to 30 canvassers going out six nights per week in Vancouver. Shortly thereafter he set up a satellite canvass in Victoria. Frank was a whirlwind of energy and soon conclusively proved to us the power of door-to-door canvassing to raise consciousness, membership and money.

Managing a canvass successfully over the long term, however, is a very difficult job. It involves continually training new canvassers, scheduling where the canvassers go, building a positive team spirit among the canvass crew members and keeping that spirit up even when they work "poor turf" (areas where they encounter little support and few donations). The turnover of canvassers is always high. Some who tackle the job are highly skilled people who have been unable to find a job in their field. Many are students working their way through college and university. The canvassing job tides them over until they land a better job. We've found that to be a really effective canvasser, you have to personally believe in, be knowledgeable about, and passionately support our cause of wilderness preservation.

Over the years I have gained a great deal of respect for door-to-door canvassers. I've made making them feel a key part of the organization one of my priorities. They are the "front line troops" in the war of ideas between the pro-development fanatics who turn their back on nature and those who believe that we must preserve nature to have any chance of a healthy planet and decent future. I truly believe there is no harder or more important job at WC².

Some people criticize WCWC for paying its canvassers. They believe that people should do this sort of work out of the goodness of their hearts. I say to these critics, "try going door to door a few nights yourself and then see if you don't agree with me that canvassers deserve every penny they earn." Volunteers who have tried to canvass regularly soon quit, not able to take the recurring rejection and negativity of those who do not agree with our activities and mission. Besides, people need jobs in our society. How many people could afford to do this hard work for free?

On average, over the long term, our canvassers make little more than minimum wage and some nights, in "tough turf" where few people are sympathetic, even the best canvassers often fail to get any donations or even one new membership. From the start we instituted a cap on the amount of commission a canvasser could get from any one big generous donation—an extremely rare but wonderful event. It amounts to a little more than one hundred dollars. Few canvassers have been skilled or lucky enough to earn this maximum.

Building public support is a slow process. Since 1989, paid canvassers have been WCWC's "educational outreach ambassadors" who distributed our education newspapers and talked to people about the wilderness issues. Changing perceptions and social consciousness takes perseverance. Several of WCWC's long term staff members started out as canvassers. They include Chris Player who did outstanding computer mapping work for us for ten years until he left in 2004 and Ken Wu who in 1999 became the campaign director for our Victoria office, spearheading from 2003 to 2005 the fight against the government's proposed "working forest" legislation that would have put an end to new park creation.

Each of the several thousand people who has canvassed for WCWC has interesting stories to tell. Over the last sixteen years our canvassers have encountered celebrities, logging executives, union bosses, lonely people, Premiers, the occasional mean dog, hostile individuals and unexpectedly generous souls and you name it.

For example, the last house Ken Wu visited one night happened to be the home of Douglas Coupland, the famous author of *Generation X*. Ken got a $7,500 donation! Coupland had just written a short story (actually more of a snippet) for a distilling company and decided to give all of the money he would receive from the company to WCWC. He only decided to write the story, which was going to be part of a company ad, after the company agreed to add a statement at the bottom saying that his honorarium was being given to Western Canada Wilderness Committee. Coupland said, "I can't think of an easier way of donating a heap of money to my favourite charity. Artists get asked to donate works to charity all the time but there's precious little writers can do or donate."

Ironically, at the same moment, just a few doors away another canvasser was having his worst day ever. An angry man physically assaulted him on the doorstep. He wasn't hurt; but he was shaken up. The police were called in; it turned out that the man was simply mentally off-balance. There are enough tales of incredible situations, interesting people and adventures that WCWC canvassers have encountered over the years to fill a large and fascinating book.

By Christmas time 1990, less than two years after Frank started our canvass, he reached his goal of 30,000 members. One of WCWC's greatest all-time failings was taking our tens of thousands of new members for granted. We were so busy campaigning, publishing and coping with our day-to-day problems, and so new to the game of building a big membership base that we simply didn't think about how to keep our members involved and happy. We had no concrete plans to service and nurture them or to renew them when their annual membership lapsed. The vast majority fell by the wayside. It took a financial crash to realize that canvass-generated members and donors—all members, in fact, regardless how they came to join WCWC—are often tenuously attached to our organization. To keep them on our membership roll we had to put more effort into continuing to educate and sustain, than we did in signing them up.

Over the next few years we also came to realize that our spectacular canvassing success of the late '80s was not the norm. It was timed with a huge surge in environmental awareness that swept across North America and the planet—the precursor to the big United Nations Conference on Environment and Development (UNCED) held in Rio de Janeiro in June 1992. It was also a unique outpouring of support for protecting Carmanah Valley, the first massively successful campaign to protect Canada's "big trees."

I kick myself now that we didn't entrench our canvasses' initial success with an effective membership renewal program. If we had, we might have avoided the worst of the financial crises to hit WCWC in the early 1990s. **It took too many years for us to realize that door-to-door canvasses are not about making money every month. They're about building long-term organizational strength. Door-to-door canvassing, used effectively, is a slow but absolutely sure way to develop and maintain a grassroots organization.**

Our canvassers dependably present our side of the issues to the large number of people—estimated to be about half the population—who don't read newspapers, listen to or watch the news. Without the face-to-face-contact with out canvassers, these people wouldn't even know WCWC existed. Besides reaching the unaware, canvassers turn sympathetic individuals into active supporters. **There is no doubt that the thing that has made WCWC into the largest membership-based wilderness preservation group in Canada is our relentless, year-after-year, door-to-door canvassing.**

WILD helps out Suzuki's Amazon Rainforest Preservation Campaign

At the same time that WCWC was recruiting members and campaigning to save local wilderness areas, WCWC's international WILD campaign was starting to tackle global issues. Like so many

Poster advertising the Amazon event on a hall wall in the school. Photo: WCWC files.

people in North America, we were aghast at the 1989 *Time Magazine* front cover featuring the Amazon rainforests being burned. In March of 1989, just a few months after the WILD campaign was approved by our board of directors, Adriane and her WILD team were asked by David Suzuki and his wife Tara Cullis to help put on a big event to help the Kayapo natives of the Amazon protect their rainforest homeland. The Kayapo were fighting to stop big hydro dam projects that, if constructed, would flood much of their aboriginal territory, destroy their traditional culture and take an inestimable toll of irreplaceable, diverse, genetic wealth. We were flattered to be asked to help, and only found out years later that we were one of Suzuki's last choices. The other local environmental organizations that he asked first turned him down.

It took us a month to plan, advertise and inform all our members about this special event. We were all blown away by the response. Over a thousand people turned out. The high school auditorium we'd booked could only hold 800 people so, due to fire regulations, we had to turn some away. Chief Paiakan wore a feathered headdress and his children wore his tribe's traditional colourful feather regalia. Those who packed in were deeply moved by the speeches of David Suzuki and plight of Chief Paiakan's tribe.

My job was to give the fundraising appeal. I drew on my childhood experiences as a Presbyterian preacher's kid listening to my dad's sermons every Sunday. I knew that most people give with joy in their hearts so I kept a monologue going while the "collection plates" were

Chief Paiakan walks to the stage in front of the packed house. Tara Cullis, David Suzuki's wife who eventually became the Executive Director of the David Suzuki Foundation is in the foreground on the far right. Photo: WCWC files.

Above top, middle and bottom: Chief Paiakan speaks to packed high school auditorium in Vancouver, raising awareness of the threat to his traditional lands and support in his fight to protect them; David Suzuki gives an inspirational speech; I present the appeal. Photos: WCWC files.

passed around, telling people that the more they gave, the happier they'd be. I'd never before witnessed people dig so deep. We raised over $11,000 that night to help the Kayapo buy an airplane so they could patrol the borders of their traditional territory. At the very end of the evening, after nearly everyone had left and we were packing up to leave, one distraught teenager asked if we could give her a bit of money for bus fare. Embarrassed, she explained that she had donated all the money she had, forgetting about her need to get home. Of course we gave it to her, with heaps of thanks for her generosity.

A few months later, WCWC-WILD worked with Tara Cullis to boost the Kayapo's campaign by producing 50,000 copies of an eight-page, full-colour, tabloid-sized newspaper titled *Canadians vow to help save the Amazon rainforest*. It featured photos taken at a conference sponsored by the Kayapo in Brazil to rally international support for their cause. David Suzuki, Tara Cullis and several B.C. First Nations' representatives attended.

This first publication by WILD asked Canadians to donate and write letters to help the Kayapo protect their Amazon homeland. I'm sure it helped, at least a little bit, in eventually getting the Brazilian government to shelve these dam projects.

Cover page of our Amazon rainforest newspaper. Document: WCWC archives.

Chapter 25

WCWC moves to Gastown and pulls off a colossal Carmanah Artists' project

Searching for storefront office space

In early '89, Joe, Ken and I began looking for a storefront office. We had definitely outgrown our West 6th premises. The little store we'd rented pre-Christmas near Granville Island was only temporarily available and too small. We had to find some place soon: our landlord at West 6th had given us three months notice to vacate. I think our increasing level of activity bothered the other tenants. After several months of unsuccessfully pursuing a location in the Kitsilano area, Joe suggested that we check out Gastown, Vancouver's old historic downtown area where he had operated his Dairy Queen. It had tons of tourists in the summer, just the kind of people who would want to take our posters home with them.

Almost immediately we found a place that fit our bill: a storefront at 20 Water Street with a double entryway and a big, three-sided display "bay" window. The owner was willing to renovate to our specifications for a tax receipt and the signing of a five-year lease. He even added a mezzanine floor for our offices. The move was a huge gamble. The monthly rent was more than five times what we were currently paying. Nevertheless, everyone agreed that we should go for it.

Raging Grannies sing at our store's grand opening

It took longer than planned to finish the renovations. Finally, on July 21, 1989, we held our grand opening. The Raging Grannies, who had only formed a couple of years earlier, garbed in colourful dresses and flowered hats, sang their bitingly satirical songs about the government's treatment of the environment and the evils of clearcut logging. Staff and directors (12 in all now) simultaneously cut the ribbon across the front entrance to our new storefront office. We were a great team, passionately engaged in a globally significant cause. The party lasted all day and into the night. Our open house drew over 200 members and friends. We reminisced about how far we had come. In

20 Water Street Storefront office with new sign installed. Photo: WCWC files.

the two-and-a-quarter years since the WC² moved from our house on Alberta Street to West 6th and now to Gastown, we had grown to have a paid staff of 12. All except one had come up from the ranks of volunteers and volunteers were still the backbone of our organization. In our new office we designated every Tuesday and Wednesday night as volunteer nights. Often more than two dozen attended, stuffing envelopes, making protest signs and planning public events.

Life was rosy. Our Carmanah campaign and door-to-door canvassing generated a steady flow of income. So did our new store. Tourists loved our posters, books and other products. I continually pushed to spend all the money as fast as it came in. I was convinced that the more papers and posters we published, and the more campaigns we started and strengthened, the greater our effectiveness would be. We spent money to make money to spend on what was needed to protect wilderness.

Rediscovery book rolls off the presses – twice

Just days after we'd made the move to 20 Water Street, WCWC's third book, *Rediscovery – Ancient Pathways New Directions*, rolled off the press. At 288 pages, with over 125 coloured photos and 39 pen and ink drawings, it was by far our biggest publishing achievement to date. To save money, the book was cleverly designed so that colour printing was only on one side of each signature sheet of the book, with black and white printing on the other. It was our first book project with Hemlock Printers, who had a new special press called a prefector that printed black ink on both sides of the sheet in one pass through it, achieving a good cost saving.

The day the 5,000 Rediscovery books were delivered was not an entirely happy one, however. We quickly discovered that many of the books had smudges on the pages. While they were minor in nature, it definitely was not a first class printing job. Hemlock was absolutely fantastic in rectifying the situation. They immediately told us to sort out the defective books and they would replace them free of charge. It turned out that nearly all of them had smudges to a greater or lesser degree.

Raging Grannies sing at grand store opening. Photo: WCWC files.

Hemlock went to press two months later, printing another 5,000 copies. The wonderful thing was that this company let us keep the defective copies as long as we did not sell them and only gave them away. We got a big stamp printed up which said "Complimentary copy – Not for sale" and for years we were able to give them away to native youth who were going to camps, and to those interested in starting up a Rediscovery camp. That printer's glitch helped boost the Rediscovery program tremendously.

The book received critical acclaim too. One reviewer raved about it saying that it was the "...*best book ever written about how to get in tune with nature.*" It sold well and steadily over the years. The money that was donated to print it was gradually recouped and recycled into other campaigns and publications. Six years later, in mid-1995, when we finally sold the last Rediscovery book, I twisted Athena's arm again and she reluctantly began to work part-time on a second revised edition. She found it very difficult to track down the active camps, get their input and update the book. Her heart wasn't in it and, with WCWC still deeply in debt, our pocketbook wasn't in it either. We didn't have the money to reprint the book and I couldn't think of a creative way of to get it. The project dragged on and Huck got more and more anxious. New Rediscovery camps needed the book.

I was completely relieved when Shane Kennedy of Lone Pine Press offered to publish a revised second edition. We turned everything over to him at no cost, including the film used to print the book.

Rediscovery – Ancient Pathways New Beginnings went on to become a classic in the field of outdoor education. It's still in print. I'm proud of WCWC's role in it.

20 Water Street office quickly becomes too small

In 1989 WCWC was growing so fast we could hardly be contained. We had to hire more staff to handle the exponential increase in membership, mailouts, campaigns, income and volunteers. We'd been in our 20 Water Street office for a few months before we outgrew it. By early fall we had to move half of our operations—our mailroom, volunteers, storeroom and accounting department—to a below-ground-level office space about a block up Water Street.

It was fairly convenient, but a constant hassle to shuffle back and forth between the two places. This went on for about a year until the back half of our 20 Water Street location became available, doubling our space there. It felt good to all move back together again. The rent was twice as expensive, but it was worth it for being under one roof boosted morale as well as increased efficiency.

Most successful three part mailer

WCWC's growth wasn't by chance. We worked hard to increase our membership and used our full collective brainpower to constantly think up clever campaign tactics and fundraising ideas. One of our Carmanah mail-out appeals was so good it worked even as a stand-alone householder mailer. That's almost unheard of in a fundraising effort. I figure it was a combination of the right appeal and the right cause at the right time.

What made this "three-part mailer" so catchy was the great cartoon on the front. Annette Shaw, an artist who volunteered to help save Carmanah, drew the cutest little young tree asking a big old tree, *"Will I grow up to be a giant like you?"* In the background was a sea of stumps in a clearcut. The trees' facial expressions were warm and friendly looking and emotionally drove the point home.

On the back of this card was our appeal to send in $25 to "adopt a tree" in Carmanah. Each of the mailer's other two postcards was different. The card to the Premier featured the ugliest possible photo of a big clearcut near Carmanah. On the back was a punchy description of the issue and a "polling question." You could check either, *"Go ahead and let them clearcut Carmanah. We need the jobs and timber"* or *"Please protect Carmanah...."* We had to offer people the choice in order to conform to the federal law that forbids registered charities from lobbying governments. No doubt you can guess which choice was checked the most! The third part of our mailer had a stunning picture of the beautiful old-growth forest in Carmanah with the same information on the back about the issue as on the back of the card to the Premier, minus the polling question. Many people told us they kept this card; and many sent them on to friends.

Nik Cuff printed the first run of 50,000 mailers for free and the following ones at cost. We sent them out to every list we could find and our canvass crews distributed them door-to-door. By the end of our Carmanah campaign we had printed 500,000 of these amazingly effective three-part mailers. At the peak of the Carmanah campaign we decided to try an unaddressed household mailing to see if enough people would respond to cover our costs. To our amazement it worked. For the next few months we mailed out tens of thousands that way. In retrospect, we should have mailed out hundreds of thousands. For never again did such a perfect opportunity to build support for wilderness preservation and membership in WCWC

Cover of Thom Henley's *Rediscovery – Ancient Pathways – New Directions* guidebook to outdoor education book with forwards by Bill Reid and David Suzuki published by WCWC in April 1989 and Reprinted in June 1989. Photo: WCWC files.

arise. Never again did a campaign so thoroughly capture the public's imagination and widespread support. It was a time when planet-wide interest in the environment was skyrocketing and the economy was booming. Never again did the media plug a campaign so hard and consistently, either. And never again did a forest company bumble so badly in its own defence.

Our colossal Carmanah artists' project

Carmanah was the most exciting and innovative campaign in WCWC's history to date. It seemed that every week in 1989 the campaign took a new turn. Arne Hansen started planning an artists' project bigger and better than the one he'd organized for our Stein Valley campaign. We leapt into another season of trailbuilding, redirecting our efforts to building a trail to the head of the watershed. Our trail in the lower valley, meanwhile, was used by some Canadian Wildlife Service bird experts to survey the marbled murrelets that flew in at dawn in springtime to nesting sites on thick moss-covered limbs of Carmanah's old-growth trees.

Celebrities also began to flock to Carmanah. One of the first was the pop singer Bryan Adams who called us out of the blue to say he wanted to help save Carmanah and would really like to see the valley. We jumped at the chance. Adriane, our children, our nanny Tina (a Bryan Adams fan from Ireland who was thrilled to be meeting this big rock star) and I went along. We all hiked down to Heaven Camp and then Key Lay took Adams up our valley trail to see some of the spectacular Sitkas and take a dive into one of Carmanah's magical deep pools. On the hike out, Adams carried our tired two-year-old son TJ on his shoulders. All of us liked his good energy. Adams later held a benefit concert to support our Carmanah campaign, and he handed out our Carmanah mailers at his other shows, too.

Bryan Adams inspects a giant Sitka spruce stump with us in the clear-cut valley adjacent to Carmanah Valley. Photo: WCWC files.

Ironic "Working Forest" sign works for us

On the Bryan Adams trip into Carmanah we exploited one of the goofiest, most-helpful-to-our-cause blunders our "enemy" MB ever made. The company had decided to erect a very big sign made out of old-growth red cedar right in the middle of a huge, ugly, recent clearcut next to Carmanah Valley. The sign sported huge letters declaring the area a "Working Forest." The only problem was that there was not a tree or even a seedling in sight! We used this sign as

Bryan Adams inspects MB's newly installed fancy *Working Forest* sign made of old-growth cedar wood in a huge clearcut right beside Carmanah Valley. Photo: WCWC files.

a "poster child" for our campaign and the image became almost as powerful as our *Big Trees not Big Stumps* poster. Stupidly, MB let that sign stand in that spot for over a year. I was amazed it lasted that long. It didn't take a genius to see how much it hurt MB's public image. At the same time MB also tried to "green-up" and groom the side of the road into Carmanah, by removing the waste piles of wood and hydro-seeding the banks with exotic grasses. It never really helped. The logging road to Carmanah still looked like a drive through hell.

The Carmanah Artist's Project book

One day in the spring of '89, just before our move to Gastown, Ken Budd of SummerWild Productions walked into our office and talked to Arne Hansen about possibly publishing a book about the Carmanah artists project that Arne was beginning to organize. Right from the start I liked Ken Budd's vision of an award-winning coffee table book of the highest quality. He had thought it out. The book would feature photos of the artwork donated by the 70 or so artists who we were planning to host on expeditions into Carmanah, and alongside each piece would be a picture of the artist, a brief biography and a personal statement written by the artist about his or her Carmanah experience.

The book would begin with short introductory pieces by David Suzuki, Randy Stoltmann, Cameron Young, Adriane and me, Arne Hansen and Sherry Kirkvold. The book would be published and launched at a huge gala affair in the fall. WCWC would sell all the art in a very upscale continuous silent art auction before Christmas! The timeline was extraordinarily tight. It would take a great deal of work, many volunteers and the incredible support of the artists. We all said, without hesitating, "Let's do it!"

Blissfully oblivious to the risk we were taking

The one thing Ken Budd didn't mention—and none of us could accurately estimate—was how much the book and artist project was going to cost us. I knew that we were dealing with big money. But I, for one, was unaware of just how high the stakes were. I'm sure if we knew, even our risk-taking WCWC might have choked. But we were driven by an overpowering will to save Carmanah and a rock-solid faith that we couldn't fail. We were completely confident the whole artists' project and the coffee table book about it, *Carmanah – Artistic Visions of an Ancient Rainforest*, would be a smashing success.

We felt pretty confident, especially with our remarkable Carmanah three-part-mailers and our door-to-door canvassing generating lots of cash, that we could handle the up-front costs as they arose: getting the artists in and out of Carmanah and buying the food to host them there. We also hired Arne Hansen to coordinate the project. Our contract with Ken Budd had him getting paid after the book was published. We scheduled the grand black tie gala opening of the art show and book launch for October 15, 1989.

That summer in Carmanah was pure magic. Almost one hundred artists hiked down our trail to Heaven Grove to paint, draw and sculpt amidst Carmanah's towering trees. They slept in small backpackers' tents, spent their days exploring and studying the rainforest and their evenings gathered around the campfire on a gravel bar beside Carmanah Creek. In our fall 1989 Carmanah newspaper we described the educational process that went on there this way:

> Campfire conversations would often turn to the stark contrast between the haunting beauty of Carmanah's virgin forest and the slash-choked, burned and blackened clearcuts that lie just outside the watershed. The artists spoke of how the distant growl of heavy

Robert Bateman, participant in the Carmanah Artists project sketching in Heaven Grove. Photo: WCWC files.

> logging equipment, carried by the wind from the next valley, affected them...a constant reminder as they sketched and painted, of why they were there.

Robert Bateman and the late Jack Shadbolt took part: both of them incorporated clearcuts into their works. Bateman told us that this was the first painting he had ever done of a clearcut. In the end, all 70 of the completed art pieces donated to us were featured in our art show and book.

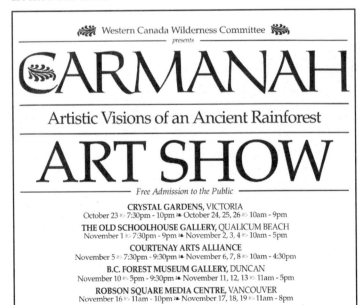

Schedule of the traveling to Carmanah Art Show. Photo: WCWC files.

Ken Budd, book producer left and Arne Hansen, project coordinator at the gala. Photo: WCWC files.

The Carmanah Gala Art Show and Book Launch

That fall we scrambled to get the book published in time. We insisted on using chlorine-free paper, which had to be shipped from Sweden because none was being produced in Canada. We hoped our decision would set a precedent that would lead to a change in the Canadian pulp industry. We decided to print 15,000 copies, naively ignorant of the fact that selling 5,000 copies would make it a B.C. best seller!

Preparations for hanging the first show went smoothly except for one unfortunate incident. A piece of artwork "disappeared" somewhere between our storage room and the hotel room where we held the show. It was a beautiful silk kimono with a stunning Carmanah scene hand dyed on it. We immediately contacted the artist, Gretchen Markle. Although immensely upset, she was kind enough to agree to create a duplicate for us as fast as she could under a paid contract.

Above: Joe Foy, WCWC Campaigner in tux enjoying himself at the gala. Below: Adriane Carr, my wife and I enjoying ourselves at the gala. Photos: WCWC files.

Frank Sloan, WCWC Canvass Manager, left and Mark Wareing WCWC Forester in tuxes enjoying themselves at the gala. Photo: WCWC files.

Within a couple of weeks, the replacement kimono joined the show and very few were ever aware of this small disaster.

Ken Budd, who performed a miracle pulling this project off in such a short time, insisted that our *Carmanah – Artistic Visions of an Ancient Rainforest* gala book and art show launch be a very grand affair. Even I had to rent a tux! We hired a string quartet to play. There were delectable hors d'oeuvres. Ken premiered a special video made about the artists' trips into Carmanah, which tacked on only a couple more thousand dollars to the huge bill for the project that none of us wanted to think about. It was truly a night to remember!

Arne Hansen, busy handling other details, left the producing of the program with information about the art on auction until the last minute. The programs arrived over an hour late. We also didn't have the regular book to sell. We only had copies of our collector's leather bound limited edition that we were selling for $250 each. Needless to say we did not sell too many of them that night. Most, even the wealthy, were willing to wait and purchase later a regular edition, which cost only $60.

But what really saved the launch and made the book and auction a success was the art. It was phenomenally good. Carmanah had worked its magic on the artists. The gala was packed with people: artists, many generous patrons and several reporters and TV cameras. We couldn't have asked for more. Someone pointed out to me a young man, relaying that he was a member of one of Canada's richest families, the Bronfmans.

That gala launch was just the first night of the art auction. Over the next two months we toured the whole art show around B.C. in a van provided at no charge by the Teamsters Union. In each gallery people had the chance to 'up' the current highest silent bid. Everyone who had placed the highest bid at each gallery was, at the end of the tour, given a chance to put in one last higher bid. It was an extremely competitive process that really worked.

One of the few paintings that did not sell, however, was Robert Bateman's *Carmanah Contrasts*. I thought, for sure, that some museum would pick it up, but none did. It was pretty unusual for Bateman. The top one-fifth of his 101 x 114 centimetre diptych was a stunning forestscape of Heaven Grove in Carmanah; the bottom four fifths of the painting was a dark, stark clearcut choked with waste wood. Bateman, who has always generously supported wilderness and environmental causes, donated $10,000 to our campaign and kept the painting himself. In his statement in the book he said: *The line has to be drawn somewhere and something as precious and unique as this [Carmanah Valley] is where the line should be drawn.* Later he featured *Carmanah Contrasts* in a powerful poster with this quote beneath it:

> The earth is not his brother, but his enemy, and when he has conquered it, he moves on. His appetite will devour the earth and leave behind only desert - Chief Seattle, 1844.

Big book promotion pays off

To help market our *Carmanah–Artistic Visions of an Ancient Rainforest* book, we published and distributed a whopping 300,000 copies of a new Carmanah campaign newspaper that fall. It was our third Carmanah newspaper in less than a year. In the tear-out coupon we offered the book to our members at a discount price of only $49 instead of the regular $60 retail price. This irritated some retailers and bookstores until their sales soared and our book became a

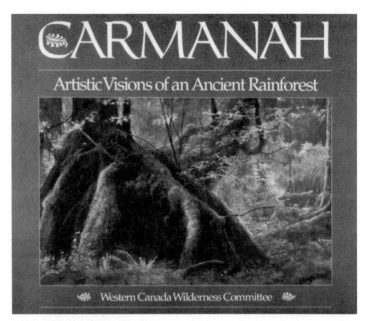

Cover of WCWC's *Carmanah Artistic Visions of an Ancient Rainforest* book.
Cover painting: Robert Bateman.

best seller for them, too.

Thanks to the courageous directors at Mountain Equipment Co-op (MEC) who gave us permission to use their mailing list, we sent out a copy of this paper to every MEC member. This was the first and last time MEC ever did this for us because—we heard later—they received a sizable number of complaints from members who were not "wilderness preservationists." At first we were quite worried about whether or not we'd make our money back on this huge mailout. We were getting back hundreds of unopened envelopes with bad addresses every day for more than a week—each one of which we had to pay postage return on. But then, the returns declined and the book orders began. Soon the orders—many of them accompanied by a donation—swamped us. We kidded about hiring an armored car to take to the credit union the daily deposit of the money coming in from the mail, our store and our canvassers.

Our massive mailings and the art show and advertising helped make our book a winner. When the regular bookstores ran out of copies a few weeks before that wonderful Christmas of '89, people wanted the book so badly that they made a special trip down to our 20 Water Street store where we had the only supply of them left in town. In one day, two weeks before Christmas, our store took in over $8,000: a record that was never beaten. Most of it came from sales of our Carmanah books at the full retail price of $60. And this day was not all that exceptional. Day after day our store receipts totaled over $5,000. Donations to help save Carmanah came in at a phenomenal pace, too. No wonder we thought that the general public had finally made the paradigm shift to putting the protection of Mother Nature ahead of everything else and that this bonanza would never end.

By Christmas day, we had sold all but a few hundred of the 15,000 copies of our *Carmanah–Artistic Visions* book we'd printed that fall. By the summer of 1990, we were completely sold out except for copies of our 500 numbered limited edition version, which we were trying to sell for $250 each.

Never again were we to see such enormous income over such a brief period of time at the WCWC.

Chapter 26

Vital projects, hectic growth and a Herculean work schedule

Helping launch a very important legal case asserting Nemiah's aboriginal right to trap

A few weeks after moving to 20 Water Street, Jack Woodward came to see me. Jack and I had become friends during the early '80s when I was consulting for several First Nations and he was fast becoming one of B.C.'s foremost lawyers in the field of aboriginal law. I always liked working with Jack. He was smart, dedicated and very funny. It was Jack who taught me the importance of culturally modified trees in proving "continuous use and occupancy" of the land by First Nations.

He convinced me that by using both "aboriginal title" and the sum total of all the aboriginal rights a nation possessed, First Nations could wield huge legal power in the fight to protect wilderness.

Jack took forward the first case in Canada to test the meaning of "aboriginal rights" after the clause affirming aboriginal rights was added at the last minute to the Canadian Constitution of 1982. Although the inclusion sounded good, there was a catch: the word "existing" was added as a qualifier. No one knew what aboriginal rights still existed, so it would be up to the courts to decide.

Jack was hired by the Huu-ay-aht, a Nuu-chah-nulth First Nation, to protect its rights to use a clam bed located beside one of its Indian Reserves along the Alberni Inlet. This band harvested this clam bed as their ancestors had done for thousands of years. The B.C. government was in the process of de facto expropriating it by issuing a licence to a company to grow mussels on top of the bed. Jack argued in court that the band had an "existing" aboriginal right to continue their "traditional use" of the clam bed and won the case.

I suspected that Jack's visit to our new storefront office that summer of '89 was more than just a friendly one. I was right. He sought our financial support for a court case that, according to him, would save a lot of wilderness. "But first off, where do you stand on trapping?" he asked me. He knew that I personally had supported the court challenge to the automatic renewal of TFL 24 with full knowledge that it had involved asserting the trapping rights of the Haida. "I support the aboriginal right to trap, but I'm against the cruelty involved just to put pretty fur coats on rich women. I know that trappers, in general, are our allies because they want to preserve wilderness in order to continue to trap, but I don't know where the other WCWC directors stand," I replied.

Jack then described the 300-member Nemiah Indian Band, the Xeni Gwet'in First Nation, that lived "a cowboy lifestyle" near Chilko Lake on the Chilcotin Plateau. The Band's chief, Roger Williams, was 20 years old and a champion wild bull rider on the local rodeo circuit. The Nemiah had been isolated without road access until fairly recently. Clearcutting was advancing towards them and they didn't want their traditional territory logged. They'd seen the negative impacts that logging had had on several neighbouring aboriginal communities on the plateau.

"It's the last of the wild west out there, Paul, where wild horses still roam. The Nemiah speak their native language and their traditional culture is strong. They have a strong legal case…the only thing they don't have is money," Jack said. He went on to outline the legal case in layman's terms. "The case is quite simple One of the band's traditions is to make squirrel skin booties for each new-born child. It's the warmest of furs. Squirrels live in the trees. If the logging company cuts the trees down, there'll be no squirrels and the Nemiah's aboriginal right to make their special baby booties will be taken away from them," explained Jack, with a twinkle in his eye.

"Even the most conservative old justice, who might not easily comprehend biologists' expert testimony about the negative impacts of logging on fur bearing animals, can understand that," I exclaimed, seeing the possibilities. "How much do you need?"

"About $100,000 to get started," he said matter-of-factly, as if it were a small sum of money.

"You must be kidding. That's impossible. You know we don't have that kind of money. What's the minimum amount you need right now?" I asked.

Jack thought for a moment and said, "$17,000."

"Anyway, we can't fund a land claim case. It's not in our mandate," I said.

"You won't be funding the case. You'll be funding the research work into the aboriginal rights to trap and how those rights relate to wilderness preservation. You'll wind up with an important document you can use," Jack replied.

I wanted to say yes on the spot. But this time I knew without doubt that I would have to get the approval of the other directors to commit that much money. "Do you need it now in one lump sum?" I asked. Undoubtedly sensing that was impossible, Jack replied that small installments over time would be fine.

It was a hard sell around the directors' table. It was a great deal of money. One director pointed out that not all our members supported trapping or native land claims. Another put it bluntly, "How do we know that the Nemiah don't just want to get control of the land so they can log it themselves?"

"What if they had a tribal park declaration like Meares Island and the Stein Valley? Could we do it then?" I asked. The answer was yes. I called up Jack and explained the situation. He invited me to come with him the next week to attend a Nemiah Band Council meeting. I was keen to go.

Chilcotin country is beautiful. The Nemiah Valley itself is dominated by a big mountain; the backdrop to the band's meeting hall. I pointed to the mountain and asked one of the natives what its name was, trying to be friendly. "Don't point!" he commanded, informing me that I had just broken a big taboo. No one points at, or climbs up, Ts'yl-os (Mount Tatlo). This sacred mountain keeps watch over the Nemiah people and their territory. What I had done offended Ts'yl-os and could cause a severe weather change.

It was an inauspicious beginning. The hall was nearly full. The men in attendance all wore cowboy hats. Many stood solemnly and silently at the side and back of the room, leaning against the walls. It was very intimidating. I was nervous as I told them bluntly about needing a tribal park declaration document in order to satisfy WCWC members' doubts as to the Nemiah peoples' true intentions regarding the use of their traditional lands. The declaration had to have a map of their traditional territory with the portion to be included in the tribal park and the rules of the land use within that park. Without this we could not provide any financial support. I gave out samples of the Meares and Stein declarations. I left the meeting with the understanding that they would think about it, and get back to us.

About a week later I got word from Jack that the Nemiah wanted to proceed. In fact, they had already drafted their own tribal park declaration! He brought me the text, a hand-drawn map and a design sketch of how they wanted their finished document to look. They wanted the declaration in their own language on one side and the English translation on the other. "There might not be an exact one-to-one correspondence in meaning between the two versions," Jack explained to me, because some of the Xeni Gwet'in's concepts did not easily translate into English. The biggest stumbling block was the idea of a park. They did not want to declare the area a tribal park because "park" was a non-native concept and parks had always barred native use.

Instead, they had come up with the title *Nemiah Aboriginal Wilderness Preserve* to more fittingly convey their long-standing traditional concept of land use stewardship. In simple, powerful words the declaration prohibited commercial logging, mining, road building and damming of rivers within their preserve.

It affirmed the Xeni Gwet'in's intention to live there forever, to practice their traditional ways, but it also conveyed their willingness to share their land with those willing to respect it and leave it unharmed. He handed me the original material and asked me to help get it printed. I felt honoured.

I went to see a young graphic artist in Victoria who one of our volunteers recommended. After briefly talking to him, I hired him on the spot and gave him the Nemiah's draft text and artwork. A week later he handed me a stunning declaration document. He suggested printing it on parchment paper (costing only a little more than regular paper). We followed his advice and printed up several thousand copies. We sent most of them to Chief Roger William, more than enough for every Nemiah Band member to have one. Our comptroller somehow found the money to pay for this printing of the declaration and to send Jack cheques every time he requested—installments on the $17,000 we promised for the legal research.

Near fatal flight aborted

A few months later we were invited to Nemiah territory for a big celebration of their declaration. Quite a few WCWC staff decided to go, joined by some members of the Friends of the Nemiah, a Victoria-based group that was also helping on this campaign. Together, we chartered a twin-engine 17-passenger plane. The round trip cost us only $1,700. We took off from Vancouver in beautiful weather. As we headed up past Whistler towards Pemberton the clouds started rolling in and the plane started bouncing. As we passed Pemberton it buffeted even more. Our two young pilots had never flown to the bush airport on the Chilcotin Plateau before (and neither had anybody else except Jack). We could see the pilots through the open cockpit door poring over maps. Jack explained that all we had to do was to pop up over the Coast Mountains and then everything would be clear and smooth on the other side. We went up one steep valley and hit a solid bank of clouds. We banked sharply, barely able to turn around.

At this point we ran out of 'barf' bags. The pilots tried one more route up another steep-sided valley looking for another way through. This time they did an even more violent manoeuvre to abort. Mark Wareing, our staff forester, freaked out! He jumped out of his seat and ran up to the cockpit and shouted, "You've got to go back! I've been in two plane crashes and know when one is going to happen!"

He made a big scene, and I, for one, was grateful. The pilots gave up trying to find a hole in the clouds and put us down on the small airstrip in Pemberton. The bleak, tiny terminal was closed, but the lone outdoor payphone worked. We discussed our options while waiting for the weather to improve.

The pilots kept asking me if we wanted to scrub the mission. I knew that if I called it quits, we would have to pay for the whole flight. But if they cancelled because of the dangerous weather, we would only have to pay a small portion...the distance we had actually travelled. A great deal of money rode on that decision. Most of our people, including Mark, who would not go back into the plane under any circumstances, wanted to call it quits. To call the pilots' bluff, Jack and I said we would go on if they felt that it wasn't dangerous. We called for cars to rescue those to be left behind.

Finally, the pilots made the decision. The weather was worsening and we had to turn back. Jack went on to Nemiah by car although he would miss the big celebration that night. I went back with the others in the plane, still feeling queasy. I think we were all very lucky that day.

The next year, the B.C. government invited WCWC to sit on its newly created Chilko Study Team. We assigned Joe to the task. Many months and many meetings later, in April 1992, the Nemiah informed us that they were pulling out of the process. They'd decided that participating in negotiations with the Ministry of Parks, Ministry of Energy, Mines, and Petroleum Resources, Ministry of Forests, Cariboo Lumberman's Association, Cariboo SHARE, and others would not achieve their goal of protecting their large Aboriginal Wilderness Preserve. We immediately followed their lead. Joe was relieved to withdraw from this time-consuming, wilderness-compromising table.

Part of the Nemiah Aboriginal Wilderness Preserve becomes Tsy'los Provincial Park

After that, the Nemiah negotiated directly with the B.C. government. A few years later, in January 1994, their efforts culminated in the creation of the 233,000-hectare Tsy'los Provincial Park around Chilko Lake. The park is nearly half the size of Prince Edward Island and it's a beauty, comprised of rugged mountains, clear blue lakes, glaciers, alpine meadows and waterfalls. The deal included co-management by the Nemiah First Nation, one of the first co-management agreements in B.C. Remarkably, the *Vancouver Sun*, B.C.'s newspaper of record, did not run even a small story about the establishment of this major park. I wrote a biting letter to the editor of the *Sun*, pointing out their snub, saying that it clearly demonstrated the paper's bias against wilderness preservation. I pointed out that the newly created Tsy'los Park was in the same class as the finest National Parks in Canada and the U.S. They didn't publish my letter; but about a week later, the *Sun* published a two-page, in-depth article about the new park.

From that time forward, the B.C. government ensured better media coverage when it created new parks by holding big celebratory press conferences in Vancouver to make the announcements. The Premier would announce the new park and everyone involved in the park-creating process was invited to attend. We were all plied with food, speeches and free posters. I liked it because it made a big thing of new parks and because WCWC's comments were usually included in the news story giving us valuable profile.

However wonderful the new Tsy'los Park was, it failed to include a critical area in the declared Aboriginal Wilderness Preserve, the home of one of Canada's last herds of wild horses, called the Brittany Triangle. Jack Woodward continued to push the litigation on behalf of the Nemiah to protect the area. Today a chillingly graphic satellite image shows a still-green Brittany Triangle surrounded by about 10,000 square kilometres of a reddish-coloured area of unprotected forestlands, nearly all of which has been recently clearcut. The Nemiah's legal fight to protect the Brittany Triangle has lasted for 17 years.

Nemiah Aboriginal Wilderness Preserve Declaration. Document: WCWC archives.

In 2001, after years of determined effort, Jack and the Nemiah Band raised several million dollars to advance a full-fledged land claims case in the B.C. Supreme Court. On March 1, 2006, after years of testimony by the elders (mostly in their own language), numerous expert witnesses, and mountains of documents, the Nemiah closed their case, turning the courtroom over to the Crown to call what evidence they might have on this issue. A decision is expected in 2007 on this landmark case. I'm proud of the small role WCWC played to help launch the Nemiah's efforts to protect their lands and to know that the Brittany Triangle still sustains its wild Chilcotin ponies, although not yet formally protected.

Walk for the Environment - September 16, 1989

Around the country every spring, thousands of people participated in Walks for Peace. Every fall, given the growing threats to the Earth's ecology, we thought that people ought to have the opportunity to take part in Walks for the Environment. Ken Lay took charge of making this idea a reality.

Organizing the first event of this kind was a monumental task. We had no grants or outside financial assistance. Ken pulled together dozens of volunteers and got all the other environmental groups in Vancouver's Lower Mainland to participate. On Saturday, September 16, 1989, several thousand people walked from Vancouver's Kits Beach to Queen Elizabeth Park, the highest point in the city. We specifically chose this long uphill route to illustrate the long, hard road we must travel to change our ways and heal the damage already done to planet Earth. This chosen route was not a particularly good idea. The walk was too hard and long for really young kids and elderly people. At the end of our walk there was not enough immedi-

Wild Chilcotin ponies. Photo: Friends of the Nemiah.

Walk for the Environment proceeding to Queen Elizabeth Park. Photo: WCWC files.

ate "reward," including fun activities and entertainment. Measured by the turnout, however, our 1989 Walk for the Environment was a success. But the money and work involved in organizing the walk was just too great for the little impact that it had, to justify us putting on another one.

Toronto Environment Fair

While folks in Vancouver were walking for the environment, Adriane and I were staffing a WCWC booth at a weeklong environmental fair at the Ontario Science Centre. We did it on the cheap, taking seat-sale flights, and staying at a modest motel room nearby. The two of us walked two kilometres along the shoulder of a freeway to get to and from the fair each day. The week was a lot of fun, especially the cold beer and $1.20 a dozen spicy chicken wings we'd have at a pub on route to our motel at the end of each long day. Thousands of Toronto elementary school classes took field trips to the fair that week and we were practically the only booth with something for sale that the young students could afford. We sold thousands of ten-cent postcards of B.C. wilderness scenes and several hundred Carmanah posters. The kids took a lot of time choosing their favourite cards, oo-ing and ah-ing over big trees that grew in the ancient rainforests of Canada's west coast. Their purchases paid for the entire trip.

At this fair I met Della Burford who was telling and acting out her story, *Magical Earth Secrets*, over and over to groups of school children, mesmerizing them with her creative costumes and her environmental message. I talked to her about us publishing it in book form. A year later Della's captivating story about an Eagle child's environmental quest became WCWC's first children's book.

Magical Earth Secrets book cover. Artwork: Della Burford.

Environment Walkers reach Queen Elizabeth Park. Photo: WCWC files.

Adriane and I also met with the local group campaigning to save a green space in the Rouge Valley near Toronto. The day after the fair they took us on a tour of this threatened area. We explained how our campaigns relied on publishing huge numbers of newspapers to get the message out to the general public and gave them some of our papers as examples. I must have convinced them that this kind of massive public education is the key to a campaign's success because a few months later a Rouge Valley tabloid arrived in the mail that looked almost identical to ones produced by WCWC.

The break in Toronto, the first long one away from our kids, gave Adriane and me a lot of time to talk about strategy and future campaigns, especially the WILD Campaign. We now had a nanny to help take care of our two young children and it was Adriane's first month on a sabbatical from Langara College working full time for WCWC.

Hectic growth and increasing pace of publications
Avant garde campaigns including Forestwatch

From 1984 through 1988, on average, WCWC doubled its income each year. With Carmanah, the pace quickened. In 1989—the first year Adriane worked full time for WCWC—we tripled our income, reaching a peak of $2.4 million. That's talking gross income. Expenses kept apace and then began exceeding income as we started to hit a wall in donations. Yet we continued to hire people, take on more campaigns and massively mail out our educational publications.

Hardly anyone at WC² knew all the different projects that were simultaneously going on. Although many were involved in Carmanah-related projects, there were many other campaigns too. Beginning in June 1989, Mark Wareing, who had become our "staff forester," launched a Forestwatch program, training local citizens on how to monitor local logging companies and their practices. He spent time that summer visiting interested groups in Bella Coola, Kamloops, Salmon Arm, Princeton and other communities.

At this time we did not have formal meetings to pass spending resolutions. The directors' meeting minutes tell almost nothing of what was really going on then. To try to bring our activities under some semblance of control, in the fall of '89 we set up a "Strategy and Finance" Committee, which acted like an executive team. Through this committee we tried to coordinate all the activities, set budgets and provide general direction for WCWC. Strategy and Finance met once per week. Once we decided to do something, the person in charge of the project just went to our comptroller and got the cheques needed or arranged to have the bills sent to WCWC. Adriane and I maintained our micro-management of the content and style of our publications, which were WCWC's major face to the world.

We hardly ever said no to any new idea that advanced WCWC's mandate in a tangible way, especially if it could be "self-funding." Self-funding meant that, at least theoretically, an activity would raise the money to pay for itself or someone else was going to pay for it. When Bruce Elkins came to us that fall with a proposal that we sponsor some two-day workshops to help people become more in tune with nature and shift their consciousness, we said sure. But first we had to try out his *Creating an Ecological Future – the Eco-Creative Shift* workshop ourselves to see if it worked. Elkins put on a free one at our house on the Sunshine Coast. Most of our Vancouver staff participated. We used the regional park that borders our property for some of the sensitivity "exercises," which were quite similar to ones described in our Rediscovery book. We all liked the experience, so we decided to sponsor a workshop on Bowen Island.

Despite inviting by mail all our members in the Lower Mainland to attend and having a very low fee, the response was poor and we ended up losing money on this experiment. The half dozen participants who did show up probably already had made the shift. Maybe it was just the wrong time. I have always believed that you'll only discover what works if you try out new ideas.

That fall of '89 we pursued another big idea: expanding WCWC into the United States. With Adriane, Joe Foy and me as directors, we registered the World Wilderness Committee as a not-for-profit society in Washington State. The registration came through in January of 1990. We dreamed of producing a U.S. endangered wilderness calendar based on the same format as our Canadian ones. It would be an advocates' tool, not just a pretty picture calendar like the one the Sierra Club produced.

We also intended to expand our fundraising by appealing to U.S. foundations that could only donate to U.S. not-for-profits. Our World Wilderness Committee, having no such restriction, could then re-donate to Canadian non-profit organizations like WCWC. One of the organizations we helped support through this route, besides our own, was the fledgling *Ad Busters*. However, producing a calendar that featured endangered wilderness areas in the U.S. remained forever an unfulfilled dream.

Baby in the dump

There were several other avant garde projects that we took on in the winter of 1989-90. An innovative group called Eclipse Productions dreamed up some far out recycling ads and produced them under our name. All we had to do was pay the expenses—which turned out to be way more than we had expected. Rental of the fancy "Steven Spielberg" 35mm slow-motion movie camera to shoot the ads (sans the film) cost more than $1,000 for the day. In March 1990 we sent copies of Eclipse's three professional-quality, 30-second ads around to over 50 major companies with high hopes that some of them would, as a public service, sponsor the airing of them on commercial TV channels. Not one did.

Perhaps it was because these ads were just too hardcore. The hardest-hitting one we nicknamed "baby in the dump." It was filmed in one long, single shot with narration only at the end. The first 15 seconds consisted of a cute close-up of a 10-month baby sitting upright, playing and cooing. Then, in slow motion, the camera panned out and you gradually realize that the baby is sitting all alone in a gigantic landfill surrounded by garbage. Finally the baby is so tiny it's indistinguishable—you just see the mountain of garbage. It concludes with the narrator saying, "What will the world be like when it is filled with garbage?" Another one of the ads focused on hundreds of seagulls soaring in slow motion. When the camera pans out, it's the same giant garbage dump below, with the same question by the narrator at the end. The last one featured a man opening up the back of a big station wagon (before the SUV era) and throwing out one full garbage bag after another for the entire half minute (the average amount of garbage one person in Canada generates annually). I thought these ads were great. Too bad they were never used.

Group writing and editing improve the quality of WCWC's publications

At the Committee we developed a peculiar "group write" style. We would always pass drafts around with the understanding that we could be perfectly honest in our comments. The goal was to make our message completely understandable to the average person. At the same time we wanted our publications to be accurate, witty, snappy, and filled with interesting facts. We all had our specialties. I was fast at 'banging out' a rough first draft, but it was usually crudely written with creative spelling, weird punctuation, misplaced phrases and run-on sentences. Joe was good at first drafts, too, and in suggesting an angle that would resonate with the "common man."

Often four or five people read the copy, adding ideas, correcting mistakes and improving grammar. Adriane could edit and polish up a written piece like no one else. Sue was fast and creative at design and layout, and usually patient with our many changes. There was a bit of everyone in the final product. The production team would always put in incredibly long hours. Most of our newspapers involved "all nighters" to finish them in time. Despite the pressure, people kept their senses of humour and joked around a lot. We'd make it through the night with pizza, popcorn and beer. We knew that together our writing and productions were much better than any one of us individually could do and that we were really making a big difference. We didn't have to make it fun; it just was fun.

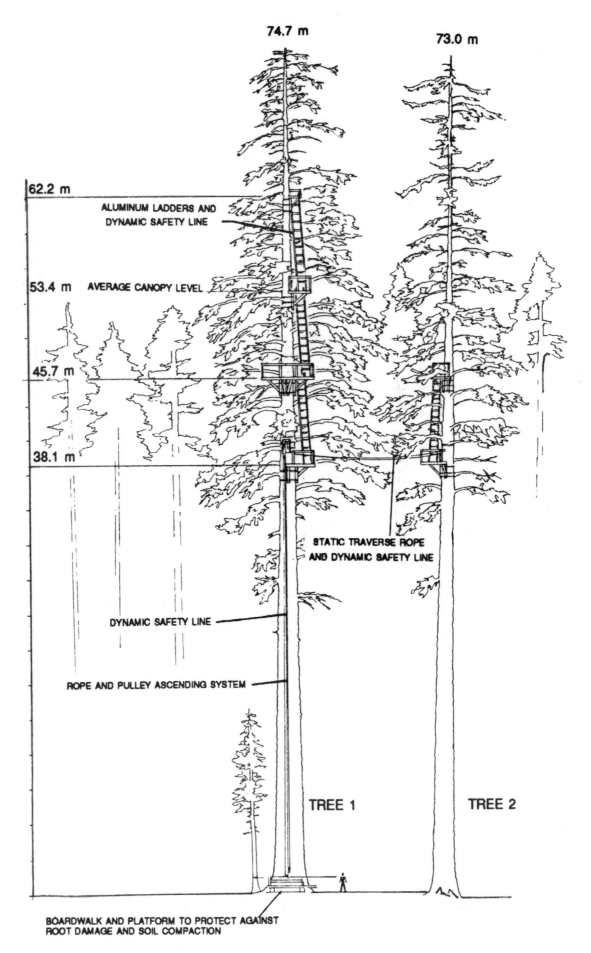

CARMANAH VALLEY CANOPY RESEARCH STATION

Chapter 27

WCWC builds the world's first temperate rainforest upper canopy research station

Shane Kennedy takes us up into the trees

That fall of '89 was busier than any before in WCWC's history. Nearly every day someone came up with a new project for us to get involved in. One of the best was a very posh *Rainforest Benefit* entirely organized by a Vancouver entrepreneur, Mary Lou Stewart, and her friends. She wanted to raise funds for our international WILD campaign and for David Suzuki's Amazon work. All we had to do was let all our members know, sell tickets and dress up in formal attire to attend.

Every other project, however, demanded massive amounts of work. Adriane and I began staying in town several nights a week to avoid the long commute every day back and forth from the Sunshine Coast. This way we could work at WC^2 late into the night and start again early in the morning when things were quieter at the office. This was only possible because we had a great nanny taking care of our kids.

Lisa Kofod, a volunteer with WCWC's WILD project, upon discovering that we were sleeping on the floor in the office, offered to let us sleep on a foamy on the floor of their basement apartment near Main and 12th, only a few minutes drive away. Lisa was living with Shane Kennedy, who was working similarly long hours at Lone Pine, a publishing company started by his father and known for its many nature guidebooks.

The four of us became great friends. Adriane and I would often pull in around midnight and, instead of going right to sleep, spend another few hours having dinner and brainstorming with them. Shane, a man of extraordinary experiences and enormous talent, was very witty and loved to brainstorm brilliant ideas.

Somehow, one night we got onto the topic of biodiversity and where to find the other 75 percent of Earth's plant and animal species that biologists asserted existed but had not yet discovered or classified by scientists.

"Lots of unknown species have to be living in Carmanah's forest canopy, just like they found out they were in the tropical rainforest canopy," I guessed.

This cue got Shane talking about his experience with some of the first canopy walkway systems in Central America. He had been part of a crew filming them and told us, "They're easy to build. You could build one in Carmanah. You use a crossbow to get a string over a limb and then pull up the first rope. You rig up a pulley system to get people up and down the tree and hoist up platform pieces and build Burma bridges made of thick rope to get between platforms on adjacent trees and aluminum ladders to go up and down to platforms at different elevations in a tree."

He explained how triangular support struts are prefabricated on the ground, hoisted into the tree, and secured in a circle around the trunk with canvas straps and a come-along ratchet—standard equipment used by truckers to hold down cargo and tarps on flat deck trucks. After the struts are in place, he explained, you deck the top

Facing page: An early scale diagram of our Carmanah Upper Canopy Research Station. Pen and ink drawing: Randy Stoltmann.

and build a railing around them. He offered to help get the project going in Carmanah. It was an offer we couldn't refuse.

From Shane's brief description I could vividly imagine what an upper canopy research facility might look like and I knew for sure that we'd find volunteers who could build it. Being rather hefty and afraid of heights, it wouldn't be me! The next morning I talked to Joe. Both of us knew we had to start focusing on the Upper Carmanah valley. Building a canopy research station there was the way to do it. We didn't worry about the cost or effort. We assumed it was do-able, legal and we could raise what it cost. We didn't ask the government for permission. Our court decision on trailbuilding would surely apply as long as we didn't harm the trees.

Earlier that fall, Joe had hiked from Heaven Camp to the top of the Carmanah watershed. There were not very many large spruces in the upper valley. In fact, he recalled only one grove with several giant trees growing closely enough together to be suitable for an upper canopy station. The grove was about a half-hour's hike down from the one clearcut that MB had made at the top of the valley a few years previous to our campaign. The only hitch was that we'd have to build a bridge over a small canyon to get to the grove.

Meeting with the Parks Minister on Carmanah

In late February 1990, Joe Foy, Derek Young (our Victoria Branch director) and I met with B.C. Forest Minister Claude Richmond in his plush legislature office in Victoria. At that meeting it quickly became evident that this minister had a closed mind on Carmanah. He was not interested in hearing our suggestions of creative and feasible win-win solutions that would provide logging jobs to compensate those that would be lost in protecting the entire valley. Richmond told us that he had already heard all the arguments. Jobs were being lost every year in the industry due to increasing automation in huge saw and pulp mills and we (the citizens of B.C.) couldn't afford any more forestland being lost to parks.

We explained to him that Carmanah Valley experienced some of the heaviest downpours in North America, with torrential dumps of over a foot of rainwater in a single day. With such heavy precipitation it was impossible to log in a way that wouldn't pose a threat to a park located downstream. We had the data to prove it: scientific studies conducted in other coastal watersheds showing that logging on steep mountainsides triggered increased erosion and landslides. The increased sediment load would ultimately fill up and widen Carmanah's creekbed and then the more frequent floods would wash away and destroy the giant spruces' flood plain ecosystem in the mid-valley.

Richmond appeared totally disinterested in our information. But he did promise to give us all the scientific information and studies that the B.C. Forest Service would be using as a basis for their recommendations to cabinet to give us an opportunity to respond before he made his decision about creating a park. We asked for this information many times, but never received anything from him. This corroborated our suspicion that we already had all the information that existed. There were no independent hydrological, wind throw or slope stability studies. The government was obviously relying on the incomplete and inadequate studies done by MB that we'd already cast doubt on. The Socreds were not going to protect the entire Carmanah Valley. For sure, however, they would soon establish a park to protect some of the lower valley—an area we'd succeeded in making famous. Now we had to make the upper valley as famous as we had the lower. We fast-tracked the work to get our canopy research station in Upper Carmanah up and running as soon as possible.

The dream becomes a reality – the world's first temperate rainforest canopy research station

We didn't have any scientists lined up to use our station before we started. All we had was faith in the maxim, "If we build it, they'll come." I was absolutely sure that researchers would find new species in the canopy. Although some scientists had inspected crowns of felled trees and not found much, these crowns were all smashed up when they fell. No one had systematically studied this temperate rainforest habitat "in situ" before. Physical barriers had held them back...especially the extreme height of the trees!

Under Shane Kennedy's guidance, a team of volunteer carpenters pre-built our research station platforms in Vancouver. They loaded

Demonstration in front of the Forest Service Office in Victoria. The letters flip over and then spell out FOREVER!. Joe Foy noticed that Carmanah and Forever! had the same number of letters and came up with this flash card idea. Photos: WCWC files.

them into a pick-up truck, took the ferry to Nanaimo and, in the dead of night, hauled them over the logging roads to the headwaters of Carmanah. On March 2, 1990, a large group of WCWC volunteers converged on the Upper Carmanah to start building the bridge, boardwalk the access trail, construct a large wooden floored research tent and begin hoisting platforms high into a huge, old Sitka spruce. They worked fast, aiming to get the research station in place before MB or the B.C. Forest Service knew what we were doing.

"How did I ever get into this?" Joe remembers thinking to himself the first time he rappelled down the immense tree. "Once I start down, there was no turning back. There was only one rope between me and the ground over 120 feet below." Soon Joe got used to this work site and he later described installing the platforms this way.

There I was, dangling from a rope 160 feet in the air in the pouring rain on a cold day in March, trying to cinch a series of four-foot-long right angle braces to the ancient Sitka spruce with canvas truck straps. It felt like I was trying to hang giant Christmas tree ornaments. It was wild.

An incredible team effort went into designing, building, transporting and installing our canopy research station. We had to ensure that the platforms did not hurt the trees, were safe and easy to remove. Even getting the first rope onto the first limb and attaching a pulley system to hoist people up and down was a difficult job.

One of our volunteers with mountain climbing experience scaled up a smaller hemlock beside one of the big spruces using boots with spurs that telephone linemen wear. He then swung over to the big tree at a height where it was of more reasonable girth to climb up further to get to the big spruce's first limbs about 10 stories above the ground. It proved to be too hard to shout instructions back and forth between the crown and the ground crews, so we got walkie-talkies to communicate.

The early volunteers on this project came from two backgrounds—sailing and mountain climbing. The equipment we initially used was a mixture of both. Early on, we had a near-fatal accident. A pulley that had formerly been used to haul people up a boat mast failed, dropping one of our volunteers a long way down before his safety rope took hold, leaving him hanging half way down the trunk of the tree. One of his hands got quite a bad rope burn in the accident, as he desperately tried to halt his fall. Somehow the cotter pin holding the shaft fell out and the pulley came apart. After that close call, we got brand new equipment designed for mountain climbing, which

Boardwalk to the WCWC's research station in the Upper Carmanah Valley. Photo: WCWC files.

Some of the first volunteer crew starting to build our Carmanah Research Station on March 2, 1990. Photo: WCWC files.

was much better suited for the job.

There was quite a bit of experimentation that went into getting the station built right. Someone, for example, had the bright idea of using a counter weight consisting of a net full of rocks to make it easier to raise and lower a person by rope and pulley. This seemed like a good idea until the first person trying it out looked up and saw the net load of rocks precariously swinging above his head over 100

Volunteer-built bridge across the Upper Carmanah with a volunteer carrying some wood across to use in constructing the research station. Photo: WCWC files.

Ascending the Sitka spruce to the canopy platforms. Photo: WCWC files.

Left top: Building a platform in one of the research trees in the Upper Carmanah. Photo: John Kelson.
Left bottom: Detail showing how triangular struts are strapped onto the tree to form the base of the canopy platforms. Photo: WCWC files.
Above: Burma bridge connecting platforms in two adjacent research trees. Photo: John Kelson.

feet above him. We quickly abandoned this innovation!

Even with a hard hat, something dropped could be deadly, like the hammer that came 'whizzing' down and sunk deep into the mucky forest floor less than a metre away from Joe. After the hammer incident we kept people away from the area under the trees when anyone was up on the platforms, unless it was absolutely necessary for them to be there. Joe himself once accidentally dropped his camera from the top platform. When he got back down he found it buried about 10 centimetres into the soft ground. Incredibly, the camera still worked! All of this pointedly illustrated how incredibly tall the trees are, how lucky we were and how important it was to work safely.

One of our first key volunteers was John Kelson. He had experience climbing big trees while hunting for marbled murrelet nests in the nearby Walbran Valley. He had natural leadership abilities and ended up taking charge of our canopy project; recruiting many of his friends to help.

He also had a lot of common sense and was able to overcome the new problems that kept cropping up. (After Carmanah, Kelson went on to perfect the technology of building upper canopy walkways and now is a director of the Greenheart Conservation Company that has built canopy walkways in rainforests throughout the world for ecotourism as well as research purposes.) We hired Kelson on contract

Above top: Looking up at a volunteer working on the canopy station from a limb below. Above: John Kelson helping construct the Canopy Research Station. Photos: WCWC files.
Right: Moving wood for a platform in an adjacent tree. Photo: John Kelson.

at $1,500 a month to oversee the construction of our canopy research facility. Besides Ken Lay, Joe Foy and John Kelson, there were several hundred volunteers who collectively put in thousands of hours of volunteer labour into this project.

By March 10th our first research platform was completed. On March 15th we held the grand opening ceremonies. On March 23rd MP David MacDonald, Chairman of the federal Environment Committee, visited the site, accompanied by Conservative MP Bob Wenman who ascended the research tree to become the first person of prominence to do so. MacDonald, in commenting about the trip, stressed the need for more research into this old-growth forest before

Neville Winchester collecting arthropods (insects and arachnids) in the canopy at our Carmanah Research facility. Photo: Bob Herger.

decision makers decided its fate.

We were right about our contention that if we built the station, researchers would come. Before it was even completed, Neville Winchester, a doctoral student in entomology at the University of Victoria, and Richard Ring, his supervisor, called us to say they were very interested. We learned years later that they asked MB officials for permission to investigate in this part of MB's TFL and they were given the go ahead, but told they probably wouldn't find much there. How wrong that turned out to be!

A volunteer's sandwich left in the upper canopy is nibbled on at night. Is it a new vertebrate species?

One day, while we were still building the canopy station, an excited message came over our marine radiophone from our Carmanah crew to our Vancouver office. A volunteer had discovered that some small animal lived high up in the tree. Inadvertently, he had left a sandwich on one of the platforms. The next morning he found the half-eaten remains of the sandwich with tiny teeth marks on it Could we have accidentally found a new species of mammal living high in the rainforest canopy? We brought live traps in and, low and behold, with a bit of peanut butter sandwich as bait, the next week we caught one. It turned out to be a mouse of the same species that inhabited the forest floor. Incredibly, these mice climb all the way up the ten-story high trunk of the tree to hunt for food. As far as I know, no one has studied this behaviour further to determine exactly what the mice were eating up there, besides our leftovers!

By June, we'd completed the construction of four canopy platforms at various heights in the main tree, connected to each other by aluminum ladders. The highest platform—a sort of crow's nest—was perched slightly over 60 metres (200 feet) above the ground. There were two other platforms, one each in two adjacent trees. Active research was now ongoing.

This project quickly became WCWC's most expensive one to date. We spent a small fortune on ropes, equipment and transportation to build and keep the station operational and safe. Our canopy station was not the "tree house" that Minister of Forests Claude Richmond dismissively referred to it as, but a serious, sophisticated research installation and the first of its kind in North America or anywhere in the world in a temperate rainforest. It stimulated a torrent of canopy research in the Northern Hemisphere and the creation of canopy eco-tourism facilities around the world. If they'd had any vision, the B.C. government could have protected the area, funded the station and taken the credit for it.

Chapter 28

B.C. government makes a half-park decision
WCWC wins prestigious award

Getting the Carmanah book into the hands of influential people and decision makers

As winter turned into spring of 1990, everyone was abuzz over Globe '90. This massive conference, held in Vancouver from March 19 to 23, was promoted as the first fully-integrated trade fair and conference that simultaneously addressed the issues of "sustainable" economic development and environmental protection. The governments of British Columbia and Canada officially sponsored this conference together with major national and international associations and financial institutions. Gro Brundtland, author of *Our Common Future* (1987), gave the keynote address.

We wanted to rent a table in the exposition hall to present WCWC's educational materials, but the cheapest one cost $2,000 so we quickly dropped the idea. Adriane, however, was invited to attend the breakout session on tourism titled, *Blueprint for Sustainable Development of Tourism*. Here she presented a scholarly paper on eco-tourism, and subsequently she was invited to participate in a working group that produced, by the end of the conference, a set of global guidelines for eco-tourism.

Other WCWC staff members attended too. Arne Hansen, who had worked so hard to put together our Carmanah book, got a media pass because he was a journalist. Sue Fox took a few of our Carmanah books and, being quite aggressive, gave one to Gro Brundtland, explaining to her that the B.C. government was not supportive of protecting this incredible place.

Arne cornered Premier Vander Zalm and gave him a book, asking him rhetorically what he was going to do about all the people who were going to blockade government offices over the issue—a tall tale he just made up on the spot to see how the Premier would react. Later, we heard that Brundtland showed Vander Zalm the Carmanah book she had been given and talked to him about Carmanah, inquiring if the government intended to protect it. We didn't hear what his answer was. The pressure was on.

When Adriane and I were in Ottawa for a meeting of the environmental groups later that same early spring, Adriane called the office of the Environment Minister to see if we could get a brief meeting with him. After explaining her way through several office staff levels she finally reached the minister's assistant and explained that all we wanted to do was to personally present him with a copy of our *Carmanah Artistic Visions* book.

"Just a minute," he said and a few moments later he came back on the line and told Adriane that if we could get there by noon we could briefly see him. A few hours later we were in Lucien Bouchard's Ministerial office talking to him while he ate his lunch. He was extremely friendly and expressed his interest in art and wilderness.

We never had the opportunity to followed up on this meeting: a few months later the Meech Lake Accord failed, Bouchard quit government and formed the Bloc Québécois. Maybe, if he'd stayed on as Environment Minister, the federal resistance to adding Carmanah to Pacific Rim National Park might have changed. You never know which politician will tip the balance in a campaign.

The day of the big announcement

By April 1990, public pressure on the B.C. government to create a park in Carmanah had reached the boiling point. On Monday April 9, 1990 someone from the media tipped us off that the government was going to announce its Carmanah decision the next day. This helped us get psychologically prepared. The next afternoon, about a dozen reporters and three TV cameras crowded into our cramped mezzanine floor office at 20 Water Street. Ken Lay, Joe Foy, Adriane and I sat behind a table facing them. On the table prominently displayed in front of us sat a bottle of made-in-B.C. "champagne," just in case a celebration was in order. Behind us, pinned to the wall as a backdrop, was our *Carmanah Big Trees not Big Stumps* poster. It was up on about 15,000 other walls across the nation at that time. We all waited patiently for B.C. Forest Minister Claude Richmond to make the big announcement

One reporter's cell phone rang. Cell phones were rare and a novelty at the time. It became obvious to all of us overhearing the conversation that he was talking to a colleague in attendance at the legislative press conference in Victoria where the announcement was about to be made. He was the only person who could hear what was going on there. Suddenly, Ken got the bright idea of asking him to hang up and have the person at the other end call us right back on our regular office phone so we could put the announcement on speakerphone so everyone could hear. He agreed.

Our phone rang. Ken pushed the speakerphone button and suddenly it was as if we were there. We could hear everything. Although I was fairly certain beforehand of what the announcement would be, I felt a deep pang of disappointment when Minister Richmond's voice boomed out that his government was establishing Carmanah Pacific Provincial Park in the lower half of Carmanah Valley. MB would be allowed to log the upper valley, but only after the company finished studies to prove that such logging would not harm the park downstream. But, I thought, at least the government hadn't gone with the leave-strip park that MB had proposed and had placed a moratorium on logging in the upper valley.

The Minister's announcement of the half-valley Carmanah Park ended abruptly and lights on the TV cameras in our room flashed on. The questions came in a flurry, "How do you feel about the decision?" "What are you going to do?"

For a second I hesitated. The four of us looked at each other, each wondering if we should pop the cork on the champagne bottle and celebrate what we'd won. I thought for a brief second that we should pop the cork, drink half the bottle and re-cork what was left, keeping it for a second celebration when the upper valley was protected. But, without saying a word, perhaps by telepathy, we unanimously decided to leave the bottle unopened and save it for the day when the upper half of the valley was also protected. Then we'd truly celebrate! In front of the cameras I called for a fat felt pen and wrote on the label "*to be opened only when the entire Carmanah Valley is protected.*"

Adriane's statement made all the news stories. Upset because she knew that a half valley park wouldn't keep Carmanah's valley bottom old-growth groves alive, she called the new park a "Solomon's decision," akin to Solomon commanding that a baby over which two mothers were fighting be cut in half. We told the reporters that we were redoubling our efforts to save the upper valley. We'd immediately finish our canopy research station in the Upper Carmanah Valley and scientists would begin the search for new species.

The government's decision, of course, included setting up a stakeholder team to plan how best to log the Upper Carmanah Valley.

Obviously we were not interested in participating. We urged all the other environmental groups to boycott it. Of course, not all did. Representatives of some environmental groups sat on it and, to their credit, they talked and stalled for years any pro-logging decision from being made.

Government threatens to remove our Canopy Research Station if we do not get a permit for it

The day after the park announcement, the B.C. government threatened to order the removal of our canopy research platforms in the Upper Carmanah unless we formally sought and got a special use permit to operate the station. We complied, of course, and retaliated with our own threat that, if necessary, we would go to court to defend our right to conduct research there. Mostly the permit application amounted to demonstrating that we had insurance coverage and could explain in detail our station's safety features.

Finally, after many months' delay and lots of correspondence, the B.C. Ministry of Forests and Lands issued us a special use permit for our Carmanah canopy research station, charging us $200 per year. Then we got a tax bill for the "improvements" we had made to the land! The Nanaimo/Cowichan office of the B.C. Assessment Authority assessed the value of the station at $32,000 (How they arrived at that figure I never figured out!) and we began paying several hundred dollars per year in annual taxes. It seemed like harassment to us, but undoubtedly it was only business as usual to the bureaucrats in charge.

Winning recognition for our accomplishments

At the end of April, soon after the Carmanah announcement, the B.C. government invited WCWC to attend an awards dinner at the Lieutenant Governor's mansion in Victoria. I went our representative. In a strange political flip flop, the guy who had hassled us so much about our research station, B.C. Environment Minister John Reynolds, presented WCWC with the "1990 Environmental Achievement Award" as the best environmental group of the year!

Reynolds handed me a framed limited addition print of a wolf with a brass plaque on it commemorating our achievement. At the dinner, I met for the first time Lieutenant Governor David Lam and presented him with a copy of our *Carmanah - Artistic Vision of an Ancient Rainforest* book, which he graciously accepted. It was the high point of the evening for me. The wolf print award now hangs on the wall in the WCWC's boardroom.

That same month a number of us from WCWC got dressed up to attend the B.C. Book Prize banquet and awards ceremonies in Victoria. WCWC's *Carmanah Artistic Visions of an Ancient Rainforest* was a finalist in the book prize competition. I was not surprised when we won the Roderick Haig-Brown Regional Prize for the best book contributing to the appreciation of British Columbia. But I was taken by surprise when our book also won the Bill Duthie Booksellers' Choice Award for the best publication of a B.C. publisher. To win this award a book not only had to be good; it had to have "legs," meaning that it had to sell well. Ken Budd got up and gave a short acceptance speech ending it with something like, "you ain't seen noth-

Carmanah Research tent cabin in March 1990. Photo: WCWC files.

Environment Minister John Reynolds (centre) presents me (left) with the Group Environmental Achievement Award for 1990, which I accepted on behalf of the Western Canada Wilderness Committee. Lieutenant Governor David Lam (right) looks on. Photo: Government of B.C. (print inside frame digitally enhanced for detail and clarity).

ing yet—wait 'till you see our *Clayoquot on the Wild Side* book next year." I think his speech jinxed us. We should have just been grateful for what we got!

In the weeks to come, we continued to raise money to fund our Carmanah research station, trailbuilding and boardwalking activities. Our best fundraising appeal was still our "adopt-a-tree" program. By the thousands, faithful Carmanah supporters sent in their $25 "adoption fee" donations with the name they had chosen for "their tree." We sent each donor a beautiful certificate with their tree name hand-scripted onto it and embossed with the WCWC's corporate seal. Preparing these certificates took time, but not nearly as much and with a lot less hassle than our first tree adoption donation program we had in the Stein.

Our upper canopy research station kept our Carmanah campaign going strong. We found that, once you explained to people how a whole watershed functions as an ecological unit, few accepted the half-a-valley park compromise for Carmanah. Anyway, almost everyone agreed that the logging companies had already taken more than their "share" of our coastal rainforest on Vancouver Island especially on the southern half. Only six out of 89 big (over 5,000 hectares) primary watersheds (river systems that empty into the ocean) remained pristine on Vancouver Island and only one was fully protected.

At a public event shortly after the government announced creating the Lower Carmanah Park, Adriane unexpectedly encountered Premier Vander Zalm. He was alone with no handlers beside him. She seized the opportunity to talk with him about Carmanah. Quickly she explained the importance of adding the upper valley to the park so that the whole watershed would be protected. "What do you mean by watershed?" he asked. She explained to him that a watershed is an area of land where the surface water flows into one stream or river drainage system.

With a puzzled look on his face he asked, "Could you explain that to me again?" She explained it in another way: that a watershed is an independent water drainage system and that all lands are naturally divided up into separate watersheds, isolated one from the other by the height of land dividing them. According to Adriane, he got it. This exchange told us that we had to do a better job of explaining to the general public key ecological concepts that we had taken for granted that they understood.

Chapter 29

Vandals leave Carmanah research station in ashes
Project Phoenix
Biodiversity bonanza

Burnt remains of Carmanah Research Station tent cabin. Photo: John Kelson.

Carmanah Research Station in Ashes

Throughout our Carmanah campaign we experienced a lot of petty harassment from the park foes. The constant removal (stealing) of the signs we positioned along the logging roads to help people take the right turn-offs to get to Carmanah Valley got to be very irritating. Someone dumped a smashed wreck of a vehicle at one logging road junction, placed a big log on top of it and spray painted WCWC on the side of the car. It remained there for a long time. We used it as one of our most permanent signposts!

Then, one weekend in early October 1990, things got ugly. Volunteers going in to work on our research station encountered a blockade set up by loggers who said they wanted to show us what it was like to be prevented from working. Our volunteer work crew protested that WCWC never participated in blockades and that there had never been any blockades involved in the campaign to save Carmanah. But this was to no avail. We were kept out of Carmanah for two days. The loggers blockade came down after the weekend.

Soon, we found out the blockade's true purpose. Behind it, vandals had completely destroyed our research station. They'd wrecked everything—except the research trees themselves and the platforms high up in them. They cut the access ropes to platforms. They sawed in half our bridge over Carmanah Creek (that had been built with considerable difficulty and with so much enthusiasm by volunteers earlier that year), dropping the two sections into the canyon. Using sledgehammers, the rampaging vandals had smashed every single one of the treads on our boardwalk and tossed many of the large stringers underneath them into the bush. (The boardwalk was more than a kilometre-long and some of the treads more than five-centimetres-thick.) Our research tent cabin had been burned to the ground in a fire so hot that the aluminum ladders stored beside it had melted and burned up. Sifting through the rubble, the only thing a volunteer found was a singed paperback book cover with the words "Heart of Darkness" on it. All the pages of the book, like everything else, had been burnt to ashes.

At first, it confounded me that the research trees were spared. Falling those old-growth spruces was the one sure way to have permanently put us out of business. Then I realized that doing so would have transformed this act of mere "vandalism" into a major crime. Destroying WCWC's property was one thing; harming MB's trees was another! If they'd done that, I'm sure the RCMP would have been much more diligent in their efforts to find and bring the culprits to justice. As it was, no one was apprehended or charged for destroying our research station. All in all, we estimated that we lost over $30,000 worth of materials and direct costs, excluding the 5,000 hours of volunteer labour that went into building everything.

Phoenix Arises

Our response to this outrage was immediate. We started planning to build an even better and more extensive research facility than the first. We called it the "Phoenix Project." It was too late to do much rebuilding work that winter. We put our energies into getting everything ready for a big coordinated effort in the spring. The response from our friends was terrific. Hundreds of volunteers signed up to help with the actual rebuilding in the spring. People contributed everything from electrical generators to spotting scopes and video cameras. American Fabricators Ltd. donated a new tent shell for the research cabin and Extreme Mountaineering Ltd. donated new climbing equipment and rope for the canopy research platforms. A tree-planting co-op donated a whole set of trailbuilding tools, cook stoves and tents.

As soon as the longer days arrived in spring, Lyn Wallace and a team of volunteers from our Mid Island Branch, undertook the task of rebuilding the bridge. They used huge amounts of wood donated from Gogo's cedar mill in Nanaimo and from Wildwood, Merv Wilkinson's eco-forestry-managed, selectively-logged woodlot in Ladysmith. Joy Craddock, WCWC's volunteer coordinator, led the team of volunteers from our Vancouver office. By the beginning of the summer '91, our Carmanah research station was up-and-running again, ready for another intense season of canopy research.

In the years that followed, as Adriane attended workshops and courses in fundraising, it was pointed out to her how foolish WCWC was in not having sent out a hard-hitting direct mail appeal for financial support immediately after the savaging of our research station and boardwalk trail. "You missed the best opportunity I've ever seen to bring in several hundred thousand dollars," one fundraising expert told her. Perhaps we did, but at the time it never occurred to

Wrecked bridge across Carmanah Creek to our research station. Photo: John Kelson.
Facing page: Smashed boardwalk with burnt out remains of our Carmanah Research tent cabin in the background. Photo: Art McLeod.

"Wood Rats" crew scavenge wood for our boardwalk trail from Gogo's Nanaimo mill's cedar waste wood pile (background). Gogo on far left. Photo: WCWC files.

us to view this criminal act as a "perfect fundraising opportunity." We were intently focused on demonstrating to our "enemies" that intimidation didn't work on us by getting our station up and running again as fast as possible. We believed that if we simply worked hard and stayed on the leading edge of the movement to save wilderness, people would recognize our efforts and donations would flow in. Later we realized it's not always that simple and our fundraising programs could be improved.

Biodiversity bonanza in Carmanah

For four years Neville Winchester and a small crew of research assistants worked out of our Carmanah Research Station. They set

Weekend Phoenix Project work party. Photo: John Morton.

Top left: Rebuilding a section of the boardwalk to the Research Station, March 1991. Photo: Joe Foy.
Top right: Phoenix work crew camping in the clearcut in the Upper Carmanah by the trailhead to the Canopy Research Station. Photo: WCWC files.
Bottom left: Volunteers building the new improved Research Cabin in the Upper Carmanah Valley. Photo: WCWC files.
Bottom right: Volunteers prepare to hoist the vinyl tent over the newly finished cabin frame. Photo: WCWC files.

traps and collected arthropods all over the research trees' canopy, including the moss pads on the big old tree limbs. They also trapped and collected specimens on the forest floor, under rocks, in the soil and even in the clearcut at the head of the valley. They knew that each different species occupies a different niche and has different behaviour patterns and survival strategies so, to find all the different arthropods (which was a main focus of their research) they'd have to sample everywhere.

All the specimens they collected were preserved, sorted into broad classification categories and saved for taxonomists (scientists who classify life forms) to identify later. The traps had to be tended and the insects removed on a regular basis. In 1992 the Research Branch of the Ministry of Forests provided Winchester with $35,000 for this research. In 1993 it dropped its support to $15,000.

The canopy of these old-growth (on average, 600-year-old) spruces turned out to be a fascinating world of huge limbs with epiphytic plants and unique micro-ecosystems. The largest limbs on these trees were bigger than all but the biggest trees in Eastern Canada. The old limbs had grown for so long that moss pads up to 30 centimetres thick had built up on them by the decay over hundreds of years of bits and pieces of plant matter dropping down from the dense foliage above. These upper canopy old-growth limbs turned out to have a unique microcosmic ecosystem populated with endemic insect species hitherto unknown to science.

Winchester figured that over three million insects crawled or flew into his traps during the four years that he collected samples in Carmanah Valley. The sheer number of different kinds of creatures makes the task of collection and subsequent classification difficult. He estimates that he found over 15,000 different species and he didn't even collect or identify them all! Not being an expert in classifying insect species (Winchester is an ecologist studying the bigger-scale interaction of species and their environment), he enlisted the volunteer help of 62 scientists around the world who are specialists in identifying specific kinds of insects, mites and spiders.

He sent each one the specimens that he was unable to identify belonging to the Families and Classes in which they had expertise. It became their job to identify them and determine which ones were new to science. These taxonomists were all working on other projects, too, so the process of identifying Winchester's specimens was slow and is still on going. It didn't take long, however, before these taxonomists were identifying one new insect species after another. In total Winchester has sent out 355,000 specimens. His lab at the University of Victoria is still filled with unidentified arthropods that haven't yet been classified.

Following Winchester's research was a real eye opener for me. Being a biologist, I always felt that our society places far too low a priority on understanding how primary ecosystems, especially forests, work. This is unconscionable considering how rapidly humans are changing (wreaking?) the biosphere. Our meager knowledge and lack of funding for essential ecological research rings alarm bells in my heart. Winchester's startling findings in Carmanah must be a wake-up call for everyone.

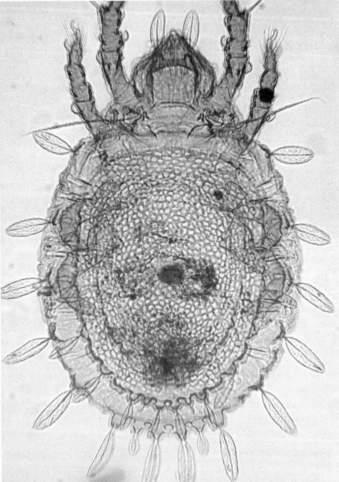

Top: Sorting specimens collected in the Upper Carmanah in a University of Victoria lab. Photo: Bob Herger.
Bottom: Unique mite found in cedar tree canopy in the Walbran Valley adjacent to Carmanah Valley. Photo: WCWC Victoria files.

A worldwide shortage of trained taxonomists is a big bottleneck in biodiversity research. In Canada, the federal government no longer trains taxonomists, and the number of taxonomists is declining. That's tragic! Without the unique skills of taxonomists, the work of Winchester and thousands of other scientists collecting samples of potentially new species around the globe will come to naught. And without identifying the players in the ecosystems, how can we ever even begin to understand how ecosystems themselves really work? I fear the consequences from insufficient money being dedicated to this kind of research: we will fail to figure out earth's ecosystems before mass extinctions irreversibly disrupt and unbalance them.

Although it's exciting to discover and name "new" species, a much bigger challenge is figuring out each species' life cycle and the role it plays in the web of life. We need to know which species are old-growth dependent and won't survive if nearly all of B.C.'s forests are converted, as currently planned, into even-aged plantations with short "rotation" cycles. There wouldn't be, for example, the thick moss pads on huge ancient tree limbs in a forest cut every 70 years and all the species endemic to them would go extinct. Retaining all the components of a natural self-regenerating forest is one of the obvious reasons why more old-growth forests must be preserved.

Winchester maintains that sustainable forestry is inseparably coupled with maintaining biodiversity. He warns that many arthropods in B.C.'s diverse rainforest ecosystems now face "Centinelan" extinction, a new term coined in the 1980s referring to the demise of a species that has never even been scientifically discovered or named. The term is linked to the real-life extinction of some unique endemic species (locally-living species that are the only ones of their kind on earth) that lived on an isolated ridge called Centinela in the western Andean foothills of Ecuador. Here, people turned a spectacular cloud forest into common farmland in less than a decade, thoughtlessly wiping out an incredibly diverse, utterly unique and barely studied ecosystem. The difference between this tragedy and the one unfolding in B.C. is that in B.C., it will be loggers, not poor peasant farmers, clearcutting old-growth forests that causes the Centinelan extinctions ...and our government is knowingly letting it happen.

One of my all-time favourite WCWC newspapers is *SAVE UPPER CARMANAH VALLEY – Home of the world's first temperate rainforest canopy research station* published in the summer of '93. It tells the story of Winchester's exciting research with many great photos and illustrations. At that time 60 new species had been identified and Winchester estimated that 200 new species would be described in all. What a conservative scientist! By 1999, approximately 300 new species had been identified but only a few named with only a small fraction of the total number of specimens having been looked at and identified. As of 2005, 26 new species had been named. The others sat in bottles with simply "new species # 1, new species # 2, etc." written on the labels. They await the time consuming task of naming. Winchester now estimated they have found at least 500 new arthropod species in Carmanah with about 30 to 40 percent being mites. Winchester's ground-breaking research has led to the doubling of the estimated number (from 30,000 to 60,000) of arthropod species native to Canada.

Despite not knowing all the species by name, Winchester has observed that a lot of predator species are present in the canopy, which he believes contributes to the dynamic stability of the ecosystem and the longevity of the trees. He explains this in an article published in the January 1999 *RPF Forum*:

"The documentation of high predator loading (e.g. Carmanah Valley, 32 species of spiders) in a functionally diverse, historically old and structurally complex ecosystem such as the ancient rainforest canopy, is in concert with previous studies. It supports the conclusion that the amount and kinds of plant material lost to herbivory are never extreme because epizootic outbreaks are prevented by a series of checks and balances, provided by natural enemies, that have developed over a long period of time."

In layman's language what Winchester is saying is that old-growth forests stay healthy and green because of the predator insects that, over time, have evolved in the old-growth's upper canopy and keep other insects in check. In the years since beginning their Carmanah studies, Dr. Neville Winchester and Dr. Richard Ring have become articulate spokespersons advocating the preservation of a lot more old-growth rainforest in B.C. As Winchester has continued to sample different canopies in different valleys across BC, including valleys adjacent to Carmanah, he has discovered unique species new to science in each one of them. He has found that biodiversity is more extensive than even he had originally supposed. Arthropods are the basis of every ecosystem and make up the bulk of the estimated 30 million species thought to inhabit earth at the present time. And we hardly know anything about them. This speaks loudly to preserving a great deal more of the old-growth forests left in B.C. The job of WCWC was to build an informed public that will pressure governments to heed the advice of scientists like these.

Famous and soon to be famous people rise into the Carmanah canopy to see this wondrous realm

During the years that our research station was up and running many important people ascended to our canopy platforms. David Suzuki and his 11-years-old daughter Severn made the trip there with Joe Foy and Joe's two young kids, Jessie and Jennifer, to film a segment for a special children's documentary that David was producing.

In July 1991, a month after federal Environment Minister Jean Charest gave WCWC the Outstanding Environmental group of the Year award, Joe Foy guided Charest and his young daughter up into the Carmanah research trees. In June of 1993, Joe Foy took Sir David Attenborough, star of countless BBC nature programs, along with a crew from the Ted Turner News Network, into our station and hoisted them up the tree. A National Geographic photographer who came along took hundreds of photos. Months later, a full-page image of the Sitka canopy taken from one of the platforms (showing part of the station) was prominently featured in that magazine, without any mention of WCWC at all.

Although she never made it up into the canopy, among our early trailbuilding volunteers in Carmanah was Tzeporah Berman, then a university student from Ontario on a summer break. She virtually camped in our 20 Water Street office for days waiting for a ride to Carmanah to help on our trails. She pitched in and diligently helped stuff Carmanah mailouts while she waited. Finally, realizing that this young woman was absolutely determined, I arranged a ride for her and she spent the rest of her summer there helping us build trails. Tzeporah went on to become one of the leaders of the blockades in Clayoquot Sound in '93 and then worked for Greenpeace and was a founder of ForestEthics, an environmental group that focuses on market pressure to motivate companies into adopting sustainable practices where she works today.

Volunteers that went to Carmanah fell in love with the old-growth forest. Some of them fell in love with each other, too, like Jay Hamburger, now head of Vancouver's Theatre in the Raw, with Atty Gell. It was a time of real excitement and adventure for the hundreds of people who worked on the trails and went up into the canopy research station. They gained a huge respect for and a gut understanding of the magnificence and complexity of B.C.'s ancient temperate rainforest. They experienced the feeling that motivated us to call Carmanah's ancient rainforest a living cathedral.

Major research at the station finished

In 1993, four years after we'd built our canopy research station in Carmanah, Neville Winchester completed his research there and our research station fell into disuse. No one else came forward to do any other major studies. Pearson College built a similar station in some old-growth Douglas firs at Rocky Point near Victoria B.C. For students attending the University of Victoria, this station was much easier to access than Carmanah and undoubtedly here they would discover a whole raft of new species uniquely adapted to that drier forest type.

Joe Foy and his two kids in the Carmanah canopy research station to be part of one of the TV shoots. Photo: John Kelson.

Premier Mike Harcourt, who was elected in October of 1991, actively sought to end the conflict in B.C.'s forests. At the end of January 1992, his new NDP government established the Vancouver Island Commission on Resources and the Environment (CORE) to come up with recommendations for what areas should be protected on Vancouver Island. Of course, the Upper Carmanah was high on all environmental groups' lists. In April of 1992, before CORE got underway, the planning team that had been established after the lower half of Carmanah became a park (to figure out how to best log the upper valley), took its recommendations to a series of public meetings in Vancouver and Victoria. We alerted our members and volunteers and they packed these meetings. As part of the "public viewing process," the planning team asked people to fill out a questionnaire. The results were never released. Many of our members asked pointed questions that the team could not answer about the impact of logging in a steep-sloped, high-rainfall valley. Government still hadn't done its research. The vast majority of people at these viewings simply said, "Don't log Upper Carmanah at all."

During the afternoon part of the public viewing in Vancouver, when hardly anyone was there except the government, company and union representatives on the team, in walked a fellow who had "streaked" several of our volunteer night sessions. As a result we had told him he was no longer welcome as a volunteer. Joe Foy, who happened to be there at the time, watched the man take off his pack and all his clothes (leaving them in a heap in a corner of the room) and calmly walk round the room stark naked looking intently at each map and chart. When the slow streaker completed his circuit he picked up his pack and clothes and left. No one in the room said a word. The IWA representative looked over at Joe and rolled his eyes. That streaker certainly didn't help our cause!

Chapter 30

WCWC branches out
WILD holds an international conference and plans to help the Penan

Branching out WCWC operations

The incredible growth in WCWC activities and profile—especially through our Carmanah campaign—brought many people to us who wanted to set up WCWC "branches." We believed in grassroots campaigning, so we nearly always said yes. In a membership paper at that time we explained it this way:

> Our head office can't possibly do everything needed to win all the issues out there! Branches make it possible to do more; to have more dedicated people working on wilderness-saving issues. The advantages go two ways. Local branches tap into the expertise of the main office, accessing the support of the entire Wilderness Committee membership for their local issue <u>and</u> the branches help the main organization by building local support for national and international campaigns. We think and act locally and globally at the same time!

In less than a year Derek Young, the founder of our first branch (WCWC-Victoria) built up a large WCWC organization with an office and store on Wharf Street in the basement of the original customs house, one of Victoria's oldest buildings. Peter Pollen, a former mayor of Victoria and a big supporter of WCWC and wilderness conservation, provided this space to us free of charge. After out-grow-

Jim Young, WCWC store manager on left and Dennis Kangasniemi, Branch Coordinator inside our new Victoria Store in Bastion Square. Photo: WCWC files.

ing this location, in 1990 Derek rented a bigger store and office in Bastion Square. Among the first paid employees there were Dennis Kangasniemi, branch coordinator, and Jim Young, Victoria store manager. WCWC-Victoria was run like an independently-minded clone of the larger WCWC body. In that first full year of operation alone (1990) their newly established door-to-door canvass increased WCWC's membership in the Victoria area more than 17 fold, from 200 to 3,500.

Besides helping on the Carmanah campaign our Victoria Branch took on the fight to save the Lower Tsitika valley. The campaign to preserve the entire Tsitika watershed, the last major unlogged water-

Facing page: Tree buttresses growing in the wild tropical rainforest in Penan territory in Sarawak, Malaysia. Photo: Thom Henley.

shed on the east coast side of Vancouver Island, was lost in the late '70s—before WCWC was born. It had been a decade-long fight by conservationists that ended up with a supposedly more sensitive logging regime. Now we aimed to save the old-growth forests left in the lower valley, near Robson Bight. Our main office helped out in this campaign. We contracted Clinton Webb, the forest technician who had helped with the pocket wilderness campaign, to analyze the "special management" logging in the Upper Tsitika. Clinton's report revealed that the Tsitika was being logged too fast to be sustainable and there weren't even key habitat areas being left for wildlife, as promised. So much for "special management!" Clinton took great photos of the logging mess and toured a slideshow around Vancouver Island to garner support for full protection of the Lower Tsitika Valley.

From left to right Clinton Webb, Joe Foy and Derek Young discuss WCWC's Tsitika campaign. Note Joy Foy wearing our Tsitika T-shirt. Photo: WCWC files.

WCWC was already growing by leaps and bounds on Vancouver Island when Laurie Gourlay, a long-time activist with SPEC (Society for Pollution and Environmental Control), stopped by our Gastown office on his way home to Vancouver Island after extended travels. We went out for a beer and I suggested that he might want to set up a WCWC Branch in Nanaimo. I explained to him that we didn't participate in any form of civil disobedience, a form of protest I knew he was attracted to. After thinking it over for a few days, Laurie decided to go with WCWC choosing Mid Island as the name for the new

Laurie Gourlay founder of WCWC's Mid-Island Branch. Photo: WCWC files.

branch, instead of opening a branch SPEC office as he had originally planned to do. In a few month's time, with the help of many volunteers, Laurie had opened a temporary store in Nanaimo and in the five days before Christmas 1989 it grossed $4,000 in sale of WCWC merchandise. In April of 1990 Laurie's Mid Island Branch opened a storefront office in the centre of Nanaimo along the Island Highway.

WCWC's Mid-Island Branch's store in Nanaimo in 1989. Photo: WCWC files.

It did a roaring business. Our Mid-Island Branch helped out greatly with both our Carmanah and Tsitika campaigns.

Joan Yardley, a middle aged woman who had canvassed for us in Vancouver, wanted to return to her hometown in the Okanagan and do the same grassroots organizing that people in Victoria and Nanaimo were doing. Soon we had an Okanagan Branch and a Vernon storefront office.

Interior of WCWC's Nanaimo Store. Photo: WCWC files.

Our three branches operated independently with very little direction or control from the main office. They all mounted their own campaigns. They all sold our posters, T-shirts, mugs, postcards, calendars and guide maps (which we gave them on consignment at a hugely reduced price) along with other products and handcrafts from local people. But what seemed like a good idea, and temporarily was a good idea, gradually turned into a nightmare. Soon, all our branch stores were losing money and the main office became preoccupied with trying to help them be good storekeepers, rather than active campaigners. We came to the sad realization that there simply wasn't a big enough market in these new locations (except in Victoria in a better location) to sell our "specialty line" of wilderness publications and products.

All of our experimentation with growth was not negative, however. In Nanaimo, because no one else was willing to do it, we signed the legal documents that made it possible for some committed people to open the city's first recycling depot. This involved accepting ultimate financial responsibility for a $200,000 operation.

Several directors on our board pointed out that we were straying far away from our core purposes and taking on huge liabilities. I argued that garbage is the antithesis of wilderness. In those days we were brash and we tackled nearly everything that came our way that would help the environment and the environment movement grow. A few years later the Nanaimo recycling depot separated from WCWC, assuming financial control of its own operations.

The Surrey-White Rock Branch forms as a completely volunteer run organization

Surrey and White Rock Branch of WCWC formed at the same time. It was different from our other branches in that it was from the start run completely by volunteers and never expressed any interest in opening up a store or an office. Its main focus was the protection of the Boundary Bay ecosystem: a globally significant natural treasure located in the Branch's "front yard." Boundary Bay had long been internationally recognized as a critical stopover on the Pacific Flyway for birds migrating north and south along the coast. Working with groups such as the Boundary Bay Conservation Committee, the Branch pushed to make Boundary Bay a UN Biosphere Reserve with special land use designations that would protect the area's precious bird habitat.

The Branch held monthly public information meetings. Liz Walker, one of the key organizers on its informally structured steering committee, would contact me or Joe and we would come down and put on a slideshow about our latest campaigns. About 70 people regularly attended. The Branch also cooperated with local naturalist

Okanagan Branch volunteers pose with WCWC merchandise. Photo: WCWC files.

clubs to hold annually a big educational event—the Boundary Bay Bird-a-Thon. The daylong event involved elementary school children and their parents and teachers. About thirty naturalists familiar with the birds took the students on field trips to several locations on the Bay and helped them identify the different species.

Besides this one big event, the chapter also offered a monthly nature walk program, helping to build public awareness of the need to preserve Boundary Bay's critical wetlands bird and wildlife habitat.

Biggest Stein Festival held in Mt. Currie

Almost all of WCWC's staff took a break from campaigning in Upper Carmanah to attend the fifth annual *Stein Voices for the Wilderness* festival in the summer of '89. This time it was held on the first weekend in August at the rodeo grounds in Mount Currie. The Mount Currie Band shared title to the Upper Stein Valley with the Lytton Band and co-signed the 1987 *Stein Tribal Park Declaration*. Bruce Cockburn, Blue Rodeo and Gordon Lightfoot headlined the show. Because of the easy road access and the big name performers, especially Cockburn, thousands of people came. All Friday night and Saturday morning, an endless stream of cars poured into the festival site. People set up camp in a huge adjacent field. They positioned camp chairs and spread out blankets to sit on for the concerts, filling the football-field-sized rodeo grounds. In all, twenty thousand people attended. It was the largest festival ever held for wilderness in B.C. The weather was perfect—sunny and hot—and the music was terrific. The mountains on the other side of the Lillooet River amplified the sound like being in an amphitheatre.

We set up WCWC's booth at the back of the field facing the large festival stage that was flanked by mountains of speakers on either side. Dozens of our regular WCWC volunteers came to help sell merchandise at our booth as well as help the festival organizer, John McCandless, with set-up and clean-up. We were one of the few booths selling anything besides food at the festival. I remember one woman, obviously a shopaholic suffering from withdrawal, tell me that she had been at the festival all weekend and she'd reached the point where "she just she had to buy something." There were lots of people like her! Incredibly, that weekend we brought in over $8,000 in sales of T-shirts, posters, cards, books and other WCWC products. Ken and Carleen slept in our canvass van with the money stuffed into their pillowcase. The chances of getting ripped off were remote, but we weren't taking any chances!

Thom Henley gets WILD involved in an international campaign to help save the Penan's rainforest

Just after the Stein festival, Thom Henley stopped by our office in Gastown on his way through town. Thom had amazing charisma. He also always had a new project he needed help with. This time he captivated us with his plea for the Penan, one of the world's last tribes of nomadic hunter-gatherers. They lived in the tropical rainforest of Borneo; to be precise, in Sarawak, the Malaysian half of the island. None of us had ever heard of them before. Thom had just got back from spending several months there and the slides he'd taken moved us to tears.

The Penan were being forcibly resettled by the Malaysian government and were suffering from diseases caused by living in cramped conditions to which they were totally unaccustomed. Logging on a massive scale was ravaging their jungle homeland, one of the world's oldest evolved forests. It was truly heart wrenching. We had to help save this unique place and the unique culture adapted to it. One of the first things we did was produce and print 5,000 copies of an avant-garde poster of the rainforest explaining the threat to the Penan in five languages including Japanese. They were distributed widely in Malaysia and we gave them out freely in B.C., too.

Thom wanted to write a book about the plight of the Penan featuring his photos and a very moving speech by a young Penan man that Thom had recorded. We all agreed that a book would be one of the best ways to bring attention to the situation and put pressure on the Malaysian government to halt the destructive logging. We decided to launch this campaign with an educational newspaper too.

Thom led an extremely busy life. The only time he had free to do all the prep work for the book was a week in January of 1990. He suggested that Adriane and I meet him in Hawaii, a convenient spot, given that he could stopover there on his return trip from Thailand. He knew a person with a place on Molokai where we could stay for free and work the long hours needed to choose the images for the book and rough out the copy. I jumped at the chance to go on this "working holiday." I'd never been to Hawaii and, although Adriane had taken several trips there as a teenager, she had only been to Oahu. WCWC's board gave us the nod to negotiate a book contract with Thom, despite the fact we really didn't have the money to publish it. The dire plight of the Penan was very persuasive.

Terry Jacks, the singer/environmentalist, gets WCWC involved in his anti pulp mill pollution campaign

Before we could even think about going to Hawaii to meet with Thom, we had lots to do, including our 1990 calendars, the Carmanah art shows, and the Carmanah book launch. As if we weren't busy enough, we had also taken on yet another project with Terry Jacks, the famous "Seasons in the Sun" pop star of the 1970s. In 1985 Jacks had started his own environmental group called Environmental Watch. It was basically a one-man crusade, assisted by insider whistle blowers, to stop the huge volumes of toxic pollution spewing out from B.C.'s coastal pulp mills. Effluent from these mills accounted for more than half of all the liquid waste dumped into B.C.'s marine waters and was, according to Environment Canada, almost three times as toxic as the combined sewage discharge from all the coastal cities.

With Jacks' help, we produced a four-page, full-sized newspaper titled *Stop the Killing of Howe Sound NOW!* The "o" in the word "Stop" had been graphically morphed into a skull and crossbones. Our lead article boldly stated:

> *While you are reading this paper these mills* [Woodfibre mill near Squamish and Howe Sound Pulp and Paper mill in Port Mellon] *are breaking the law—exceeding their permitted pollution limits. Our B.C. government lets them get away with it, saying that the companies promise improvements, which will provide adequate environmental protection. Meanwhile, Howe Sound is dying.*

The photos turned your stomach, especially the one of the fish found in Howe Sound with cancerous tumors. Several years earlier, Jacks had requested the maps and statistics on cancer mortality in B.C. and in particular, Howe Sound. The table he got back from the Cancer Control Agency showed that the Howe Sound region had the highest rates of both male and female lung cancer deaths in all of B.C. We printed unaltered copies of this letter, map and statistical table in the paper, simply adding the title: "*According to the Cancer Control Agency, the School Districts of Howe Sound rank number one per capita in British Columbia for both male and female lung cancer deaths.*" Alongside this material we featured a huge picture of the Woodfibre mill shrouded in the smoke that was roiling out of their stacks. Nowhere in our paper did we say that the mill effluent caused this high rate of cancer.

We printed 100,000 copies and mailed them to every household around Howe Sound. It caused a huge uproar. Someone from the Cancer Control Agency called me to complain, explaining that nothing in their statistics specifically blamed the pulp mills' air pollution for the high rates of cancer. "It could mean that there are more recent immigrants from Eastern Europe (who are heavy smokers) living in the area," he said.

"Yes, that could be due to that," I agreed. "But the particulate pollution and other stuff spewing out of the pulp mill stacks could be the reason. We've left it up to the readers to decide," I continued, standing up for our paper.

Tricked by unethical reporter

A few days after this conversation I got a call from a freelance reporter writing a story for the *Vancouver Magazine* about our *Stop Killing Howe Sound* paper. He was very aggressive, nasty and got my goat right away by insisting that our use of the cancer statistics in our paper was misleading. I explained over and over again that we just put two things together, visual evidence with statistical evidence. It was up to the readers to draw their own conclusions.

Finally in exasperation I said, "You're just trying to get me to say that we did it [used the cancer statistics in a misleading way] on purpose to scare people and further our cause. And I'm not going to say that." The magazine came out a month later. In the article was my quote: *"we did it on purpose to scare people...."* I was furious at the intentional distortion, using this quote out of context. I got our lawyer Don Rosenbloom to threaten a libel lawsuit. The reporter was extremely cooperative, telling Don that he had taped our conversation and would send a copy of the tape to him. It took several more months and many more phone calls before the reporter finally sent it.

When it arrived the quality of the tape was so poor you could hardly make out the words. And my qualifying phrase, "You're just trying to get me to say that..." was missing! It had to have been edited out, but the tape quality was too poor to prove it. I couldn't afford to take it any further. I just had to let it blow over. What I learned from this incident was never again to lose my cool or make facetious comments when talking to reporters. Some industry-biased journalists will stoop to anything to discredit environmentalists.

We also learned by this "one-shot" Howe Sound newspaper (and other one-time one-issue papers) that they were of limited effectiveness. It takes several educational newspapers spread out over time—often years—to win big issues. Each publication has to repeat the basic ideas, explain the problem, provide the solution and offer the newest possible information and arguments to refute industry's counter propaganda.

With so many environmental issues crying for more exposure, we published as many newspapers as we could. If a local environmental group had already done the research, gathered the facts and photos together to make production easy, we nearly always put one out, even if we knew it was only going to be a one-shot effort because that is better than nothing.

Tricked again by a photo-journalist who provoked me into providing an angry look

It actually took another blunder for me to learn my lesson regarding biased media and the use of clever tricks. Shortly after the *Vancouver Magazine* article, a photographer came by our office wanting a picture of some WCWC campaigners at work for an article being written about our campaigns for new parks. He had a twist in mind. Instead of WCWC being shown in a forest setting and the

Set-up by a photo-journalist who provoked me into looking angry.
Photo: Alex Waterhouse Hayward.

MB forest company executive in his office, he wanted it reversed. We complied. Adriane, Ken and I posed in our boardroom table looking over some maps. The photographer took about an hour to set up his fancy coloured lighting, puttered around, taking sample Polaroid shots to see what the photo was going to look like. He brought each one over and gave it to us. Finally, exasperated, I glared at him and snarled, "We've got work to do." He immediately said, "I'm done," packed up and left. Guess what picture was featured in the magazine article? Of course, it was the one with me angrily looking directly into the camera! That's what the magazine wanted for this pro-logging article, and that's what he'd tricked me into doing.

Sue Fox comes to the rescue; publication deadline met

1990 was a record year for WCWC newspaper production. We published 14 different papers, printing over one million copies in total. Staring at the towering stacks of newspapers forced us to revamp our distribution strategy. We were already dropping off bundles of papers at friendly stores, but many places wouldn't take them unless we had a fancy distribution box to keep the papers tidy.

The solution was to hire one of our volunteers with carpentry skills to construct 50 "Ecoboxes" out of rough-sawn cedar with our WCWC logo branded into the top. Stores and health food restaurants loved them. From then on we nominally paid someone to keep our ecoboxes filled with our papers and free from weird publications and sales flyers that other people stuffed into them. It was a great grassroots distribution system.

Producing every one of our papers was a marathon affair. I was particularly bad in terms of making last minute text changes. Somehow, I couldn't get the copy right until after I saw it pasted up. One of our volunteer graphic artists, Bev Deausalt, who worked at layout and paste-up professionally, did a fantastic job. Like me, she was a perfectionist, making every change herself and staying until the job was done. If her name was on the paper as the layout person, she had to be sure there were no errors in it.

One night, while finishing up a Carmanah paper, we kept finding typos and editing to improve the text. At 2 a.m. there were still two brief articles unfinished. We were trying to edit them down to fit the small space still available. As time rolled on, Bev, who had to work the next day, got more and more tired and impatient. Adriane and I hurried up as much as we could, but we could only work so fast.

Newspaper production team gathered around a mock up on the light table. From left to right - Adriane Carr, me Joe Foy and Sue Fox. Photo: WCWC files.

Finally at 4 a.m. Bev said, "That's it! I can't work with you anymore!" She cut her name out of the credits, walked out the door and never came back to volunteer for us again.

The press had set 10 a.m. as the firm deadline for getting the layout flats to them. If we missed that press deadline, it would mean a delay of more than a week to get the paper printed. Our campaign couldn't afford the delay. I knew I was incapable of finishing the paste-up myself: I'm just not neat enough to keep the lines straight. Adriane suggested we call Sue Fox, her WILD campaign volunteer, who had a studio nearby and had done some work for us before. With some trepidation I phoned her at about 6 a.m., woke her up and begged her to come and rescue us. She did. Arriving by taxi a half hour later, she did a great job of finishing the paper, putting some of her own design flare into it.

From then on Sue did most of the design and paste-up for WCWC's publications. She worked on a contract basis, at a very low non-profit rate. Over time, we became increasingly dependent on her and eventually hired her on full-time. Only rarely, under extreme provocation, would Sue get impatient and snappy. At the same time Sue took on more and more responsibilities with Adriane's WILD campaign and maintaining our media and external contacts list. The Penan book was the first big project that Sue worked on with us.

Trip to Molokai to plan the Penan book

Finally, the day arrived when Adriane and I left on our January '90 working holiday in Hawaii. It rained hard every day we were there, which must have been an omen that reminded us that we were there to work, not play. The apartment Thom Henley lined up for us to stay in was very nice, but fairly small. Day and night we clicked through tray after tray of Thom's slide images, selecting the most compelling ones and sorting them into a story sequence for the Penan book. We talked endlessly about the plight of the Penan and brainstormed about what WCWC-WILD could do to help them, besides publishing the book.

Here we were in Hawaii and we hardly got outdoors. It was more or less OK with me because, as we learned from our host, there wasn't much wilderness left on Molokai: its natural ecology had been trashed. Excessive land clearing, introduced species and pineapple plantations using mercury fungicide had despoiled the land. The sandy beach and shallow lagoon in front of the condo where we were staying must have been beautiful at one time. Now they were completely silted in with red volcanic mud. Every time it rained hard, rivers of silt flowed down the denuded mountainsides, coating the sandy bottom of the lagoon behind the coral reef. Much of the reef itself was smothered in silt, too. It was the first time I'd seen tropical eco-damage that matched the eroded clearcut slopes of B.C.'s coast.

Our one touristy day-trip to the wettest end of the island nearly turned into a disaster. It rained harder and harder as the day progressed. On our way back, our host decided to drive our rented car across a ford in the road where the water had risen to a torrent since we crossed it earlier in the day. I yelled at him to stop, pointing out that the line on the sign beside the stream that warned it was unsafe to cross if the water level had risen above it was no longer visible. "Don't worry. I've lived here for five years. It'll be OK," he said as he gunned our car forward. The whole car bumped sideways several times as the brown surging water, which came half way up the side of the car doors, nearly succeeded in sweeping us off the road and into the gorge below before we reached the other side. My extra pounds probably kept us alive. It was a very close call!

During that week we hammered out a book contract with Thom. Aware of our financial limitations, we put in a clause that if WCWC-WILD were unable to come up with the money for printing it, Thom would owe us nothing and he would be free to take all the work we had done elsewhere to another publisher. Even though we couldn't guarantee that we could complete the project, at least we could help get it started!

The three of us spent hours talking about the plight of wilderness planet-wide. Thom, who did a lot more traveling than we did, knew from firsthand experience how fast wild places were going down. Adriane explained to him the WILD vision and her dream of holding an international conference to launch a worldwide effort to map the remaining wild places left on earth and help save them.

Dawat Lupung, Penan hunter and gatherer "fires" his blowpipe. An image selected for inclusion in the *Penan - Voice for the Borneo Rainforest* book. Photo: Thom Henley.

She wanted it to be a "working conference" where participants would bring their research and actually begin the mapping work. The maps would respect traditional cultures and support aboriginal rights and title to lands. Aboriginal peoples occupied almost all the wilderness areas left. In fact, wilderness without people only existed on one continent: Antarctica.

Trip to Honolulu to find a perfect site to hold our "Mapping the Vision" International WILD Conference

Thom suggested that the East-West Center on the University of Hawaii campus in Honolulu might be a great place to hold our WILD conference. He offered to check it out with us before we all departed from Honolulu airport at the end of the week. He also knew a wonderful woman who taught Hawaiian Studies at the university who might help us.

A few days later, with the slides selected for the Penan book, the outline of the book settled, and a contract signed, we flew back to Honolulu and visited the University of Hawaii. We met with Thom's contact and she connected us to a group called Kumu Honua (very roughly translated, she said, means "Hawaiian Archipelago Wilderness Society"). We went to Kumu Honua's tiny off-campus office and the Hawaiian sitting at a desk there said he'd be glad to help. Their group was actively opposing several mega-projects that would threaten native Hawaii vegetation and culture. One in particular, a proposed geothermal electrical generation project on their sacred volcano Mount Pele, was the focus of an ongoing protest.

Next, we went to check out the East-West Center. It was set in lovely tropical flower gardens and was just the right size. Its largest lecture room sat 350 and there were several breakout rooms of various sizes. The only weeklong slot they had left for a year was in June, just six months away. The university cafeteria was just a five-minute walk from the centre and we could rent the unused rooms in the dormitory (June was the break between spring and summer semesters) at a very reasonable price. The total rent for the centre for our weeklong conference was incredibly reasonable—just under $4,000 US. There was no immediate deposit needed, so we booked it. The East-West Centre would hold the June period and give us two months to decide and pay a 20 percent deposit. We could withdraw our booking at no cost if WCWC's directors decided against holding the conference. It was almost too good to be true!

WILD's visionary goal

We came back with contagious enthusiasm for holding a "Mapping the Vision" convention in Honolulu, and nearly everyone at WCWC got swept up in it. The board approved the $800 U.S. deposit (a little more than $1,000 Canadian). Of course, we were all counting on the conference being self-funded. "The participants' fees should cover all costs and perhaps even make us some money," Adriane predicted optimistically. It was a bit of megalomania...not just the WILD conference itself, but the idea that our tiny group could actually accomplish its goal of compiling the first ever comprehensive global inventory of threatened wilderness areas and head-up a globally-integrated wilderness-saving campaign. But, as they say, if you don't aim for the moon you'll have absolutely no chance of even reaching it. At the beginning, the WILD maps were expected to identify:

1. *Ecologically self-sustaining areas where natural flora, fauna and habitat have not been substantially altered by human activities, including areas where indigenous people live in an essentially traditional way, in harmony with nature*
2. *Areas of various sizes and states of wildness, ranging from large tracts of wilderness to small natural areas important for significant landscape features or biota*
3. *Wild areas that are unprotected or endangered as well as areas protected by government*
4. *Areas of natural significance, which must be rehabilitated in order to stabilize the natural ecology*
5. *Wild "hotspot" areas that are particularly threatened*

The goal of the WILD maps was to put more detailed information about wilderness into the hands of people campaigning to save it. We knew that the mapping would be hard. Information was scarce and some grassroots wilderness groups and First Nations' people were reluctant to put lines on a map. But our experience was that, without defining the boundaries of proposed protected areas, it is difficult to get widespread public support or political action to protect a place. Maps ground the fight for wilderness protection in reality.

Back home, our Hawaii WILD Conference takes shape

Adriane's WILD Team, which now included over 30 volunteers, started meeting weekly. Adriane instructed the group in their tasks as if they were taking an advanced college course. They gathered data and found addresses for over 2,500 prospective environmental groups, research institutes, government agencies and First Nations around the world that might be interested in the mapping of wilderness. Sue Fox designed a wildly colourful poster to promote the conference. Perry Boeker, the sales representative for Hemlock Printers, the company now printing our calendars, got his company to donate the printing of our "Mapping the Vision" conference poster. Adriane worked on the conference schedule and materials.

After a month of regular phone calls with Kumu Honua, Adriane realized that, while sympathetic to our cause, this activist group was very busy with their own campaigns and could only provide us with minimal volunteer help. If we wanted our conference to be run smoothly and be sensitive to the indigenous Hawaiians of our host "country," the smart thing to do was to hire someone from Kumu Honua on a full-time contract for the next five months to organize things there. We did and it made our conference truly remarkable.

As soon as our new conference organizer and Adriane hammered out the conference details, we printed up the rest of the promotional materials. It took two days for all the WILD volunteers to stuff and label the 2,500 piece mailout. The stamps, especially for international airmail, cost a fortune. By now WILD had spent over $10,000 of WCWC's money and was committed to spending a lot more. It was our first international conference and we had to do it right.

World Wildlife Fund leader comes to visit

In early spring of 1990, in the middle of our angst over funding the WILD conference, Monte Hummel, President of World Wildlife Fund Canada, paid an unannounced visit to our Gastown office. None of us had met him before. Hummel wanted WCWC to support WWF's new Endangered Spaces campaign. I could hardly object to their goal of protecting 12 percent of Canada's land base by the year 2000. In our beginning days WCWC had set a much lower target, first asking for only a one percent increase in the amount of B.C. lands protected in parks. But we soon learned that setting too low a limit was a mistake, and that made us nervous about the 12 percent goal.

A few years earlier WCWC has endorsed the "wish list" of proposed parks published in map format by the Valhalla Wilderness Society. All together (including existing parks) they totaled about 13 percent of B.C.'s land base. Since then, scientific studies had indicated that even 13 percent was not enough to do the job of fully protecting natural biodiversity over time. Today, the best scientific estimate by conservation biologists is that, on average, between 30 to 40 percent of lands, including interconnecting corridors between larger wild areas must remain in a natural state to sustain the full range of existing biodiversity indefinitely.

Environmentalists constantly "move the goal posts" (as our detractors put it) and pursue an "unfinishable agenda" of wilderness protection. We aren't participating in a sport like hockey or soccer. We are trying to save nature, and, as scientists learn more, the goalposts naturally have to shift. There is and was nothing covert about it. Our changing agenda simply reflected how little we knew about

Cartoon that graphically portrays the inadequacy of 12 percent wilderness preservation. It features the "faceless one," a character we used in a series of cartoons to represent the industry's power brokers. Artwork: Geoff Olson.

our biological world and our honesty in asking for more wilderness preservation once science revealed we needed it. I made sure that Hummel agreed with us that 12 percent was just another goal, not the final goal. He did, understanding that the total of the areas already proposed for protection exceeded 12 percent of B.C. lands. We both agreed that the 12 percent would get governments to focus on meeting a specific target. The way Hummel put it, WCWC and WWF should work together on areas we both agreed should be protected and not work against each other when we independently pursued protection of different areas. In retrospect, it's easy to see how governments seized upon the 12 percent as a firm upper limit of what they would ultimately protect.

The Harcourt NDP government elected a year-and-a-half later in B.C. adopted WWF's "Endangered Spaces" goal wholeheartedly. Harcourt set up planning processes that facilitated a torrent of new protected areas in the 1990s that more than doubled the size of B.C.'s park system. But "12 percent" also became the upper cap on B.C. parks for the NDP, for industry and for the B.C. Liberals who formed government in 2001, making protection of the areas left out much more difficult to achieve.

After we agreed to endorse WWF's Endangered Spaces Campaign, Adriane told Hummel about our WILD conference in Hawaii, explained its goal and invited him to come. He not only accepted, but he also said that he would see what he could do to help out financially. He came through to the tune of $20,000 from WWF Canada and another $20,000 donation from WWF in the United States. It was a very successful meeting!

Money crisis puts Hawaii WILD conference in jeopardy

As the weeks went by and planning proceeded, letters of interest in our WILD conference kept coming in. By the end of March, over a hundred wilderness activists and academics from all over the world had expressed interest in coming. Late one afternoon, only two months to go before the date of the big conference, Adriane came to me in tears. "We can't do it. Most of the people who really want to come and we know would be key at the conference can't afford their airfares, let alone our conference fees. We have only 17 paying participants, and nearly all of them are from Canada and one of those is from Share BC (an anti-wilderness group). There are fabulous people from all over the world who say that they really want to attend but can't—unless we find some way to subsidize their travel and other conference expenses."

Our "Mapping the Vision" Conference was a bargain. The $795 (CDN) fee for the seven-day conference included all meals and accommodation at the University dorm and cafeteria. We had reduced NGO rates of $595 and even slightly lower rates for students. However this helped only a few because most wilderness advocates are notoriously poor and academics and activists from "developing" countries are equally poor and cash-strapped.

We had already spent more than $30,000 on getting the conference going. Everything was lined up; expectations were building. We'd reached the point of no return. "There's no way we can send back the money we've already raised—like the $20,000 Monte Hummel so generously granted from World Wildlife Fund Canada (WWF Canada) and the $20,000 from WWF USA—because most of it is already spent! Besides, this is too important," I said. My final rational: "We'll pay for it with the money we'll get from Rainforest Benefit II in September."

Then I asked Adriane the big question, "How much is it going to cost to pay for all the key participants' to attend?" trying to get a handle on just how much money it would take to pull it off.

Adriane had the figure already worked out. "About $60,000 more than we've already raised," she said in almost a whisper. Her sympathetic and hard working travel agent had hunted far and wide to find the very lowest fares for everyone. "Oh, that's not as bad as I thought," I said. "Do we have to pay for the plane tickets up front?" I asked.

"No," Adriane explained, "we just have to book them now and pay the bill later." She continued, "The East-West Center is not asking us for any more money up front either."

Realizing that we were already slipping rapidly into financial difficulties at WCWC and couldn't come up with any cash, it was reassuring to know that our credit was still good! Overtaken by an overpowering urge to do what's right for wilderness and not let our pocket book be our master, I replied, "Hey, that's not much money at all. Let's just do it!" I said confidently to Adriane. "And worry about paying for it later," I whispered less confidently to myself. And that was that. We never again questioned whether or not we could afford to do it. We couldn't! But we were holding it anyway!

WILD Conference proceeds — full steam ahead

Adriane immediately contacted all the people she had identified as key participants from the "south" (no one used the term "third world" anymore) and told them the good news that we had lined up scholarships for them. The conference was on in full swing, although scaled down to 150 from the 350 participants we had originally hoped for when we began.

During the next two months following this "no turning back" decision, WCWC's financial situation worsened considerably. Our payables were now $350,000 and growing. Garry Ulstrom, our comptroller, constantly blew the warning whistle. In his end-of-April report, Gary requested a special directors' meeting to deal with these two main questions: *Do we write cheques in excess of our overdraft limit or not?* and, *What is the maximum debt that you, as directors, are willing to let the Committee take on?*

As far as I can remember we never held that special meeting. Already we were writing cheques above our line of credit limit, counting on revenues coming in from the store and canvass to cover them in the following day or two. We were unable to grapple with the situation rationally. Our collective belief was that the tide would surely turn for us soon, and revenues would increase sufficiently to stem the deficit and pay back the debt. Meanwhile, we kept thinking positive and big, and continued to expand our campaigns.

Spencer Beebe, head of newly formed Ecotrust, visits Vancouver and consults with WCWC

A short time after meeting with Monte, we got a call from a big-time American environmentalist, Spencer Beebe. He had just founded a new conservation group called Ecotrust. Most of Ecotrust's founders, including Beebe, were former employees of Conservation International, one of the largest environmental groups in the U.S. Beebe told us that Ecotrust was interested in focusing its efforts on protecting the temperate rainforests along North America's West Coast. He had come to Vancouver to talk to key activists, including us, about the extent of the remaining primary rainforest in B.C. Adriane and I spent a whole evening briefing him on issue after issue. He impressed us: he had the smarts and financial backing to conduct major campaigns. He wanted our help in identifying the two biggest unprotected, pristine wilderness areas for his group to work on in B.C.

We didn't know what the biggest area was on the mainland coast; but we did know that Clayoquot Sound was the biggest wild area left on Vancouver Island. We told him that no one really knew how many watersheds along the coast were left untouched by logging. It would take a lot of research to find out this vital information. He asked us whom we would recommend to do the research. Immediately, both Adriane and I thought of Keith Moore. Adriane had been in graduate school with Keith and I knew him from pre-WCWC days in Haida Gwaii. He had recently left working for the Ministry of the Environment as a habitat protection officer on the Queen Charlotte Islands (a frustrating job) and set up his own consulting business. He had a master's degree in geography, had passed his Registered Professional Forester exams and was looking for interesting work. We gave Keith's contact information to Beebe.

A short time later we heard that the needed survey was underway. Keith Moore's 54-page report titled *Inventory of Watersheds in the Coastal Temperate Forests of British Columbia*, conducted for the B.C. Endangered Spaces project and published by Earthlife Canada in cooperation with Conservation International/Ecotrust, appeared less than a year later. It verified that out of 89 large (over 5,000 hectare) primary watersheds (river systems that drain directly into the sea) on Vancouver Island only six had never been logged. It also revealed that of the twenty-five primary watersheds in B.C.'s coastal temperate rainforest that are larger than 100,000 hectares none were in an entirely pristine condition. One of them, the 275,100-hectare Kitlope watershed, located 100 kilometres southeast of Kitimat, was pristine except for a tiny logged 30-hectare island in the lower valley. It was the largest undeveloped wild watershed left. Kitlope and Clayoquot became the two focal points of Ecotrust's conservation work in B.C.

Domestic campaigns expand too

In the spring of 1990, while Adriane focused on preparing for the WILD conference and Joe and Ken worked full time on our Carmanah research station and campaign, I spent most of my time producing more WCWC publications. Ric Careless had formed a new organization called Tatshenshini Wild to take the lead on protecting "the Tat." It didn't take much convincing by him to get us to wholeheartedly join the campaign to protect this million-hectare watershed in the far northwestern corner of B.C. I immediately began working with Tatshenshini Wild to co-publish 100,000 copies of a four-page tabloid we titled *TATSHENSHINI – Ice Age Wilderness –Help Protect North America's Wildest River*. Immersing myself in the issue, I read everything available about the proposed monstrous mine that threatened this wilderness. I came to the conclusion that the best way to stop it was to focus on the terrible effects of the titanic amounts of sulfuric acid the mega-mine's waste rock would generate for generations to come.

Knowing that a focus on the ugly consequences of industrial development does not fully sell wilderness conservation, we also produced a dramatic poster titled *Tatshenshini - Ice Age Wilderness*. We

Our Tatshenshini *Ice Age Wilderness* poster. Photo: WCWC files.

sold many copies via a tear-out coupon in the paper. The poster had an icy northern mood that made people looking at it shiver. According to Ric, who had been there, it perfectly depicted the area's glacial grandness. A raft full of people in colourful raingear provided a cheerful focus in the middle of the gray-blue river highlighted by mountains with massive white glaciers in the background. No one could tell that we had digitally removed from the image the name of the rafting company, *Canadian Rivers Expeditions*, which had been prominently displayed in huge letters on the side of the raft. I didn't want the poster to look like an ad and knew (without asking) that Johnny Mikes, who had donated the image and owned the rafting company, wouldn't mind. He was passionate about protecting "the Tat" and worked tirelessly at thwarting Geddes' proposed Windy-Craggy copper mine to keep this big river wild.

That spring I also worked with the Tetrahedron Alliance, a group on the Sunshine Coast just north of Vancouver, to co-publish 75,000 copies of a four-page, two-colour newspaper titled *Protect the TETRAHEDRON*. The paper called for a 6,000-hectare provincial park to protect this high elevation wilderness recreational area.

Staff dissension grows

All spring, as the money situation at WCWC deteriorated, the grumbling against WILD's international work grew. The griping wasn't just over money. Some staff felt that by spending so much time on international work we were short changing our efforts to save local wilderness. But for Adriane and me, and many others at WCWC, WILD's international work was a natural outgrowth of WCWC's intense and effective local work in B.C. It was about thinking and acting both locally <u>and</u> globally.

In my mind, trying to help save Earth's oldest tropical rainforest in Sarawak perfectly complemented trying to preserve the world's tallest Sitka spruces in the ancient temperate rainforest in Carmanah. I felt that both campaigns were vital to the conservation of natural biodiversity. Both forests were needed to sustain the long-term health of our living planet.

At that time we didn't really perceive any limits on what we could successfully undertake. We also obviously didn't have a firm grasp on our finances. We didn't have project accounting, budgets or spending resolutions for each big outlay of money. All the incoming money was jumbled up into one big pot—or so it seemed—and we made payments to the squeakiest wheel. To handle the growing number of "squeaky wheels" we kept getting an increasingly larger line of credit at our CCEC Credit Union. We used our growing inventory of publications and products as collateral and then used our credit line to the max. It's interesting how you can suppress financial worries when you are so busy doing things that you strongly believe will, however unrealistically, ultimately help solve those worries.

Chapter 31

The WILD Conference — an international effort to map Earth's remaining wilderness

After a 24-hour nightmarish journey we finally arrive and our Hawaii WILD Conference starts on time

Excitement grew as the flurry of activities snowballed into a frenzy in preparation for our WILD conference. In the last week we were still writing and photocopying huge volumes of conference materials. We were also making banners, signing up last minute participants and solving transportation problems. Adriane and I worked around the clock the last two days without any sleep at all. "We'll catch up on the flight," we told each other.

Our travel agent's cheap tickets for our WILD party of 14 involved leaving Vancouver at dawn, flying to Seattle and transferring to another plane bound for Honolulu. It took a long time to check in all our stuff through customs. We had over 30 boxes of conference materials. Each person had to check in two of these boxes through customs along with his or her own single personal suitcase so we didn't have to pay any extra baggage charges. On top of this we had 30 cases of Shane Kennedy's video equipment, needed to document the entire seven-day conference. Finally, all checked in, we boarded our small plane for Seattle. We chatted boisterously as the plane engines warmed up. We sat for a long time. Finally I asked the stewardess what was going on. "It's foggy in Seattle," she replied. As soon as it cleared—"any minute now"—we'd be taking off.

After waiting for over an hour with the engines running the entire time (to power the plane's air conditioner the stewardess told me), our flight was scrubbed. All our gear was unloaded and we had to take it back through customs. We had already missed our connecting flight to Hawaii. There was no way that all 14 of us could now fly out together. There weren't enough empty seats in any one plane. Heather Souter, a WILD volunteer who had worked as a travel agent, worked feverishly with the airline to figure out how best to get us all to Hawaii. Luckily, we had allowed an extra day for set up before the conference began, so there was no immediate panic.

We split up into smaller groups destined to Seattle, San Francisco or Los Angeles to pick up connecting flights to Honolulu. A few people went home to bed to wait to catch a direct flight much later that day. Adriane, Shane Kennedy and I and some other core conference planners waited in the airport to catch the next flight to San Francisco via Seattle to connect with a flight to Honolulu that would supposedly get us there the quickest. We checked all our baggage through U.S. customs again.

After several hours, the fog lifted and we were on our way. We touched down in Seattle, where the captain announced there would be a short stopover to pick up passengers before we continued our flight to San Francisco. They allowed people to get off the plane to stretch. I talked Adriane into going to the nearby lounge to relax. All of a sudden Adriane frantically said, "We've got to go," sensing that we had been in the lounge too long. We rushed back to see our plane taxiing away. What a disaster! Adriane was in tears. She did, however, convince the check-in people to radio the plane. Meanwhile, inside the plane Shane was pleading with them to re-dock for us. They just couldn't depart without their WILD leaders, he declared.

It was a different era back in 1990. The airline captain had compassion and more freedom. He backed up the plane and they reattached the ramp. We rushed on board and sheepishly sunk into our seats to the cheers of everyone on board!

Our troubles got worse. In San Francisco our plane to Honolulu was late too. Finally after many hours—it was now early evening—we boarded our plane. We were to arrive in Hawaii about midnight. But this plane, too, just sat there. Eventually the captain informed us that they had a slight electrical problem. The emergency cabin lights didn't work. We all de-boarded and waited in the terminal for it to be fixed. After a two-hour wait they informed us that "the problem was fixed" and we again boarded the plane. On the final pre-take-off flight check, the trouble light came on again and we had to get off again. This time they sent for a new plane. They told us it would only take a couple of hours to arrive.

By now, people were really upset. It was well past midnight. Adriane led a peaceful passenger protest demanding that the airline at least feed us. All the restaurants in the terminal were closed. The protest was successful and the airline got us sandwiches from the crippled plane's catering trolley. I curled up on the carpeted floor, like most of the others in our troupe, and "crashed." Adriane couldn't sleep.

In the wee hours of the morning our new plane arrived. We staggered on and finally arrived in Honolulu in mid-morning, 30 hours after we had departed from Vancouver. Adriane had been without sleep for more than three days. It was the day of our conference, which was scheduled to start at 4 p.m. that afternoon. We had only six hours to get everything together. Ironically, those in our party who had gone home to sleep and came back to the Vancouver Airport later that day got to Hawaii first!

Overjoyed at finally arriving in Honolulu 24 hours after departure from Vancouver. From left to right — Adriane Carr, WILD Conference leader; Heather Souter, volunteer conference organizer and me. Photo: WILD files.
Facing page: In small group sessions participants worked on mapping wilderness areas, some highly threatened in their respective countries. Photo: WILD files.

Thanks to our great volunteer team, everything was pulled together really efficiently and was ready in time for early registration. Nearly everyone who was booked to come to the conference arrived that afternoon. There was only one big disappointment. At the last minute, the delegation from China could not make it. Red tape prevented a top Chinese mapping expert and another expert on China's natural areas from attending our conference. Africa was barely represented, too. There was one representative from the Zambezi Society in Zimbabwe, a world-renowned bird expert, Ian Sinclair.

We had failed to find grassroots organizations in Africa that championed wilderness preservation for its own sake. The hunting-safari

Traditional Hawaiian greeting opens WILD's *Mapping the Vision* conference. On right in white with head bowed is Kumu Hula John Keolamaka'ainana Lake (Kumu John Lake), our cultural guide and mentor. Photo: WILD files.

types so prevalent at the 4th World Wilderness Congress in Denver in 1987 did not attend our conference.

Traditional Hawaiians' welcome launches and sets the tone for the WILD conference

That first evening Kumu Honua hosted a welcoming party for the participants with traditional hula dances, the finest Kona coffee and a kava ceremony. An elder passed around a coconut cup filled with the kava—a thick white liquid that, with one small sip, numbed your mouth and gave you a warm mellow feeling all over. It was our initiation into Hawaiian culture and the kind of ceremonies that were to guide us through the whole conference.

Kumu Honua had engaged Kumu Hula John Keolamaka'ainana Lake to be our cultural guide and mentor. He was one of the most renowned keepers of Hawaiian traditional knowledge and a leader in expanding the use of the native language and spiritual practices. Every morning at 8:15 a.m. sharp, Kumu John Lake led us in a fifteen-minute presentation of a spiritual message drawn from Hawaiian culture. Each one was especially designed to guide us for that day of the conference. The themes were: *"Ancient Hymn of Creation," "the Earth is a Sanctuary," The Earth Endangered," "In the Spirit of Healing the Earth,"* and *"Reverence for the Planet."* The messages were presented in both dance and chant and helped us focus and keep on track with the tasks we were undertaking. These early morning ceremonies were very formal. No one was allowed to enter the room late. Almost everyone attended all of them. Every lesson included a traditional Hula performed by a half a dozen dancers. We quickly learned that this kinetic art form was at the heart of Hawaiian culture. We all also learned that here, too, as in B.C., native cultures and natural ecosystems were inextricably interconnected.

On the first full day of the conference I woke very early. On the spur of the moment I drove into town in the hopes of finding some flower leis for the WILD women volunteers who had been working so hard. I found a small flower shop open that was in the process of making them and got them at a very reasonable price. Thus began my own little conference ritual. Every morning I got fresh leis. They smelled and looked beautiful, contributing to the ambiance of this incredible conference. I believe that small touches like this underlie the success of great campaigns and events.

As the first two days unfolded, it became evident (to our surprise) that our Hawaiian hosts were deeply involved in the Hawaiian sovereignty movement. They were seeking independence from the U.S.! Understanding that aboriginal people's support was key in protecting specific wilderness areas, we had made a special effort to invite First Nations people from Canada to attend. Some of the aboriginal British Columbians, like some of the aboriginal Hawaiians attending the conference, blamed all white people for their current problems. An aboriginal from Estonia attending our conference was personally affronted by this assertion. He was a white! Things started to go off the rails the second day of the conference as the radicals from Hawaii "teamed up" with the more radical aboriginal people from Canada and tried to shift the conference's focus away from mapping wilderness to what they believed was a more pressing issue—justice for First Nations.

They complained that "wilderness" is basically a "western" concept. One native from B.C. said that her culture did not have a word for "environment" because they didn't need it. "They lived it." That day, by consensus, the working group on North America decided that First Nations people as indigenous inhabitants are included in the concept of wilderness which they defined, by consensus, as "a place maintaining natural systems and native biodiversity."

Another spokesperson representing native people in Northern Canada complained about "green fascists" who love animals but didn't respect the rights of First Nations' people to kill wildlife. He was particularly hostile to the representative from International Fund for Animal Welfare who was attending the conference. Many indigenous people questioned how the WILD maps would be used; noting how maps had been used by early explorers to guide other Europeans to gold and other exportable resources and are still used by outsiders today to rip-off the wealth of aboriginal lands.

Adriane rescues the WILD Conference from being hijacked by a few with different agendas

On the third day after the morning's opening ceremony, Adriane was urged by everyone on our WILD team to take the podium. Many had been in tears at the daily "de-briefing" session held the night before because of the mounting tensions between the First Nations and other participants. Everyone on our team wanted Adriane to pull people together and put the conference back on track. Acknowledging the First Nation's agenda and the importance of social justice, Adriane made a passionate case for the conference to remain focused on mapping and protecting the natural living fabric of Earth that underpinned all cultures. *"We are here to share a life-protecting dream to identify and protect Earth's wild places before they are degraded and their special wild character and biodiversity are lost."* That was what the conference was about. As she pleaded with everyone to *"work things out, and think not of what other people can do for you but what you can do for other people by protecting our Earth,"* she broke down in tears. Everyone gave her a standing ovation. The conference was back on track.

All that day Adriane went from group to group and encouraged participants to share their experiences and goals and help each other. Both the native and environmental activists attending had much in common. Nearly everyone who attended condemned the multinational corporations and governments that ran roughshod over natural and cultural diversity. We all shared a sense of reverence for the spirit of nature so well presented to us by our native Hawaiian hosts.

Adriane Carr, conference leader, addresses the plenary on the morning on the third day of the conference. Photo: WILD files.

The significance of the WILD Conference

Our WILD conference was packed with notable leaders in the environment movement. Those from B.C. included broadcaster David Suzuki, Tara Cullis (then with a group called Our Common Ground), Vicki Husband of the Sierra Club of B.C., Bristol Foster of the Friends of Ecological Reserves, Bill Wareham from EarthLife Canada, the late Grant Copeland of the Valhalla Wilderness Society and Thom Henley of Rediscovery International. Monty Hummel from WWF Canada attended, as did renowned ethnobotanist and writer Wade Davis who was then with the U.S. Smithsonian Institution. Well-known environmental leaders from other countries who attended included Dave Foreman, founder of Earth First; Frank Yong of the Environmental Protection Society in Malaysia; Bill Duvall, a founder of the deep ecology movement in the U.S.; Yoichi Kuroda of JATAN (Japan Tropical Forest Action Network); Stamatis Zogaris of the Hellenic Society for the Protection of Nature in Greece; Hernan Verscheure of CODEFF (Comité Nacional Pro Defensa de la Fauna y Flora) in Chile; Mark Bellingham of the Royal Forest and Bird Protection Society in New Zealand; Michael McCloskey, Executive

Renowned Penan researcher Bruno Manser (on right) and ethnobotanist Wade Davis discuss the possibility of a Penan world tour on the lawn outside the EastWest Conference Center. Photo: WILD files.

Director of the U.S. Sierra Club; Kailash Sankhala, founder of Tiger Trust India; and, Tirso Maldanado from Fundacion Neotropica de Costa Rica.

Most them had never met each other before. They were approaching the problem of wilderness conservation from many different perspectives. An amazing exchange of ideas occurred in the meetings, hallways and after-hour events. One interesting contact of special note was between Monte Hummel, then head of the conservative World Wildlife Fund Canada and David Foreman, then head of the radical Earth First! movement in the U.S.

Wade Davis, Monty Hummel, David Suzuki, Michael McCloskey, Dave Foreman, Kailash Sankhala, and Tirso Maldanado gave evening addresses to the plenary. Hummel acknowledged that a wide range in tactics was needed to save wilderness: *"Sometimes it takes sand in the gas tank* [referring to park rangers fighting poachers in African parks] *and sometimes it takes a prince"* [referring to Prince Philip, honorary head of WWF]. His statement was tape-recorded by Brenda Armstrong, wife of Patrick Armstrong of Share BC, a group that had been opposing our efforts to save wilderness in B.C. ever since the early days of our Stein Valley campaign. When she got home she tried to use his statement out of context to discredit Monte Hummel, saying that he supported ecotage (damaging and destroying machines and equipment to stop ecologically damaging developments) tactics. Her sleazy efforts didn't work.

Most of the vital work of the conference occurred in the small group sessions held every morning and afternoon. Experts from each "eco-realm" worked diligently together to start mapping the remaining wild areas that needed protection in their region. The Central and South American realm was the most active and best represented of the eco-realms. Experts attended from Costa Rica, Columbia, Ecuador, Chile, Brazil and Peru. Their sessions were carried on in Spanish. The walls of their room, the largest of the breakout rooms at the Center, were plastered with maps. Every table in their room overflowed with maps too. Almost all of the participants had brought big tubes of their own maps with notes and back-up research. WILD volunteer Guadalupe Jolicoeur, a vivacious young university student

Monty Hummel, president of World Wildlife Fund Canada, addresses the plenary and states that saving wilderness sometimes requires a Prince and sometimes extreme tactics. Photo: WILD files.

who spoke fluent Spanish, led that group. Her enthusiasm and sense of humour were key factors that made the group work so well together. Participants in this group formed a real bond and vowed to continue working together after the conference.

In most of these Latin American countries there was a lot of wilderness left with a growing will to preserve it at both the local and political levels. The Latin American participants attending offered to host WILD's next conference in Brazil the next year (1991), the year before the UNCED (United Nations Conference on Environment and Development) was scheduled to be held in Rio de Janeiro.

South and Central American leaders working on their wilderness maps in their breakout room. Photo: WILD files.

The Luau

To wrap up the conference, we arranged to have a real luau with traditional hulas and Hawaiian foods in the University's beautiful outdoor amphitheatre. By this time we were starting to think about going home to the bills, so we took a chance on the weather and didn't rent any tents or canopies. The gods were with us. It didn't rain. We extended invitations to all of Kumu Honua members and their families as well as those who had presented the inspirational "services" every morning. About 300 people attended. The food was great, although few non-Hawaiians liked the poi, the starchy taro paste that was a traditional food staple.

The highlight of the luau was a fabulous "show" on how the hula had evolved from pre-contact times through the missionary period to contemporary expression. A Hawaiian orchestra played songs from each time period. Kumu John Lake gave an extremely humorous historical explanation of each hula "phase" as the troupe of hula dancers changed regalia for each dance. Near the end of the show my sixth sense told me that he was going to invite members of the audience onto the stage to learn how to hula. I didn't want to be among them. I unobtrusively slipped away into the back of the audience leaving Adriane alone right near the front. Sure enough, that was the next agenda item. Kumu Lake invited Adriane specifically to come up. Of course she couldn't refuse. She was great at catching on to the basic movements. At the end of the lesson she was instructed on how to perform the "naughty" dance movement (called the "pelvic thrust") that was vigorously suppressed by the early missionaries. She caught on quickly, to the hoots of laughter and cheers of all.

The dramatic final two days

The last two days of the conference were packed with reports on the small groups' mapping work. There were also emergency resolutions from activists who wanted international expression of support for their local campaigns. Bruno Manser of the Society for Threatened Peoples, emerging from seven years of living with the hunter-gatherer Penan people in Sarawak, arrived late to the conference, but just in time to give a passionate speech about the need for action to save the Penan. Thom Henley gave a similar plea, illustrating his talk with his slideshow about the Penan and their "jungle" homeland. The conference then endorsed the world tour asked for by Penan leaders so that the Penan themselves could take the message of their plight directly to the world and put pressure on the Malaysian government to stop the logging of their traditional rainforest home. WCWC-WILD agreed to take on the job of helping coordinate the tour.

On the last day, Kailash Sankhala, founder of Tiger Trust India and the person most instrumental in saving the tiger from extinction in India, gave his report concerning the state of wilderness in India. His first map showed the protected areas in India. On it were just a small number of dot-sized parks scattered around the country. Many of them were recently created as tiger reserves—the result of his work. Then he flipped this map over to reveal the next map underneath which showed the remaining wilderness in India.

This map was identical to the first! There was no unprotected wilderness left in his densely populated country of then nearly a billion people. Finally, he flipped this map over to reveal the map of threatened wild areas in India. His third map was identical to the first two! Every single one of the currently protected areas was threatened by human encroachment. It was a shocking foreshadowing of the state towards which the whole planet was heading.

Kailash Sankhala, "father of tiger conservation" and founder of Tiger Trust India displays his maps at a plenary session. Photo: WILD files.

In Adriane's closing address she reaffirmed WILD's belief that there is power and utility in mapping endangered ecosystems. *"It's time for us to make our own maps. They are a tool in our fight to save this planet."* She urged everyone to complete the task we had just started.

Exhilarated, exhausted and relieved that the conference was over and successful, Adriane and I went for a few days to Kauai to relax and recuperate. Then it was home to WCWC's headquarters, and back into the thick of our financial problems.

The full impact of our *Mapping the Vision* conference has is hard to assess. It certainly spurred some people on to work harder to save the wilderness left in their countries. It put many powerful environmental and conservation leaders in touch with each other and it helped the Canadians who attended better understand the global situation and provide our work at home with a planetary perspective.

In his weekly editorial column in the *Vancouver Sun* published July 21, 1990, David Suzuki had this to say about our conference:

> *...As environmentalists focus on the struggle to save small fragments of wilderness, it is often difficult to remember the whole planet of which the contentious bits are a part. That's what made the weeklong WILD conference held in Honolulu in June so important.*
>
> *"Sponsored by British Columbia's Western Canada Wilderness Committee, the conference brought people from 26 countries of the world together to begin a monumental task—to map all areas of wilderness left on the planet. In an age of satellite imaging and super-computers, it is surprising to realize that we don't even know what there is in the way of untouched areas on earth...*

Some of the plans that were made at the conference didn't pan out. My proposal to create an *International Redbook of Endangered Wilderness Areas* was a great idea, but we never had the time or the financial resources to follow through. Sadly, from the several hundred videotapes of conference proceedings, no presentable video of the highlights of the *WILD–Mapping the Vision Conference* was ever produced. The raw material was there; but the money, experience and technology to do the follow-up work were not. We accessed a government "top-up" grant to hire a young student on unemployment insurance to produce a half-hour video of the conference. She tried hard for seven months but was not able to put together a workable documentary. In this pre-digital video age it was not easy to edit the reams of tapes. Every suggestion we made to improve the video meant a complete manual re-assembly of the segments—a task that took countless hours.

Globe hopping to Switzerland, WILD campaigners are all over the map

Before settling back into the work at home, Adriane set out on another international trip. A rich conservationist in Switzerland paid all the expenses for her and our forester Mark Wareing to fly to Switzerland and attend "Sol 3." This two-day international conference on deforestation brought together forestry experts from around the world. Wareing presented a paper on the state of B.C.'s forests. Both he and Adriane learned a lot about the dire threats to most of the world's forest ecosystems, including the killing impacts of acid rain on eastern Europe's forests and the "drying up" of U.S. interior forests because of large-scale clearcutting on the west coast.

One nice tangible outcome of this conference was a contract for WILD's cartographer, Robyn Sydneysmith, to map the state of the world's remaining ancient temperate rainforests. We ended up publishing this map on the cover of our 1992 WCWC-WILD four-page, full-colour, full-sized newspaper titled *British Columbia's Temperate Rainforest - A GLOBAL HERITAGE IN PERIL*. The map showed that Chile was second only to B.C. in having the largest extent of still-wild temperate rainforest left in the world. It inspired our WILD campaign team to work more intensively with several Chilean environmental groups and, a few years later, to develop a partnership project between WCWC-WILD and CODEFF (who attended our WILD conference in Hawaii), which was eventually funded by CIDA's Environment and Sustainable Development Program.

WILD goes all out to help the Penan; but to no avail

Upon our return to Vancouver, Adriane, Sue Fox and I put enormous effort into finishing up our book about the Penan in time to distribute at Rainforest Benefit II in September. We had arranged with Mary Lou Stewart, the generous patron who organized the first Rainforest Benefit and now was organizing the second, that everyone attending the event would receive a copy of our Penan book. We agreed that $17 from each $100 ticket to the benefit would automatically go towards paying for the printing. Given WCWC's dire financial situation, this was the "ticket" that made publishing the Penan book possible.

Our WILD team also started organizing the Penan world. The tour team comprised Thom Henley, two young Penan men and a Kelabit native (another tribe in Sarawak also hurt by the rampant logging). Organizing this tour was probably one of the most complicated and stressful projects WCWC ever undertook. In the midst of handling the logistics, we also rushed to produce an educational newspaper for the team to take on tour with them. It was another mammoth effort. Adriane and I stayed up for 48 hours straight to finish the newspaper in time for the group's arrival in Vancouver.

The Penan World Tour event in Vancouver, thanks to Thom's creativity and the compelling innocence of the young Penan participants, was one of our best events ever. The high school theatre was packed with an audience of over 600. After everyone was seated, the traditionally dressed Penan, made a dramatic entrance stalking from the back of the room, as if hunting, down the isle towards the stage. One of them raised a blowpipe, their traditional hunting weapon, to his mouth and propelled a dart at amazing velocity into the top of the stage curtain where it stuck firmly. From this startling start to the end, the Penan show had people in rapture. Everyone was moved by Thom's slides and the natives' detailed descriptions of how the logging was destroying their rainforest and impacting their lives.

After the show there was a Penan fundraising party for wealthy supporters in a fancy West Vancouver home. Adriane gave a brief fundraising pitch and then, exhausted by her two nights without sleep, fell fast asleep on a couch. Apparently people came over to her throughout the evening and tucked cheques into her hand. When she finally woke up several hours later, she discovered she had raised $8,000. It was money we desperately needed to keep the tour going. Adriane still says that her snooze was her most unique fundraising effort ever. The next day before the tour troupe left, I asked Thom if they had retrieved the blow dart stuck in the high school stage curtain. He said no. Then he informed me that he thought it might have had deadly poison on its tip! I was pretty sure he was kidding me about the poison, because, otherwise, he surely wouldn't have left it there.

The Penan tour ran through October and November of 1990. It encompassed 28 cities in 10 European countries, the U.S., Australia, Canada and Japan. The tour group met with Prince Bernhard of the Netherlands, who had phoned Adriane early one morning while we were planning the tour to say that he would be delighted to help the Penan and would host them when they were in the Netherlands.

Penan father with child suffering from conjunctivitis (pink eye). This contagious infection was common and untreated among the children in the crowded longhouse re-settlement villages that the Malaysian government had established for the Penan as their forest was being destroyed by logging. Photo: Thom Henley.

They also met with Maurice Strong, Noel Brown, who was the head of the United Nations Environment Program, U.S. Senator Al Gore, representatives of UNESCO, the International Human Rights Commission, the International Red Cross and the World Council of Churches. All promised to help find an ecologically and socially just solution to the Penan's problem. One solution favoured by the Penan was the creation of a large United Nations Biosphere Reserve. Adriane and Thom even had a meeting with the Secretary General of the United Nations in New York asking for his help. He said he'd use "quiet diplomacy" to resolve the situation.

After the tour, WILD published a 100-page tour report and then kept on with the campaign because logging in Sarawak increased and the lives of the Penan got worse. In our next Penan newspaper we said, *If we are unable to win this campaign, what chance is there for the rest of the world's tropical rainforests?*

WILD sponsored MP Svend Robinson, then the federal New Democratic Party's foreign affairs critic, on a fact-finding trip into Sarawak. We also sponsored Dr. Ron Aspinall to assess the Penan's state of health and health care. He reported back that basic medical care was lacking in the resettlement "longhouses" that the Malaysian government had built to house the displaced Penan. We fundraised to send several ceramic water filters to try to help purify drinking water because all the rivers had turned muddy with the logging-caused erosion. But, all in all, these efforts were only token and certainly weren't the solution. WCWC-WILD was not mandated to provide humanitarian aid. Our mission is to save wilderness and that was what the Penan wanted us to do—save their rainforest homeland.

At the United Nations Conference on Environment and Development [UNCED] held in Rio in 1992, Adriane and Thom met with the Malaysian delegation, including Prime Minister Mahathir of Malaysia. When Adriane presented Mahathir with a copy of our *Endangered People – Endangered Places 1993 UN Year of the World's Indigenous People Calendar* which featured some young Penan children on the cover he said, "Oh, they're my favourite people!" Given what his government was doing to them, Adriane was aghast. He insisted that his government was treating the Penan well, bringing them out of their primitive status. When Adriane returned to Vancouver, she organized a picket of the Malaysian tourist bureau.

Out of the blue, a Malaysia company reprinted our *Penan – Voices for the Borneo Rainforest* book. We simply got the publisher's request in the mail, along with a contract and a $200 cheque. We cashed the cheque, signed the contract and sent it back, expecting to receive a request for the film to make the press plates. We never did.

Our book appeared in Malaysian bookstores and at the airport in Kuala Lumpur a few months later. They had simply scanned a copy of our book to produce press plates for theirs. The quality of the colour images suffered, of course, and we never got any of the promised royalty payments. "Oh well," I thought, "at least it's keeping the pressure up on the Malaysian government." But despite all the pressure, which continued to mount globally, the Malaysian government wouldn't budge. They were bent on allowing full-scale logging until the trees ran out.

During the two years following the Penan world tour, Adriane traveled to Ottawa regularly. She sat on a CIDA (Canadian International Development Agency) advisory committee for their new Environment and Development Support Program. She was also a member of the federal government's national advisory committee on implementing the Biodiversity Convention in Canada. The airfare for these trips cost the government a lot—over $2,000 if she didn't stay over a Saturday night—and no government meetings were ever planned for weekends. Adriane frequently offered to stay longer to get an $800 reduced fare; saving the government money and giving her time to meet with other people in Ottawa.

The most concrete pressure that Canada could apply to try to convince the Malaysian government to halt the logging of the Penan's homeland would be to impose trade sanctions. Adriane arranged many meetings with officials in External Affairs in Ottawa. After several diplomatic meetings where she was "nicely talked to," Adriane finally asked a senior bureaucrat point blank, "Tell me the truth. Will you ever consider putting sanctions on trade with Malaysia over this issue?"

"No," he said. "Malaysia is a preferred trading partner and we want that to continue." At least someone in External Affairs was finally honest with her.

This truth hit Adriane hard. "It's money, not principle, that drives Canada's foreign policy," she told us at a staff meeting when she returned home. The money that Canadians were making through trade with Malaysia blocked the Canadian government from doing anything to help the Penan.

About this time an article appeared in a newspaper in Sarawak claiming that environmental groups, including ours (which was named), were just using the Penan to raise money to fund ourselves. It alleged that we were only exploiting the Penan and did nothing concrete to help them.

The truth was that every penny of the money we raised (not a huge amount) specifically for this campaign (and a lot more) went directly into the campaign, the Penan tour and educational publications about the issue. But some people believed these accusations. I suppose it was beyond their comprehension that people like us would do what we did only because we believed in the cause.

However, nothing we did helped. Not only was logging accelerating, but it also became politically ugly in Sarawak with the government arresting and threatening to arrest all those who were trying to stop the logging, including foreigners who were coming to assess the situation and see what they could do to help. It became harder and harder to communicate with the Penan. It was frightening to know that people might get killed in this campaign. Naively, we had been thinking that all we needed to do was raise awareness and global pressure and eventually the Malaysian government would do what the public wanted. But it didn't work that way at all.

Mutang Tu'o Penan, and Mutang U'rud, Kelabil translator, are interviewed by Rafe Mayer, talk show host on CKNW, during the Penan World Tour. Photo: Thom Henley.

Our Penan Campaign quietly dies

Three years after the Penan World Tour, ethnobotanist Wade Davis visited the Penan again. On his return home to Washington D.C., Wade stopped by to talk with Adriane at our office in Vancouver. Extremely upset, he informed her that the logging had destroyed the last intact unprotected rainforest of any size in the Penan's traditional territory. There wasn't a place you could go where you wouldn't run into a logging road within a few hours hike. The rainforest was completely fragmented. On his trip there two years earlier he had to travel for several days by riverboat through pristine roadless jungle to get to the Penan's territory. This time, he drove the full way via logging roads. "There is not enough rainforest left to sustain them," said Wade. Both he and Adriane broke down crying. Our Penan campaign was over. Despite everyone's best efforts, we had lost. It was tragic. The Malaysian government ultimately established a tiny Biosphere Reserve for the Penan, but it was far too late and too small to mean anything.

WILD gets involved in saving wild places in Greece

Despite the setback in Sarawak, Adriane and her WILD team never lost heart or enthusiasm for their overall work. WILD's mandate to map and help protect wild places around the world opened our doors to virtually any campaign and one of them came from a keen volunteer, Stamatis Zogaris, who'd been coming by our offices since 1988. Stamatis was a naturalist who had been involved in wildlife habitat protection in his native Greece. As a foreign student attending Capilano College, he got active with a group campaigning to save the last large, still-natural wetland left in Burrard Inlet—Maplewood Flats—located just east of the Second Narrows Bridge in North Vancouver. Kevin Bell, head of the Lynn Canyon Eco-Centre and a future chair of WCWC's board of directors, worked on this campaign with Stamatis. Together they wrote an 18-page brief for WCWC to present to government: *Maplewood Flats - Reasons for Protection and Enhancement*. The successful conclusion of this campaign—protection of Maplewood Flats—cemented the involvement of both Stamatis and Kevin in WCWC.

Stamatis became one of our WILD campaign's first volunteers. He'd spent most of his life in Greece and knew many field naturalists and conservationists through the Hellenic Society for the Protection of Nature and the Hellenic Ornithological Society. Excited about the WILD mapping campaign, he mentioned that there was no map of Greece showing all the nature reserves, the few still relatively wild places and the threatened wilderness "hot spots." In fact, Stamatis said there weren't any detailed maps of Greece because the government didn't want enemies to have maps that might help them invade their rugged country! He offered to gather the information needed to produce a WILD map of Greece—if we helped pay his expenses. He was sure it would help protect these threatened wilderness areas that increasingly vulnerable species like raptors needed to survive. This was a unique opportunity we just had to take advantage of. In the spring of 1990 WILD gave Stamatis $1,100 in expense money, including $600 for 40 roles of slide film, to visit, map and document the hot spots in Greece prior to our WILD Conference in Honolulu. We also sponsored Stamatis to come to our Hawaii conference to present his findings.

After the Hawaii conference we wanted to publish the information he gathered in newspaper format, but we didn't have the funding to do it. Fate positively intervened. A woman walked into our office representing a family foundation, the Cundill Foundation. Adriane presented the *Save Wild Greece* newspaper project to her and the foundation gave us a grant to do it! The results: our spring 1991 16-page *Save Wild Greece* newspaper. The production into camera-ready flats was another all-night marathon affair. Before the work session began, Sue

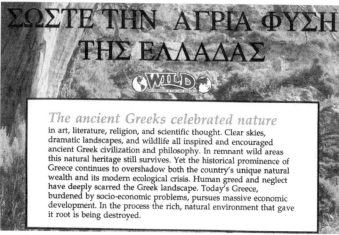

Part of the cover of our Greek newspaper. Artwork: WCWC files.

WCWC posters of watersheds in the proposed Cariboo Mountains Park. The Penfold was eventually clearcut while the Niagara was protected. Posters: WCWC stockroom.

Fox went to a nearby used bookstore and purchased a very old textbook on ancient Greek art with lots of black and white illustrations. We used a number of ancient pottery designs depicting wild animals and birds, which added a great graphic flare to the paper.

It was an intense night. Putting together sixteen pages is like doing four of our regular four-page papers at once. Thank goodness Stamatis worked all night with us. He handled lots of last minute details, edits and questions. "What does liming mean, Stamatis?" I asked. He explained it was putting sticky stuff on tree branches so when wild birds alight on them they get stuck fast. It was illegal, but often used by rural people to poach small birds to eat as a delicacy. Greece, by virtue of its location was "at the biological crossroads" between Europe, Asia and Africa. It was by far the most biodiverse country in Europe. There are over 6,000 different plant species, 760 endemic plants and 26 different raptor species in this one country. Our paper's centrefold map was truly a masterpiece; the first graphic documentation of natural areas left in Greece. It featured 79 numbered "hot spots," with a brief description of the ecological importance and threats to each. In this paper we blew the whistle on the uncontrolled, ecologically-unsound development that had taken off in Greece since it entered into the European Community in 1981.

Many of the 35,000 papers we printed were shipped to non-profit naturalist and conservation groups in Greece. Stamatis told us that they were exceedingly grateful and used our paper extensively. We also handed out papers at the Greek Orthodox Church in Vancouver, encouraging people in B.C. to help the conservation work "in the old country." In less than a year all, except a few, copies were gone. It was a great feeling to be a part of doing something to help protect this birthplace of western civilization.

Lending a hand to another campaign – saving the Cariboo Mountains Wilderness between Wells Gray and Bowron Lake Provincial Parks

During '91, Doug Radies and his wife Ocean Hellman came to our office regularly to keep us updated on the Cariboo Mountains Wilderness Coalition and its efforts to connect Wells Gray and Bowron Lake Provincial Parks by protecting a 160,000 hectare Cariboo Mountains National Park Reserve that bridged the gap between them.

This coalition also proposed to append an additional 60,000 hectares onto Bowron Lake Provincial Park. Doug and Ocean were taking the lead in a full-out campaign and had gathered together many persuasive arguments and statistics. These were great proposals that made ecological sense. Although I had never visited these areas, I came to know them through looking at hundreds of diverse slide images and talking to Doug who had explored the area extensively.

In the summer of '91, Doug and I wrote and produced a full-sized, four-page full-colour newspaper titled *Save the Cariboo Mountain Wilderness*. Doug lined up significant financial support, including a big donation from Mountain Equipment Co-op, to help pay for the printing. He also helped distribute the paper, arranging for its insertion in the *B.C. Sierra Club Report* newspaper. Doug was not always easy to work with because he was as strong willed as I am and he had his own ideas on how best to say things. But we worked through our differences and the final product turned out to be great.

Several years later the B.C. government established a Cariboo-Chilcotin land use planning process to identify and recommend which areas should be protected in the region. To provide pressure to get protection for the bridging wilderness WC^2 published two large, full-colour posters of watersheds in the proposed Cariboo Mountains park area. One was of the Niagara Valley, which was successfully saved. The other was of the adjacent Penfold Valley, which we'd heard was going to be sacrificed. Our poster, which I though was the more spectacular of the two, was unsuccessful in getting the Penfold protected. On November 24, 1994 The B.C. government announced the protection of part of the Cariboo Mountains Wilderness—the Mitchell and Niagara watersheds linking Wells Gray and Bowron Lakes Provincial Parks. Right after the government decision was made, loggers moved in like gangbusters to foreclose forever the option of the Penfold watershed being added to the park at a later date.

Chapter 32

Desperate financial shape
WILD moves out
Drinking water defended

Adriane leaves teaching to work full time for WCWC

Gaining Adriane as a full-time permanent employee of WCWC was a spin-off that I count as one of the biggest successes of our WILD campaign. The WILD work hooked Adriane completely on environmental activism. Right after the WILD conference in Hawaii, she requested and got an unpaid one-year extension to her leave of absence from teaching at Langara College. When that leave ended in August 1991, she made the final plunge. Despite her love of teaching and the constant financial pressures at WCWC, Adriane resigned from her position at Langara Community College and continued to work full-time for the Committee.

The desperation of our financial crisis deepens as our activities expanded

In blissful ignorance of our true financial condition, WCWC's branches—Victoria, Mid-Island in Nanaimo and Vernon—charged ahead. They all were running stores and campaigns with varying diminishing degrees of success. The branches brought us lots of complications and cause for concern. None of them made money for the main office, and they all contributed to soaring administration costs. Our comptroller Gary Ulstrom, wrote this in his financial report to the Strategy and Finance Team, a month after we had returned from the WILD Conference in Hawaii:

> The crisis has arrived. The phone keeps ringing with suppliers wanting payment; individuals or small companies that are already really tight for cash are hurting because we are not able to pay them. Relationships with people who have given us great deals are being strained to say the least. We are losing a lot of good will. It sucks.
>
> Two cheques bounced last week, one to Hemlock Printers for

Gary Ulstrom in the accounting area at 20 Water Street. Photo: WCWC files.

$4,500 and one for All Destination Travel (to pay for flights to the Wild Conference) for $5,000.... Spending and expenses are totally out of control.

The only thing that saved us from immediate collapse was our Gastown store. It was averaging gross sales of over a thousand dollars per day. Since most products sold were our own posters, books and T-shirts we already had in stock, the cash coming in was almost pure "profit." We decided to do everything to maximize store sales while the summer tourist season lasted. Ulstrom also recommended a 40 percent across the board cut in wages, which would save us $12,000 per month. At that time we had about 24 full-time employees working out of 20 Water, not including canvassers. We all took cuts. Those who were paid so little that they already were barely able to cover rent and food were not required to cut as much.

Ulstrom, like all of us at the time, was still infused with unrealistic optimism that we could power out of our financial slump. In his report printed in our summer 1990 members' paper he predicts: *I'm confident that with the sacrifices made by the staff* [the voluntary pay cuts of 20 to 50 percent], *the implementation of our management plan and with your* [referring to our members and donors] *support the Committee will get back on a solid financial footing within the next few months*. It didn't turn out that way!

The big split — WILD goes its semi-separate way

After adding up all the bills, WILD owed $90,000 to WCWC. It was a big part of WCWC's debt, but, as we learned a few months later, not by any means the biggest. WCWC was actually in the hole over half of a million dollars! Some people blamed WILD. Some people blamed our Branches. Some people blamed consensus decision-making for our inability to lay off any staff. But we all agreed that the WILD campaign should become a semi-independent special project of the WCWC with *"financial obligations and decision-making privileges similar to a Branch of the Committee."* From then on WILD had to be financially self-sufficient. It had its own bank account and ran its own affairs and did not go further into debt. The agreement gave the directors the power to shut down WILD on short notice if it ran financially amuck. At our September 1990 AGM this arrangement was formalized in WCWC's bylaws.

At the 1990 AGM we also limited the number of WCWC directors to 12 with overlapping two-year terms; with six directors elected every year. The staggered election of directors was a move to stem the possibility of an "outside takeover" (although I couldn't imagine who might want to take over such a heavily indebted group). These new rules wouldn't kick in until our next AGM. Meanwhile, almost all of the staff were re-elected to our current 18-member board including Adriane and myself. Following the AGM, the WILD campaign team elected Adriane, Laurie Gourlay and me to be "directors" of the new WILD 'branch." Adriane concentrated on fundraising for WILD so it could pay its own way, fund new projects and start paying back the money owed to WCWC central for the Hawaii Conference.

Before the formal passage of these bylaws, we had what was probably WCWC's biggest internal fight. On the afternoon before our July 1990 evening directors' meeting, our Strategy and Finance Team, which included Adriane, Joe, Derek Young from Victoria and myself, held a special meeting to decide which projects should continue and which projects and publications should be shelved to save money. We easily agreed on most of the cutbacks. One controversial issue, however, was the publication of our third annual (1991) *Canadian Endangered Wilderness Calendar*. I strongly argued that this calendar already had a good market amongst our members and was profitable, although we didn't have the financial tracking to prove it. Our Canada calendar was also a great educational tool to feature more threatened wilderness areas. Reluctantly, the other members of the Strategy and Finance Committee, which at the time performed a role similar to that of a multi-headed executive director, agreed to keep it on the list of projects that would go ahead.

Neither Adriane nor I were able to attend the directors' meeting that evening. We found out the next day—July 20, 1990—that the 1991 Canadian calendar project had been killed, with the other members of our Strategy and Finance Team arguing against the decision we'd agreed upon together. I was furious. Feeling betrayed and unable to trust the other members of the Strategy and Finance Team, both Adriane and I resigned from it. We continued to sit as directors of WCWC, but put the bigger part of our energy into WILD.

Another casualty of our rocky financial times was the loss of our single biggest donor, the one who took me on the yacht trip in South Moresby shortly before the area was saved. He sat in on one of our directors' meetings in the fall of 1990 to see how things were going. He was so disgusted with the haphazard money management, preoccupation with merchandising and inept decision-making he witnessed that he decided to drop us from the portfolio of conservation groups he funded. He told me we spent far too much time and effort "trying to make good little shopkeepers" out of people (referring to our focus on our stores in Nanaimo, Penticton and Victoria) rather than working to save wilderness. He was right. Things were wonky.

We were failing as entrepreneurs and if we did not get our finances in order quickly, we'd soon be out of business. What started out as a great way to get our products and message out, modeled after the successful Wilderness Society chain of stores in Australia, had turned into a nightmare. The donor said he would gradually decrease his support to zero over a couple of years. I appreciated his gradual gentlemanly way of quitting, but it was a big blow to have someone like that lose confidence in us.

WILD rents separate office quarters

Soon after that blow-up over cutbacks, WILD got a federal government UI top-up grant (the unemployed participants got extra money from the government to supplement their UI payments for working at a job where they were learning new skills and getting employment experience) to hire four people at no cost to us for six months. There simply wasn't enough room for everyone at our 20 Water Street location. WILD rented cheap office space on the corner of West Cordova and Homer, about three blocks away. This move, along with separate accounting for WILD, helped mollify those who felt that too much money was being spent on our international campaign. I moved with WILD and started working mostly on WILD publications, leaving the running of WCWC to others. Ken Lay took over WCWC's day-to-day leadership. The anxiety and unhappiness were palpable.

Adriane and her WILD team focused on preparing a grant application to CIDA's Environment and Development Support Program for a project to map and conserve natural ecosystems in Latin America. WILD proposed to work with two experienced and enthusiastic cartographers who had attended our Hawaii conference: Clayton Lino of SOS Mata Atlántica in Brazil, and Tirso Maldanado of the Fundacion Neotropica in Costa Rica. The goal was to get the maps done before UNCED, being held in Rio in 2002.

Our outspoken comptroller abandons ship

In October of 1990 WCWC's new board of directors appointed Joe Foy, Ken Lay, Allan McDonnell (one of the new directors) and Lisa DeMarni to the Strategy and Finance Team, which now called itself "WCWC's *Survival and Management Planning Coordinators.*" It was a tough assignment.

Sadly, Gary Ulstrom, our witty and over-worked comptroller, gave us his notice the following month. He was the first of many staff to go. He had a good reason: returning to school to get his CA papers. It made all of us feel even more worried. After advertising and interviewing quite a few candidates, we hired Mike Rodgers. Although extremely competent, Mike faced an impossible task of trying to fix our financial crisis. At the same December 1990 directors meeting that we hired Mike, we passed a motion to hire an "Organizing Administrator" at $2,500 per month to oversee and make sure "Team Leaders" (the staff who headed up each Team [department] of WCWC) stayed "within budget and on track." It passed six to four with Adriane and I voting against it mainly because we didn't have the money. I insisted that our board of directors had to do the hiring. The job was never posted.

At another board meeting during this time period we accepted the donation of 50 percent ownership in a house in Ucluelet (on the west coast of Vancouver Island near Tofino), the town where there was great resistance to any wilderness protection in Clayoquot Sound. The donors wanted us and co-owner Friends of Clayoquot Sound (the group the owner gave the other half of the house to) to use this house as an "environmental outreach centre." This seemingly good decision caused us lots of grief in the years to come.

WCWC forester destroys the myth that watershed logging doesn't harm drinking water quality

Despite our differences and financial difficulties, WC^2 publications continued to roll off the presses. Foundation grants and financial support from other environmental groups helped fund quite a few of them.

Mark Wareing, our staff forester, embarked on several new campaigns during this time period. He wrote a newspaper about "New Forestry" called *Crisis in the Woods* that clearly explained why selection cutting is superior to clearcutting. Then he started work on another newspaper to discredit clearcut logging in the three watersheds (Capilano, Seymour, and Coquitlam) that provided Greater Vancouverites with drinking water. Mark wanted to present concrete evidence that logging was causing the muddying of Greater Vancouver's water. The simplest way to get this evidence would be to travel up the watersheds' logging roads during torrential rainstorms and collect samples of stream and runoff water. But we couldn't do this because the watershed lands were off limits to the public.

Wareing, when he worked for the B.C. Forest Service Lower Mainland District, had such access. In fact, part of his duties had been to monitor silvicultural activities in these watersheds. Shortly after Mark quit his Forest Service job in late 1988 to work for us, he tried to enter the Coquitlam watershed. He was denied access and his "green card" that permitted him to enter the watersheds was confiscated at the gate. The Greater Vancouver Water District (GVWD) officials didn't want public scrutiny of their logging activities. Despite this setback, Mark kept doggedly working on this campaign.

Will Koop, researcher, founder of the B.C. Tap Water Alliance and stalwart campaigner for drinking water protection, extols the importance of Mark's early efforts in a report called *Silty Sources*. In it Koop writes:

> As a direct result of public criticism in late 1988 by professional forester Mark Wareing with the Western Canada Wilderness Committee, the Water District contracted consultants to conduct an

Ken Lay posing beside a load of giant logs clearcut out of one of the watersheds supplying Greater Vancouver's drinking water. Photo: Mark Wareing.

internal audit of the watershed logging program. The audit, which began in March of 1989, was not presented to the GVRD's Water Committee for information until four months later in July 1989. During this process of investigation by consultants who were hired without public consultation, the public was never asked to provide input for this process, nor were politicians given meaningful status reports during the inquiry. The final report was intended to be submitted to the GVRD Board at the end of 1990, without public scrutiny and review, had it not been for a large landslide in a Seymour watershed cutblock in late November 1990, which shut down the Seymour supply for a number of weeks. This event triggered enormous public concern and media attention, which caused the internal report to take a sudden turn.

In November of 1990 several of Mark's friends who still worked for the B.C. Forest Service tipped him off that a big landslide had occurred in the Jamieson Creek tributary of the Seymour watershed. Mark knew that this could be the convincing evidence we were waiting for. On his own, Mark sought out a big donor who was equally passionate about protecting Vancouver's drinking water and got a donation to hire a helicopter to go up and take photos of the slide.

His photos (fortuitously highlighted by a thin veneer of new-fallen snow) showed clearly that the slide had begun in a clearcut; in fact, in the very clearcut where Doug Golding, a UBC forest hydrology associate professor, authorized the logging in a formal study about logging and water quality. Mark coupled this photo with another photo taken during a flight over the watersheds the preceding summer, showing active logging road building on incredibly steep slopes for

added visual evidence that showed it was logging-induced landslides, not natural "slope failures," that muddied Vancouver's water when heavy rains fell.

Mark did all the writing and research for what became the first of a series of WCWC publications against logging in Vancouver's watersheds. It was a four-page, full-colour, tabloid called *Halt Watershed Logging*. The sub headline read: *Evidence mounts linking your dirty tap water to CLEARCUT LOGGING*. The paper drew public attention to the fact that since 1961 the Greater Vancouver Regional District had built over 300 kilometres of roads and clearcut over 5,000 hectares (much of it on steep slopes) in the three watersheds that provided Greater Vancouver with drinking water.

It's hard to believe, but true, that the GVWD and the UBC Forestry professor who were studying Jamieson Creek denied that logging had anything to do with the Jamieson slide or the dirty drinking water. Professor Golding explained that he was studying small particle erosion, not mass wasting; so the big landslide didn't count! The public outcry we helped create forced the GVWD to hold two days of public meetings in May 1991 at the Robson Square Media Centre. Many citizens spoke out against continued logging in the watersheds, including Mark Wareing and myself. But the forest industry was there in full force and scoffed at the idea of a logging moratorium. Their "experts" provided about half the input, which, of course, supported continuation of the logging. The Water District consultants, who reviewed the public's presentations, recommended that the Board continue to log the watershed under a "pro-active" management philosophy framed as "improving water quality." So log-

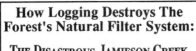

A three-panel brochure co-published with the Citizens for Clean Water after the Jamieson Creek slide. Document: Nik Cuff's files.

ging continued, but not for long.

Over the next few years WCWC's forester Mark Wareing, researcher Will Koop, lawyer and environmental activist Paul Hundel and others continued to keep up the pressure to halt the logging. Driving loggers out of Vancouver's watersheds turned out to be a long, hard fight.

WCWC helps launch the Sierra Legal Defence Fund

In the early fall 1990, a young lawyer, Stewart Elgie, stopped by our office. He had been working for the Sierra Legal Defence Fund (SLDF) in Alaska and thought that it would be good to have a similar organization like that here in Canada. Anticipating my first question, he explained that it would not be duplicating what the West Coast Environmental Law (established in 1974) does—empower citizens to participate in forming policy and decisions about protecting our environment and provided free legal advice, advocacy, research and law reform services. SLDF would be more like a law firm dedicated to enforcing and strengthening the laws that safeguard Canada's environment, wildlife and public health.

It would focus on taking actual court cases forward for clients like WCWC. In fact, he had already recruited Don Rosenbloom to be on their board of directors. I always believed that in the principle of "the more the merrier" or more accurately "the more on your team the greater chance of success." We did not have enough organizations actively fighting to save the environment!

Then he asked if he could use our membership list to send out an appeal for people to join this new organization. I explained to him we never sold or lent out our membership or supports' list. In a veiled quid pro quo request, I asked him if a case against the Alberta government for unauthorized spending to establish another Forest Management Area (FMA), which would destroy a great deal of wilderness was the sort of case SLDF would take on. A few days before, Gray Jones, who had recently launched our Alberta Branch, had called and explained this legal case he would like to pursue. Stewart said yes that was exactly the kind of case it would take forward...if the fact that I outlined to him were correct.

In a quid pro quo gesture, I then offered to do the next best thing to giving him our members' list. I told him that in our next mailing to our members, I would write a personal letter encouraging them to join the newly formed SLDF and enclose a postpaid envelope with the address of their new office on it (if they provided it).

I never found out exactly how successful this appeal was. I only got a verbal report from Stewart that it was "very successful." About a year later, the Sierra Legal Defence Fund, acting for our WCWC-Alberta Branch, successfully won our case against the Alberta government. The justice ruled that a cabinet committee did not have the statutory authority to guarantee paying a logging company millions of dollars in compensation if its application for a FMA in the boreal forest of northern Alberta (which the committee had invited it to submit) was ultimately turned down.

It was the beginning of a long string of legal battles that the SLDF championed for us, many of them not as successful!

Chapter 33

WCWC publishes Penan and Clayoquot books
Fights dams and mines

Rainforest Benefit II – an educational and entertainment success but a financial flop

A few weeks before the October 1990 Rainforest Benefit II, over 30 WILD volunteers showed up at Hemlock Press for a big work party to collate our *Penan – Voice for the Borneo Rainforest* book. To save money we had used two different kinds of paper: more expensive glossy coated for the pages with colour photos; less expensive recycled matte for the text only pages. They were interspersed in the book, which made it impossible for a machine to do the job, so Hemlock Press let us put the pages in order by hand on their pressroom floor. Starting in the early afternoon, we worked in four-hour-long shifts, ordered in pizza for dinner and got the job done by midnight. Someone taught us the trick of using "rubber fingers" to help pull off only one sheet of paper from each pile at a time.

There were 80 different piles stacked around a big set of tables. By the time we were done we had made 6,000 trips around the tables gathering up nearly a half a million sheets of paper. It went off with only one hitch, which was my fault. I put the wrong page on one of the stacks when I was refilling it. Luckily, a sharp-eyed volunteer caught the mistake after less than 100 circuits and we repaired these misassembled books. After they were bound, every volunteer got a free book for helping. The book finally arrived from the bindery in the afternoon of the big Rainforest Benefit.

Rainforest Benefit II, held in the Vancouver hotel, was an amazing event. David Suzuki, Wade Davis, Miles Richardson, Thom Henley and Adriane spoke. The music was outstanding, including Doug and the Slugs, Jim Byrnes, Long John Baldry, Ann Mortifee and Sarah McLachlan. The profits were to be shared equally between David Suzuki's Amazon Fund and our WILD Campaign.

The only trouble was that there were no profits! The benefit grossed about $98,000: 770 in ticket sales at $100 a pop, a fabulous silent auction and some unexpected generous donations. But the expenses were a little over $100,000. The hotel bill alone, including the hospitality suites for the musicians, was over $30,000! The newspaper ads to get people to come was the biggest expense. Our hope that Rainforest Benefit II would generate the funds we needed to cover the cost over-runs on our WILD Hawaii conference evaporated that night. At least the benefit paid off most of the printing bill for our Penan books. Wisely, we had insisted beforehand that, whether the event made money or not, $17 from every ticket sold would go to WILD in exchange for each Benefit attendee getting a copy of our book. The event's sponsor, Mary Lou Stewart, absorbed the deficit and personally honoured that commitment.

There was no Rainforest Benefit III. Mary Lou was definitely burned out and so were we. Forever after, any time anyone suggested putting on a big event to raise us big money, Adriane and I groaned. Big events are, without a doubt, big energy drains, and the riskiest way to fundraise. We feel lucky that at least everyone had a great time at Rainforest Benefit II and, thanks to Mary Lou, we got to publish the Penan book.

Image of two young Penan men hunting with blowpipes featured on the cover of our *Penan – Voice for the Borneo Rainforest* book. Photo: Thom Henley.

In the morning of the evening Rainforest Benefit II even, WILD held its *Penan – Voice for the Borneo Rainforest* book launch in the MacMillan conservatory at the top of Queen Elizabeth Park in Vancouver. David Suzuki who wrote the forward and author Thom Henley hold up a copy of the book with Sue Fox, who designed the book standing between them. To the right of Thom is ethnobotanist Wade Davis, then Rock and Roller Doug Bennett of Doug and the Slugs, Mary Lou Stewart, sponsor of Rainforest Benefit II and Red Robins, well known disc jockey. To the left of David Suzuki is me (mostly hidden), Adriane Carr, head of the WILD project, two unknown people and on the far left Tara Cullis, Executive Director of the newly formed David Suzuki Foundation. Photo: WCWC files.

Launching our *Clayoquot on the Wild Side* book in style in the Vancouver Planetarium

We had no time to mourn the failure of Rainforest Benefit II to rescue us financially. We were too absorbed in the final production of our new coffee table book, *Clayoquot on the Wild Side*, the sequel to our hugely successful *Carmanah – Artistic Visions of an Ancient Rainforest*. Ken Budd had been working for a year with Tofino-based photographer Adrian Dorst, whose images were featured in the book, and Cameron Young, who wrote an incredible text to accompany Dorst's images. We spent weeks deliberating how many books to print. In the end, despite being broke, we decided to print 10,000 copies. Any fewer and the cost of printing per copy would have been exorbitantly high. This number was down from our originally planned 15,000 because we knew that Clayoquot wasn't as hot an issue as Carmanah had been the year before. We were right!

We launched our Clayoquot book in the Vancouver Museum's H.R. MacMillan Planetarium in November. Adrian Dorst gave a great talk and slideshow, ending with a stunning sequence of 360 degree panoramas of Clayoquot, shot using a special tripod loaned to him by the Planetarium. Simon Lucas, an eloquent Hesquiat chief who strongly advocated protecting wild fisheries and wild forests in his traditional Clayoquot home, also gave a moving speech. Cameron Young was there to sign the books. But, despite extensive advertising, the planetarium was less than half full for the launch. Of the 150 people who did attend, many were our volunteers and friends who didn't have much money. Few bought the book.

After the Planetarium launch, Joe toured Vancouver Island with a slideshow to promote the book. In every location the attendance and book sales were low. Joe's tour ended up costing us money. We were counting on this book to pull us out of debt. Instead, it made things much worse. The 10,000 copies cost us more than $170,000 to produce. By year's end we had sold less than 1,800 of them even though we had reduced our selling price to $39.95 from the original $60 price. The book was obviously ahead of its time. When the Clayoquot issue finally heated up in 1993, our *Clayoquot on the Wild Side* book proved to be a very smart investment and made us solid money that we used to pay down our debt. But in 1990, with WCWC sinking, it seemed like a very stupid decision.

The last of our hopes for turning around our financial fortunes was our Christmas appeal and catalogue sales. Ken Lay hired an outside professional he had met at a fundraising workshop to design a new innovative fundraising package called "Deep Roots." Donors were given many choices of activities to support, including different campaigns and publishing ventures. They were also given several choices regarding how much they would donate, with different incentives for each level of support. For example, a thousand-dollar sponsorship of a chapter of a new hiking guidebook would garner a hardcover copy of the book. No one took up this option. I thought the whole thing was ridiculous and was proved right when "Deep Roots" flopped.

Too many choices! Most of our supporters liked all our publications and campaigns and couldn't decide which one they wanted to fund. They either gave a general donation as they had always done, or felt confused and didn't give anything at all. No single project got overwhelming support. Most of the projects didn't bring in enough money for us to embark on them. So ultimately there were disap-

Cover of *Clayoquot on the Wild Side* book. Photo: WCWC files.

pointed donors who didn't get the results they felt they'd paid for. Although I liked to think that my "send in $25—it's only the cost of a large pizza" style of fundraising letter would have done better, years later I realized that no appeal would have been successful in 1990. The economic recession had us beat. The wave of support for wilderness preservation had collapsed, at least for the time being.

During these tumultuous times we continued to be divided into two WCWC "camps" each with our own headquarters in Gastown. Adriane, the WILD staff and I occupied the Homer Street office and Joe Foy, Ken Lay and the rest of WCWC's staff held the fort at Water Street. Although only three blocks separated us, I seldom went over to Water Street even to visit, still miffed at the way the decision was made not to produce a 1991 Canada calendar. I let Ken and Joe run things. That fall I had nothing to do with either the 1991 Christmas appeal or the product catalogue until late in November when Joe asked me if I would mind giving the catalogue a final check before it went to press. I went right over to take a look.

I was pleased with the design and the photos of the products featured. Then I started reading the text. It was full of typos, grammatical errors and inconsistencies in prices and product numbers. It was a mess. It took a few days for Adriane and I to do a complete edit and a thousand dollars to make the corrections and reset the type. It was a waste of money and a delay we could ill afford.

A minor campaign ploy to help save Boundary Bay has unforeseen major consequences

Although we were preoccupied with trying to raise money through most of the fall of 1990, our campaigns kept building and expanding. The WILD campaign team was not only immersed in managing the Penan World Tour and pursuing more mapping work with enthusiastic partner groups in Latin America, but it also was involved in two campaigns to stop gigantic hydro-electric projects. While the Penan tour was wrapping up, we produced and published a new four-page tabloid titled *HELP STOP JAMES II & Kemano II – Two proposed nature-destroying hydroelectric megaprojects*, that compared the devastating effects these two projects would have on the First Nations people and the natural environment in their respective areas. WILD got involved because two women active on the James Bay campaign walked into our Homer Street office and volunteered to help in the research and production of this campaign publication. The paper was partially financed by Ken Kirkby who donated several of his original Inuksuk paintings to support the cause.

Nik Cuff, meanwhile, continued to volunteer for WCWC and involved us in key campaigns in the Greater Vancouver area. One creative idea he had was to use an upcoming provincial by-election in South Delta to further the campaigns to protect Boundary Bay, a major stopover for birds migrating along the Pacific Flyway, and nearby Burns Bog. Nik persuaded John Ball, who lived in Delta to run for the Social Credit Party's nomination and use his nomination speech to explain to right-wing-thinking Socreds, who were not normally very accessible to us why these areas should be preserved.

A geologist turned conservationist, Ball had walked into our office a few months earlier and told us that he had personally worked on the Geddes' Windy Craggy mine exploration site in the Tatshenshini wilderness and on mineral exploration sites elsewhere where he had witnessed the wholesale slaughter of grizzly bears. He volunteered to prepare a map of the Tatshenshini area showing the proposed park and planned Windy Craggy mine and mining road. He also helped us estimate the number of ore concentrate trucks that would travel that road daily if this mega-mine got the go ahead. His information graphically illustrated the devastating impact the proposed mine would have on the wilderness and wildlife. Impressed by his volunteer work, we hired him on a small contract to help with the Tat and our ban grizzly bear hunting campaigns.

With Nik's promise that he would be Ball's personal campaign manager and find someone to write his 20-minute nomination contest speech, John Ball volunteered to run. He joined the Socred Party and went door-to-door successfully gathering the few signatures of party members he needed to get his name on the ballet. A few weeks before the big meeting, Ball threatened to pull out because Nik hadn't done a thing as his campaign manager. Nik promised to do more: in particular, to make big posters for the nomination meeting and fill out the Socred party's lengthy background questionnaire in order to qualify John to stand for nomination. Reminiscing years later, Nik said that he simply made everything up on the qualification form, fancifully filling it out to make Ball look good. "After all," Nik explained to me, "it was only a lark, one small ploy to advance the Boundary Bay and Burns Bog campaigns."

At the last minute, a few hours before the meeting, Nik used a feature on his print shop's colour photocopier that automatically scanned a 35mm slide, enlarged it, and split it up into 16 partial images and printed them on 11" x 17' sheets of paper. Using a slide of Ball's head and shoulders, he made four sets. He taped the pieces together to make four giant posters, put in large letters *A New Commitment to the Environment* across the top of each and put them through his big laminator. (We used this same technique to make many large displays of clearcuts and threatened ancient forests for protests and rallies in the years to come).

Nik arrived late to the meeting. Hundreds of Socreds already packed the hall. The only wall space not already plastered with posters was high up above the doors. He borrowed a stepladder from the janitor and scurried around putting up the giant four-foot by six-foot blow-ups of Ball's face above every fluorescent red exit sign. These posters were much, much bigger than the other candidates' posters. Someone snidely said to Nik as he was putting them up, "That guy sure has no problem with ego!" The Socred party faithful listened respectfully as John Ball gave a good strong presentation of his prepared speech about Boundary Bay and Burns Bog (ghost written for him by environmentalists involved in these campaigns). Although his speech was well received, he got only a few votes.

Months later, long after John Ball had left the employ of WCWC, he won the Socred nomination in Richmond, the seat that the former Premier Bill Vander Zalm had held. It was the safest Socred seat in the province. Immediately after the election was called on September 19, 1991, accusations surfaced in the press that Ball was a Holocaust denier. We were shocked. The entire first week of the Socred campaign was dominated by the controversy it kicked up. Ball ultimately withdrew and the Socreds nominated another candidate, but the damage was done and thereafter the Socred campaign never picked up much steam. During the six months Ball had worked and volunteered for WCWC, no one had any inkling about this side of his character.

By convincing Ball to run for the nomination in the Delta by-election and by "creatively" filling out the qualification papers, Nik believes he changed history. Nik undoubtedly made party officials believe that Ball had passed their screening process. If they had checked him out, his bizarre beliefs and past activities with German-Canadian Holocaust denier Ernst Zundel, that were revealed in the press during

Next two pages: A cartoon-style map shows the devastating environmental effects of the proposed Geddes' Windy Craggy mine in the Tatshenshini wilderness. The other side (not shown) illustrated the benefits of a park. Artwork: John Ball.

One Day in the Life of the Tatshenshini Area if
(Environmental hazards which would occur if Geddes 105 km/65 mi long by 25 m/80 ft wide road and develop

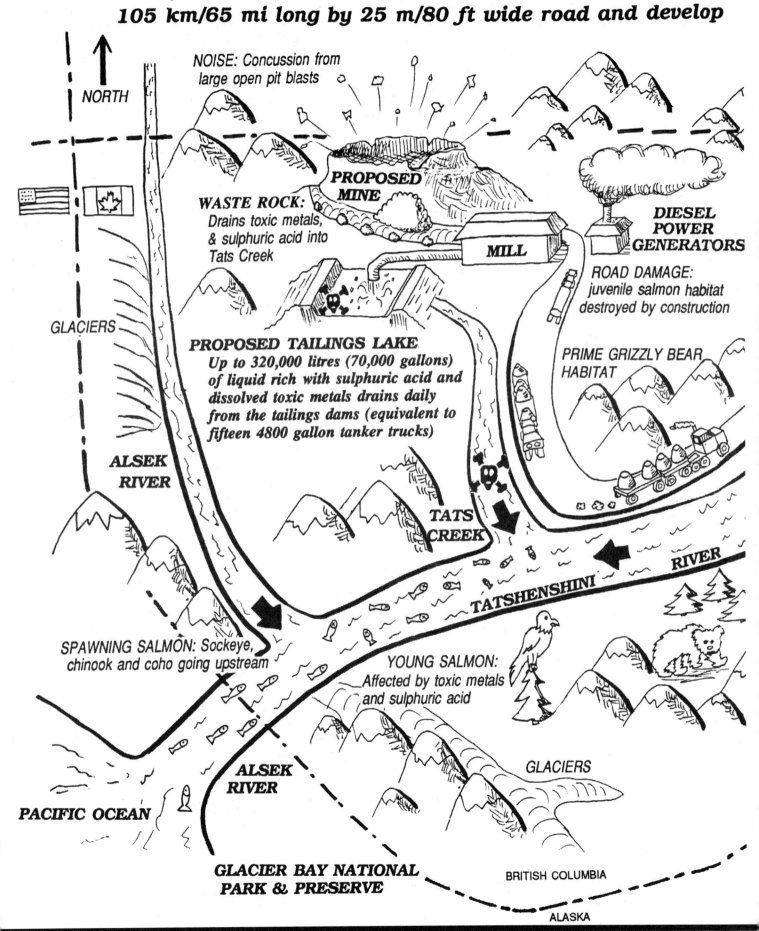

the Proposed Windy Craggy Mine is Built

Resources is allowed to construct a the Windy Craggy open pit copper mine)

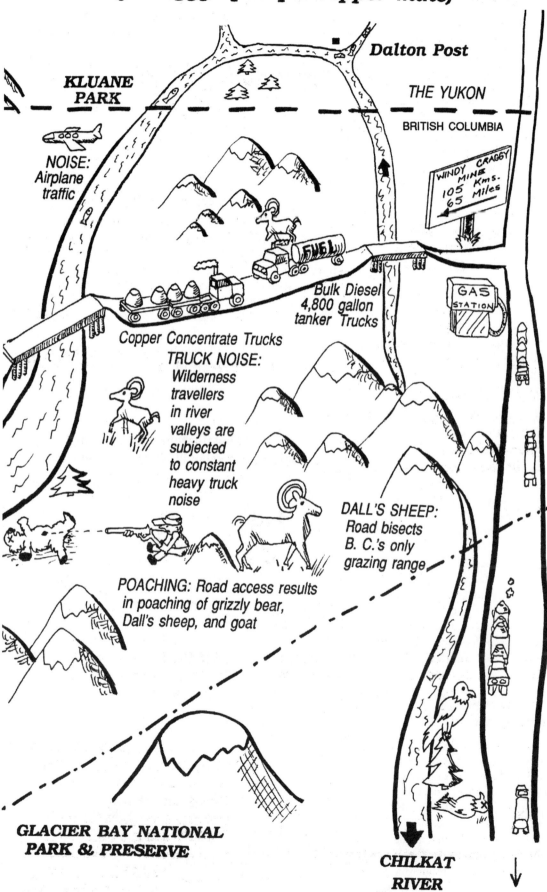

the first week of the election campaign, would certainly have come to light. Thus a minor Boundary Bay campaign ploy helped pave the way for the Harcourt NDP win in the October 1991 election!

Lower Tsitika-Robson Bight becomes a big campaign for WCWC Victoria and Mid-Island Branches

Derek Young, founder and executive director of our Victoria Branch, wanted WCWC-Victoria to act like it was a completely independent mini-WCWC that made all its own decisions and conducted its own campaigns. That was mostly OK with us because we were not a centralist organization. Our only squabbles were about money and over maintaining the high standards in publications and campaigns that had won WCWC its good reputation.

WCWC-Victoria's first big campaign (after Carmanah) was the Tsitika. Along with our Mid-Island Branch, it was one of a number of groups advocating for a large Lower Tsitika Provincial Park to protect the old-growth forest that grew adjacent to the killer whale rubbing beach near the mouth of the river at Robson Bight. In the summer of 1990, WCWC-Victoria set up a research camp in the Lower Tsitika. I was a bit skeptical of some of the scientific work—one of the studies was on the movement of slugs—but our camp certainly helped draw attention to the issue. That summer, roads and clearcuts did not advance any closer to Robson Bight.

During this time a protest camp sprung up along the proposed roadway. In late October 1990 people there began blockading to halt the logging of the Lower Tsitika Valley. In fact, the movement to save the Lower Tsitika became quite hot. Laurie Gourlay, founder of our Mid-Island Branch who was on the side of the Tsitika logging road video-documenting nine blockaders getting arrested, was himself arrested. His name, along with the fact that he was a WCWC director, was plastered all over the media.

This was the first and only time a WCWC director was arrested for civil disobedience. Laurie immediately called me to explain his innocence, but offered to step down off our board knowing we eschew such illegal action. The board accepted his resignation.

Laurie was later cleared of the charges and re-appointed back to our board. Soon afterwards he moved to Ottawa and established a WCWC Branch there.

Inspired by our big Carmanah art project, our Victoria Branch also launched a series of Tsitika–Robson Bight artists' expeditions in the summer of 1990 and started working on producing a book. Later that fall it held an art show in Victoria's Crystal Gardens featuring the artists' donated works where the art was silently auctioned off, like we had done the year before with the Carmanah art. Not being able to afford to get all the art works photographed, colour separated and reproduced as we had done for our *Carmanah Artistic Visions* book. WCWC Victoria's Tsitika book, instead, told the story of the artists' expeditions.

Quality of publications becomes an issue with a fight over "group orgasm" with WCWC-Victoria

Bickering about the quality of writing was part of the scene at WCWC. We had a publications team that was supposed to check over every press release and newspaper, including those of our branches before they went public. But it was not practical to do it all the time. In the late fall of 1990 this bickering came to a head with the publication of our WCWC-Victoria's *Beyond the Gate: artists' journeys to save the Tsitika Valley and Robson Bight*. This book was the first major WCWC publication not produced at our main office. Derek Young resisted editorial control, refusing to let our publications team even proofread their book before it went to press.

Finally, I made a special trip over to Victoria to attempt to see the manuscript. After much coercion, Derek agreed to let me take a look at it. I spent the afternoon reading it through, finding many typos, inconsistencies in style, capitalization errors and poorly written sentences and I was not the best proofreader at WCWC.

There was one sentence that I really objected to in the book. It boldly asserted that the feeling the artists felt when they saw their first killer whales was "...*a group orgasm.*" I pleaded with Derek to change that analogy, because people would either laugh at it, or find it offensive. He said that he couldn't change it because it interfered with the rights of the author. I insisted that the book go through a final copyedit, pointing out that, besides this offensive phrase, there were many careless errors that detracted from the book's message. Finally he told me the truth. It was too late. *Beyond the Gate* was already "out of the gate." It was being printed as we spoke. "What a waste of my time," I thought, as I boarded the ferry back home. I was most angry about the fact that so much money was being wasted on a mediocre book that could damage WCWC's precious reputation. While the donated Tsitika art itself and the commitment on the part of the artists were definitely on par with that of the Carmanah artists, our Victoria Branch's *Beyond the Gate* book did not do them justice.

Besides the conflicts over editorial control, there was constant pressure from our Victoria Branch for us to return to them a fair share of the money collected by the main office from WCWC members who lived in the Victoria area. (Victoria was financially strapped, too.) We all agreed, in principle, that this was a good idea. In practice, however, money was so tight at our main office that it was impossible to do. Every penny we received went to pay pressing bills. The financial crisis and the fight regarding head office control over branches came to a head a year later, at the end of 1991 when the new "outside" WCWC board took over.

Preventing a possible hostile takeover: realizing our employees' board can't solve WCWC's financial crisis

Although we didn't realize it until our 1990 AGM, WCWC was very vulnerable to takeover by hostile outside forces. At our annual general meetings we always allowed nomination of directors from the floor and usually only a handful of members besides staff attended these meetings. A week before our 1990 AGM someone tipped us off that some Share BC members (who opposed any new parks in B.C.) had clandestinely joined WCWC and were planning to come en masse and elect themselves to our board. We sent out an emergency call for support from our members, resulting in over 100 of our most faithful coming to that AGM. A few people who seemed like they might be SHARE BC types attended, but not nearly enough to be of any threat. It was sobering, however, to realize that those who opposed more wilderness protection could easily have taken over WCWC if they had made a more concerted effort.

Following that AGM we called a special general meeting of members and adopted bylaw changes that required persons running for our board to gather 10 signatures of endorsement from other WCWC members and submit them to our office 15 days before the AGM. This would, of course, alert us to any hostile takeover. At a later extraordinary meeting we went a step further, adopting two-year director terms and staggering board elections so that only half of the board members came up for election at each AGM. Back in 1990, however, we failed to confront the bigger problem of not having an "independent board." Almost 100 percent of our board members still made their living working for WCWC.

Chapter 34

WCWC and WILD reunite
WCWC gets top Canadian award

WILD and WCWC make up and WILD moves back to 20 Water Street office again – long term employee Ken Lay quits

Early in 1991, after the UI top-up grants to WILD ran out and the extra staff were laid off, WILD and WCWC reconciled and WILD moved back to 20 Water Street. Soon after we reunited WILD and WCWC, we received some very good news: WILD had been awarded a CIDA grant to mount its Latin American WILD Mapping project. This amounted to $73,000 from January to May with an additional $50,000 to support mapping work in Latin America from June to April 1992. Most of the money would go to our Southern partners, but some of it would stay in Vancouver and help cover our administrative costs on the project.

It was good for all of us to be back together again. We tried to work friction-free, but relationships were strained because there simply wasn't enough money to go around. We held meeting after meeting about how to save money and how to increase revenues. At one of these meeting Ken Lay, our appointed fundraiser, proposed to mailout another appeal in March 1991 to follow-up on his "Deep Roots" Christmas appeal.

Our new comptroller, Mike Rodgers, recommended against it. Our pockets were truly empty. We couldn't even afford to pay the postage. The staff voted Ken's proposal down and the next day Ken quit. He was the first big casualty of our financial crunch and everyone really missed his energy, enthusiasm and hard work.

Remarkably, our internal turmoil did not affect how the world viewed WCWC. In fact, only the environmental movement insiders knew how bad a state WCWC was in. In 1990, despite our financial struggles and internal wrangling, we published more issue papers and carried on more effective campaigns than ever before. Among the 14 papers produced that year was one to help local people in West Vancouver beat a referendum to cut down an old-growth forest and put in a golf course in West Vancouver on Hollyburn Ridge adjacent to Cypress Bowl Park. We also published our first members' paper, outlining all the projects we had on the go. Our Victoria Branch produced its first newspaper about protecting the Lower Tsitika.

To save money, a few years later we rented cheap warehouse space on the fifth floor across the street from 20 Water (55 Water) and let go of the back end of our 20 Water Street space that fronted on Blood Alley. It gave us twice the area at half the cost. The canvass, mailroom and product storage moved to 55 Water.

"Outstanding Group of the Year" award from the Canadian government goes unreported by the media

In May 1991, just a few months after Jean Charest became Federal Environment Minister, we got an official letter informing us that WCWC had been chosen as one of the finalists for Canada's Outstanding Environmental Group of the Year award. I was designated to go to Montreal to attend the ceremonies on June 7th on behalf of WCWC. Of course, the federal government paid for everything, the flights, hotel and food. I booked a stopover in Ottawa afterwards so that I could meet some MPs (we never lobbied; only educated) regarding our campaigns to protect wilderness to make the most out of this free trip.

Believing that we had no chance of winning, I made no preparation for an acceptance speech. After the state dinner, I spent the evening in the pub with the leader from Paulatuk, N.W.T. This Community was also up for the same award for its work in shaping northern environmental policies, learning about environment problems in his territory. The next morning at the ceremonies held in the Montreal Botanical Gardens, I was genuinely shocked when Jean Charest announced that Western Canada Wilderness Committee was the Outstanding Environmental Group of the Year! I walked up to the podium in a daze and accepted the heavy glass and rock trophy with our name engraved on it. Not having prepared an acceptance speech was OK, for there wasn't an opportunity to give one. There was just enough time to briefly shake Charest's hand, say "thank you," and walk off the stage. Charest gave out many other environmental awards that morning.

Winning this award was a huge honour for our organization. No media, however, was there to report on it. There was only an official government photographer who took pictures primarily of Charest giving various awards to companies. This photographer didn't take a picture of me accepting the award; he only took one afterwards of the two other finalists in our category and me holding our awards. The government sent us a copy of this photo a few weeks later. There was nothing like the fanfare that greets winners at music, movie and TV award ceremonies! Nothing about the awards appeared in any newspaper in B.C. or anywhere else as far as we could tell.

We finally got someone in MP Lynn Hunter's Ottawa office to fax us a copy of the government's press release a week later. It stated why we won the award:

> The WCWC has worked 10 years to preserve B.C. forest, increase awareness of forest ecosystems, and encourage sustainable forestry through appropriate logging practices. In addition, the organization has helped give global prominence to the need to save wilderness elsewhere on earth through its sponsorship of the Wilderness is the Last Dream (WILD) conferences.

No media in B.C. seemed to have received this press release when it was sent out on the awards day. We found out later that it was the

Photo taken moments after the Honourable Jean Charest, federal Minister of the Environment presents the 1991 Canadian Environmental Achievement Award to WCWC. I am flanked by the two runners up. Photo: courtesy of the federal government.

responsibility of our Environment Canada's regional office to send out the news release and the B.C. office failed to do so. Arne Hansen investigated for us and concluded that the failure to decimate the news was not due to some sort of conspiracy; it was simply a case of bureaucratic incompetence.

Being deemed Canada's top environmental group that year didn't give us any financial boost. I rationalized that perhaps if it had become news, it might have even hurt us. People might have thought that, since we were so successful, we didn't need their small donations (the major source of our support was thousands of small donations). The one concrete benefit we did get was the hefty trophy which has been on display in our main office ever since to impress all who visit.

Modified motion to preserve and protect at least 12 percent of each of Canada's natural ecosystems passes unanimously in the House of Commons

On Monday, June 17, 1991 the Parliament of Canada unanimously passed an amended Motion (M-330) titled *Protection of Canada's Ecosystems*. The original M-330 motion introduced by MP Jim Fulton and seconded by Charles Caccia read:

> That in the opinion of this House, the government should consider the advisability of preserving and protecting in its natural state at least 12 per cent of each of Canada's ecosystems.

According to Fulton's speech in the House of Commons in support of the motion, Liberal environment critic Paul Martin amended the original motion *"with the assistance of Kevin McNamee of the Canadian Nature Federation"* ostensibly to clarify the shared responsibility involved. The revised motion that passed unanimously read:

> That, in the opinion of this House, the government should consider the advisability of preserving and protecting in its natural state at least 12 percent of Canada by working co-operatively with the provincial and territorial governments and assisting them to complete the protected area networks by the year 2000.

We were duped! Martin's amendment had little to do with clarifying responsibilities and everything to do with letting the federal government off the hook in terms of protecting a representative 12 percent of every natural ecosystem in Canada including the increasingly rare old-growth forests on Canada's west coast.

Fulton's speech also revealed some other interesting facts, including a rundown on how much of Canada's land base had been protected to date:

> Canada's total lands are 9,922,00 square kilometres and our 34 national parks total 182,272 square kilometres. Our provincial and territorial parks, ecological reserves and wilderness areas of which at this moment there are 1,160, total 138,000 square kilometres, substantially less than what is presently in our national parks system.
>
> Of those 1,160 such protected sites at the provincial and territorial level, there are only 43 designated as wilderness areas for a total of 8,680 square kilometres, If you add up all the so-called protected areas, which is 6.3 per cent of Canada or just over 320,000 square kilometres, and take out the national, provincial and territorial parks which allow logging, hunting, mining and other kinds of activities, that total drops to just 254,000 square kilometres or 2.6 per cent of our national land base.

At that time no one knew the percentage of protection that was afforded to each unique ecosystem in Canada. The figures weren't calculated and, as far as I know, they still aren't today! But it was

Bumper sticker. Graphic artwork: WCWC files.

common knowledge that mountain tops, rock, ice and alpine were greatly over represented and valuable forestlands with commercial grade timber growing on them were highly under represented in Canada's parks.

Fulton sent WCWC a letter with Motion M-330 and his speech enclosed. In his letter Fulton said, *This Motion is unanimously endorsed by the House of Commons. Let's run with it!* We at WCWC certainly followed up on his suggestion.

WILD goes to Brazil

Under deadline pressure, our WILD team worked feverishly on its Latin America mapping project. Unlike the Hawaii conference, most of this conference's costs were covered by our CIDA grant. About 70 percent of the funding was designated for our "Southern Partners" but that still left a good amount for our overall coordination of the project. The WILD Team now included two coordinators working under contract in South and Central America. Participation in the Brazil conference was by invitation only for those actively involved in mapping wilderness in Latin America.

Clayton Lino, head of our Brazilian partner group, SOS Mata Atlántica, was working on protecting the Atlantic Rainforest in Brazil as a UN Biosphere Reserve. This tropical rainforest, a strip of forest along the east coast of Brazil, the most highly developed and densely populated region in the country, was packed full of rare, endemic and endangered species. Much less well known than the Amazon, it was much more threatened. Clayton arranged to hold our Latin American mapping conference in Picinguaba, a tiny village in the heart of this endangered forest. It was quite remote, located in a beautiful coastal bay within the Mata Atlántica National Park about half way between Rio de Janeiro and Sao Paulo. Our weeklong Brazil Conference was strategically timed for May 1991—a year prior to UNCED (the United Nations Conference on Environment and Development) scheduled for Rio in June 1992. Already the UNCED "Earth Summit," billed as the biggest gathering of heads of states the world had ever known, was generating great excitement.

Our Picinguaba mapping conference caught the eye of the Brazilian national media and it was covered daily in the news. The conference went well with about 50 in attendance including academics, environmental activists and indigenous people. The participants adopted a standardized methodology to use in mapping the remaining natural ecosystems and indigenous homelands in Latin America. This was quite an achievement in and of itself as Latin American academics had been fighting for years over which mapping methodology was best. At the conference another unique effort was launched to draw up a comprehensive map of the Andes mountain region and develop a peoples' proposal for a system of national Parks and biosphere reserves to protect the region and the indigenous homelands of the people living there. This came to fruition years later.

Adriane would call me about every other day to touch base and check up on our kids. There was only one public phone in Picinguaba and it always had a lineup of people waiting to use it, so she couldn't talk very long. The day before the end of the conference she said, "I've been asked to go to Brasilia to talk to government officials there. They are paying for everything. I don't know what it's all about," and then she had to hang up.

When Adriane arrived home five days later, after hugging me and our kids, she described her "official" visit to Brasilia. "It was so unreal. They had no idea that we were a tiny group that had taken a $100,000 government grant and stretched it to the max. They thought

Clayton Lino, right and Adriane Carr, left inspect a map of the Pacific Flyway at the Brazil conference with Ganga Jolicoeur, centre and Sue Fox, right behind. Photo: WCWC files.

we were loaded with money and could fund their national mapping offices. For two days they toured me around, pointing out one thing after another that they needed, especially computers for their GIS lab. I kept explaining that we were just a small environmental group and couldn't help. But they persisted. I was so happy to leave."

During the year following our Picinguaba conference, participants continued mapping their regions using the methodology adopted at the conference. WILD published the results in both English and Spanish in soft cover books titled *Mapping Natural Ecosystems in Latin America – Mapping Natural Ecosystems for Sustainable Communities*. This 47-page publication with full-colour maps and information about remaining wilderness in 10 countries in Central and South America culminated WILD's two-year CIDA-funded project.

In our members' report for 1992, Adriane wrote:

The really good news is that our work to date has actually contributed to increased indigenous rights and wilderness protection in Latin America by helping our partners in their local campaigns. Three new national parks (identified on the WILD map) have been designated in Venezuela. Traditional land rights have been guaranteed for the Guarani people (whose homeland was identified by our partner group SOS Mata Atlantica).

It was a great feeling of accomplishment knowing that in a small way our efforts helped save some fantastic wilderness in Latin America, including Brazil's Mata Atlántica ecosystem, which our Picinguaba conference had spotlighted. Our project had focused our Brazilian partners on mapping those parts of the Atlantic rainforest that remained relatively wild and provided them with some much needed financial assistance to develop their biosphere reserve proposal. The timing was fortuitous. In the lead up to the 1992 Earth Summit, the Brazilian government wanted to demonstrate that it cared about the environment and so, just months before the Rio conference, it protected its Atlantic rainforest—one of the most endangered and impacted tropical rainforests in the world. Today, Clayton Lino, the Brazilian organizer for our Picinguaba WILD conference, is the director of the Mata Atlántica Biosphere Reserve in Brazil.

After that CIDA grant was finished, we no longer advanced WILD's grand wilderness mapping vision. In large part this was because the detailed wilderness mapping we had pioneered was taken up by much better funded, bigger organizations like Conservation International. In addition, the World Bank had begun funding national governments to undertake national conservation strategies, including mapping work. We simply couldn't compete; but remained hopeful that the mapping work we knew was much needed was going to be done.

There was only one more global mapping project we embarked on: to help with the base-mapping needed for the Andes Protected Area network project that was talked about at our Picinguaba conference. Our major partner was based in Peru and the project was funded by Canada's International Development and Research Centre (IDRC). Coordination was a nightmare. Deadlines were constantly missed. I'd never seen Adriane so frustrated. She wrote up the problems and then wrote off the project. Later we learned that the work we'd started continued and a new Andean Biosphere Reserve was established. Our WILD campaign went on to work under its next CIDA grant with several partner environmental groups in Chile on a campaign to

Cover of our Brazil WILD Conference newspaper. Document: WCWC archives.

Group photo of the participants at WCWC WILD's Picinguaba mapping conference. Photo: WCWC files.

protect their ancient temperate rainforest. These Chilean conservation groups had many organizational and on-the-ground similarities to WCWC. It felt good to focus our international work on campaigns similar to those at home.

The Kitlope Declaration

Ecotrust's survey of big B.C. wilderness areas by Keith Moore revealed that the Kitlope was the largest undeveloped temperate rainforest watershed left in coastal B.C. This did not mean it had the most old-growth timber in it (about three percent of this steep mountainous watershed has commercially valuable timber growing in it); only that it covered the largest area. The Kitlope's timber was strung out in a thin strip for many kilometres along the narrow valley floor, which made it "inoperable" (meaning it cost more to log it than could be made through the sale of the logs). Yet years earlier the B.C. government gave West Fraser Timber the rights to log this magnificent valley.

Ecotrust picked the Kitlope as one of two areas they would work on with First Nations to help gain protection. (The other area Ecotrust picked to focus on was Clayoquot Sound.) The Haisla, traditional landowners of the Kitlope, were already actively opposing logging in the Kitlope, having seen the effects of logging in the Kitimat Valley.

In 1991, the Haisla First Nation issued its Kitlope Declaration, which stated:

> To those who would despoil our land: we will oppose any proposals or acts that threaten the lands, waters, and living creatures of the Kitlope. You will find us implacable, for we are protecting the very core of our existence as a people.
>
> To those who would approach us in friendship and harmony; who would join us in wonder and respect for this place: our laws require that we make you welcome, and share our most precious gifts with you. You are welcome here; we know that once you have seen and felt this place, you cannot leave here unmoved and unchanged.

As part of its campaign, Ecotrust brought influential people from the U.S. to see the spectacular Kitlope valley, including the owner of Patagonia, one of the world's foremost outdoor clothing companies. We were in the depths of our debt problems when Haisla Chief Gerald Amos approached me about producing a paper about protecting the Kitlope. Fortuitously, his request came with an attached grant that Ecotrust had lined-up from Patagonia. Otherwise I would have had to say no to the request.

The Kitlope paper was a joy to work on. In the spring of 1991 we published the full-colour, full-sized, four-page newspaper titled, *Save the Kitlope – Protect the Kitlope*. It featured the Haisla Nation's *Kitlope Declaration* to protect from industrial development their ancestral lands to which they had never relinquished ownership. We went with a different web press because it cost half as much as the one we had been using, enabling us to print 100,000 copies, stretching our Patagonia grant to the max. But we paid for it in lower quality. The colours turned out all muddy: a trade-off that we did not repeat again. The good thing was that people in BC were eager to hear about a new, huge wilderness area under threat on B.C.'s remote central coast. The paper's massive distribution really helped plant the issue in the public's consciousness.

We didn't do too much more on the Kitlope campaign, except featuring the area every year in our endangered wilderness calendars. In August of 1994, in a media release titled *WORLD'S LARGEST COASTAL TEMPERATE RAINFOREST PROTECTED*, the B.C. government announced to the world it had reached an agreement with Haisla Nation and West Fraser Timber to save the Kitlope. Years later, Chief Gerald Amos told me he felt that our Kitlope paper played a key role in their successful campaign.

Front cover of our Kitlope newspaper. Document: WCWC archives.

Chapter 35

New B.C. NDP government is pro-parks
WCWC struggles to stay solvent

The lure of big money from U.S. foundations

For more than a year leading up to spring 1991, rumours circulated that a handful of major U.S. foundations wanted to help fund the front line environmental organizations fighting to save B.C.'s wilderness, especially the ancient temperate coastal rainforest. The U.S. had literally run out of big wildlands to protect. The last large chunks of unprotected and threatened wilderness left in North America were all in Canada. For months, no money materialized.

According to those "in the know," this was because the U.S. foundations wanted to make sure the large sum of money they planned to give—we'd heard it was one million dollars—would be spent wisely; that the diverse B.C. groups would coordinate their efforts and not peck at each other. I was skeptical about this manna from mama America. Sure, the foundations there had the money, but did anyone really influential in the U.S. really care about saving wilderness outside U.S. borders?

Our experience indicated that trying to interest U.S. conservation groups in B.C. wilderness issues was a hopeless cause. Over the last ten years Adriane and I had attended at least half a dozen events in the U.S., mostly in Washington State, including several at Western Washington State University in Bellingham. We had a heck of a time getting people there to even take a copy of one of our educational newspapers about B.C. wilderness.

Our threatened wilderness areas were larger, more spectacular and closer, than all of the areas in the U.S. that they were campaigning to save. Most of our proposed parks, including Carmanah Valley and the Stein, were only a few hours drive away by car from where these people lived in the U.S.

Driving back from every one of these events, I would rail on about the xenophobic Americans and their myopic vision of the world. The one exception was the radical Earth First! organization, with its clenched fist logo and monkey-wrenching ways whose illegal actions I knew, if deployed in B.C., would do a lot more harm to our wilderness conservation causes than good. Yet they were the one group that seemed the most interested in B.C.

Of course, I had the right to be so critical of the U.S. because it was my "old country." I was born in Wisconsin and lived in the American Midwest for the first 28 years of my life. I knew from personal experience that my "new country," Canada, which granted me citizenship in 1974, had more and better wilderness than the U.S. It was one of the reasons I came to Canada and stayed. In fact, Western Canada had more than a dozen unprotected, on-the-brink-of-destruction wilderness areas on par with the finest national parks established in the U.S. generations ago.

Although we were unable to sell many WCWC posters at these events in the U.S., we were always able to give them away to leaders and employees of environmental groups. But selling our Endangered Canadian Wilderness Calendars in the U.S...forget it. There was no market for them, except for the few Canadians working in the U.S. who were homesick.

Seeking support from American foundations
Question: "Where's the Money?"

Packing this load of prejudice, I traveled to Vancouver Island to a big meeting with some American foundation representatives. I don't know who exactly organized it, but Vicky Husband of the Sierra Club of B.C. played a major role. This meeting was held in a historic old schoolhouse turned into a community hall somewhere in the countryside of Central Saanich. Just one person per environmental group was invited to attend. Everyone else at WCWC was busy, so I reluctantly went. At the very best of times I'm not very good at meetings. I get bored, fidgety, and impatient. Worse still, I sometimes get disruptive and even angry when things seem blocked.

At WCWC we had focused very little of our fundraising efforts on garnering grants from foundations. In the past, several applications that took me considerable effort to write (one to the Vancouver Foundation asking for support for our South Moresby campaign) were summarily turned down. That had soured me on this kind of fundraising. I had to admit, however, that Adriane was exceedingly good at it. I stayed focused on the tried and true method of bringing in new members and $25 to $50 donations from individual citizens to keep WCWC's financial heart beating. It took more work to garner this grassroots funding, but it was secure, steady and came with no strings attached except to campaign hard, something I really liked to do.

"It's going to be worth your time," Adriane said to me as she prepped me before I left. She insisted that I dress up for the meeting, so I put on an old suit jacket (with no tie of course). I even combed my hair. I arrived on time. Of the approximately 30 people there, over half I had never seen before. Most of them, I soon learned, were potential funders. The meeting started slowly with the usual round of introductions. I forgot most of the names and affiliations by the time the round was completed. Then representatives from the foundations gave general statements about what an important era we were moving into and what great opportunities existed in the present political climate to protect more wilderness in B.C. They were referring to the new NDP government's commitment to increase park protection in B.C. to 12 percent. It was all great general stuff, but nary a word was mentioned about money or funding to help us. "Just part of the dance," I thought to myself, as I sat there waiting patiently.

Then came a hotshot consultant from the U.S. who began to give us a special presentation on how to more effectively use the media. "More diversion," I thought, as I continued to sit quietly. But as she lectured on, I started to get angry. The woman was extremely condescending. She spoke to us like we were children or country bumpkins that knew nothing about modern media and TV technology. At WCWC we were ahead of her ideas in many areas. But it was her continual smug attitude and air of superiority that really set me off. She was obviously implying that our failure to be successful so far in B.C. was because we were ineffectively using the media. Her talk went on and on.

Finally, she finished and there was a question period. I couldn't hold back any longer and ranted, "I came here to talk about funding. WHERE IS THE MONEY?" I talked on for a few more seconds about dangling a carrot in front of us and not coming through and then walked out of the meeting. It was a beautiful spring day outside and I really didn't care that I had been so rude. Thankfully, I never again was sent to represent WCWC at funders' meetings.

That evening Vicky Husband hosted a party for everyone at her house. As I usually do at formal parties, I felt out of place, not being good at small talk and meeting new people—that's Adriane's forte. Also, my outburst had alienated people. Most of the funders shunned me like the plague. During the course of the evening a man came over and struck up a conversation. He asked me bluntly what would help WCWC the most. I explained to him about the lack of knowledge about how much wilderness was still left on Vancouver Island. I told him we needed to map the remaining wild areas there and figure out which ones were most threatened by overlaying a map of the approved and proposed logging cutblocks. I knew that a comprehensive map like this would be invaluable information for Vancouver Island's CORE planning table. I suspected we'd all be shocked at how fast our wilderness was being logged and it would motivate all the conservation groups on Vancouver Island to fight harder for more parks. Making a big "pitch," I told him we knew a qualified person who would do the work on contract dirt cheap—a forestry technician named Clinton Webb. All we lacked was the $10,000 needed to pay him.

The person I was talking to, I found out later that evening, was Emory Bundy, a granting officer with the Seattle-based Bullitt Foundation (a foundation I'd never heard of before). Just recently Bullitt had expanded their mandate to include the entire Pacific Northwest, not just the U.S. Its primary interest was preserving wild forests and wild salmon and it was not afraid to take a chance on a radical, not-so-diplomatic organization like ours.

The Bullitt Foundation selects WCWC to be its first grant recipient outside the U.S.

Remarkably, despite my boorish outburst, WCWC became the first organization outside the U.S. that the Bullitt Foundation funded. Soon after that infamous funders' meeting in Saanich, we were encouraged by Emory Bundy to make a formal application and, in late 1991, were granted $10,000 to do the research and publish a report on the *Status of Vancouver Island's Old Growth Temperate Forests*. We hired Clinton Webb to do the research. He had to go to every district forest office on Vancouver Island to find the maps of planned cutblocks. Information about future logging plans was no longer being kept in the Forest Service's Regional Office. It took a lot longer to complete the project than we'd estimated. We had to push and beg Clinton to get it done so we could report back to the Bullitt Foundation on time. It took Clinton until the summer of 1992 to finish the map and his analysis. Adriane and I pulled another all-nighter, with Clinton hovering over our shoulders, as we edited and finished up the report to Bullitt at the final deadline hour.

The centrepiece of our report was what ended up getting called "the measles map." It showed Vancouver Island's intact wilderness in

Media Conference to release the our *Status of Vancouver Island's Old Growth Temperate Forests* research report. From left to right WCWC-Victoria campaigner Denis Kagasnenni, Joe Foy, and Clinton Webb presenting. Photo: WCWC files.

green with red dots representing approved cutblocks for the next five years of logging. The red dots were everywhere. It even shocked me how fast logging was penetrating the last large chunks of wilderness on Vancouver Island. We sent a copy to every conservation group on Vancouver Island (about 75 groups), to every media outlet and to all B.C.'s cabinet ministers. We personally delivered a copy to CORE Commissioner Stephen Owen, who had been appointed a few months earlier to head the land-use planning table on Vancouver Island, but had not yet got the process going. Our report revealed that 70 percent of Vancouver Island's remaining wilderness was threatened with logging or fragmentation during the next five years. It galvanized the environmental community by destroying the myth that there was endless wilderness left.

Bullitt's generous donations to WCWC and to many other B.C. conservation organizations in the years following, and its insistence on detailed planning and reporting, which was especially helpful in keeping our organization strategically focused) made a huge difference in the preservation of wilderness in B.C.

A generous donation makes a trip down the Tatshenshini possible

Another issue we were working on during 1990-1991 was protecting the Tatshenshini watershed from the intrusion of an access road and the development of a huge copper mine. Ric Careless kept urging us to join a ten-day rafting trip down the Tatshenshini that he was planning for July 1991 with leaders of other environmental groups. Tatshenshini Wild made an offer that was difficult for Adriane and me to refuse. Johnny Mikes of Canadian Rivers Expeditions had donated half the cost of the trip and Peter Pollen (a former mayor of Victoria and longtime WCWC supporter), donated the other half so both Adriane and I could go. All we had to do was to pay for our return airfare to Whitehorse. At first we weren't going to go because of all the financial problems at WCWC, but we finally decided we just had to go. It was a really smart decision. We re-discovered the fact that there is nothing like wilderness to rejuvenate ourselves and our commitment to the cause.

This was a budget trip, and a "who's who" of the environment movement. All of us had to help prepare the food...we didn't have the full complement of guides along to pamper us like full-paying rafting clients have. But we were all experienced outdoors people. On the trip were Ric Careless and Donna Reel of Tatshenshini Wild, Vicky Husband of the Sierra Club of B.C., Sally Ranney, co-founder of American Wildlands, and other leaders from across Canada and the U.S.

Adriane and I were gung-ho rafters. On our first big day down "the Tat," I volunteered us to sit in the back of the raft and help paddle through the rapids that we would encounter a few hours downstream. It turned out to be quite an adventure. We ended up getting "stuck" in the middle of a standing wave. The raft slid backwards under it and remained stationary while tons of cascading water poured down over Adriane, slowly drowning her. Quickly, following the brief instructions our rafting guide gave us before we left on how to deal with such a situation, the rest of us scurried to the front of the raft and successfully "popped it out." When we pulled in for a rest stop a short time later, someone in another raft who had not seen what had happened said to the dripping wet Adriane, "Your rain gear certainly doesn't seem to be working!"

Adriane and I also volunteered to prepare the first day's breakfast for the group. Not having an alarm clock (all of us were told to leave our watches behind), I woke with the sun. We made a fire,

Camping on a sandbar in the Tatshenshini River. Photo: WCWC files.

got the coffee going and everything else prepared and then waited... and waited... and waited for the others to get up. Finally, someone stumbled out of their tent and asked "What are you doing up? It's 5 a.m." We had forgotten that we were near the land of the midnight sun. Adriane and I had been up cooking since 2 a.m.!

That ten-day trip was remarkable. "It really is an ice age wilderness," I said to Adriane near the end of our trip as I put some glacial blue ice in a glass with some Bailey's Irish Cream we'd carefully saved. We raised our glasses and toasted "the Tat" while we sat on the shore of Alsek Lake and watched miniature, centimetres-high tsuna-

Riding in the front after shooting the worst rapids on the Tatshenshini Rafting trip, July 1991. Photo: Ric Careless.

mis ripple onto the shore from the icebergs calving off the glacier several kilometres away at the head of the lake.

On our "Tat" trip we got to know many key leaders in North America's wilderness movement. We also bonded to this wilderness area, coming home determined to do more to help win this campaign.

Trying to make better environmental protection a 1991 provincial election issue

In late summer 1991, I came down with a case of election fever. From mid-September to mid October, I took an unpaid leave of absence from WCWC to run on the Green Party ticket against Social Credit Premier Rita Johnson in Surrey-Newton. The Green Party's leader, Stuart Parker, talked me into it by convincing me I would have the opportunity to debate the Premier at all-candidates' meetings and raise the case for better environment and wilderness protection. It didn't work out that way. Premier Johnson didn't attend a single all-candidates' meeting. The whole campaign was a nightmare. Garnering a mere 0.8 percent of the vote, I vowed never to run for political office again—and kept it.

WCWC, too, was active during this pivotal election. We were trying to make the environment a major election issue, in a non-partisan way, of course. Several months before the election, Terry Jacks produced a four-minute video for WCWC titled *The Faceless Ones*. It featured a poem recited by its West Vancouver author about the "faceless" people in big office towers who make decisions that wreck the environment. The poem concluded with the assertion that we have the power to stop them, because *we have the power to vote.*

We contracted Lisa Jaeger, a keen volunteer, to conduct a candidate questionnaire campaign. She worked tirelessly, sending out copies of Jacks' video to every candidate along with a questionnaire to ascertain their position on important environmental issues. By making persistent follow-up calls she got many candidates and most of the parties to respond—except for the governing Socreds. We published the results in a soft cover report and distributed it widely right before the election. Whether or not it made much difference is hard to say, but a new "log on the fire" never hinders the cause.

Harcourt wins and places logging moratoriums on some key wilderness areas without consulting WCWC

A few weeks after Premier Mike Harcourt and the NDP took office in November 1991, we read a front page *Vancouver Sun* article that said he had placed logging moratoriums on the contentious areas that B.C. environmentalists wanted protected. The article told of a meeting between the new Premier and the BCEN, listing WCWC as a member group, giving the clear impression that we had participated in this meeting. We hadn't. In fact we did not even belong to BCEN's Forestry Caucus, the sub-group that met with the Premier. The BCEN was not supposed to be a lobby group. Supported mostly by federal government money, it was supposed to be a networking organization that helped environmental groups exchange information and get their message out to a wider public. To get around this restrictive mandate, some BCEN members formed "caucuses" that gave them more liberty to lobby. Members of the BCEN's Forestry Caucus worked hard to get into the backrooms with the NDP to lobby for increased park protection.

The reason we did not belong to any BCEN caucus was that we did not want other groups lobbying or speaking on WCWC's behalf. What if they supported an action that included civil disobedience or in some other way compromised our reputation? We also could not afford to be part of a coalition that subsumed us and took credit for our actions. We needed to get the credit for the work we did so our members would keep financially supporting us. The *Vancouver Sun* article about the logging moratoriums did what we feared most. The deal that the BCEN's Forestry Caucus struck with Premier Harcourt left out several of the key wilderness areas that WCWC was campaigning to protect. It now looked to the public that we agreed to the list printed in the paper and had given up on the areas not on the list. I was furious.

A couple of days later, by chance, I ran into Jim Cooperman, an environmental campaigner from the Shuswap area, who was very active in the BCEN and its Forestry Caucus. I expressed my displeasure with what had happened. It wasn't his fault, he said. He had not claimed that WCWC was part of the Forestry Caucus. "It was the media's fault," for including us on the list. The problem was that the media did not distinguish between the BCEN's membership as a whole and the members of its various caucuses. We solved the misinterpretation problem by quitting the BCEN.

Having a pro-park Premier helped us get more parks. A side effect that few environmentalists talked about was that a sympathetic government depressed donations. Some people simply thought that a supportive government would protect wilderness of its own volition, and our demanding voice did not need to be heard as loudly; in other words, the crisis had abated. Of course, that wasn't true at all. The NDP, with its strong IWA ties, needed as much of a public push on wilderness protection as the Socreds before them did. Many of the NDP parks also came through costly trade-offs.

Reward offered for conviction of "faceless" lawbreakers

Shortly after the NDP took office we heated up our campaign to stop the clearcut destruction of B.C.'s old-growth forests. Someone pointed out that, since we had a $5,000 standing reward for informa-

Cartoon of a "faceless one" lawbreaker on our WANTED poster.
Artwork: Geoff Olson.

tion leading to the arrest and conviction of tree spikers, we ought to have a similar reward for clearcutters who break the law. We had Geoff Olson, a great local cartoonist, do up a cartoon of a "faceless one" character (a corporate executive), sitting at a desk situated in the middle of a clearcut. We made up a poster that looked like a Wild West "WANTED" poster. The text below the "faceless one" mug shot read as follows:

> I am authorized by the **Board of Directors** of the **Western Canada Wilderness Committee** to offer a **reward** of up to **$5,000** leading to the first **conviction and imprisonment** of any forest company executive, government official, professional forester or any person or persons responsible **for unlawful logging-caused damage to fish-bearing streams anywhere in British Columbia** pursuant to the Sections of the <u>Fisheries Act of Canada</u> pertaining to offenses.
>
> The disbursement of all or any part of this reward will be decided by the Board of Directors of the Western Canada Wilderness Committee.
>
> Only those persons who come forward and volunteer information to **Western Canada Wilderness Committee** are eligible for this reward...
>
> **THIS OFFER EXPIRES DECEMBER 31, 2001**

The poster was authorized and signed by me on behalf of WCWC's board of directors. We distributed over a thousand of them. No one ever took us up on this offer.

WCWC's Chilcotin Art project scrapped; Chilcotin book project transformed and rescued

Arne Hansen, a stalwart WCWC employee from the Carmanah hey-days, had a hard time adjusting to our hard times. He kept trying to get us to undertake another moneymaking art project. He started out proposing to take 100 artists to the Chilcotins, the stunning wilderness area we planned to feature in our third coffee-table book to be produced by Ken Budd. Arne projected the cost at only $54,000, with estimated revenues of $300,000. However we just didn't have any start-up funds to get it going.

Arne kept persisting, reducing the scale of the proposed project until all he said he needed was $22,000 in total and only a couple of thousand dollars up front. He had already lined up artists who were keen to participate and donate their art to us. WCWC's Board rejected this proposal, too: we didn't even have an extra $200. The Carmanah, Clayoquot, Chilcotin book trilogy as originally conceived was dead. We had exhausted our ability to print on credit.

Several months earlier, however, before pulling the plug on the Chilcotin book project, we had contracted Terry Glavin to research and write the text for this book. We'd already invested several thousand dollars. Glavin was writing the fascinating story of the Chilcotin war of 1864 between the European newcomers and the Nemiah First Nation from the Nemiah point of view. He understood our financial bind and came to our rescue by lining up New Star Publishing to take his manuscript and publish a more modest soft cover book.

We turned over all the rights to New Star. I promised to help market the book through our catalogue. I felt it was important to get this story out, both to promote the protection of the Chilcotin area and to broaden our understanding of the Nemiah's cultural heritage. In 1992 New Star published *Nemiah: the Unconquered Country*. It was a great book, illustrated with photos by Gary Fiegehen, Rick Blacklaws and Vance Hanna and was short-listed for the Duthie book prize that our Carmanah book had won a few years before.

Artists who contributed to our *Carmanah: Artistic Visions of an Ancient Rainforest* book project prepare to autograph their page in the numbered collector's limited edition book. From left to right—Jack Wise, Graham Herbert and Ron Parker. Photo: WCWC files.

Autographs add value to our *Carmanah Artistic Visions* collectors' edition book

Arne continued to hunt down projects that would bring in needed money. We were completely out of hard cover copies of *Carmanah: Artistic Visions of an Ancient Rainforest*, which we could have easily sold. We had, however, about 350 copies of the 500 original collectors' edition books left. Each one was numbered on a special insert page, leather bound and sleeved in its own box. They were priced to sell at $250 each. They hadn't ever sold very fast and, since Raincoast Books' soft cover edition of our Carmanah book came out, now they weren't selling at all. Arne Hansen came up with the brilliant idea of getting each artist to autograph these books on the page featuring their artwork. He convinced us it would make the books more valuable and saleable. We finally said OK to this project.

It was a huge amount of work. Even with two people assisting—one opening the book to the right page and putting it before the artist and another one closing it and putting it in its box after it was autographed—it took several hours for each artist to sign all of the books. Some of the artists could come to our Gastown office to do it; but many couldn't. So Arne had to haul the books to them. All the artists (except a couple who had left the province) signed them, including the famous Jack Shadbolt, Robert Bateman, and Toni Onley. It took Arne several months to complete this task. Today many of these books are still stored in WCWC's product "cage." Over the years we have donated dozens of these "collector edition" books to other charities to auction off and help them out.

Arne went on to design another art-based fundraiser, which involved producing 93 folios of limited edition prints of 18 famous artists that would only be available in that one unique package. Each package would cost $5,000. Again we balked at the project, having no money to invest in it. With our blessing, Arne left our employ and took this project with him to the Wildlife Rescue Society.

WCWC presents a proposal to Nuu-chah-nulth

Despite our money woes, we kept pushing on our key campaigns. In the fall of 1991 we got permission to present Randy Stoltmann's proposal to protect an area he called the "West Coast Trail Rainforest" to the Nuu-chah-nulth Tribal Council's annual meeting. Our campaign team felt that we had to work much more closely with the Nuu-chah-nulth on issues of wilderness preservation within their traditional territory on the west coast of Vancouver Island to have any chance of success. This was particularly true regarding Randy's proposal to expand National Park protection along the Lifesaving Trail from Port Renfrew to Bamfield. The proposed park expansion lay entirely in Nuu-chah-nulth territory and the Nuu-chah-nulth were notoriously touchy about parks.

Clinton Webb put together a slideshow showing wilderness areas within Nuu-chah-nulth territory, contrasting them with huge clearcuts on steep slopes and the damage they were causing and pointing out that a similar fate awaited most of their lands if the status quo continued. We put together a cerlox-bound brief with colour photos and samples of Tribal Park declarations made by other First Nations to protect special places within their territories. We asked the Chiefs at the meeting if we could start an ongoing dialogue with them and meet on a regular basis in the future to exchange viewpoints.

Having worked for the Nuu-chah-nulth Tribal Council ten years earlier, I was apprehensive of this approach. Delegations such as ours consisting of "white do-gooders" usually got a rough ride. I also knew that there was a prevalent feeling among the Nuu-chah-nulth leaders that environmental organizations had abandoned them when they went to court to protect Meares Island.

They had raised hundreds of thousands of dollars through raffles, bake sales, lahal (a traditional west coast gambling game) tournaments and through digging deep into their own pockets using their limited discretionary funds (most money they received from the federal government was earmarked for specific projects or programs). The environmental groups, including ours, had given practically nothing to help. Many Nuu-chah-nulth believed that all environmental groups were wealthy and could have helped out a lot more although it was certainly not true for WCWC.

We got to the meeting early in the afternoon of November 22, 1991, about a half-hour before we were scheduled to give our presentation. Proceedings were running late and we did not get to present for a couple of hours. Clinton gave his slideshow. Joe handed out a copy of our cerlox-bound brief to all the chiefs around the big table. I just sat in the audience. Everyone watched the slideshow intently, listening to Clinton, but there was no way to block out the sun from the windows high in the gym room so the images were not as effective as they would have been in the dark. When Clinton finished he asked if there were any questions. One Chief got up and started talking about an entirely different subject and the meeting went on as if we had not even made our presentation.

As we all drove back in WCWC's canvass van, Clinton and Randy expressed great disappointment, wondering where we had gone wrong. I told them not to be discouraged. It could have been a lot worse. They could have outright turned down our proposal. Anyway, I explained, it was better to work with individual bands on specific issues and areas that were within that band's territory, rather than try to work on the tribal council level. Each band was a separate First Nation and made its own land use decisions regarding its territory.

With Carmanah we had to work with the Dididahts who had a village on Nitinat Lake. With the Walbran we had to work with both the Dididahts and the Pacheedahts who live near Port Renfrew. With our West Coast Rainforest Trail proposal we had to work with those two and the Huu-ay-aht. Our brief did one good thing. It let the Nuu-chah-nulth clearly know where we stood on aboriginal title and rights: we recognized and respected them.

Celebrate Wilderness Gala is a fundraising failure

All the while our financial situation continued to deteriorate. Mary Lou Stewart, who sponsored our gala Rainforest Benefits I and II and paid all the bills they generated, decided not to put on a Rainforest Benefit III. We decided to risk putting on our own event to be held in November 1991, hoping against odds that this one would raise money. We convinced ourselves that we knew all the pitfalls and could avoid them. Our event, which we decided to call *Celebrate Wilderness Gala*, would be different. We made it less formal—a "black tie and sneakers" event. We served inexpensive (but good) pizza to save money. We picked a really up-scale venue, the prow at the front of Canada Place that had a spectacular view of Vancouver harbour. We organized it entirely "in house," although we contracted a full time person to handle all the details and garner donations for our silent and live auctions.

Our Celebrate Wilderness Gala had all the ingredients of being a success except the most important one—paid attendees. Our tickets, priced at $125 per person or a "bargain" at $200 per couple, put us out of our own class. We should have known better. The vast majority of WCWC members did not have that kind of coin. The weather, however, dealt it a final, fatal blow. On the night of our gala the winds gusted in excess of 100 kilometres-per-hour and were accompanied by torrential rains. All the power on the North Shore was knocked out. The miserable weather kept away many supporters we counted on coming, including some who had prepaid and reserved tickets. It was the worst storm in a decade.

Our gala morphed from being a fundraiser into a fundloser. Instead of making tens of thousands of dollars, we lost several thousand. Our loss was this small only because almost everything was donated. On an upbeat note, those who attended the event loved it and there were some very happy people who got incredible bargains in the auctions. Monte Hummel, Adriane and others gave great speeches. There was a fabulous slideshow by Jeff Gibbs of the Environmental Youth Coalition. Then, impromptu, John Cashore took the podium and gave his first public speech as the newly appointed Environment Minister in the just-elected Harcourt government. He spoke cautiously but optimistically and heavily hinted, to the delight of those in attendance, that many new parks would soon be created.

WILD applies its mapping methodology in B.C.

Just because we were in financial crisis and our fundraising efforts were failing, this did not mean that we slacked off in our campaigning. We plowed ahead with publications, most of them covered by specific grants. WILD had garnered a grant to apply some of its mapping expertise to B.C. Early in 1992 WILD came out with its full-sized newspaper titled *BC's Temperate Rainforest – A Global Heritage in Peril*. It popularized Ecotrust's survey of rainforest watersheds that had just been completed by Keith Moore showing all the remaining undeveloped primary watersheds in B.C. over 5,000 hectares in size.

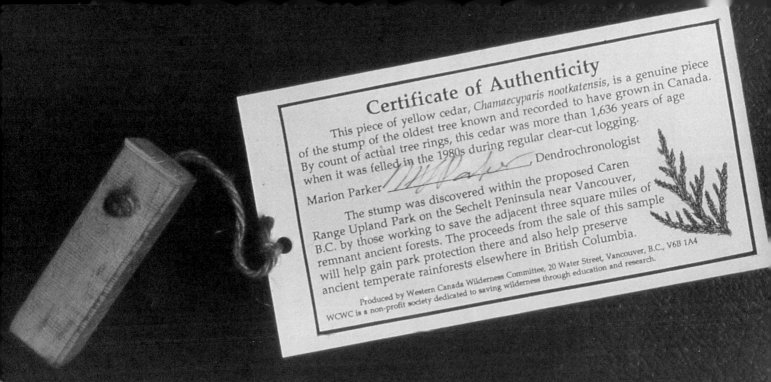

A small block of wood from the stump of the then oldest known tree (yellow cedar) in Canada. We sold these in our store for $10 each as a novelty item to raise money for the campaign. Photo: WCWC files.

Besides showing their boundaries, it also presented conservation values for each watershed. A comprehensive table showed the salmon species that spawned in each river, as well as waterfowl and grizzly bear ratings and whether or not the area was protected. In many ways it was a forerunner of the "Great Bear Rainforest" campaign that came nearly a decade later.

WC² attends the IUCN's 4th World Parks Congress

A month later, in February 1992, WCWC-WILD took its new *Temperate Rainforest* paper to the International Union for the Conservation of Nature's (IUCN's) 4th World Parks Congress held in Venezuela. The IUCN was, and still is, the world's largest environmental knowledge network focused on conserving the integrity and diversity of natural ecosystems. It has helped over 75 countries prepare and implement national conservation and biodiversity strategies. Headquartered in Gland, Switzerland, the IUCN today has over 1,000 people on staff located in 62 countries.

The IUCN holds a Parks Congress only once every 10 years. This one brought together 1500 people from around the world—scientists, representatives from government agencies and non-government organizations all interested in parks and protected areas. The conference was an ideal place to distribute WCWC's publications and embarrass Canada and British Columbia into protecting more wilderness. Also many friends from Latin America came who were contributing to the WILD mapping work in South and Central America and who had attended our mapping conference in Brazil six months earlier. These connections proved to be a great asset.

We loaded down Adriane and Ian Parfitt, director of WILD's mapping work, with an incredible amount of WCWC materials including, besides our recent temperate rainforest paper, 1500 copies each of our Kitlope, Tatshenshini and Penan newspapers—one for everyone attending the meetings—plus many of our Endangered Wilderness Calendars and a large assortment of ancient temperate rainforest posters. But, of course, we only sent samples of our posters and calendars—we couldn't afford to give a copy to everyone and they would fill up a whole plane. While Adriane and Ian arrived in Caracas fine, unfortunately, the boxes of WCWC materials didn't.

They were immediately seized by customs in Venezuela. It looked like there was no way Adriane or Ian, who worked at the problem diligently, could get them cleared. Adriane called me to give me the bad news. Fortunately, she also told all of our WILD mapping partners who were in Caracas, too.

Almost magically, one of our contacts worked through some local connections and got everything through customs without any duty charges. The ten big boxes were delivered straight to the conference centre. Our papers were a big hit. Ian managed to get into the conference delegates' mailroom and delivered our papers into every mailbox. No one else was distributing information this way. Our posters and calendars were an even bigger hit! Adriane nearly precipitated a riot when she put them out on a table and people scrambled to get them. She told me afterwards that she felt they did more good than did the scholarly paper she presented on the need to expand the world's park system and on the role that NGOs (WCWC-style) can play in public education and conservation work.

Canada's oldest trees found to date are located in the proposed Caren Range Provincial Park

In February of 1991 Paul Jones forwarded a proposal to the provincial government asking it to create a park on the Sechelt Peninsula. It was an area he was familiar with that was rapidly being liquidated. A forester by profession and a naturalist by avocation he felt that this forest was extremely old and that marbled murrelets nested there. It turned out to be the oldest known forest and it became the first forest where the nesting behaviour of this incredible seabird was observed. (*The Marbled Murrelets of the Caren Range and Middlepoint Bight* by Paul Jones. Published by WCWC in 2001) Together with some other concerned citizens on the Sunshine Coast he formed a group called Friends of the Caren. Every weekend in that summer they took interested people up the logging road and into the old-growth still remaining along the top ridge of the Caren Range, which runs down the spine of the peninsula. Paul gave me a copy of his Caren Park Proposal soon after he presented it to government and imme-

Display round in WCWC Gastown Store from the stump of Canada's oldest known tree, a 1,835 year old cedar that grew in the Caren Range on the Sunshine Coast of B.C. Photo: WCWC files.

diately I said, "Let's co-publish a newspaper about it!" Paul agreed it was a good idea. But first I had to visit the area. Our family went with other interested people on one of the Saturday tours of the area and it (the Caren forest) lived up to its reputation. It was beautiful and the trees dripping in moss did indeed look ancient. Already one was found (a stump of a cedar that had been cut nearby the proposed park) to be about 1717 years old. Just before spring '92 we published our joint paper titled *Canada's oldest tree on Sunshine Coast...Chainsaw found it first* with a picture of some children playing on the stump and we distributed a copy to everyone on the Sunshine Coast.

Paul Jones was convinced that there were older ones there. It was just a matter of finding them. In the summer of '93 the Friends of Caren took dendrochronologist Marion Parker up to the Caren to check out some stumps. While they left to search for one promising stump they'd seen in one of the clearcuts they left Marion on the road. Casually looking out over the clearcut, he saw a coke can on top of a stump a fair way into the clearing.

Thinking it was one the Friends had previously marked, he struggled over to it and counted the rings. He found over 1,800 of them. It was older than any other found up there by a hundred years.

When the others came back he raved about the stump they had marked for him. "What stump?" they asked. They hadn't put the old pop can there and didn't know anyone who had. Perhaps a logger had eaten his lunch there when the area was logged and carelessly left it behind.

Eventually, the Friends got two slices cut off this yellow cedar stump. We contracted Ken Lay to sand and mount one of them. Marion Parker dated it; the final careful count turned out to be 1,835 growth rings. We added interesting historic dates, (including 395 AD – the fall of Rome) and made it into a display for our Gastown Store. It's still on display in our store today. It's still the oldest known (verified by ring count) tree in Canada, as far as I know.

Chapter 36

WCWC's new "outside" board tackles our internal problems and imposes a solution

A period of hope

In the last few months of 1991, the only reason all of us were not completely freaked out about WCWC's financial situation was the hope that when our new "outside" board was elected in December, things would dramatically improve. I was one of those who strongly championed the concept of having a board of directors made up wholly of volunteers, none of whom worked for WCWC. The fundamental problem with our staff-dominated board was that it wouldn't lay off staff (board members) in order to reduce the payroll to fit reduced revenues. We fixed this flaw in another extraordinary meeting of members in August of 1991, adopting bylaw changes that made only WCWC members who received less than 10 percent of their annual income from WCWC eligible for a directorship on our board.

The period between the decision to create an "outside" board and the election of the new board members was a time of hope. Joe Foy and one other person stepped down from our board to "smooth the transition." Two non-employees, Bryan Evans and Christine Lattey, were appointed to fill these vacancies on the board, and were subsequently elected at the December AGM.

A few months prior to the "outside" board election, we had held a day-long staff planning session conducted by Lattey, who was a practicing psychologist. She volunteered her time and effort to help us develop a "Mission Statement" and establish a new way of making decisions through a "Team Management System" structure. She also led us in a two-day staff retreat at Manning Park during that "fall of hope" before the December AGM.

Internal turmoil helps us define ourselves

One positive accomplishment of our retreats and internal "navel gazing" was the development of a Mission Statement, a Values Statement and two fundamental Policy Statements: one on Civil Disobedience; the other on the Recognition of Aboriginal Title. They have stood the test of time and remain firmly in place today.

- *WCWC Mission Statement* – Western Canada Wilderness Committee (WCWC) is a registered non-profit environmental society with federal charitable status working through research and education for the preservation of wilderness in Canada and around the world. WCWC promotes ecological principles that achieve sustainable communities. WCWC works only through lawful means.

- *WCWC Values* – Life in all its abundance, complexity and mystery; wilderness with all its natural biodiversity, as absolutely vital to the continuing health of the planet and human survival; accuracy in all our information and excellence in all our endeavours; and the diversity of individuals and groups whose combined non-violent strategies give the environment movement its strength.

- *WCWC Policy on Civil Disobedience* – We do not participate in any form of civil disobedience. We do not condemn non-violent civil disobedience by others. We condemn activities that threaten life, property or the environment.

But these measures, well-intentioned as they were, did not ease the tensions at WCWC. This is because our primary problem did not stem from our inability to get along with each other or work together efficiently. Our bickering was rooted in the stark fact that, no matter how hard we tried, we could not generate enough money to pay our debts, or support our current overhead. Several solutions were floated during these difficult times. Some thought that we were trying to do too much and should reduce the number of campaigns and publications. A few, including Adriane and I, thought that we had too many employees and that staff should be radically reduced and more use made of volunteers...at least temporarily. Most (including myself) thought that if we just worked harder, smarter and hung on, eventually things would take a turn for the better.

With an air of relief and celebration, in December 1991 we held the election of our new board in the basement of the Kitsilano United Church. The new board included Ric Careless and Nik Cuff. Our kids (along with some other children) ran around uncontrollably, especially TJ who was now five years old. He climbed up on the stage and hid behind the curtains while the acceptance speeches were being made by the new board members who all won by acclamation. I felt that we turned a new leaf and WCWC had finally "grown up."

Group photo taken at the al-day staff workshop in early 1991 that developed WCWC's Mission Statement, Values Statement and Policies on Civil Disobedience and Aboriginal Title as well as established the Team Leaders management style. Photo: WCWC files.

Recession was a major contributing factor

We spent years, after the fact, wondering why our income dropped over a million dollars in 1991 and what we could have done differently to have averted the crisis. Some blamed the fact that we let too much of our fundraising ride on our Carmanah campaign and after half the valley was made a park in April 1990, it was harder to garner support for this campaign. It's true that our meteoric rise in income was in great part due to "Carmanah fever." It rose from $350,000 in 1987-1988 before this campaign started (WCWC's financial year goes from May 1 to April 30), to $590,000 in 1988-1989 the year our Carmanah campaign began. As Carmanah gained momentum and our door-to-door canvass took off, our gross income rose to $2,400,000 in 1989-90 and peaked in 1990-91 at $2,800,000. Then, in 1991-1992, a half-Carmanah-valley of big trees saved, our income plummeted to $1,700,000, 1.1 million dollars less than the year before.

Although we didn't realize it at the time, our troubles weren't all due to the collapse of our Carmanah campaign or other mistakes we made. The Gulf War and the global recession of 1990-91 took their toll and undoubtedly accounted for the lion's share of the big decline in revenues at WCWC.

We were caught in a situation over which we had no control. Other environmental groups in B.C., Canada and the U.S. were experiencing the same crunch. Some even thought that the election of Premier Harcourt and President Clinton who were more "wilderness and environment friendly" also contributed to our woes. Regardless of blame, the wave of environmental consciousness crested in the late 1980s, and then collapsed. We had to adjust, wait to catch the next wave and, in the meantime, try to create big waves ourselves.

Another retreat fails to solve the problems

In January 1992, a little over a month after being elected, WCWC's new board sponsored a special retreat for staff and directors. The goal of this two-day affair was to get us all working together. At it we talked about campaigns and priorities, then broke into smaller working groups and drank beer together in the evening. Adriane again made a strong case that we had no choice but to reduce staff. She pointed out that large groups in the U.S. (the Sierra Club and Greenpeace) had already reduced their staffs by half. Her advice fell on deaf ears.

This big retreat solved nothing and maybe even made things worse. The chairman of our board, Allan McDonnell, wrote in his board report that not much was accomplished because there were *...too many sacred cows and pet projects.* I think Adriane and I were the "sacred cows" he alluded to and the WILD project was the biggest of the "pet projects."

Things drifted along for another month with no substantive action to correct our situation. Our comptroller reported that monthly expenses in February outstripped monthly income by $17,000. The idea of bringing in an outside executive director to pull everything into shape was vigorously promoted by some board members and staff. The idea of spinning off our book and poster production and sales into a separate company that would feed its profits to WC^2 was another idea that had its supporters.

The upstairs/downstairs split grows

The "outside" board failed to provide a magic cure. In fairness, the new "outside" board inherited a much worse situation than any of its members had imagined. As our financial crisis deepened and the red ink continued to flow, the "upstairs/downstairs" split that began prior to the new board taking over grew and deepened. The WILD Team, Adriane, Joe and I were "upstairs" and the accounting, marketing, shipping and store staff were "downstairs."

Another serious problem that no one had expected landed on our shoulders. The couple who had donated half ownership in their Ucluelet house to us about eighteen months earlier now wanted the money from the sale of the house to go to the David Suzuki Foundation, not us, because we had not fulfilled our obligation to use their house as an outreach centre. The board agreed to do that, thinking that our Victoria Branch held our share of the sale money in their bank account. But the money was gone. Derek Young, head of the Victoria Branch, without the boards' approval, had used it—about $28,000—to pay off a loan he had personally taken out to pay for the *Beyond the Gate* book about the Tsitika.

Derek resigned from the board over this controversy and quit as head of our Victoria Branch. On behalf of WCWC, I attended an intense session mediated by the Victoria Dispute Resolution Centre and reached the following agreement with the elderly couple who donated the house. WCWC would make donations of $500 a month to the David Suzuki Foundation until the entire amount we realized from the sale of the house was transferred. It was just another obligation piled onto the mountain of debt repayments we'd already committed to making.

The staggering pile of bills we'd racked up peaked at nearly $700,000 in January 1992. Every plan to rescue us had dismally failed. The outside board became a lightening rod that attracted internal dissent. Now some staff members, when they disagreed with a Team Leader's decision, brought their complaint directly to a sympathetic board member's attention. Some board members championed a "wronged" employee after hearing only one side of the story. Some spent fruitless hours trying to mediate. Often I was the one who "wronged" the employee by being outspoken in my criticisms or in opposition to their plans.

WCWC sends me to an anger management workshop

With the tensions rising and bankruptcy looming, my bad habit of blowing up and screaming at people when things didn't go right got worse. My outbursts now happened almost on a daily basis. When I became frustrated with staff members whom I felt did stupid things, I'd simply yell. At the height of WCWC's problems, someone got the bright idea that I should go to "The Haven," a new age touchy-feely retreat on Gabriola Island, to take its four-day anger management workshop. I protested that we couldn't afford it. But everyone at WCWC, including Adriane, thought that, afford it or not, I must get help. Somehow our financially-challenged organization came up with the registration fee. (I suspect that one of the board members donated it.)

This anger workshop experience opened my eyes. I got "cured" in two days instead of four! The feedback from others, especially women, who listened to my "Vesuvius" eruptions (controlled reenactment of an angry outburst) made me realize that my verbal rantings evoked real fear that I'd be physically violent. I reassured the other participants that I never did get physically violent and realized I had to change.

Most of the men attending this anger workshop were not loud and obnoxious like me. They were quiet and calm, suffering from repressed anger. When they finally did let go, they really blew their stack. Some did get physically violent. At one of the sessions where each person did their "Vesuvius" I went on and on in my passionate (angry) voice complaining how wasteful I thought it was to be spending so much money to attend this workshop. For the same price I could have flown to Hawaii and hiked in the Na Pali Coast wilderness instead of wasting my time with such neurotically screwed up people! Not surprisingly, the participants in my workshop voted me the angriest person in the room. I didn't convert anyone to the cause of wilderness preservation there either!

Regardless of my putdown of the programme, the therapy sessions helped me stop constantly yelling at everyone at WCWC. I realized how counterproductive it was. But my stopping yelling really didn't solve the overriding problem at WCWC. We still lacked money. We were on the edge of bankruptcy. Something radical had to be done... and soon.

All along we could not tell our members in our appeal letters the true state of our financial affairs. We were advised that donors only give to winners and groups that manage their affairs prudently. An appeal to "help us pay down our debt" was an admission that we had been reckless, poor money managers and a risky investment for their conservation dollars. So we had to think of catchy, inventive appeals that stressed our positive attributes and campaigns.

We were trapped in a giant catch 22 treadmill; constantly scrambling to find the money to campaign and produce more educational newspapers and send them out (with an appeal, of course) in order to get more support and campaign even more. In the summertime we were buoyed up by the cash flow that came in from our 20 Water Street store. In the fall we would go into debt to produce our calendars and catalogue, with the firm belief that improved Christmas sales would produce a profit that would keep our financially-challenged ship afloat until the summer. All the while, hope sprung eternal in us that somehow there would soon be a bigger wave than Carmanah and we would catch it.

Foolish decision not to mail out surplus calendars

We made some stupid decisions during these trying times. One of the worst concerned the mailout of 5,000 surplus 1992 Western Canada Wilderness Calendars. In early 1992 we still had stacks of unsold calendars, more than usual due to the recession and drop in consumer spending. In previous years when we had leftover calendars, we'd sent them out early in the new year to lapsed members. It proved to be a good renewal strategy. In the letter accompanying them, I'd explained that we did not want to let our calendars go to waste and decided to send them gratis to all members "not currently in good standing" knowing that they cared about wilderness and hoping they would renew. I clearly explained in the accompanying letter that they were under no obligation to rejoin or send in any money for the gift calendar...but we would really be pleased if they did because *the threatened wilderness needed them.* In previous years every one of these mailouts returned revenues to cover all the mailing costs and the cost of the calendars, which would otherwise have been a dead loss. Business wise, it was cost effective.

It was hard to get the Team Leaders' approval for this mailout. But ultimately everyone agreed it was worth it. I wrote a dynamite appeal letter and helped Nik print it. Monalisa Amirsetia, who replaced Carleen Lay in the job of managing our database and accounting information, output the labels. Our Wednesday night volunteers stuffed, sorted, bundled them by postal code and bagged the bundles in mail sacks for a bulk mailing in order to save big money on postage. Adriane and I worked late into the night right alongside the volunteers to complete the job.

A couple of months later Nicholas Spears, a stalwart volunteer who was then putting in as much time as any employee did, told me that I should have a look under some tables in the back of the mail room. Poking around, I found a bunch of mail sacks hidden under a table in a far corner against the wall. "I wonder what's in them," I puzzled as I opened one up. "What the hell?" I cried out in astonishment. There were the calendars that I thought had been mailed out months earlier. Angry at the stupidity of not sending them out, in a controlled voice I asked George Yearsley, who worked in our mailroom. "What the hell are these still doing here?"

"Oh, those. We didn't have money for postage so I put them there." I knew (anger workshop) it was not worth yelling at or blaming George. I fumed off and soon found out that it was Mike Rodgers, our comptroller, who had pulled the plug on the mailout. "There simply wasn't enough money to do it!" he explained to me.

"Why didn't anyone tell me?" I cried. "Somehow we'd have found the money. It is so stupid. We can't afford a mistake like this." We'd invested hundreds of dollars to print the letters and envelopes now all of it was wasted.

"Why don't we send them out now?" someone asked at the next TLC (Team Leaders Committee) meeting.

"It is way too late," I said, explaining that sending out a calendar four months after the year had started would look like a joke. We had missed the critical window of opportunity.

This incident brought home to me how dysfunctional WCWC had become. We were beyond the point of no return. We were so broke that we couldn't even afford to do what was absolutely necessary to continue to survive.

My six-page letter to the board suggests solutions

Realizing that something had to be done, with the help of Adriane, I wrote a six-page proposal to be read out at the Board's special meeting on the last day of March 1992. In it I detailed how best to reduce staff–*the only sure way to reduce overhead.* I explained that under the current team management structure and consensus decision-making, there was no way the staff itself could decide to let people go; therefore it had to be a board decision. I recommended that the board direct the TLC to reduce staff by at least 25 percent. I also suggested specific people, which, in hindsight, was not a smart thing to do. This obviously got back to the staff, further fueling the "upstairs, downstairs" split in the organization. In this letter I stated succinctly the central philosophy that had guided the Committee since its inception 11 years earlier. It remains essentially intact today. Here is the text:

> *The style that has been the hallmark of the Wilderness Committee; action oriented; quick response to changing conditions, fighting for specific areas, building public attachment to issues by getting the public to know why specific areas must be preserved using high quality publications including single issue free newspapers must continue to be our focus. Producing and selling educational products will be secondary to campaigning for wilderness. We will spend most of our time directly contacting the public and not get bogged down in the numerous government and company sponsored planning processes.*

Regarding adopting an executive director management structure, I explained that if the board chose the executive director route: *...as the cofounder and the person that has carried out, in a de facto way, many of the duties of an executive director, my hat would be in the ring for the job. If I was not selected, it would be a clear message to me that it was time (for me) to move on.*

Fleeing from a supper of silly solutions

Obviously, what I had to say was not what the board wanted to hear, for the day after it met, board member Ric Careless arranged with Adriane and I to have a special supper meeting the following week at a restaurant near where we lived. Although Ric said that he just wanted to float a few ideas and was not put up to it by the board, I was not convinced.

At this dinner party the small talk did not last long. Before the food arrived, conversation turned to serious matters. Ric outlined his idea of a special role for me in a "newly restructured" WCWC. He thought that hiring an outside director to run WCWC would free me up from the terrible burden of trying to work with all the difficult staff members. Under the new structure, I'd fill a special, vital role as "Creative Director," where I could do what I did best—"brainstorm good ideas." Ensconced in my own special room with "Creative Director" painted on the door, I could dream up brilliant campaigns and work on publications and let someone more skilled at working with people and managing a mid-sized organization make wise business decisions. This was Ric's (and I suspected was the board's, too) vision on how to turn WCWC's fortunes around.

It was a screwball plan, totally out of whack with reality; and it made me furiously angry. I knew that WCWC had no money to hire someone at an executive level wage to go this route and, even if there was the money, there was no miracle-working person that could take over and easily fix our organization, given the dire financial straits we were in.

Knowing that I couldn't argue with Ric without exploding, I just got up and said something like "I can see you've already made up your mind," and quickly left the restaurant. I walked around Gibsons for hours, cooling down and thinking things through before finally walking home. It was the lowest point in my life with WCWC.

I arrived home in the wee hours to a tearful Adriane who had remained behind and finished the dinner with Ric before coming home many hours earlier. She comforted me and we stayed up talking for the rest of the night, considering our options. In an epiphany-like turning point, we realized we had the drive and experience to create another organization. We worked out our "Plan B." It was a turning point for me.

A board member interviews staff and volunteers seeking a solution

At the next board meeting the board appointed director Bryan Evans to investigate the situation at WCWC and come up with a set of recommendations on how to fix the problems. Bryan spent the next week in our office separately interviewing every one of the staff and regular volunteers. Well suited to this job, Bryan was a graduate student in resource management at SFU and sat on the boards of several other conservation groups active in gaining protection for alpine areas in B.C.

Rational and well spoken, he was well respected by everyone at WCWC. He had been brought onto our board to fill a vacancy prior to the last AGM by way of a recommendation by Randy Stoltmann.

During my interview with Bryan, I repeated my opinion that we needed a firm decision-making body to take charge and lay off some of our employees. We should not close down half of our activities or publications, because to do that would mean that even less money would come in and we would go into a downward spiral. I felt the "Team Management System" we'd adopted was a complete failure. We were too small an organization to handle this bureaucratic corporate structure of dividing staff up into separate teams each having its own leader and then having those leaders meet as a TLC. It was obvious to me that this structure contributed to the bickering and divisions between staff members.

Bryan then explained to me Joe's proposal to form an Executive Team (E-team) consisting of Joe, Adriane and I and asked me what I thought of his idea. "I'll back that," I replied immediately, despite it being the first I'd heard of it. It made more sense than bringing in an outside person or appointing someone on staff or from the board to run WCWC. Joe was easy to work with. I knew the three of us could work together and make the hard decisions needed.

By then, Adriane I had decided we'd leave WCWC if the board decided to hire an executive director to run the show. We'd thought out our "Plan B." We would take our international WILD campaign, which was already operating semi-independently as a branch of WCWC and register it as a new society and put our energy full-time into building it and making it a success.

The night of the big decision

Finally, April 27, 1992, the night of the big board decision regarding the fate of WCWC, arrived. At the beginning of the board meeting, before the directors began their deliberations on restructuring, Adriane made a brief presentation regarding funding WILD's new big "pet project"—to produce and publish a *1993 International Year of Indigenous Peoples Calendar*. Adriane was prepared to go into personal debt and provide a loan of $20,000 for this project (all her personal savings)...if the board approved. She had interested partner groups willing to co-publish and buy into the press run and assured the board that she'd make the project a financial success.

After Adriane made her presentation, she and I retreated to the Jolly Taxpayers Pub a few blocks away to have a beer and wait for the board's decision on her proposal and on WCWC's future direction. As we drank down our beers, Bryan Evans delivered his report to the directors at their in-camera session. We had arranged with Nik Cuff (who was a member of the board) to meet us at this pub after the in-camera meeting ended. We needed a "heads up" so we could prepare to handle the fallout following the board's decision announcement to the staff the next morning.

While we waited, Adriane and I talked more about our "Plan B." We were confident that, if the board opted for an outside executive director and we left, we could build WILD into a powerful conservation force. Most board members didn't want WILD. Yet WILD had a good reputation, lots of supporters, great international partners, dedicated staff members and loads of volunteers and supporters. Most importantly, WILD was debt-free. It had been running financially independently and in the black for more than a-year-and-a-half.

As a "separation package," all we wanted were two things: a fair share of WILD's unsold merchandise and a one-time mailing to all WCWC members and supporters to explain in positive terms the parting of the ways and to ask them to continue to support both WCWC and WILD. Although it seemed like ages, less than a month had passed since the "dinner from Hell" with Ric Careless (and the epiphany that Adriane and I had during our all-night brainstorm on what we would do if we left WCWC). We waited and waited for Nik to arrive. Why was it taking so long? Whatever happened, it would be a relief to have the decision made.

The good news and the bad news

Finally Nik arrived. I could tell by the look on his face that all was not right. "You want the good news or the bad news first?" he asked, as he sat down. Without giving us a chance to respond, he started right in.

"You got what you wanted—the E-team. You two, Joe and the comptroller—to make sure the financial matters are considered in all decisions—will run WC². That's the good news"

"Then...," he paused, "the whole Board resigned."

"What?" I asked, totally dumbfounded.

"Yes. After deciding to let you guys run WC², everyone wrote out a resignation letter and handed it to McDonnell."

"You, too?"

"Yes."

"That sure sends a message that none of you have faith in your decision to go with the E-team! Whose idea was that?" I asked, feeling completely betrayed.

Nik either couldn't remember or didn't want to tell me. Undoubtedly those details would really be breaking the confidentiality of the board's in-camera discussion. Telling us the big-picture decision was a more forgivable breach. Nik explained that everyone was so burned out with the problems at WCWC, that they all just wanted to wash their hands of it, including him. The resignations all just sort of happened spontaneously.

I could tell this line of questioning wasn't going to reveal anything more. I suspected that the resignations had nothing to do with spontaneity. Recently, there'd been talk about the extent of each board members' liability for a society's debt if they had not acted prudently and with due diligence to deal with it. Obviously, getting out before WCWC sank—a fate that many believed was inevitable—was prudent.

Trying a different tact, Adriane asked, "What was Bryan's report like?" Nik's face lightened up and so did the mood around the table.

"It was hilarious. There are enough good stories to fill a couple of Peyton Place novels!" Nik wouldn't elaborate with specifics.

He then explained that the board really didn't have any other option. Adriane and I had run the WCWC as a "mom and pop shop" for so long that our personalities and style were inseparable from that of the organization. The board reasoned that it would likely be fatal if Adriane and I left at this crisis point. Much of the credit that printers had advanced was because they trusted and had faith in us personally. Realistically, there was no way anyone else could take over WCWC and turn it around. If anyone had a chance to succeed, it was Adriane and I along with Joe Foy who had become such an outstanding campaigner, spokesperson and solid balance to our personalities.

The three of us had the commitment and will to do it, as well as the goodwill of the members. Our new E-team had the power to do whatever it took to deal with the crisis he asserted.

Then, all of a sudden it dawned on me. "If there's no WCWC board of directors, does the WCWC still exist as a society? Wouldn't that be ironic? The good news is we get WC².... the bad news—there is no WC²!"

It was a good question. Adriane had another one. "Were all the directors present at the meeting?"

"No, a few were absent," replied Nik, rattling off three names.

"I think that the remaining directors have the power to appoint new directors to fill the vacancies. As long as there is at least one director remaining, there's a valid and legal WCWC board of directors. We'll have to call those directors first thing in the morning and make sure they stay on at least long enough to act as a bridge. They can appoint new directors to fill the vacancies until the next AGM," Adriane stated.

"I'll call one of our lawyer friends tomorrow to verify that this is how it works," I added.

As the evening wound down, I chided Nik. "You should withdraw your resignation." Nik didn't bite. He had only stood for the board because I had "twisted his arm." He never liked attending meetings. It didn't mean he was no longer going to be my friend. In fact, he'd be a better friend not being on the board. Finally, he relented and said that if it were absolutely necessary, he would stay on briefly until a replacement for him was appointed. He had no intention of seeing us fail because there was no board. He hadn't thought through the implications of the whole board resigning. I wondered if others had. I hardly slept that night thinking about the next day.

The E-team assumed control of WCWC with quelling the "insurrection" its first job

Adriane and I arrived at 20 Water early the next morning. Joe arrived early, too, and we told him the news. Then Allan McDonnell called and said he would like to meet with the three of us somewhere outside the office as soon as possible. We suggested a coffee shop two doors away. We went over there and a few minutes later Allan arrived in his lawyer suit and looking very formal. He began by telling us that the board had chosen our proposed E-team option. Then he said, "WCWC is yours. Here are the resignations of the directors," holding out a bunch of papers toward me. Somehow I knew instinctively the right way to respond.

"Don't give them to me. I'm not a director. Give them to one of the remaining directors to deal with," I said, refusing to take them. Why Allan tried to give me, a staff member, the resignation letters, I'll never know. It wasn't at all proper. He was a lawyer and must have known the rules about how director resignations are handled. Was it some kind of trick?

After I refused to take the resignation letters, Allan asked us to tell the rest of the staff members the news. We refused. It was his board's decision, and since he was chairperson when it was made, he had an obligation to convey the message to our staff, despite the fact that he had subsequently quit.

Allan reluctantly agreed to do it. It was not a happy job. Together the four of us went back to the office where the staff were anxiously waiting. Allan called everyone together. In a solemn authoritative voice he quickly announced the board's decisions. There was a gasp. Mike Rodgers, our comptroller, who had already put in his resignation angrily shouted out to Allan, "That's not what you told me was going to happen!"

Someone else shouted out, "There's no board of directors. They (referring to Joe, Adriane and me) can't be in control! Don't go anywhere! Lock the doors! Wait until a receiver is appointed. It's not over yet!" Allan coolly replied that everyone would have to deal with the E-team and walked out.

This was not the last we saw of Allan McDonnell. He went on to become a director of the Sierra Club of B.C. and served as president of Ric Careless's Tatshenshini Wild organization. A few months later a new organization formed to accept large donations from American foundations and provide a coordinated, strategic approach towards achieving the 12 percent protected areas in British Columbia by the year 2000. Allan eventually emerged as the executive director of this organization that eventually called itself BC Wild.

As Allan was leaving, Adriane told everyone that there was, indeed, a board of directors because not all WCWC board members had resigned. She explained that the E-team would be talking to each staff person individually as the day progressed to work things out. Adriane, Joe and I then went upstairs and immediately got to work. First, we called the directors who weren't at the meeting and urged them to stay on. They all did. Then we called some of the others who had quit the night before who we felt might be sympathetic to our situation and asked them to stay on for at least for one more directors' meeting when appointments could be made to replace them. Several

Volunteers stuffing a mailout at a regular Volunteer Night meeting. Photo: WCWC files.

said that they would. We now had secured our legal structure.

As we worked away we could hear muffled voices of the staff that had remained downstairs, but I didn't venture down there. As the day wore on, things thawed a bit. The "upstairs" and "downstairs" factions really weren't winners and losers. Everyone knew deep down that there wasn't enough money to keep on going the way we were. It was a relief to have the conflict over. Everyone who was unhappy with the decision chose to get laid off, rather than to quit, so that they would immediately qualify for unemployment insurance benefits. In one fell swoop we lost half our staff including our volunteer coordinator, store manager, marketing person and accounts receivable collector, shipper and a few others. With their departure, one of our most acute problems disappeared—the struggle every month to meet payroll.

The business world has since coined the word "downsizing" to describe the painful process we went through. Debt restructuring and payment schedules were the next steps in WCWC's recovery.

Randy Stoltmann among those laid off

It was hard to see good people, many of whom had worked for several years for WCWC and had been good friends, leave. I especially felt bad about Randy Stoltmann's departure.

Randy wanted us to work more on preparing technical briefs for government processes; but those did not bring in donation dollars. We had had our share of conflicts, especially over editing his writing. I knew, however, he would continue to work full-time to save B.C.'s wilderness.

The energy and drive that got him out into the wilderness, his passion to save big trees, and his photographic, writing and drafting skills made him a powerful force on the side of wilderness preservation. Our parting of ways made sense, but saddened me all the same.

We let Randy retain all the rights to re-use the material we had published in the *Hiking Guide to the Big Trees* book as well as all the photo images and other material he'd generated while working for us, knowing he'd put them to good use for the cause.

However, his West Coast Trail Rainforest campaign fared poorly. It floundered without Randy to spearhead it. We sent the newspaper out to our members, but got very little return in donations and interest in that campaign. Even that small expression of interest we did not follow up. Those of us left could only do so much.

Chapter 37

Restoring our financial credibility; new personnel and campaigns revitalize WCWC

An old friend and employee shows up to help out

On the day after the E-team took over, Ross Muirhead dropped by our office by chance. He had first started to work for us as a canvasser under Frank Sloan in 1990 when, according to Ross, "the average middle-class person was just becoming aware of what an old-growth forest was." Over the years Ross had worked on and off as a canvasser, as a canvass field manager, and at various other tasks including the mailroom. This is how Ross describes the encounter:

> After being away for almost a year, I stopped in to visit on a beautiful summer day and found the store closed. I went around to the side door and there was Paul, the founder and environmental crusader of WCWC, counting out a T-shirt order to mail to make a few bucks—not a good scene when clearcuts were still wiping out precious valleys. There had been a failed coup on the executive and everyone had quit. Needless to say, I'm glad I missed out on that one! Down the stairs came Randy Stoltmann with his boxes, tape measures and maps. They shook hands and Randy wished Paul good luck. I thought, 'holy smokes', how is this organization going to survive with one guy trying to run everything and Mr. Big Tree who had led them to Carmanah walking out the door? It was a sad day in relation to the bustling time and good camaraderie that I had known a few years earlier.
>
> Paul came back to the mailroom and without looking up from packing the Stein Valley T-shirts said, 'Do you want a job running the canvass? But just to let you know I can't pay you anything. It would have to be on a percentage of what you raised.'
>
> How could I say 'no' at this moment of seemingly desperate time? I said, 'Sure. Let's run a few ads for canvassers and see if we can raise some cash flow.' In a few weeks we were bringing in $10,000 a month and in three to four months up to $25,000 a month. The current paper we had to pass out at the door showed the remaining intact valleys left on Vancouver Island.
>
> I always told my canvassers to seek out the salient concept and present something at the door to shock the person awake like; 'Did you know that only two percent of the original ancient forests in B.C. are protected?' That would usually strike at the heart of most long-time B.C. residents and get them to sign up. After getting the canvass going again from the ashes, it was time for me to get a 'real job'. John Minty, a former Greenpeace canvass manager, took over and ran a real good operation for several years.

Canvassing off and on during the four years from 1990 through 1993, Ross estimates that he personally signed up at least 2,000 members. One dark rainy night in North Vancouver, at his last house (a long climb up at the top of a dead-end street), a man wrote out a cheque to WCWC for a thousand dollars. Like most who have canvassed for WCWC for any length of time, he has a book's worth of colourful stories about his experiences. Ross and many others who came forward and helped us get through those hard times assured WCWC's survival.

WCWC's situation becomes public

A few weeks after the big blow up, the news of our downsizing reached the media. The *Vancouver Sun* and *Vancouver Courier* ran stories. Joe handled the probing questions. He told the *Courier*'s reporter in response to a question about it decreasing our effectiveness, "*It is difficult to tell if the drop in funding and manpower has affected WCWC's work. It's always pretty hairy around here. Like all environmental organizations we tend to hold things together with a lot of bubble gum and shoestrings. We want the word to get out there that although we've gone down in staff, we want to go up in volunteers.*" It was a brave face. We were wounded and hurting, but not to the point of being ineffectual.

A young rebel comes through the door, picks up loose ends and helps hold things together

Just a few days after the big shake-up at WCWC, a young woman named Andrea Reimer showed up on our doorstep. She found her way to us via a government programme that helped "street people" re-enter the work force. She had never been active in the wilderness preservation movement, but had been an activist in the fight to get housing for the homeless, having lived in "squats" herself. Years later she told me that she had originally asked her social worker to find her a training job in landscaping because she liked the outdoors and thought, because she liked to travel, it was a "portable" skill she could take with her and always get a job. Somehow, her caseworker thought she would be better suited working for an outdoors-activist group like ours. Not yet twenty years of age, Andrea bragged about never having held a job longer than a few months.

Andrea brought with her a "pre-packaged" job offer. The deal was too good for us to turn down. If we provided her with a job that included office training, the government would pay her full wages for four months: we had no obligation to keep her on after that. We were in desperate need of help and she was pleasant and seemed interested in what we were doing. We assigned Ramona, who worked with Adriane's WILD Team, to be her "boss." On Andrea's first day of work, Ramona gave her the sink-or-swim job of calling all the volunteers who had filled out an information sheet in the last couple of years—there was a huge stack of them—and finding out who was still interested in volunteering for us. Andrea knew nothing about the internal conflict that had been raging for the last while at WCWC and nothing about us, yet she tackled the job systematically and competently with enthusiasm.

Andrea was the kind of person who liked challenges. We simply assumed that if she didn't know how to do something, she would tell us. We kept piling more and more work on her plate, also assuming that, unless she complained, she could handle it. Years later we found out that she hadn't done any office work previously, didn't know how to type, and the only time she'd used a computer was in her fifth grade gifted class!

On Andrea's second day at work Ramona put Andrea in charge of working with the "bike people" and the Environmental Youth Alliance (EYA) to help organize a second Environment Walk like the one Ken had organized two years before. But this time it had morphed into a "Walk and Ride (bicycles) for the Environment." The EYA had a grant from Environment Canada to put this event on and we had promised to help. The EYA was working with the "bike people," a sort of anarchistic group that promoted the greater use of bikes instead of cars in Vancouver. Andrea poured herself into this job, too. When the day came for the actual Walk and Ride, there were more volunteers at first aid stations and checkpoints along the way than there were participants! She had organized nearly 300 vol-

Andrea Reimer takes charge and handles all tasks assigned to her. Photo: WCWC files.

unteers to supervise the less than 200 walkers and riders that showed up. Despite nice weather and lots of lead-time, the other groups failed to bring out the masses. This fiasco reinforced our belief that these kinds of events, although not a total waste of time, were very close to it.

Andrea had been on the job less than a month when Ramona asked her to type out a long report on the Vancouver Watershed that Mark Wareing had hand written. It had to be done by the next day. Unbeknownst to us, Andrea went to our bookkeeper and asked him to give her a quick lesson on how to use the word processing program. She pecked away all night long with two fingers and successfully completed the job.

We never suspected that she was just winging it. But in our own way all of us were just winging it, too, working as hard as we could, learning new skills and figuring out new strategies and tactics as we went along! From that time on, Andrea got more and more work piled onto her plate. When her four-month-long training period was up and it came time to let her go, we really didn't want to. But because of our precarious financial situation, we couldn't afford to keep her on. Undaunted, she went back to her caseworker and somehow got an extension to the programme for a few more months.

Adriane took over her supervision and became the person who wrote the monthly reports on what skills Andrea was learning, to prove that the taxpayers' money was being well spent. When this extension ran out, Andrea found another program that she was eligible for where WCWC paid her minimum wage and the government topped it up by a couple of dollars more an hour. Andrea has to be one of the biggest success stories ever of these types of programs. By the time the top-up program was exhausted, Andrea had become indispensable to WCWC and our financial situation had improved enough to hire her, albeit at a very meager wage.

Key positions filled

Mike Rodgers, our financial comptroller who put in his resignation several weeks prior to the board's decision, was very gracious and professional about leaving the job. He helped us find and train a new comptroller to take his place. We immediately advertised for the job and after interviewing several people hired Margaret Halsey. We also had to find a new store manager as our entire store staff had left and we found it impossible to keep the store open using other WCWC staff. We lucked out in hiring Chantelle Desharnais, a young woman with lots of positive energy and willingness to work hard. She invested a tremendous amount of creative energy into making the store a success and in helping coordinate our volunteers.

One of the biggest ticket items in WCWC's unsold inventory was our *Clayoquot on the Wild Side* book. The 8,000 we had in stock in 1992, even at $30 wholesale, represented a whopping $240,000. I insisted that we hold firmly to that price and go no lower. I had confidence that eventually the books would sell. It turned out to be a very wise decision and over the next three years, especially after the Clayoquot issue heated up in 1993, brought in consistent revenues, especially through our store.

Diplomatic skills of new comptroller rationalize our debt and sets up payment schedules

Our new comptroller, Margaret Halsey, made the conversion of our current debt into long-term debt a top priority. During the first few months with us, Margaret Halsey went to every one of our creditors, explained our situation and worked out realistic repayment schedules varying from one to three years. She got some to reduce the amount we owed and some to donate it outright. She lined up meetings between me and a few creditors whom I knew really well, so I could make similar arrangements.

In the end, between the two of us, we reduced WCWC's debt by several hundred thousand dollars. How Margaret got everyone to agree to the new terms still seems to me a miracle. She had a wonderful, straightforward way of explaining our financial position very much like an educated Forrest Gump might have. For example, in her financial report to the 1992 AGM she summed it up this way. *The main financial difficulty facing the Committee is consistent with prior years, that being the lack of available cash resources.*

In looking back, it's remarkable that none of our creditors pulled the plug on us. They must have believed in our cause and did not want to see us fail. I'll always be grateful for their understanding, patience and their faith that we would ultimately pay them off. We always printed in B.C. and bought from local suppliers. Quality was one of the reasons we did it, the other was our belief in the importance of supporting our local B.C. economy. However, letting us remain in business was also a smart business decision on the part of our creditors. We had no assets other than posters and books, which would have fetched very little in a fire sale. Ultimately, every one of WCWC's debts from this era got paid off.

Debt repayment schedule—$9,000 a month—gets us out of the situation we spent our way into

When Margaret was finished refinancing our short-term debt, we had to pay back about $9,000 a month...every month for the first year. Then we had to come up with a little less—still over $8,000 a month for another year and so on for the next four years. We never really got rid of all our debt because new debt was incurred. But the short-term debt—or line of credit—eventually became manageable after five years of hard slogging.

The main reason that our financial statements of the time and in earlier and later years never looked as bad as the situation really was, was due to the potential worth of our huge unsold product inventory. Theoretically, even at a greatly discounted wholesale value, the stuff was worth hundreds of thousands of dollars. If we could only sell it—a big if—then voila, our financial crisis would be over.

Our biggest single creditor, by far, was Hemlock Printers. This company was especially understanding. Our past due bill, at its highest, was well in excess of $200,000. Since we began printing with Hemlock three years earlier, working with Perry Boeker, their extremely helpful and competent sales representative, we had established our credit by always paying our printing bills on time. They never suspected that, all of a sudden, we'd be unable to pay for the Clayoquot book printing and our 1991 calendar printing.

During the previous year, WCWC's comptroller Mike Rodgers and I met with the owner and his chief financial officer several times and made promises to meet a certain payment schedule and then failed to do so. Paying down the debt to Hemlock was crucial because we needed this company to extend more credit so we could print the next year's calendar. We had to publish our next annual calendar to keep in business. Hemlock allowed many adjustments in our payment schedule. We always squeezed out the maximum we could to pay them. It took us five years to finally pay off our Hemlock debt in full. After our last payment cheque, the owner of Hemlock held a big celebration party for us and donated back to WCWC all the interest money we had paid on our debt to them over the years!

Despite our financial woes, WCWC launches another new campaign

In taking over the reins of WCWC, our E-team assumed a big responsibility. At our meeting we decided that we had to do more than just pay down debt. We had to keep campaigning, despite our greatly reduced staff. The previous summer Joe Foy, who always had a keen interest in local wilderness areas, went on an exploratory expedition into the about-to-be-logged the Boise Valley west of Pitt Lake to check

"Cedar Spirit Grove" in the Boise. From left to right John Gillespe, Volker Bodegom, Mark Haddock. Photo: Joe Foy.

up on rumours of a legendary cedar grove there. He found it.

On his return we launched a campaign to save the Boise Valley but, nine months later, we still had not done much. Joe wanted to kick-start this campaign with a newspaper about protecting the Boise (and the adjacent Burke Mountain and Pinecone Lake area) and mail it to our members and all the hiking and naturalist clubs in the Lower Mainland.

The E-team creatively came up with a super low-cost solution; to produce our first half-tabloid—a two-colour, two-page paper. We did this by taking a four-page two-colour tabloid with two identical pages and cutting them in half and having volunteers separate them by hand. So with a press run of 30,000 we ended up with 60,000 *FOOLS GOLD REGION – BRITISH COLUMBIA – Share in a Wilderness Adventure of a Lifetime – Help Rebuild the Fools' Gold Heritage Trail* papers. It came off the press at the beginning of summer 1992. The back page was devoted to selling shares to fund the re-opening of the "Fools' Gold Trail", a trail that existed more in the imagination than in reality. The Burke Mountain Naturalists helped immensely in the distribution of this paper.

Foy's loaded canoe putting up the Fraser River near Port Mann towards the Pitt River and Pitt Lake. Photo: Joe Foy.

Trail marker fancier than the actual trail which in places is only a roughly flagged hiking route. Photo: Joe Foy.

Joe goes on a foolhardy journey to the Boise Valley, the fools' gold wilderness

Joe is a hands-on person. Just writing about a campaign area and distributing papers asking people to help save it would never keep him satisfied. Perhaps to get away from the constant pressure of debt repayments, Joe decided to take an "ultimate wilderness trip," departing civilization from his front door in New Westminster and trekking into the coastal wilderness by canoe, bike and foot.

Joe had talked WCWC's forester, Mark Wareing, into going along with him. But when Mark saw how low in the water their fully-loaded canoe sank, Mark emphatically declared, "I'm not going" and took his stuff out of the canoe. Joe, stubbornly refusing to abandon the expedition, decided to go it alone. If I'd been there, I'd have insisted on Joe aborting the expedition. I knew that Pitt Lake, the largest fresh water tidal lake in North America, was notorious for killing people. Most of its shores are lined with steep walled cliffs with no place to pull out. Outflow wind suddenly kicks up huge waves and many much larger boats than Joe's canoe have gone down without a trace in the Pitt's deep, treacherous waters, taking the foolhardy souls with them.

Powered by a tiny outboard trolling motor, Joe canoed down the Brunette River and up the Fraser River to the Pitt River, then up the Pitt into Pitt Lake and on to the head of it. From there he planned to journey by bike to the end of the logging road and hike into the Boise Valley wilderness. Everything went according to plan...until he reached the top of Pitt Lake. The winds suddenly came up and he decided to go up the upper Pitt River a ways to pull out. He got through the waves at the mouth and entered the river but got only a little ways against the swift current before he reached a point where, with the trolling motor that propelled his canoe going full speed, he came to a stand still. There was no place to pull out along the bank. His only option was to swing around and go back out into the lake. He barely managed to make the turnaround without capsizing, only to find when he reached the lake that the current hitting the bigger chop was now generating a series of gigantic waves nearly two-metres-tall. He had no choice but to run through them. He hit the waves

Joe's canoe pulled up on one of the few small beaches half way up Pitt Lake. Photo: Joe Foy.

Crossing Boise creek. Photo: WCWC files.

rough, steep terrain. This trip was not the only close call Joe had in our Boise campaign. He later had to stare down a big black bear that made its home in our trailbuilding camp when we temporally abandoned it because our trail boss's foot got crushed under a big boulder that rolled over it as he crossed the swift Boise Creek.

Crossing Boise creek. Photo: Joe Foy.

Record tree book project rescued

A few months after Randy left WCWC, a "Green Gold" grant we'd applied for earlier to cover the printing costs to publish his *Record Trees of B.C.* book came through. None of us wanted to squander this opportunity. Randy and I amicably worked out a contract in which he would get a lot of books for the time he'd spend getting his text ready for publication. His book was the first one that compiled a list of the biggest known specimens of native tree species, their dimensions and their locations in B.C. Randy's book undoubtedly created public pressure to preserve the groves where these trees grew. This is probably why the forest industry never gathered this kind of information or made it public prior to the publication of this book.

head-on. The canoe went near vertical going over them. The angels were with Joe; he made it through. He then managed to pull safely onto the shore. "It was really close to a disaster," Joe understated when recounting to us his "fool's" adventure on his first day back in the office. Joe made it to the Boise Valley and brought back photos we used in our next newspaper.

Although the Boise Valley was quite close as the crow flies to Vancouver, it was one of the harder-to-get-to wilderness areas we'd campaigned to save. Trailbuilding there was difficult because of the

Tragic death of a quiet unassuming conservationist prompts WCWC to create an award

That fall of 1992 the news of Eugene Rogers sudden death came as a shock. Eugene was a stalwart figure in the B.C. conservation movement. He'd worked countless hours to protect the Stein and first put Joe Foy in touch with WCWC. He'd worked with the Steelhead Society for years to protect wild salmonids. Someone in our office said offhandedly, as we discussed all that Eugene had done, especially behind the scenes, that he should have received an award.

That lit the light bulb. Compared to most other fields of endeavour that people pursue—music, sports, film, the arts—there were hardly any awards for outstanding environmentalists. I felt we especially needed an award to honour unsung heroes like Eugene Rogers: people who gave our movement the backbone and strength to succeed. Ramona Tabando, who besides working on our WILD campaign was also a volunteer with the Vancouver Branch of the United Nations Association (UNA), suggested that WCWC create a joint annual environmental award with the UNA. She arranged it all and in a few months time we had a beautiful commemorative plaque—the Eugene Rogers Environment Award—featuring two spawning salmon carved by Squamish artist William Watts. On the bottom there was lots of room to put engraved brass plates with the names of the yearly winners in the years to come.

"Scary Crossing" along the Boise trail. Photo: WCWC files.

Eugene Rogers Environmental Award

In memory of Eugene Rogers who passed away suddenly in 1992, to honour his tradition of hard work and dedication in the conservation movement in B.C., WCWC annually presents this award to a person who has worked exceptionally hard to protect nature. The plaque features spawning salmon carved by Salish artist William Watt, and the names of the recipients:

1992 - Terry Jacks for his work to stop pulp mill pollution in Howe Sound.

1993 - Joe Martin a Nuu-chah-nulth tribal member for his work to protect Clayoquot Sound.

1994 - Ocean Hellman and Doug Radies for their unrelenting work to protect the Cariboo Mountains.

1995 - Danny Gerak commercial fisher and fishing lodge owner, for his work as a tireless defender of the fish and wildlife habitat of the Upper Pitt River Valley.

1996 - Maureen Fraser for her work on a community-base level to help in the progress towards preservation of the ancient rainforests in Clayoquot Sound.

1997 - John Clarke for his work as a wilderness educator and as an advocate for the preservation of Coast Mountain wilderness areas.

1998 - Marion Parker for his documentation of culturally modified trees and his pioneer efforts to protect BC's thousand-year-old trees.

1999 - Will Koop for protecting Vancouver's Drinking Watersheds from logging.

2000 - Will Paulik for his unfaltering work to help protect the Fraser River.

2001 - Mae Burrows for her outstanding work in the environmental and labour movements and for bringing activists from both movements together to work on common issues of interest.

2002 - Ruth Masters for her work over several decades in resisting mining in Strathcona Provincial Park, the hunting of bears, as well as logging in the Walbran Valley, Tsitika Valley and Clayoquot Sound.

2003 - Clint Marvin for his effort's to preserve and protect the Fraser Valley's tallest trees in the Elk Creek Rainforest.

2004 - Betty Krawczyk for her courageous work to conserve ancient forests and defend a citizen's right to protest unjust laws without undue government harassment.

2005 - Telálsemkin Siyám (Squamish Nation Chief Bill Williams) for his tireless efforts to protect the Squamish Nation's Kwa kwayex welh-aynexws (Wild Spirit Places).

Cartoon we commissioned to expose the true make-up of the pseudo-environmental group B.C. Forest Alliance shortly after it was formed. Artwork: Geoff Olson.

The first year (1992) we awarded it to Terry Jacks, famous for his three #1 pop hits including one of the best selling records of all time *Seasons in the Sun*. In 1985 Terry had founded Environmental Watch and during the next seven years had spent nearly all his time persistently trying to get B.C.'s coastal pulp mills to stop polluting the ocean and the air. The next year we awarded it to Joe Martin for his work to protect Clayoquot Sound. We have continued to award the Eugene Rogers Award annually to outstanding environmentalists ever since.

Paul Watson almost awarded the prestigious Eugene Rogers Environmental Award

A couple of years later our Eugene Rogers trophy was awarded to Paul Watson and then withdrawn and given to someone else. This embarrassing situation occurred because the UNA and WCWC took turns deciding who would receive the award. The group in charge each year was to inform the other of its decision and get their approval before awarding the award. In 1994 the UNA decided to give it to Paul Watson of the Sea Shepherd Society for his work in protecting whales. Unfortunately, they informed him of their choice before talking to us. We first heard about Watson getting our award when we read about it in the *Vancouver Sun*. The media-savvy Watson did not wait until the awards ceremony to crow about winning this award. After getting the phone call from the UNA, he immediately put out a press release making the announcement himself.

I was firm in my opposition. An award with WCWC's name on it couldn't go to someone who advocated and practiced civil disobedience, including destroying property like sinking rogue whaling ships. Besides, the Eugene Rogers Environmental Award was to honour people like Eugene, unsung heroes who did not work in the media spotlight. We explained this to the UNA and they saw our point. They consented to withdraw the award; but told us we had to inform Watson of the decision. I was the one elected to bear the bad news. To say the least, Watson was not pleased; and to this day, I believe that he still thinks poorly of WC^2 because of it.

Honeymoon with mainstream press comes to an end

The whole time that WCWC's office was located behind the press club on West 6th Avenue, we enjoyed positive coverage in the *Vancouver Sun* and *Province*. We got especially sympathetic coverage on our Carmanah campaign. But beginning in 1991 things began to change. The "good" reporters were squeezed out and the *Sun* started to give less and less coverage to the environment. That year the *Vancouver Sun* hired Burson-Marsteller, the American-based multinational public relations company, to advise them on how to market their newspaper. It was the same PR firm that set up the B.C. Forest

Demonstrating in front of the B.C. Forest Alliance's office, shortly after it formed holding up the *Real Forest Alliance* cartoon by Geoff Olson. Photo: WCWC files.

Alliance earlier that year for B.C.'s 13 biggest forest companies to convince people of a big lie—that these companies had changed their ways and were now doing a good job. From that time on, environmental coverage in these two papers rapidly declined.

In 1992, the *Vancouver Sun* eliminated its forestry reporter position and told the remaining reporters to concentrate more on Greater Vancouver area issues and less on B.C.-wide environmental issues like the disputes over clearcutting and wilderness preservation. In mid-October 1992, it came to a head with the publication of an editorial in the *Sun* that was unqualified forest industry propaganda. It urged B.C. forest companies to be more aggressive in defending their clearcut logging practices, claiming that enough old-growth forest was already protected in parks to ensure the survival of ancient-forest dependent species.

Throwing caution to the wind, Joe Foy issued a blistering press release titled "*Vancouver Sun* goes anti-environmental - WCWC cries foul." It was the first time we took a newspaper to task. There is an unwritten rule amongst non-profit groups not to complain about media coverage because the response will be even less or worse coverage for your group or issue. *The Sun's editorial stance flies in the face of the facts and simply mimics forest industry propaganda,"* said Foy in our anti-*Sun* press release.... *I guess the Vancouver Sun's policy of no B.C. environmental news is good news as far as the major logging companies and Burson-Marsteller are concerned. But this is no way for a major B.C. newspaper to act. It's bad enough that the forest companies are mucking around in our drinking water source, now they have got their fingers in our daily newspaper.*

Eventually the *Vancouver Sun* relented and hired a full time forestry reporter and began to report more B.C.-wide environmental news. But neither the *Vancouver Sun* nor the *Province* newspapers ever returned to the level of coverage we'd seen in the late 1980s. It became absolutely vital to be our own press and publish our educational papers to get our message out. The major press, now corporately concentrated, wasn't ever going to do it. When hearings went on a few years earlier about the concentration of newspaper ownership in Vancouver (the two major dailies and most of the weekly newspapers are now owned by the same company) people worried about the monopoly driving up advertising prices. What they really should have been worried about was control and manipulation of the news content in the papers.

Chapter 38

We maintain our independence
UNCED hopes dashed
UN uses WILD's unique calendar

WCWC doesn't get a share of the $1 million per year grant to help save wilderness in B.C.

After my "Where is the money?" outburst at the first big funders' meeting in Victoria, the E-team decided, by consensus, that I would not be attending these kinds of meetings in the future. We chose Adriane to be our "negotiator" regarding how B.C. environmental groups would share windfall U.S. foundation funding, which was by now looking very promising. Almost immediately after being elected in October 1991, the NDP government began setting up land use processes to consider additions to the provincial park system. Harcourt made a commitment to increase the size of the provincial park system to cover 12 percent of B.C. lands by the year 2000—WWF's *Endangered Spaces* goalpost.

Although the conservation movement in B.C. was big compared to other provinces in Canada, compared to the conservation movement south of the border, it was tiny. When it came to money, B.C.'s environmental groups were very poor country cousins indeed. We all knew that we did not have the resources to fully take advantage of this opportunity to establish more parks.

The Pew Charitable Trust, the largest private charity in the world, was the lead organization weighing in to help protect B.C. wilderness. Its style, we were told, was to pick a cause and put a big chunk of money in for a few years to have a major impact and then leave. The catch was that Pew had its own ideas on how its money was to be spent. Our loosely bound consortium, five major groups (Sierra Legal Defence Fund, Sierra Club, Valhalla Wilderness Society, WCWC and BC Spaces for Nature), picked Ric Careless, because of his skills at negotiation and his inside connections with the new NDP government, to represent us in informal negotiations with a Pew grants officer. Ric went back east to find out how we could make a successful application that would satisfy the "strings" Pew attached.

Ric came back bearing the message that Pew would only fund a centrally-run coalition consisting of all the groups involved in protecting wilderness in B.C. working together in a tightly coordinated way. We at WC² had a different idea. We thought that the best way to use this money was to strengthen the existing groups and overall movement. Each group had its unique strengths and existing campaigns that could benefit from increased funding. We felt strongly that this big chunk of new money should not be used to create a new super-group, new structure or to skew the movement. We could continue to coordinate our efforts and ensure that groups didn't work at cross-purposes informally, as we were currently doing. In other words, we should fine-tune our efforts but not re-build from scratch.

We argued first and foremost that the existing groups, not some outside funders, knew best what was needed to win campaigns in B.C. and each group knew better than anyone else what they needed to increase their effectiveness. Since this new money would eventually dry up, when it did, the B.C. groups should be left strengthened with greater membership and capacity.

Diverting money, time and effort into a new structure could leave B.C.'s existing groups weaker. Furthermore, the set-up and maintenance of a new structure would undoubtedly waste money on creating a new bureaucracy within our movement.

There was another, strategic reason why we were so opposed to a new super group being formed. We feared that the opposition could more easily counter one big pro-park voice with one big "no." Most of the others in our loosely organized "Pew money" consortium believed, however, that we should jump through whatever hoops the funders wanted us to.

Greg McDade of the Sierra Legal Defence Fund, who was devoting a great deal of time to devising a plan that the funders would find acceptable, explained to us how it all might work over a friendly beer in our Gastown hangout at that time (Greek Characters' neighbourhood pub). Greg outlined a plan that involved each group getting a share of the expected $1,000,000 US roughly equal to its current level of activities. We, being by far the largest group, would get a fairly big chunk, but not the lion's share as would be strictly dictated by our membership and current budget. Our share of the $1 million would be $180,000. We had to write up projects to cover that amount, complete with budgets. Each other group would write up its own projects independently. Then there would be a big meeting where all the participant groups would gather, listen to each other's proposals and vote to fund them all. Then we'd package them all together as one integrated strategy in the grant application. Greg asked us if we would agree to these gentlemanly rules and vote for the other groups' proposals—even if we might think that it was not the wisest use of the money—because, after all, these were the individual groups' own priorities. Joe, Adriane and I agreed.

Adriane spent more than a month putting together our package of proposals, making sure the total budget equaled our pre-arranged share of the grant. She talked to the other groups to make sure that no one else was proposing to do something similar. It wasn't hard. Everyone knew that WCWC would focus on educating and mobilizing people, in the firm belief that a broadly-based, well-informed public is the prerequisite to government action—even if there was already a commitment to create new parks. We had honed our skills as the "public educators" of the wilderness preservation movement, but knew we could always expand and improve. With the new money we planned to mass-produce and widely distribute a series of educational newspapers on issues related to the new park-creation processes that we could not otherwise afford to do. We included slide-show tours and membership-building activities, making sure the projects would leave WCWC stronger when the U.S. money ran out.

It's easy to imagine how good that money looked to us. It was no secret to other environmental groups that the WCWC was in dire financial straits. Our E-team, in power for only a few months, had stemmed our upward-spiraling deficit, but had only made a tiny dent in bringing down the debt. We were surviving; but it was a struggle. Finally, the big day of the meeting came. Adriane returned late in the afternoon. Immediately, I could tell by the look on her face that things had not gone well. I had never seen her so distraught. Joe and I sat her down and tried to calm her down.

"What happened?" I asked.

"They voted our proposals down!"

"What?" I exclaimed, not believing what I heard. "That's impossible! It was pre-arranged that everyone would support each other's proposals. You voted for all their proposals, didn't you?" I asked, somewhat accusingly.

"Yes, of course I did," replied Adriane.

"Well, what happened then?" I asked more gently, seeing how distressed she was. Seething with anger and barely holding back her tears, Adriane explained how the meeting had progressed. They had considered the proposals alphabetically by groups, with WCWC presenting last. Everything went as planned. Groups kept their proposal budgets to the pre-arranged amounts. There were friendly questions, friendly explanations and everyone approved the projects unanimously as expected. When Adriane's turn arrived, she presented WCWC's proposals and suddenly the sniping began.

The first question came from John Broadhead (JB) of Earthlife Foundation, a small group that he had founded on the Queen Charlotte Islands after South Moresby was protected and the Islands Protection Society had folded. "You want to spend all that money on public education? Where has that gotten you?" he probed.

"What do you mean?" asked Adriane.

JB outlined our Carmanah campaign as an example, commenting that we'd put out a book, five newspapers and four posters and we'd saved only half a tiny valley. "Obviously WCWC's methods don't work!" he exclaimed to everyone around the table.

Colleen McCrory, who might have defended our work but had attended the meeting by speakerphone, had hung up earlier. No one came to our defence. Around the table people were saying that maybe we didn't need that kind of public education anymore... "The movement has moved beyond this kind of work," someone asserted.

Vicky Husband, Lloyd Manchester of Earthcare Society, John Broadhead and Ric Careless proceeded to vote down most of our proposals. At this point, Adriane told us that she had "lost her cool and yelled at all of them in rage." She walked out of the meeting before it was over.

"I felt so betrayed," Adriane sobbed. She continued explaining that the meeting had begun with Ric reporting that Pew had rejected the approach we'd all agreed on and wanted to pump its money into one central organization with a "fully coordinated strategy." It was exactly the opposite of what we'd envisioned and what I had feared might happen.

"We don't want their money! We're better off building our own membership and funding our own campaigns. Somehow we'll make it on our own," I said resolutely. Joe wholeheartedly agreed. Adriane hugged us both with a sigh of relief. "That was the most horrible meeting I've ever been to," she said. That ended our involvement with the Pew-funded coalition efforts. WCWC never received a penny of the millions of dollars Pew pumped into B.C.

Adriane, who had spent so much time and effort during the preceding couple of months to make the co-operative effort work, felt completely deflated. That night she had the first and only migraine headache she has ever had. It was the feeling of betrayal, not the loss of this source of money that upset her the most. In the long run, far from being bad, it turned out to be a positive turning point for WCWC. We abandoned the chase after so-called easy dollars. With the utmost determination we re-focused our efforts on building a strong membership base that provided us with our own independent funding. We became fiercely independent and feistily creative in our efforts. Refusing to be part of the Pew funded coalition did not mean the end to all foundation funding for us. We continued to get Bullitt grants and other foundation support that played a vital role in funding many of our publications and campaigns. But we never let the foundation grants and support overwhelm our organization. Grant funding hovered at between 10 to 15 percent of our total income—the same proportion it is today.

A new group is formed to spend Pew's grant money

In the fall of 1992, soon after our withdrawal from the Pew money talks, the consortium of Pew fund-seekers formed a new super group called Ancient Forests and Wilderness Campaign (ANFOR) to ...*provide a coordinated strategic approach towards achieving a representative protected areas system in B.C.* Its directorship was the "who's who" of the wilderness movement: John Broadhead, Ric Careless, Maureen Fraser, Vicky Husband, David LaRoche, Colleen McCrory, Greg McDade, Lloyd Manchester and Bill Wareham. When I read the newspaper article announcing the formation of this new organization, WCWC was named as a member group. It infuriated me, because we had quit the consortium months earlier. We learned later that ANFOR had left our name on the proposal that they submitted to the Pew Charitable Trust in order to convince Pew of the movement's solidarity and get the funding. No doubt it helped, for we, despite all our financial troubles, were still by far the largest wilderness preservation group in B.C.

Soon after, Allan McDonnell, the former chair of the Wilderness Committee board who resigned with the others en masse in April 1992, became the new executive director of ANFOR and the group changed its name to BC Wild - Earthlife Canada Foundation—known as BC Wild for short. According to an article written by McDonnell about this organization published in the BCEN news, the bulk of the work of BC Wild had to do with ...*developing and delivering important research information—including economic analysis, mapping and communications—and assisting grassroots environmental organizations.* He went on to explain that, *BC Wild is not a sub-funding agency. It is an independent group that focuses its resources on delivering information that will help the environmental movement as it pushes government to protect remnants of the province's diminishing wilderness.*

While we continued our public education role, BC Wild primarily funded people (at much higher salaries than ours) to do research and attend government public planning processes set up around the province to work out consensus agreements as to which wild areas would be protected and which ones committed to industrial use, using the 12-percent-protection yardstick as the guide. BC Wild set up an office only a few blocks away from our 20 Water Street office. During the eight years of its existence from 1992-2000 (when it disappeared with hardly a trace) we only heard bits and pieces of gossip about what BC Wild was doing and who was joining or leaving its ranks. The Pew money mostly flowed through the Earthlife Foundation. BC Wild operated a bit secretively because it was always afraid of being exposed as an agent of U.S. conservationists, wielding undue influence over B.C. land use decisions.

There was then, and still is today, a double standard in the Canadian mind-set. Massive intervention of U.S. dollars to control, own and exploit B.C.'s natural resources, destroying wilderness in the process, is OK No one even raises an eyebrow. But U.S. money spent to save wild areas from being industrially destroyed raises cries of foul.

At WCWC we sometimes envied the high salaries BC Wild paid its executive director and others working on contract. We paid such pitifully small salaries in comparison. But working for BC Wild wasn't the greatest according to Ken Lay, who worked with them for a year on the mid-coast region's issues. He left frustrated because he could never get money to publish educational material, the style of campaigning he'd learned while working at WCWC.

BC Wild totally bought into the NDP's land use planning processes. It funded participants, some of whom were inexperienced and naive and no match for the industry and union participants who were well-versed in hardheaded negotiations that characterized the CORE and LRMP (Land and Resources Management Plan) tables' processes. Vital areas not included in newly created protected areas were often designated "special management areas" which, we suspected at the time, were meaningless designations. "Special management" proved to utterly fail to provide any additional on-the-ground protection. It was a good negotiating tactic from industries' viewpoint.

Without the BC Wild's money, there probably would not have been any full-time "stakeholders" representing the environment and preservation sector at many of the government-run land use allocation negotiating tables. As time went on, the B.C. government got more and more clever at designing the terms and conditions of each successive land use negotiation process in different parts of the province. The Greater Vancouver RPAC table was the worst. The meetings were held behind closed doors and environmentalists at the table had to sign protocol agreements that stated, after the conclusion of the process, that they would defend the decisions and not criticize them. It was inevitable that WCWC would come into conflict with these enviro-compromisers, especially those who decided to trade off and sell-out at their negotiating table some key areas we were fighting for—like the Upper Elaho Valley.

When in mid November '98 the federal Reform Party and the IWA put out a press release saying that that U.S. foundations were donating millions of dollars to B.C. environmental groups fuelling the wilderness preservation movement in B.C., we were able to deflect the heat. John Duncan, Reform MP for Vancouver Island North, spearheaded this anti-environmental group campaign. He stated it had become clear to him that a lot of what had occurred in B.C. in recent years was being driven by outside interest and those funds might even be donated because of trade instead of environmental interests. He didn't like the idea of Americans bankrolling the B.C. environment movement and influencing forest policy in B.C.

People generally believed WCWC accounted for most of the money being spent by the movement in B.C. so the media naturally came to us for comments about Duncan's allegations and we were legitimately able to deny them as far as WC^2 was concerned. Less than five percent of our annual funding came from U.S. sources and Joe Foy told the reporters that we were extremely grateful for our U.S. grants. They amounted to about $51,000 of our $1.5 million annual budget that year with contributions from ordinary individual Canadian members making up most of the rest. Joe also noted that the forest industry in B.C. was driven primarily by U.S., Japanese, and European dollars.

Few suspected that BC Wild accounted for the lion's share of money being spent to promote the establishment of new parks during that era and that B.C. Wild was almost completely dependent on U.S. foundation money. Despite the fact that we had our differences with BC Wild, we did not want to see them or the wilderness movement discredited.

We did occasionally cross swords, however. When a new funder, the Brainerd Foundation, held an orientation meeting in Vancouver for environmental groups, the first question the head of BC Wild pointedly asked the presenter was whether or not Brainerd funded environmental groups that publicly criticized other environmental groups. The answer was no. Brainerd wanted B.C. environmental groups to work cooperatively together and therefore wouldn't fund divisive groups. We had just criticized the Sierra Club of B.C. in an op-ed piece in the *Victoria Times Colonist* for supporting the Vancouver Island Land Use Plan that did not include all the remaining pristine, unprotected watersheds over 5,000 hectares on the Island. Needless to say, BC Wild didn't like our criticisms of their compromising negotiators, either. We never got a grant from the Brainerd Foundation.

While attending a Christmas party several years after BC Wild had been active on the B.C. scene, a plaque on this law office's wall caught my eye. It was produced by BC Wild to congratulate itself on all its successes in protecting six percent of B.C., listing some areas that WCWC had played the major role in campaigning to save long before BC Wild even existed. Their plaque was beautifully done and totally self-serving. At the time I was incensed. Later, I thought to myself that even BC Wild had to show its donors that it was an effective, successful organization in order to keep getting funding. It wasn't the first and likely won't be the last time a group takes credit where it wasn't due. At WCWC we always tried to fairly credit all those who worked on an issue. But I'm sure we, too, have been guilty to a lesser extent of the same thing. Most people personally involved in winning a campaign don't really care who gets the credit. Their reward is seeing the area protected and knowing it's going to be there for their children, grandchildren and generations of wildlife to come.

WILD's 1993 international *Endangered Peoples, Endangered Places* calendar takes shape

WCWC's new E-team worked hard on taking care of daily business. We also fit in long hours working on what we loved best: WCWC's campaigns and educational publications. Our top priority in the spring of 1992 was finishing our 1993 *Endangered Peoples – Endangered Places* calendar in time to take it to the Rio Earth Summit in June. We wanted it to be ready well in advance of 1993—the United Nations International Year of the World's Indigenous Peoples. UNCED was a perfect place to get those early sales we always dreamed would make our calendars a financial success.

Ande Axelrod, a young graphic designer who had walked through WCWC's door a few months earlier looking for work, took on the job of designing and helping produce our *Endangered Peoples – Endangered Places* calendar. She put an extraordinary amount of talent and effort into this project. Each month of our folio-sized (extra large) full-colour calendar featured a portrait shot of a different indigenous people, a map of their traditional territory and a large photo image of the natural "wilderness" where they lived. The calendar's purpose was to raise global awareness of the deep-rooted connection between Earth's biological and cultural heritage. It illustrated the fact that most of Earth's remaining wild places are also the homelands of indigenous peoples who still try to live sustainably following traditional ways and values.

Knowing international pressure helped us in our own local battles, we included a big emerging B.C. issue—the native people living in Clayoquot Sound and their situation—in this calendar. On the front cover we featured an emotive Thom Henley photo of three Penan children. On the back cover we featured the David Suzuki Foundation's *Declaration of Interdependence* that was written especially for the upcoming Rio Earth Summit. Brilliant, but still unheeded as of today, the statement concludes with:

At this turning point in our relationship with Earth, we work on an evolution from dominance to partnership, from fragmentation to connection, from insecurity to interdependence.

Gathering the information and producing our international calendar turned out to be more difficult than we'd ever imagined. It took hundreds of phone calls and faxes to locate the highest quality images and check the accuracy of the information with partner groups. We produced four different editions, one in each of the UN's official languages: German, Spanish, French and English. Faithfully and accurately translating and typesetting the texts proved almost as difficult as gathering the images and the original information. Many people remarked that it was the most beautiful—and profound—calendar they'd ever seen: a true masterpiece. I still regret the fact that we never submitted it for any awards.

International Year for the World's Indigenous People — *Año Internacional de los Pueblos Indígenas Mundiales*

ENDANGERED PEOPLES 1993 ENDANGERED PLACES
POBLADORES EN PELIGRO / LUGARES EN PELIGRO

Cover—1993 *Endangered Peoples – Endangered Places* calendar. Cover photos: WCWC-WILD files.

We start producing our own line of greeting cards to raise funds and awareness

At the same time that she was working on our international calendar, Ande Axelrod helped us produce WCWC's first set of greeting cards. She had designed similar cards for the Sierra Club in the U.S. when she lived in California. She convinced us that greeting cards were a lot more marketable than postcards: a sure bet to raise income. Her suggestion turned into a new "product line" that has made money for WCWC ever since.

Picking images for each new set of cards, while simultaneously picking the images for the upcoming calendars, became a regular part of our late summer routine. Since those first cards in 1992, WCWC has produced and sold over half a million greeting and holiday cards featuring more than 200 different photo images.

Most of our cards featured threatened species and endangered places, but some are simply stunning images of more common birds, flowers, trees and landscapes. It's been a great way to showcase Canada's wilderness and wildlife and build up more support for wilderness preservation.

Preparations for the 1992 Rio Environment Summit

UNCED–the June 1992 United Nations Conference on Environment and Development in Rio de Janeiro, Brazil–was a beacon of hope. We fervently wanted this face-to-face meeting of the world's leaders to be *the* big "turn around" event where the critical environmental crises facing the world would finally be seriously addressed.

In the two years between our Hawaii conference and UNCED, Adriane and other WCWC staff participated in several UNCED lead-up activities. To some extent, our whole WILD mapping campaign and conference in Picinguaba were UNCED lead-up events! Certainly our southern partner groups hoped that their WILD mapping work would translate into more nature protection under the UNCED pressure for governments to be more environmentally responsible. The official UNCED lead-up meetings were called "Prep Coms" (Preparatory Committee meetings). They focused on government delegations negotiating the wording of conventions to be passed in Rio on forestry, climate and biodiversity. Adriane and her WILD Campaign assistant, Ramona Tibando, got their way paid to several of these meetings.

We also added to the pile of pre-UNCED documents. In November 1991, Adriane and I completed a brief commissioned by the Canadian Council for International Cooperation, that would be part of the NGO input to the Biodiversity Convention negotiations. Our brief was titled *GLOBAL BIODIVERSITY: PROTECTING THE HERITAGE ESSENTIAL TO THE SURVIVAL OF THE PLANET*. It called for an International Convention on Biological Diversity that:

1. *Enshrines the principle of the inalienable right of all species to exist,*
2. *Initiates a massive effort to catalogue species and ecosystems and monitor the state of protection of natural biodiversity in every nation,*
3. *Establishes measures, and elaborates on obligations to save traditional and ancient varieties of cultivars; and,*

4. *Provides both incentives and disincentives that are significant enough so that all countries maximize their conservation of biological diversity.*

Even though we were doubtful that our brief would make even the tiniest impact on the government-to-government negotiations, we liked working on this document because it clarified our own thinking about what was needed in our campaigns to save biodiversity throughout Canada.

According to Adriane, the most exciting pre-UNCED meeting had nothing to do with governments. It was a meeting of non-government organizations (NGOs) held in Paris in late December 1990, paid for by the UN and the French government. Over a thousand NGOs from around the world gathered there and worked on *Ya Wananchi*, a manifesto that both outlined a shared commitment to sustainability and urged governments to take action on specific global problems. Adriane was successful in getting sections on temperate and boreal forests added to this document.

On arriving home after that meeting, Adriane raved about a state reception put on by the French government for all the participants, that served the most incredible wine and food, including real caviar. Then, she noted, despite the incredible generosity of their hosts, the delegates' first resolution was to condemn the French government for its sinking in July 1985 of Greenpeace's Rainbow Warrior, the ship that had been protesting French nuclear testing in the South Pacific!

I still wondered to myself whether or not such resolutions and manifestos really do any good.

Further preparations for UNCED

Three months before the big Rio meetings, Adriane and Ramona flew to New York to take part in Prep Com IV at the UN. At the first session Adriane brought up the fact that there were people attending from B.C. who were from an anti-environment NGO: the B.C. Forest Alliance. "Everyone should be careful in terms of what they say," Adriane warned a meeting room full of people. She explained that the goals of the B.C. Forest Alliance were to defend clearcut logging and fight against the establishment of new parks and protected areas. (This pseudo-environmental organization was formed a year earlier in B.C. by the forest industry with the help of the infamous PR firm Burson-Marsteller.)

Adriane pointed out to everyone in attendance the B.C. Forest Alliance's delegates: Patrick Moore, the former Greenpeacer who had sold out to industry, and Claude Richmond, a former Socred environment minister. Her statement and request that the B.C. Forest Alliance be barred from the meeting threw the room into turmoil and caused a heated exchange. Defending his right to be there, Patrick Moore said he would be happy to talk to anyone, including Adriane, about the B.C. Forest Alliance's position. When Adriane approached him in the hallway immediately afterwards, he simply swore at her. "I don't take verbal abuse," said Adriane as she walked away.

At this Prep Com the NGO delegates did a lot of wordsmithing, trying to tighten the concepts and avoid loopholes and weasel words on a document called *Agenda 21* (UNCED's global sustainable development action plan leading into the 21st century). This was the big document that would be finalized and passed at the Global Forum, the NGOs' parallel summit to the United Nations' Earth Summit (UNCED) in Rio. Working with a number of other NGO representatives from around the world, Adriane and Ramona successfully lobbied for improved language in the biodiversity, forests and oceans sections. Again, thousands of WCWC newspapers got distributed to delegates.

As the NGOs at Prep Com IV negotiated *Agenda 21*, in another

Our booth in Rio at the Global Forum. Photo: WCWC files.

part of the building government delegates were negotiating the wording on a proposed forestry convention. For a few hours, Adriane sat in the gallery and watched the Malaysian delegation resist wording that might give some clout to the convention. The Malaysians specifically refused to include the term "forest ecosystems" in the convention language because, they said, "such a term implies soils, waters and species other than trees, detracting attention away from the management of forests."

After that, Adriane expressed as much skepticism as I did regarding the whole UNCED and convention-negotiating process. "Countries like Malaysia are too stubborn," she told me. "Their whole effort is directed towards watering down the convention with the most non-compelling words and definitions so that they can continue to log their rainforests as rapaciously as ever." Then she heard that the Canadian delegation was also aiming for language that sounded good but wouldn't compel our government to require on-the-ground changes by industry. "What's the point of it all?" she asked our staff back at home in Vancouver. "They're arguing over whether to put 'may' instead of 'shall' in sentences about taking action when they really mean 'won't'."

Participating in UNCED in Rio

Sometimes, even if you have grave doubts, you have to see a plan of action through to its conclusion. Five people from WCWC-WILD flew to Rio to attend the big June 3-14, 1992 meetings. WCWC sent our staff forester, Mark Wareing, to focus on the final forestry convention negotiations. Our WILD campaign representatives included our cartographer Ian Parfitt and Latin American coordinator, Ganga Jolicoeur, who aimed to wrap up WILD's CIDA grant with one final meeting of WILD's Latin American mapping partners. Adriane and Ramona went to follow the negotiations on the biodiversity and climate change conventions and to sell our hot-off-the-press *Endangered Peoples - Endangered Places* calendars.

There were really three "summits" in Rio—sort of like a three-ring circus—going on simultaneously: the NGOs' Global Forum, an indigenous peoples' Earth Parliament, and the United Nations' Conference on Environment and Development (UNCED). UNCED, also called the Earth Summit, was the largest gathering of heads of state that the world had ever seen to date. The real power brokers were all at UNCED, engaged in final hot and heavy negotiations regarding the adoption of the biodiversity, forestry and climate conventions. Their conference was across the city (over an hour away by taxi) from the NGO hoards at the Global Forum.

Our team had an outdoor booth in a park at the Global Forum. Everyone took turns staffing the booth, handing out newspapers, selling calendars and talking to people who came by over the ten-day-affair. I thought our staff's overriding goal was to line up other NGOs to buy in bulk our *1993 Endangered People - Endangered Places Calendar* to take back to their home countries to sell. They failed to line up a single group to help sell the calendars! Needless to say, when I found this on their return, I was choked. I felt that if I had been there this would not have been the case.

According to Adriane, all the NGO delegates, except those from the very largest and best funded environmental groups (who walked around the air-conditioned Earth Summit centre in suits, talking to their colleagues on fancy mobile phones), were completely frazzled trying to cope with the masses of people, conflicting events, logistical nightmares in traveling from one summit to another, frustration with feeling ineffective and the heat. Every day the temperatures rose above 40 degrees Celsius and the air humidity was 100 percent. The NGOs' meetings weren't air-conditioned and some people passed out from the heat. Although Adriane tried to talk groups into bulk purchases of our calendars, no one there wanted to deal with business transactions.

It wasn't as if our team members weren't busy. Mark participated full-time in the NGO forest treaty process—producing yet another lofty-worded document that, in my mind, meant no real change in the real world. Ramona did the same in the NGOs' ocean treaty process. She also helped out the Penan delegates who attended the meetings and played a role in organizing a big rally in support of indigenous peoples. Ian and Ganga organized WILD's public workshop on mapping at the Global Forum and at the Earth Parliament Ian and Adriane gave a well-received presentation of WILD's mapping work. Our staff spent most of the time at our booth explaining the dismal record of the Canadian and British Columbian governments in destroying our old-growth forests and natural biodiversity. People from the developing world loved to get the environmentalists' side of the story and were astonished to learn about the mismanagement of Canada's forests. They'd thought that bad forestry practices were a problem found only in developing countries.

The B.C. Forest Alliance delegation gave us some competition, manning their own booth at the Global Forum. They distributed B.C. forest industry propaganda, claiming that B.C. had the best forest practices in the world. Much to the dismay of our forester Mark Wareing, they also handed out tree-growing kits with exotic Douglas fir seedlings to everyone who came by their booth. This demonstrated its blatant disregard of the fundamental principle of biodiversity conservation not to introduce plants and animals (exotics) into ecosystems where they are not naturally found.

Our WCWC-WILD booth quickly became the meeting place for many Canadian environmentalists, including Colleen McCrory and Vicky Husband. Watching the B.C. Forest Alliance in action, we decided to hold a joint WCWC-Sierra Club-Valhalla Wilderness Society press conference at the official UNCED conference site to make the media and conference delegates aware of the threats to Canada's forests. Near the end of the 10-day extravaganza, Jean Charest, Canada's Environment Minister, briefed all the Canadian NGOs on how things were going at the Earth Summit—not so good.

UNCED's failure brings environmental hard times

As the Rio conference wound down, its tragic failure became starkly apparent. The Convention on Forests was squashed. The Convention on Climate Change failed to be meaningful because all the specific targets for greenhouse gas reductions were removed. The Biodiversity Convention was weakened by the refusal of President George Bush Sr. to sign on behalf of the U.S. Adriane came back from Rio filled with disillusionment. The problems facing our Earth were so enormous and the lack of political will was so evident that it was impossible not to be discouraged. She felt that most environmentalists' efforts in Rio, including our own, were either scattered and unfocused or directed to creating nicely-worded documents that would likely have no real impact. I reassured her that the large, well-organized environmental groups she described to me with such envy had just spent much more money achieving little or nothing, too. She was desperately worried that peoples' hope would vanish after all these good intentions and lofty commitments had failed to be put into action.

She told the rest of us at WCWC her conclusion: the best we could do internationally now was to work closely with a single environmental group with a style and goals similar to ours, in a country where there was some possibility of having a positive result, like

Extinction Sucks bumper sticker. Artwork: WCWC files.

Chile. But locally, all of us had to hunker down and fight to save what we could. Adriane turned her energies more towards working on issues in Canada.

Canada is the first western industrialized nation to sign on to the biodiversity convention

Canada was the first industrial nation in the world to actually follow-through on one big commitment made at UNCED: to save Earth's biodiversity. On December 4, 1992, with great fanfare, Prime Minister Brian Mulroney ratified the *International Biodiversity Convention*, pledging to conserve our country's biodiversity. Adriane and I were invited to attend the signing ceremonies at the George C. Reifel Migratory Bird Sanctuary in Delta. It was a warm, sunny day, beautiful for bird watching and for watching Mulroney's publicity-generating political gesture. At the time, of course, we had no inkling that it would take more than 10 years for Canada to enact a Species at Risk Act to protect endangered and threatened species; an Act that was so ineffectual that it ended up being condemned by every major environmental group in Canada. In retrospect, all the work that went into the crafting of a well-worded international biodiversity convention seems to have been so useless.

My parents lend WCWC the money to print both our Western and Canadian calendars

Knowing that we did not have enough credit to publish our 1993 endangered wilderness calendar, and wanting to publish our Canadian Calendar again, I asked my parents for a short-term loan to pay for the printing of both calendars. I was really afraid that they would turn me down because my dad, a Presbyterian minister, was retired and they were not rich. What a relief it was to find out they could. My dad's cheque came when Adriane was at the Rio Summit. In the accompanying letter he wrote: *"By now Adriane is down in South America attending the Conference. Our thoughts and prayers go with her. Of course we cannot pin all our hopes on one important world-meeting, but it does make a wonderful beginning effort to save our little blue-world planet."*

By February 1993, I had paid off the loan from my parents. It had top priority at WCWC. Without the help from my family at this critical time, it would have been very difficult, perhaps even impossible, to publish our Western calendar and certainly impossible to publish a Canadian one again.

UN uses WILD's calendar to launch the International Year of the World's Indigenous Peoples

Shortly after the Rio Summit, Dr. Noel Brown, head of the United Nations Environment Programme (UNEP), visited our offices. He was in town for some other meeting and phoned us "out of the blue" to see if we could meet with him that afternoon. Of course, we jumped at the chance. At our meeting Dr. Brown told us that he really liked our international *Endangered Peoples – Endangered Places* calendar and asked if we could use the same images, expand the text and publish it in booklet format for UNEP to hand out at its launch of the International Year of the World's Indigenous Peoples at the UN headquarters in December. It was a huge honour. But, of course, we told Dr. Brown that we couldn't afford to do it on our own. He said that he would find the funding.

Right away, Adriane got to work with Ande Axelrod, who'd designed the calendar, to produce a mock-up of the booklet that UNEP wanted. It disturbed me that Dr. Brown kept offering direction and encouragement on the project, but kept giving us the runaround on funding. I was right to be wary. Ultimately, Dr. Brown came up empty-handed and we had to cancel the project. Ande Axelrod was pretty choked, and so was I. Realizing that the booklet was a no-go, Dr. Brown decided that in substitution, he would pass out our *Endangered Peoples – Endangered Places* calendars. It became obvious to us that UNEP had no other printed material prepared to commemorate this special year! Clearly getting the message from Dr. Brown that UNEP did not have much money, and attracted by the prestige of the deal, we arranged to sell UNEP 1,500 of our calendars, charging just the cost price and the rush shipping to get them there in time.

Adriane flew to New York to attend the UN's launching of its International Year of the World's Indigenous Peoples on UN Human Rights Day, December 10, 1992. It was a major event with leaders and delegates from indigenous groups from around the world gathered at the UN headquarters for the occasion.

At the end of the General Assembly's plenary meeting that morning, the UN Secretary General officially declared 1993 the Year of the World's Indigenous Peoples. He then adjourned the meeting and announced that it would reconvene in the afternoon to hear statements by the world's indigenous leaders. To Adriane's great surprise, and to the great insult of the many indigenous people who had traveled long distances to be at these ceremonies, hardly any members of the General Assembly attended that afternoon to listen to what the prestigious indigenous leaders had to say. One after another, they addressed the nearly empty assembly hall. They poignantly described their tragic situations and pleaded for help. Adriane recorded many of them by whispering into a tiny tape recorder the words of the translator she was listening to via her earphones.

One of the most powerful speakers was a Yanomami leader. He started out by saying: "We were the last people in the Amazon to be invaded by the whites. I am part of nature. We are one of the last people to live in the forest. Our lands have been demarcated, but not guaranteed by that demarcation. They say our lands are too big for us."

He went on to explain that gold miners had recently invaded their lands and now were dredging for gold and using mercury to separate it. He had never seen so much destruction. "We have only one land in the whole world. We, the indigenous people, say nature must be preserved. Many other places have already been destroyed…. We want the United Nations to make sure nothing bad happens to us." He informed those present that in recent months over 8,000 miners had invaded their territory. Their situation was desperate. It was another situation that we at WCWC could do nothing about.

There was little, if any, local, national or international press coverage of their pleas or any of the other proceedings at the launch of the International Year of the World's Indigenous Peoples. It was shocking. The UN was not making a big thing of it. It wasn't because there was insufficient time to prepare. The UN had been designating International Years—not every year but most years—since 1959 in order to draw attention to major issues and to encourage international action to address concerns that have global importance and ramifications. The UN had designated 1993 to be the Year of the World's Indigenous Peoples back in 1990!

That evening at the reception in the UN, copies of our calendar were handed out to everyone present. That same night at Robson Square Media Centre in Vancouver we (WCWC-WILD) sponsored the world premier of the Endangered Peoples Project's video *Cry of the Forgotten People* about the plight of the Moi people of West Papua. West Papua (Irian Jaya) is the largest and easternmost province of Indonesia, occupying the western half of the world's second largest island, New Guinea.

The independent state of Papua New Guinea occupies the other half of the island. New Guinea is the most culturally complex island on Earth, having the world's greatest number of languages and the largest intact tropical rainforest left in all Southeast Asia.

The video, made by Ian Mackenzie, a director on WCWC's board, eventually went on to win 10 international awards. It exposed the extreme threat of logging, which was opposed by the local tribal group that owned the land. The show was a big success. Despite the fact that this video brought the Moi's desperate fight to international attention, it failed to stop the logging.

UN never pays for calendars - WILD gives more away to worshipers in the world's largest cathedral

In the months to come, we experienced first hand the UN's desperate financial problems which everyone heard about in the media. Despite our repeated billings and diligent phone calls, and despite UNEP's repeated promises to pay, we never received a single cent for the calendars we sent them, not even the money to cover the courier costs. We finally wrote off our account receivable to UNEP. It was over $5,000—undoubtedly small potatoes to UNEP—but to us, deeply in debt, it was big money and their failure to pay was a bitter disappointment.

A month into 1993 we still had over a thousand copies of the English-French versions of our international calendar left. We were looking for a creative way to use them and decided that we should try to raise some money for the Penan campaign with them. We figured the cover shot of the Penan children would work. Adriane came up with the idea of handing them out along with a request for a donations in Saint John the Divine Cathedral of the Episcopal Diocese of New York. St. John the Divine was, and still is, the world's largest Gothic cathedral and the congregation had been very supportive of the Penan issue when Adriane, Thom Henley and the Penan delegation had gone to New York in the fall of 1992. Everything was arranged. We dreamed of raising tens of thousands of dollars. The calendars were handed out to the congregation with a table set up in the vestibule for people to donate at the end of the service. But the amount of money collected was only a few thousand dollars. Adriane later heard that the donation pitch made during the sermon was "very low key." All in all, the gamble didn't cover our costs to produce and ship our calendars, but it was a lot better to have the calendars used than to have them shredded. Ultimately, we had to accept the fact that our one and only international calendar was an artistic and educational success, but not a financial one.

Left: Declaration of Interdependence printed on the back of WILD's 1993 *Endangered Peoples Endangered Places* **calendar. Document: WCWC archives.**

Declaration of Interdependence

This we know

We are the earth, through the plants and animals that nourish us.
We are the rains and the oceans that flow through our veins.
We are the breath of the forests of the land, and the plants of the sea.
We are human animals, related to all other life as descendants of the firstborn cell.
We share with these kin a common history, written in our genes.
We share a common present, filled with uncertainty.
And we share a common future, as yet untold.

•

We humans are but one of 30 million species
weaving the thin layer of life enveloping the world.
The stability of communities of living things depends upon this diversity.
Linked in a web of community, we are interconnected-
using, cleansing, sharing and replenishing the fundamental elements of life.
The stability of communities of living things depends upon their diversity.
Our home, planet Earth, is finite; all life shares its resources and the energy from the sun,
and therefore has limits to growth.
For the first time, we have touched those limits.
When we compromise the air, the water, the soil and the variety of life
we steal from the endless future to serve the fleeting present.

•

We may deny these things, but we cannot change them.

This we believe

Humans have become so numerous and our tools so powerful
that we have driven fellow creatures to extinction, dammed the great rivers,
torn down ancient forests, poisoned the earth, rain and wind and ripped holes in the sky.
Our science has brought pain as well as joy;
our comfort has been purchased by the suffering of millions.
We are learning from our mistakes, we are mourning our vanished kin,
and we now build a new politics of hope.
We respect and uphold the absolute need for clean air, water and soil.
We see that economic activities that benefit the few
while shrinking the inheritance of many are wrong.
And since environmental degradation erodes biological capital forever;
full ecological and social cost must therefore enter all equations of development.
We are one brief generation in the long march of time; the future is not ours to erase.
So where knowledge is limited, we will remember all those who will walk after us,
and err on the side of caution.

This we resolve

All this that we know and believe must now become the foundation of the way we live.
At this turning point in our relationship with Earth,
we work for an evolution from dominance to partnership,
from fragmentation to connection, from insecurity
to interdependence.

The David Suzuki Foundation

Chapter 39

Campout at the Legislature gets meeting with the Premier
Rally turns into a riot

WCWC joins a government process team

I must admit, I was a bit surprised when the B.C. government invited WCWC to send a representative to sit on its Burke Mountain-Pinecone Lake Planning Team. The team was being established to consider park status for the region above Pitt Lake, which included the entire Boise watershed that we were campaigning to protect. We were told that the team's meetings would be open to the public, that it could initiate studies and that it did not have to come up with a "consensus" plan at the end. It could hold open public meetings to take public input on the options it developed and present to government the results.

In January 1993, we joined the planning team. It was WCWC's chance to show it could work together with industry representatives and the government towards reaching an agreement on protecting more wildernesses when the ground rules were fair. Joe Foy, who initiated and was leading our campaign to protect the Boise Valley, became our representative. For the next year-and-a-half, Joe regularly attended the team meetings and made the most out of our participation. It was his "flagship" campaign. When the time came in the spring of 1994 for public input on the options, WCWC's volunteers and supporters shone.

Clearcut perception problem - Premier Vander Zalm discovers the "Black Hole"

But our main focus during this time period continued to be on the west coast of Vancouver Island. In the early 1980s, while working for the Nuu-chah-nulth Tribal Council and attending the Meares Island Planning Team meetings, I had hitchhiked back and forth between Tofino and Port Alberni many times. In transit, I learned a great deal about people's perception of Clayoquot Sound and what a beautiful, wild place it was. At that time, almost the entire active logging taking place there was out of sight in remote areas.

People unfamiliar with the local landscape needed someone to point out the clearcuts on the distant horizon and explain what they were before they could actually "see" them. I regularly pointed out the small-looking, white, rectangular patches on the distant mountains and explained, "Those are actually huge clearcuts." People didn't know that the crisscrossing white lines were logging roads and the brown v-shaped areas were massive landslides caused by the road building and accelerated runoff from clearcut slopes. But once they were made aware, the clearcuts, roads and landslides jumped out at them. No one needed to be convinced that clearcuts were bad. They were horrified by what they saw and became part of the movement opposed to clearcut logging.

The one famous exception to the fact that there weren't any recent clearcuts close to the road to Clayoquot Sound was the clearcut appropriately nicknamed "the Black Hole." This huge roadside clearcut was located only a few kilometres from the t-junction turn-off to Ucluelet and Tofino. In keeping with the general forestry practice of the day, it had been slash-burned (deliberately set on fire to burn up the debris and waste logs) after being logged. Upon seeing the "Black Hole" on his first visit to the Sound in 1989, when the damage was fresh with charred stumps and bare blackened soil, Premier Vander Zalm blurted out, "Who the hell did this?" It prompted him to launch yet another planning process for Clayoquot Sound in a vain effort to find a solution to the conflict over the clearcutting. This process lasted only a year. Local participants representing the environment and the tourism industry quit in disgust because the areas under consideration for protection were being logged while they talked. In the summer of '93, although it was already four years old, the "Black Hole" clearcut had not yet "greened up." It still looked like hell.

Forest industries' PR forest tours

I often wondered how people could be so blind to the forest industry's destructive impact on nature and have come to realize that it's due mostly to lack of familiarity—not knowing what details to look for. This naiveté was exploited by the industry when they took politicians, dignitaries and their customers on forest tours to try to allay concerns about B.C.'s logging practices. Most came from Europe, where environmentalists' claims about the catastrophic impacts of clearcut logging on B.C.'s old-growth forests were swaying public opinion and causing cancellation of contracts for B.C. paper and pulp.

On the typical "deception tour" of B.C. forest practices, foresters and public relations officers would take visitors to see a 10-year-old plantation forest on level, low-elevation, fertile ground to show them how fast the second-growth trees grew on the coast. Then they would walk them into another forest somewhere near Duncan on Vancouver Island where there were second-growth trees, especially Douglas firs nearly one metre in diameter and 15 stories high, much larger than anything the "tourists" were familiar with back east or in Europe. The guides took delight in the fact that most people jumped to the conclusion that they were experiencing a genuine B.C. old-growth forest as they commented on how beautiful it was and how big the trees were. Just as the foreign "suckers" were deep in rapture (yes, some older second-growth forests are beautiful) and right before people started to look more critically, suddenly, the guide pricked the illusionary bubble.

"Look there," he'd say, pointing out a big old rotting stump covered in moss with its unmistakable notches for springboards on which the lumberjacks stood above the butt swell to whipsaw the tree down. "This forest has already been logged!"

The next game the foresters played was, "How old do you think these big trees are?" The guesses were wildly wrong.

"No, they are not 150 to 200-year-old trees, they are only 80-year-old Douglas firs."

This completed the indoctrination process. The innocent visitors were duped into concluding that clearcut logging isn't really bad after all. The clearcuts look bad only for a short while; they're really only "temporary meadows."

In B.C. it's true that in some places along the coast the forest grows back incredibly fast (although not nearly as fast as the tropics). But it's only in a few places that the trees in an 80-year-old second-growth forest have grown so large that they might be confused with old growth. Less than one percent of B.C. has the rich agricultural-grade soils and long growing season needed to produce such verdant second growth. The vast majority of forests on B.C.'s coast grow on steep slopes with thin soils. On these higher elevations and erosion-prone sites the trees grow very slowly.

During the '90s, many "fact finding" delegations came to B.C.

to check out the "horror stories" that environmentalists told about clearcutting. The forest industry, because they hosted these delegations, had the lion's share of their whirlwind tour time. Some, including European parliamentarians, stopped by our office at the end of their visit, spending a couple of hours with us before they caught their flights back home. We gave them a brief slideshow and then showered them with posters of old-growth forests—the magnificent forest many of them had come to see, but hadn't. Many regretted that they hadn't seen the majestic ancient forests or the horrible clearcuts we showed them on the screen. "Next time," they vowed.

Refusal of our request for a meeting with the Premier launches us on our biggest campaign ever

Over the years, WCWC had always fought "valley by valley" to save wilderness. The notion that all the remaining ancient forest in the entire Clayoquot Sound should remain unlogged came from the locally-based environmental group, the Friends of Clayoquot Sound. It was an idealistic and seemingly impossible goal, but we jumped right on board with the campaign. With wilderness, bigger is definitely better! The mostly-still-wild Clayoquot Sound, at over 350,000 hectares, about the size of Prince Edward Island, was a very big and very special place.

Since the beginning of WCWC, we had worked on issues in Clayoquot Sound, particularly Meares Island. We had already published several educational papers including one calling for the whole Sound to be declared a UN Biosphere Reserve. The succession of government land use planning teams for Clayoquot had all failed. Local campaigns were heating up. Now the new NDP government was bound and determined to make a "final" decision to end the endless quarreling.

At WCWC, we felt compelled to take our concerns about this special area to the top. In our entire 12 years of existence we had never been able to get a meeting with a B.C. Premier. After every general election, we dutifully wrote a letter of congratulations and requested a meeting. Every time, we were politely refused and referred to the new Environment Minister as the appropriate person with whom we should meet. Adriane vowed she would get a meeting about Clayoquot with Premier Harcourt. We had heard through our contacts in government that his cabinet would be making its decision regarding the fate of Clayoquot Sound in the upcoming spring session of the Legislature. Being by far the largest membership-based environmental group in B.C. and having done solid work on this issue, we felt we had a right to be heard by the Premier.

Adriane sent off a well-worded letter to Premier Harcourt requesting a meeting and followed it up three weeks later after we'd received no response with a phone call. Talking her way through several secretaries who tried to give her the brush-off, she finally got through to Andy Orr, the Premier's press secretary. He said he would see what he could do. A couple of days later Orr called back. Harcourt's answer was a firm "No." The Premier had all the information he could possibly need to make the right decision on Clayoquot. "The place had been studied to death," said Orr, and the Premier was "too busy" to meet with a delegation from WCWC. We received the official rejection letter a few days later. It was a rude awakening as to how little power we wielded. Were all our efforts really for naught?

"We can't take a flat "No" for an answer, can we?" I asked Joe and Adriane as we held a special E-team meeting to deal with Clayoquot and the Premier's rejection letter. We were worried that our failure to get a meeting meant that the NDP's soon–to-be-made decision was going to be bad news.

Cold and rainy "campout" on the Legislature lawns forces a meeting with Premier Harcourt

"Let's go to Victoria and wait outside Harcourt's office until we get our meeting!" exclaimed Adriane.

"That could be a long time—days, even weeks. We'd have to camp out there," said Joe.

"Once we start, we can't give up. We couldn't wimp away without getting a meeting. Then it would do more harm to the cause than good," I added.

"We simply have to do it," said Adriane. "I can't think of any other way to force the issue. We'll ask Sylvia if she's willing to take care of our kids full time," she added.

Sylvia, my niece who had graduated early from Grade 12 in Wisconsin, had come to live with us and be our nanny for the spring and summer before entering university.

"Can you stay in Victoria for as long as it takes, Joe?" asked Adriane.

Joe said he would have to check it out at home. He got the OK and the rest of the staff said they could hold the fort at 20 Water Street while we were gone.

In many ways, it was a crazy idea. It was winter and it was sure to be wet, windy and cold. We couldn't have a campfire to keep us warm. Our "action," of course, was not going to involve any civil disobedience. We planned to comply with the rules of the Legislature regarding protests on the lawn. But it would make a strong statement anyway, because we had never done anything like this before.

On February 24, 1993, the three of us sent a letter to Premier Harcourt putting him on notice that starting with an all night candlelight vigil on Tuesday, March 2nd, we would remain camped on the Legislature lawn *...as long as it takes to speak with you about this issue.* We ended the letter with, *If you order your security guards to remove us, we will, of course leave peacefully. But we believe it is our democratic right to meet with our elected officials and your democratic duty to meet with us on an issue of such grave importance.*

The next week we packed our tents, sleeping bags and WCWC's laptop computer. We brought along a table, three lawn chairs and a big display featuring photos of the clearcuts and beautiful old-growth forests in Clayoquot. Of course, we also brought along copies of our Clayoquot petition, posters, newspapers, and our big "Save Clayoquot Sound" banner.

Joe brought along his favourite "prop"—a large Canadian flag. What could be more patriotic than fighting for the protection Canada's irreplaceable natural heritage? We also rented a cell phone for a month. We sent out a press release announcing our intention of camping out on the Legislature lawn for as long as it would take to get a meeting with Premier Harcourt.

The day before we left for Victoria someone came to the store, handed us a plain brown paper envelope, urged us to read it right away with the words, "It's important!" and then quickly left. It <u>was</u> important. It had just what we needed to make our campout at the Legislature more meaningful and reinforce our case to have a meeting with the Premier.

Inside the envelope was a copy of a "top secret" tourism study of Clayoquot Sound authored by Brian White, a Capilano College tourism instructor. His study was done under contract for one of the recent planning processes for Clayoquot. The pro-forestry chairperson of that planning team had not released it to the public because of its conclusion that protecting the ancient rainforest in Clayoquot would generate more dollars and jobs in the tourism industry than logging would generate. This "new information" might be just what was needed to persuade the government from making the pro-logging decision that rumours indicated was about to be made.

A "palace guard" orders us to pull down our tents

Joe, Adriane and I arrived at the Legislature early on February 24th so we could set up camp and prepare for the candlelight vigil that night. We positioned our tent in front of the main legislative building on the walkway below the paved ceremonial driveway beside the stone wall. We were out of sight of the front entrance, but highly visible to people driving by or walking along the street in front of the Legislature. We worked quietly and quickly, aiming to avoid detection by the guards before we were completely set up. We had just finished putting up the tents and were stringing up our banner when two guards approached us.

"You're not allowed to camp on the grounds. You'll have to leave," said the senior guard.

"Oh, we're not camping here. This is a demonstration. We are simply protesting until we get a meeting about Clayoquot Sound with Premier Harcourt," responded Joe.

"The tents will have to go," insisted the guard, as he looked inside one of them and saw the sleeping bags inside.

"They are part of our props. We are a wilderness preservation society and tents are one of our identifying symbols," I tried to explain.

"Well, we'll see about that," said the senior guard and the two walked away.

The three of us waited around, kidding about the fact that the Legislature grounds were far better protected than our province's ancient forests were. About an hour later, a more senior member of the security staff came to talk to us. We could keep our tents up, he said, but they and everything in and around them, had to be removed each night. We could put them back up in the morning, if we so desired. He also informed us that no sleeping was allowed at anytime, day or night, in the tents or outside of them on the Legislature grounds.

Our first media interview after setting up "camp" in front of the Legislature to get a meeting with Premier Harcourt to discuss the fate of Clayoquot Sound. Photo: WCWC files.

We comply with the rules

We agreed to the terms presented to us. It was a no-brainer: we were there to save the ancient forests of Clayoquot and draw attention to this issue, not to challenge the rules governing demonstrations at the Legislature. It would be more work for us to pack up and set up every day, but the fact that we could be there from dawn to dark maintained the illusion that we were camping there. We'd figure out where to sleep later.

All day we continued to organize for a rally late that afternoon and candlelight vigil that evening. Thanks to the help of other environmental groups a crowd of about 250 people showed up, including some drummers who pretty well continuously drummed in front of the Legislature the rest of the days we were there. There was TV coverage too. I gave a passionate speech, imploring people to get more active to save Clayoquot from the clearcutters' chainsaws. As evidenced by the clip used on TV, I got quite carried away, gesturing with my fist up in the air and rallying the troops to take action. I was impassioned by my fear that we were losing this pivotal campaign. So much for my anger management workshop.

After the rally, with darkness descending, Adriane and I took down our tent and put all our stuff into Joe's van where he would sleep parked at a WCWC staffperson's home. Before the rally started, Adriane and I had been offered an old beat up 1973 VW camper van nicknamed "Rossie" to sleep in. Rossie belonged to Alison Spriggs, who had begun working for WCWC-Victoria as a canvasser in late 1991. Full of positive energy, Alison was now running our Victoria door-to-door canvass and helping with campaigns. Alison parked Rossie beside the Legislature earlier in the evening after the hourly parking restriction ended. She gave us the keys and said we could use the van as long as we needed.

About fifty hearty souls remained after the rally to participate in the candlelight vigil. There was no shortage of candlepower. That morning we had purchased all the candles we could find in the local thrift stores. That March night was very long, very wet and very cold. Adriane and I lasted until about 2 a.m. and then trundled off to bed in Rossie. Joe continued the vigil.

He told us the next day that people huddled together with their candles on the Legislature steps, until gradually heads began to droop. Immediately, one of the vigilant security guards would come over and gently prod the "head nodder" to awaken. There was zero tolerance for sleeping on the Legislature grounds. Joe gave up a few hours after we did. Several people remained.

When I got up very early the next morning, no one was there. Exactly what time the vigil ended, only the last diehard to leave knows. The only indication that there had been a vigil the night before were several large, and a few smaller, puddles of hardened wax on the stone steps leading up to the Legislature building's ceremonial entrance. Not wanting to leave any mess, I worked hard to scrape them off. As soon as Joe arrived—it still was very early, before any of the workers arrived or before the traffic began to flow—we put up our tents and other paraphernalia, thus succeeding in projecting the illusion that we had been there all night. So far so good, I thought.

A mad rush to get the suppressed tourism study into the cabinet ministers' hands in time

Day two of our campout was a busy one. We focused on getting the leaked Clayoquot Sound Sustainable Development Strategy Steering Committee's suppressed tourism study into the hands of sympathetic government cabinet ministers. It was an 11th hour effort. We contacted MLA Tom Perry who was sympathetic to the cause and told him about the tourism study. He was excited and said that he needed copies immediately for an important cabinet meeting that morning at 11 a.m. Clayoquot was on the agenda.

We got one of the volunteers from our office in Victoria to take the document to a high-speed photocopy shop and make copies for all the cabinet ministers. Tom sent a messenger down to pick up the packages just in time for the meeting. We were really excited: perhaps we had saved the day for Clayoquot. We later heard that our copies had made it to the meeting in time, but learned nothing about what

Adriane types out a press release on our laptop in out tent as I pass time outside. Photo: Joe Foy.

transpired at that meeting. Of course, it was a cabinet meeting and they are always held in camera.

In the afternoon we released the tourism study to the media, but there was no great rush of interviews. The weather got wetter, and the publicity we hoped to generate did not materialize. We sat around and paraded with our signs and talked to the few people who happened to come to see the buildings that day—mostly tourists. A few people who had heard through the environmentalists' grapevine that we were out there specifically came by to see us. Late in the afternoon one nice elderly lady brought some cookies and hot tea. While this was nice, I knew that we had to do something more to attract public attention.

"'Till the cows come home"

Day three. Adriane called Andy Orr and asked again for a meeting with Premier Harcourt. This time he was adamant. "You can stay out there 'till the cows come home', but you're not going to get a meeting!" Things looked bleak. Joe, Adriane and I quickly huddled and agreed that we had to stick it out. The Legislature was going to open in another two weeks and we resigned ourselves to being camped out there until then. Adriane was busy on the phone calling everyone she knew, telling them about our "campout" on the Legislature lawn and why we were doing it. But there were no TV cameras and no reporters. Despite the bad weather and no publicity, more and more supporters were coming to visit us at our "camp in." We were becoming quite a show. People brought fresh baked muffins, sandwiches, steaming coffee and piping hot homemade soup.

There were even a few cash donations. We told everyone that we were not going to give up. We had a right in a democratic society to a meeting with our Premier about the most important decision his government was about to make on the fate of the ancient forests of Clayoquot. Everyone was shocked that Harcourt refused to meet with us, the biggest wilderness preservation group in the province.

To make good use of our time, we decided to organize a giant rally for Clayoquot and the ancient forests of B.C. on March 18th, the opening day of the spring session of the Legislature. We walked into the Sergeant-at-Arm's office and applied for an opening day rally permit. It undoubtedly helped that, during the last three days we had been on the lawns, there were no negative incidents.

We had followed the rules to the letter and made sure we made no mess. No one else had applied for a demonstration permit that day, so we immediately got our permit. With all the time on our hands to plan it, we intended this rally to be the biggest the Legislature had ever seen!

That afternoon on the same day that we got our Legislature rally permit, our cell phone rang. Adriane took the call. "You have your meeting," the voice on the other end said.

"Who is this? Is this a joke?" replied Adriane not believing that it was, indeed, an official call.

"I'm Andy Orr and I'm in the Premier's office looking down at you right now. Take a look up." Adriane looked up and sure enough there was Andy standing in front of a window on the right wing of the buildings—the Premier's suite of offices—holding a phone to his ear. He gave us a wave.

March 16th was the earliest the Premier could meet with us. We would have only a half-hour to make our presentation, but we were invited to attend a press conference the Premier was holding afterwards that day. Having accomplished our objective—getting a meeting with the Premier on Clayoquot—there was no point in continuing our "camp." It took us less than an hour to pack up and leave.

Although we were excited about the meeting, it was a bit of an anticlimactic end to our campout. We had started to really get into it; enjoying the people who came to talk with us and brought all their goodies. We issued a media release thanking the Premier for finally agreeing to meet with us, noting both the time and place of our meeting as well as the Premier's press conference afterwards.

As we were heading to the ferry, Adriane had a brilliant idea. She had seen a T-shirt in one of the stores in Victoria with cows on it—big white ones with black spots—with the statement "The Cows Came

Home." We went and got it and gave it to Andy Orr as a gift, thanking him personally for getting us the meeting. He laughed.

Organizing a big rally to save Clayoquot Sound on opening day of the Legislature

During the days that followed Adriane, Joe and the folks in our WCWC-Victoria Branch office continued organizing and publicizing our Save Clayoquot Rally. In the week before the rally we passed out thousands of flyers throughout Victoria and put ads in the local papers. We planned the rally to begin at 1 p.m. with a "silent candlelight vigil—a message of hope!" This was intended to avoid any confrontation, like the throwing of pinecones and pine needles at the Premier's entourage that some young Walbran Valley logging protesters did during the Legislature's opening ceremonies the year before. Yet, at the same time, it would send a powerful message during the cannon-blast salute and the military band music that always accompanied the arrival of the Lieutenant Governor. It didn't quite happen that way!

We also planned that after Premier Mike Harcourt and Lieutenant Governor David Lam walked up the steps and through the Legislature's ceremonial doors, we would set up a microphone system so prominent speakers could talk about Clayoquot on the steps outside while the Throne Speech droned on inside. This didn't go as planned either.

The Sergeant-at-Arms would not let us have electrical power while the Legislature was in session. The earliest we could set up the mike and use our loudspeakers was 5 p.m. We agreed to the terms (we really had no choice) and consequently planned and advertised a two-part rally: a very long candlelight vigil during the opening ceremonies and after the throne speech, speeches and music. We called it the *Rally to Save Clayoquot Sound – A Turning Point for the Protection of B.C.'s Ancient Forests*. It certainly ended up being a turning point!

Everyone knew this had to be a big event to make the Harcourt government wake up to public pressure and protect Clayoquot Sound. The Sierra Club, Friends of Clayoquot Sound as well as some very active students at the University of Victoria and Camosun College all helped get people to come. Some local radicals had been demonstrating (mostly drumming) at the Legislature in support of Clayoquot Sound almost continuously for the last several weeks.

These noisy protesters discovered that if they stood inside the cupola-roofed formal entryway, pounded a syncopated beat on their drums and chanted over and over again "Save Clayoquot Sound" they created a naturally amplified sound that reverberated and throbbed throughout the whole Legislature buildings. It was mesmerizing and, we later heard, extremely annoying to everyone inside. It certainly got everyone within earshot thinking about Clayoquot, although, I suspect, mostly negatively. A columnist for Victoria's *Times Colonist* wrote about these noisy protests saying that the rhythmic noise that penetrated the building made it hard for those inside to function. Subsequently, the Legislature guards banned people from drumming inside the cupola, putting a ribbon across it with handwritten signs saying, "keep out."

Meetings with Harcourt unleash First Nations' fury

The day of our meeting with the Premier, Tuesday, March 16, 1993, came far more quickly than seemed possible. Joe, Adriane and I put in lots of extra effort to be fully prepared for this hard won meeting. We printed copies of a detailed agenda that outlined what we hoped to accomplish in our precious 30 minutes. We planned to start with a five-minute video that we had taken pains to have our video team shoot just four days earlier in Clayoquot Sound and had edited especially for Harcourt. It clearly showed that logging practices had not changed, contrary to statements made by him during his recent European trip in support of the B.C. logging industry.

Arriving at the Premier's office a half-hour before our 2:30 p.m. appointment, we learned that we were not having our meeting there, as we expected. Instead, Harcourt would meet with us in one of the big formal committee rooms on the east side of the building. I was disappointed. I was looking forward to seeing Harcourt's office and its décor. I was sure all the major company and union bigwigs got to meet in Harcourt's office. But alas, not us. We were also informed by his aide that we were not the only ones meeting with Harcourt that afternoon. He was already meeting with the Sierra Club and after us he would be meeting with IWA representatives.

While Harcourt met with Vicky Husband and several other Sierra Club representatives, we waited outside in the hallway nervously discussing with each other about how best to approach Harcourt. When Vicky came out, we were quickly ushered in without having a chance to even exchange a meaningful glance. We sat down and Harcourt began. He was very, very friendly and overly cordial, engaging in small talk far too long. He asked Adriane and I about our kids, mentioning them by name (he had obviously been briefed) and Joe about his kids and how things were going in general at the WCWC.

We knew this ploy for what it was: a time-killer used by politicians when they want to avoid a sensitive topic. We answered his personal questions as briefly as possible without being curt—we now had only 25 minutes left—and then launched right into the core of our business, presenting our strongest case and appeal for him to protect all of Clayoquot Sound. He brushed off seeing Joe's video show saying he had seen lots of pictures of Clayoquot and knew the issue very well, maintaining that it would be better if we just talked. So, we gave copies of our video to his aides.

None of us mentioned our three-day campout on the lawns in front of his office that got us this meeting. We tried to convey how precious, marvelous and rare on a global scale the ancient rainforests that remained in Clayoquot Sound were. We tried with words to describe its huge moss and fern-draped cedars and the spiritual feeling they evoked in visitors. From Harcourt's comments and body language, I got the distinct impression that he had never experienced such feelings himself. So I asked him outright, "Have you ever personally spent time in an old-growth forest in Clayoquot?" Harcourt did not answer with a straightforward yes. He gave a convoluted answer that made the three of us conclude, independently, that he had not only never experienced the magic of Clayoquot, he had not experienced any coastal old growth close up and intimately.

Adriane continued with this line of questioning. She asked him if he would like to spend a half-day on a guided tour with us to special places in Clayoquot's old-growth forest, or even to our Carmanah Research Station before he made his decision. This would help him understand what was at stake and the strong spiritual inspiration that drove people to be so passionate and committed to protecting Clayoquot's ancient temperate rainforest. Harcourt politely declined, saying he would like to, but he was too busy. At one point near the end of our meeting, Joe, either out of frustration with Harcourt's passivity, or simply because he cared so much about Clayoquot and realized he couldn't find the right words to move Harcourt, broke down and started to cry. I had never seen him like that before or anytime after.

At the conclusion of our meeting we gave Harcourt a pile of our educational publications. We included our Clayoquot newspaper publications, a copy of our *Clayoquot on the Wild Side* book, posters, maps and a copy of our three-part opinion poll mailer card with which he had to already be familiar. We had sent out tens of thousands of the mailers over the previous six months and thousands of concerned citizens had sent them back to him. Harcourt graciously thanked us for coming and said he found our input "very useful" and explained that it was going to be a difficult decision he had to make. His job was to find the "proper balance."

All in all, it was one of the most frustrating meetings I've been in. We failed to penetrate through what seemed like a plastic mask that shielded the real Harcourt from us.

Only months later did we learn some facts that explained why all three of us experienced such a surreal feeling at this highest-level meeting we had ever had with government.

Upon emerging from the meeting, the IWA representatives were ushered in and we immediately faced several TV cameras and a couple of reporters. We had sent out a media release the day before reminding everyone of the time and place of our meeting with the Premier. Ever since our campout, and the sending out the dates of our meeting with the Premier and the big Clayoquot rally, the media was reporting everything about Clayoquot. In front of the cameras we tried to put the best spin possible on our frustrating meeting, saying that it was friendly and that we hoped Harcourt had listened.

We waited in the hall along with the TV cameramen and the reporters for the last of Harcourt's half-hour Clayoquot meetings to end. We wanted to hear what Harcourt had to say. In less than 20 minutes the IWA representatives came out and hurried past the cameras. Then Harcourt emerged from the room. They immediately cornered him in a scrum and the first thing he said was, "Now I've met with everyone and we can make our decision." I can't remember what else he said. It was non-committal. We knew now for sure that the decision would be coming down soon, but not before the opening of the spring session of the Legislature.

The three of us drove to the ferry and traveled back to Vancouver that evening. Later that night Adriane got a call from a good friend in Tofino who was well connected with several of the Nuu-chah-nulth chiefs. He told her they were watching the six o'clock news and saw and heard Harcourt state emphatically that he could make the Clayoquot decision now because he had consulted with everyone. That made the Nuu-chah-nulth furious. Premier Harcourt had not consulted with the First Nations owners of Clayoquot Sound who comprised over 50 percent of the population living there!

WCWC rally at the Legislature turns into a riot

Two days after our meeting with Premier Harcourt—Thursday, March 18th—the spring session of the B.C. Legislature began. The weather was blustery but not rainy. We left on the first ferry from Vancouver, went to our Victoria office to help with last minute planning for the rally, and arrived at the Legislature well before the "vigil" part of our rally was to start. Before noon there was already a sizable crowd of Clayoquot supporters gathered on the lawns.

People had come from all over Vancouver Island. Many carried signs to save the wild forests and make more parks in B.C. from other protests from years past. Someone had a big cardboard gravestone with "Tsitika R.I.P." on it. Someone else held a sign that said "No Jobs on a Dead Planet." The crowd kept building, although it wasn't nearly as big as I hoped it would be. We unfurled our huge Save Clayoquot Sound banner at 12 noon. It took eight strong people holding the poles to keep the banner upright in the gusty wind. A silent candlelight vigil was obviously a foolish idea—there was no way candles would stay lit—and no one showed up with them! There was a scattering of drummers but nothing like the powerful drum beat a couple of weeks earlier at our Clayoquot campout rally.

It was getting on past 1 p.m. and the crowd was still thin when all of a sudden I heard faintly in the distance the familiar syncopated percussion drum beat. It grew louder and louder. Finally I saw a parade of people coming down Government Street. The placard-carrying protesters behind the drum band kept coming and coming. There were hundreds and hundreds of people following behind, many of them young students, carrying a variety of signs and chanting in unison over and over again "Save Clayoquot Sound... NOW!" The "NOW" followed a half-second after the rest of the chant, perfectly timed to the syncopated drum beat. It was mesmerizing.

"Where did all these people come from?" I asked Alison who was standing beside me. She explained that, as planned by the student activist groups, they had started out at 11 a.m. at the University of Victoria, picked up people at Camosun College and others all along the way as they marched to the rally.

The crowd gathered was by far the largest that I had ever seen at a pro-environment rally at the Legislature. The entire front half of the lawn was filled with people right up to the concourse roadway that ran directly in front of the building. A rope was strung next to the roadway to keep people back because Premier Harcourt and then the limousine bearing Lieutenant Governor David Lam would arrive via this roadway to bring them to the ceremonial steps into the Legislature. Many environmentalists present, including me, had been through it all before. It would be the same ritualistic play of pomp that occurred on every opening day of a new legislative session.

With media always present, it had always been a good day to stage a protest. The army honour guard in dress uniform with rifles sporting fixed bayonets arrived and stationed themselves in front of the crowd of protesters.

The Premier arrived first and waited on the steps for the Lieutenant Governor to arrive. Harcourt stood stoically facing the crowd of protesters pretending not to see us and giving no indication of being fazed at all by the loud rhythmic chanting and persistent heckling. His disregard fuelled the frustration that people felt about the Clayoquot situation and the chanting became more urgent and more demanding.

Finally, the Lieutenant Governor arrived. The chanting grew even louder and everyone, including me, tried extra hard to drown out the military band, the bagpipes and especially the 19 gun (cannon) salute, that hurt the eardrums, shook the buildings and symbolized the power wielded by lawmakers who would soon be sitting inside. When the cannonade concluded, Harcourt, Lam and their entourage slowly ascended the steps and disappeared into the building.

The army honour guard marched away and the crowd milled around for a couple of minutes. I saw someone, an older man that I had never seen before, untying one of the ends of the rope that held the crowd back from the steps and the building. The rope, basically a symbolic barrier, dropped to the ground. A few people ran across the concourse and started to mount the steps of the entry-way towards the now closed doors. Immediately others, including the percussion band, followed. Meanwhile, Joe, Adriane and I and a few of our volunteers remained back on the lawns holding up our big banner where we had originally centrally positioned ourselves in order to be highly visible in the TV coverage. We watched and heard the drumming and chanting by the protesters packed into this confined cupola entryway space. We carried on a casual conservation commenting on the fact that they must be making an awful racket inside. We also discussed our plans for our 5 p.m. rally—the music and speeches; who should speak first—not paying much attention to what was happening in front of us.

Protestors in front of the B.C. Legislature on March 18, 1993. Photo: John Yanyshyn of the Vancouver Sun.

"What the hell's happening down there?"

Suddenly we noticed it was much quieter and we looked at the Legislature and saw a few demonstrators walk up the steps and disappear through the ceremonial door. There were only a few hundred demonstrators left on the lawns. I thought—but only fleetingly—that it was very strange that they would let demonstrators in through the big ceremonial doorway. We continued to talk when our rented cell phone rang. Adriane answered. It was a woman from the media.

"What the hell's happening down there?" she asked.

"Nothing much, the rally's over," said Adriane.

"Nothing much!?" replied the caller incredulously. "Right now I'm looking at my TV screen and there's a riot going on inside the Legislature buildings!"

"Oh my God, that's where all the people are," cried Adriane, as it finally dawned on her what had happened.

"You better do something to calm it down. They're pounding on the stained-glass window in front of the Legislature chamber. It's getting very ugly in there," said the reporter who was sympathetic to our cause.

"OK, thanks for calling. We'll do what we can. Goodbye." Adriane hung up and yelled to me, "Paul, we've got a big problem! Come over here now!"

Obviously agitated, she quickly began to explain the situation. Before she'd said much, I got it. I realized that there had to be hundreds and hundreds of people protesting inside. Listening carefully, I could faintly hear the drumming and chanting emanating from inside the building.

Joe goes in to calm the crowd

"Where the hell is Joe?" I asked, sending out several volunteers to go find him. In less than a minute Joe came rushing over. "One of us has to go in there and try to calm things down and get the people to leave!" I told him.

"You should go, Paul," said Joe.

"No you should go," I countered. "They're mostly young people and they'll more likely listen to you. You're younger and have a much calmer personality."

So we decided Joe was the man. He quickly strode off towards the Legislature, walked up the steps and disappeared into the building.

There was nothing Adriane and I could do but wait and worry. After what seemed like an incredibly long time, the cell phone rang. This time it was Nicholas, our regular volunteer who was alone back at our 20 Water Street office. He was freaked out! The phones were ringing off the hook. Dozens of members had called in and angrily quit WCWC for leading a mob into the Legislature, seriously hurting a guard and destroying property. Adriane explained to Nicholas that we didn't do it. Other people we didn't know, or have any control over, stormed the Legislature. Nicholas told her people said they saw us do it on TV.

We found out later that the initial TV coverage showed me extolling people to get more active to save Clayoquot (the rip-roaring speech I had given at the rally three weeks earlier at the launch of our camp-in at the Legislature) and then immediately flashed to the people storming into the Legislature and disrupting the Throne Speech. It made it look like I was the one instigating the "riot." Subsequent coverage did not have the clip of my earlier speech. I saw a videotape of this first CTV news story about the incident a few days later and it was certainly a case of deliberate distortion—not that we wanted to dodge our legitimate responsibility for what happened. We were the lead organization, the group that obtained the permit for the rally, and we ultimately were accountable for what occurred.

I took the phone from Adriane and talked to Nicholas. I told him to remain calm and explain to everyone that he was just a volunteer

Second part of WCWC's "infamous" rally on March 18, 1993. Photo: WCWC files.

and did not know what had happened. I asked him to be sure to get everyone's name and phone number and tell them that I, personally, would call them back and explain what happened as soon as possible.

Finally, after waiting to hear from Joe for over an hour, I could see a constant dribble of people coming out of the Legislature building. Shortly thereafter Joe emerged looking very tired. He said that it was really tough in there. "It was like going to meet a lynch mob," he told us. He explained that it took him quite a while to get through the packed crowd past the percussionists who were drumming in a circle in the rotunda. Finally he got into the anteroom right before the chambers of the Legislature and tried several times to speak, but there was too much noise and no one was listening. Finally one young guy shouted out "SIT DOWN. LET JOE SPEAK! SIT DOWN AND GIVE HIM A CHANCE!" People gradually quieted down a bit. Then, one by one they sat down. Joe tried to talk again, but they still couldn't hear him. Someone else who wanted to hear what Joe had to say yelled out "GIVE JOE THE MEGAPHONE!" The person who had been pounding it against the locked stained glass windowed Legislature chamber door handed it to Joe.

Somehow, Joe told us, he instinctively knew what to say. He started a dialogue by saying that people should leave, that what they were doing was not helping save Clayoquot Sound. Joe couldn't remember the details, but whatever he said, it worked. (He is a genius at thinking on his feet and speaking plainly and passionately.)

The mob calmed down and eventually left peaceably. Only a few panes of stained glass were smashed. We had not yet heard about the elderly guard who had been bowled over and broke his hip when the crowd first burst through the front doors.

No mention of riot in Legislature's Hansard record

When you read the Hansard record of the proceedings that day in the Legislature you'll find no mention of this whole incident. The only trace record of this momentous event—the first time in B.C.'s history that an angry mob stormed the B.C. Legislature (or any Legislature in Canada for that matter)—is found on Page 4703 in Hansard of the first day of the second session of the thirty-fifth Legislative Assembly of the province of B.C.

The House began the session at 2:04 p.m. and the Lieutenant-Governor was just giving the preamble to his throne speech when:

> [Interruption.]
> **Hon. M. Harcourt:** Your Honour, it's my suggestion that we adjourn briefly.
> The House recessed at 2:16 p.m.
> The House resumed at 3:37 p.m.

It was during this one-hour-and-twenty-one-minute adjournment that it all happened.

Relieved that things had calmed down and there wasn't too much damage, the first thing I asked our group was whether or not we should hold the second part of our rally with the speeches scheduled to begin at 5 p.m.

"Will they turn on the power for us after what had just happened?" I rhetorically asked out loud. All of us agreed that the show must go on—if we were allowed.

Apparently, the government wanted to put this incident behind it as quickly as possible just as badly as we did. We asked the Sergeant-at-Arms if we could hold our rally after the throne speech rally and he said of course we could and that we would have electrical power for our sound system as pre-arranged.

Our late afternoon rally takes place as planned

All the speakers showed up. About 300 of the demonstrators attended. David Anderson, MP for Victoria, whom we had hoped would pledge his support for protecting Clayoquot as the core piece of a UN Biosphere Reserve, used his entire time to shame us and admonish the unruly demonstrators.

He didn't say a word about Clayoquot. Adriane, Joe and I used our opportunities at the mike to try again to put pressure on the provincial government that was giving every indication that it leaned towards a pro-logging decision to protect Clayoquot. We also expressed our disapproval of the riotous behaviour and asserted that it was counterproductive. Some people booed us for criticizing the protestors, but we held our ground.

John Cashore, then Minister of Environment, Lands and Parks, spoke for the government. He gave the standard noncommittal speech, saying that the B.C. government would make the best possible decision—a balanced one—because we needed to protect the forest industry too.

This second part of our planned opening day Save Clayoquot rally concluded without a hitch and the crowd was rather subdued. The news that an elderly guard had been knocked over in the rush into buildings and in the fall had broken his hip had a sobering effect on everyone.

Joe, Adriane and I met afterwards at a nearby restaurant with leaders of the Sierra Club, Friends of Clayoquot Sound and other organizations that had helped with the rally. We decided to hold a press conference in front of the Legislature the next day at 11 a.m. WCWC, as the lead organizer of the rally, would acknowledge full responsibility (even though we had nothing to do with organizing or participating in the riot) and apologized profoundly to the B.C. public for what had happened.

Chapter 40

Infamous April 13, 1993

Meeting with the Honourable Tom Perry the next morning before our press conference

A week earlier we had arranged to meet with Dr. Tom Perry, who was the government's Minister of Advanced Education, in his Legislature office at 9 a.m. on the day after the opening of the Legislature. Tom had been one of the environmental movement's leaders in the campaign that had successfully thwarted U.S. plans to build the high Ross dam, thus saving the Skagit Valley from being flooded. The first thing Tom did at our meeting was try to impress upon us how scary the events of the previous day had been for some of the MLAs. He repeated several times that he was fearful that one of them would have a heart attack and die from the stress. The MLAs had no idea how far the anarchistic mob would go as they chanted and pounded hard on the stained glass windows of the chamber's ceremonial doors. Several glass panes were broken. Some MLAs feared that the angry crowd might break into the Chambers and do worse.

I admonished him saying, "Come on. They couldn't have been that afraid. This is Canada and people are not that violent." But Tom kept insisting that from the viewpoint of many of MLAs within, it was a scene of terror. His assertion was hard to believe. Later we heard from another person who had been inside the Chambers that afternoon that some of the MLAs picked up pieces of the broken stained glass and took them away as souvenirs. Obviously not every MLA was terrified!

Next, Tom took us to task for our extreme position on Clayoquot Sound. While we insisted that people supported full preservation of the ancient forests in Clayoquot and would react vehemently against any government that allowed it to be logged, he insisted that we had spent too much time in our downtown Vancouver office and were out of touch with the B.C. public at large whom he knew wanted a "balanced" decision.

Tom was absolutely sure that the citizens of B.C. had moved on from the Carmanah craze of a few years before when protecting ancient trees was popular. Today, they weren't as interested in preserving B.C.'s old-growth forests. I got so angry at his analysis and his wiser-than-thou attitude that I walked out and left Joe and Adriane to finish the meeting. I knew that if I'd stayed, I would have undoubtedly blown up and yelled at him.

Even if Tom was right, which I highly doubted, I felt that he should be championing the preservation of these incredible ancient forest ecosystems because it was the right thing to do. He, of all people, should not be gearing his political position to "balanced" compromise. I was more convinced than ever that the Harcourt government was going to sell-out Clayoquot. I was morose at the thought that we had no champions for the preservation cause at the government's cabinet table.

About a half-hour later Adriane and Joe emerged. I'm glad I didn't stay. According to both of them, Perry went on and on about how being an elected representative meant that he must look at things from a bigger perspective and represent all his constituents' points of view. The "balance" he said he sought was code for "selling out" as far as I could see.

Later that morning, Joe, Adriane and I held our press conference in front of the Legislature. It went off well. We announced that we would pay for any property damage done to the Legislature buildings and, in the future, promised to take precautionary measures and have peacekeepers and marshals for crowd control to make sure that nothing like that ever happened again at one of our rallies. The government obviously made a similar vow to strengthen security measures. Thereafter, strengthened security measures at opening day ceremonies eliminated the possibility that anything like that could ever happen again.

One WCWC director did break in with the crowd

We also asserted at that press conference that none of our directors or staff had participated in any of the unruly events inside the Legislature buildings that day. This turned out to be untrue. We found out the next day that one of our directors, Silvaine Zimmermann, did enter the building during that time period.

In her letter of apology and resignation from WCWC's board, Silvaine explained that she had followed the tail end of the crowd into the building after the initial surge. While inside she had participated in the loud chanting of *"Save Clayoquot Sound"* and *"Save our Planet."* After a while she became concerned that so many people were packed into the small entryway to the Legislature chambers and that a few had begun pounding on the door. Feeling powerless to do anything about it, she had left.

At the end of her letter she said that we might have averted the mounting of the steps if there had been some good entertainment to capture people's attention right after Harcourt entered the building. Silvaine did not know that we had originally planned it that way and were forced to have a three-hour gap because we weren't allowed to have electrical power until the government session inside was over.

Legislature riot puts Clayoquot on the map

Bad as this incident was, it put Clayoquot on the map and into national consciousness like nothing else had ever come close to doing before. The riot was front-page news in every paper in Canada and was repeated at the top of the news on TV around the world including in Europe. Coupled with the images of the rioters inside the Legislature building was footage of the beautiful lush Clayoquot rainforest and its big trees and views of the ugly clearcuts—making vivid to everyone what propelled the protesters' passionate plea to "Save Clayoquot Sound." It was a turning point in the campaign. From then on, Clayoquot was big news. The campaign to save this tiny part of the planet took off like a rocket.

We returned to 20 Water Street late on the day after the riot. The damage to WCWC was not as great as we had feared. After the first flurry of members who quit after seeing the misleading TV coverage (which was not repeated after the first video news story), the number of upset members dwindled to a trickle. One member called in to contribute $1,500 to help us pay for the damage. Many people thanked us for apologizing and taking responsibility even though we were not the ones who instigated the civil disobedience.

Never had a crime scene in progress been so documented. All the major TV stations were there for the opening of the Legislature. All the evidence needed was contained in their hours of film footage. Every channel covered the event live. There was footage covering every angle and detail of the incident as it unfolded. Eventually, four people were charged and convicted as ringleaders. One of them was a 14-year-old boy.

I do not believe that there was a big conspiracy that planned the riot. I think, by and large, it was a spontaneous outburst of frustration by environmentally-concerned people who felt ignored by the Harcourt government.

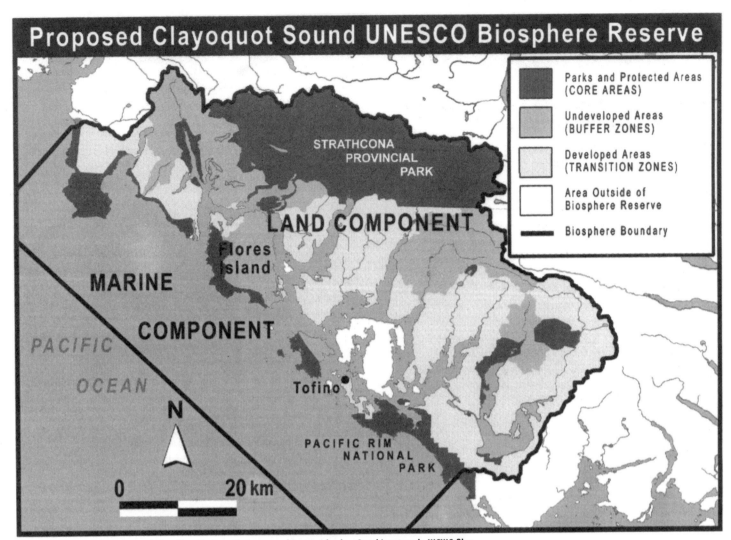

Map of Clayoquot Sound proposed Biosphere Reserve that appeared in our calendar. Graphic artwork: WCWC files.

Who were the riot leaders and how did the mob bust through the massive oak doors?

We knew only one of the four people charged with instigating the riot, the 14-year-old boy who calmed down the mob so Joe could speak. The year before, he had helped WCWC as a volunteer on one of our trail projects. He asked Joe to testify as a character witness at his trial. In 2003, he came up to me at a WCWC table I was helping staff at a summer music festival and told me his side of the story. The "riot at the Legislature" trial had resulted in his conviction and sentencing to 200 hours of community service. While at the time he "wasn't too concerned about it," the whole thing really upset his parents. Here is his description of what happened:

"Someone loosened the rope barrier that kept the protesters off the front steps of the Legislature. I ran up the steps along with the others, stopping at the closed doors through which the Lieutenant Governor and the Premier had just passed a few minutes earlier.

I was at the front right against the big doors when someone behind me started shouting to the people below, 'The door is open! Let's go in!' The crowd surged forward with crushing force. I tried to get out of the way but I couldn't. The breath was being squeezed out of me and I couldn't breathe when the big doors finally gave way and broke open. All the guards had retreated except an elderly one who, I heard later, got his hip broken when he was bowled over as we burst through." [Note: In English parliamentary tradition, the front doors of all commonwealth parliaments and legislatures, including the B.C. Legislature buildings, open inwards to symbolize that these democratic institutions are always open for commoners to bring forth their petitions for consideration. If the doors opened outward, as they do in most public buildings for safe egress in case of fire or a panicked exit, the crowd would not have burst in. But then some people might have been crushed to death.]

The most amazing thing that this young man told me was that he raced ahead and was the first to arrive at the big stained glass doors behind the rotunda directly opposite the ceremonial entrance the crowd had just broken through. He had no inkling that the stained glass doors were the ceremonial entryway to the Legislature chambers. He opened them and there was the Legislature chamber with all the MLAs seated inside. Shocked, he closed the doors and locked them before the rest of the mob arrived, preventing, according to him, a much worse situation.

B.C. Forest Alliance propaganda infuriates us

In the aftermath of the riot in the Legislature, we began to get stronger and stronger hints of a negative NDP government decision on Clayoquot. One source of rumours was the B.C. Forest Alliance, the industry-sponsored, pseudo-environmental group formed to counter conservationists' efforts to protect B.C.'s old-growth forests. This organization, formed two years earlier, was now in full swing. In March the B.C. Forest Alliance published a slick publication titled *Choices – Issues and Options for B.C. Forests – Public Participation 'putting your two cents in.'* This publication confirmed all our suspicions that

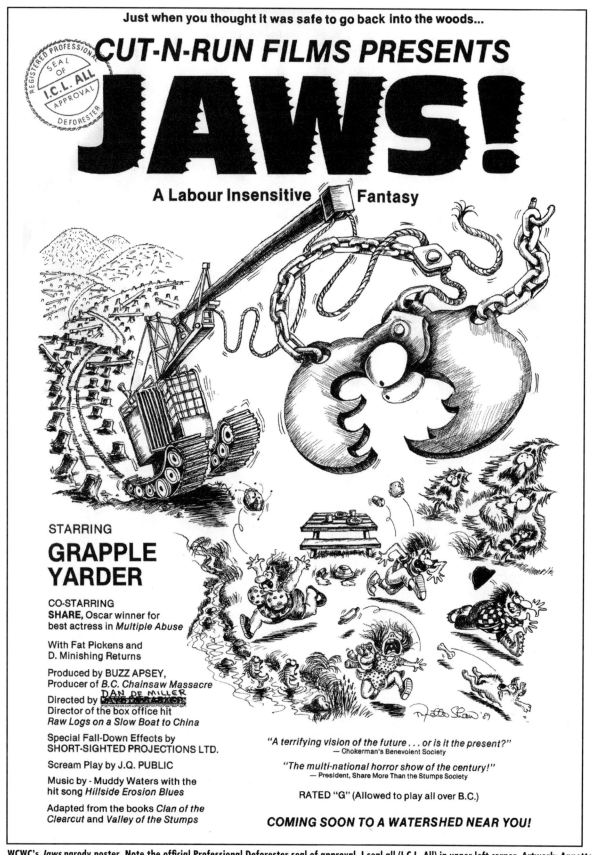

WCWC's *Jaws* parody poster. Note the official Professional Deforester seal of approval, I seal all (I.C.L. All) in upper left corner. Artwork: Annette Shaw.

the real intentions of the government's public involvement processes were to co-opt and waste activists' time, assuage public concerns and cleverly manipulate compromise outcomes. Otherwise, why would the forest industry laud these processes so effusively? We published a spoof duplicate *Choices* cover, stapling it over hundreds of the B.C. Forest Alliance's newspapers. Our cover read, "Choices – Either the Multinationals own the Forests or the People of British Columbia Do – Your opinion is worth more than two cents.

On the backside of our spoof cover we printed a film poster parody titled *Cut-N-Run Films Present JAWS! – A Labour Insensitive Fantasy* decrying the increased use of grapple yarders, a new way of clearcutting that used more roads and caused more erosion. It was fun to put out this biting satire. It was cheap, too, because Nik Cuff printed it up for free. But it didn't do anything to change the situation.

Increasingly, grapple yarders were being employed, regardless of the environmental damage they wrought, because this logging method employed fewer people and therefore saved corporations money and increased their profits.

Behind our humour was real anger. In a press release dated March 29, 1993, we slammed the B.C. Forest Alliance's *Choices* publication as *untruthful propaganda*, especially about Clayoquot Sound. In it we quoted Joe as saying:

> 'It is written to appear as if it comes from an impartial point of view, but through omissions and totally ignoring the ecological reasons why people want to preserve the whole area, including the last remaining pristine watersheds in Clayoquot, it ends up being totally misleading.
>
> 'The main article in the slick Alliance paper lauds democracy, but the kind of democracy they want is rigged,' continued Foy. 'Their article on Clayoquot Sound completely glosses over the reasons for the failure of the Socred designed [Clayoquot] planning processes. It attempts to make the industrial option to log over 85 percent of the ancient forest there look reasonable by colouring the future clearcuts green and the proposed preservation of bog forest and alpine forest a darker green.
>
> 'The Wilderness Committee believes that the Alliance newspaper was written when industry was confident that they had succeeded in getting through the back door of the government to get the Clayoquot region excluded from the Commission on Resources and the Environment (C.O.R.E.) process for Vancouver Island. It would never have been written in such a blatantly slanted way if it had been written giving the public real choices,' added Foy.

This last paragraph turned out to be prophetically true.

Harcourt announces his infamous Clayoquot logging decision atop Radar Hill

On April 13, 1993, a couple of weeks after the release of the *Choices* newspaper and a month after our meeting with Premier Harcourt in Victoria, Harcourt, his entourage of cabinet ministers and advisors, and the press corps flew to Pacific Rim National Park Reserve. Steven Owen, the appointed Vancouver Island CORE Commissioner who was supposed to be independent from government interference, was among those there. On Radar Hill, a dramatic hilltop setting a few kilometres south of Tofino, Harcourt announced his government's Clayoquot land use decision. Only one-quarter of Clayoquot's magnificent old growth would be protected: the Megin Valley and some other smaller areas, most of which had forests of low commercial value. The NDP opened up 74 percent of Clayoquot's productive old-growth forests to industrial logging. The government land use designations were almost identical to those shown on the land-use map published a month earlier in the B.C. Forest Alliance's *Choices* newspaper. The only place in Clayoquot Sound left "undecided," was Meares Island because of the pending unresolved court cases.

I've never seen such a unanimous, negative reaction to a government decision from the environmental community. Everyone was upset and angry. We were angry with Premier Harcourt for ignoring our pleas for orchestrating this so-called "balanced" decision without even taking the time to visit the old-growth forests it would destroy. We were angry with Stephen Owen for selling out his integrity by being on the podium, thereby providing a de facto endorsement of the

The TV news covers the government's Clayoquot announcement on Radar Hill. From left to right: John Cashore, Minister of the Environment, Michael Harcourt, Premier, and Minister of Forests Dan Miller. Photo: WCWC files.

NDP's Clayoquot plan. Ironically, for an hour or so, when we got the leak that Owen was with Harcourt in Tofino, we thought that maybe the NDP had listened to our plea to include Clayoquot in the CORE process. How wishful thinking we were.

WCWC had advocated the inclusion of Clayoquot in CORE for two reasons. First, we thought that it would include government-placed logging moratoriums over all the areas under consideration for protective status there, which would give us additional time to build unequivocal public support for them. Second, CORE's original mandate was to protect 12 percent of each ecosystem type. Since there was so little low-elevation old-growth forest left on Vancouver Island, we figured that its inclusion would mean full protection for all the key old-growth areas in Clayoquot Sound that we wanted protected, including the remaining pristine watersheds.

Because WCWC had consistently stated in the media that if the B.C. government decided the fate of Clayoquot Sound outside of the CORE process we would boycott CORE, we proceeded immediately to make good on that threat. Joe, Adriane and I took steps to make sure that all our representatives working within the CORE process quit and started working with us to make Clayoquot Sound our "flagship" campaign. We vowed to work more closely with the First Nations there, who had been ignored by the NDP in this land use decision. Our E-team decision to shun CORE did not sit well with certain folks in our Victoria and Mid-Island Branches who were already deeply immersed in this process. Joe and I had a heavy-duty verbal argument with Russ Forester, who worked at fundraising for our Victoria Chapter in the evenings and spent most of the rest of his time participating on our behalf in CORE. Joe and I contended that CORE would drain away energy in time-wasting meetings. We would be more effective by working outside the CORE process, developing our own protected area vision for Vancouver Island and "selling" it to the public who in turn would pressure CORE to adopt it. "Let the Sierra Club be on the inside. Our outside pressure will help its negotiators get more at the CORE table," I insisted.

Russ left WCWC over this decision. George Gibson, who worked so hard to get our Mid-Island Branch up and running after the financial meltdown of 1991-92, also ultimately decided to leave. He joined the Sierra Club and continued to work on CORE.

Chapter 41

The Tatshenshini gets saved
Other campaigns expand

The Tat goes fast track

At the same time that we were "at war" with the Harcourt government over Clayoquot Sound, we were pushing hard to bring the "Tat" campaign to a successful conclusion. A few years earlier, for a short period in the spring and summer of 1990, the Tatshenshini had been one of WCWC's hottest campaigns. We had motivated our members and done what we could at that time. Since then, Ric Careless had organized an international campaign, wisely understanding that pressure from the U.S. would provide a powerful push to save this Canadian wilderness area, which bordered directly on the U.S., especially because the proposed industrial development inside the area would hurt U.S. interests.

The umbrella group Tatshenshini International, which included most of the big environmental and conservation groups in the U.S. and Canada, really put pressure on the B.C. government. Despite our differences of opinion with Ric over other matters, we continued to work with him on the Tat. We co-published a second very powerful eight-page, full-colour, tabloid newspaper on this issue. On the cover we featured this quote from U.S. Vice President Al Gore:

> The (proposed) development of a huge open-pit copper mine in the midst of one of the world's most rare and pristine regions is an environmental nightmare that threatens the river and every living thing in the region.

Actually, we did little to make this publication happen. Ric arranged the funding, which included a grant from WWF's "Endangered Spaces" program. He lined up all the writing and production work. All we did at WCWC was to put our name on it as the co-publisher, have our volunteers mail them out to everyone on our mailing lists, both to WCWC members and supporters and to other environmental groups around North America to insure a rapid distribution of all 75,000 copies.

I didn't really believe that the Tat decision was imminent, even though the headline on the cover of our newspaper stated so in a blazing big red-bannered headline *B.C. Government to decide immediately. Your help needed now!* Such big decisions always took more time! But the pressure was on. Harcourt had a choice of either being a hero by creating the park, or facing insurmountable opposition from the U.S.

Due to the harm that the massive amounts of acid mine drainage arising from oxidation of the sulfide-rich waste rock and the heavy metals dissolved in that drainage could inflict on Glacier Bay and an American fishery, the U.S. steadfastly refused to allow the export of the ore concentrate from the proposed mine through a U.S. port. With no way to ship out its metal-rich concentrate, the proposed mine was going nowhere. With the outcry over the April 13th Clayoquot decision still reverberating, Harcourt moved quickly to be the environmental hero on the Tatshenshini issue.

Ric, who was pulling strings behind the scenes, wanted to make sure that when Harcourt made the Tatshenshini park announcement, WCWC would join in the celebration and not hold back because we were so upset by the Clayoquot issue. I assured Ric that there was no problem with WCWC simultaneously being of two minds about the B.C. government's environmental policies. Of course we would do what we could to get the government the kudos it deserved if it saved the Tat. Acting on Ric's urging to make this

Group photo of those present at the founding meeting of Tatshenshini International. Ric Careless on the far right holds onto the banner. Photo: WCWC files.

clear to the powers that be in the NDP, we put out a special pre-announcement press release titled, *WCWC Poised to Praise Harcourt Government When Tat Protected*. This was designed more to reassure Harcourt that his government would get heaps of credit for protecting nearly one percent of B.C., than it was to get picked up by the media. Situated in the uppermost northwest corner of B.C. between Kluane National Park in the Yukon and Glacier Bay and Wrangell-St. Elias National Parks in Alaska, the Tatshenshini watershed was the keystone that knit them both together and made the combined protected wilderness (including the Tatshenshini) the largest protected area in the world, approximately 8.5 million hectares. Who wouldn't praise a Premier for doing that!

Cheering down the miners' jeers

Despite our fawning press release, I was caught off guard when Ric called me in a panic early Tuesday morning on June 22, 1993, to tell me that Harcourt would be holding a press conference in a couple of hours at Canada Place, only a few blocks away from our 20 Water Street office to make the Tat Park announcement. His excuse for not informing us about this earlier was that the government had been keeping it quiet to limit the protest—this time from the mining industry. Now he had just learned that the mining industry had been tipped off about it and would be there in force to jeer and boo and make it a really bad scene for Harcourt. He asked me to get as many people as I could to the media conference to cheer for the decision. I told him I'd do my best, but it would have helped a lot if I'd been given more notice. It was obvious to me that if there weren't a protest a brewing, we would have been excluded.

I had only about an hour to get people to go and no time to make any signs. We could only muster a little over a dozen people, most of them our staff members, so we had to make up in enthusiasm what we lacked in numbers. We set out on foot for the press conference in time to get there fifteen minutes early.

Already there were a couple of dozen or so mining executive types milling around the back of the room. I told our people to hurry up, spread out and grab seats up front so that the cameras would see us clap and cheer at the top of our lungs when they panned the crowd as the announcement came down. More people kept pouring in, mostly mining executive and speculator types, along with a few fellow environmentalists from other organizations. Soon all the chairs were filled and the standing room at the back packed. Just as Harcourt was about to begin, some members of the pro-mining faction started catcalling and chanted, "Recall! Recall!" As Harcourt lauded the attributes of this wild area in his preamble one guy in the back heckled, "Can I get a job there?"

The mining lobbyists continually waved professionally made signs that proclaimed *Envirosaurs Rule Again* and *Windy Craggy - We Will Never Forget* and others saying they were heading to Chile to mine (which in some cases was true). There was no doubt this decision was unpopular with them!

Finally, when Premier Harcourt announced the establishment of the 958,000-hectares Tatshenshini-Alsek as a Class A Provincial Park, all of us wildly cheered and clapped as loudly as we could. Although the mining lobby outnumbered us by about four to one (25 to 100), we "envirosaurs" expressed our appreciation so loudly that we almost drowned out their anti-park boos.

Because there was such a kafuffle, it became a major news story. Every point of view was covered. But because the conservationists present were leaders representing different large groups, we yea-sayers got more TV and newspaper coverage than the nay-sayers. George Smith of the Canadian Parks and Wilderness Society (CPAWS), Mark Angelo of the Outdoor Recreation Council (ORC), Ric Careless of Tatshenshini Wild, and our WCWC spokespersons all got extensive statesmanlike "sound and print bites." Adriane's quip was widely quoted: *This is a day to celebrate. We are losing so much of the natural fabric of this planet that when we actually have the courage to set aside an area like this, it is extraordinarily significant.*

Although some will claim that Clayoquot had nothing to do with it, the "Tat" good news announcement came just 69 days after the disastrous Clayoquot decision. Harcourt created the park in time to receive a major environmental award in the U.S. and to get the good news reported in a National Geographic magazine article about the "Tat" which came out shortly thereafter. It also garnered the NDP a high year-end grade from WWF for "conserving wild spaces." Whatever backroom negotiations drove this deal, it was a great conservation achievement of global significance—and the B.C. economy thrived despite the "loss" of this colossal copper mine pipe dream.

Tat campaign wrap-up press release goes out before the champagne flows at WC²

As we walked back to our office, we made a not-so-hard decision to take the rest of the day off to celebrate. But first we typed up and faxed out a celebratory press release titled: *WCWC heaps highest praise on NDP Government for making the right decision in protecting the Tatshenshini Wilderness as a Provincial Park.*

In it we covered all the bases:

> *It is not too often we get to celebrate…the decisions to save wilderness have been few and far between,' said Paul George, anticipating the bubbly taste of champagne he's going to uncork to celebrate the NDP government's decision today to save the Tatshenshini wilderness from industrial development.*
>
> *George remembers first hearing about the area eight years ago when the Committee included it in their annual Western Canada Endangered Wilderness Calendar. 'We couldn't even find one great picture of the area then. Now there are so many that people who may never be able to go there can still see what this vast and wild river feels like,' continued George.*
>
> *George and his wife Adriane Carr, the Wilderness Committee's Clayoquot Campaign Coordinator, took a two-week rafting trip down the Tatshenshini in the summer of 1991. 'It is the wildest country we've ever been privileged to experience and I can't tell you how overjoyed I am to know that it is now protected,' exuded Carr. 'I can't praise the government enough for this decision. We are asking all our members and supporters to thank Premier Harcourt. If he gets enough praise, I am sure that he will make more ecologically sound decisions which future generations will thank him for.'*
>
> *George is worried however that this large area takes up much of B.C. government's commitment to increase the parks to 12% of the land area. 'The 12 percent has no validity in science. It was just picked out of a hat. Places like B.C. which has the greatest biogeoclimatic diversity in North America, must protect a lot more,' commented George.*
>
> *The Wilderness Committee believes in fair market value compensation for the mining interests. 'We don't want industry fighting every park proposal in the future,' quipped George, who said the Committee was not against all mining in B.C. as he headed down to the WCWC staff room to join the celebrations and thank everyone from all organizations who fought so hard to win protection for the Tatshenshini.*

WCWC no fan of the new Forest Practices Code

In early November, a few weeks after the Tat announcement, the B.C. government launched its long awaited Forest Practices Code. While formulating this Code, the B.C. Forest Service sent a delegation to WCWC to pick our brains as to what we thought should be in it. They spent a whole afternoon in our boardroom with Mark Wareing, our staff forester, Joe and I. What a waste of our time. Not one of our suggestions was incorporated in the final product! We wanted more conservation of old growth and rare forests coupled with a much-reduced rate of cut, based on true sustainability. We recommended watershed-by-watershed plans to protect soils, fish and wildlife and water. We suggested that planning horizons be based on a full rotation (the length of time it takes for trees to grow to full maturity—hundreds of years in the coastal rainforest). This was because we knew that the current short rotations, based on clearcutting the forest right after the rate of young tree fibre acquisition peaks, was producing low quality and low value wood.

We wanted full rotations coupled with commercial thinning to maximize the quality and value of the wood produced. To maintain the full range of forest ecosystem functions, we recommended that single stem selection generally replace clearcut methods of logging. Overall we aimed to make ecological considerations, and thus sustainability, paramount in forest management.

I spent a whole day reading through the fine print of the copy of the new Code. I was disgusted. In almost every case where it would cost the company more money to meet the Code and protect the environment, a B.C. Forest Service District Manager had the discretionary power to wave that requirement. The huge number of "weasel words" in the Code drastically weakened what little environmental protection the Code might have provided.

Every time we checked out forestry practices on the ground (or from the air) we couldn't tell the old "chainsaw massacre" from the new "up to Code" logging, except for the accelerated pace of road building penetrating the wilderness and slightly smaller cutblocks. Leave strips to protect streams were still inadequate. Clearcutting continued. The volume of timber being logged was just as great.

In my opinion, the Code was more about public relations than about real change. The huge volume of rules didn't bring about ecologically sound logging practices. In fact, we sympathized with the logging companies who complained that the most tangible impact of the Code was increased red tape and paperwork. I knew, however, that the major forest companies' cacophony of complaints was more 'crocodile tears' than a genuine expression of real hardship. The bureaucratic burden of the Code hit smaller companies much harder than the 'majors'.

On the coast, the smaller more dispersed clearcuts gave companies the perfect opportunity to target the best groves and most valuable timber, especially red cedar. It also paved the way a few years later for the greatest forest industry scam in the scandal-ridden history of B.C.—grade setting. Although there were many infractions of the Code over the years, the million-dollar fines set out in the Code were never imposed. These fines were just the false paper-mâché teeth of this paper tiger law.

To add insult to injury, the Code also attempted to take away WCWC's hard-won right to build trails through provincial forestlands. Under the Code we had to get permission from the government first before we could build any trails anywhere in B.C.'s publicly owned forests. We knew this was not just to avoid new trails in ecologically sensitive areas; we already avoided those places. The obvious point was for the B.C. Forest Service to deny us permission to build a trail in a politically contentious area where the local logging company opposed our objective of protecting the area from logging... and that meant everywhere we wanted to build trails! At least the Code's authors were loophole happy enough to leave an escape clause in trailbuilding, too, that we could ingeniously weasel through. They didn't want the Code requirement to be so strict that every time timber cruisers flagged access routes into the forest, they had to obtain permits. So, from the time the Code came in force in 1995, WCWC did not build any new trails. We only re-opened old traditional trails

WCWC bumper sticker promoting trailbuilding. Artwork: WCWC files.

and flagged and clipped access and egress routes for reasons of safety, so that our forest researchers could get in or out quickly in case of an emergency situation. The Code didn't stop or slow us down. But there's no doubt that the anti-trail rule was mean-spirited.

At the same time that the new Forest Practices Code came out, we came up with our own solutions to improve B.C.'s forest industry in an eight-page paper titled *How to Save Jobs in the BC Woods*. In it we advocated commercial thinning of second growth coastal forests that were over 40 years of age to improve the quality of timber, and periodic thinnings thereafter. At the time there were about half-a-million hectares of these "thrifty mature" forests on Vancouver Island.

We also advocated only selective logging in all "mature" second growth forests on Vancouver Island. We hired eco-forester Ray Travers on contract to analyze how many jobs the practice of commercial thinning and selective logging would yield over and above the jobs generated when second growth forests are clearcut. It turned out to be more than a thousand permanent jobs. To write the paper WCWC contracted Ken Drushka, who had written several books about logging and who at the time was an advocate of small scale logging companies, before he became an apologist for the big forestry corporations. We printed up and widely distributed 50,000 copies. It made no discernible difference in improving forest policy in B.C.

Cover of WCWC's *How to save jobs in the woods* newspaper. Document: WCWC archives.

The "on record" dinner with MB's PR brass

Out of the blue during the same time period, I got a call from Scott Alexander, head of MB's public relations. It was about something he disagreed with in one of our newspapers. At the end he casually said, "You know, we really should go out for lunch sometime. We're both professionals and, like all professionals, we share things in common and should get to know each other socially."

Personally, I was not in the least bit interested in socializing with MB's PR man. If fact, it was one of the last things I would ever think of doing! So, to brush him off I replied, "Sure, that's a great idea, but it would have to be 'on the record'," explaining that I would only agree to such a meeting if we had some representatives of the press there to record the momentous occasion.

Scott thought for a moment and then agreed saying, "...but you'll have to arrange it." He also suggested that I bring Joe Foy along and he'd bring Dennis Fitzgerald, another member of MB's PR staff.

"OK, I'll check it out with others here and get back to you."

The whole idea was really off the wall. Neither Joe nor Adriane were keen on it. Both asked me, "What will we gain in doing this?" "Lots of publicity for our cause, I replied."

They were both skeptical about being able to get reporters to come out to such an affair. Joe was willing to come to the luncheon, but he suggested that, if it really was to be a "friendly" meeting, it should be a supper date because then it could last into the evening and not encroach on the working day.

We had no trouble coming up with the place for the dinner: Kosta's Greek Characters, the neighbourhood pub and restaurant right near our office that was WCWC's new hangout that had replaced our old office favourite, the Press Club. We knew that there would be plenty of room in the restaurant section during a weekday to give us privacy. The food would be great, even if the conversation wasn't. It had to be "Dutch treat" with each of us paying for our own meal and drinks. We wanted it clear that no one was hosting this affair, currying favours or creating obligations.

After getting the ground rules squared away, I called around to see if any reporter was interested in attending and found two: Glen Bohn of the *Vancouver Sun* and Ms. Lush of the *Globe and Mail*. I made the call back to Scott and the date was set. I personally was looking forward to this upcoming on-the-record dinner.

The night came. Both Joe and I wore suit jackets, but no ties. We got to the restaurant a bit early. I ordered a beer; Joe declined. Soon the others arrived and we exchanged pleasantries. Then we went over the ground rules again. Everything we said was going to be 'on the record' and could be reported in any way our journalistic observers saw fit. We ordered our food. Then we got into it.

They pointed out that we undoubtedly shared many of the same values including making the world better for our kids. We all agreed on that one!

I insisted on making one point very clear from the start; this was not a meeting of equals. I was not talking about the big spread between what Joe and I got paid and what they got paid for doing what they did. I was referring to the fact that Joe and I were CEOs of our organization with the power to change things if we came up with an agreement or new insight this night. They were not CEOs. They were merely employees who took orders. They had little power at MB to change things.

"Get your CEO and the other decision makers in MB to come to a dinner with us and then maybe we could work things out. But I doubt it. Your company, when push comes to shove, is in the game to make as much money as it can," I jabbed. "Ultimately what we want—the withdrawal of licensed cutting rights from MB to make more parks—will cost your company money."

Our conservation went on-and-on, back-and-forth with the two journalists busy scribbling away in their notebooks. Joe and I used the opportunity to explain how strongly we believed in the cause of wilderness preservation. We were fair players and believed in the democratic game. One thing for sure, we were not motivated by money to work for our cause. While we personally respected each other's differing viewpoints, it became abundantly clear that Joe and I could never relate to the two MB employees as "professional friends working in public relations" whose sole difference from us was that they worked for a different outfit. Our work was about public education and mobilization, not PR. Our work world was not nine-to-five. It was 24-7. We could never go out after work with them for a friendly beer, for there never was an "after work" or "off the record" time in our minds.

I don't think Scott and Dennis fully appreciated the difference between them and us nor the passion that drove us to work so hard for the cause. **Just as it is virtually impossible for someone who has never fallen in love to empathize fully with someone who has, it's equally impossible for those who have never "fallen in love" with a wilderness area to understand the behaviour of those who have.** Long hours of work, low or no pay, are not hardships. They simply come with the territory.

The reporter from the *Globe and Mail* left first, then Joe left. I stayed on for a while longer, continuing to drink beers and chat with Glen, Scott and Dennis. We broke up about 9:30 p.m. On the way back to the office, where I was going to sleep on the couch that night (like I often did when I worked late) I thought to myself that things had gone pretty well. I was looking forward to seeing the articles. When they appeared about a week later, both were good and I felt readers would conclude that we had won the debate. But Bohn ended his full-page feature article in the Sun with the fact that I accepted a cigarette from Scott and we both lit up, showing that we both were equal in terms of sharing the same vice. I did it on impulse as a departing friendly gesture. I'd quit smoking 15 years earlier. This stupid act diminished my credibility—I should have known that it, too, was going to be "on the record."

Meares Island Big Cedar Trail project

There's far more to campaigning than waging word battles in the media. We always grounded our campaigns in concrete activities that would make a lasting positive difference in protecting wilderness. With over 70 percent of the First Nations people living in Clayoquot Sound unemployed, we knew that any lasting wilderness preservation must be tied to jobs that would continuously employ First Nations. According to Brian White's tourism study, protected wilderness would generate more jobs in tourism than logging would generate.

The challenge was to make sure the First Nations got their fair share of them. Meanwhile, the forest companies had figured out the same social dynamics. Following the government's April 13 decision, the major logging companies vigorously courted the First Nations in the Clayoquot, offering them various roles in industrial forestry development including "partnership" deals if they agreed to logging in their territory. Historically, the First Nations had been more involved in fisheries. Few were employed in the logging industry. Most believed through first hand observation that logging negatively impacted the wild salmon. But jobs in a jobless society were a big pull.

Circumstances were conspiring against us. The Big Cedar Trail on Meares, which had created some employment for Tla-o-quiaht band members who used their small boats to ferry tourists between Tofino and the Island, was closed. As an easy-access way to experience Clayoquot's ancient temperate rainforest, the Big Cedar Trail had simply been loved to death. Meares, the third largest island off the west coast of Vancouver Island and the second largest in Clayoquot Sound, was renown for its wild forests and awe-inspiring big trees. Thousands of people annually hiked the trail. The heavy use had turned it into a big, wide mud wallow as people trampled ever-wider routes through the undergrowth vegetation to try to avoid the muck. To let it heal, the elders of the Tla-o-quiaht First Nation had declared the trail closed to the public, asking water taxi operators not to ferry people from Tofino to the trailhead and back. Most people respected this closure and complied. First Nation water taxi jobs were down.

We had to do something. Re-opening the Big Cedar Trail was the easiest way to increase tourism jobs for First Nations as well as get more people directly experiencing Clayoquot's rainforest so they could be inspired to help protect it! WCWC's E-team decided to use $17,000 of the $80,000 we received in 1993 from the Bullitt Foundation to fix up the trail. We costed out the project and then added in another $7,000 of our own donation money. We talked at length with the Tla-o-quiaht, especially the Band Manager Howard Tom, and drew up a contract between WCWC and the Band to fix the trail and construct a boardwalk through the muddy sections.

The B.C. Supreme Court injunctions would not stand in our way as they did when we roughed out the Meares circuit trail in 1988 because, in this case, we were not actually creating anything new. We were simply maintaining the status quo by fixing up an existing trail and keeping it in good condition. The Meares Big Cedar Trail had been scouted out by Adrian Dorst and built by local environmentalists in 1982, before the 1984 Meares confrontation and ensuing court cases.

In addition to our contract with the Tla-o-quiaht, we hired Robinson Cook, a creative young carpenter with experience in building beautiful cedar boardwalk trails, as WCWC's liaison with the First Nations crew doing the work. The amount of money we put towards this project really wasn't enough. It was a big job involving the construction of over a kilometre of boardwalk including several steep sections that required stairs.

Through dedication, hard work and good craftsmanship, the crew did a beautiful job completing the boardwalk to the Hanging Garden Cedar, the second largest known cedar in B.C. and one of the largest in the world. They used lots of wood scrounged from beaches and a lot of ingenuity in incorporating unique pieces into the design. Near the end of the contract, in order to get the job done when money and time ran out, the crew actually donated their labour and put in unpaid overtime. They also got lots of support from local Tofino businesses. A number of businesses donated services including a package trip to Clayoquot and a stay in a bed and breakfast, which we raffled off to our members to raise more money for the project.

Near the end of the summer, when the new boardwalk was nearly completed, Arnold Schwarzenegger visited it. This led to a LA film company using the location in a film and giving a contribution to the Tla-o-quiaht, which helped pay some of the overrun costs of the boardwalk construction.

In October of 1993, after the long summer of blockades and mass arrests was over and the boardwalk completed, the Tla-o-quiaht hosted a grand opening of the trail. About fifty people attended including dignitaries from Tofino. The ceremonies began at the trailhead

Meares Island Big Cedar boardwalk trail under construction. Robinson Cook, who helped with the design, is standing on my right. Photo: Adriane Carr.

with speeches by Tla-o-quiaht hereditary chiefs who thanked everyone involved, especially the trail crew. Then everyone hiked along the new boardwalk, holding onto driftwood rails as they climbed up and down the beautiful, sturdy staircases, to the Hanging Garden Cedar. Adriane and I walked along slowly with the Opitsaht elders as our two young children ran on ahead. As we progressed I got a wonderful feeling that we were experiencing a performance piece of artistic carpentry that fit perfectly into the ancient rainforest that it wound through. The smell of the fresh cedar planks permeated the air and enhanced the enchantment.

While we were admiring the Hanging Garden Cedar, both of our children were guided into a three-metre-long passageway through the heart of the tree where the wood had rotted out. Afterwards TJ and Kallie could not stop talking about this marvelous journey through one of the oldest living creatures on our planet. Although it is impossible to know how old the Hanging Garden Cedar is (the early growth rings in the tree's centre rotted away long ago), an estimate of 2,000 years old is not unreasonable.

The trail's opening day ceremonies ended with a feast in Opitsaht, the Tla-o-quiaht village on Meares Island. The newly boardwalked trail was a real source of community pride. All the crew members were publicly recognized for their work. All of us celebrated Meares itself, too. Almost entirely cloaked in its original primary forest; it is a truly special place.

Chapter 42

Clayoquot conflict heats up
A *Conservation Vision* for Vancouver Island

Ombudsman confirms that First Nations were not consulted on Clayoquot land use decision

A black cloud hung over Harcourt's April 13th Clayoquot decision. In the days following the announcement, several First Nations lodged complaints with the B.C. Ombudsman that they were unfairly treated in not being consulted by the government.

The Ombudsman accepted the case. His investigation took months. Finally, in November 1993, he tabled *Public Report No. 31* titled *Administrative Fairness of the Process Leading to the Clayoquot Sound Land Use Decision*.

This strongly worded report found that the government's simple assertion in the press release accompanying its Clayoquot decision, that this decision would not in any way prejudice the First Nations' land claim, was not good enough. In response to this Ombudsman report, the B.C. government entered into negotiations for an Interim Measures Agreement (IMA) prior to a treaty settlement with the Nuu-chah-nulth.

When the IMA was concluded in March 1994, it established a framework within which people had to work to solve the ongoing land use conflicts in Clayoquot. It also created the Clayoquot Central Region Board (CRB), a unique decision-making body with representatives from all the communities in Clayoquot, both native and non-native. In an amazing act of power-sharing, the IMA gave the First Nation Bands of the Central Region of the Nuu-chah-nulth Tribal Council the opportunity to vet all future resource development proposals for Clayoquot Sound and the power to veto their go-ahead.

Nothing like this had ever been tried before in B.C. But then, nothing on the scale of the Clayoquot conflict had ever erupted before, either. I felt a twinge of satisfaction that our "tent-in" at the Legislature had got the ball rolling by forcing a meeting with Harcourt, which in turn provoked him into asserting that he had "consulted with everyone," which of course he hadn't. First Nations had seized the opportunity and negotiated a profound new agreement.

WCWC was invited to send representatives to the IMA ratification ceremony. We were the only Canadian environmental group to be honoured there with a gift from the chiefs. (They also recognized "Bobby Kennedy's group," the Natural Resources Defense Council in the U.S. that had powerfully taken up their case of aboriginal injustice. This organization had helped in many ways including embarrassing the B.C. government into beginning to right the wrongs.) The IMA ceremony included the most extensive native singing, dancing and feasting I'd ever witnessed.

As the night wore on, all of the other environmentalists slipped away, each one confirming with Adriane that she would stay as their overall representative to the end. Well past midnight Adriane and John Cashore, Minister of Aboriginal Affairs, joined in the final circle dance, stepping in time to the drum beat.

Facing page: Cartoon lambasting Harcourt for going to Europe to promote B.C.'s forest industry. Artwork: Geoff Olson.

Harcourt's consultation with us was phony

In late summer of 1993, I discovered a most amazing piece of information vital to understanding the chain of events that occurred during March and April of that year. A visitor to our office casually asked me if I had read the report released July 28th by Commissioner Justice Seaton. Seaton was appointed by the B.C. government to look into whether or not there was a conflict of interest in the B.C. government buying a big block of MB shares in February, just three months before the government made its Clayoquot pro-logging decision. (MB shares went way up in value immediately after that decision.) In his report Seaton concluded that no conflict of interest had occurred.

"No, I haven't read it, but I know about it…the share purchase was for some government employees' retirement fund," I responded off-handedly. Even before the inquiry, I thought the conflict of interest charge was a "red herring." I didn't believe that the share purchase was part of a grand conspiracy connected to the decision. It had to be purely coincidental.

"No, I'm talking about the testimony in the report that reveals when the actual cabinet decision regarding Clayoquot was made. It was a couple of weeks before the big riot at the Legislature!"

"Is that really true?" I asked, incredulously.

"Yes. Harcourt waited for weeks until he thought it was the right time to make the announcement."

Suddenly it all made sense to me. The Clayoquot decision was made at the beginning of March, on the first day of WCWC's Legislature "tent-in." The important cabinet meeting that we'd rushed to get the copies of the repressed tourism report to was actually THE Clayoquot decision-making meeting! The meeting we had several weeks later with Harcourt was all for show. The decision was already made. No wonder it seemed so surreal and Harcourt didn't even bother to watch Joe's video. It was all totally phony! How could anybody, let alone a Premier, be so duplicitous? I'd never respect Harcourt again!

Maybe the frustration expressed by those who crashed into the Legislature on opening day was fueled by a psychic sixth-sense that the big Clayoquot demonstration that day would not make any difference. It was too late! No wonder Tom Perry, Minister of Advanced Education, told us during our meeting with him the day after the "riot" that we should not be so fixated on Clayoquot and that the public was not behind us. What he was really saying to us in so many words was that the government had made its decision and moved on, and we should move on too.

I just couldn't get over it. We had been suckered. It wasn't a nice feeling. But it sure motivated me and everyone else us at WCWC to campaign even harder to prove Harcourt and the NDP wrong and build the public determination to win protection for Clayoquot. It definitely woke us up to the fact that the NDP government would not make ecologically-sound decisions regarding creating new protected areas. They were committed to "balance," the new code word for trade-offs. We would always be pitted against the IWA and forest industry who dictated the NDP's bottom line. Already, Harcourt had backed down from its "land use strategy" commitment to protect a 12 percent representative sample of each and every one of the diverse ecosystems in B.C. The new goal was simply to protect 12 percent of B.C.'s land base. We all knew this meant that there would be a much higher percentage of rock and ice than old-growth forest protected.

Harcourt hated the heat he got over Clayoquot. Environmental groups, especially Greenpeace, launched international 'market' campaigns to get major customers to cancel big contracts to purchase wood products that might come from Clayoquot. They especially

Cartoon we had drawn when we first heard about the B.C. government's recent purchase of MB shares. Artwork: Geoff Olson.

targeted MacMillan Bloedel (MB), as a major forest licensee in Clayoquot. Industry was hurting. Harcourt hopped on a plane to Europe trying to sell Clayoquot as a "balanced decision." It was not the first time he had made the trip there. During his first trip in 1992 to defend the B.C. forest industry and its clearcutting methods, we commissioned local cartoonist Geoff Olson to create a cartoon titled *Mike's trip to Europe*. It depicted Harcourt flanked by two giants, one labeled *IWA* and the other *Forest Multinationals*, with their arms around his shoulders in the foreground. In the background was a totally clearcut mountain with tiny protesters standing on it holding up signs saying *Save the Ancient Forests, Stop clearcutting*, and *Big Trees not Big Stumps*. On the bottom it quoted Harcourt as saying, *And, on behalf of the people of B.C., I'll be going to Europe to counter the misinformation campaigns of environmentalists....*

Work on our Vancouver Island Vision Paper continues

Although Clayoquot Sound became our "flagship campaign," we kept working hard to protect the rest of Vancouver Island's wilderness, too. With the big boost of our second Bullitt Foundation grant, we began work in earnest on developing a "Conservation Vision" for the entire island. Our Victoria Branch hired Misty MacDuffee as a researcher. Immensely talented, she soon became our branch's lead campaigner. Along with our Victoria canvassing director Alison Spriggs and other Victoria volunteers and staff, Misty also started a major campaign to protect Victoria's drinking watershed from logging.

Our Victoria-based research team collected and analyzed existing maps and data. They used this information, together with conservation biology principles, to design a conservation biology-based protected areas system that would fully conserve Vancouver Island's biodiversity over the long haul. It meant nearly 40 percent of the island needed to be protected in a network of interconnected parks.

They included areas called "Proposed Restoration Protected Areas" where second-growth forests would not be logged. These were important to rebuild the biodiversity in ecosystems that had already been nearly destroyed by past logging, especially in the dry belt Douglas fir ecosystem in the Nanaimo Lowlands Ecosection. At the same time, they proposed that the rest of Vancouver Island's land base be managed in a truly sustainable way under locally held community forest tenures. WCWC's vision served as an alternative vision to the CORE

Misty MacDuffee working in WCWC-Victoria's Bastion Street office. Photo: Alison Spriggs.

One of the giant red cedar trees in Clayoquot Sound. Photo: Mark Hobson.

(Commission On Resources and Environment) recommendations. At our Vancouver office, WILD's GIS staffers Ian Parfitt and Andrew Boldt also worked on this Conservation Vision map. It was a great exercise in science-based planning. I always wanted WCWC to do the same thing for the entire province, but it was just too big an undertaking. We never had the funding to do a credible job. We have, however, worked on pieces of it and it still remains a WCWC dream. Most of all, I wanted real science to be the backdrop for political decision-making. Too often political decisions and trade-offs are made without fully knowing what's at risk.

Knowing that our *Conservation Vision* map had to have the backing of all the environmental groups on Vancouver Island to be really credible, Adriane and I took our draft map around Vancouver Island and sought their input, including their "ground-truthing" of our data. We incorporated the information that they provided, ultimately ending up with a document that virtually all the conservation groups endorsed. Although we had the Bullitt grant funding in the bank to print 300,000 copies and distribute them to all the households on Vancouver Island via inserts in community newspapers, we held off on publishing. To be maximally effective, we put our *Conservation Vision* paper out just prior to CORE's public input meetings being held in communities around Vancouver Island. We knew that people's attention spans and memory spans are very short.

Chapter 43

Building a Witness Trail
Making a stand at Sutton Pass

We named it the "Witness Trail"

Besides advocating a major preservation package for Vancouver Island, we needed a focal point for our Clayoquot Sound "flagship campaign." We chose the imminently-threatened 7,680-hectare Clayoquot River Valley, which had one of the best salmon streams left in the Sound. Slightly larger than the Carmanah Valley, it was difficult to access and still completely pristine. MB had plans to start logging there soon.

We reasoned that a hiking trail into this valley would work like the one we built into Carmanah. We sent Joe Foy and Mark Wareing to scout out the best route. Their trail surveying work quickly revealed that Clayoquot Valley was as spectacular and, in its own way, as beautiful as Carmanah. However, we had to build a longer trail over more difficult terrain to get into the valley. I don't know who came up with the name Witness Trail, because there was never any debate about it. It was an apt name: hikers were able to witness both the beauty of the valley and the red flags signaling MB's intention to build roads and log.

Making our stand at Sutton Pass, the entrance to Clayoquot Sound

Our trailbuilding activities in Clayoquot give us little profile. We had to do more. But what? Nearly two months had passed since the B.C. government made its awful Clayoquot decision. Hard as I tried, I couldn't come up with anything new to do.

Weekly staff meetings were one of our few office routines. At a memorable meeting in early June, Joe, Adriane, Andrea, I, and other staff members and volunteers—about 15 of us in all—were sitting at our big worktable at 20 Water Street. These meetings often turned into brainstorming sessions with staff and volunteers trying to come up with creative campaign ideas, to solve problems, and to plan the week's work, including volunteer nights. They were lively, fun and full of good energy...when we weren't talking about our debt.

"You know, people haven't got a clue really where Clayoquot Sound starts or how to get to the Clayoquot River Valley," commented one of our new volunteers who was about to head off to work on the Witness Trail. "They're not on any map, are they?" We all knew that the seven-kilometre-long abandoned logging road that branched off from the Port Alberni-Tofino highway at Sutton Pass was partially hidden; an unmarked access to the trailhead of our Clayoquot Valley Witness Trail. We also knew that the height of Sutton Pass marked the entry to Clayoquot Sound. About a 40-minute drive from Port Alberni, Sutton Pass is the highest point along Highway 4 to Tofino and Ucluelet, the only road access into Clayoquot Sound. The pass defines the boundary between Alberni and Clayoquot Sounds.

"You have a really good point," said Joe. "It's even hard for the volunteers we send up to work on our trail to find the turn-off. Our signs keep getting stolen. There ought to be a big sign saying, 'You are entering Clayoquot Sound.' How can we protect the place if people don't even know where it is?"

Facing page: Boardwalk on the Witness Trail to Spire Lake has nearly a thousand names of sponsors routered into its treads. Photo: WCWC files.

"Most people think that Clayoquot Sound begins at Tofino or Long Beach. They don't know the clearcuts they see after Sutton are Clayoquot clearcuts," I chimed in.

All of a sudden someone said, "We should put up an information booth at the pass with a big banner saying, 'You are entering Clayoquot Sound'!"

"Nuu-chah-nulth First Nations' Territory," added Adriane, completing the wording for the banner.

"We need another banner that says 'Protect Clayoquot's Ancient Temperate Rainforest,'" someone else added.

It was a fabulous idea. If tourists stopped at our booth, we could enlist their support for the campaign. Even for those who just passed by and didn't stop, our Clayoquot message would get into their heads. "Do you think we can get permission from the Nuu-chah-nulth chiefs to put up a temporary kiosk there?" I wondered out loud.

There was no question about it. Only with their permission would we do anything that involved even the smallest physical change in their unceded territory. **This policy of respect for aboriginal people's rights was one of WCWC's founding principles. Our working with First Nations was a prime reason for our success in achieving so much wilderness protection in B.C. so far. If a First Nation wanted to keep industrial activities out of specific areas in its traditional lands to preserve the ecological foundation of its traditional culture, and if it was willing to stand on the line to stop unwanted development, it had the moral and legal clout to succeed.** Adriane, who was already working as our Clayoquot Campaign Coordinator and had earned the respect of many First Nation leaders, said she would seek the necessary permission for the kiosk.

I had another question that needed a yes answer. "Is it legal?"

"It's got to be," said Joe. "People put up stands like that to sell fruit all over B.C."

"Maybe they're breaking the law and the RCMP just aren't enforcing it," I countered.

We knew that if it were illegal, our kiosk would swiftly be shut down. In the eyes of our opponents we weren't peddling something akin to fruits or vegetables. They'd certainly complain to the RCMP and have us removed. If there wasn't a law specifically making it legal... well, we'd call it legal, until proven otherwise. Joe said he would find out. Sue Fox said she would handle making the banners.

We already had a kiosk tent we'd purchased a couple of years earlier and used at several Stein festivals. We had no shortage of posters and thousands of *Clayoquot on the Wild Side* books to sell and lots of free WCWC Clayoquot educational newspapers to give away. We especially needed more signatures on our Clayoquot petition. I really hoped this kiosk plan panned out.

Our roadside Sutton Pass kiosk not illegal, but not exactly legal either

Joe quickly found out that our kiosk idea was in a legally "gray" area. In discussions with the RCMP, he learned that as long as we were far enough off the road so as not to be a hazard, and as long as we did not put up a permanent structure, what we proposed to do was not against the law. If we set up our educational booth on the edge of the road allowance area on crown land, the RCMP would not make us remove it.

However, we had to remove our information kiosk every night to qualify as a "temporary structure." The Sutton Pass Summit site couldn't have been a more perfect location. There was a big pull-off on the right hand side of the road—a large, flat area where tractor trailer trucks often over-nighted and a stand-by snowplow parked

Sign informing people they were entering Nuu-chah-nulth territory beside the highway near our kiosk at the highway entry point to Clayoquot Sound. Photo: WCWC files.

during raging winter storms that brought metres of deep snow and treacherous icy road conditions. We could set up our eight-by-ten-foot, three-sided vinyl kiosk far off the road with enough room for cars to pull off, park and pull out safely back onto the road.

We put casters on the bottom of the corner poles of our kiosk so we could roll it intact down an abandoned logging road and stash it out of sight about 70 metres away from the pull-out. Our volunteers could put up a tent and sleep in a tiny patch of old growth nearby. To top it all off, there was a long embankment that faced east perfectly situated to display our huge welcome banners where they would be highly visible to all passersby heading into Clayoquot Sound and easy to take down at night.

Adriane got permission from the Nuu-chah-nulth. It certainly helped that we had already obtained permission for building the Clayoquot Valley Witness Trail. The Tla-o-quiaht were completely supportive of this effort to gain public support for keeping logging out of this best remaining sockeye river within its territory. They wanted to fully protect this river and didn't want any roads or clearcuts there. They knew from experience elsewhere in the Sound that logging inevitably harms the wild salmon. It probably helped that our latest Clayoquot educational newspaper clearly stated our support for the Nuu-chah-nulth's aboriginal titles and rights to Clayoquot Sound.

Finding the right person to "man" the kiosk

Finally, we had to hire a good person to run our kiosk. It was the toughest of jobs. People were polarized and emotional. Conflict and confrontation were inevitable. Clayoquot loggers, most of them living in Ucluelet and some in Port Alberni were vehemently against everyone who was "trying to take away their jobs." Every day angry loggers would be driving by the kiosk. On the positive side, there would be tens of thousands of potential new preservationists, tourists coming to the west coast seeking summer surf and sand. Many of them wouldn't even know that a "war in the woods" was in progress or even be aware of the fact that they were visiting Clayoquot Sound. Through our efforts, many would be educated, convinced and converted to the cause. That's what our campaigning and our kiosk would be all about.

We could only afford to pay our kiosk manager $1,000 a month plus food and travel expenses. For that, we wanted more than an arm and a leg: 24-hour duty all summer long with only a few days off here and there. It was not the kind of a job that you could advertise through Canada Manpower! It was a job that required someone who passionately believed in the cause of saving Clayoquot Sound and in our educational tactics who had people skills and tenacity. I couldn't think of any person I knew—besides Joe Foy, whose plate was already overflowing and couldn't possibly do it—who could fill these shoes.

I asked everyone on staff if they knew of someone who might be suitable and willing to try it out for a few weeks. Andrea Reimer suggested James Jamieson, a new volunteer who was helping out in our mailroom. James was interested and came in for a job interview with the E-team. He was short and solidly built, sort of like a barrel with a bit of fat that rounded out his face. He looked scruffy, like me, and had a firm air of friendliness and assurance about him. He believed in what we were doing and was eager to help.

During his job interview he demonstrated that he knew the issues backwards and forwards. He had managed the Sierra Club's door-to-door canvass in Victoria for a while. I didn't even ask him about his educational background. It made no difference for this job. The big question was: could he stand the heat? Would he be intimidated by threats from our opponents and quit right away?

James assured us that he was not afraid. He understood that angry loggers would yell at him. I assured him that at least one volunteer would be with him at the kiosk at all times so he would not be alone at the site. James was also positive that he could sell a lot of posters and books and get donations for our Witness Trail work. He was available right away. He was willing to work the entire summer.

"You're hired!" I said to James as we shook hands. "How much money will you need to buy supplies?" We gave him a couple hun-

WCWC Kiosk with our big Save Clayoquot Sound sign behind it. Note the large rock placed in front of it to prevent vehicles from coming too close. Photo: Adriane Carr.

dred dollars from petty cash and off he went with Mark Wareing, who agreed to help James set up the kiosk and stay with him for the first week until we could lineup volunteers to replace him. They left on Tuesday morning with everything they needed, including our old radiophone that worked on marine frequencies. It had come in handy at our Carmanah Valley research station.

We told James to call us after they got set up. Early that evening I got the call. Everything worked out fine except the radiophone: it didn't work at all! Sutton Pass, located in the centre of Vancouver Island, was too far from the ocean. He had to drive nearly all the way back into Port Alberni to get reception. Living in the pre-satellite phone age, we were stuck with having no direct contact with our kiosk. I told James that Adriane and I would bring additional supplies and check out everything on our way to attend a Tofino meeting on the upcoming Saturday; in four days.

Witness Trail project takes off

However puny a defence our Witness Trail was against the juggernaut of industrial forestry, it was enough of an impediment to thwart MB's plans to put in a road and begin logging in Clayoquot Valley that summer... and the next. We hired Glen Hearnes to coordinate our Witness Trail project and sent out an appeal to all our members and supporters to help by donating enough to cover the costs of our trailbuilding campaign. The response was good. A team of eight volunteers started work in the clearcut at the trailhead, clearing a route over a hill to a lake we named "Spire Lake." It was an odd lake, created when a big natural landslide in 1970 dammed the river, leaving hundreds of dead trees standing as silvery spires throughout the new lake.

In an article titled "The Clayoquot Witness Trail - Pathway to Understanding" written for the BC Environmental Network newsletter and published that summer of 1993, Joe put it this way:

Western Canada Wilderness Committee (WCWC) volunteers are constructing a trail into the pristine Clayoquot Valley in order to allow the public to learn for themselves what is at stake in the Clayoquot Sound controversy. People are invited from everywhere, who are interested in learning more to come and hike into the endangered ancient forest of Clayoquot Sound. We believe most people, once they have actually experienced this magnificent area, will work to protect it. The Wilderness Committee is also calling for volunteers to help construct this Wilderness Trail.

Threatened but not driven out — "self funding" works

About the same time we set up our kiosk, the Friends of Clayoquot Sound and Greenpeace opened up a "peace camp." Their camp was located beside Highway 4 in the "Black Hole" clearcut just before the turn-off to Tofino and Ucluelet about 35 kilometres down the road from our kiosk. It attracted hundreds of protesters who began to mobilize for soon-to-start "direct actions" (non-violent civil disobedient protests) against the ongoing clearcut logging in Clayoquot Sound at the nearby Kennedy River Bridge, the major logging road into the heart of Clayoquot Sound. While the threat of blockades didn't slow down the rate of logging, they enraged the loggers who feared that the government would give in and stop the logging. I was a bit apprehensive about the impact of all of this on our Sutton Pass kiosk staff. We hadn't heard from them since the first night when they reported that our radiophone didn't work. I could only assume that everything was OK.

On our way to Tofino that next weekend, Adriane and I talked for hours about the campaign. Along most of Highway 4, there was little evidence of recent logging. The clearcuts, done when the road was built several decades earlier, had "greened-up" and didn't look very alarming. Rows of robust second-growth trees lined one stretch of the highway, the results of a silvicultural experiment in pre-commercial thinning that made the trees grow bigger sooner. It was a clever corporate move to locate these plots right beside the highway to give the false impression that B.C.'s forests were well managed. It made it tougher to mobilize public sympathy for our cause if people couldn't see the devastation that clearcutting caused.

We knew that our Witness Trail project and our educational kiosk were tame stuff, as far as media interest goes, and would garner little publicity compared to the blockades and arrests that would soon be happening down the road. **But most of the essential work in all successful social movements has no glory and little recognition. I always felt our educational efforts and work with First Nations on tourism-type projects were the critical underpinnings for the big shift to a more conservation-minded society.**

We arrived at Sutton Pass at about 11 a.m. With the big banners, the whole set-up looked great. Already we could see that it was working. There were several cars parked with people inside the kiosk talking to James. Before we even came to a complete stop, however, Mark Wareing rushed out to meet us.

"You've got to shut this thing down! Someone is going to get killed! You have no idea of the hostility and threats we are facing!" he shouted at me.

Taken by surprise and angered by his outburst, I shouted back, "No way! We're staying. That's exactly what the loggers want to do—scare us away. We can't let them succeed!"

"You haven't been here! You'll be responsible for getting someone killed! It's not safe!" Mark shouted back.

"We can't wimp away. The bullies would love it and they'd only get worse. Besides, this is Canada. No one will be that violent!" I yelled. But despite what I said, Mark was adamant that it was too dangerous to remain.

"These guys are really violent. They aren't just threatening!" Mark said vehemently.

"Well, if they kill one of us, that person will become a martyr and do more to save the old-growth forest than anyone else alive! WCWC is not going to leave and that's it! You don't have to stay here. If James doesn't want to continue, and no one else will do it, I'll take over. We can't back down cowardly in the face of threats from the loggers." I replied.

Without saying another word, Mark walked away down the abandoned logging road to the their hidden campsite.

I could tell that Mark truly believed that some of the loggers would carry out their threats. After James finished with the customers, Adriane and I talked to him. James took the threatening incidents much more in stride. He was not the kind of person you could intimidate. He recounted several shocking incidents: a pickup truck driving by at high speed, kicking up and spraying gravel at them; a guy throwing a big rock from a passing truck that just missed them; one night finding a huge dump of human excrement on the front seat of our truck parked out of site by their hidden campsite.

But it did not reduce my resolve to keep our kiosk up and running. Nor did these incidents succeed in scaring off James. He liked the job and was more than willing to stay on. James was one of the key people in the effort to "Save Clayoquot."

Both James and I optimistically projected that as the loggers got used to us being there, this kind of behaviour would diminish. Adriane wanted to be more proactive. She went to see the local bosses at MB and asked them to talk to their guys and de-escalate the situation. They were cordial and concerned, but assured her that it was not their company's loggers who were doing it. It had to be contractors' loggers, over whom they had no control.

On the upside, James told us that there was a steady stream of people who supported our cause. They stopped, got information, bought

The Witness Trail passes by "Spire Lake" located on the upper Kennedy River. Wintertime Image. Photo: WCWC files.

our campaign products and donated generously. He was already bringing in, on average, over $125 per day! With money still very tight, we had an unwritten rule that every one of our new campaigns had to be "self-funded." No matter how good an idea or campaign, if it meant generating new debt or jeopardized meeting our payment schedule to reduce our old debt, we couldn't do it. We projected—albeit optimistically—that we would fund the position of kiosk manager through donations and selling T-shirts, books, posters etc. at the kiosk. For once our "self-funding" projections were proving true.

Sue Fox, who designed the display and helped select the merchandise for inside the kiosk, had done a great job. There was a big map of Clayoquot Sound on the back wall. Framed posters decorated the sidewalls and our Save Clayoquot T-shirts were hung by the entryway. There was a big table with our Clayoquot petition, three-part mailers, free newspapers and lots of campaign merchandize including our portable plastic mugs with the Margaret Mead quote about how it took only a small number of dedicated people to change the world. There was a big donation jar filled with bills and coin. It was a regular mini WCWC store. It was a happening place!

Kiosk and Witness Trail key campaign components

Outside, beside the kiosk, James had set up a table with a Coleman stove and a constantly fresh pot of coffee. There were also several lawn chairs to sit on. He poured the coffee, sat down with us and continued to talk about our Witness Trail (which was coming along nicely) and the campaign. He was most excited about relaying the latest gossip from the Black Hole Peace Camp. He reported that the cold 'War in the Woods' was about to become very hot. The first blockades would soon begin. In fact, the first arrests occurred on July 5, 1993 when 15 protestors, including MP Svend Robinson, refused to obey a court injunction to get off the road and let the loggers go to work.

Then he told us of his plans for the kiosk. With the help of volunteers, he was going to keep the kiosk open long hours every day. There was lots of daylight now and some of his best "customers" came late in the day—like people driving up from Vancouver and Victoria after work on Friday for a weekend at Long Beach. The roughest time of the day was late afternoon when the loggers working in Clayoquot passed by on their way home to Port Alberni. Most just rolled down their windows, gave him the finger and swore and yelled. A couple of loggers had pulled over, got out to argue and threatened him.

One thing that James wanted for protection was a camcorder so that they could capture on video the threats that Mark was talking about. He reasoned that pointing a video camera at someone driving by or who was angrily ranting away would act to cool that person

Volunteers building the boardwalk on the Witness Trail in Clayoquot Valley. Photo: WCWC files.

Shelter built on the landing at the end of the logging road where the Witness Trail takes off through the clear cut towards the Clayoquot River Valley. Photo: WCWC files.

down and get him to tone down his rhetoric. It would also document and provide evidence in case something really ugly happened. It was easy to fill that request. We had an old video camera that someone had donated to us a couple of years earlier somewhere in our office. I found it, got some batteries and tapes for it, and sent it out right away. This turned out to be an auspicious move. It was put to good use a few weeks later.

The next week we hired Dai Roberts, an older man who had carpentry skills and lots of practical experience and common sense, to lead a second group of volunteers to build the more difficult sections of the trail. His crew began to build stairs in a steep section using wood gleaned from Gogo's cedar sawmill in Nanaimo. Gogo again generously donated waste wood just like he had for our Carmanah Valley trail. Boardwalking was needed to protect all wetter sections of the delicate forest floor from hiker's boots. Without such protection a well-used trail quickly turned into a quagmire.

George Gibson, chairperson of WCWC's Mid-Island Branch, led teams of volunteers to get the wood for the boardwalk. We called them "wood rats." In the evenings, after Gogo's mill shut down for the day, the "wood rats" combed the mill's huge pile of waste wood and pulled out usable cedar planks for boardwalk treads. We rented a flatbed truck to take the loads of wood from the mill to the trailhead. By the end of July we were feeding and supporting Witness Trail crews of 10 to 30 volunteers. They had constructed one-and-a-half kilometres of cedar boardwalk with stairs and railings to get people up and over the steep hill to our "kitchen camp" nestled in the forest beside Spire Lake.

Drainbows

Most trailbuilding volunteers worked really hard and more than earned the free food we provided. But occasionally we'd get some who wanted to get stoned, meditate or party rather than pound nails, carry wood or clip shrubs. They would stay up, sleep late and take advantage of the free food without doing any significant work. They totally deflated the energy of the rest of the crew. Some were cast-offs from the civil disobedience Peace Camp down the road where, we later learned, they had also taken more than they contributed. Someone coined the word "drainbows" to refer to these malingerers. (I guess it came from the fact that some of them referred to themselves as members of the "Rainbow Tribe"). The trail boss learned quickly to ask those who drained too much of the crew's energy "to move on."

Among the hardest working volunteers were young German tourists who stopped in for a few days or a week or two to help out. They

A group of trailbuilding volunteers at the eastern Clayoquot Witness Trail trailhead near Sutton Pass. Photo: WCWC files.

would get up at dawn and work hard for 10 to 12 hours every day and hardly ever take a break. They could not understand why Canadians were so laid back and did not work as hard as they did.

We had a few close calls but never any serious accidents or incidents. There were wasp stings and one bout with the lower intestinal runs that made the rounds of the camp. It helped that we had strict rules about cleanliness. We also never allowed volunteers to run our chainsaw, the most dangerous tool in our trailbuilding kit. We had some weirdos pass through, though. For example, one of our volunteers who had been acting strangely gave away most of his possessions one day, including his precious *I Ching* book, and set off down the unfinished trail barefoot towards the mouth of the Clayoquot River, never to return. This gave us quite a scare. We sent out search parties, informed the RCMP and alerted our friends in the Peace Camp down the road. A week later, to everyone's relief, he showed up in Tofino. He had no clue that anyone was worried about him.

Mark Wareing puts up paper roadblock to stop logging road construction into Clayoquot Valley

We sent Mark Wareing to lead an exploratory trip deep into the uncharted heart of Clayoquot Valley to scout for the best trail route from the Lower Clayoquot Valley back to a logging spur road in the Kennedy River watershed further to the west. This was the road that MB was proposing to extend into Clayoquot Valley. Although Mark was not so good at handling the rednecks at the kiosk, he was superb at handling the roughest of wilderness bushwhacking. Mark, along with Glen Hearnes who was taking a well-earned break from supervising the trail crew, left on the first week of August and returned a week later. To their great surprise, deep in the heart of this wild rainforest, they came upon an MB crew. They exchanged a few pleasantries and then noticed a tree with some flagging tape on it. After parting ways, Mark and Glen doubled back later and discovered that this crew was not timber cruising. They were engineers laying out and flagging a route for a logging road. Mark and Glen decided to follow the road survey ribbons back up the mountainside. They discovered that the proposed roadway was going to cross a very steep (45-degree) slope right above a salmon-bearing stream.

On his way back to Vancouver, Mark followed up on their discovery. He stopped in at the B.C. Forest Service's District Forest Service office in Port Alberni and found out that MB had made a preliminary application a week before to build that logging road. MB did not yet have final permission to build it, but was "on the fast track" to getting it. Mark quickly wrote a detailed report on what he had found and presented it to the Forest Service. The points he made about slope stability and the potential for increased runoff, landslides and siltation that could smother salmon eggs in the stream below the proposed road were legitimate concerns that had to be addressed before MB could get government approval. Mark forwarded copies of his report to federal fisheries officials as well as to MB and the media.

MB's accelerated logging plans kicked us into higher gear. We immediately hired a second trail boss and pulled together a crew to start clearing the Witness Trail at the end where MB planned to build their road. We routed the lower end of our trail right beside the proposed logging road. It wouldn't interfere with MB's planned road, so the company couldn't object. Of course MB didn't like the fact that it spotlighted the real threat. As people hiked up or down the steep sloped trail they could vividly imagine how damaging a logging road would be in the stunningly beautiful Clayoquot Valley.

Without a doubt, Mark's exposé of the potential hazards of MB's proposed road delayed MB's invasion of Clayoquot Valley. With blockades and arrests a daily occurrence at the Kennedy River Bridge, it was unlikely that MB would want to open another "front" in the Clayoquot war. They knew efforts to build a road into Clayoquot Valley would be confronted by blockaders from the Black Hole Peace Camp who would likely be joined by Tla-o-quiaht Band members. The Tla-o-quiaht had outspokenly expressed opposition to any roads or clearcuts and resolved to protect the Clayoquot Valley. It was their last pristine salmon-rich valley.

Evidence of murrelet nesting found

During the same week that Mark and Glen discovered MB's proposed road, a team of researchers from Germany hiked down the completed eastern section of the Witness Trail to Norgard Lake in the heart of the Clayoquot Valley. Here they planned to set up a base camp from which they could conduct rainforest plant studies. En route, one of the researchers noticed a two-centimetre-by-four-centimetre eggshell fragment on the trail. Bird expert Irene Manley later determined that this piece of eggshell was from a marbled murrelet egg. Marbled murrelets are a remarkable sea bird that nests on the mossy limbs of old-growth trees. Because so much of their nesting habitat had been destroyed, their numbers had greatly declined and they were listed as a threatened species. A member of the German research team climbed several 70-metre-tall trees near where they found the eggshell in a failed attempt to find a nest.

Marbled murrelet nests are notoriously hard to find. In fact, as of 1993, only five marbled murrelet nests had been discovered in B.C. A WCWC research team headed up by John Kelson and Irene Manley found the first one on the moss pad of a big tree limb 46 metres up the trunk of an old-growth Sitka spruce tree in the Walbran Valley a couple of years earlier. The most remarkable thing about this discovery in the Clayoquot River watershed was the distance of the nest from the ocean, over 20 kilometres as the murrelet flies. Several times every day the murrelet chick's parents would fly back and forth to the ocean, catching fish and bringing them back to feed their hungry nestling. That is an incredibly long way to go. This and other discoveries helped build our case for the preservation of Clayoquot Valley.

The best discovery to potentially block logging would be to find cedar trees that were culturally modified centuries ago. Such evidence of "traditional use and occupancy" was the cornerstone of the successful counter injunction that was keeping Meares Island from being logged. The best place to look for such trees was in the lower valley where the upper Clayoquot River emptied into Clayoquot Lake. A portion of this area was protected in Harcourt's April 13 Clayoquot land use decision. Further up the valley, beside this new protected area was a near-sea-level red cedar forest that would have been readily accessible to Nuu-chah-nulth peoples in pre-contact times. We contracted Steve Lawson and Susan Hare to look for CMTs (culturally-modified trees, in this case cedars) there. They got to the area first via the ocean on their herring skiff, then by hiking along the lower river to a cabin that had been built on Clayoquot Lake by a new research group called the Clayoquot Biosphere Project, then by canoeing up the lake to its head. There they began their search for CMTs. Steve strained his back during the expedition, which limited the extent of their explorations. Despite this hindrance, they found quite a few CMTs. But we were unsure as to whether or not they were located far enough up the valley to be beyond the small protected area the government established in their April 13th announcement.

Assault at our Sutton Pass Kiosk caught on Video

All the while, harassment at our kiosk at Sutton Pass did not abate. During the first few weeks it was particularly intense. Several times vehicles pulled into the lot, turned around so the back end of their vehicle faced the kiosk tent entrance and backed up close to it. Then the driver revved his engine up full bore and popped the clutch. The spinning rear wheels pelted the kiosk with gravel flung at high speed. Fortunately, no one was ever hurt.

One afternoon, during the third week of being there, a couple of guys pulled up and got out of their truck. They were very abusive. One of our volunteers pulled out our video camera and started shooting. This angered one of these guys so much that he put his hand over the camera and gave a big push, swearing, "Shut that f~ing thing off...." They then proceeded to get back into their truck, which was a MB company truck. The volunteer's videotape of the man and his hand being thrust in front of the camera clearly showed the MB logo on the door of the pickup as they drove off, proving that these were not contractors working for MB, but actual MB employees. Immediately after receiving the phone call from our volunteer about this incident, Joe, Adriane and I got together and discussed our options. We dismissed the idea of going to the RCMP. It wasn't really that much of an assault. We could go to the media with the story but we felt that this might just inflame the situation out there. We finally decided the best thing to do was to call the head of MB's Kennedy Lake Division and tell him what we had on video. We'd inform him that we weren't going to go to the media with this tape because we trusted that MB's management would handle it by talking to their loggers and making sure that their men would not bother us in the future. But if any kind of incident like this happened again, we'd pull out all the stops.

I was elected to make the call. I made it early the following morning. When I told the receptionist who I was, I got through right away to their head person. I explained to him exactly what had happened and that we did not want to make a big thing of this. I told him what we really wanted was an end to further harassment. He was very polite and understood the seriousness of the matter. He assured me that he would talk to all their employees and make sure that we were no longer bothered. At the end of the conversation he asked, "Is there anything else I can do for you?"

Impulsive request instantly granted

"Yes," I replied, shooting spontaneously from the lip with a bold request, "We would like a Special Use Permit to take some of the waste cedar left in the clearcuts near our trail project to use for stringers and treads on our boardwalk. The guys pulling out cedar shakes have already been there and salvaged all the good stuff and we would just be taking what has been left behind. We would be exceedingly careful not to trample any of the seedling trees growing back,"

To my surprise he replied, "Yes, we can do that. I'll fax a permit over to you right away. What's your fax number?"

I gave it to him, thanked him and hung up. I couldn't believe this fortunate turn of events. It transformed the loggers' negative energy into a bountiful windfall for us. We really needed to use that waste wood. There was so much of it; thousands of splintered cedar logs and hundreds of discarded blocks of cedar left behind by the cedar salvage contractors because there were too many knots to make roofing shakes and shingles. The leftover wood was perfect, however, to

Volunteers packing in the cedar treads along the Witness Trail for the boardwalk being built about a kilometre away. Photo: WCWC files.

split up for thick treads on our boardwalk. In the weeks before we got this permit, we had resisted the temptation to use any of this waste wood. Despite how tempted we were, especially knowing this wood had absolutely no commercial value and had been left behind to rot, by the terms of MB's TFL agreement, it all belonged to MB. If they caught us stealing any of it, they would have grounds for shutting down our trail project.

Two hours later, the promised permit *"to salvage cedar for your project"* rolled out of our fax machine. The signed permit specified exactly which cutblocks we could enter to get the waste wood and clearly stated that we must not harm or damage the new forest regenerating in these clearcuts while we were retrieving the waste wood. We danced around the office in glee knowing how much of a godsend this was and how great a difference it would make in our trail project, especially in saving us lots of money.

From that day on, until we permanently closed our kiosk, MB employees no longer hassled our staff and volunteers there. We found out through the grapevine that MB management had laid down the law. If any of their employees did anything to threaten or intimidate Clayoquot logging protesters, they would immediately be "sent down the road." MB had zero tolerance for such violent behaviour from its employees.

Harassment continues – Joe catches a thief

But the efforts to drive us away from our Sutton Pass kiosk stand did not end. The worst incidents involved contract loggers. One week

Two Witness Trail volunteers carry a cedar stringer for the board walk salvaged from MB's cut block. Photo: WCWC files.

Volunteers splitting cedar treads from "defective" cedar blocks left as waste wood in the clearcut which we could now legally salvage under our permit from MB. Photo: WCWC files.

when Joe was on duty at our kiosk with James, relaxing in front of the kiosk on a lawn chair, he heard a vehicle approaching up the winding road from the west. Its engine was revving like a drag race car preparing to accelerate to the max. Suddenly a pickup truck careened around the corner, crossed the centre line into the oncoming lane, and kept going onto the gravel area towards where Joe was sitting. The driver locked up his brakes and did a sideways skid, stopping only a few metres from Joe. Then he fishtailed back onto the road and disappeared around the next corner.

For many days one log truck driver would delight in slowing to a crawl every time he passed the kiosk with a new load of old-growth logs. He held a baseball bat out his side window and beat it slowly on the side of his door as if to say, "You're next." One day as Joe, James and the volunteers were all speaking to a group of tourists who had stopped at the kiosk, a logging truck pulled over and parked near by. It wasn't uncommon for large trucks to pull over there to cool their engine and check their brakes, so after an initial glance, Joe and James did not keep a watchful eye on it.

When the truck's diesel engine kicked to life, it seemed as if the driver was pushing the truck harder than normal, jerking quickly through the gears. As the truck picked up speed, they could see to their horror that the driver had tied a rope from the back of the truck to our 15-metre-long "Protect Clayoquot's Ancient Temperate Rainforest" banner. With a pop, pop, pop all the ropes holding the banner in place broke. In a flash the truck was headed towards the west coast with our banner flapping behind it.

Just before the truck disappeared around the bend, the banner whipped across the windshield of a small car coming the other way, blinding the driver for an instant. Thank goodness it snapped back again. The car's occupants must have been scared witless. Joe called us from a pay phone in Port Alberni to relay the bad news. We dug into our meager funds and had another banner made right away.

On the following day, another loaded logging truck drove slowly past the kiosk. At the rear, nailed to a butt end of one of the old-growth logs was another one of our signs! Joe hopped into his van in hot pursuit and followed him down the highway towards Port Alberni.

Not wanting to lose him, Joe did not stop at the first phone on the outskirts of the town, but instead kept trying to reach the telephone operator using our marine band radio. Earlier, he'd rigged up an antenna on the roof racks of his van and used his official call-sign, which was "Clayoquot Defender." Somehow he actually got through to an operator, while driving down the road and soon he was put through to Adriane at our office and told her the story. She contacted the RCMP in Port Alberni while Joe continued to follow the logging truck through town and out, over the pass and through Cathedral Grove. The RCMP intercepted the truck near Coombs. Joe, who was still right behind the truck, stopped and explained the situation to the officer.

The culprit, a contract logger, got charged. He pleaded guilty and was fined $100. He wrote a letter to us offering an apology for his rash behaviour, which we accepted. The upshot was that, from then on, logging truck drivers never stole another one of our signs or banners from Sutton Pass.

Harassment continues – someone tries to stink us out

Next, we faced the onslaught of someone trying to stink us out. Mysteriously, late at night somebody dumped water with rotting fish guts all over the kiosk site. The foul odour was horrendous. Scraping off the fish remains and the surface gravel did little to stem the stench. I figured that the only thing that might work, besides a heavy rainfall to wash it away, was to sprinkle a lot of powdered quicklime over the area. James went into Port Alberni and got some. It worked... to a degree.

At least once per week, always sometime in the middle of the night, more rotting fish offal was dumped and James and the volunteers scraped it away and covered it with more lime. When I stopped in to visit in late August I could hardly stand the smell, but James did not even seem to notice it. He had obviously become habituated to it. I asked him if the odour affected sales or donations. "Only positively," he quipped, telling me that when he revealed the origin of the terrible smell, people felt sorry for him and gave even more money to our cause!

Like the harassment, the weather at Sutton Pass was something fierce. Regularly, strong winds would blow up in the late afternoon and evening, kicking up volumes of dust. It got very cold at night, but our valiant crew persevered. We didn't keep track of the exact numbers, but figured we educated thousands of people that summer. Virtually everyone who stopped at our kiosk and talked to James and our volunteers became part of the growing tide of public opinion to save all of Clayoquot Sound.

Robert Kennedy Jr. and Chief Francis Frank visit the Witness Trail

On July 31, Chief Francis Frank of the Tla-o-quiaht First Nation brought Natural Resources Defense Council (NRDC) attorney Robert Kennedy Jr. of the famous U.S. Kennedy clan to the Clayoquot Witness Trail. A host of Nuu-chah-nulth chiefs and elders accompanied Chief Frank and "Bobby," along with an entourage of media. Adriane and I spent that evening talking to Bobby Kennedy and answering the dozens of questions he peppered at us. He and the NRDC made protection of the Clayoquot rainforest and justice for the aboriginal peoples living there one of NRDC's top priority campaigns. In the following years Bobby Kennedy and Liz Barrett-Brown, another lawyer working for the NRDC, played key roles in helping resolve the Clayoquot conflict in favour of more preservation and greater control for First Nations.

Holly Arntzen entertains in front of WCWC Sutton Pass Kiosk. Photo: Adriane Carr.

Donors immortalized in the Witness Trail boardwalk

The Witness Trail construction was a costly project. In a major appeal that spring of '93 we had asked our members and supporters to sponsor a trailbuilding volunteer. If they donated $50 or more we promised to carve their name into a plank on the boardwalk. Several

Names of financial supporters of the Witness Trail project routed into the boardwalk treads. Photo: WCWC files.

months of trailbuilding elapsed and we still had not started work on the name-carving. Several donors had come by to see their name and, of course, found nothing. Finally, in early August, John Gellard, a man with the right skills, volunteered to do it. He spent weeks carving names into planks, using a router and a portable generator. When he started there was a backlog of over 800 people who had contributed over $50. By the middle of August, John had carved the names of 500 of them into our boardwalk.

By the end of the summer Glen and his team of volunteers had cleared about 12 kilometres of trail deep into the heart of the Clayoquot River Valley. When we completed the trail the following summer it turned out to be 20 kilometres long. Andrea Reimer and her boyfriend rolled a bicycle wheel device with a counter on it to measure the distance between every camping spot and place of note for a guide map to the Witness Trail that we published that fall. Before we completed the trail, hundreds came, hiked and witnessed part of one of the largest expanses of lowland temperate rainforest left on Earth.

Peace Camp and then our Kiosk close

All that summer of 1993 our Sutton Pass kiosk was a busy place. Hundreds of tourists stopped by every day. We'd recruit trailbuilding volunteers and encourage people to sign our "save Clayoquot Sound" petition and donate. Activists stopped to get the latest news on the Clayoquot campaign. Even the occasional reasonable forestry worker stopped to get our side of the story. The donations and sales revenues covered everything including most of the food for the trailbuilding crew. Trail crew workers who wanted a break often spent time volunteering at the kiosk. It all worked out incredibly well.

In mid September the weather deteriorated into torrential rain and the tourists diminished to a trickle. We finally shut down...but only until the following late spring. The Friends of Clayoquot and Greenpeace had already closed their "Black Hole" peace camp a couple of weeks earlier. Summer is the time of protests. The weather is just too foul in the fall and winter to do it effectively. Although there were many arrests—more than any other protest in Canadian history, nearly 1,000—there was no sign that the government was going to give in and preserve even one more hectare of Clayoquot.

Craig Delahunt, whom we contracted to make a video about our Witness Trail, understandably got caught up in documenting the largest civil disobedience campaign in Canadian history, occurring just down the road from us. It was a story too big to resist! He incorporated a bit of footage about our trail and kiosk, a lot of footage of the protestors' blockades and arrests, as well as some footage about the Clayoquot Biosphere Reserve project sponsored by Ecotrust into a one-hour documentary titled *Clayoquot Summer*. We made 500 copies of the video to sell to our members, supporters, libraries, and the general public through our 1994 Wilderness Catalogue.

We sent out complimentary broadcast quality tapes to over 30 cable stations in Canada and the U.S. for their use free of charge. Because this project was paid for out of our Bullitt grant, we kept good track of the results for our annual report to this foundation. Our video was, at the time, the only comprehensive explanation of the summer of '93 events at Clayoquot Sound. It was aired more than 130 times on 25 public TV stations; most of them were on Vancouver Island and in Greater Vancouver. In addition it was also shown in Toronto and Calgary. Major TV stations used segments of our Clayoquot Summer footage in many newscasts. Our *Clayoquot Summer* video was nominated for five different international awards, but never won.

Trees along Witness Trail. Image used in Clayoquot poster. Photo: WCWC files.

Chapter 44

Stumpy's extraction
Stumping to Ottawa seeking a Clayoquot Biosphere promise

Morphing a "big log tour" into a "big stump tour"

With our kiosk at Sutton Pass closing and Clayoquot Valley trailbuilding over for the season, we had to do something to maintain our Clayoquot campaign's momentum until we resumed our on-the-ground activities in the spring. One option was to get a big log from a thousand-year-old tree in Clayoquot Sound and tour it across Canada. We'd floated this idea in our *1992 Members Report* newspaper (along with a proposal that never went anywhere to build an ancient forest museum in Gastown to bring the virtual experience of an old-growth forest to people who would probably never go to see a real one).

Our old-growth log tour, patterned after one successfully done in the U.S. several years earlier, failed to get off the ground for several reasons. Big logs were expensive. It would cost us $10,000 or more to buy one…if we could find one for sale. Since two big companies "owned" almost all the trees in Clayoquot and logged them to feed their own mills or sell them as raw logs to foreign mills, getting one for a pro-preservation tour was near impossible). Even if we were able to purchase a log, we'd then have to lease a big semi truck with a flat bed trailer and hire a driver, which we couldn't afford, either.

Then our Carmanah poster slogan popped up in my mind, "Big Stumps not Big Trees! That's it! We'll tour a big stump, instead of a big trunk." We can haul a stump around on a small flatbed trailer. There is no shortage of stumps in Clayoquot. We can salvage one!

The wording of the special use permit that MB had granted us a couple of months earlier made salvaging a stump for our tour possible. Undoubtedly because MB did not want to specifically acknowledge our trailbuilding activities in their TFL, the wording of the permit stated that we could take waste wood from the two of MB's clearcut blocks *for the purposes of your project*. A stump is a kind of wood waste; we'd be taking it for our project to raise money to complete our Clayoquot Valley Witness Trail, which was part of our bigger project to save Clayoquot's remaining ancient temperate rainforest! Best of all, the stump wouldn't cost us a penny.

I enthusiastically talked up the stump tour idea with all the staff at WCWC. After dispelling the usual worries of how much it would cost, everyone thought it was a great idea. Everyone, that is, except my good friend Nik Cuff who quipped, "Who in hell wants to see an old stump?"

"Lots of people," I replied. "It's the novelty of it. You'll see!"

At our E-team meeting that week we decided to go ahead if we could find a suitable stump. Our permit also stated clearly that we couldn't damage any of the young seedlings during our salvage operations. That meant the stump would have to be situated right beside the road with no seedlings growing near it. Since our permit expired on December 31, 1993, we had to act quickly. It would be next to impossible to salvage one after mid-November when the snow started accumulating in Sutton Pass.

I dreamed of getting a three-metres-in-diameter stump from a giant tree that was over a thousand years old when cut down. I could vividly imagine the monstrous stump's huge impact on the public. Fantasizing the fantastic eventually got me thinking about what was practical and within our sphere of the attainable. I figured out that even if such a huge cedar stump existed in our salvage area, it would be impossible to take it on the road. Even with its rootwad cut way down, the stump would still be over seven metres wide at the base and weigh tens of tons. We'd have to hire a huge flatbed tractor-trailer, a driver, and a pilot car with a flashing "WIDE LOAD" sign. It would be impressive, but prohibitively expensive—costing even more than a "big log" tour would.

"Baby" red cedar stump found

We had to settle on a "baby" stump, relatively speaking, a small one that would fit on an eight-foot wide flatbed trailer, the widest trailer we could legally tow without special permits. It had to be big enough to be impressive, but not so big as to be unmanageable. We sent Mark Wareing and a volunteer on a "wild stump chase" into the clearcuts along the logging road leading to our Witness Trail trailhead; the area covered by our salvage special use permit. There wasn't a whole lot of choice.

Only one stump came close to fitting the bill. The more we looked at it, the more we realized it was perfect. We immediately nicknamed it "Stumpy," and got to work extracting it.

Joe couldn't find a flatbed trailer, new or used, that we could rent or purchase that matched our specs. But he did find an outfit that could build us one for around $2,000. They'd have the trailer completed in less than 10 days if we put down a substantial deposit to get them started.

By this time, everyone at the office had caught "Stumpy fever." Reasonable expenses were not going to stop us. Wearing our familiar rose-coloured glasses, we projected that by fundraising on the tour we would easily make up the two grand spent on the trailer, cover the stump's extraction costs and all the tour expenses. We'd bank the rest for our continued Clayoquot campaign!

Volunteers extracting Stumpy aided by old-timer with construction know-how

We had to hurry. We wanted to get Stumpy to Ottawa before the end of the federal election which had just got underway. Our goal was to use this icon to help extract a promise from Prime Minister Jean Chrétien that, if re-elected, he would help protect Clayoquot and support UN Biosphere Reserve status for the area. We asked Dai Roberts, the handyman who, a few months earlier, had supervised the building of the difficult boardwalk and stairs sections on our Clayoquot Valley Witness Trail if he would, with the help of volunteers, extract the stump and load it on our trailer. It was a challenge he couldn't refuse.

Our stump was situated on top of a big boulder about two metres back from the road and a metre higher than the roadbed. With a chainsaw, jacks, shovels and several come-a-longs, our crew of eight volunteers worked from dawn to dusk for four days straight to carefully cut the roots around Stumpy, shovel out the dirt, get the jacks under it and lift it free. It took another day to move it slowly and easily, metre-by-metre, onto the roadway, jack it up higher and get the trailer underneath it.

The stump's location—a long way up an abandoned dead end logging road—made detection by the opposition unlikely. We all prayed that no MB employee would discover us nabbing Stumpy. If caught in the act, MB might very have tried to stop us. Luckily, no one came by to put our legal interpretation of MB's special use permit to the test!

Above: Easing Stumpy onto the waiting trailer. Photo: WCWC files.
Below: Clowning for the camera before Stumpy is eased onto the trailer. Rob, John, Mona, Jack. Dai Roberts centre, Andrew Kotaska right. Photo: WCWC files.

Stumpy tour launched in front of the B.C. Legislature

We planned to launch our stump tour to Ottawa from the most obvious place, the steps of the B.C. Legislature. Because we chose a Sunday, we had no problem getting a permit to drive up right in front of the main entrance and park our flatbed trailer there for the afternoon.

We thought that we had allowed enough lead-time to get Stumpy ready to roll when we got our special event permit several weeks before the planned launch. But we hadn't. The crew finally managed to lower Stumpy onto the trailer in the late afternoon on the day before the rally. As it settled down on the flatbed, it became painfully obvious we had a problem. The bulging tires provided mute evidence that we'd greatly exceeded the trailer's load limit. It was hard to believe. Red cedar wood is known to be a quite light, low-density wood.

We discovered, to our surprise, that wet red cedar stump wood isn't light at all. It had to be one of the heaviest softwoods in the world! Since we had not yet lined up a suitable towing vehicle for the trip to Ottawa, we had no choice but to use our old canvass van to tow Stumpy to the Legislature. It was a risky thing to do, but a "no show" wasn't an option. So Joe headed off to Victoria, less than two hundred and fifty kilometres away, late that afternoon with the heavy stump load in tow.

Before the crew left the extraction site, they carefully replaced all the soil and gravel they had removed and put back moss over the rock where the stump once stood. Looking at the spot after they'd finished, it was hard to envision that, only five days earlier a stump had sat there, and five years before that a young and thriving 380-year-old western red cedar grew, 1.8 metres in diameter at breast height, in the midst of an old-growth forest.

"Preaching" beside Stumpy at the B.C. Legislature of the Stumpy tour to Ottawa to seek government support for Biosphere Reserve status for Clayoquot Sound. Photo: Joe Foy.

The first thing Joe did when I met with him before the rally the next day in Victoria was vividly describe his hair-raising trip. Taking it very slowly, he traveled well into the night. He was barely able to make it to the summit of the Malahat (a small mountain near Victoria) in low gear. Then, he was barely able to keep from running away as he crept down the other side. The trailer's brakes were either inadequate or not working properly. The big problem was Stumpy's excessive weight. We both agreed that we had to do something about it! We couldn't tow Stumpy across the country the way it was.

Only about 50 people came to our launching rally. Misty MacDuffee, Adriane, Joe, several others and I all got a chance to stand on the bed of the trailer, put our hand on Stumpy and pontificate. It was preaching to the converted, but fun nonetheless. Thanks

Stumpy on the last stop in Vancouver before going on the road to Ottawa. Photo: WCWC files.

to the sheer determination and hard work of our motley crew, we'd succeeded. Now we all shared the vision that Stumpy was going to do yeoman's service in helping save Clayoquot Sound's ancient temperate rainforests.

Right off the bat, a clever forest industry PR hack trying to put us down characterized our campaign as a "publicity stump." When a reporter asked me if that was true, I said, "Of course it's true! We're stumping Canada to get as much publicity as possible. We're letting everyone know that MB is still cutting down the big old-growth trees in Clayoquot Sound and asking them to help us stop them."

At the rally one of the volunteers who had helped extract the stump, Dr. Andrew Kotaska, made an unexpected announcement. He and his wife were "putting their life on hold" so they could drive the stump to Ottawa. Andrew had become involved only a month earlier as a volunteer helping build our Witness Trail. Realizing that not having a decent tow vehicle for the stump trailer was a big weak link in our campaign, Andrew sold his car and borrowed the money to buy a brand new four-wheel drive vehicle big enough to tow it. It was fabulous news! He and his wife's generosity made this first leg of our Stumpy tour a success. Andrew was a reasonable person, not against all logging. In fact, he had paid his way through medical school working in a sawmill. On top of the money, time and energy, he contributed credibility to our Stumpy tour too.

Stumpy gets a "root canal"

Joe had a great idea about how to lighten up Stumpy. "Let's give it a root canal," he said, suggesting that we flip Stumpy on its side and get Teri Dawe, our eco-logger friend, to chainsaw out the inside wood on the underside of it. Teri agreed to do the job as long as he got no publicity. He had to continue to make a living in the forest industry and that meant getting occasional contracts from the large companies. He was rightfully fearful that he would be blackballed if the "timber barons" knew he was assisting us.

The Stumpy team found an equipment yard in Vancouver with a Hi-Ab crane truck that, for a small fee, lifted the stump off the trailer, tipped it on its side and let Teri work on it. Teri fired-up one of his big chainsaws with a metre-long bar and did the "dental work" on Stumpy. It took him several hours of precise carving to complete the job. Mark Wareing carefully weighed the chunks and chips as Teri cut them out. A huge pile of material accumulated on the ground beside the stump weighing nearly two thousand kilograms—one third of Stumpy's original weight.

"Root canal" completed, the Hi-Ab plunked Stumpy back onto the trailer and our crew chained it down. Looking at it, you'd never know it was hollow inside. From that day on, although still a mighty heavy load (4,000 kilograms), it towed much easier, reducing tire wear and gas consumption.

Ignoring our press releases, the *Vancouver Sun* did not report our Stumpy tour launch or anything about Stumpy at all. Early in the morning after the root canal—on the tour's round of Vancouver before heading to Ottawa, we parked Stumpy it in front of the *Sun* and *Province* press building, which was then situated at 6th and Granville. We marched around the front of the building carrying signs protesting the poor coverage of environmental issues in these two major papers. The happy result: a picture of Stumpy in the *Sun*. The next day we parked it in front of the CBC building, and on the following day in front of the Vancouver Art Gallery where it stayed all day and late into the evening for people to gawk at and ask questions. On the way out of town the next day Stumpy protested at a wharf in Vancouver where raw logs were being loaded on a ship for export. We videoed the event, acting as our own media, thinking we'd probably use the footage sometime in the future.

Misty MacDuffee, who had started volunteering for WCWC-Victoria in 1991, took charge of the Stumpy campaign. She recognized its potential as a "market awareness tool" to inspire customers of MB's paper products to tell MB that they would take their business elsewhere if MB did not change how and where it logged. This tactic proved to be hugely effective in achieving greater protection for Clayoquot Sound, despite the B.C. government's refusal to budge from its pro-logging decision. Misty presented slideshows at events organized by local environmentalists in all the major cities along the way. Her slides and talk transformed people who knew nothing about the issue into dedicated Clayoquot Sound preservationists.

Everywhere Stumpy went it got publicity. Photos of Stumpy appeared in nearly every newspaper along with its *Save Clayoquot Sound*, *Stop Clearcutting & Save Jobs* and *Clayoquot Sound NOT Clearcut Sound* banners. In Edmonton our WCWC-Alberta Branch organized a big rally and used it as a prop to promote protection for Canada's boreal forest. The further east Stumpy went, the more impressed people were with its size. Even the largest of the original giant red and white pines still left in Ontario (almost 1.4 metres in breast-height diameter) were smaller in diameter than our baby red cedar stump's trunk had been.

Along the way to Ottawa, Stumpy visited 33 towns, 22 university campuses, 20 schools and 18 malls. It was featured in the media everywhere. If the media didn't cover our Stump tour of its own volition, our stump tour crew brought the star of the show to them, parking Stumpy outside TV stations and newspaper offices. It was parked in front of five legislature buildings of every province it passed through on the way to Ottawa. We have photos and news clippings in WCWC's archives to prove it.

Stumpy makes it to Parliament Hill - Liberals promise to protect Clayoquot Sound

Our Stumpy caravan arrived in Ottawa on Friday, October 22, 1993—three days before the federal election. An *Ottawa Citizen* newspaper photographer took a great picture of Stumpy with the House of Commons' clock tower in the background which was featured in next day's (Saturday) edition of the paper with a short write up about our cause—saving Clayoquot Sound.

Unfortunately, Stumpy failed to make its debut at the big Clayoquot rally on "The Hill" the day before the election. That morning the trailer broke an axle just five blocks away from the Parliament Buildings and the crew could not get it fixed in time. Fortunately, Stumpy had

Stumpy rests at the outskirts of Ottawa on its way to the rally. Photo: WCWC files.

Chantelle and Monalisa posing with our Stumpy T-shirt sold through our catalogue and in our store. Photo: WCWC files.

already gotten the needed publicity when they rolled into town two days before. We had to consider ourselves very fortunate that the axle didn't break when they were rolling along at 80-kilometres-an-hour down the Trans-Canada Highway on the way there!

Stumpy's journey to Ottawa and the ever-increasing numbers of people getting arrested on the logging road in Clayoquot Sound had a major impact on the federal Liberals. They were aiming to take over the reigns of power from the Progressive Conservatives who had imploded in the last few months under the leadership of Brian Mulroney and then Kim Campbell. Aiming to pick up green-leaning voters, the Liberals' environment critics, Charles Caccia, Marlene Catterall and Paul Martin, issued a media release committing the Liberals to pursue protection for Clayoquot Sound. We were ecstatic. On October 25th, the Liberals won the federal election. The prospects for Clayoquot's ancient rainforest brightened.

B.C. government creates a blue ribbon panel to recommend how to best log Clayoquot

Undoubtedly feeling the pressure of the protests, on the same Friday that Stumpy reached Ottawa, the B.C. government announced the establishment of a *Scientific Panel for Sustainable Forest Practices in Clayoquot Sound* (Clayoquot Sound Science Panel). The government mandated the scientists and First Nations representatives it appointed to this panel to review the existing forest management standards, recommend changes to them appropriate to the ecological conditions of Clayoquot Sound and develop *world-class standards for sustainable forest management* for the area by combining traditional and scientific knowledge. The Panel was not allowed to recommend preserving more forest in Clayoquot, which gave us little hope that it would help solve the conflict. But we were curious about how the First Nations and forestry members of the panel would mesh. We'd heard from Keith Moore, a friend of mine from the Queen Charlotte Islands, that when one First Nations elder explained how they viewed the forest from a holistic seven-generation perspective it radically changed the thinking of the whole group.

The Clayoquot Scientific Panel didn't table its final report until July of '95, two years after it started work. In a move lauded by the environmental community, the B.C. government accepted all of its 128 recommendations. These included a moratorium on logging in pristine watersheds so inventories of all forest values could be conducted (a godsend for us preservationists) and the introduction of "variable retention" logging where patches of old-growth trees would be left standing in the clearcuts. B.C.'s Environment Minister exclaimed to the media that the adoption of variable retention logging was "the end to clearcutting in Clayoquot."

The coastal forest industry quickly realized the potential of "variable retention" as a newspeak concept. Soon most coastal logging companies were no longer clearcutting, a practice the public had come to revile. Now they were doing more ecologically-friendly logging, leaving patches of scrub old-growth trees here and there in their "variable retention cut blocks." The trees they "retained" were almost invariably worthless from a corporate sales point of view. So, it ended up that the companies were looking better media-wise and looking better, too, in their financial bottom lines.

Chapter 45

Stumpy goes to Europe
Stumpy tours the States
Stumpy lost in Montreal

Back home we finally publish *A Conservation Vision for Vancouver Island* and give 3000,000 away

We were busier than ever in the fall of 1993. The Vancouver Island CORE table, finally getting down to business, was on a self-imposed tight timetable. Although WCWC had stopped participating in CORE ever since the April 13th Clayoquot decision, we still wanted to spur the CORE table on to make the best recommendation it could.

To this end, we had been working almost two years on a Vancouver Island land use *Conservation Vision* newspaper. We finally went to press with it in October, printed 300,000—tied for our largest press run with *Carmanah – Canadian Rainforest Deserves Protection*. I believe

Conservation Vision newspaper cover. Photo: WCWC files.

it was one of WCWC's best papers ever, with lots of information packed into its eight pages. It featured two very cutting cartoons especially drawn for it. *Stephen in Plunderland* depicted Stephen Owen, Chair of the CORE process, as the Mad Hatter at an Alice in Wonderland-like tea party. He was cutting and handing out pieces of the Vancouver Island pie; 88 percent going to the big logging companies and the leftovers going to all the others around the CORE table—mostly cute caricaturized animals. The second cartoon was titled *The Faceless Ones* and depicted a faceless forest company CEO leading the *Share BC Boys* into thinking that environmentalists were their biggest enemy. The centre two pages of the paper featured our scientifically researched *Conservation Vision* map.

It was a zinger, showing a utopian land use plan with an interconnected system of protected areas that included restoration areas where the forest had already been cut, but would never be cut again, where there wasn't the needed old-growth forest left to protect.

Facing page: Cartoon lampooning the Vancouver Island CORE process deciding wilderness preservation on Vancouver Island. Artwork: Geoff Olson.

Applying scientific methods and conservation biology principles, we determined that to sustain Vancouver Island's natural biodiversity would require protecting about 40 percent of the Island's land base. This was science, not politics, talking. It was a far cry from the 12 percent parks goal fixated upon by the Harcourt government.

We had a really tight deadline to get this paper out. The Vancouver Island CORE table was still seeking public input into its process, but the end date was fast approaching. We had to get our paper from the printing press plant in Vancouver to a distributor's warehouse in Nanaimo that evening so the papers could be inserted into all the weekly freely distributed newspapers on Vancouver Island. This would reach over 200,000 families, all the households on Vancouver Island. I called around and found out that it was going to cost a small fortune—more than $500—to rush courier them there in time. We couldn't afford that so I rented the biggest moving truck I could get and took them over myself, taking the last ferry at night to avoid traffic. I delivered the skids of papers around midnight, slept in the truck and brought it home safely the next morning saving over $350.

An Official Petition sums up our goals

On the top half of the back page of this *Vision* newspaper we included, under the headline *Let's make Vancouver Island's temperate rainforest work for local residents not multinational corporations*, a petition titled *Official Petition to the British Columbia Provincial government asking for an end to clearcut logging, reform of the forest tenure system, a just land settlement with Native Nations, and full protection for natural biodiversity on Vancouver Island*. At the bottom of it there was room for only five signatures. We requested that people fill them and send them back to us so we could get them presented to the spring session of the B.C. Legislature. We made an extra effort to get the wording right.

> *To the Legislative Assembly of the Province of British Columbia in Legislature Assembled this petition of British Columbia voters and other concerned citizens humbly showeth:*
>
> **WHEREAS** *the current forest land use system of Vancouver Island clearly doesn't work, with the ancient forests being cut at an unsustainable rate, with less and less forest-based employment every year, with natural forest ecosystems being damaged by current logging practices, with soil erosion and salmon stream damage, and with long term productivity of the land being sacrificed for short term corporate gain;*
>
> **WHEREAS** *a park and protected area system that is arbitrarily limited to only 12 percent of the land base is inadequate and will fail to protect the Island's natural biological heritage;*
>
> **WHEREAS** *it is impossible to establish an adequate protected area system and reform forest tenures without addressing and settling the valid claims of First Nations to the land so that they also have a land base and resources for their development of sustainable communities;*
>
> **THEREFORE BE IT RESOLVED** *that we the undersigned British Columbia voters and concerned citizens, humbly pray that your Honourable House may be pleased to pass the necessary legislation to phase out multinational corporations' forest tenures and clearcut logging on Vancouver Island, to enable First Nations to acquire a portion of these lands and to establish both a protected wilderness area system that fully preserves Vancouver Island's natural biodiversity and a system of local, community-controlled forest reserves where only ecologically-sound selection logging methods are allowed, providing a sustainable future with more permanent jobs, more value-added manufacturing and more benefits and profits to the people of Vancouver Island.*

Besides doing the mass drop into all the Vancouver Island households, we distributed copies of our Vision paper directly to the media, B.C.'s MLAs, and all the participants at the CORE table as well as to all our WCWC members. VINE, the Vancouver Island Network of Environmental groups that provided coordinated direction for the conservation "sector" representatives at the CORE table, endorsed our Vision map. It was satisfying, too, to see our *Conservation Vision* treated as one of the legitimate land use alternatives by the biggest newspaper on the Island, the *Times Colonist*.

Despite the science backing our 40 percent *Conservation Vision* goal, I knew it was going to be tough to exceed the NDP government's CORE commitment to 12 percent protection. From the start of the CORE process, the 12 percent goal was clear, although before the CORE process got underway participants were told that the goal was to protect a representative 12 percent of each and every unique ecosystem on Vancouver Island.

The forest industry and IWA must have got to Harcourt right away, because this quickly degraded into protecting a straight-up 12 percent of the land base, discarding any representational requirements for specific ecosystem types. This made us at WCWC really mad (we predicted a lot of rock and ice and not too much old growth would end up as new parks) and doubly happy we weren't sitting at the CORE table legitimizing the process.

Protecting 12 percent of B.C. became something like a quest for the Holy Grail for the Harcourt government. Twelve percent was mandated into every parks planning process and established as an upper limit to the amount of land those participating in these processes could recommend for protection. As time went on, I began to rue the day that we had signed onto the WWF Endangered Spaces Campaign. But we didn't know better at the time. Ten years earlier, while fighting to save South Moresby, WCWC had advocated that just 6 percent of B.C.'s forests be protected. The visionary Valhalla Wilderness Society's map had asked that only a little more than 13 percent of B.C.'s land base be protected in parks.

Just as the forest industry decried, we'd kept moving our goal posts higher and higher. But until our Vancouver Island *Conservation Vision* map, we'd never done the science to know how high the goal posts should really be to conserve biodiversity! By endorsing the WWF goal in '89 we had helped reinforce the myth that 12 percent preservation was sufficient. Now we knew it was more like 40 percent of the land base that needed to be kept natural. Anything less would mean the loss of species and ecosystems.

I often wondered where the 12 percent goal originated. I knew it was popularized in the Brundtland World Commission on Environment and Development Overview Report titled *Our Common Future* published in 1987. This report is also infamous for coining the fuzzy concept "sustainable development," a label now affixed to nearly every development proposal to enhance its acceptability no matter how ecologically unsound it is. We tracked down the origins of this 12 percent goal and discovered that Jeffrey McNeely, a Canadian and a chief scientist at the World Conservation Union (IUCN) in Switzerland was the author.

In a personal conversation with McNeely about how he picked this figure (when she and McNeely were both attending the first Conference of the Parties to the international Biodiversity Convention held in the Bahamas in 1994), Adriane learned that the 12 percent goal had no basis in science whatsoever. McNeely arbitrarily fastened on this figure in an off-the-cuff remark he made during informal discussions at a previous conference about protected areas around the world. At that time protected areas comprised about two percent of the earth's land base. McNeely told her that he guesstimated that 12 percent might be a politically achievable protected areas goal for the planet. He also told her that, although many countries would find it hard to protect 12 percent of their lands, he never thought that a rich country like Canada with an environmentally aware electorate and lots of wilderness would ever stoop to establish a protected areas goal as low as 12 percent.

1994 calendar celebrates "clearcut sports"

Despite all the intense Stumpy and *Conservation Vision* campaign work going on, we still had to focus in the fall of 1993 to produce our '94 calendars and catalogue. Deciding which threatened wilderness areas to feature, selecting the photo images, getting the maps produced and writing the 150-200 words that accompanied each month never got any easier for me to do. In our 1994 Canada calendar we featured a full page cartoon inside the front cover titled *1994 Commonwealth Games Special Events – BC Clearcut Sports*. Victoria, B.C. was going to be the host of these games. Some Clayoquot campaigners proposed to picket and make a stink during the games.

We felt that this would be in bad taste, and most likely would turn more people off than on to our cause. We chose to use humour instead. Annette Shaw, who had a delightful imagination and had drawn the cartoon of the anthropomorphized old-growth trees with a young one asking an old one, *Will I grow up to be big like you?* for our successful Carmanah Adopt-a-Tree fundraiser, drew this satirical cartoon in full colour. It included these trees in the background with commonwealth "athletes" competing in a "Biodiversity Assessment Slash Dash" and a "Mass Landslide Replanting Scramble" in a huge stump-filled clearcut landscape.

Our efforts obviously are having some effect

B.C.'s NDP government loathed our Stumpy road tour and our *Conservation Vision* paper. Leonard Krog, NDP MLA for Parksville-Qualicum Beach, sent us a stinging letter in late October '93 decrying our criticism of the government's "compromise" Clayoquot decision. He warned us that if we kept *pissing in the government tent...somebody is going to close the flap and you can howl in the wilderness*. Krog was one of the few NDP MLAs who strongly supported more parks.

He threatened us, not very subtly, with a negative government backlash that would slow the pace of new park creation if we continued our heavy campaigning. We leaked Krog's letter to the media, knowing it would keep Clayoquot in the news. It made quite a big hit. Premier Harcourt defended Krog commenting to the media that he understood his frustration. Harcourt was a stubborn man and there was no way he would ever back down or reverse his Clayoquot logging decision. WCWC wouldn't back down either.

During the month following the October 23rd federal election, Misty and her crew continued to tour Stumpy around Ontario "self-funding" their tour through the donations they got along the way. Their efforts helped build greater and greater national support for protecting Clayoquot. We put out media releases almost daily, similar to a political leader's election tour. News articles about Stumpy appeared in virtually every media outlet, even college newspapers. At Union Station in Toronto, Stumpy met up with the transcontinental protest train organized by Elizabeth May and the Sierra Club of Canada that was traveling west across Canada to support Clayoquot Sound.

At this Union Station rally, outspoken author Farley Mowat stood beside Stumpy and spoke to the crowd saying, "*It's all part and parcel of the desecration of the planet. The destruction of the fish in the Atlantic and the forests in B.C. fit hand and glove.*" Stumpy also attracted support

Joe's kids practice "stump ball toss" a possible candidate for calendar use (it didn't make the cut). Photo: WCWC files.

amongst rock stars, sitting on its trailer outside several benefit concerts for Clayoquot where REM and Spirit of the North played.

While over-wintering, Stumpy gets a Greenpeace European tour contract

In late November Misty left Stumpy and its trailer in Caledon, Ontario, near Ottawa. Sympathetic supporters Josie and Bradley Carmichael agreed to display Stumpy in front of their Caledon Inn for the winter.

By now, Stumpy had become a worldwide celebrity, certainly within environmental circles. In early spring I got a call from Greenpeace International. They were interested in buying Stumpy and touring it in Europe. I explained that, under the terms of our wood salvage permit, we could not sell the stump, and anyway, we needed it to do more touring to raise greater awareness in North America during the next year. However, they could lease it for a few months.

We worked out the terms. They would pay a nominal monthly rental fee and cover all the costs to ship Stumpy back and forth to Europe from Canada. They didn't want the trailer, just the stump. (We had measured it up and it fit snuggly into a standard shipping container with no centimetres to spare on the sides.) They agreed to use our display board that explained our "project," the Clayoquot Witness Trail, wherever they took Stumpy.

Phytos nearly quash Stumpy's European tour — but a determined volunteer solves the problem

A volunteer named Peter (no one can remember his last name), a Ph.D. student from York University, offered to help ship the stump. He had to take it from the Caledon Inn and deliver it to a Greenpeacer in the U.S. who would take it to New Jersey to put on a boat headed for England. However, when he went in to get an export permit (as essential a document for cargo as a passport is to a person) he ran into a snag. Since Stumpy was a raw piece of wood and not a finished product, he was told that it might be harbouring unwanted "bugs." In order to pass British Customs it had to have a certificate from Agriculture Canada stating that it was "phytosanitary." Peter thought that perhaps by removing the bark, all bugs would be gone, too. Without removing Stumpy from its flatbed trailer by Caledon Inn Peter stripped off every centimetre of bark. He contacted Agriculture (Ag) Canada and they informed him that this debarking was not good enough. Either Stumpy had to be kiln dried or autoclaved.

Kiln drying Stumpy would wreck it. Stumpy would shrink and split. Besides, it would take weeks to dry out such a huge chunk of wood and it would cost a lot of money, even if we could find a kiln owner willing to do it. And there probably wasn't an autoclave anywhere large enough to do the job. So, it looked like we were stymied. Peter went back to Ag Canada, found a sympathetic person and explained our problem to him. Based on the fact that autoclaving involves heated steam, they came up with a creative solution: to steam clean Stumpy using a portable commercial steam cleaner that people normally use to power-wash engines, walks and walls. But it could not be a regular steam-cleaning job. To ensure Ag Canada that all the bugs were gone, it would have to be an exceptionally protracted steaming—10 straight hours—to simulate the immersion of an autoclave. Peter rented the equipment and proceeded to clean Stumpy vigorously. We heard that visitors to the Caledon Inn, where the steam poured for hours in huge volumes around Stumpy, were mightily impressed. The whole process ending up aging Stumpy considerably, giving it a weatherworn look. The final result: Stumpy passed the Ag Canada test and was issued its Phytosanitary Certificate.

Stumpy stumps Europe

With the certificate in hand, Peter delivered the trailer with Stumpy on it to the Greenpeace driver who was going to haul it to New Jersey. The Greenpeacer hitched up his 4x4 to the trailer, climbed behind the wheel, put it into first gear and popped out the clutch.

His front wheels lifted off the ground! "No way!" he said, unhitching Stumpy's trailer. Undaunted, Peter rented a flatbed truck with his own money and got a crane to lift the stump onto it and chained it down. He drove it down to New Jersey by himself getting there just in the nick of time to catch the boat. Peter was very apologetic when he submitted the bill to us for his out-of-pocket expenses, which totaled over a thousand dollars. In his letter accompanying it, he said he'd understand if we couldn't pay, since he was not authorized to spend the money. Of course, we gladly reimbursed him. It's heroes like Peter who never give up and overcome seemingly insurmountable obstacles who win campaigns.

From the moment Stumpy was lifted out of its container in Liverpool, it continually made the news. If our baby cedar stump seemed big in Eastern Canada, it was huge in Europe where all the large old-growth trees had been cut down centuries earlier. During its stay in Europe, Stumpy was shared between Greenpeace Great Britain and Greenpeace Germany. The English Greenpeacers made up a Canadian passport for it that looked like the real thing, except that it was jumbo sized—four times bigger than normal. As Stumpy toured around England, the Greenpeace campaigners got the mayors of the cities they visited to stamp Stumpy's passport and com-

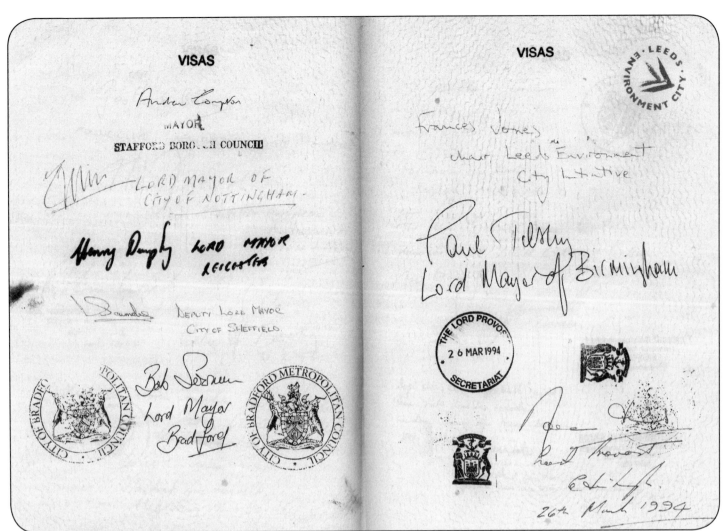

Pages from Stumpy's passport stamped by mayors who committed to using only old-growth-free paper. Document: WCWC archives.

mit to buying only old-growth-free paper in the future. Stumpy also made a guest appearance on a popular kids' show called *Blue Peter*. In Germany Stumpy "locked down" on a Canadian Embassy by being parked up tight against the front doors so that no one could open them. It was not our style of protest and was against our basic tenet of not being involved in illegal activities. We couldn't protest, however, because we had failed to put into the leasing agreement any prohibition on using it in such a manner, never having imagined using it in a civil disobedient way was possible.

Share BC wants us to "Bring Stumpy Home"

The B.C. forest industry eventually assigned several PR lobbyists to follow Stumpy around Europe and counter its message. Then someone in Share BC, the so called "grassroots" organization financed by the forest industry to counter environmentalists' campaigns to create more parks, got the brilliant idea of launching a "Bring Stumpy Home" campaign. In what was probably the most bizarre PR battle in B.C.'s "War in the Woods," Share BC actually published a large, full-colour poster featuring a photo of a beautiful 50-year-old second-growth forest with a large moss-covered rotting stump, presumably a Stumpy look-alike, in the foreground.

Above the image of the stumps in bold print was emblazoned *Stumpy in its Natural Habitat*. Below that was a short paragraph about the ecological importance of stumps. I don't know who would display such a poster, except for us, where one still hangs today on our Gastown office wall as a reminder of one of the most successful of all WCWC's campaigns. **It's true for all campaigns: if the opposition is focused on responding to your initiatives, you're winning.**

Tony Eberts, then the outdoor columnist for the *Province*, wrote a piece on the Share's Bring Stumpy Home campaign titled *It's a rotten way to go*. Poking fun at it, he mused, *All those who believe Share BC are experts in talking rot will no doubt see the sense in getting Stumpy back with the other 247 trillion stumps the timber industry has so generously provided*. Eberts was referring to Share BC executive director Michael Morton's assertion that Stumpy would do more good for the forest ecosystem by providing nutrients for regeneration of a healthy new forest than by being used to *misrepresent the facts about improved forest practices in B.C.*

When a *Victoria Times Colonist* reporter asked Adriane what she thought about Share's Bring Stumpy Home campaign, she retorted, *At first I honestly thought it was a spoof—that they couldn't be serious. When you contrast the biomass represented in our stump with the biomass that's been removed by clearcuts, it's laughable that they should suggest that we have been ecologically irresponsible*. In the same article Alison Spriggs, our WCWC-Victoria Branch spokesperson said, "*I think personally their concern would be perceived as a lot more sincere if they focused their attention on the massive disruption that clearcuts cause rather than on the removal of one stump.*"

We held a press conference in the clearcut at the site where we had salvaged Stumpy to illustrate the point of how misguided and ridiculous this Share BC campaign was. Here, Joe promised the forest industry we'd *bring back the stump when you bring back the trunk*.

Greenpeace Europe returned Stumpy to us in midsummer 1994,

delivering it back to the Caledon Inn. By now Stumpy had logged over 50,000 kilometres. Misty immediately began an ambitious road tour through Ontario, eastern Canada and then down through the U.S. Helping her organize the tour and accompanying her on it were Misty's brother Cameron, Cheri Burda, and Greg Dowman. They kicked the tour off on Saturday, July 23, 1994 at the St. Lawrence market in Toronto.

1994 was a very busy year. WCWC's Clayoquot efforts in September

THE 400 year-old red cedar tree stump that Greenpeace dragged around Europe this spring was carelessly uprooted from its natural habitat in British Columbia's Clayoquot Sound rain forest. Even though the tree was felled to make furniture, cabinetry and other building materials, Stumpy could have contributed far more to the forest eco-system. Decaying tree stumps and fallen trees provide vital nutrients for the regeneration of a healthy new forest. Many become nurse logs, with second generation trees setting roots in their decomposing material. Stumps and fallen trees left in place after harvesting also provide critical habitat for a host of forest animals. So when Greenpeace rips a four-tonne stump from its native soil to promote an overseas fundraising campaign, do they really have the best interests of British Columbia's forests and wildlife at heart?

SHARE B.C.

For more information call:
(604) 726-2002
P.O. Box 639, Ucluelet, B.C.,
Canada V0R 3A0

The ironic and hypocritical Stumpy poster published by Share BC, the pro-logging "grassroots" lobby group funded by the big forest companies. Photo: WCWC files.

included organizing a special primo-ticket holder party with Bonnie Raitt after her Clayoquot benefit concert at the Orpheum Theatre in Vancouver as a fundraiser for WCWC. Both the concert and the fundraiser were fabulous! We also reopened and ran our Sutton Pass kiosk and trailbuilding efforts during the summer months.

This time we were the only large environmental group active on the ground in Clayoquot. There were no blockades that summer and the harassment of our volunteers and kiosk staff by loggers was not nearly as intense.

Stumpy shames the New York Times for contributing to the destruction of Clayoquot's ancient rainforest

After an appearance in front of the Parliament Buildings for the opening of the fall session of the House of Commons (part of our continuing efforts to get Prime Minister Jean Chrétien to honour the Liberals' promise to help protect Clayoquot Sound made right before their election as government nearly a year earlier), the Stumpy tour headed eastward to Quebec and the Maritimes. Eventually the caravan traveled south, arriving in New York City in October.

They parked Stumpy in front of the *New York Times* building in Manhattan and joined a rally to urge the *Times* newspaper to stop buying newsprint from MB. At a fundraiser in the evening after the rally, with Stumpy outside, Robert Kennedy Jr. gave a powerful speech condemning B.C.'s logging practices and its failure to consult with the Nuu-chah-nulth regarding the fate of their forests in Clayoquot Sound. Kennedy credited his personal efforts with helping ensure that the First Nations had a say in the matter. This prompted Premier Harcourt to say in the press that Kennedy was *no longer welcome in B.C.* after making such *crummy remarks*. First Nations' leaders agreed with Kennedy.

Stumpy journeyed on from New York to Washington DC. Here Misty's statements about B.C.'s subsidies to the forest industry through low stumpage rates that the B.C. government charged them for cutting the publicly owned timber kicked up a hornet's nest. Some in Canada construed Misty's comments to mean that we were backing the move on the part of the U.S. forest industry to have duties imposed on Canadian softwood imports.

But we weren't. We simply wanted a higher stumpage collected in B.C. so we could afford to have a bigger Forest Service and enforce more environmental protection and at the same time stop the U.S. from pocketing this money.

Stumpy then went across the U.S. visiting dozens of communities, including Denver and San Francisco. Our Stumpy slideshow and caravan crew's strong message about the continuing battle to save Clayoquot Sound and ban clearcut logging in B.C. garnered plenty of verbal support, but their "pass of the hat" barely took in enough money to keep them going. Disaster struck in San Francisco. On the first night there someone broke into their van and stole the projector, the entire slideshow and all their gear.

We put together another show from our slide collection in Vancouver and couriered it to them the next day, and the show went on. Patagonia, a big outdoor clothing and equipment store, generously gave our crew new gear.

Still in San Francisco, Stumpy put the pressure on communication giant, Pacific Bell to stop using paper from MB that contained Clayoquot Sound old-growth fibre in their phonebooks. After a final few stops including Seattle, Stumpy and our road tour team arrived back home in the early fall. We stored Stumpy and waited for another opportunity to put it to good use. Misty got right back to work in our WCWC-Victoria office, focusing on our Clayoquot and Sooke Hills campaigns. At a large 1993 enviro year-end party, Karen Mahon, Greenpeace campaigner, summed up some noteworthy number-related achievements and said: 365; *"the number of WCWC media releases all of us got this year detailing the escapades of Stumpy."*

Taking the fight to save Clayoquot to the Bahamas

Not long after Stumpy returned home, Liz Barrett-Brown, a lawyer working for the Natural Resources Defense Council (NRDC), urged WCWC to attend the first Conference of the Parties to the International Biodiversity Convention. The Biodiversity Convention, adopted at UNCED in Rio a year-and-a-half earlier, was one of the few accomplishments of this eco-political extravaganza. The UN was holding this 10-day conference at the end of November and early December 1994 in Nassau in the Bahamas. Liz painted a powerful picture of how we could build international pressure at this conference to help preserve the ancient forests in Clayoquot. She offered a free place for us to stay; sleeping-room on the floor of her hotel room near the conference.

"Bring Stumpy along," Liz urged. We checked it out, but it proved to be atrociously expensive as well as logistically impossible. Convinced it was still worthwhile to attend, we came up with airfare money to send Adriane and Misty MacDuffee. They arrived—sans Stumpy—but with the Stumpy slideshow and all our printed educational material about Clayoquot Sound the day before the conference. Liz, who really "knew the ropes" of how such conferences operated, had already devised a detailed game plan.

"Most people are stuck in meetings all day," she told Adriane and Misty. "They don't have any time for a break from their intense sessions. They are all hungry at the short noon break. The way to entice them is to rent a room during lunchtime right next to where they're meeting, serve them free sandwiches and put on the Stumpy slideshow while they eat. We'll change locations every day to make sure we cover all the different break-out sessions.

Adriane and Misty followed her advice to a "T." Every day they produced and printed a little flyer that announced the time and place of that day's slideshow. The title highlighted how Canada was allowing the destruction of biodiversity in the ancient temperate rainforest in Clayoquot Sound. Misty and Adriane would stand by the exit doors of the morning plenary sessions and hand out the flyers advertising the free lunch show. As they handed out the flyers they called out, "See how Canada is destroying its biodiversity!" They got people's attention!

Liz worked mostly at a senior level urging delegates to see the show. Over the length of the conference, she got virtually every member of the U.S. delegation, as well as many senior people from other delegations, to attend one. NRDC also helped cover the cost of the room rentals and the food. One morning Liz roller bladed down to a shop in town to pick up the food, only to discover the boxes were too big to carry back by roller blade. Juggling all the boxed food, she hitched and got a return ride back in the back of a pick-up truck!

Delegates from the "southern" countries loved our exposé. They were the ones most often in the hot seat for destroying biodiversity. Canada had always proudly presented itself as a world leader in biodiversity protection. In fact, we were considered a "Boy Scout" of the biodiversity movement for being, in December 1992, the first industrialized nation to ratify the International Biodiversity Convention. Delegates "just lapped up our information" according to Adriane. Of course, the official Canadian delegation just hated it. They were on the phone to the B.C. government several times per day trying to figure out ways to counter us. At one point a member of the Canadian

Collage of news articles about Stumpy: Artwork: WCWC files.

delegation came up to Adriane and, more seriously than kiddingly, asked, "How much would it cost to send you home?"

At another point during the conference, Misty brazenly ejected the videotape running a perpetual loop of pro-Canada propaganda at the un-staffed Canadian government information table in the exhibition hall and replaced it with *Sulphur Passage*, a very emotional five-minute Clayoquot video directed by Nettie Wild. The video featured Bob Bossin's rousing song sung by him and a host of well known B.C. musicians Valdy, Stephen Fearing, Raffi, Roy Forbes, Ann Mortifee and others accompanied by images of clearcut logging, incredibly beautiful ancient temperate rainforest and peaceful blockaders being arrested during the summer of '93 at the Kennedy River bridge in Clayoquot Sound.

Dozens of people immediately crowded around to watch this protest video and listen to *Sulphur Passage*, the anthem of the Clayoquot preservation movement. Needless to say, when the B.C. government delegation discovered Misty's tape caper they were not amused. Adriane told Misty to cool it or else they would probably get kicked out of the conference.

"Being there was brilliant," Adriane enthused when she returned. "More than one-third of all the official delegates to the conference saw our show. Usually at a high-level conference like this we have no real impact. But this time we weren't trying to get the wording changed from 'may' to 'shall' in a document that ultimately would mean very little. We were actually showing the world that Canada and B.C., while hypocritically promoting their Boy Scout enviro-image, were actively destroying biodiversity. The embarrassment we caused had to have helped move both governments closer to doing something to protect Clayoquot."

Stumpy stirs up controversy in the House of Commons

Our impact at the international biodiversity conference in the Bahamas was likely one of the reasons for the attack on WCWC four months later during a question period in the House of Commons by Bill Gilmour, the MP that represented the Clayoquot Sound area. Shortly after this attack, Revenue Canada again audited WCWC. Here is the text from Hansard of Thursday, March 30, 1995:

Mr. Bill Gilmour (Comox-Alberni, Ref.): *Mr. Speaker, my question is for the Prime Minister.*

Last November I asked the Minister of Foreign Affairs why his department was funding environmental groups whose main purpose was to actively discredit B.C. logging practices in North America and in Europe. Since 1990 the Western Canada Wilderness Committee has received over $754,000 from foreign affairs, environment, human resources development, heritage and natural resources.

11298 [page number]

Does the Prime Minister support providing federal funding to groups whose main purpose is to undermine our number one industry in Canada, forestry?

Right Hon. Jean Chrétien (Prime Minister, Lib.): *Mr. Speaker, these groups have been helpful because most of the provinces have improved, quite dramatically, the way they harvest forests. It is helping make our case when we are abroad.*

We have to be careful. For example, we are asking people to help protect the fishing environment of the sea. We have to do the same thing in Canada. When we have groups in Canada that want to harvest trees in a proper fashion, they are not causing a disservice to Canada, they are helping us to do the right thing.

Mr. Bill Gilmour (Comox-Alberni, Ref.): *Mr. Speaker, some of us may recall seeing Stumpy, that large cedar tree stump from the Clayoquot in my riding on Vancouver Island, around Parliament Hill last fall.*

The Western Canada Wilderness Committee not only took this stump across Canada at taxpayers' expense, but is planning to take it to Berlin next month to discredit, again, the Canadian forest industry.

My question is for the Prime Minister. Will he act immediately to cancel all funds to this group, which has not only cost taxpayers $754,000 but continues to cost the Canadian economy millions in lost revenue?

Hon. Sheila Copps (Deputy Prime Minister and Minister of the Environment, Lib.): *Mr. Speaker, the Prime Minister has underlined how important it is for Canada to have sustainable forestry practices. The work of organizations such as the Western Canada Wilderness Committee reinforces the notion for an international forest practices code where all countries can be judged on a level playing field.*

The member would be the first to criticize the government if it did not have a proactive international stance by which it shares Canada's sustainable forestry practices with those of other parts of the world. That is what we are attempting to do, working with environmental groups, rather than damning them all as the Reform Party would do.

Cartoon lampooning the unsustainable overcut. Artwork: Annette Shaw.

This large amount of money that Gilmour touted we got came as a complete shock to me. Except for a couple of grants to do a poster and the Hot Spot brochure in the early 80s and the few CIDA grants to do international work recently, I could not think of any other money the federal government had given us. It was a fact that WCWC didn't use one government dime to fund our Stumpy tours.

I asked our new comptroller, Brian Conner, to look into it. It turned out that Gilmour was adding in all the money that passed through our accounts from government programs that helped employ people including UIC "top-up grants" and youth employment grants. Our chapters and branches had used most of these employment programs and they helped many people. Andrea Reimer, who became WCWC's executive director in 2002, started out at WCWC on a provincial government grant to help unemployed youth get job skills. Many others were trained in a variety of office duties so they could more easily re-enter the workforce. After their training period with us, nearly all moved on to other work.

Stumpy's final journey

In late February 1995, Greenpeace International called. They wanted to host Stumpy in Berlin for the opening of the first meeting of the signatories to the 1992 Climate Convention, being held from March 28th through April 7th. Greenpeace offered to pay for everything; reimburse us for the cost of transportation over and back and return Stumpy in good shape in four months' time. This time, however, they couldn't afford to pay WCWC any rent. We thought the idea was great. However, the timeline was tight. The convention was only five weeks away.

We checked with Magnacargo, the shipper we had used before, and they said they could do it. The trip across Canada took about a week; ships left Montreal every week and the trip across the Atlantic took about two weeks. They urged us to hurry, however. While theoretically there was plenty of time, a rail strike was looming that might "slow things down a bit." We wasted no time in getting Stumpy to the terminal in Vancouver and waved good-bye as it started its next journey. Beyond our control, things soon went "off the rails." Caught in slowdowns and rotating strikes, Stumpy's container crept across Canada. Upon reaching Montreal it encountered a longshoreman's strike. We asked Magnacargo what we should do and they suggested we take it by rail to Halifax and ship it from there. On the way there, Stumpy encountered more slowdowns and more delays. By the time it reached Halifax, there were only four days to go before the start of the climate change meeting in Berlin. Greenpeace didn't want Stumpy if it couldn't make the opening of the conference. There was no way it could get there in time. We told Magnacargo not to ship it to Europe and to hold it in Halifax while we figured out what to do next.

Stumpy becomes "abandoned cargo"

Magnacargo started charging us $20 per day for storage. By the time Stumpy got to Halifax the shipping and storage expenses had already mounted up to $3,800. We contacted some environmental groups in Nova Scotia and, a couple of months later, their arrangement with the City of Halifax to allow us to prominently display Stumpy in a city park came through.

When I asked Magnacargo to deliver Stumpy to the park, I found out that several weeks earlier, without asking or informing us, they had shipped Stumpy back to their main warehouse in Montreal. There sat Stumpy, continuing to rack up $20 per day in storage fees. Magnacargo explained that they had to pay some other shipper for the storage in Halifax, so it was cheaper for them to keep it in Montreal. We now, they said, had to pay for the freight back to Montreal too. The bill was in excess of $5,000 and growing daily.

In a year's time, storage charges alone would amount to over $7,000. To bring Stumpy back to Vancouver would cost nearly $1,000 on top of what we already owed. No stump, no matter how famous, was worth that!

Furious, wanting to "sue the bastards," I called a lawyer who specialized in commerce law. Quickly, he let me know the score. There wasn't a thing we could do. The fine print in every shipping contract everywhere in the world contains the clearest and tightest legalese on earth. It boils down to this; no matter if it is entirely the carrier's fault, the shipper always has to pay. The only way we could stem the tide of red ink was to pay up and get Stumpy back or "abandon cargo" and let them have Stumpy. Reluctantly, I "abandoned" Stumpy, pleading in my letter to Magnacargo that they not chop it up for firewood, and, of course, we let the media know what had happened. Instantaneously, Stumpy belonged to Magnacargo and we were no longer liable for any more storage charges. But we were liable for the all charges up until then, no small chunk of change! Magnacargo could sell Stumpy to recover as much of what we owed them and then they could still go after us for the rest in court.

We generated quite a few media stories about the plight of Stumpy. Soon thereafter I got a call from CBC's famous *As It Happens* radio show. They wanted to feature Stumpy's story live on air with a discussion between a representative from Magnacargo and me. They gave me an exact time to phone in.

Adriane, Joe and I rushed back on our return trip from our research cabin in the Upper Carmanah trying to get to a place where our cell phone would work by airtime. Finally, at Shawnigan Lake, just minutes before I was to go on air, we got in range.

This spot on *As it Happens* turned out to be a godsend. I got the opportunity to explain our side of the sad tale. Michael Enright, then host of the show, grilled Magnacargo's representative who said, on air, that the company was not going to go after us to collect the outstanding money we owed and that they were not going to destroy Stumpy. In fact, he said that they would find an appropriate home for it. Obviously, the company didn't want anymore bad publicity. We never heard from Magnacargo again: we never contacted them again either.

A year later we learned that Stumpy still lived. A consultant working on developing exhibits for the new Museum of Natural History in Ottawa, which was going to open in a couple of years, contacted us. He thought Stumpy would make a great display. We explained to him that we no longer owned Stumpy and gave him Magnacargo's number. A few days later, the consultant called me back to let us know the good news. He had called Magnacargo and found out that Stumpy was available.

He said that he would be working on purchasing it for the museum. But his effort fizzled out. Instead of purchasing Stumpy, he told us later that the museum would be commissioning the construction of an artificial stump, a composite made out of chunks of wood from every province glued together—a typical bureaucratic Canadian solution that averted any possible inter-provincial rivalries and avoided angering the logging industry.

In the summer of 1995, Andrea Reimer, who was holidaying in Montreal, searched a whole day around the transport yards in the city looking for Stumpy, but she came up with a blank.

Where is Stumpy today? The answer is out there somewhere. It may still be in Montreal, collecting dust in a dark corner of some cavernous warehouse.

Bumper sticker. Artwork: WCWC files.

Chapter 46

Parts of Paradise saved
Clayoquot campaign continues
Victoria watershed logging nixed

Losing our forester Mark Wareing

At WCWC we always ploughed every bit of income we could back into our campaigns. At the end of January 1994 we had to make some hard decisions. We didn't have enough money to make our scheduled debt payments for the next month. With great regret we laid off our forester Mark Wareing as well as Adriane's longtime assistant Ramona Tribando. Ramona went back to school to take up marine biology, something she was planning to do anyway. But Mark was tainted by having worked for us and because he had campaigned vigorously against clearcut logging and shamed foresters for their unethical support of bad forestry practices. The forest industry does not forgive those who break rank. Although we let Mark take with him all the files and photos he'd collected while working for us, he did not find it easy to get work. Suffering from periods of depression, Mark took his own life in June of 1996. At WCWC we cried. He was family.

Mark had joined WCWC because of his passion to protect the watersheds that provided Greater Vancouver with its drinking water. He felt it was his duty as a professional forester to protect the public interests he had sworn to uphold as required by the B.C. Professional Forester's code of ethics.

During his time with us he documented and exposed the ecosystem-destroying logging he'd always felt was unethical and dealt a fatal blow to the industrial apologists who stubbornly rationalized logging in Vancouver's steep-sloped drinking watersheds.

A few months after Mark left, WCWC finally published a four-page full-colour tabloid-sized newspaper titled *STOP ALL LOGGING in GREATER VANCOUVER'S WATERSHEDS NOW...before it's too late!* (Educational Report Vol. 14 No. 6). It made an undeniable case based on Mark' research that halting logging would improve water quality because the ancient temperate rainforest is *"...nature's unbeatable water purification system."* It took eight more years of constant pressure to completely shut down logging in Vancouver's watersheds (November 1999). The credit for this win is shared amongst many who diligently kept up the pressure including Will Koop through his in-depth research and Paul Hundel through his work with SPEC. But most of all, the success in halting logging in Vancouver's drinking watersheds is Mark Wareing's legacy.

Vancouver CORE report tabled

On February 9, 1994, the CORE Vancouver Island table presented its final report. According to one newspaper account, the reaction was *"howls of discontent from both environmentalists and loggers."* Owen recommended that 13 percent of Vancouver Island be protected including 7.8 percent of the island's forest. Our *Conservation Vision* campaign, including mailing our newspaper calling for 40 percent protection to every household on Vancouver Island, probably helped CORE exceed the 12 percent limit on protection... by one percent.

Facing page: Meares Island Tribal Park in Clayoquot Sound as seen from Lone Pine Mountain on Meares Island. Photo: Adrian Dorst.

When it came right down to it, the 13 percent didn't give us much. It actually involved protecting only 1.7 percent more of the island's land base over and above that which was already protected. This was nowhere near enough to protect 12 percent of each ecosystem type. There was far too much alpine, rock and scrub forest in the island's existing parks and another 1.7 percent couldn't make up for the imbalance. As we'd expected all along, Vancouver Island's valuable old-growth forests would mostly end up logged. CORE also recommended that another eight percent of Vancouver Island be designated Regionally Significant Lands to act as buffers and as corridors between protected areas where conservation needs would take precedence over industrial wood fibre needs.

Without a firm government promise that logging practices would be totally different there, we concluded that these lands were mostly designed to garner positive PR. CORE brought neither peace nor ecological integrity in the woods.

The environmentalists who participated at the CORE table explained that they had achieved as much as they possibly could and even expressed fear that the Harcourt government wouldn't accept all their recommendations. Meanwhile, we at WCWC were choked at how little they got. The logging community, true to form, rose up in arms against more parks and began organizing a massive protest rally on the Legislature lawns in Victoria.

Wilderness Defender bumper sticker. Artwork: WCWC files.

Paradise lost or saved

When we heard that the loggers' rally was going to be held on March 21, 1994, we decided to quickly pull together a science-based response to the CORE recommendations. It was maybe a bit of wishful thinking, but we hoped we would look reasonable compared to the logging community's knee-jerk reaction. We also hoped that our thoughtful explanation of the need for biologically-based conservation goals integrated with forest tenure reform and settlement of First Nations' land claims would spur on every environmental group to demand more.

We published our eight-page full-colour tabloid newspaper titled *Vancouver Island – Paradise Lost or Saved* one week before the big loggers' rally. Joe Foy, Misty MacDuffy and I wrote the text, Adriane edited it and Sue Fox designed and laid out the paper. Ian Parfitt and Andrew Boldt produced a masterpiece of a map for the centre two pages. They superimposed CORE's recommended protected areas on our *Conservation Vision* map showing just how much area needing protection was left out by the CORE table. On page two, we featured another shocking map showing Vancouver Island in 2040 with tiny, scattered patches of green, all that would remain of Vancouver Island's old growth if only the CORE recommended areas were protected from logging. Several articles in this paper have stood the test of time. The back page article *Debunking the seven great myths of industrial clearcut forestry* ripped apart the myths still in circulation today, including *there's no difference between a mature second-growth and old-growth forest* and *clearcutting is safer for workers*.

The hard-hitting article we wrote for page two, *Eight Thousand New Jobs*, exposed the real reasons why forestry workers justifiably fear the loss of their jobs. These reasons are as true today as then: mechanization, depletion of high-value old-growth timber and subsequent closure of mills, failure of forest companies to diversify manufacturing and the shipment of raw logs out of their region.

We printed 150,000 copies of our *Vancouver Island – Paradise Lost or Saved* paper, only half as many as our *Conservation Vision*. Although we didn't have the cash to deliver a copy to every household on Vancouver Island as we did with our *Vision* paper, we had enough to insert our *Paradise* paper into the Saturday edition of the *Victoria Times Colonist*, just two days before the big rally.

The biggest rally ever held on the Legislature lawns

I decided we had to attend the loggers' anti-park rally. We may not have agreed with their main message, but we sure sympathized with their fear that forestry was going downhill on Vancouver Island! "We can make the point that increased raw log exports are having a much bigger negative impact on eliminating jobs than the proposed parks would," I said. A few WCWC staff members thought that our presence might provoke violence. I was sure it wouldn't. I countered with my standard line, "This is Canada."

We got to Victoria early on Monday morning, March 21st, and "set up shop" about two-thirds of the way back on the Legislature lawn. When we arrived I immediately noticed that at the corner of the lawn, next to the major intersection by Victoria's inner harbour,

One of the most ignorant signs at the logger's rally. Photo: WCWC files.

were a couple of rough looking guys with a huge sign that proclaimed in large letters *Screw the Environment. We need jobs!* I was sure that the rally organizers were not too pleased with that. Of course, we had a number of our own signs, including our big WCWC banner, a big sign that Nik Cuff had made for us using his image-splitting photocopier and laminator featuring our cartoon *Stephen in Plunderland* cartoon, and another sign saying *Save Jobs – Stop Raw Log Exports*. As soon as we arrived, several IWA peacekeepers came over and stood by us. They, as much as we, didn't wanted any ugly incidents to occur.

A steady stream of loggers, mill workers and their families from all over Vancouver Island and the Lower Mainland flooded in. They arrived by the busloads. Some came in loaded logging trucks and slowly circled the Legislature buildings. There was even a self-loading log barge in the harbour with a huge anti park banner strung between its cranes. Most of the logging companies and mills had given their employees the day off with pay. They eventually packed the entire front lawn and spilled out onto the streets. According to the media,

Someone flies a sarcastic banner over the Legislature during the big logger's rally. Photo: WCWC files.

over 15,000 people rallied at the Legislature that day. I believe the actual number was closer to 27,000 as claimed by the rally organizers. The huge crowd completely engulfed our tiny contingent of about a dozen that included Joe, Andrea, Alison, Adriane and I and our two young kids and a few stalwart WCWC volunteers. As far as I could see, we were the only ones there that day who were supporting the CORE recommendations and asking for more!

A few people passing by made disparaging remarks and one guy grabbed a sign away from one of the volunteers and stomped on it (he was quickly restrained by the rally peacekeepers). That was it as far as ugly incidents. The rest tolerated us. It probably helped that an organizer spoke to the crowd from the podium and asked everyone

At the March 21, 1994 Loggers Rally, WCWC's small group of park supporters engulfed by protesting loggers. Photo: WCWC files.

to respect those in the crowd who may have some different points of view.

After the first few anxious moments, something amazing happened. The huge crowd of loggers began chanting over and over again, "TWELVE PERCENT AND NO MORE." Adriane couldn't get over it. Just a year earlier the IWA's position was no more new parks. In fact, they even maintained that current parks should be opened to logging! She saw this chant as proof positive that our 300,000 *Vision* and 150,000 *Paradise lost or won* newspapers had shifted the balance.

We stayed to listen to all the speeches of union leaders decrying the CORE proposal for more parks. Meanwhile, during question period inside the Legislature Chambers, Liberal forestry critic Will Herd asked how many jobs and how much revenue to the Crown would be lost if CORE's recommendations were implemented. In his preamble to this question Herd stated: *There are 20,000 people on the lawns of the Legislature today who have been waiting for two and a half years for that study to be tabled.* The government's answer was that it was in the process of determining that. Herd was extremely displeased that neither the CORE report nor the NDP had the figures.

Standing in the midst of the largest rally in the history of the B.C. Legislature, I would like to believe that we tempered the loggers' zeal a tiny bit. **Many of them must have realized in their hearts that environmentalists were not really their enemy. If there was an enemy, it had to be the big multinational companies that took too much wealth out of the public's forests for their excessive executive salaries and their shareholder dividends, not leaving behind enough for workers, nature and our children's security.**

Wilderness Committee bumper sticker, a publication for a campaign that's still ongoing against raw log exports in B.C. Artwork: WCWC files.

Strange meeting with the new Leader of the Opposition Gordon Campbell

That spring of 1994, shortly after Liberal leader Gordon Campbell won a by-election and took his seat in the Legislature as the Leader of the Opposition, we met with him. It was a strange encounter. Joe, Adriane and I arrived at his newly opened Vancouver office bearing gifts of our posters. We were one of the very first delegations he received. We quickly figured out that he was completely unfamiliar with WCWC.

In fact, from the bent of his initial questioning, it was obvious that he mistakenly assumed we were one of the Share groups that opposed the creation of new parks. After a few awkward moments during which he quickly caught on that we were a parks advocacy group, he asked intelligent questions about our activities and looked at our posters and diplomatically said he liked them.

As we walked back to our office, the three of us compared impressions. We all thought he lacked good advisors and were disappointed in his lack of knowledge about park issues. Although he feigned interest, we doubted that he had any real sincere commitment to wilderness conservation. We had a sinking feeling that, while the NDP government was not great for wilderness, a Liberal government under Gordon Campbell might be a whole lot worse.

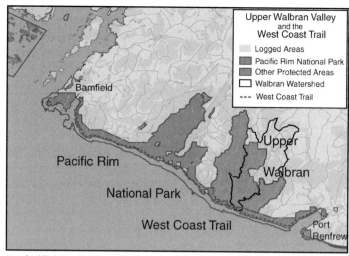

Map highlighting the Upper Walbran that failed to get protected in the CORE decision. Map: WCWC files.

Upper Carmanah almost gets the axe

On June 22, 1994, Premier Harcourt held a media conference to announce his government's response to the CORE's recommendations. WCWC was not invited. Although we were pretty sure that the government would accept most of its recommendations, we knew that park protection for the Upper Carmanah was by no means assured, because of the strong IWA and corporate opposition to protecting it. We waited in our office to hear the news. In contrast to the crush of media gathered with us when Vander Zalm announced the half-valley Carmanah park four years ago, a lone CBC TV reporter and his cameraman came to our office to record our reaction. This time we heard the good news of the acceptance of the entire package recommended by CORE, live on radio. The new parks included the 3,343-hectare Upper Carmanah and 9,500 hectares of the adjacent Lower Walbran (additions to Carmanah Pacific Provincial Park now renamed the Carmanah Walbran Provincial Park) and 23 other parks, many of which we had campaigned to protect over the years.

Bubbleless celebration of second Carmanah win

We brought down from the shelf the aged bottle of champagne on which we had written after the announcement four years ago, *Do not open until all Carmanah is protected*. "Here's to Carmanah!" I exclaimed as I loosened the wire around the plastic cork (it was inexpensive B.C. bubbly).

There was a tiny feeble "poop" as the plastic stopper rose a few centimetres off the top of the bottle before tumbling back down onto the table. We all had a good laugh. I poured a small amount in my glass and took a sip of it to make sure it was OK, before pouring glasses for the rest. I spat it out. "You'd have to be desperate to drink that!" I declared, hamming it up for the rolling video camera. "There are very few drinks worse than champagne without bubbles," I continued, making light of the situation. I resisted the temptation to make the analogy that the decision was weak, too, for there were a number of other worthy areas on Vancouver Island that should have been recommended for protection and protected.

We continued to joke around, but the TV cameraman and reporter were not amused. They argued that the feeble bottle opening wouldn't make good TV coverage of such good news and insisted we get a new bottle and do it all over again. We complied (for they were right) and repeated the scene with a loud popping of the cork, clinking of glasses and Carmanah toasting. Friends who saw the brief TV news clip said we looked great.

While in our hearts we celebrated this as a big victory for nature, there remained unfinished preservation business on Vancouver Island, the campaigns to save Upper Walbran, Clayoquot Sound, East Creek, Nahmint and other unprotected areas on our Vision map. After the media left we vowed to fight on to save them, despite the odds. In fact, we went straight out and bought five bottles of good champagne (that would age well) labeling one for each of the big still-wild watersheds unprotected by CORE. We'd only drink them when they were saved.

In the weeks following the decision, Ric Careless, Vicky Husband and other prominent B.C. environmentalists heaped praise on Harcourt for his CORE decision and said nary a word about key wilderness areas that were sacrificed. WC^2, taking a hard line, criticized these environmentalists for settling for too little. Long after the ink was dry, Ric Careless, the "inside man" who helped "finesse" many of the park deals in B.C. in the '90s, told me that the government almost choked at protecting the Upper Carmanah. It was only because Ric and others agreed to wholeheartedly support the decision publicly that the whole CORE package went through. Being the first of many such land use planning tables decisions to come, it was absolutely crucial that government implement the whole package without modification. To get that, those environmentalists at the table who supported the process had to agree to pile on the praise.

Park victory spells doom for canopy research station

This good news announcement made our Carmanah research station irrelevant as a campaign tool, although useful for continued scientific research. The station was in desperate needed of expensive repair, including replacing the climbing ropes and the strapping that held the platforms in place. We simply couldn't afford to keep it up and running. "Now that it is in a park, let's sell it to the government at its assessed value," I suggested. We wrote to the government, but our offer was turned down. Then I wrote to the University of Victoria, offering them the station for a dollar and suggesting that their ecology department could use it to undertake longitudinal studies of the upper temperate rainforest canopy. Having no funding for such a facility, UVic turned us down too.

With liability considerations foremost in our minds, we decommissioned the canopy platforms by removing all the ropes needed to ascend to them. Later that rainy fall of 1994, Adriane, our two kids and I went out to check up on our station and have a two-day mini-holiday in the research tent cabin. Our kids remember the multiple leaks in the roof and the mice that ran rampant. Nature re-conquers quickly in the temperate rainforest. The station sat idle for another year. Our special use permit ran out and we did not seek to renew it. We no longer paid the tax assessment, although the bills kept rolling in. There were so many other hot active campaigns that we hardly gave the place a thought.

The Carmanah Research Station's final day

A couple of years later, on the spur of the moment, Joe decided to visit the station to see how it was faring after being abandoned for so long. Incredibly—a case of synchronistic serendipity—he arrived there on the very day that a crew of provincial park employees were dismantling the last of the canopy platforms and burning them up in a fire they had set to consume our tent cabin. They had also dismantled most of the boardwalk, leaving only the first hundred metres intact out of respect for the hundreds and hundreds of donors whose names were routered into the cedar treads to commemorate their donations to fund our trail work and research station. B.C. Parks wanted to keep this part of the valley a wilderness and discourage people from hiking there. The only trails maintained in Carmanah today are in the mid valley.

The demise of our research station is actually not that sad a tale. The station fully succeeded in fulfilling its prime purpose, saving Upper Carmanah Valley! It also proved, through Neville Winchester's research that led to the discovery of so many new arthropod species, that this temperate rainforest ecosystem was a lot more biodiverse and complex than anyone had previously dreamt and that plantation forests with short cutting rotations could never conserve all the old growth's biodiversity. Now, with the station gone, Carmanah's canopy continues to evolve as true wilderness—hopefully, for eons to come.

In the months after its CORE decision, the B.C. government made it clear that the land use and parks plan for Vancouver Island was complete. In a speech in Port Alberni, Premier Harcourt assured the forest workers that there would never be another park created on Vancouver Island. The rest of the land base outside of the established parks was open to logging and industrial development forever! This, we learned, was the deal that bought industry and union acquiescence to the CORE and other subsequent land use decisions in B.C. The logging industry gave up a tiny percent of its land base in exchange for secure access to all the rest. It didn't matter how sound the ecological or economic merits were to protect other places, the NDP's mind was closed. Only our faith that a huge, loud and growing public outcry would eventually sway even the most closed-minded politicians kept us going.

WCWC refused insurance

Soon after the loggers' big rally, our longtime insurance broker, the Co-operators, refused to renew our insurance. This insurance company had made a lot of money off us. During our many years with them, we had made only two small claims for broken window glass when our store in Gastown was burgled. Our new comptroller, Brian Conner, told me not to worry: he'd find us another company. But Brian soon discovered that no other company would touch us, not even Lloyds of London. Having insurance was not a luxury for us. It was a necessity. We had to have general liability insurance. It was a requirement in our 20 Water Street storefront lease agreement.

Finally, we complained to the insurance industry's professional organization and went public, informing the media that the insurance industry had blackballed us. Mark Hume wrote an article about our dire situation which ran in the *Vancouver Sun* on April 27, 1994. In it, a spokesperson for the Co-operators confirmed what we had suspected; the growing militancy of the forest workers and the threats of violence we had received at our kiosk during the summer made them believe that our store could easily be the target of a firebomb.

I asserted again my belief that Canadians were not violent people and that the likelihood of such an incident occurring was infinitesimally small. To prove it, I cited the fact that we had been counter demonstrating in the midst of the crowd at the loggers' rally a month earlier with no violent incidents.

I'm sure that the media exposure shamed the insurance industry. Shortly after Hume's article came out, a company came forward and agreed to insure us. But our premium took a big jump and our deductible was raised to two thousand dollars, effectively no longer providing coverage of nuisance break-ins.

Over the years we had quite a few break-ins and lots of shoplifting at 20 Water Street. Four break-ins occurred while I was on the premises, peacefully sleeping on a foamy in the upstairs mezzanine level. In each case I was wakened in the wee hours of the morning by a loud crashing of glass. My hollering was enough to scare two of the intruders off. Another time the determined thief scampered in and snatched a handful of our T-shirts and then rushed out.

The T-shirts were worth way less that the cost of fixing the window. Dealing with the situation after the police had gone; sweeping up the glass, finding a piece of plywood and nailing it up to cover the open window was always aggravating. You had to do it well or another thief would tear it off and do a second burglary. This happened to me. A person broke in at 4 a.m. through the temporary patch I had put on the window a few hours before after the earlier break in. By chance, an off duty policeman who happened to be going by apprehended the thief as he was inside our big display window stealing merchandise. Three break-ins at 20 Water Street happened shortly after the NDP announced that only those who had lived in B.C. for at least three months could apply for and collect welfare.

There were lots of desperate people on the streets in the downtown eastside, shoplifting and stealing during that time period. Many other store owners in Gastown like us paid for this NDP shortsighted and hardhearted decision. Eventually, like all the other businesses in the area, we got iron bars that we installed inside our windows and door each night.

Even though our windows were fairly flimsy, they were an effective deterrent. The deteriorating street scene of drugs and crime in the lower part of Gastown hurt our business. Revenues from our store declined steadily from 1994 on.

WCWC rebounds financially with the Clayoquot campaign

Despite the decline we started to experience in store revenues, our May 1993 to April 1994 fiscal year was a good one. Our hot campaigns gave us the financial boost we needed to recover from the brink of bankruptcy. We reduced our old long-term debt by $100,000 and paid off some of our more recent bills, too, so that by the end of February 1995 we owed only $132,000 in short-term debt. We now operated in a much more business like manner—especially in our tracking of revenues and expenses—thanks to our new comptroller Brian Conner. Our membership bounced back from 15,000 to 22,000. We paid a lot more attention to keeping our WCWC members happy through better communication. Our membership renewal rate more than doubled from the troubled years of 1991 and 1992. It now stood at over 50 percent.

Our decision to automatically send out calendars to all our members every fall (including those who had failed to renew) worked. Revenues in the fall of 1993 showed the success. During that time period, 25 percent of our income came from sales of educational products; 19 percent from membership fees (for which we did not issue tax receipts); 42 percent from individual donations and 13 percent from foundation and government grants (the government grant at that time was a youth employment grant in Clayoquot). One thing for sure, money was still tight; there never was enough to do all we wanted to do. The big blessing that fall was a large bequest of over $100,000. We used it to pay off our most pressing bills. It was truly a gift from heaven.

1994 and 1995 were our busiest campaign years ever with, on average, about one new campaign paper coming out every month. We published 15 different "educational reports" in 1995: a record year for newspaper production. We were also resuming our old habit of using every opportunity we could to publish and promote wilderness through art posters. Choosing the right image, being at the press to adjust the colour during printing, and seeing our posters up on peoples' walls was always a thrill for me.

The federal government funds a WC² publication!

One of my favourite publications during this very active period was titled *Protect Canada's Biodiversity* (WCWC Educational Report Vol. 14 No. 4). We printed 120,000 copies of this full-colour tabloid-sized paper. Teachers and students especially loved it and many schools ordered class sets. We ended up having to put it back on the press and ran an additional 50,000 copies. This paper was the only concrete good that I could see arising from Adriane's sitting on the government-struck multi-stakeholder Biodiversity Convention Advisory Committee. It came about because she and Elizabeth May of the Sierra Club of Canada, both of whom attended the advisory committee's monthly meetings in Ottawa, organized the Canadian Coalition for Biodiversity, an alliance of eight non-profit conservation and environmental groups.

Besides WCWC and the Sierra Club, membership included the Canadian Nature Federation, Friends of the Earth Canada, and the Canadian Biodiversity Institute. The coalition got Environment Canada funding to help publish some educational materials about Canada's biodiversity and the need to conserve it. Adriane offered to put together one of our educational newspapers for the coalition. The aim was to enlighten the public about what biodiversity is, why we need it, and convince readers that the biodiversity crisis was real and that Canada needed to enact strong federal legislation to protect Canada's rich natural biodiversity heritage. It took a huge amount of Adriane's work and much of back-and-forth communication amongst the coalition partners to get a product everyone was happy with.

Andy Miller, spotted owl export working for WCWC dons a spotted owl costume to fight for protection of the habitat for this endangered species. Photo: WCWC files.

This eight-page newspaper was packed with statistics and interesting information about threatened and endangered species in Canada. It featured this quote by the Honourable Sheila Copps, then Deputy Prime Minister and Minister of the Environment, *The greatest legacy we can leave our children is a world rich in its diversity of life. For it is that diversity that will allow them to meet the challenges of the future and realize their hopes and dreams.* Copps had also on November 17, 1994, stated that she would be introducing a new law to protect endangered plants and animals in B.C. She did. But the weak law never went anywhere.

It eventually died on the government's order papers when the next election was called. It wasn't until July 2002 that the House of Commons passed Canada's first act to protect endangered species in Canada. This Species at Risk Act (SARA) is extremely weak and only specifically protects species on federally-controlled land. So far it has proven to be ineffectual in providing any real protection—prime example: the spotted owl and marbled murrelet—for Canada's endangered species.

Canadian government gives WCWC's largest ever petition a shrug

In January of 1994, an Angus Reid poll found that only 14 percent of Canadians surveyed nation-wide felt that the B.C. government's decision to allow clearcut logging on approximately two-thirds of the land in Clayoquot Sound was a good one. This same poll found that 67 percent of Canadians opposed the practice of clearcut logging. It was obvious that our efforts, and those of the brave souls who had put themselves on the line and were arrested to save Clayoquot's ancient rainforest, had changed Canadian consciousness. But it wasn't enough to move the B.C. government to revisit its decision.

In less than a year's time, with the help of other organizations and many hundreds of volunteers, we had gathered over 120,000 signatures on our "Official" Clayoquot Petition. Our petition asked the federal government to initiate negotiations with Clayoquot Sound's First Nations and the B.C. government with the goal of terminating the logging companies' tenures and protecting Clayoquot's irreplaceable wild forests. Signed sheets came in from the Stumpy tour, from our Sutton Pass kiosk and from countless volunteers.

Every day for months on end our "mail bag" had envelopes with petition sheets full of signatures. A week before the first anniversary of Harcourt's Clayoquot decision, volunteers counted the number of signatures on the sheets we had collected—three apple boxes full—in preparation for Adriane to take them with her to Ottawa on her next regular trip there.

Adriane felt it was important to have someone on the government's bench present the petition in the House. This created some bad feelings. Svend Robinson, the NDP's environment critic, was "royally ticked off" when Adriane told him that he wouldn't get to do it. But the insider advice we'd been given was to give the petition directly to a Liberal so government would take ownership of the issue. Charles Caccia, Liberal MP and chair of the House Standing Environment Committee, was diplomatically the right choice.

On April 12, 1994, one day shy of the anniversary of the Clayoquot logging decision, Adriane held a press conference in the House of Commons' media theatre and gave the 120,000-signature *Official Save Clayoquot Sound Petition* to Caccia. Robinson spoke at the media event, too, as did the Sierra Club of Canada's Elizabeth May who'd helped make Clayoquot a national issue. Adriane phoned me later to talk over the meetings she'd had that day, including one with Sheila Copps' political staff who told her that Copps, Canada's Environment Minister as well as Deputy Prime Minister, was "receptive and wanted to help." At the same time, however, they insinuated that our petition, with by far the most signatures we'd ever gathered, was not a strong enough show of support to get the Liberals to act!

"How many more signatures would it take?" I asked Adriane rhetorically. We continued to gather signatures and by that summer we had another 20,000.

MB holds its annual shareholders meeting on the first anniversary of the Clayoquot pro-logging decision

On the next day, April 13, 1994, MB held its annual shareholders meeting in a downtown Vancouver hotel. Was it mere coincidence that this day was also the first anniversary of Harcourt's infamous Clayoquot decision? Was it just happenstance that the hotel MB chose had a myriad of entrances so shareholders, board members and their CEO could easily slip in and out without being confronted by protesters? We held up our huge Clayoquot banner *Save Clayoquot Sound – Nuu-chah-nulth First Nations' Territory* on the sidewalk outside the hotel (the one we had displayed all the previous summer at Sutton Pass) and passed out our latest Clayoquot newspapers to passers by.

We had barely set up when a pickup truck pulled up and a person dressed in a bear costume jumped out and started shoveling a load of manure onto the sidewalk in front of the hotel entrance. Needless to say this Earth First "direct action" eclipsed our protest and made the news! It didn't faze me: I was still convinced that the less showy process of mass public education is the backbone of social change.

Tzeporah Berman of Greenpeace, who was one of the leaders at the blockade site the previous summer, said hello as she went inside with some shareholder proxies so she could ask questions and vote. She presented a report to shareholders listing all the logging violations and toxic spills that MB had been convicted of over the past few years; making the point that MB was the real criminal. Passions were still running high as hundreds of protesters were now being convicted and sentenced to fines and jail time in mass trials of criminal court contempt for their non-violent civil disobedient road blockading to stop MB's Clayoquot logging the previous summer. We thought her tactic was smart and in the years to come we got shareholder proxies, too. Joe Foy attended many forest company shareholders' meetings to further the cause of wilderness protection with pointed questions that never got truthfully answered.

Victoria's big court case needs a boost

Early in 1994, I got a call from Alison Spriggs. She informed me that a key court case we had launched was in jeopardy. The case was against the Greater Victoria Water District Board and aimed to establish that it was illegal for the Board to authorize any logging in the watersheds that supplied Greater Victoria's residents with drinking water. The Sierra Legal Defence Fund lawyer taking forward the case on behalf of us as well as the Sierra Club of Western Canada and six individuals needed a letter to present to the court certifying who would pay the court costs if we lost the case.

Without the letter, the case wouldn't proceed. The costs could amount to several thousand dollars if we lost. Alison went on to explain that none of the individual plaintiffs or the Sierra Club were willing to assume the risk. Could we do it?

I immediately said yes, of course, pending approval of the E-team and directors, which I was sure would be forthcoming. I knew some of the particulars of the case. We contended that logging and the road building associated with it seriously jeopardized the drinking water quality by increasing sedimentation, the possibility of contamination and the need for chemical disinfecting. More to the legal point, the 1922 statute that created the Greater Victoria Water District (GVWD) clearly stated that it did so for the sole purpose of supplying water for the area's inhabitants.

Logging wasn't mentioned as a permitted activity. The GVWD board justified its logging in the watersheds—which had been going on since the 1950s—by saying that it made money doing it which made the water for the residents of Greater Victoria less expensive because the money made from logging, in effect, subsidized the GVWD's operations. They obviously were not logging to improve water quality!

Winning this case was important and the chances were good that we would. However, there is no such thing as a sure court case. WCWC's board just wanted to make sure the risk was not too great. Since there was no injunction involved and therefore no compensation due to a company and its workers if we lost, our lawyer assured us that at the very most we would be out five to six thousand dollars. Our board unanimously approved WCWC assuming the potential liability. I wrote our guarantee letter, sent it to the lawyer and the case was on.

Logging in the watershed that provides Greater Victoria with its drinking water is ruled illegal

In a landmark decision in March of 1994, B.C. Supreme Court Justice Melnick found the Greater Victoria Water District guilty of illegally logging Greater Victoria's drinking watershed. The GVWD, however, dragged its feet in canceling its logging contracts until the Sierra Legal Defence Fund obtained a B.C. Supreme Court order immediately stopping all commercial logging in the Greater Victoria watershed. It was a great win that inspired us to redouble our efforts to halt logging in Vancouver's drinking watersheds too. In Vancouver the logging had been legally entrenched by an amending indenture that essentially turned the drinking watersheds into a mini-TFL run by the rules in the Forest Act and managed by logging engineers working for the Greater Vancouver Water Board.

Court win provides stimulus for a big Sooke Hills wilderness campaign

Shortly after we'd stopped the logging in Victoria's watershed, Ray Zimmerman, a passionate community activist who had joined WCWC as a co-petitioner in the court case, stopped by our WCWC-Victoria office. Zimmerman loved to hike in the hills around Victoria. He wanted Alison and Misty to launch a full-blown WCWC campaign to establish a Sooke Hills Regional Park that would protect the Greater Victoria Water District's "off-catchment" lands for the recreational use and enjoyment of Victoria residents. These lands were called "off-catchment" because the water from them did not drain into a reservoir that supplied drinking water. They were "surplus" because they would never be required to provide drinking water for Greater Victoria. They totaled 4,900 hectares, nearly five times the size of Vancouver's Stanley Park.

Alison and Misty jumped into the campaign with full enthusiasm. Since our court win, these lands were more threatened than ever. The GVWD was now contemplating selling parcels of its off-catchment lands to real estate developers for housing subdivisions. It was also considering trading some of them to a logging company in exchange for private lands it held within the catchment area. The Association of B.C. Professional Foresters (ABCPF) also had their eyes on them, proposing to develop a "demonstration forest" like the one in the Lower Seymour Valley in Vancouver where logging would take place under the guise of being an educational experience.

WCWC argued that a park was the best land use for the 35 percent of the landbase under the jurisdiction of the Victoria Water District that would never supply drinking water to Victoria (off-catchment lands). Besides, Environment Canada had already categorized the Nanaimo Lowlands Ecosection (in which the Sooke Hills off-catchment lands lay) as *one of North America's most endangered ecosystems*. Ninety nine percent of this ecosection had already been logged or developed and less than one percent was protected. The CORE process did little to add to the protection because nearly all of it was privately owned. Although most of the Sooke off-catchment lands had been logged at one time or another in the past, there remained pockets of ancient coastal Douglas fir, Garry Oak and Arbutus that had escaped the axe and chainsaw. These pockets were among the last stands of their kind left in the 342,000-hectare Nanaimo Lowlands area. In short, the Sooke Hills area that we were campaigning to save was the largest undeveloped wilderness of this unique ecosystem left on southeast Vancouver Island.

In June 1994 our Sooke Hills campaign got underway. Its goal was to create a huge protected *Sea-to-Sea Green-Blue Belt* from Sooke Basin to Finlayson Arm of the Saanich Inlet. Our proposed park included added protection for marine areas at either end and the addition of some private lands and some other Crown parcels, besides the GVWD's surplus, off-catchment lands. It stretched along the western horizon of Victoria, about 15 kilometres away from the steps of Victoria's City Hall and our small WCWC office. No other major urban centre in the world had such a wonderful opportunity to protect a wilderness so large, close and magnificent.

WCWC-Victoria ran this campaign and financed it through its

Sooke Hills campaign team. Misty MacDuffee campaigner far left, next to her Simon Spencer, Canvass Director, Joyce Dalgleish, store volunteer, Alison Spriggs far right, kneeling left and right Russ Forester, fundraiser, and Greg Boweses, store volunteer. Photo: WCWC files.

WCWC Sooke Hills campaigners picnic in the proposed park with the Juan de Fuca and the Olympic Mountains in the background. From left to right Zane Parker, Chris Genovali and Misty MacDuffee. Photo: Alison Spriggs.

door-to-door canvass and campaign newspaper tear-offs. Soon after the campaign got underway, Misty took off on the Stumpy tour across Canada and the U.S. leaving Alison Spriggs in charge. Alison started working for WCWC as a canvasser and took over managing our Victoria office when no one else wanted the job after our near fatal financial meltdown in 1992. Her real love, however, was campaigning to save wilderness. She constantly struggled to get enough money to run our Victoria office and often went without her own meager wages to pay the rest of the staff. Her sympathetic parents helped her out to make ends meet during the toughest times.

WCWC-Victoria research helps to save the Sooke Hills

One of the first things Alison did was hire co-campaigner Zane Parker to organize and publish a preliminary biological inventory of the off-catchment lands, recruiting top-notch scientists who were willing to donate their time and effort to the project. Dr. Richard Hebda, Curator of Botany at the Royal BC Museum, Dr. Nancy Turner, renowned ethnobotanist and professor in the School of Environmental Studies at the University of Victoria, and 10 other researchers took part. After considerable pressure, in April 1994 the GVWD Board finally gave WCWC-Victoria's research team permission to access the area. It was a special exception to the Board's policy of "no trespassing" on GVWD lands regardless of whether or not the water draining from them entered into Victoria's drinking water system. Beginning in May and continuing on into the fall, this research team made nine trips into the area to collect specimens and take field notes.

As the campaign heated up, and the workload grew, Chris Genovali joined the Victoria team. Through relentless campaigning to keep the issue in the local newspapers and force the backroom dealings of the Board out in the open, our Victoria WCWC campaigners along with Ray Zimmerman and others mobilized the vast majority of local citizens in favour of protecting the area. At WCWC's main office in Vancouver we assisted the campaign by helping publish two four-page, full-colour newspapers—*Save the Sooke Hills* (Winter 1994) and *Protect the Sooke Hills – Wilderness now or never!* (Summer 1995). We printed 60,000 copies of each and helped distribute them in the greater Victoria area.

WCWC-Victoria hires a high powered campaigner

After Misty got back from the Stumpy tour in 1994 Victoria hired a new campaigner, Chris Genovali to help. He was a skilled communicator and a big part of his job was to write opinion pieces and letters to the editor and do radio and television interviews. This gave a much higher public profile to our Branch's campaigns. As our Victoria forest campaigner he primarily worked on the Sooke Hills campaign. But he started getting more and more involved in the emerging campaign that became known as the Great Bear Rainforest campaign. His involvement started out through friendships with the Victoria-based McAllister family.

While the majority of environmentalists focused on Carmanah and Clayoquot, Peter McAllister, who originally was with the Sierra Club of B.C., founded a new organization in 1990 to focus on B.C.'s

Dr. Chris Pileou identifying a plant on a Sooke Hills research expedition. Photo: Alison Spriggs.

Dr. Richard Hebda and Adriane Pollard on research expedition in the Sooke Hills. Photo: Alison Spriggs.

vast and remote mainland coast, an area that the rest of us knew virtually nothing about. Peter got his son Ian, a superb photographer, and daughter-in-law Karen involved.

McAllister's Raincoast Conservation Society worked in partnership with scientists, First Nations, local communities and other non-governmental organizations to steadily build support for protection of the marine and rainforest habitats in this "Great Bear Rainforest." It was destined to be the big campaign of the late '90s and the first half-decade of the 21st century.

This is what Chris has to say about one of his campaigning adventures with WCWC:

The Sooke Hills campaign was the most rewarding and enjoyable campaign I had ever worked on. It was particularly satisfying because it was a true grassroots community-oriented effort, as well as being an initiative that benefited the entire region. In addition, we had such a great team of creative people at WCWC- Victoria at that time and it was a pleasure working with all the other grassroots activists around Greater Victoria who were also advocating to protect the Sooke Hills. We spent a significant amount of time on the ground, in the Sooke Hills, documenting and inventorying the area, and maybe even more importantly fostering a close bond to the land itself.

Chris went on to recount one of the more of their most exciting adventures during the Sooke Hills campaign.

At one point the Great Victoria Water District (GVWD), who held the lands we were trying to protect, was considering trading a key valley (the Waugh) in the Sooke Hills to a private logging company in exchange for lands the company had already trashed within the catchment area. This potential land swap, if it went through, would have essentially cut the Sooke Hills greenbelt in half. The GVWD Board was being typically secretive and refused to reveal what kind of backroom dealings they were engaged in.

When we were tipped off that the GVWD was going to be holding a secret meeting regarding the land swap, we were determined to track down the location of this meeting and inform the local media in order to shine a light on the board's shadowy land swap dealings. At 5:30 a.m. on the day of the meeting, Misty MacDuffee, Alison Spriggs and I parked down the street from one the GVWD board members. We were on a stakeout. We slouched down in my car so as not to be seen and waited for the board member to emerge from his house. About an hour and a half later and consuming numerous cups of tea, he came out, got in his car and started to drive. We followed him throughout Victoria, dodging around the traffic to keep him in our sights. It was like a scene out of a spy movie or a TV cop show, and it was hard not to laugh at ourselves playing detective. When the guy parked his car downtown we did the same, then getting out and following him on foot through the back alleys of Victoria, trying to keep ourselves from being seen but not losing his trail. He entered an office building, but soon came back out - too fast for any meeting. He got back in his car and eventually led us to the location where the meeting was to take place. We tipped off the media and the subsequent coverage publicly exposed the proposed land swap and the threat to the Waugh Valley.

After two short years Chris Genovali left our employ and went to work for the Raincoast Conservation Society where today he is its Executive Director.

Chapter 47

WCWC fights for Boise
WILD fights for Chilean wilds

The public gets to decide (comment) on park options

In the year leading up to February '94, the Pinecone Lake-Burke Mountain Study Team met frequently. But the team could not come to consensus, so it developed four different land use alternatives. The Study Team presented all four options in its final report aptly titled *Options Report of the Pinecone Lake - Burke Mountain Study Team*. They ranged from no park at all to complete park protection for all of "Vancouver's 38,000-hectare Backyard Wilderness" as we called the study area.

Joe, who sat as WCWC's representative on the Team, made sure that the full-preservation option was backed up with compelling reasons to choose it. Among them were that the Boise and DeBeck Valleys—the heart of the region—comprised the closest, largest old-growth forested wilderness to Vancouver outside the few remnants already protected in provincial parks; that 31 old-growth dependent species lived in this unprotected forest; and that a unique species of hybrid char (a trout-like fresh-water fish) inhabited Boise Creek.

Boise bumper sticker. Artwork: WCWC files.

The *Options Report* also said that this wilderness contained a 50-kilometre-long hiking route (the trail we'd worked on building over the last two summers), which connected the top end of a logging road near Coquitlam to the top end of another one near Squamish. The code word here was "route." The rugged terrain between the two logging roads was steep and extremely difficult to transverse. Much of the route was merely sketched out with flagging tape to prove that one could bushwhack their way through—if they had enough determination and skill.

The clincher, however, was a tourism study that optimistically predicted that over the long term, nearly twice as many jobs would be created under the complete park protection option as compared to the full-out logging option. It also concluded that complete park protection would "cost" only a four percent reduction in the current rate of logging throughout the Fraser Timber Supply Area, thus making the new park affordable.

The Team held four public input meetings to gage public opinion: two in logging communities and two in urban communities. It also sought additional public input through a short questionnaire that asked people which option they supported. As soon as we found out the schedule for the meetings, we sent out a special notice to all our members in the Lower Mainland urging them to attend.

Besides WCWC, the Burke Mountain Naturalists, a very large and active local environmental group based in Coquitlam, also sat on the Study Team and built public support for the large park. In mid

Facing page: Image of the Alarcé Ancient Rainforest in Chile with trees over 3,000 years old used for WILD's poster titled "Cathedrals of Life." Photo: Daniel Dancer.

February '94, the Team held its first public meeting in Coquitlam's Dogwood Pavilion. Over 1,200 people showed up and tried to crowd into a space designed to hold 300. Speaker after speaker supported the maximum-sized park to thunderous applause. On the following night at the meeting in downtown Vancouver in Robson Square's Judge White Theatre, nearly 500 people packed into the 300-person capacity room and several hundred more had to be turned away.

Again, the overwhelming majority favoured the full park protection. When the Study Team introduced themselves from the stage, Joe got a standing ovation. And again thunderous applause greeted Adriane when she spoke in favour of the full-park option from the audience microphone.

We knew that the tables could be turned at the last two meetings in Chilliwack and Squamish, both loggers' strongholds. To have at least some pro-park people at the Chilliwack meeting, Joe suggested that we bus in our supporters. So we rented a bus and took 50 of our volunteers, many of whom had worked on our Boise trail, from our Gastown office to the meeting and back afterwards. We decided not to bus people to the final meeting in Squamish, because that could be seen as "overkill." The loggers in Squamish were not going to be directly affected by the government decision. Joe, who was part of the Study Team, would be our only emissary there. As it turned out, only about 150 turned out to the Squamish meeting and among them were a few local "voices for the Boise wilderness."

Our bus arrived early in Chilliwack and we promptly set up informational pickets outside the entrances to the hotel, and tried to hand our latest Boise newspaper to people coming to attend the meeting. Few would take them. I put a stack of them on a hall table beside the door to the meeting room. I noticed when I came back about 10 minutes later that they were all gone. That just wasn't right. I checked every wastepaper receptacle and found all of our papers stuffed into the last one at the far end of the hallway. "They're not too smart," I thought to myself as I rescued them. "This meeting is going to be a hot one!"

I went in and stood at the back of the room now filled to near capacity with about 400 people. During the pro-park presentation by Joe, some of the loggers started to heckle. I couldn't help myself and started heckling back. I like to heckle. Adriane came over and gave me the big nudge to silence me. After the Team members' reports, several brave souls who supported the creation of the park (not from our bus) stepped up to the mike. We cheered them loudly. But a lot more people got up and spoke out against it. After it was over, on the way out of the meeting, I saw several loggers pick up a few extra comment forms from a table in the back, presumably to take them home and get their friends to fill them out. I took a few extra myself.

On the bus ride home I told Joe about the "enemy" taking home extra comment sheets and we brainstormed up the idea of mailing out a copy of the questionnaire to all our members in the Lower Mainland. This time (I believe it was the first time ever) we decided not to include an appeal for money in the mailout. Everyone at WCWC agreed that this was a great idea. We had been hitting our members so frequently and so hard that we were generating "donor fatigue." We were sure that our members would welcome this non-monetary call to action.

I began the cover letter by making it clear that this was not another WCWC appeal for money. It was an appeal for people to use their pens, not their pocketbooks to save wilderness. I asked every member to fill out the enclosed public input comment form and send it right back to us in the enclosed pre-postpaid envelope. Receiving the questionnaire meant that we could count the responses before forwarding them on to the Pinecone-Burke Planning Team. We rushed to get the pieces of the mailout printed in time for our next volunteer night two days later. The Study Team's meeting to discuss the public comments was only five weeks away. As usual Joe, Adriane and I worked alongside our faithful volunteers to stuff, label, seal and sort the mailout. We were motivated! Everyone worked late into the night to get the job done.

With the thought, "We'll show them what real public input is!" running through my mind, I dropped off the ten mail sacks filled with bundles of letters on Vancouver Main Post Office's dock the next morning. We waited with great expectations for the returns. I always liked to take a quick look at the mail every morning when it arrived to get a pulse on what was coming in from our various mailouts. It was sort of a ritual. I was sure the response from this one would be spectacular.

It takes about a week for the first returns to start coming in. By the end of the second week, given that it was a local mailout, I expected a flood. But only a trickle came in. At our E-team meeting we speculated as to why the low return. "Perhaps the people are sending them directly in to the Study Team. The address is on the bottom of the form," suggested Adriane.

"Not too likely. We provided a postpaid envelope. It's got to be apathy," I cynically quipped back.

"We've got to find why," said Joe, whose campaign was on the line.

We decided to hire a phone canvasser who had successfully solicited donations for our Clayoquot Witness Trail, to call everyone and urge them to fill out this questionnaire and send it back to us ASAP. After a couple of nights of calling, he had some startling news for us. More than half of the people he had talked to hadn't even bothered to open our letter yet, thinking it was just another appeal for money. Some had already thrown it away unopened! Many of our supporters also supported other environmental groups and they were flooded with appeals. When he explained what was in our envelope and how important it was for them to register their opinion on making the entire Pinecone-Burke Mountain area a park, nearly everyone was eager to help out. We sent out a second questionnaire to quite a few people who had "lost" or "misplaced" the original mailout. The phoning resulted in generating hundreds of pro-park public input forms, greatly out numbering the pro-logging ones. We'd learned a good lesson. From then on, we took into consideration "donor fatigue," limited the number of appeals we sent out per year, and printed something catchy on the outside envelope to entice people to open it.

We expected the Harcourt government to make a quick decision after the Study Team submitted its report. But we had to wait until June 1995, more than a year for it. Grafted onto the NDP's Pinecone-Burke Park announcement was another one—a big ugly surprise.

WILD Campaign focuses on helping Chilean conservation organizations and preserving global biodiversity

During this same time period WCWC-WILD took off in a new direction. After completing the project to map as much of Latin America's remaining wilderness as possible with the limited funds available through our CIDA grants, our WILD team decided to focus its energy on Chile, a country with similar temperate rainforest ecosystems to those found in B.C. We "partnered up" with Fundacion Lahuen, a non-profit non-government environmental group registered in Chile with goals to protect the surviving primary forest habitats in that country. Together with this partner we designed the *Cani-Clayoquot Project: Protecting Temperate Rainforests in Chile and British Columbia* and got CIDA funding for it. At that time less than 10 percent of the Earth's primary (original) temperate rainforest re-

Cover of the newspaper we co-published with our southern partner groups (English version). Document: WCWC archives.

mained and by far the largest fragments of it existed along the west coasts of Chile and Canada.

Fundacion Lahuen, among other things, ran their own privately-owned 400-hectare Cani Forest Sanctuary where they conducted education workshops for local people. Both of our groups were committed to using "education as advocacy," i.e. public education as a means to achieving conservation of what little remained of the old-growth temperate forests. Our CIDA funding enabled Fundacion Lahuen, which prior to this project was completely volunteer run, to hire people and greatly expand their educational work.

Adriane and our GIS mapper, Ian Parfitt, visited our conservation partner in late February 1994. Besides her descriptions of remarkable forests of ancient araucaria ("monkey puzzle" trees—one of the most ancient conifer species) in their natural high elevation habitat, Adriane returned with a tale of almost being run over by a herd of tarantula spiders. This is what happened. While taking a nap during the lunch break on a rigorous hike to one of the remote areas that our partners wanted to protect, she was startled awake by fellow hikers shouting at her in Spanish. A troupe of huge tarantulas was moving towards her, only a few feet away in the grass.

Almost as remarkable (according to Adriane) was her experience the next day when a teacher led the group through some wilderness sensitivity exercises. In his hand he held a well-worn copy of Thom Henley's *Rediscovery* book that WCWC had published five years before!

Adriane returned to Vancouver with a plan to continue the partnership project with Fundacion Lahuen and expand it to include some national campaign work by CODEFF (Comite Nacional pro Defensa de la Fauna y Flora), a conservation organization in Chile very similar in scale and operations to WCWC. Adriane told us all about CODEFF's involvement in the global efforts to establish the Forest Stewardship Council (FSC): a grassroots process to certify eco-friendly forestry. It was the first time we'd heard about FSC.

Randy Stoltmann brings us a new wilderness proposal

Around the same time that Adriane and Ian returned from Chile, Randy Stoltmann stopped by our office to give me a copy of a brief he had just completed titled *The Clendenning/Elaho/Upper Lillooet Wilderness -"Stanley Smith" Wilderness*. He had prepared it as input to the provincial government's Protected Area Strategy (PAS)—the name the NDP gave to its overall provincial parks planning process (including its regional CORE tables) that aimed to protect 12 percent of B.C.'s land base. Randy exuded the same infectious enthusiasm for the wild Clendenning/Elaho/Upper Lillooet area as he had for Carmanah Valley. He let us know that this was the next big one. He wanted us to get involved.

Randy told us that he was about to go on an extended ski traverse in the Kitlope area with John Clarke. Clarke (whom I'd not yet met) was an accomplished mountaineer who knew more about B.C.'s Coastal Mountain Range, including the proposed Stanley Smith Wilderness Area, than anyone else because he'd personally explored most of it. We agreed we would talk more about our involvement when he got back. After Randy left I studied his 32-page proposal, which was sponsored by Mountain Equipment Co-op. It advocated the preservation of a 260,000-hectare wilderness area encompassing the Upper Elaho, Sims, Clendenning Valleys (all within the Squamish Watershed and TFL #38) and the Upper Lillooet Valley.

Randy's reports always had a wealth of information in them that compelled sympathetic readers to want to help save the areas. This particular wilderness was first traversed and described only a century earlier by the legendary explorer Stanley Smith. Randy quoted from Smith's original report on Clendenning Valley: *Close to the summit we crossed a glacier* [now named Havoc Glacier], *which filled the valley. The scenery in many places is very fine, and we found luxuriant vegetation close to the snow. There are falls on this route rivaling Shaffhausen or Yosemite...*

Randy also pointed out in his report that the B.C. Ministry of Forests, through the Parks and Wilderness for the 90s process, had proposed that part of the Elaho/Upper Lillooet region be designated "wilderness," which provided *an official recognition of its special value.* Over the last 12 years Randy had systematically explored the mountainous country within 200 kilometres of Vancouver and felt that he was *...qualified to rank the Elaho/Upper Lillooet wilderness as the most important unprotected wilderness area in the Eastern Pacific Ranges in terms of variety of special features, backcountry recreation opportunities, remoteness and overall wilderness conservation values.*

Randy also noted that there were only three large wilderness areas remaining within easy reach of Vancouver: the Garibaldi complex, the Stein/Upper Nahatlach watersheds and his proposed Stanley Smith Wilderness. Not one of them was fully protected. The Stein system remained threatened by logging. Additions to Garibaldi to give it low elevation continuity were pending. All the major valleys of the Stanley Smith Wilderness were already scheduled for road building this year and logging the next. There was already logging road access to the Lower Elaho on the southern boundary of the wilderness. Interfor, the logging company that now held the cutting rights to the area, was about to build a major bridge across the Elaho near its confluence with the Clendenning River to access the wilderness on the west side.

Renaming a wilderness in memory of Randy Stoltmann

I finished the report, excited about getting involved in this campaign and looked forward to talking to Randy in a few weeks when he returned. But he didn't return. On May 21, 1994, Randy Stoltmann, only 32-years-old, died in a mountaineering accident, swept off a cliff by a small avalanche as he was skiing across a slope in the remote ranges west of the Kitlope River in B.C.'s central Coast Mountains. The news hit like a bombshell. It shocked and saddened the whole extended family of wilderness advocates and campaigners in B.C. John Clarke, who was the leader of this tragic expedition, said, "*For me, sharing a tent with him on his last trek sharpened the tragedy of his death, but also gave me a sense of his vision for the future of wild places. His knack for being liked by people on both sides of this argument* [whether or not to preserve wilderness] *grew from his belief that this was one divisive issue that ultimately involved everyone. Randy said to me, `In the long run, we're all in this one together'.* "

For a long time I thought about Randy and how important he was to the movement and how great a loss his death was to us and our work. At WCWC we felt that the best way to honour his life was to win protection for the last area he asked us to help save.

The first thing we did was to reprint several hundred copies of his *Stanley Smith Wilderness* brief and widely circulate it. By consensus everyone agreed that we should rename the Stanley Smith Wilderness the Randy Stoltmann Wilderness in honour of his work and proposal. Before we did, Joe Foy asked for permission from Randy's parents and brother, Greg and got it. That summer we were tied up in Clayoquot Sound completing the Clayoquot Valley Witness Trail. It wasn't until January 1995 that we got going with the Stoltmann Wilderness campaign. It turned into the toughest, meanest and most disheartening campaign we had ever tackled.

Art posters defend wilderness

In August of 1994, a young artist who just graduated from the Ontario College of Art, Drawing and Painting named Dominik Modlinski, walked into our office. He showed us a folio of watercolour paintings he'd just created during a solo hike he'd taken in the Lower Stein Valley the week before. He offered to donate one to us to help save the Stein. We hadn't done anything for the Stein campaign in a couple of years and here was an opportunity we couldn't pass up. We chose our favourite one and produced a beautiful poster titled *"Stein Valley – Protect its Natural and Cultural Heritage Forever."* Sue Fox picked a purple background to frame the image and placed the famous Stein pictograph beside the title on the bottom. In total, this art poster cost less than $2,000. Modlinski used it as a promo for his art show in Toronto in September. We hoped we'd recover our cost when the original painting sold there; but it didn't. The original, still hanging in the entryway to our office in Gastown is going to be worth a lot of money someday. Regardless, the poster gave the Stein campaign a good boost when it needed it.

Over the years we produced and published over 25 different art posters. Our most popular series featured the paintings of Linda Frimmer, some of which were even purchased by BC Ferry Corporation for their passenger lounges. I helped pick a few titles including *Trees of Life, Earth Gardens* and *River Eternal*. Linda's generous donation of her art posters raised thousands of dollars for WCWC's campaigns over the years.

I love art that reaches into peoples' emotions and tugs hard on the bonds to wildlife and wilderness that I believe are deep-wired into all humans. Among my favourite art posters is *Scorned as Timber, Beloved of the Sky* featuring a 1935 painting by Emily Carr of a tall spindly tree that loggers left standing in a clearcut. One of our volunteers with connections to the Vancouver Art Gallery suggested that we use this painting for a poster and arranged for us to meet with the curator of the museum. There was no problem getting permission and there was no royalty charge because we were publishing it to raise money for our charitable society's cause of protecting nature and not for our personal profit. It didn't take much imagination to view the painting as a contemporary anti-clearcutting statement.

The idea of publishing art posters was not so much to make money as to marshal the power of art into the service of saving wilderness. But, like all our other publications, the goal was also to raise enough money so we could continue to publish more images and educational materials. Active campaigns constantly need refreshment.

Poster published to celebrate the successful completion of the Ahousaht Wild Side Heritage Trail. Poster image: Graham Osborne.

Chapter 48

Walking the Ahousaht Wild Side

A "Working Holiday" on Flores Island

In the summer of 1994, Adriane and I decided to holiday with our children in Clayoquot Sound. We had been campaigning intensively to protect the place for over a year and a half and we really hadn't seen much of it ourselves. It had been nine years since our last trip into Clayoquot's wilderness—our canoe expedition up the Megin River in the summer of '85 when Kallie was 18 months old.

The place we had to see, I convinced Adriane, was a small Indian Reserve the Ahousahts called Utla-Cutla on the northwest coast of Flores Island. When I was working for the Nuu-chah-nulth Tribal Council in the early '80s, I was captivated by a story told to me by the late John Jacobson, a noted Ahousaht historian, about this special place where his ancestors lived. He explained to me in detail how they had practiced "salmon enhancement" there.

Every fall they would camp near the mouth of the Utla-Cutla River by a pool below a ten-foot waterfall. This falls was so high it prevented most of the salmon from reaching the excellent spawning gravels in tributaries feeding the big lake above it. They'd catch the returning coho swimming in the pool and carry them live in cedar baskets to the lake above and let them go. "It's a beautiful place," he said, describing a protected cove perfect for landing a small boat and a trail to a nearby sandy beach where you could camp next to a grove of enormous cedars. I promised myself then, that some day I would visit Utla-Cutla.

To camp there, we had to get the permission of the Ahousaht Band Council. Adriane called the then elected Chief Councilor of the Band, Louie Frank, and asked him how we should go about asking for permission. He explained the protocol: write a formal letter to the Council and they would consider it. I thought that was probably a polite way of saying no to us, but we sent off the letter anyway and waited. Three weeks later the reply came: permission granted!

We began our preparations for an August trip. I called Ron Aspinall who, although no longer a WCWC director, still supported our work. He was one of the doctors serving the Ahousaht community and he offered us the use of his house there. "Be sure to see Susan Jones, the wife of the new store manager, while you're in Ahousaht. She's quite active in the local effort to protect Flores Island," Ron advised me as he hung up.

In order to get to Utla-Cutla we caught a scheduled run of the Seabus water taxi to Ahousaht and there hired another water taxi to take us to the remote reserve. We planned to hike back along the outer rocky shoreline to Cow Bay and from there across beaches and over headlands back to Ahousaht so we didn't arrange a boat to pick us up.

Utla-Cutla was just like Jacobson had described. There was a steep 20-metre-long path from the pool at the base of the falls to the lake at the top. The rocks in the path were as deeply worn as the steps in the Parthenon in Greece (according to Adriane who'd walked both). It was an ancient-feeling place. For days we saw no one as we explored and relaxed. We got revitalized and re-inspired. Flores Island was indeed a magical place—the heart of Clayoquot Sound. After a little bushwhacking with our two young kids we realized that hiking back would be too arduous. So, after hanging out at Utla-Cutla for 10 beautiful days and with a big storm approaching, we "flagged down" a small fishing boat and hitched a ride back to Ahousaht.

Beginning of WCWC involvement in "Walk the Wild Side" trail with the Ahousaht First Nations

We walked to the house that Ron said we could use, found the hidden key and "made ourselves at home." After taking showers and doing our laundry, we decided to try to find Susan Jones. It wasn't hard, Ahousaht is a small place. We started at the only grocery store in town, the one that Mike, her husband, had established. He was in the process of training Ahousahts to take over and manage it. Mike immediately invited us to supper.

That evening was the beginning of a whole new dimension to WCWC's Clayoquot campaign. Susan, who came from Calgary with a background in property management, was as full of ideas as she was of energy. She was a natural born campaigner.

She told us the story of how a few months after she arrived there (a year earlier) she and several Ahousaht women had formed "Walk the Wild Side," a First Nations women's eco-tourism initiative, to secure steady jobs and income for their community. They dreamed of opening up a world-class hiking trail on beautiful Flores Island. On a cold morning in March '93, they put out a call over the local VHF (the marine two-way radio channel that everyone there listens to) for help in clearing the traditional route from the Ahousaht village to the wild side beaches of Flores Island. Fifteen volunteers in total including men, women and children came out that day to help clear the brush and fallen trees from the path. That first summer, with just sweat equity and a $158 investment in a brochure and posters, which they put up in all the Tofino tourist spots, the Walk the Wild Side women pulled in $19,000 worth of art sales, seabus rides and trail guide wages.

By summer 1994, after just one year's tourist use, the trail had degraded into a series of huge mud wallows. Susan estimated that over 8,000 people per year were walking, kayaking and boating to the beaches on Flores. Ahousaht elders were preparing to shut down the muddy trail unless measures were taken to protect the environment. She asked for our help and involvement. She took us down and showed us a new building beside the grocery store. It was not quite finished and she thought that WCWC might be able to rent the place quite reasonably and set up an office in half of the space, with Walk the Wild Side using the other half to sell crafts to the tourists that came through and hiked the trail.

It would be the first time that an environmental group had opened an office on an Indian Reserve in B.C. or Canada, as far as we knew, and the idea immediately caught our imagination. The Ahousahts' traditional territory, located in the centre of Clayoquot Sound, encompassed over 70 percent of the remaining wilderness in the entire Sound. They were making strong moves to protect their lands and their salmon bearing watersheds from being damaged further by clearcut logging.

Our proposal to open the office was sidelined in August 1994 when we got "caught in the crossfire" of a dispute between Greenpeace and the Nuu-chah-nulth Chiefs. Greenpeace arranged to take Nuu-chah-nulth Chiefs and MB officials on an educational visit to see eco-forestry operations in Europe, and then abruptly cancelled the tour. The reason that Greenpeace gave for doing this was that MB was using correspondence between MB, the Nuu-chah-nulth and them in a public relations package they were sending out to their customers in Europe, claiming that it showed that the conflict in Clayoquot was over. This "dirty tactic" being deployed by MB, whose clearcutting continued to destroy the ancient forests in Clayoquot, was undermining Greenpeace's markets campaign aimed at getting MB's customers in

Europe to cancel their contracts to buy MB's "Clayoquot-destroying" paper. Besides canceling the trip, Greenpeace retaliated by demanding that MB immediately cease all clearcutting in Clayoquot. They withdrew from further joint meetings until the clearcutting stopped. MB continued to log. The Nuu-chah-nulth, who were miffed by the abrupt cancellation of the trip, accused Greenpeace of not respecting First Nations' rights and decided that they did not want anything more to do with environmentalists, including us. Hence, our negotiations for the Ahousaht office were off.

Truth be told, WCWC exacerbated the tense situation between the First Nations and environmental groups by excitedly talking about our "new office" in Ahousaht as if it were a sure thing. It got reported in the media, which incensed the First Nations, because they had not even given approval yet. Adriane, who had been chosen by the Nuu-chah-nulth to be the liaison to all the environmental groups, had to do a great deal of diplomatic work to get things back on track.

Susan Jones made arrangements for Adriane to meet with Cliff Atleo, who was a powerful figure in Ahousaht and also was one of two Nuu-chah-nulth designated spokespeople on the whole Clayoquot issue (the other was Chief Councilor Francis Frank of the Tla-o-quiaht First Nation).

The meeting between Adriane and Cliff was one of the most intense in the whole Clayoquot campaign. It took place during the lunch break in one of the big treaty negotiating sessions. There were over 100 people there. Cliff took Adriane to the far corner of the room. Ahousaht chiefs sat in a row facing her. Adriane only had Susan there for support. Cliff started by angrily berating Adriane for the whole Greenpeace affair. He would not let up, accusing her of not doing her "liaison" job which he felt meant pulling Greenpeace in line. Adriane countered by explaining that the various environmental groups were like the various First Nations—the Ahousaht, Tla-o-quiaht, Hesquiat and other 11 nations that comprised the Nuu-chah-nulth Tribal Council. Just as it was near impossible for the Ahousaht to tell the Tla-o-quiaht what to do, it was near impossible for WCWC to tell Greenpeace what to do. But she'd talk with them nonetheless, and certainly reiterate protocols in communications.

Cliff said that wasn't good enough. Adriane was so frustrated she was near to tears. Just when it looked like the whole meeting was going to end in an impasse, Cliff dramatically changed his tune, commended Adriane for her work and assured her that the Ahousaht would approve the opening of WCWC's office in their village. Adriane was absolutely shocked and relieved. She and Susan left, elated with the result. Then Susan explained to Adriane what really happened. Everyone in the negotiating room was covertly eyeing Cliff and Adriane's intense discussion. At one point well into the discussion, the women of Ahousaht came to Adriane's defence. Adriane couldn't see it, but they began to pull up chairs and quietly, one by one, sit in a row behind her. They were the women of power in Ahousaht. Cliff could see them, and when their support coalesced behind Adriane, he gave in. In late September 1994 the Wild Side store and WCWC's office in Ahousaht opened for business.

From Pretty Girl to an emergency trip to the Ursus

Our office in Ahousaht provided WCWC with unique opportunities to help protect Clayoquot. Right after we opened, Susan, who made dropping into the Ahousaht Band Office a part of her daily routine, called us and said that the Band was concerned about MB's activities in the northern part of their territory. Upon the urgings of the Ahousaht Chief and Council, Susan organized an expedition with Joe Foy, Alison Spriggs and several young natives. They explored

Easter Lake expedition crew. Photo: Joe Foy—delayed timer.

the region that was threatened with logging—Young Bay, Easter Lake and Pretty Girl Lake. The expedition resulted in a report outlining this area's outstanding features and its tourism potential. We presented our report and a slideshow about the expedition to both the Ahousaht Council and community.

Another nearby wilderness, the Ursus Valley, was particularly important to the Ahousaht. Susan suggested we hire an Ahousaht native guide and go on an exploratory mission there in September of 1994. It was a good idea and Joe Foy was keen to head up our team. Joe had great bush skills and always took advantage of every opportunity to get out into the wilderness. The Ursus watershed is only 6,567 hectares in size. But, because of the gradual gradient of the river from source to mouth, it has a huge expanse of spawning gravels and is exceptionally rich in salmon. All five salmon species spawn in the Ursus. This valley is also known for its abundance of bears, hence the name, and for its resident herd of Roosevelt elk.

It is so special that, while being included in the logging zone in the government's April 13th decision, it was designated a "Special Wildlife Management Zone." This was supposed to mean that full environmental impact studies had to be undertaken before any industrial activity would be allowed in that watershed in order to make sure that such activities wouldn't harm the valley's wild life.

During this first trip into the Ursus, Joe and the crew happened upon a Ministry of Environment (MOE) helicopter on a sandbar in the river. Unbeknownst to the Ahousahts, these MOE researchers were undertaking scientific studies. When Joe and crew got back to Ahousaht, they informed the Band Council about this activity. Joe returned to Vancouver with an incredible set of 35-mm slides and couldn't stop talking about the place. He succinctly summed it up: *"The Ursus is one of the ecological jewels of Clayoquot Sound."* He was looking forward to going there on an extended exploratory expedition the following spring.

But that was not to be. Later that same fall some Ahousaht elk hunters came upon some MB researchers in the Ursus who had helicoptered in to do wildlife studies. The hunters learned from them that MB was seeking approval to build a road up the Ursus Valley to access the timber in the Upper Bulson Valley, another contentious area in Clayoquot. MB obviously aimed to complete the studies quickly so it could fast track this development. There had been no consultation with the Ahousaht Band regarding any of this research activity. MB's total disregard of their aboriginal rights in the Ursus angered the Ahousahts.

Roselie Thomas pulling boat of equipment up the Ursus River for our CMT research camp with Kurt John assisting. Photo: Joe Foy.

In December of 1994, the Chief and Council of Ahousaht, alarmed by possible road building and logging in the Ursus—rumoured to begin as early as springtime—decided to launch an emergency expedition to hunt down and document evidence of aboriginal use there. They asked us to participate. It was the worst possible time of the year to do field work. December is cold, wet and often snowy and the days are short. But, since the Ahousaht Chief and Council wanted us to go and the logging threat was imminent, we went.

Three Ahousaht natives along with Joe and Susan Jones went in first to establish a research base camp consisting of a big wall tent that we had purchased, at the lowest "cost" price, from American Fabricators, the same company that had donated our Carmanah research tent. It was not easy getting the tent and equipment in there. The expedition team took turns wading up the river towing a small boat piled high with all the gear. Joe, never one to complain, told me years later that his feet got so cold in the icy water that it took weeks for the numbness to go away. Eventually, they found a good spot and set up the tent. With an airtight cook stove stoked up with wood, it became nice and warm inside and steamy as all the wet clothes began to dry.

After establishing the camp, Joe helped Marion Parker, the dendrochronologist who was an expert in dating culturally modified trees (CMTs), hike in. Marion had worked for us in the Queen Charlotte Islands during the South Moresby campaign. His expertise was vital to the success of the expedition. Getting him in and out alive, however, turned out to be a real challenge. Marion failed to tell us that he had a deteriorating heart condition. He was weak, and constantly short of breath but he was game. He frequently downed nitroglycerin pills to dull his heart pains. Steve Lawson, who had a great deal of experience in finding CMTs and his son Matt also joined this winter expedition. Our mapping expert Ian Parfitt rounded out the crew.

Although the crew spent a week in the Ursus, the short daylight hours meant that they only had time to survey a tiny area. Still, they found a remarkable number of CMTs including some related to two canoe building sites—one dated sometime prior to 1869 and another prior to 1895. The 30 CMT sites recorded and mapped included cedars with test holes and scars where bark had been stripped. In

Marion Parker looks at the growth rings in a core sample taken with an increment bore from the healing lobe on the side of a CMT while on the Ursus expedition in December 1994. Photo: WCWC files.

With the help of WCWC director Bob Broughton, we also put this paper up on WCWC's new world wide web site: the first paper we published electronically.

Our CMT report effectively put an end to MB's efforts to fast track road building in the Ursus. There would have to be a thorough CMT study, which would take more than a year to complete, before the company could even contemplate barging in with a bulldozer. Such a study would surely document extensive aboriginal use, much of it prior to the time when British sovereignty over British Columbia was established by the Oregon Boundary Treaty of 1846. Like other types of archaeological sites in British Columbia, forest utilization sites with CMTs modified prior to 1846 are protected by the B.C. Heritage Conservation Act. The B.C. government arbitrarily chose the signing of the Oregon Treaty as the "cut off" date separating archaeological sites that would be protected or unprotected when it revised the Heritage Conservation Act in the 1980s. Besides possible heritage protection, the Clayoquot Sound Interim Measures Agreement (IMA), signed in March 1994, would also undoubtedly help stop MB. The IMA, through the Clayoquot Central Region Board (CRB), gave Clayoquot Sound's First Nations the right to review all proposed resource developments in Clayoquot and the power to stop those they believed were harmful to the environment or native cultural heritage.

Tongue-in-cheek comparison between the Hawaiian Islands and Clayoquot Sound

In the summer of 1995, WCWC published and distributed 30,000 *Clayoquot Sound –the Hawaii of British Columbia* two-part mailers. It made a "tongue in cheek" comparison between the Hawaiian Islands and Clayoquot including size, shoreline, and statistics on mild temperatures. One of the two large colour postcards featured a beautiful aerial photo of pristine Clayoquot, the other of recent massive clearcut destruction in Side Bay to the north of the Sound as an example of what could happen to Clayoquot.

The ugly postcard addressed to Prime Minister Jean Chrétien contained one opinion poll question: should he keep the promise the Liberals made during the last election to help save the ancient temperate rainforest in Clayoquot Sound or not. We never did find out how many people checked the "keep the promise" box and mailed the card in (no postage required). The important thing in a campaign is to always have something new for supporters to do. In tough campaigns, humour especially helps.

The *Ahousaht Wild Side Heritage Trail and Eco-Tourism Project* is launched

During 1995, Adriane and Susan worked closely with elected Ahousaht chief Louie Frank, to develop a joint Ahousaht Band Council-Wilderness Committee proposal—the *Ahousaht Wild Side Heritage Trail and Eco-Tourism Project*—to submit to Youth Services Canada (YSC). We sought funding to employ native and non-native youths to upgrade and complete a 16-kilometre-long trail, following the route of an ancient traditional trail, to the farthest wild side beach on Flores Island that everyone called Cow Bay (although it had no cows). Louie Frank, a powerfully gentle and visionary man, completely backed the joint project and agreed with the wisdom in having native and non-native youth work together on the project, even though unemployment rates were so drastically high amongst the First Nations people.

It took almost a year to get YSC approval. YSC, a federally funded program, required letters of support from MB as well as ones from all

Test hole cedar. Checked out for soundness, years ago by aboriginal "loggers" this cedar failed and thus was left standing. Photo: WCWC files.

January 1995, the Ahousaht First Nations and WCWC co-published a 45-page report about our joint expeditions into the Ursus titled *Preliminary Investigations of Culturally Modified Trees (CMTs) by Aboriginal Use of the Ursus Valley in Ahousaht Territory of Clayoquot Sound*. A few months later, in the spring of '95, we co-published PROTECT URSUS VALLEY – AHOUSAHT TERRITORY printing 100,000 copies of this four-page, full-colour, tabloid-sized newspaper, which gave a popular account of the expeditions and findings. Both publications went through an extensive review process by the Ahousaht First Nations before they went to press.

We worked hard at distributing the PROTECT URSUS VALLEY newspaper. Besides getting copies to all the MLAs and MPs, we delivered one to every household in Ahousaht and mailed copies to every First Nation Band Office in B.C. and to every environmental group and media outlet included on our extensive list of North American contacts for our Clayoquot Sound campaign.

Ursus bumper sticker. Artwork: WCWC files.

Clayoquot Sound, is on the central west coast (open Pacific Ocean) side of Vancouver Island. The watershed basins which flow into the Sound cover 272,000 hectares of land. More than 70 percent of this region is natural wilderness, never roaded nor industrialized by man. It is a place of adventure, mystery and beauty beyond description. The lush jungle-like rainforest and high energy coastline have inspired some people to call the area the "Hawaii of British Columbia".

Clearcut logging, which has destroyed much of the natural temperate rainforest along the entire west coast of North America now encroaches upon Clayoquot. The Nuu-chah-nulth natives who hold traditional title to the land, local residents, and people around the world are joining together to prevent further clearcut logging destruction and save Clayoquot Sound. Will you join them? Address your letters of concern to the Premier of B. C., Legislative Buildings, Victoria, British Columbia, Canada V8V 1X4.

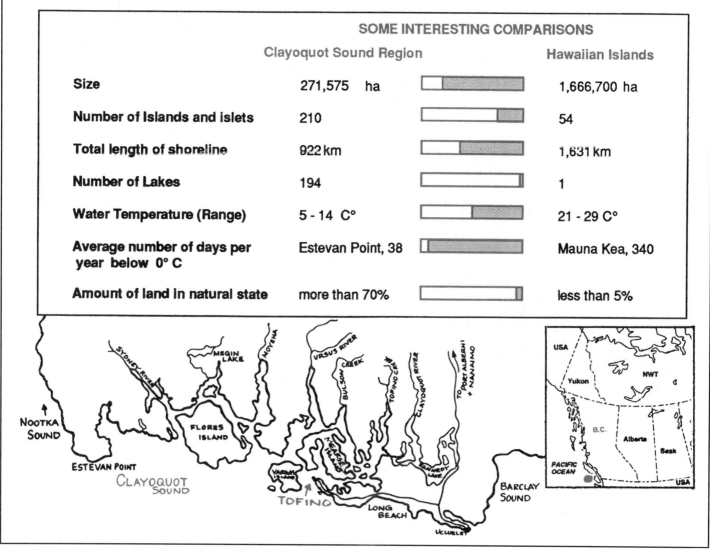

Humorous statistical comparison featured in our *Clayoquot Sound – the Hawaii of British Columbia* two-part mailer. Graphic art: WCWC files.

the surrounding towns and villages in Clayoquot, including Ucluelet where most of the Clayoquot loggers lived who were vehemently opposed to any more preservation in Clayoquot. No easy feat! The biggest hurdle was the letter of support from MB. Susan, who never let anything stand in her way, called MB's new Vice-President of Environmental Affairs, Linda Coady, and made the request. Linda offered to provide not only the letter of support but also the assistance of an experienced faller who would saw through the deadfalls that crossed the trail and train the youth participants on how to handle chainsaws. The $194,000 Youth Services Canada funding came through in January 1996. Adriane heard months later that it took some champions amongst the sea of skeptics in Ottawa to get the final approval. Although it was still winter, we had to get started on the project right away.

This is how Susan Jones reported on the project, midway through, in a WCWC membership paper:

> *In January of 1996, at the tail end of one of the biggest storms I've ever seen (the one that generated over 100 landslides in Clayoquot Sound), Louie, [Louie Frank, elected Ahousaht Band Council Chief], Adriane and I journeyed to every community in the Clayoquot-Alberni district to tell people about the project and encourage youth to apply. By March we had received over 70 applications and hired five staff and twenty youth—ten natives and ten non-natives—who began to salvage wood from an old mill site and construct a boardwalk over the mucky Wild Side trail.*
>
> *It was a profound moment on June 25, when I stood on the beautifully crafted boardwalk and joined the Ahousaht community celebration marking completion of the first four kilometres of the trail. 'A powerful shared vision, hard work and hope,' I thought, 'that's what it takes to make things happen.'*

We haven't for a moment been able to stop fundraising for this project (trailbuilding is expensive!) But we found lots of great donors [including among many local business, MB and the Keg Steak House]. Most importantly, we found a source of strength in the people and especially the elders of Ahousaht, who offered advice and help every time it was needed.

For me, all the tough times are worth it to see the positive energy, dedication and inspiration in the faces of the youth working on the project, and to know that the community of Ahousaht will be able to use this trail to secure steady employment and protect their Island.

During the months the project was underway, Susan Jones acted like a mother hen to the 20 participant youth. She took personal interest in every one of them, making sure that they successfully completed their tasks. It took seven months of hard work, battling storms, mud, personal doubts and fatigue for the crew to accomplish

Susan Jones on the way out to Cow Bay to see the trailbuilding crew. Photo: WCWC files.

what it set out to do. When finished, the trail crossed headlands and linked eight beautiful white sand beaches which ranged from serene coves to surf-pounded bays that faced the open Pacific Ocean. During the seven-month-long project, the crew boardwalked about three kilometres of the trail, the boggiest and wettest places. Joe Foy helped a team of youth participants demarcate a four-kilometre-long trail route, an extension to the Wild Side Trail, from Cow Bay to the top of 970-metre-high Mt. Flores. It's a rough trail and a rugged climb to a breathtaking view that's still in use today.

We produced two reports during the project: the *Ahousaht Wild Side Trail Survey Options Report*, which the Ahousaht people used to select which route they wanted the trail to take and the *Ahousaht Wild Side Heritage Trail and Ecotourism Development and Management Options Report*, which WCWC contracted ecotourism consultant James MacGregor to produce for the Ahousahts. This report outlined some of the potential tourism employment opportunities the trail could generate.

We also co-published with the Ahousaht First Nations a stunning poster titled *Ahousaht Wild Side Heritage Trail, Flores Island - Ahousaht First Nations' Territory - Clayoquot Sound, Treat with Care and Respect* in time to hand out at the celebration marking the completion of the trail project held in Ahousaht on August 24, 1996.

It featured a Graham Osborne large format photo of the beautiful newly constructed boardwalk wending through the ancient cedar-hemlock forest (see first facing page of this chapter). Adriane took hundreds of these posters to Ottawa, handing them out to government staff and, of course, to every MP, with a letter thanking the federal government for its support of the project. People on "The Hill" in Ottawa loved this Clayoquot poster best of all.

One of the youth participants, Regan Thomas from Port Alberni, had to say about his experience.

When I came out here I didn't know what to expect. What I found was a beautiful island. I have learned a lot, seen a lot and worked a lot. I had the opportunity to hike the entire trail route and I am quite confident that it ranks with the best in Canada and probably the world.

During the year following the completion of the trail project, Ahousaht elder Stanley Sam Sr. wrote a guidebook for it. An Ahousaht artist made beautiful signs that were installed at the historic locations referred to in Sam's guidebook. In 1997, with financial assistance from the National Resources Defense Council, we published "Grandpa" Stanley Sam's 92-page *Ahousaht Wild Side Heritage Trail Guidebook*, which connects the amazing history of this area with the geographic features and wonders of nature found there. Robert R. Kennedy Jr. wrote the foreword to this book. In it, he related the memorable time he had hiking this trail with his eight-year-old

A B.C. Provincial Parks official reviews the portion of the Ahousaht Wild Side Heritage Trail Project boardwalk route that goes through White Sands Provincial Park on Flores Island with youth participants in April 1996. Photo: WCWC files.

Ahousaht Wild Side trailbuilding crew at Cow Bay, Flores Island. Photo: WCWC files.

Above: Stanley Sam Senior holds a partially completed sign. Photo: WCWC files.
Right: A completed sign. Photo: WCWC files.

daughter, Kick. Kennedy concluded:

> I want to thank the Ahousahts for sharing their land and history so generously and for being such wonderful hosts to me and my family. They have made themselves vigilant guardians of their resources on behalf of all humanity.

WCWC's first and last official visit to Carmanah Provincial Park

In the summer of '95, WCWC marked the end to our Carmanah campaign. B.C. Parks invited WCWC to send representatives to attend a special ceremony in Carmanah Provincial Park that was

Sue Fox, Joe Foy and me beside the sign at the entrance to Carmanah Pacific Provincial Park. Photo: Bryan Evans.

intended to honour the role Randy Stoltmann had in launching the public efforts to protect the valley. It was a small gathering for Stoltmann family, friends and invited guests. Joe, Sue and I went. None of us had been to the mid-valley since it had become a park. The new park access trail followed the road route originally laid out by MB to the mid-valley's cathedral-like trees.

It was a far cry from the original steep trail our volunteers built only seven years earlier, all traces of which were gone. The new trail was wide and gentle enough to accommodate wheelchairs. After a short stroll along the boardwalk in the valley bottom that WCWC had constructed during the heated campaign and commenting on the need for more boardwalks in muddy places where there was obviously heavy trail use, we reached the edge of Heaven Grove. There on the edge of it was a large new sign covered up with a cloth. The dedication was a somber affair. After saying a few kind words about Randy, an official from B.C. Parks unveiled it. On it, the history of Randy Stoltmann's discovery of the grove and his work to have it protected was written and illustrated. It renamed Heaven Grove the *Randy Stoltmann Memorial Grove*. On the way back to Vancouver Joe, Sue and I agreed that Randy would have much preferred that B.C. Parks memorialize him by protecting the huge wilderness north of Squamish that he had raved about to us just before he died.

Us beside new sign erected in Carmanah Park naming the spectacular Sitka spruce grove in the mid-valley the Randy Stoltmann Memorial Grove. Photo: Bryan Evans.

Chapter 49
Stoltmann Campaign launched
Spotted owls doomed
RPAC sell-out begins

Joe goes on a mid-winter expedition into the Elaho

In early 1995, Joe Foy, Kerry Dawson, our new volunteer coordinator, and Simon Waters, one of the founders of a more radical preservation group called Forest Action Network (FAN), embarked on a late winter expedition into the Stoltmann Wilderness. Simon made the expedition possible by arranging the loan of a snowmobile. It was a good time to go because the extremely low flow made fording the Elaho River possible. In the spring, summer and fall the Elaho rages with glacial melt waters, making it virtually impossible to cross.

The goal of our expedition was to check out an area that Interfor planned to access and log in early spring. The previous summer this logging company had built the footings for a bridge over the Elaho. Once the decking was installed in spring, it had access to the west side of the Elaho Valley and the Clendenning Valley. The Upper Elaho and Clendenning were two of the big wild watersheds that Randy Stoltmann had proposed for protection.

The snowmobile could only carry two people at a time so it took a long time to go back and forth repeatedly to ferry three people up the 80-mile snowed-in road to Interfor's river crossing site. They reached the bridge site in the late afternoon.

According to Joe, it was so cold there that even standing around a big fire they couldn't keep warm. "I can't take it any more," he said as he jumped up from the fire and began to run around and around the parked snowmobile to get warm. The others took turns running the same circuit until they finally crawled into their sleeping bags inside their tents. That night Joe left his socks up against the side of his tent and when he went to put them on in the morning he discovered that they were frozen solid. He had to put them under his armpits to soften them up enough to get them on his feet.

Later that same beautiful blue-skied morning, after the sun warmed things up, they picked a place where the water was moving slowly and paddled across the Elaho. They explored the other side on snowshoes and took many photographs. That night when Kerry crawled into her tent to go to bed she discovered that her sleeping bag had turned into "a big solid block of ice." For some strange reason, while they were gone during the day, the moisture inside her sun baked tent condensed on the tent top and rained back down onto the bag completely soaking it. After the sun set, her bag had frozen solid. All three of them slept together in Simon's tent that night to keep warm and survived!

Interfor did not finish building the bridge in the spring, summer or fall. The area our team had explored became protected as part of the Clendenning Provincial Park (October 1996). Interfor built its bridge further to the north. But that's jumping far ahead in the story. The following winter Joe, with a few volunteers, journeyed back up and held a New Year's celebration at the same site.

Road building along Sims Creek makes a mockery of the new Forest Practices Code

WCWC started its on-the-ground Stoltmann Wilderness campaign in the spring of 1995. We decided to concentrate on Sims

Above: Simon Waters and Kerry Dawson paddle across the Elaho river. Photo: WCWC files.
Facing page: Douglas fir in the Upper Elaho Valley. Photo: Joe Foy.

Creek, the southernmost major watershed within the proposal area. Here Interfor was road building as fast as it could and had several cutblocks along the new road scheduled to be clearcut that summer. We planned to start trailbuilding (euphemistically called hiking route surveying to make it legal under the NDP's new Forest Practices Code which made trailbuilding illegal) as soon as the snow melted.

We wanted to beat Interfor to the area and bring to public light the beautiful cedar forests that were about to be destroyed. On Joe's first weekend reconnaissance mission, he discovered that Interfor was building its new road into the area beside Sims Creek (which was really a small river). In the process, Interfor's road crew was side casting the waste rock directly into the river in direct contravention of the new Forest Practices Code and the Federal Fisheries Act, since this was a fish-bearing stream containing trout, although this point was above a falls that blocked salmon from reaching it.

At our Monday morning staff meeting Joe relayed how ironic it was that at the very same time he was watching Interfor's road builders dump rock into a fish-bearing stream, Premier Harcourt was in Europe telling everyone that we had the best forest practices in the world under his new Code. It was balderdash propaganda. We put out press releases about the situation, circulated photos and took John Werring of the Sierra Legal Defence Fund there to see what we could do about it.

Joe went up to the Sims every week. He was astounded at how fast the road went in. "They had it down to an art. There were more machines on the site than men," explained Joe. "They had way fewer guys than we had on our trail scouting crews. All it took was two men to build the road. They ran from machine to machine. That's what technology can do," Joe observed. Interfor's aggressive road building thwarted our Sims campaign plans. It would be foolish to build a trail right alongside a road being constructed. There was no way to get to the other side of the Sims to build a parallel trail there. Sims Creek was too big to build a footbridge across easily.

The Elaho Giant

Giving up on developing a hiking route into the Sims, we moved on to scout the Upper Elaho. It had by far the largest tract of old-growth forest in the Stoltmann Wilderness proposal. In July we began surveying a possible trail route to access the wild upper valley. Almost immediately, our surveyors discovered a huge eight and a half-metres-in-circumference Douglas fir, which they aptly name the "Elaho Giant." At 2.7 metres in diameter and 47.5 metres tall, it turned out to be B.C.'s third largest known Douglas fir tree.

In August during B.C. Day weekend, we held our first camp-out to introduce WCWC members and supporters to the Stoltmann Wilderness. It was WCWC's fifteenth anniversary and our big launch of our Stoltmann campaign. Over 100 people attended the gathering, held in a large burned-out clearcut at the end of the logging road on the edge of the remaining wilderness in the Upper Elaho Valley. At night, using a rented generator, Kerry Dawson and John Clarke put on slideshows against a big white sheet strung up between stumps and draped over the road embankment.

Despite the setting, people were inspired by the slideshows, the talks given by Joe and others, and by the guided nature walks along our newly surveyed hiking route to the Elaho Giant by Kerry, our volunteer coordinator, who was an experienced naturalist guide. At the end of the campout as they left, the oldest couple there (who had to be in their late 70s) handed me a cheque to help with the campaign. I took it and thanked them, putting it in my pocket without even looking at it. As their car rolled away someone asked me, "How much did they donate?" I looked. It was a thousand dollars!

John Clarke, renowned mountaineer, adds authority and credibility to our Stoltmann Wilderness campaign

At that campout I met John Clarke for the first time. A dedicated mountain man, he had spent every summer of the last 35 years exploring the Coast Mountains of B.C. from Vancouver to Alaska. His

John Clarke speaking at a WCWC rally at the Legislature to preserve the Stoltmann Wilderness. Photo: WCWC files.

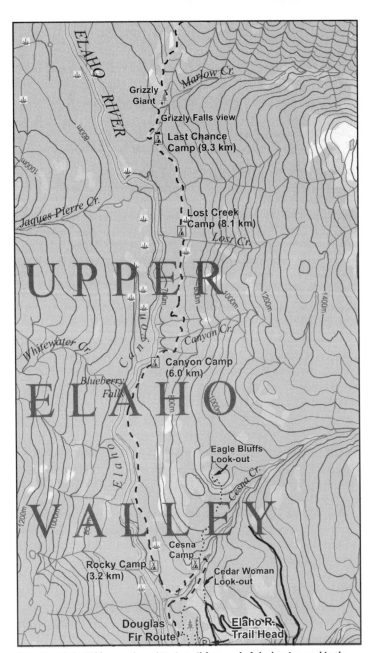

A portion of the Wilderness Committee's trail from end of the logging road in the Upper Elaho to a logging road near Meager Creek Hot Springs. Map: WCWC files.

first hand knowledge of this range and its valleys was vast. He told me that the timber companies and the B.C. Forest Service had been challenged to name another Douglas fir/western red cedar forest like

Summer snow in mid-June greets WCWC volunteer trail-surveying "lopper crew" along a high elevation portion of the Stoltmann Hiking Route. Photo: WCWC files.

that found in the Upper Elaho anywhere else on the mainland coast. They couldn't! "The Stoltmann Wilderness is the last of them," he said. John impressed upon us how little of the big old growth was left along the B.C. coast. Here is what John Clarke said in an interview with just retired *Province* outdoors columnist Tony Eberts, who wrote our first newspaper about the Stoltmann Wilderness. Titled *Randy Stoltmann Wilderness Area – Save it Now* (Educational Report Vol. 14 No. 7), of which we published and distributed 80,000 copies in the spring of '95.

> Many people think that there's one great expanse of untouched forest all up the coast from the Vancouver area. They couldn't be farther off the mark. As you fly up the mainland coast from Vancouver to Bella Coola, stretched below you is one of the world's grandest and most intricate coastlines. Your eyes search the fiords for an unlogged valley, but until you reach Kwalate Creek in Knight Inlet, every valley running in from tidewater has its system of roads and clearcuts. What really makes an impression is the number of logged-out valleys that could still be providing employment if they had been harvested on a sustainable basis. Historically, the classic, valley-bottom stands of giant fir and cedar were removed during the railroad and steam-donkey era of the 1920s and 1930s. What followed was 50 years of truck logging that used zig-zag roads to log far up the sides and into the headwaters of these valleys. What remains today in most mainland valleys are patches of forest on benches above cliffs and a fringe of high elevation sub-alpine trees that weren't considered valuable before. Now, even these remnants are being helicoptered off, removing the last trace of integrity that these watersheds still have.

In our paper Eberts explained that the mainland coast of B.C. is rapidly going the way of Vancouver Island, where only about one-half of one per cent of the once great Douglas fir forests have escaped logging. Dean Channel, some 450 kilometres northwest of Vancouver, is the northern growth limit for all but stunted, bonsai-like Douglas fir. Within this 450 kilometre long coastal zone, Clarke asserted that the Douglas fir stands in the Stoltmann Wilderness are unique.

Clarke ended his interview with this statement:

> It's a miracle that these fine old-growth Douglas fir trees still exist, so close to logging operations and so easily reached from Vancouver. It would be a tragedy to see them destroyed. The Stoltmann Wilderness mix of forested valleys, high meadows, small lakes, soaring mountains, swift water and ice fields add up to incredible beauty. In places you seem to be walking through a series of Japanese gardens. The lakes show an unbelievable range of colours. Some of the animals, such as mountain goats, have little fear of people. When you consider that it's close to 57 percent of B.C.'s population and has so much to offer in terms of recreation and conservation values, preservation is the only logical course.

By late October '95 when winter began to close in and our volunteer trail surveyors had to cease work for the year, they had completed 10-kilometres of a hiking route up the Elaho Valley.

Harcourt sells out the spotted owl's old-growth habitat, dooming the bird to extinction in B.C.

The Harcourt government, praised far and wide for its record of making new parks, did so at a terrible price as far as conservation of species and old-growth forest ecosystems. Nothing illustrates this better than the NDP's sacrifice of the highly endangered spotted owl. The whole situation slammed home to us on June 8, 1995.

Ric Careless, who always knew what was happening in the back rooms of the NDP government, relayed to us by phone the good

Trailbuilders beside the "grizzly giant" Douglas fir tree near last chance camp along the Elaho to Meager Creek trail. Photo: WCWC files.

news that Harcourt was going to announce the creation of the Pinecone/Burke/Boise Provincial Park several days before the official announcement. In the phone conversation he told me that WCWC was getting everything it wanted—the whole 38,000 hectares. The deal also included the signing of a joint management agreement with the Katzie First Nation, who had also fought to protect this area and whose support had made the government decision in favour of preservation possible.

"But we're really afraid that there is going to be a big protest by the loggers at the media conference when Harcourt makes the announcement," said Ric. "Can you get out a big crowd to cheer and support the Premier in this difficult decision?" he asked me. "The announcement is going to be made at a recreation centre in Coquitlam next Thursday morning at 10 a.m." Delighted by the news, I assured Ric that we would get at least one full bus load of supporters there to enthusiastically greet the announcement. But, being a working day, I couldn't guarantee a huge crowd.

Joe Foy and our staff worked hard calling everyone who had been part of the campaign to inform them of the celebratory event. We had a large pool of volunteers who had worked hard flagging and clearing the Boise hiking route, attending public hearings, spending hundreds of hours stuffing Boise mailouts and doing the work needed to win protection for the area. Our campaign had spanned nearly five years. We rented a bus to take the celebrators from our Gastown office out to the media event and back.

It was the first time we'd played "by the book," fully participating in a government planning process. The Pinecone Lake-Burke Mountain Study Team was the first to come out of the government's *Parks and Wilderness for the '90s* initiative and predated Harcourt's CORE processes. It allowed for a range of final options and full public consultation. We considered this land use decision-making process a showpiece of democracy in action. We had won fairly in the court of public opinion and it was gratifying to see this translated into a positive government decision.

Planning out what we'd do to capture the media's attention in case the decision's opponents got too rowdy, we decided to pop the cork on a champagne bottle—our traditional way of celebrating—in front of the cameras. Knowing it would be in a public place, we got several bottles of non-alcoholic champagne from the grocery store and some plastic champagne glasses. They looked exactly like the real thing. We also made up thank you banners and posters. We were set to celebrate. Occasions like this come all too infrequently in the environmental movement!

Early in the morning of the appointed day, I woke up feeling great. We had been the lead player in this campaign. By our hard work and persistence, we had won one for nature. It seemed too good to be true! And it was! Just as we were about to drive off in our bus, I got a call from Ric. He told me he had some bad news. He explained that the NDP decided it couldn't just create a big park without appeasing the IWA and logging industry. Harcourt would immediately follow his announcement of the new provincial park with an announcement that government was opening up the nearby Spotted Owl Conservation Areas (SOCAs) to some very limited logging. Ric said he had tried to stop it, but to no avail.

I went ballistic. "How could Harcourt do such a thing! It will ruin the announcement. Instead of getting positive reinforcement Harcourt will get boos," I shouted at Ric, accusing him of concealing this information when he originally asked me to get out supporters to the announcement event. He assured me that he did not know about it until just now. But I didn't believe him.

Ric reasoned that it was better to go along with the owl announcement and fight to have it changed later. I insisted that it was a death sentence for the spotted owls, that it would be impossible to get reversed and that there was no way we could or would be quiet about this sell-out.

When we got to the hall in Coquitlam, all of us were fuming. Adriane, our most diplomatic envoy, saw Harcourt and his entourage enter the building. She went right up to the Premier and begged him to back off. "Please let us celebrate this park. Don't link a wrong decision with a right one," she pleaded. But Harcourt was resolute. He told her it was about being "balanced." She said, "there's no balance in condemning spotted owls to extinction."

There were no angry loggers in the room. The Burke Mountain Naturalists, who also had been orchestrated into attending, were holding up a huge banner with the words "Thank You, Premier Harcourt." Obviously, they had not been informed about the tradeoff, which opened up to logging 175,000 hectares of proposed Spotted Owl Conservation Areas (SOCAs). In Harcourt's supposedly "balanced" deal the industry gained access to tens of thousands of hectares of prime timber in exchange for giving up a few thousand hectares of marginally merchantable timber in the new 38,000 hectare park that was mostly high alpine, rock and non-commercial forest. Harcourt's decision also included giving new directions to the Wildlife Branch to prepare a "management plan" which would not have more than a 10 percent impact on forestry's AAC within owl management zones. It was, indeed, a sure death sentence for the spotted owl.

As we waited for Harcourt to speak, Ric Careless sidled up to me and whispered in my ear, "If you protest, you won't get any more parks from Harcourt." That was enough to ensure I'd make a scene. But I wasn't the only one who catcalled and booed when Harcourt announced his tradeoff deal. All of the conservationists were angry. The TV cameras caught a shouting match between Ric Careless and me in which I accused him of selling out the owls. I held the closed champagne bottle in front of the cameras and turned it upside down stating in disgust, "I won't celebrate the Boise over the bodies of dead spotted owls." I then went outside, popped the cork and poured it out on the grass.

The WCWC contingent departed on our bus thoroughly depressed. We vowed to initiate a campaign to save the spotted owl and appeal to our environmental friends in the U.S. to help. I felt really sorry for Joe who, as WCWC's lead Boise campaigner, deserved to celebrate the Boise's preservation and see this long-awaited celebration day turn into a wake.

Thrust into the campaigning to protect B.C.'s endangered spotted owls

We had obviously not done enough to help protect the beautiful and highly specialized spotted owl. The spotted owl was southwest B.C.'s "canary in the ecosystem cage." It was also an "indicator species" whose presence or absence in the forest where it historically lived indicates the relative health of countless other species that share its natural habitat. While we were winning specific battles to save special wilderness areas, we were losing the larger war. We were not achieving enough protection of un-fragmented wilderness ecosystems to maintain natural biodiversity over the long term.

Some people argued that since the B.C. spotted owl population represents the very northernmost range of the owl and there were never many of them here, it didn't matter if they went extinct in B.C. But conservation biologists assert that it is the individuals living on the outermost parts of their species' range that offer the most hope for that species' survival. The individuals in peripheral populations have a genetic makeup that often differs from that of individuals in the centre of their species' range. These genes provide the raw material for rapid adaptation through natural selection when the species is stressed.

In the early '90s, the huge fight over the fate of the remaining old-growth northern spotted owl habitat on the United States' west coast resulted in much of these birds' remaining old-growth habitat being protected. We thought that this U.S. move, coupled with the move that COSEWIC made in 1986 to list this species as endangered in Canada, would give us a leg up on our fight to save them in B.C. But neither meant much, for without any federal or provincial endangered species protection laws, there was no legal clout to protect them.

In 1990, the B.C. government established a Spotted Owl Recovery Team (SORT) which was mandated to ...*examine the current status of spotted owls in Canada* [they are only found in the southwest corner of mainland B.C.] *and develop a national recovery plan for the species.*

SORT began as a serious attempt to save the spotted owl. The B.C. government put some of their top biologists to work on it. But due to pressure from the forest industry in the Chilliwack and Squamish Forest Districts (the districts that would be impacted by any withdrawal of land to protect the owl's habitat) the newly elected NDP government soon began to impose limits on what SORT could consider. For example, it could <u>not</u> come to the obvious conclusion that nearly the entire remaining old-growth habitat in the spotted owl's range in southwestern B.C. had to be protected. Nor could it recommend the establishment of corridors of second growth that would never be cut again to connect the few remaining patches of old growth between spotted owl recovery areas. These two conservation measures are crucially needed to enable fledgling owls to search successfully for a mate and thus keep the owl's tiny gene pool from fragmenting into isolated pockets destined to extinction.

Spotted Owl bumper sticker. Artwork: WCWC files.

This team did, however, map Spotted Owl Conservation Areas (SOCAs). These represented some of the owl sites where the 50 or so pairs of owls still surviving in B.C. at that time were known to nest. Scientists estimated that there were over 500 active spotted owl nest sites before European settlement and logging took place. The decline of this bird towards extinction had hit a crisis point. The destruction of its habitat and its shrunken population meant that, even with a complete halt to all logging in its habitat and protection of corridors of second growth forest for habitat recovery, the owls' survival in B.C. still could not be absolutely assured.

What transpired after the initial good work of SORT is a sorry tale of sell-out science and political compromise. Conservationists often refer to the professional biologists who stretch their research and interpretation of facts to justify industrial development as "biostitutes." Biostitutes, for example, rationalized ongoing fisheries on the East Coast of Canada to the point of the cod's collapse. Biostitutes, in the case of B.C.'s spotted owls, affirmed that logging, with minor adjustments, could continue without negatively impacting the owls.

Political compromise sealed the owls' fate. In 1993, to appease its IWA supporters, the NDP government changed the terms of reference of its Spotted Owl Recovery Team. There were loggers' trucks in the Fraser Valley sporting bumper stickers saying "Shoot a Spotted Owl" and "Eat a Spotted Owl for Supper." No doubt they were inspired by Jack Munro, President of IWA Canada, who had actually told a New York Times reporter, "I tell my guys if they see a spotted owl to shoot it." Instead of preparing a scientifically-based recovery plan for this endangered species, SORT was re-mandated to simply prepare an "options report" spelling out different management strategies. They were asked to guesstimate the chance of success of each strategy to avert the owl's extinction. Coupled with this redirection of focus was a reduction in funding for the Team's work.

As I saw the owl sell-out unfold, I felt guilty that WCWC was doing little to educate the public about what was happening and the need to immediately take action to save this endangered bird. At the time it was easy to make excuses; the foremost being that there was simply no money to do a newspaper and that we were completely overloaded with the Boise, Clayoquot and other campaigns. The government was also quite secretive about the location of the owls and our regular campaign formula of building trails to provide public access wouldn't work, because these areas should not be disturbed. Consequently we could not build a campaign team of dedicated volunteers who personally experienced and bonded to the threatened area. Our spotted owl campaign was thus more philosophical than on-the-ground.

Harcourt's announcement at the Pinecone Burke Provincial Park event spurred us into action. Within a few months we wrote and published 100,000 copies of our first spotted owl paper and distributed it widely. However, the returns, as measured in donations from our members and the public, were poor. We rationalized that our other large wilderness campaigns may be creating "donor fatigue" and this campaign was a much tougher one to explain because it involved complex ideas about ecosystems and species' needs.

A year later, when the B.C.'s Chief Forester set the Annual Allowable Cut (AAC) for the next five years, for one region with spotted owls, the small reduction in the cut we had anticipated to accommodate the owls wasn't made. Acting on our behalf, the Sierra Legal Defence Fund launched a judicial review of the Chief Forester's decision regarding the Squamish Forest District AAC. We lost this case on a justice's bogus ruling that a reduction of the cut to protect the owls was a political decision, not one that the Chief Forester could make. The only thing that would constrain the Chief Forester was a law requiring him to put the needs of a species ahead of the needs of the forest industry. We needed an Endangered Species Protection Act in B.C. to do that!

Our E-team subsequently met with Moe Sihota who, as Minister of Environment and Parks, had managed to get many new parks established. He told us that he would be introducing endangered species legislation during the next session of the Legislature. It was great news. But when we tried to contact him about this again, we could not get through to him. A government insider told us that the IWA had strenuously objected to such legislation and that Minister Sihota had given up on introducing that law.

We temporarily gave up, too, on trying to get a provincial endangered species law and concentrated instead on pushing for a federal endangered species act with teeth. We held up the spotted owl as an example of one of the most obvious endangered species that such an act must protect. Adriane, in fact, had started the ball rolling on this through her participation in the national Biodiversity Convention Advisory Committee that met in Ottawa every month beginning shortly after Canada signed on to the International Biodiversity Convention in December, 1992. She got all of the environmental groups sitting on that advisory committee on board to make the committee's main legacy a federal endangered species law. She insisted, however, that if the federal act didn't fully protect B.C.'s spotted owl and require mandatory implementation of a biology-based recovery plan for it (i.e. no logging in its critical habitat) then that law would be inadequate. Here, too, we failed. The Species at Risk Act eventually passed by the federal government in 2002 after two earlier failed attempts didn't require protection of endangered species' habitats (unless they lived on federal lands such as national parks, post offices or ports) and failed to require implementation of recovery plans.

Our dire predictions about logging leading to the demise of the spotted owl sadly came to fruition faster than anyone had predicted. In 1997, the government adopted an owl management plan that included clearcut logging of up to a third of the owl's so called protected areas, including the old-growth forests with trees over 250 years old that everyone who had any credibility knew were absolutely critical to the owls' survival. As the owls' prime habitat continued to be logged and fragmented, the owls' population continued to decline.

During the following years the Sierra Legal Defence Fund, acting on behalf of WCWC, tried every legal avenue to get the endangered spotted owls the protection they needed. The courts would not rule against the forest industry's sell-out scientists, nor make up for the lack of political will or decent species-protecting laws. The truth is, the die was cast, the spotted owl's death sentence was pronounced, with Harcourt's sell-out decision in June 1995. Nearly all were gradually executed, pair by pair, over the next decade. In the years to come, WCWC constantly blew the warning whistle. In September 2002, WCWC and Sierra Legal Defence Fund published a 53-page report titled *Logging to Extinction: The Last Stand of the Spotted Owl in Canada*. It featured this dire warning: *Without an immediate halt to industrial logging of the Spotted Owl's old-growth forest habitat, the Spotted Owl could be extinct in Canada within a decade.* Logging did not stop. Three years later in December 2005, WCWC and Sierra Legal Defence Fund published a 34-page report titled *In Defence of Canada's Spotted Owl*. It noted that only six known pairs of spotted owls were left in B.C. and it predicted that this bird would be extinct in the wild by 2010. BC Timber Sales, the B.C. government Crown Corporation in charge of selling timber off Crown land not tenured by forest companies, was the biggest culprit in hastening the owl's demise.

Harcourt government opts to adopt an undemocratic land use planning process for the lower mainland

Despite our disgust with Harcourt's pairing of the spotted owl and Boise park decisions, we continued to laud the Pinecone-Burke Study Team process as a model of democratic and participatory public land use planning. It was the last time the government used this model. Instead, it instituted processes that let it escape the political minefield of choosing between options. The new processes shifted the hard decisions onto the backs of the appointed planning team members. They were mandated to make tradeoffs and come up with a single ion—a "consensus" plan for their region—for the B.C. government to adopt. In true Machiavellian style, they also made sure the environmental players at the table were moderate or weak enough to consense on a plan that government and industry would like.

In spring '95 the NDP government initiated a land use planning process for the Lower Mainland region based on this new model. It quietly appointed thirteen members to its Regional Public Advisory Committee (RPAC). Besides various government representatives, it included those who represented the interests of Interfor, the IWA, the B.C. Wildlife Federation, B.C. Wild and the Canadian Parks and Wilderness Society. Even though RPAC was going to be recommending the creation of new parks, the government did not include a representative of the Provincial Parks Branch. Instead, it appointed a representative from the Lands Branch, notorious for its pro-development stance, to represent the Environment Ministry. This was particularly bad because the government's parks planners had a long list of proposed "pocket parks" around the Lower Mainland that they wanted to see protected. They would have greatly enhanced the quality of life in the Lower Mainland in years to come, but they were never adequately considered. Of course, there was also the problem of the cap on the amount of land that could be considered for protection

WCWC was not invited to sit on RPAC, even though Joe Foy, our National Campaign Coordinator, was well known as a leading environmentalist, knowledgeable about Lower Mainland wilderness protection issues. He had founded the Pocket Wilderness Coalition eight years earlier and had been campaigning for more park protection in the Vancouver region ever since. He had also recently successfully served on the government's Pinecone Lake /Burke Mountain planning team. But even if WCWC had been invited to have a representative sit on RPAC, we would have refused. Our bottom lines of a fair, open and democratic process were simply not met. The RPAC process was designed to be undemocratic, coercive and secretive. Their meetings were closed, the table was stacked, and the ceiling on protection too limiting, especially given the huge population served by the parks in this region. Unbeknownst to us at the time, RPAC members also had to sign a document at the conclusion of the process promising to publicly support the "consensus" decision reached by RPAC and never criticize it in public!

"It has to be this way," one of our environmentalist friends appointed to RPAC told us at the beginning of the process. He explained that since the NDP's mandate was running out and an election would have to be called within eight months, this was the only way to expedite decision-making and get the Lower Mainland's needed new parks established before the election call. "The Liberals are almost certain to win and then we won't get any parks at all!" he concluded.

"But what about the limit of only allowing a total of 12 percent of the landbase to be in protected area status?" I asked him. "We already have over nine percent consisting mostly of alpine, rock and ice protected in the Lower Mainland. Twelve percent doesn't allow enough room to protect the Stoltmann Wilderness, let alone all the 'pocket wilderness' areas in the Lower Mainland that need park protection,"

"Oh, we'll be very smart in what we select. Get out a map of your proposed Stoltmann Wilderness and I'll show you. Here," he drew his finger along the dark green shaded area beside the Elaho outlining a long narrow park that would only encompass the commercial forests along the valley bottom, "we'll select this area. The Stoltmann is mostly rock and ice too and we don't have to protect that. We'll just go for the lowest elevation areas with commercial timber and de facto we'll get it all. If we do that everywhere, we can save every area you want even with the 12 percent cap on the amount of protection," he asserted with an air of confidence.

He failed to convince us. There was no way that RPAC, stacked with industry and union representatives well-versed in adversarial negotiations from years of contract bargaining, would ever agree to such ribbon parks. Naive environmentalists would be no match for these professional negotiators.

"If all the environmental groups boycott RPAC, the government will have to change it. The ground rules and committee membership are stacked against us," Joe countered. He contended that there was enough time before the next election to complete a process like the Pinecone Lake-Burke Mountain one. "If it's an open process with options being developed and full public involvement, there'll be the opportunity for us to build huge public support. We would get a lot more protected than you will in a secret trade-off process," asserted Joe. We vowed to put the full machinery of WCWC behind getting the needed public support.

But our arguments fell on deaf ears, as did all our other efforts over the coming months to convince our colleagues not to participate in RPAC and the B.C. government to open RPAC up.

In August 1995, RPAC began its deliberations. Thus began a sorry chapter in the history of the environment movement in B.C. Frustrated by our failure to get a more fair process, we felt compelled to go public in our opposition. This inevitably led to us publicly criticizing the environmental groups that chose to participate. The NDP had successfully divided the environment movement with their new

WCWC volunteers protest the RPAC process outside a RPAC closed, secret meeting in the winter on Vancouver Island. Photo: WCWC files.

undemocratic process. While differences between groups were not something new, this schism was. There had always been moderate groups and more strident ones that boldly criticized governments. Now, however, for the first time environmental groups not actively campaigning to save a particular wilderness area were willing to trade that area away, against the wishes of the environmental group campaigning to protect it, in order to save another area. It was sad.

Hard as we tried to convince the chosen few environmentalists on RPAC not to play the trade-off game, the fix was on. Nothing would persuade them. Kevin Scott, a man in his mid-twenties who had assisted Ric Careless in the Tatshenshini campaign, sat on RPAC representing B.C. Wild. He was the youngest member on committee and he was particularly stubborn and arrogant. There was no reasoning with him. He was lauded in a 1996 *Vancouver Sun* newspaper article titled *standout individuals* as being *...a new breed of environmentalists, one that seeks compromise instead of confrontation.*

On February 22, 1996, just a few hours before Glen Clark's swearing in ceremonies as B.C.'s 32nd Premier, we decided to protest the RPAC process outside the community centre in East Vancouver where the ceremonies were taking place. We had the picket signs and people, but we didn't have the transportation to get there. I suggested we hire a limousine. It cost less than two taxis and everyone—all seven WCWC protestors—could fit in with their protest signs. This idea was a stroke of genius. The limo was the ticket to getting right up to the front steps of the centre. The security guards thought it was some celebrity coming to attend and let our limo bypass all the barriers. When Joe and the other protestors stepped out, Clark's handlers were shocked. They were hoping to keep protesters, if any, confined to the sidewalk far away from the event.

Many Clark supporters expressed their displeasure at our protest, believing it was an inappropriate occasion for such activities. I believed that it was appropriate. It was a public event. But our protest was ineffectual. From the very beginning of the RPAC process, we protested everywhere we could. About 20 WCWC volunteers led by volunteer coordinator Kerry Dawson camped out in the winter in protest all weekend long in front of Dunsmuir Lodge outside Victoria where inside RPAC members were having one of their first weekend-long in camera think-tank sessions. It garnered little media attention. It was a horrible feeling to know that through this process the Stoltmann Wilderness was going to be sacrificed.

With Glen Clark at the helm, the window of opportunity to save wilderness in B.C. rapidly closed. Clark had no commitment to wilderness protection and parks. He was only committed to finishing up the initiatives in progress started by Premier Harcourt.

Tragedy strikes our Mid-Island Branch

In the wake of our withdrawal from CORE after the Clayoquot decision and George Gibson's abandonment of WCWC, Annette Tanner took over the leadership of our Mid-Island Branch. She stored all the office equipment and the posters and other products from the branch's closed store in her basement. The Branch began to concentrate on local wilderness campaigns and hold regular issue nights, which were well attended. At the main office we had little to do with these efforts besides occasionally presenting campaign up-

Photo of Spirit Bear (Kermode, white coated black bear) donated by Myron Kozak to WCWC to be used in a WCWC poster. Photo: Myron Kozak.

dates at 'issue night' public meetings. Joe, Adriane, Andrea and I all gave campaign talks.

After the CORE decision, there was a great deal of follow-up work to do—especially on picking up more small parcels of land to protect in the next phase called "Goal Two," follow-up process (which WCWC did not boycott). On the east coast of southern Vancouver Island there were only small pieces of wilderness left to protect. Less than one percent of the dry belt Douglas fir ecosystem was protected and nearly all of the remaining unlogged forests were held privately, part of the original E&N Railroad grant lands. But over the years a few scattered blocks of forested land had reverted to the Crown and Annette and the Mid-Island Branch believed that all of them should be protected.

To see what some of these blocks looked like and view some controversial logging on a local mountainside (Mt. Benson) overlooking Port Alberni, Annette organized a reconnaissance fly-over. She contacted LightHawk, a non-profit organization in the U.S. that provided aerial support for conservation efforts. They were happy to help. She also got Myron Kozak, a very talented and generous outdoor photographer who had provided many images to WCWC, including one for a Spirit Bear poster, to come along and take photos. Mike Humphries, a retired pilot who was the first person to attach a remote TV camera on the wing of his small Cessna plane and who had flown tens of thousands of kilometres documenting the logging destruction along B.C.'s mid coast for the Raincoast Conservation Society, was the volunteer pilot. On the afternoon of August 31, 1995, Mike, Myron, Scott Tanner (Annette's husband, who was also active in WCWC's Mid-Island Branch) and John Nelson, a conservationist who was volunteering to help identify and protect areas for Goal Two took off. Something went terribly wrong. The plane didn't make it out of a steep, narrow valley and crashed into the trees, killing the pilot and Myron Kozak. John was only slightly hurt, but Scott was gravely injured and it took him months to recover.

This was another terrible loss to the conservation movement. Both Mike Humphries and Myron Kozak had contributed greatly to the cause of wilderness preservation. They were irreplaceable.

WCWC's persistent efforts to obtain government digital mapping info fail

Lost on that disastrous flight, as well, was the only set of maps with the patchwork of forested public lands demarcated on them. This loss slowed the efforts to capture all of these precious remnants in the protected area system for Vancouver Island.

Generating our own maps with critical campaign-related information had become a powerful tool in WCWC's wilderness saving tool kit. In the mid '80s the B.C. government began assembling a set of digital base maps for the province on which various layers of geographic information were overlaid. These TRIM (Terrain Resource Information Management) digital files represented state-of-the-art capability in B.C. mapping.

They quickly became indispensable to resource companies, land use planners and conservationists. When finished, about 7,000 of them would cover the entire province. At WCWC we got a grant of a SUN Station computer and software specifically designed to handle GIS mapping and the government's new large TRIM files. But right away we ran into a snag. In March of 1993, the Ministry of Environment, Lands and Parks rejected our request to obtain, free of charge, three TRIM map files covering Clayoquot River Valley without commenting on the merits of our research proposal.

A long legal battle ensued in which the West Coast Environmental Law (WCEL) Society did everything it possibly could to obtain for us access to these digital maps. The Ministry insisted that we had to buy each digital map at $600 per map data set. The government had established a "cost recovery" policy regarding its mapping. We argued that when we bought a printed map from the government with the same data on it (not in digital form) it cost only $5 or $6 dollars. Therefore, the price for the data should be about the same, as long as we were not using the information to produce some product that we were going to eventually sell. It was evident to us that the government's high price was an effective way to deny groups like ours access to government information that we had a right to have. The exorbitant price tag violated, at the very least, the spirit of the Freedom of Information Act.

We subsequently tried again, asking for the 42 TRIM digital map sheets for all of Clayoquot Sound for free, or at a reduced cost price, so we could provide input to the Clayoquot Scientific Panel. The Ministry again refused to provide them to us at any price lower that the full retail price, which came close to $30,000 when taxes and everything else were included. The Ministry claimed, after looking at our financial statements, that WCWC could afford to buy them because we stated in our annual report that we had $200,000 worth of assets. We pointed out that this wasn't cash in the bank but the cost value of posters, books and other products we had in storage that we couldn't sell immediately (if ever).

After a lengthy exchange of letters with the Ministry, WCEL took our case to the Information and Privacy Commissioner of the Province of British Columbia who heard it in late '95 and early '96. There were a number of intervenors supporting our side, including the Canadian Union of Public Employees (CUPE), the BCEN Forestry Caucus, the Sierra Club of Western Canada, the head map librarian at SFU, Ecotrust Canada and the B.C. Freedom of Information and Privacy Association. But to no avail. In mid March 1996, Privacy Commissioner David H. Flaherty issued his 15-page report (Order No. 91-1996) on our request that he overturn the decision by the Ministry of Environment, Lands and Parks to charge WCWC the full price for its TRIM Digital Map Data information. Flaherty's report makes for fascinating reading. It reveals the tight bond between the resource industry sector and the government.

Resource companies most often got the digital map sheets free in different exchange deals. The Ministry of Environment, Lands and Parks made its money by "selling" maps to other government departments at its "cost" of $150 each. This created the illusion that this project was "recovering" money. The Commissioner ultimately ruled against us because he said that he was not empowered by the law under which he was operating to order what we wanted. However, as consolation he wrote, *While I do not intend to make an order requiring the Ministry to reconsider its decision under section 20(1)(a), I am of the view that there should be a general reconsideration by the Ministry and other government departments about the application of its overall pricing policies for TRIM map sheets.* He went on to point out the extensive list of intervenors who supported WCWC's side.

Our WCEL lawyer claimed that this was a win, of sorts, for us. He assured me that Commissioner Flaherty's suggestion that the government review its policy had a lot of moral force and the NDP government would heed it. I was skeptical. The government never did heed the commissioner's hint. The Environment, Lands and Parks Ministry never relented. But we did get some of the critical digital map information we needed under the table through e-mail transmissions from friendly bureaucrats. E-data is hard to contain and keep secret.

Chapter 50

Stein is finally saved
WCWC's prolonged protest at B.C. Legislature sets record

A Park in the Stein – Harcourt's parting gift

The Stein Valley finally got its long overdue protection as a provincial park in late 1995. It was ultimately due to the sheer determination and hard work of Aboriginal Affairs Minister John Cashore and his executive assistant Bob Peart. In October '95, when Harcourt announced that he was stepping down as Premier, Peart, who had worked for Canadian Parks and Wilderness Society (CPAWS) before working for Cashore, noted that one outstanding wilderness area—Stein Valley—had not yet been protected even though it had solid First Nations' backing.

The Stein had been the focus of intense efforts over many years, but the campaign was on the back burner because a former government Forest Minister had promised the Lillooet Band that no logging would occur there without the Band's agreement, which they vowed never to give. Cashore and Peart sold Harcourt on the idea of protecting the Stein as Harcourt's parting gesture of environmental good will. There was only one catch. They had just a few weeks to put the whole deal together.

Not being privy to government's backroom planning (I picked up all this information at Bob Peart's retirement party many years later), we were taken by surprise when WCWC got a formal invitation to come to the beautiful longhouse at UBC on November 22, 1995, to witness the Stein Park signing ceremonies. At that event Premier Mike Harcourt and Chief Byron Spinks of the Lytton First

From left to right in foreground Colleen McCrory, Valhalla Society, unknown, Adriane Carr, WCWC, Vicky Husband, Sierra Club of Western Canada and George Smith, CPAWS applauding the announcement. Photo: WCWC files.

Nation announced, with great fanfare, the creation of the 107,000-hectare Stein-Nlaka'pamux Heritage Park, to be jointly managed by the Lytton First Nation and B.C. Parks. It was a triumph of the many who campaigned for 25 years to save the area, to keep industrial developments out of this last large completely unlogged watershed in southeastern B.C.

At the celebratory event, Harcourt honoured Chief Ruby Dunstan who had put so much time into protecting the Stein when things looked bleak in the late '80s. But John McCandless, who was Dunstan's "Stein Coordinator" and had also worked tirelessly to save the Stein, including organizing seven annual Voices for the Stein Festivals, was not personally thanked. I went over and talked to him. As we ate the sandwiches and drank fruit juice, the free lunch provided by the government, we reminisced about the campaign.

Chief Byron Spinks of the Lytton First Nation and Premier Mike Harcourt about to sign the agreement establishing the 107,000 hectare Stein-Nlaka'pamux Heritage Park. In the background from left to right Adriane Carr, John Cashore, Moe Sihota, and Ruby Dunstan. Photo: WCWC files.

At a meeting in Victoria several years earlier where Environmentalists discussed their issues with His Royal Highness Chief Ruby Dunstan presents Prince Philip, Honorary head of WWF with a ceremonial drum and tells him about her First Nation's efforts to save the Stein Valley. John McCandless (behind) looks on. Photo: WCWC files.

WCWC fights on for the Stoltmann and establishes record for the longest continuous protest in front of the B.C. Legislature

About a hundred kilometres away from the Stein Valley and Chilcotin Mountains, WCWC's flagship Stoltmann Wilderness campaign was flagging. There were compelling reasons why the Lower Mainland region, where more than half the people in B.C. lived, should have many more parks than allowed under the NDP's policy of an approximate 12 percent limit. We argued that more parks create higher quality living and, in the new economy, it's a country's quality of life that attracts investment capital and talented people. But neither this argument, nor anything else we said or did had resulted in any special consideration for the Stoltmann. None of this convinced the government to open up its behind-closed-doors RPAC planning process now underway that was creating a "final" park package for the Lower Mainland region.

James Jamieson, stalwart campaigner. Photo: WCWC files.

We both agreed that the real reward for all the effort the two of us and thousands of others had put in over the years was not accolades, but seeing the Stein protected. Like all other major wilderness wins, a huge number of people deserve credit and only a few get the limelight.

This Vancouver celebration was the first of several held that day. Harcourt and the chiefs left for Lytton by helicopter to celebrate at the entrance to the new park and later with the Lytton First Nations people in their community hall. It was a great day for the earth's fragile ecosphere and the announcement got the extensive media coverage it deserved.

I vehemently disagreed with those who later asserted that accompanying the Stein Park decision was an unannounced end to further park protection in the Lillooet region because, with the protection of the Stein, the government's 12 percent cap had been exceeded.

I never accepted that the Southern Chilcotin Mountains, another outstanding wilderness in the Lillooet region with a whole set of features not found in the Stein, could not be protected. The Southern Chilcotin Mountains had been a park candidate-in-waiting since 1937—even longer than the Stein. The 12 percent cap made no sense for the diverse, wildlife and scenic-rich Lillooet region that had marginal timber and such a tremendous potential to expand its tourism economy. If we built enough pressure, the government would surely see the sense in protecting it.

We'd featured the Southern Chilcotin Mountains in nearly every one of our endangered wilderness calendars. Now we had to ramp up our campaign to protect it. I'd hiked the area once for ten days in the late '70s with Brenna, my teenaged daughter, the youngest from my first marriage. Spruce Lake, Mt. Sheba (the first mountain I'd ever climbed), Hummingbird Lake, Warner Pass and all the other features of this incredible wilderness area that I saw, were indelibly imprinted in my mind. It was a joy in the spring of 1996, a few months after the Stein Park was announced, at last to produce and publish our first newspaper about this wilderness area, a four-page, full-colour, tabloid-sized newspaper titled *Southern Chilcotin Mountains – B.C.'s "secret, gentle wilderness"* (Educational Report Vol. 15 No. 9 with a press run of 80,000 copies. Heavy with stunning photos, it made a persuasive case that this longstanding park candidate must now become a Class A provincial park.

One afternoon, while writing an article about the Stoltmann Wilderness, I became extremely disconcerted by our campaign's ineffectiveness. Out of frustration I yelled over to Joe, whose desk was separated from mine only by a bank of four-drawer filing cabinets, "Do you think that we should campout on the Legislature lawn to get the RPAC meetings opened up like we did to get the meeting with Harcourt about Clayoquot?"

"It's a great idea, but not for us," he retorted (meaning himself, Adriane, and me).

"How about asking James Jamieson, if he's willing to do it? He could put up a tent every morning and take it down every night like we did. Our Victoria volunteers could picket along with him," I continued. We both knew that Premier Clark was really stubborn and it was highly unlikely he would cave in under this kind of pressure. But, at the very least, the campout would create greater public awareness of the issue.

Without opening up RPAC and getting the public involved we knew the writing was on the wall—most of the Stoltmann Wilderness would be lost. We asked James if he would take on the job under a contract similar to the one we gave him for running the Sutton Pass kiosk in Clayoquot and if he would stick it out for as long as it would take. He said yes to both questions and thus our second 'siege' on the B.C. Legislature began.

Since the Clayoquot "riot" at the Legislature opening in March '93, the rules regarding demonstrations on the Legislature lawns had tightened up greatly. James took his camping gear, Stoltmann signs and headed for Victoria, hoping for the best. After a few days of pseudo camping, security staff told James that he was no longer allowed to pitch his tent even just temporarily during the day on the Legislature's lawn.

"Let's put our tent on rollers and then James can tow it around in front of the buildings," suggested Joe. So James attached casters to the four corner poles of our old exterior-framed backpacker's tent and towed it behind him. Covered with signs, it was more effective than the stationary tent had been. For the next few weeks the tent rolled along behind him around the Legislature like a dog on a leash.

A heap of credit goes to James for sticking it out through the wet and nasty winter weather. He undoubtedly was bolstered by his firsthand experience of the Stoltmann Wilderness, having spent much of the previous summer with our trailbuilding crews in the Elaho Valley's tall-treed, old-growth forest. Still, it takes a special kind of person to continue this sort of endless protesting for only paltry pay. At times volunteers came out to lend James support. But most of the time he was traipsing around the Legislature on his own.

After five weeks of continuous protest, security guards told James that he could no longer tow the tent around the buildings because it was causing disruptions on the walkways. We all knew that this was a bogus excuse. People could easily get around the tent. Feeling the campaign profile building, there was no way we would let them banish our prize prop from the Legislature!

My first reaction was to fight this decision by making a big thing

Right: James "camping out" in front of the Legislature to get RPAC meeting open to the publics so we would have a chance of saving the Stoltmann Wilderness.
Below: Lone protester continues the vigil at night across the street from the Legislature to keep up the pressure. Photos: WCWC files.

in the media about the new rule and resort to the courts if need be. Instead, we brainstormed late over a few beers that night and came up with a creative way to circumvent the new rule—we'd float our tent in the air! Quite a few years earlier we had purchased a large weather balloon, filled it with helium and "flew" it at one of our demonstrations at the Legislature. It attracted a good crowd. Tethered by a long, heavy nylon fishing line, it soared almost as high as the top of the dome. Unfortunately, some didn't like our prop. An annoyed seagull swooped down and pecked it. It burst and its shredded shards, still attached to the line, dropped straight to the ground.

The floating tent. Photo: WCWC files.

Andrea Reimer comes over from our main office in Vancouver to spend a day "on the line." Photo: WCWC files.

We didn't intend to fly our tent so high. Not knowing how many balloons it would take to lift the tent, we brought some large balloons including a dozen four-foot-in-diameter weather balloons at a novelty shop in Vancouver and rented a large helium-filled tank. I must admit I felt a twinge of conscience using this rare and absolutely un-renewable resource. The whole thing was a bit expensive (a couple of hundred dollars in total) but well worth it.

Floating tent at the Legislature in an effort to get RPAC to hold open public meeting to give us a chance to save the Stoltmann Wilderness. Photo: WCWC files.

With a couple of the balloons tied to the outside poles and a few inside the tent, it lifted off and floated. Now the "palace guards" policing the Legislature had no grounds to object. Our tent was well above the sidewalk. We loved the fact that it attracted hundreds of people and media attention to our protest. But it, too, totally failed to sway the government to open up the RPAC process. Several more weeks went by. Then one morning James called Joe to tell him that our tent had inadvertently "escaped."

I took the phone from Joe. "What happened?" I asked, annoyed at the loss and worried about where the tent might go. James explained that he had temporally tied the tent down to a pole while he was doing something else when all of a sudden a big gust of wind came up and tore it loose.

A few hours later we got another report: our tent had almost hit a helicopter as it ascended. The stunned helicopter pilot said over his radiophone, "I can't believe it, but I think I just saw a tent fly by, heading north!"

It was no laughing matter. Joe, Adriane and I had an emergency

Flying tent looks like a ghostly flying saucer at night. Photo: WCWC files.

E-team meeting. "We'll have to put out a press release right away and somehow put a positive spin on it," I said.

Everyone agreed.

"It's got our banners on it. If some logger finds it in a clearcut somewhere on Vancouver Island They'll claim we are 'environmentally irresponsible' and we'll get bad publicity," said Adriane.

"There's not much chance of that. It probably has already dropped into the ocean and is long gone by now," I replied.

"If that's the case, let's offer a reward for its return," said Joe, coming up with a practical solution.

We decided our reward had to be substantial enough for people to take it seriously, but not so rich as to be ridiculous. We picked $500: impressive, but not exorbitant. Our press release was humorous and upbeat. The whole incident made a great fluffy news story and we thought that we had heard the end of it.

Two weeks later I got a call from a man living on Galliano Island. "Are you the people who lost the tent and are offering a $500 reward for its return?" he asked.

"Yes," I answered.

"Well, I think I have it here in my garage." He went on to explain that he had found it on a beach a few days earlier. Besides the reward, he asked if he could keep the balloons for his grandchildren. I said sure. His discovery meant that we had to shell out $500. But it also meant that we got another news story and another opportunity to explain our protest and why we wanted to have RPAC meetings made public in order to have a chance to save the Stoltmann Wilderness. Our protest now moved over to Vancouver's Trade and Convention Centre, sans tent, where Premier Clark was spending most of his time. About a week later we called it quits. We'd lost this round.

Artist camp in a recent clearcut in the Upper Elaho, explore the old-growth forest and create art they will donate to help raise money to save this forest while Interfor is clearcutting and blasting further into the Upper Elaho Valley. Photo: WCWC files.

WCWC launches a Stoltmann Wilderness Art project and a court case to judicially review a Forest Service decision to fast track logging in the Elaho Valley

In the spring of 1996 we launched a Stoltmann Wilderness Artists' Project, hosting over 80 artists on hiking and camping trips into the area. Some went into Sims Valley, but most visited the Upper Elaho. Just like those who participated in our Carmanah project, the artists experiencing the Stoltmann were as shocked by the huge clearcuts

Artist Jack Campbell in the Stoltmann Wilderness with his painting created for our fundraising art auction. Photo: WCWC files.

Craig Delahunt documents for WCWC the new road Interfor is rapidly bulldozing into the Upper Elaho Valley. Photo: WCWC files.

they saw on their way in as they were inspired by the ancient rainforest when they got there. Some had participated in our Carmanah project eight years earlier, but most were new. Meanwhile, our hiking route surveyors pushed deeper into the Elaho Valley, discovering new groves of big trees.

Our campaign was having an impact. Public support was growing and Interfor changed logging plans to thwart us. In June of '96 Interfor rapidly worked to extend its logging road deep into the Elaho Valley. It came very close to the renowned Elaho Giant (B.C.'s third largest known Douglas Fir tree) that WCWC trail surveyors had discovered the year before. According to the logging development plan that Interfor had presented for public viewing a year earlier, this road was not supposed to be built until 1997 with clearcutting to follow in 1998. Upon inquiry at the Squamish Forest District Office, we learned that the B.C. Forests Service had speedily approved Interfor's application for its accelerated pace of road building. This flew in the face of a moratorium on development that was supposed to be in place pending the government's decision about new parks in the Lower Mainland. It also contravened provisions in the new Forest Practices Code, which require public notice and sixty days for public review and comment before such a proposed amendment to a forest development plan could be approved.

We were so sure that Interfor and the Forest Service had broken the new law that we instructed our Sierra Legal Defence Fund (SLDF) lawyers to seek an injunction in the Supreme Court of B.C. to stop road construction. The Justice hearing this case refused us the injunction on the grounds that greater environmental damage would result from stopping construction on the partially constructed road than if the road was completed. Go figure that one! The Justice's reasoning seemed like Alice in Wonderland logic to me. Despite not having been granted an injunction to stop the road, we continued with the legal proceedings to establish that we were right.

Four months later, the Attorney General attempted to have our case thrown out of court. Our SLDF lawyer successfully rebuffed the Attorney General's application, arguing that the integrity of the Forest Practices Code's public participation process was at stake. In a press release after this "win" SLDF asserted that, *WCWC will have their day in court later this winter.* We did—about a month later. In the meantime, the B.C. NDP government amended the Forest Practices Code, no longer requiring public notice and review of changed forest road building and logging plans.

The Court, therefore, ruled our case moot and dismissed it. It was one more damning piece of evidence that the government and the forest companies were completely in cahoots.

On a happier note, working hard all that summer of '96 our WCWC trail crew successfully completed surveying a 30-kilometre-long hiking route up the Elaho Valley, over the Hundred Lakes Plateau and down into the Meager Creek Valley to hook up with a logging road there. Their work was done by fall and the route was set up to become a real trail through heavy public use the following summer.

WCWC continues to promote a global shift in consciousness through applied Deep Ecology

From time to time WCWC experimented with unusual tactics to build public sympathy for wilderness preservation. On June 21, 1996, WCWC hosted a Council of All Beings Solstice Celebration on the front lawn of the Legislature in Victoria. It was a "new age" kind of gathering with Bill Duvall from California, one of the fathers of the Deep Ecology movement, leading us through the ceremonies. At an All Beings Council people wear animal costumes and assume the animal's identity and worries. I contacted Evelyn Roth, a flamboyant artist whom I had first met in 1977 on the Queen Charlotte Islands. Inspired by the native culture of the Pacific Northwest, she had sewn a giant salmon out of light nylon parachute fabric and colourfully-painted it for a salmon festival with the Haida people there. It inflated to about 50-feet-long and 10-feet-high powered by a small electric blower fan.

As an added feature, a flap in the side of this salmon unzipped and people could go inside and meditate or listen to someone give a talk. I thought this huge work of art would be perfect to attract people to our "all beings" celebration. Evelyn was pleased to participate. All she wanted was to have us pay for her travel costs from Vancouver to the event and back. As an added bonus, she volunteered to bring along her wardrobe of over a hundred animal costumes she had created, including birds with flowing wings that people, especially children, could borrow and wear.

I had to be the animal that I best personified, so I splurged and rented a full-bodied grizzly bear suit complete with realistic head for the day. Adriane and I arrived on the Legislature lawn about noon and I put on my bear suit in the car to surprise those already there setting up for the event. Everything went smoothly until Evelyn started the gas-powered generator, which was hidden underneath a tree, and turned on the blower. Before the salmon was half filled with air a security guard stormed down the steps of the Legislature and yelled, "You can't have that here! Take it down! There is nothing in your permit that mentions it."

I tried to reason with him explaining that the nylon salmon couldn't possibly hurt the grounds. The whole thing weighed less than 20 kilograms. But he wouldn't listen. Finally I declared, (emboldened by the grizzly persona I had assumed), "We are not taking it down! Go ahead. Call the cops and have them arrest me!" I could see the headline "Grizzly bear arrested with illegal salmon on Legislature lawn." This wasn't an act of civil disobedience on my part. I was sure that we

Evelyn Roth's salmon with grizzly bear standing in front at WCWC sponsored Solstice Council of All Beings on the B.C. Legislature lawn. Photo: Kerry Dawson.

were not breaking the law and that the jumbo salmon was harmless. The security guard stalked away and nothing more came of it.

About 60 people, a few decked in animal garb, sat around a big circle on the ground that afternoon. Each one of us voiced our concerns for what was happening to our species' niches on planet Earth and how we (speaking from the point of view of the animal persona we had assumed) felt about it. Shortly after everyone had spoken and we had broken the circle, an elderly man in military dress with lots of medals on his chest came over and asked me if I would pose with their group. I said sure. I walked over to the other side of the Legislature lawn and stood with a group of retired legionaries from the U.S. in full military garb as someone snapped lots of pictures. As the sun beat down on my grizzly bear costume and the sweat poured down my face (which no one could see), I wondered to myself, "Is this really helping save wilderness?"

A friend of Bill Duvall's told me later that he, too, found the whole gathering to be surreal. Councils of All Beings are normally held in wild places, not in cities. Years later we hosted another one in a remote old-growth-forest near the Stoltmann Wilderness led by John Seed, the guru of the Deep Ecology movement from Australia. It was a lot more appropriate.

Bill Duvall leads a Council of All Beings meeting in front of the B.C. Legislature. Photo: WCWC files.

At one time grizzly bears roamed the western coastal mountain ranges of North America from Mexico to Alaska. Two hundred years of habitat destruction, hunting and poaching have killed off the great bear in the southern half of its range. Today the grizzly bear's southernmost front line on the Pacific coast is the Stoltmann Wilderness in British Columbia, Canada. It is located within the traditional territories of the Squamish and Lil'wat First Nations, 200 kilometres north of Vancouver, B.C.'s largest city. A 260,000 hectare region of pristine rainforested valleys, soaring mountains, and massive glaciers, it is called the Stoltmann Wilderness in memory of Randy Stoltmann, a young wilderness conservationist and author, who tragically died in an avalanche shortly after he proposed that this area be protected in 1994.

The living heart of the Stoltmann Wilderness is the lush temperate rainforest of the Elaho Valley. Some environment groups including B.C. Wild, the Canadian Parks and Wilderness Society, and the Sierra Club of B.C. have backed a proposal to allow InterFor to clearcut log throughout most of the proposed park area, including the Upper Elaho. They endorse protecting only 20 percent of the area in two small separate parks. They have accepted the current provincial government's policy of limiting forever the total protected parkland in British Columbia to only 12 percent of the province's landbase, believing that this is the best we can achieve.

We say FULL PRESERVATION – NO COMPROMISE with the clearcutters. The *entire* 260,000 hectare Stoltmann Wilderness must be preserved as a Provincial or Tribal Park. The grizzly bears' range must not shrink any more – their front line of defence must be held.

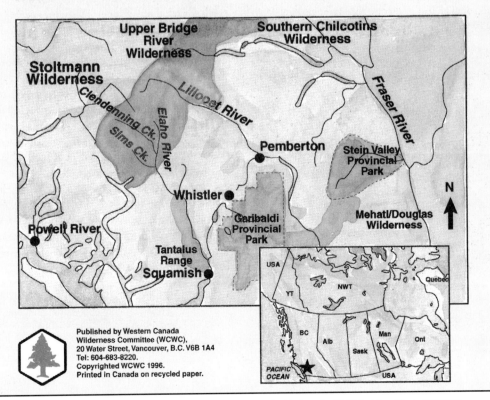

Information printed on the side of our Stoltmann Wilderness poster published before the RPAC decision. Photo: WCWC files.

Chapter 51

Stoltmann Wilderness' blackest day

"Old men's" boreal road show

Jeers greet the B.C. government's Lower Mainland wilderness sell-out announcement

We knew the news would be bad before we'd even left for the media conference in Vancouver's Trade and Convention Centre early Monday morning, October 28, 1996. Besides all our staff, about 20 WCWC volunteers came and stood at the back of the room to witness Premier Glen Clark's Lower Mainland parks announcements. We'd brought along our big Save the Stoltmann Wilderness banner and unfurled it so that when the TV cameras panned the crowd it

WCWC protests at the government press conference at the Pan Pacific Hotel announcing the RPAC decision. Photo: WCWC files.

would stand out. Clark stood behind the podium with Bob Sitter, President of Interfor, then B.C.'s second largest logging company, at his right hand side. Also assembled on the platform were the members of RPAC who were there to praise the government for adopting their consensus recommendations. While Premier Clark preambled, Sitter smirked. The reason for his facial expression became clear after Clark announced that his government was creating 23 new parks and protected areas totaling 136,000 hectares. The dent in Interfor's cut was miniscule. Prime areas of valuable old growth, especially in the Stoltmann, were earmarked for logging.

RPAC deal allocates an unfair share of parks to the Sunshine Coast

The Sunshine Coast definitely got the short end of the RPAC deal, leaving less than three percent of the Sunshine Coast's land base protected, one of the lowest levels of protection in all of B.C. Among the newly protected areas were three tiny parks on the Sunshine Coast's Mt. Elphinstone and a small park in the Caren Range, located just north of Mt. Elphinstone on the Sechelt Peninsula and site of Canada's oldest known trees.

Only a mountaintop island of old growth was protected in the Caren. The campaign to protect the Caren Range was launched by Sunshine Coast resident and forester, Paul Jones. WCWC co-published two newspapers with Jones' group, Friends of the Caren: one in 1991 and the second, *Create the Caren Range Provincial Park to Protect Canada's Oldest Known Forest*, in 1996. We called for an ecologically sound 8,500-hectare park that would include the forest that descends down to the Sechelt Inlet as well as the mountain top area of ancient trees. One key reason the whole ecosystem gradient needed protection is because the Caren provides nesting habitat for marbled murrelets, a species at risk that nests in the old growth but lives the rest of its life at sea. In 2001 WCWC published *The Marbled Murrelets of the Caren Range and Middlepoint Bight* by Paul Jones. The book contains his remarkable detailed field observations—the first ever—of active nesting marbled murrelets. Unless a lot more old-growth forest along the coast of B.C. is protected and preserved, this species is doomed. It can't reproduce in industrial plantation forests.

Mt. Elphinstone was also shortchanged by RPAC. Sunshine Coast residents had been lobbying for a 1,500-hectare park on Mt. Elphinstone since 1989. WCWC produced a newspaper in 1995 to help them out. The RPAC process ended up protecting only 140 hectares: three small parcels of 70, 50 and 20 hectares each. Mt. Elphinstone, where aboriginal jurisdiction is split between the Squamish and Sechelt First Nations, has fabulous forest-growing lands that were obviously too valuable for the forest industry to give up.

Instead, industry representatives proposed protecting a large area in the northern part of the same forest district—the 20,000-hectare Bishop River, which has less valuable timber and is for all practical purposes inaccessible to the public. The nice thing about this area from industry's point of view was that it took up most of the new parkland allotment for the Sunshine Coast region of the Lower Mainland, so the district's forest cut would stay the same.

Following RPAC, WCWC took up the Elphinstone campaign in a more concerted effort to build public support and save the entire 1,500-hectare proposed park. We hired a local campaigner, published another newspaper in 2000 and offered support to local citizens, some of whom camped in areas scheduled for clearcutting; some of whom, like Rick O'Neill, guided regular free public walks for people to experience the unique habitat on this gentle-sloped mountain. Elphinstone has one of the most diverse and complex mushroom flora in all of North America. It is heartbreaking to see it whittled away, cutblock by cutblock, as the B.C. Forest Service issues cutting permits in the most valuable areas, foreclosing preservation options despite widespread local support from residents and regional, municipal governments and First Nations representatives asking for more protection.

WCWC livid with RPAC for selling out the Stoltmann Wilderness

Although we were upset with the miserly meting out of little parks on the Sunshine Coast, we were angriest about how RPAC dealt with our flagship campaign area, the Stoltmann Wilderness. Only the Clendenning and Upper Lillooet valleys were protected: 20 percent of the total area that we were campaigning to save. The 80 percent of the Stoltmann left unprotected for Interfor to clearcut included the biggest, oldest, most wildlife-rich and most spectacular ancient forests in this wilderness area.

At the big media event, Premier Clark gushed that the new parks brought the amount of land protected in the most densely populated region of B.C. to 14 percent, more than any other area in the prov-

The Stoltmann Wilderness art show and silent auction, thanks to the donations of works from over 80 artists and the participation of hundreds of WCWC's members and supporters, raised over $20,000 to help support our continuing battle to save the Stoltmann Wilderness. Photo: WCWC files.

ince at that time. What Clark didn't say was that much of the new, and most of the old, parklands were alpine, rock and ice. He also left out that his decision failed to provide protection for the endangered spotted owl that lived in this region. Nearly all of its habitat was scheduled to be destroyed by logging.

The RPAC announcement was a crushing blow to WCWC.

Right after Premier Clark spoke, Kevin Scott, a young man representing the enviro-sector on RPAC, got up and spoke. He began by saying, "This is the happiest day of my life." That's when I lost it! I yelled out, "You sell-out!" "Shame on you!" and other barbs expressing my disgust. The WCWC contingent all booed and made quite a negative scene. When coverage of the announcement hit the TV screens that night and the papers the next day, everyone knew that WCWC was bitterly disappointed with the decision.

We were pretty morose back at our office. Our opponents were armed with signatures of environmentalists saying that they agreed with the decision and would never criticize the plan in public. The logging industry, the IWA and the provincial government hailed the new parks as a wonderful compromise and *"the final solution."* From then on it would be a big uphill battle for us to save more of the Stoltmann wilderness.

Only two and a half years had passed since Randy Stoltmann proposed that this wilderness area, located 200 kilometres north of Vancouver in the headwaters of the Squamish and Lillooet River systems, be protected. When Randy died accidentally only a month later, we vowed to campaign to preserve the area as a memorial to him. We weren't about to let this short-sighted compromised RPAC decision defeat us.

WCWC opens an art gallery to show off the Stoltmann Wilderness art show and auction

One project we had to complete was the Stoltmann art project we'd started in the summer. We'd tried unsuccessfully for several months to find an art gallery that would host our Elaho Expedition art works and host an extended silent auction on our behalf. In the end, we decided to rent our own space and run our own show. We found a perfect place, a storefront right on Granville Street near Broadway along "Gallery Row" and rented it for three months. A former gallery space, it already had all the lighting in place to illuminate the works to their best advantage. With a huge amount of help from

volunteers, Sue Fox set up the show and we held a grand opening in the fall of 1997. For two months we displayed the donated works of the 100 artists who had traveled to the Stoltmann Wilderness the summer before. It was a fabulous show.

Unfortunately, there was no way we could afford to produce a coffee table book featuring this art, although the art was as worthy as that produced to save Carmanah. We photo-documented each piece,

Opening night at the Stoltmann Wilderness Art Show in the gallery we rented on Granville Street. Photo: WCWC files.

just in case a huge donation came along and made such a book possible. It never did. On the final night of the show we hosted a big party, invited all the artists and the people who had bid on the art or had purchased art at our other shows. Our gallery was packed. After some last minute frantic bidding for some of the pieces it was all over. We raised over $20,000 and a priceless amount of new awareness of the need to protect the Stoltmann Wilderness.

The "old men's" boreal forest tour

Around this same time Gray Jones, our Alberta WCWC director, approached me to firm up an idea that had been percolating for months for the two of us to tour across western Canada all the way to Toronto to raise awareness about the ongoing destruction of Canada's boreal forest. Gray called it "the old men's tour." Gray had come to WC^2 in the late fall of '91 during the darkest days of debt driven conflict at WCWC wanting to get a job with us. The only thing available was canvassing, so he went door to door for about a week and, although successful, told me that he really wanted to campaign. I knew he had experience campaigning a few years earlier at SPEC and I had wanted for years to get WCWC more active in Alberta. I suggested to Gray that he go to Edmonton and establish and build up a WCWC Branch there. He expressed real interest in the idea. Knowing he had no money, I gave him $60 out of my own pocket to help him get there. Gray set off as the snows set in.

All winter long he went door-to-door in Edmonton building membership and financial support. By spring he had rented a small office and WCWC was up and running. I flew out for the formal meeting where the members voted unanimously to establish our WCWC-Alberta Branch. Over the next few years the Branch grew rapidly as Gray got talented people to work with him on campaigns and publications. It was self-funding and ran pretty well autonomously. Gray lived by Edward Abbey's famous motto: *The idea of wilderness needs no defense. It only needs more defenders.*

A brief history of WCWC-Alberta Branch's early campaigns and accomplishments

Gray's first major WCWC-Alberta campaign was to help stop the logging in Wood Buffalo National Park. He worked with other environmental groups gathering information and keeping the issue in the news. In the spring of '92 WCWC-Alberta published its first newspaper: an eight-page, three-colour tabloid titled, YOU MUST HELP US STOP THE CLEARCUT DESTRUCTION OF WOOD BUFFALO NATIONAL PARK (Educational Report Vol. 11 No. 3 with a press run of 60,000 copies). On the front cover Gray featured a clearcut inside Wood Buffalo Park with a pick-up truck in the foreground sporting a Parks Canada logo on its door to shame the federal government. People were shocked to learn that 60 percent of the park's old growth was already logged and that only 400 hectares of prime white spruce forest, one of the most rare and awe-inspiring forest ecosystems in Alberta's boreal forest, remained in the park.

The paper hammered home the point that it was time to stop the logging. Of course, Gray mailed a copy to every MP. Shortly thereafter CPAWS was successful in its legal effort to bring a permanent end to the logging in Wood Buffalo National Park. We counted it as Gray's first win, too.

Right from the beginning Gray focused primarily on campaigns to protect the boreal forest. He had the skilled help of Claire Ashton, who soon became his wife. She provided invaluable support as his "Executive Assistant" keeping the business side of the campaigns in order and getting grants to fund the campaigns. In the fall of '92, WCWC-Alberta published its second newspaper, an eight-page full-colour tabloid-sized newspaper titled *Save our Boreal Forests – The Mystery & The Heritage* (Educational Report Vol. 11 No. 7 with a press run of 60,000 copies). Mountain Equipment Co-op helped fund the cost of printing.

The paper featured a centre-spread article by Dr. Jim Butler, professor of Parks, Wildlife and Conservation Biology at the University of Alberta. Butler was a world-renowned expert on the boreal forest and author of the well-respected book *Canadian Boreal Forest: The Great Unknown.* In it he related the results of a study done by several of Alberta's leading wildlife biologists on the impact that logging would have on 309 species of vertebrate animals known to live in Al-Pac's (Alberta Pacific Forest Industries') Forest Management Area near Athabasca. These scientists concluded that 59 species would benefit, 29 would be neutrally affected and 229 (74%) would be negatively affected. Forty of the species negatively affected (13% of all the species) were of special concern because they were rare. Crude as this measure was, it forecast that there would be disastrous consequences from the massive boreal clearcutting that continued to accelerate.

Dr. Butler also encouraged his ecology students to get active and volunteer for WCWC. In 1993 he helped Gray establish a WCWC Boreal Forest Research Station 800 kilometres north of Edmonton near LaCrete on the banks of the Peace River. Here biologists and students from the University of Alberta conducted research for that summer and the next. The major focus of the research was the examination of old growth and riparian communities on the islands in the middle of the Peace River, including their avian components. The researchers found much higher species diversities in these places than they found in industrially disturbed areas and immature forest communities in the surrounding area.

During the research project, Dr. Butler coined the phrase *Boreal Coves* after the *Appalachian Coves* of old-growth hardwood forests found in the U.S. Protected from fire by dampness and isolation, the *Boreal Coves* had both mature and "overmature" spruce, aspen and birch trees which provided maximum feeding, breeding and nesting habitats for wood warblers, a group of songbirds that winter in central and South America and summer in the boreal forest where they nest and raise their young, and other songbirds.

This finding, coupled with his worry that migrating bird habitat was in alarming decline, led Gray to write up a proposal to protect an *International Boreal Flyway*. His idea was a unique one: to protect the linear flyways along the Peace and Athabasca River systems comprising the older growth forests adjacent to both rivers and the islands in them.

These linear parks would protect the nesting habitat for approximately 22 species of neo-tropical wood warblers (a teaspoon-sized bird considered by many to be the most colourful and interesting of all the songbirds). Although it was a great idea, Gray never followed it up with a campaign. It was probably just too ambitious and too big for one campaigner working on his own. It's still waiting for some keen, young, dogged environmental campaigner to run with it.

I had the privilege of meeting Jim Butler only once—a memorable afternoon and evening. Gray, Jim and I drank beer and talked non-stop about a wide range of subjects. When we touched on the Penan in Malaysia and their way of life, Jim brought out a traditional blow-pipe and darts he'd acquired years earlier. "Do you want to try it out?" he asked us. Of course we did! We went out into his backyard and target practiced into some bales of hay placed against his garage. I put the light down-tufted dart into the end of the tube (pointed end first!), lifted the long tube to my mouth and puffed out a blast of air from my lungs. It was incredible how fast, far, hard and accurate I shot the dart and what a thrill it gave to do it. I guess the hunter in every man runs deep. We also talked about crop circles. He had recently been on a scientific expedition to investigate them in England. Many of them were so intricate and complex and completed so quickly and accurately he had to conclude that they were not a hoax.

We talked about various national parks around the world, their design and interface with tourism. Then about slowing and halting the human-caused mass extinction that is underway. Butler was truly a brilliant renaissance man.

Andy Miller, who later came to work for WCWC and head up our campaign to save the spotted owl, was a Master's Degree student under Dr. Butler. He was steered to WCWC by him. Andy had spent two summers doing bird research out of our boreal forest research station. Besides being a remarkable naturalist and birder, Andy also turned out to be an incredible door-to-door canvasser who still holds the record for night-after-night getting the most new WCWC members. I ask Andy how he did it. He told me that after explaining to a person how important protecting wilderness was, when people offered to give a $10 or $20 dollar donation, he would urge them to become a member for $30. If they resisted, he would pleasantly refuse the donation and tell them to give it to another charity because what WCWC really needed was members to give it strength. As he turned to walk away, Andy explained, nearly everyone called him back and joined up. No one else, as far as I know, has ever even tried such a bold move, let alone pulled it off successfully.

Not being against all development in the boreal forest, WCWC-Alberta proposed some solutions. In 1993 Karen Baltgailis filmed and helped produce for WCWC a 55-minute video titled *A New Leaf: Real Sustainability for the Boreal Forest* which presented an alternative way to develop and "harvest" Alberta's northern forest. It shows how current forestry clearcutting and replanting is unsustainable, both environmentally and economically. *A New Leaf* proposed a different kind of economy for northern Alberta: integrated, small scale and labour intensive, including nature tourism, horse and machine selective logging, small sawmills, value added wood products manufacturing, and small clean pulp mills. It also featured examples of real working alternatives like using crop fibres such as flax to manufacture paper. Although *A New Leaf* was shot in Alberta, the principles outlined in this video apply everywhere.

Like many good solutions, those proposed by WCWC in *A New Leaf* did not take hold. In fact, development went in the opposite direction. The rate of boreal forest logging kept increasing. Each Canadian province went at it as fast as it could. **A major part of the problem is that no one is in charge, looking at the big picture and projecting the consequences of such massive changes to this huge ecosystem over space and time. I believed then, and still do today, that this enormous deforestation is unhinging earth's natural homeostatic mechanisms (compensatory changes in response to destabilization in order to maintain balance) that is opening up a real Pandora's Box of yet unforeseen climatic calamities.**

WCWC-Alberta published its third newspaper, about the boreal forest destruction by the largest multinational corporation in the world, in the summer of '94. It had been less than a year since Al-Pac, then the largest single-line bleach Kraft pulp mill in the world, began producing over half a million air-dried metric tons of pulp per year. (The mill, located 200 miles northwest of Edmonton, fed on wood clearcut from a vast region around Athabasca, Alberta). Titled *Al-Pac: Mitsubishi's Attack on Alberta* (Educational Report Vol. 13 No. 7 with a press run of 60,000 copies), it told the story of how the Al-Pac mill was built with the greatest taxpayers' dollar subsidies ever given to a private company, totaling hundreds of millions of dollars.

Gray's Al-Pac paper pleaded with the Alberta government to give back control over Alberta's boreal forest to local communities. Al-Pac threatened to sue WCWC over the paper, but never did. We had the perfect defence. Everything in it was true. If the company had tried to take us to court, our lawyer informed us that we would be able to gain access to more "sensitive" information about how bad the Al-Pac mill deal was for the Albertan taxpayers and the Alberta environment—just what we wanted! A year later I went and took a look at the mill with Gray. Outside the mill stacks of logs five stories high went on for over a kilometre. There were only a few cars in the parking lot. It didn't take many people to operate this mill. With its belching stacks it looked like something out of Danté's Inferno.

From August 23 to 27 of that same summer of '94, our Alberta Branch hosted the Taiga Rescue Network's second international conference, "People of the Snow Forest," which brought together boreal forest people from around the world in an effort to find ways to protect the boreal forest. Over 200 people attended from more than 12 different countries. At the conference the Lubicon joined the network noting in their press release that indigenous peoples all over the boreal region have suffered from forest destruction long before the international environmental movement even recognized the problem. They have fought against oil and gas exploration and destructive logging for decades.

One presenter warned that if the global demand for primary pulp (pulp that contains all virgin and no recycled fibre) continued to increase as it has in the past decade, four to five new Al-Pac sized mills would have to be built every year on our planet to meet this ever increasing demand!

A year after the conference (the fall of '95) WCWC-Alberta published a four-page, full-colour tabloid-sized newspaper titled *Take a Stand with the Last of the Lubicon Cree* (Educational Report Vol. 14 No. 12 with a press run of 50,000 copies). It asked readers to support the Lubicon's efforts to get a just land claim settlement that gives them the right to control developments on their traditional lands and stop the wilderness-destroying logging and oil exploitation there.

The "old men's tour" takes off

Andrea Reimer, our main office assistant and Branch liaison at the time, was a big promoter of the idea of the "old men's tour" to save Canada's boreal forest. The idea originated in a big brainstorm session at our Alberta Branch's AGM that Andrea and I had attended in Edmonton in the fall of '95. Our Branch's board was fully in support. I agreed that we had to do more to protect the boreal forest and thought the idea was a good one, too, but didn't push it. Andrea told me years later that she continued to promote the idea in large part because it was a way for her to keep in touch with Andy Miller who was volunteering for the Branch and had struck her fancy when she met him at the Alberta AGM that fall. Her covert plan worked. Andrea eventually married Andy. However, Gray and our low budget mid-November boreal forest slideshow tour wasn't as great a success.

Except for our launch in Vancouver where well over a hundred supporters came out, the audiences we spoke to were tiny: 50 at most and as few as 25 in Calgary, Saskatoon, Winnipeg and Toronto. It looked like the local environmental groups who sponsored us did little to promote our shows. But the truth was that these groups were small and our subject matter was not a hot environmental issue and they had done the best that they could.

I slept outside on porches in sub zero temperatures on the tour because nearly every home that billeted us had cats and I get deathly asthma attacks from them. I got the impression along the way that there wasn't much hope for Canada's boreal forest. Already nearly all of this vast northern woodland, the largest ecosystem in Canada, was "controlled" by multinational timber companies under various logging licence agreements with the provinces. Very little was protected. And in Manitoba and Ontario the boreal forest "protected" in provincial parks was not really protected because logging was, and still is, a permitted and accepted use for many of them!

WCWC alleges WWF grading system is corrupt

Gray and I both harboured a smoldering resentment over WWF's tactic of annually awarding each province a letter grade based on the progress it had made towards the goal of protecting 12 percent of its land base as protected areas. The grades for B.C. were always high because B.C. was establishing some large parks. But this kind of grading glossed over the bad behaviour throughout the rest of our province. It gave the public the false impression that the B.C. government was doing a good job in protecting the environment. From my point of view, giving B.C. high grades fostered complacency—both in the public and in government itself. It could get away with destroying a great deal of wilderness and endangered species if it made just one big park at the same time.

In 1993 when the largest act of civil disobedience involving police arrests in Canadian history occurred in B.C. in reaction to the government allowing two-thirds of the ancient forests of Clayoquot Sound to be logged, WWF gave B.C. an "A" grade because it had protected the Tatshenshini. In 1995 WWF gave the outrageous grade of "B+" to Alberta. Alberta was destroying wilderness faster than any other province in the country. Alberta's Special Places protected area program was a sham. Their existing protected areas were being opened up to the oil and gas industry. This industry was cutting seismic lines through Alberta's boreal forest, ecologically fragmenting the wilderness. There was nothing that could rationally account for such a high grade.

Old man's protest at WWF's Toronto headquarters

Our ten-day boreal forest tour culminated and terminated in Toronto. Only 35 people attended our last show. A third of them were WCWC door to door canvassers. To make up for our lack of success, Gray insisted that we hold a protest demonstration in front of WWF's headquarters before we left town to express our opposition to the "B+" grade that they had just given Alberta in their yearly Endangered Spaces grading exercise. If there ever was an un-together protest, this was the one! We had no volunteer base and no other environmental groups that would support us in this protest. WCWC's operations in Toronto consisted only of a paid canvass director and canvassers. We invited our canvass crew to participate in the protest but all were busy doing something else that day.

The day before our protest I called Monty Hummel and told him that we were going to demonstrate in front of his headquarters the following noon and requested a meeting with him. He said he wasn't going to be there then, but if we came over right away, he would meet with us. Gray and I immediately flagged down a taxi and arrived at WWF's headquarters in less than an hour. Monty gave us a brief tour of WWF's operations and then ushered us into his office to talk turkey. Gray and I tried to convince him that WWF's high grades were giving the public a false impression that we were winning our campaigns to protect wilderness and conserve biodiversity.

Monty was unable to justify to us why Alberta received a B+ grade that year. He denied that it was because Alberta's oil patch players had given WWF a big chunk of cash. Monty's main argument was that his organization used "the carrot and stick approach" to get governments to protect wilderness knowing that governments had to be rewarded for making even tiny steps in the right direction. He asserted that our protest wouldn't hurt WWF even a little bit. In fact, he told us that our action would only reflect badly on WCWC and might even hurt our funding base. Just as we failed to dissuade him from continuing to give out misleading grades in the future, he failed to dissuade us from carrying out our protest the next day.

That evening Gray and I went to see one of Gray's friends who owned his own small avant-garde graphics firm. "Can you take WWF's logo, open up the panda's mouth and put WCWC's old-growth tree logo in it and print it out at the top of a few sheets of paper for our press release?" I asked. (This was before PC software was available that makes such a task kindergarten play today.)

His answer: "No problem."

Opening up the WWF panda's mouth unintentionally changed this cute and cuddly mascot into a mean and menacing looking bear. We put the computer doctored-up logo onto letterhead and Gray and I typed up a press release announcing the time and the purpose

WCWC-WWF protest logo used only once on a press release. Artwork: WCWC files.

of our protest picket outside WWF headquarters the following day. Early the next morning, I faxed out this media release to Toronto's ten top media. At noon John Yates, who had started WCWC's Toronto canvass in November '93 and successfully managed it ever since, dropped us off at WWF's office with a big sign showing what was happening to Alberta's forests and some protest signs saying WWF had sold out to the oil and gas industry. He picketed with us for a short time and then left for another appointment. I went up to WWF's office and delivered a copy of our press release to the front desk. I'll never forget the look on the face on the secretary as she exclaimed in horror, "What have you done to our panda?"

Picketing outside WWF was no fun. It was windy and bitterly cold. No one from the media showed up and the few people walking by were merely puzzled by what we were doing and not in the least bit interested.

Just as we were about to pack it in after fruitlessly walking back and forth in front of the building for more than an hour, up drove a City TV truck with a cameraman and Bob Hunter of Greenpeace fame. Bob's first question was: "Where are the demonstrators?"

"We're it!" Gray and I replied simultaneously.

Hunter commented disparagingly that this was the smallest demonstration he had ever gone to during his entire career as a TV journalist. And perhaps, because he felt sorry for us, he decided to cover us. Afterwards, we went with him to a nearby pub, had a few beers and filled him in on the latest news and gossip about the wilderness preservation movement in the west. I never saw his TV piece but, according to accounts by people who did, it was great.

WWF grades continue to grate

Our tiny protest failed to deter WWF from continuing to issue its annual report card. From then on, however, they called us and got our "input" and sometimes even lowered the grade a bit to blunt our criticism before releasing their grades to the media. In 1996, shortly after the B.C. government decided that 80 percent of the Stoltmann Wilderness would be clearcut, WWF called us to say they planned to give the B.C. government a high "B" for recently making areas including the Empire Ranch at Churn Creek a provincial park to protect its rare grasslands. I was furious. A high grade would make the Stoltmann sell-out acceptable to the general public and greatly undermine our ongoing campaign to save the rest of the area the government had excluded in its RPAC decision.

We complained. WWF listened to us and instead of giving B.C. the B+ that they had originally planned, they decided to award B.C.

Grey Jones and myself with our protest display board outside WWF's Toronto Headquarters. Photo: WCWC files.

a C+. We felt that a C+ grade, while not stellar, was still a respectable passing one in the eyes of most people and would hurt our efforts, but we would have to live with it. Unwinding in Greek Characters, our after-working-hours hangout, on the evening of the day before WWF was going to release its grades for the year, we lamented about what a drag it was to be sold out by "moderate" organizations in the wilderness movement and how powerless we were to prevent it.

If we complained to the media, we would lose foundation funding because funders insist that we work together with other environmental groups and never air our inter-organizational conflicts or dirty laundry in public. Regardless of that disincentive, we agreed that we shouldn't do it anyway because in the grand scheme of things it was

Bob Hunter interviews me on CTV. Photo: WCWC files.

counterproductive to complain publicly. Publicly squabbling environmental groups would play right into our "enemies" hands. Our environment movement was so tiny in comparison to the huge crushing forces of the corporate sector; a divided environment movement meant that they'd conquer all the remaining wilderness for sure.

Then all of a sudden Andrea got a cheeky idea. "Let's grade WWF and give them an F for failing to support campaigns to save big wilderness areas like the Stoltmann and for giving high grades to governments who are not really protecting enough wilderness to preserve biodiversity."

Emboldened by several beers, we went back to the office and typed up the condemning press release and faxed out only two copies of it: one to WWF headquarters in Toronto and one to Ric Careless, who, at that time, was acting for WWF in B.C. It worked.

The next morning both recipients thought we had sent our press release to the media and were extremely angry with us. It was a great practical joke. Going home on the ferry that evening with Andrea, who was temporarily staying at our house and helping babysit our kids, we walked by Ric Careless who was sitting in the ferry's cafeteria with his back towards us. As we passed by, in a loud stage whisper, I said "F." Ric reacted furiously, berating our organization and me personally. All the while I kept telling him it was just a practical joke and not to take it so seriously.

When WWF's Endangered Spaces campaign came to an end in the year 2000, with B.C. being one of the few provinces that had actually achieved the 12 percent goal, WWF abandoned its practice of handing out annual grades to the provinces.

Chapter 52

WCWC adopts Marr's BET'R Campaign

WILD takes a new turn

In November 1995, WCWC-WILD hired a new campaigner—Anthony Marr. A year earlier Anthony, a Chinese-Canadian, had launched his BET'R Campaign dubbed by the media as a "one man crusade" to save the world's endangered bears, elephants, tigers and rhinos. Anthony had chosen to focus on these four species groups because they were the ones most devastated by the Asian tradition of using animal parts for medicinal purposes. Anthony explained to us that he had launched his BET'R campaign not in spite of his being of Chinese descent, but because of it. BET'R's goal was to save these species from extinction by reducing the demand for their body parts,

Anthony Marr shopping for medicines containing bones of endangered tigers. Photo: WCWC files.

BET'R Logo. Artwork: WCWC files.

which had made poaching lucrative and was causing rapid population declines in these animals. He was also in tune with WCWC's mandate: to protect these species by protecting their natural habitats. But the main reason Anthony came to WCWC for help was because he had run out of personal money to fund his campaign.

The first thing Anthony did after we hired him was to stir up a media storm by exposing the sale of Chinese medicines containing tiger bone in the Chinatowns of Vancouver, Toronto, and Ottawa. Since these medicines were still being sold openly, WCWC hoped to shame the Canadian government into enforcing legislation that outlawed all trade in endangered species parts in Canada that had already passed through parliament. For some unknown reason the government had never got this act proclaimed by the Governor General, the final step in making legislation into law.

To address the supply-and-demand nature of the $6 billion global endangered wildlife trade, Anthony explained that his BET'R campaign employed a "Yin/Yang" strategy. The Yin track used education and media to reduce and eventually eliminate the global demand for endangered species body parts. The Yang strategy used laws and their enforcement to stop poaching of endangered species and the trafficking and smuggling of their body parts.

The following is an example of Anthony's BET'R's Yang track at work. Up until June of 1996, it was illegal to import Chinese patent medicines containing endangered species parts into Canada. But once smuggled in, incredible as it may seem, because of some loophole in the law they were legal to sell! "Medicines" containing these ingredients were prominently displayed and copiously available for sale in most Chinatown apothecaries. In March of 1996, Anthony received a reply to a letter he had sent a few months earlier to the Federal Environment Minister Sergio Marchi asking him to do something about this situation. In the letter Marchi said:

> The federal government's capacity to deter this trade will increase with the proclamation of the Wildlife Trade Act this spring. Under the Act, poachers and smugglers will be liable to penalties of up to $150,000 and 5 years imprisonment. Corporations are liable to fines of up to $300,000. The maximum fine can be doubled for a second offence....

In May of 1996, four years after parliament passed Canada's Wild Animal and Plant Protection and Regulation of International and Interprovincial Trade Act (WAPRIITA or shortened to Wildlife Trade Act), it was finally proclaimed. In June of 1996 the Minister made good on his promise and these products were gone from the shelves in Vancouver's Chinatown. In 1996 alone, enforcement of-

Illegal medicines purporting to have tiger bone in them purchased by Anthony Marr in Vancouver's China Town. Photo: WCWC files.

ficers for WAPRIITA seized 39,000 individual tiger containing products in B.C.

Employing his Yin strategy, Anthony gave presentations to thousands of high school children in the Greater Vancouver area, many of Chinese heritage. The youth were very supportive. However, when he appeared on Chinese language open line talk shows, Anthony caught a lot of flack for his attack on these traditional medicines. Cultural change, he told us, does not come easily.

BET'R campaign concentrates its efforts on the "T"

In 1996, with the help of CIDA (Canadian International Development Agency) funding, we established a partnership with Tiger Trust India (TTI), an organization founded in 1989 by Kailash Sankhala, India's "father of tiger conservation." Pradeep Sankhala, Kailash's son, had taken over the organization after his father died in 1994.

The main reason we decided to partner with TTI was because we got to know and respect Kailash Sankhala when he attended our WILD Mapping the Vision Conference in Hawaii in 1990. In a plenary at that mapping conference, Kailash had presented three identical maps in quick succession showing remaining wilderness, protected areas and threatened wilderness in India. It was a stunning glimpse of what is undoubtedly destined to be the predicament everywhere on Earth sooner than people expect.

At the turn of the 20th century, India had approximately 50,000-80,000 Bengal tigers. At the beginning of World War II there were still about 30,000. But by 1972 only an estimated 1,800 tigers remained. The year following the '72 survey, Kailash Sankhala launched Project Tiger, which succeeded in getting tiger hunting banned outright in India and in establishing over 25 protected tiger reserves. The tigers in India began to rebound. By 1990, there were over 4,000. It was a spectacular success story.

But then, following a change in government, the tiger conservation effort lost momentum. Meanwhile, the East Asian economies began to boom, bringing about an increased demand for tiger bone medicinal products, causing tiger poaching to increase. Tiger populations again were falling.

Under WCWC-WILD's partnership with TTI, Pradeep Sankhala established a pilot educational facility with a medical clinic and conservation centre and a medicinal plant garden at his eco-tourism lodge next to Kanha National Park, one of India's key tiger reserves. TTI also developed women's craft programs, arts programs and a Nature Guide Training Program to help local people living in villages surrounding the tiger reserve to derive some benefit from tiger-related eco-tourism.

The WCWC-TTI partnership efforts to save India's endangered tiger expanded the next year with CIDA's agreement to provide three more years of program support. We co-published with TTI 120,000 copies of a newspaper titled *Tiger, tiger burning dim*...and 5,000 copies of a poster entitled *Mother Earth With No Wild Tigers?* Both publications featured a series of maps showing how the tiger's range had rapidly dwindled in the 20th century—a trend that had to be reversed or the tiger would be extinct in the wild before 2010, the next Year of the Tiger in the Chinese calendar.

To raise awareness of the tiger's dire situation, in 1997, we organized and held on the third Sunday in October our first annual *Save-the-Tiger Walk* in Vancouver. We specifically chose this late fall date so our walk would be after all the other charity fundraising walks for the year were over. Over a hundred hard working volunteers, unseasonably sunny weather and WCWC staff made this event a huge success.

Diana Vander Veen with her tiger face painted son Ainsley wearing our tiger T-shirts at WILD's Vancouver *Tiger Walk*. Photo: Greg McIntyre.

We contracted Evelyn Roth, whose salmon enlivened our Council of All Beings celebration on the Legislature lawn a year earlier, to make us a giant inflatable tiger cub. When the blower was turned on, its head grew four-metres-tall and its body nearly 15-metres-long. Like the salmon, we could unzip a panel and people could go inside while the tiger was inflated. About 1,000 children, parents and teachers participated in the walk, raising nearly $18,000 in donations.

We held four more annual *Save-the-Tiger Walks*. At our second annual *Save-the-Tiger-Walk* our tiger escaped! A gust of wind whipped out the stakes that tethered it to the ground and it soared up at least 25 metres in the air and like a kite and headed off towards Burrard Inlet less than 100 metres away.

Then, just as suddenly as the strong gust arose, it withered, and calm returned. Our tiger floated gently down, landing on the grass about 25 metres from the shore. This taught us a good lesson. From then on we always put sandbags inside our tiger and made sure it was staked down firmly when we inflated it outside.

Anthony and Tim Murphy, a young university student we hired through another grant, inflated our tiger in many primary school gyms and put on "tiger shows." Anthony finally took our tiger to India where he gave it to our partner group TTI who used it extensively at school rallies to increase public support to save the tiger. The school children there nicknamed it "*Bara Bacha*"—"Big Baby." To replace *Bara Bacha* we had Evelyn make us another one, a smaller "tiger cub", that better fit in elementary school gyms.

In 1998, during Anthony's second two-month-long trip to India to work with our partner group, he created quite a stir. He was determined to accomplish more than we had to date. To combat the local peoples' dependence on firewood for cooking, which was causing the illegal deforestation of tiger preserves, he brought with him several homemade models of simple solar cookers made by a "backwoods"

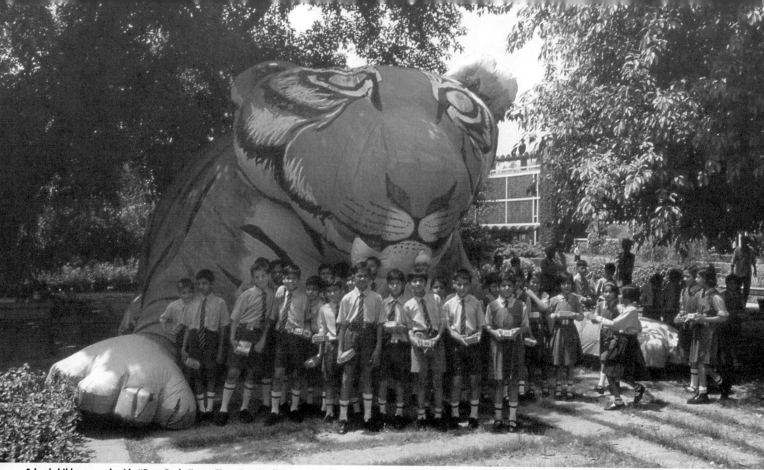

School children pose beside "Bara Bacha" at a Tiger Trust India's sponsored event to raise awareness about the need for tiger conservation. Photo: Anthony Marr.

solar energy activist on the Canadian prairies. It was a simple device that could be constructed in India out of locally available materials. Anthony put on many demonstrations and reported that the local people were amazed when he opened the solar oven to reveal the fluffy sun-cooked rice, which they then happily ate. Anthony also bought a Haryanna bull that local people had been longing for, because its offspring would be high-yield milk producers.

He also spent time talking with local people, including leaders in 120 out of 178 villages in the "buffer zone" region surrounding the Kanha Tiger Reserve. He discovered that the general sentiment of these villagers was that Kanha was little more than a rich peoples' playground that generated no financial benefit for them. Their concerns included getting no compensation for their cattle that tigers sometimes killed. Anthony introduced the idea that the park fee (which then was only $2.50 per day for a foreigner to visit the Tiger Reserve) should be greatly increased and that the additional revenues generated be kept and used locally to help these villagers. Entrance fees for wildlife parks elsewhere in the world, Anthony pointed out, were much higher—up to ten times more. This was really popular with the village leaders, but it was not so popular with our partner group TTI. According to Pradeep Sankhala, the fees would be raised, but undoubtedly the money would not come back to the local people. It would simply disappear into the government's general coffers.

Anthony also thought the money we provided to our partner group could go much farther in creating positive change than it did. Anthony was impatient and had a hard time accepting the cultural differences that existed between the slower pace in India and the swifter western approach to change.

Tiger Walk morphs into Endangered Species Walk and then quietly goes extinct

For various reasons our Vancouver *Tiger Walks* never lived up to our dream of how big we felt they should be and needed to be. Each year it took a lot of work to organize and promote the walk. Perhaps because we held them so late in the fall, the attendance was always disappointing. The weather, even when sunny, was always windy and chilly. They never attracted huge crowds, despite lots of advertising, mailouts to elementary schools seeking their participation, fancy posters, special mailouts to all our members in the Lower Mainland and listings in the *What's Happening* sections of free weekly and all the major daily newspapers.

In 2000 we expanded the concept of our walk to include all endangered species. The next year we changed the venue to Trout Lake Community Centre, thinking that a less windy place with use of an indoor space that was closer to family housing would attract more participants. Our Trout Lake *Walk for Endangered Species* was successful in one respect—those who came out had a good time. Unfortunately, only a couple of hundred people attended.

No matter how hard we tried over the years, we did not attract sufficiently big crowds to justify all the effort. We had to be realistic. Our walks were fun but they weren't building a great wave of people committed to helping save endangered species. When our CIDA funding ended, which paid for the organizational work behind our *Tiger Walks*, so did our annual tiger and endangered species walks in Vancouver.

Chapter 53

A concerted effort to ban bear hunting

Ban Bear Hunting Initiative – brainchild of an impromptu brainstorming session

In March of '96 during one of our evening brainstorm sessions at Greek Characters Pub, we thought up a new way to get the BET'R campaign into high gear. "Let's be the first ones to use B.C.'s new Recall and Initiative Act to get a ban on trophy hunting of bears, like Anthony suggests," I blurted out as we unwound over a few cold pints. The B.C. government passed this unique act in 1995 and it had come into force four months earlier. I had the general idea of how an initiative worked in other jurisdictions. It involved circulating a petition to gather signatures from people who supported a proposed piece of legislation.

If the proponent got the required signatures, citizens got the opportunity to vote on the proposal in a referendum and if the majority voted yes, it became law.

"Good idea," was the general consensus around our table. Someone offhandedly said, "Why don't you check it out?" That was enough of a green light for me.

I quickly learned that advancing an Initiative under B.C.'s new Act was an extremely complicated and convoluted process. For starters, it was impossible for WCWC to launch an Initiative to ban bear hunting. The law only allowed an individual to be the proponent of an Initiative. At the same time it allowed an unlimited number of people and organizations to register to oppose it.

I also found out that we had already lost our chance to be the first one to launch one. At the end of 1995, right after the legislation was proclaimed, three citizens had tried. But they'd all quit after barely getting started. Not one of them handed in a single signed petition sheet. So, if we launched a ban bear hunting initiative and actually collected signatures and handed them in, we'd be the first to actually try out the new law.

The question was: if I personally became the proponent, could WCWC help fund the effort? I checked it out with Elections BC, the independent government agency that administered this new law, and found out that it was OK. However, I had to be personally responsible for keeping the process within the law, including making sure that none of the people authorized to canvass were paid for collecting signatures on the petition sheets.

Furthermore, to conform to rules regarding federally approved charitable societies, WCWC couldn't accept donations from individuals earmarked specifically for this Initiative. Donations to the Wilderness Committee had to be general in nature, leaving us the option of choosing how to spend the money. Sound convoluted? That's only the tip of the iceberg of bureaucratic rules regarding Initiatives.

Every canvasser had to be authorized by Elections BC. There was a different petition sheet for each of B.C.'s 79 different electoral districts. All the signatures had to be collected within a 90-day period. All the money I personally spent on the Initiative had to flow through a special bank account. I had to have a Registered Financial Agent who took charge of the account. The agent had to complete a detailed report on the spending for the Initiative and file it with Elections BC no later than 28 days after the signature-gathering period was over.

Our comptroller, Brian Conner, graciously volunteered to do the financial agent job and this made the campaign feasible. The amount of money I could spend on pursuing the Initiative was limited—up to a maximum of twenty-five cents for each registered voter, which added up to over half a million dollars. This limit on spending was not a critical factor in determining our success! I spent less than a tenth of that. Little did I know at the exciting beginning of this effort that I'd become an expert in an obscure "made to fail" law that was theoretically supposed to expand democracy but instead ensnared those involved in a bureaucratic nightmare.

It did not seem that obtaining the required 10 percent of the voters' signatures on a petition during the 90-day period would be an impossible task. That total, only about a quarter of a million signatures, which averaged out to only about 2,500 per day!

An Angus Reid poll taken in 1995 showed that 78 percent of the people in B.C. supported the banning trophy hunting of black bears. Polls showing support for banning the hunting of grizzly bears ran even higher. The vast majority of people are against wantonly killing of these intelligent, beautiful wild animals for sport.

Having just received a large donation from a person who wanted us to campaign to protect bears, we decided an Initiative campaign would be an effective way to spend it. While we were at it, we'd also campaign to establish grizzly bear sanctuaries. For without even a shot being fired, scientists predicted that the threatened grizzly bear would be driven into extinction in B.C. if all large tracts of wilderness in B.C. were roaded and logged or mined.

I contacted Greg McDade, head of the Sierra Legal Defence Fund and he offered to help. SLDF would look into the legal requirements and help me draft the anti-bear hunting bill—the proposed legislation that we would be seeking petitioners to support. After examining the Initiative legislation, David Boyd, then a staff lawyer for SLDF, wrote a letter to me with the following advice:

> Despite the major hurdles [a huge understatement] this law imposes, you might still want to give it a try. As you know, Paul, this will be a real uphill battle. As Greg [Greg McDade] said, "How are you going to get 5,000 signatures in the Peace River country?" The final decision is up to you. However, I still like the idea—it will raise the profile of bear hunting, inform and educate the public, increase pressure to end the hunt and, if your efforts ultimately fail, show that this law needs an overhaul. If you have any further questions, please do not hesitate to contact me.

By this time I got SLDF's letter we were so psyched up about this campaign at WCWC that even if Boyd had said it was a foolish idea we would have done it anyway. It was a new government-sanctioned window of opportunity and, even if the opening was tiny, we were going to try to crawl through it. With the E-team's formal approval, I gave SLDF the go ahead and within about a week they drafted a piece of legislation that would ban bear hunting in B.C. After a few minor modifications, we were happy with it.

Our proposed law was reasonable. It allowed wildlife officers and their agents to dispatch problem bears and allowed First Nations who traditionally hunted bears to continue to do so. In retrospect perhaps it would have been wiser to have only "targeted" grizzly bears, which were listed as a "species at risk," rather than all bears including black bears, which, relatively speaking, are plentiful and not endangered in any way. However, we wanted to attract the help of animal rights advocates who had never really jumped on board in support of our wilderness habitat-saving campaigns. They believed that none of these

Facing page: Image of a grizzly bear from the Great Bear Rainforest featured on a WCWC poster. Poster image: Ian McAllister.

June 5, 1996, press conference announcing the launch of the ban bear hunting Initiative process. Right to left, Anthony Marr, the proponent (me), and Sierra Legal Defence Fund lawyers Greg McDade and David Boyd. Photo: WCWC files.

remarkable mammals, that some argue are smarter than dogs, should be shot for fun or profit. And I agreed.

With the draft legislation in hand I, the Proponent, accompanied by representatives of WCWC and SLDF, held a joint press conference in the Park Pacific Hotel to launch the Initiative. It was well attended and we got good coverage in the major media.

It took time to get things rolling. First off, there was a delay in getting the proposed bill through the hoops of approval in Elections BC. The wording of our proposed legislation had to be fine-tuned to pass the government lawyers' test of "workable legislation." Then there was the requirement that notice of our Initiative be published in the BC Gazette (the government's publication of official announcements) allowing 60 days after the notice for opponents to register. During that 60 days I garnered 107 opponents. They included practically every rod and gun club in the province. These organizations were "gunning" for us right from the start. So instead of getting started on our 90-day signature-gathering period in mid-summer as we'd planned, we were delayed until early September.

The star foot soldier in my WCWC-backed Initiative campaign was Anthony Marr, our BET'R campaigner. From the moment we conceived the idea he was keen on it. In fact, he was fearless in his support. In June and July, before the 90-day signature gathering process began, he toured on his own around all of B.C., even the north, to raise awareness, educate people and recruit canvassers who would gather signatures when the Initiative officially got underway. On this road trip Anthony visited over 50 cities and towns and traveled 12,000 kilometres to present slideshows at public meetings and pre-sign up canvassers. Everywhere he went he also stirred up hornets' nests of angry hunters who opposed the Initiative.

At least a couple of enraged hunters showed up to protest at each public event. Due to the conflicts that ensued, our ban bear hunting initiative got extensively covered in local media. Anthony was relentless in his pursuit of media coverage and he kept meticulous track of it. According to his count, he was interviewed by newspaper reporters more than 150 times, and was featured on TV and radio more than a dozen times during that pre-Initiative trip.

Anthony could not be intimidated. He believed that his Chinese Canadian heritage gave him a unique opportunity to take on this racially charged conservation issue. He had tirelessly crusaded against poaching and the illegal trade in bear gall bladders. He had already gone directly into Chinatown stores in Vancouver and exposed the illegal sale of tiger bone products and been chased out by angry store owners. This did not endear him to a certain segment of the Chinese community that profited from this trade.

He quickly became the pariah of B.C.'s hunting fraternity, too.

The following excerpt from a July 5, 1996, newspaper article in *The Prince George Citizen* by Gordon Hoekstra about a meeting held in Prince George during the first week of July provides a snapshot of what Anthony's public meetings were like. The article, entitled *Fur flies at meeting to ban bear hunts*, began as follows:

> *It was barely civil and sometimes downright ugly. In the end, it took a representative of the Western Canada Wilderness Committee close to two hours to deliver a plea for help to ban bear hunting in B.C. Anthony Marr was interrupted, shouted down, and generally abused by hunters in an audience of more than 100 that spilled out of the conference room at the Civic Centre Thursday evening...*

Anthony couldn't see any difference between killing a bear for its gall bladder and paws and killing a bear for its head and pelt. As he put it *Hunters go for the head and hide and poachers go for the gall and paws; they're all after bear parts.* Of course, physically and morally there is no difference, but under the current law in B.C. there was and still is one.

One activity is legal under licensed circumstances and the other one is illegal at all times. According to Anthony, the discrimination against Asian uses was taken into account by judges who consequently gave convicted bear poachers and merchants who dealt in illegal bear gall bladders low fines. They never got jail sentences. Because of the growing market for these "medicinal" products, the minimal enforcement of wildlife laws and paltry punishment of convicted offenders, bear poaching was increasing.

At the packed meetings one of the questions hunters frequently asked was, "Why are we being persecuted when the real culprits are the poachers?"

Anthony replied, "Both are culprits. The difference between a hunter and a poacher is irrelevant if you're a bear." Pro-hunting advocates constantly harassed Anthony on this trip and throughout the whole campaign. Pickup trucks routinely tailgated his car and he received anonymous threats of violence by phone. His blunt unflappable style infuriated the opposition.

"Deep down inside, it's a moral issue," Anthony asserted repeatedly. "It's immoral to kill for entertainment. And abominable that adults teach their children to kill for fun." He estimated that at least

90 percent of hunters shoot bears for the trophy and that 65 percent of those hunters actually come from urban areas. Although our position at WCWC was clear that we were not against people hunting animals for food when wildlife populations were strong, Anthony never stressed this point. Some in the hunting lobby thought that our opposition to bear hunt as the thin edge of the wedge. They claimed that our hidden agenda was to stop all hunting and then take away their guns!

Above and below: Bear Initiative bumper stickers. Artwork: WCWC files.

In the fray, we lost a few WCWC members, but not that many. I received several letters eloquently defending hunting and chastising us for getting involved in the animal rights movement. We also gained a few members who thought that we were courageous in fighting for the rights of wild animals.

Elections BC issued me my Initiative Petition on September 9, 1996. There were actually 75 different petitions (a unique one for every electoral district)—and from that day on we had 90 days to sign up 10 percent of each district's registered voters. To become a canvasser, a person had to be a registered B.C. voter and had to have lived in B.C. for the last six months. Elections BC had to approve of every canvasser and issue each one an identification card (that came to me as the Proponent and then I had to forward it on) before the canvasser could gather even one signature. Canvassers had to personally witness each signature they collected and make sure the signature was on the right petition sheet, the one bearing name of the electoral district where that person was registered to vote.

I personally did little of the paperwork or canvassing. My biggest responsibility as the proponent was to oversee the campaign and sign every canvasser application form before it was sent in to Elections BC. Just before the 90-day Initiative began, I made sure WCWC published a compelling four-page tabloid titled BEARS Need More Protection Now (Educational Report Vol. 15 No. 13) that outlined the reasons bear hunting should be banned in B.C. I hoped that this publication would help recruit more canvassers and bring in additional money.

Most of the bureaucratic details were handled by Bonita Magee (nee Charette), a young WCWC volunteer who I ended up hiring on a full-time contract for the duration of the Initiative. The first thing she did was organize volunteers to stuff our BEARS need more protection paper into a mailout appeal to all WCWC members. Through the whole 90 days she worked long hours to help canvassers get registered and make sure they got the right petitions for their area. She sent out the canvasser cards, organized events and sorted the incoming filled-out petition sheets by electoral districts.

The response to our WCWC appeal was lukewarm. In all, less than five hundred WCWC members volunteered to circulate the petition. Together with the canvassers recruited by Anthony, only about 1,500 people made the commitment o help collect signatures. After people agreed to be a canvasser, there was quite a time delay in getting them registered by Elections BC. Once they had their official Initiative Canvasser ID card they faced other hurdles. We quickly found out that it was really hard to sign people up on the right petition sheet when canvassing on the street. Very few people knew the name of the electoral district in which they lived. People in public places in the cities came from all over. Canvassers in Vancouver had to carry 20 different petitions and a map of all the electoral districts in the Lower Mainland.

Star ban bear hunting Initiative canvassers. Evelyn Kirkaldy, centre and Doug Carter, left. Photo: WCWC files.

Then there was the harassment. Instead of intelligently relying on the unworkable rules of B.C.'s Initiative legislation to defeat us, the BC Wildlife Federation (BCWF) went "ballistic" in attacking us. They launched an intense fundraising campaign saying that they needed a $40,000 war chest to counter our efforts. They told people that poaching was not out of control, that bear populations (including grizzly bear populations) were healthy and could support hunting, and that hunting was a valid way for wildlife officials to manage wild animal populations.

We could handle debating these opinions, but the BCWF did not stop there. In the fall issue of its magazine Outdoor Edge, BCWF executive director Doug Walker breached the boundaries of free speech

Ban bear hunting Initiative office in WCWC's space at 55 Water Street. Bonita Charette in centre behind the desk. Photo: WCWC files.

with his rhetoric. In his editorial column he passionately urged his membership to donate money and actively oppose my Initiative, likening WCWC to *terrorists groups who threaten human lives, burn houses, send razor blades in the mail or kill family pets to get attention....*

This was an outrageous and deliberate act of defamation. It was also kind of ironic because BCWF President John Holdstock had been quoted in the press about a week earlier as saying, *even if all the logic is on our side, it is hard to counter emotion*, referring to our moral arguments against killing bears for sport. Our organization was well known for always being law abiding. We never even once used peaceful civil disobedience as a tactic.

Angered by BCWF's libelous tactic, I immediately sought the help of one of Vancouver's best law firms familiar with this complex area of law. One of the senior partners took the job. I dreamed of funding our whole anti-bear hunting campaign with the tens of thousands of dollars in compensation that the courts would award us for this unlawful character assignation.

Our lawyer immediately prepared to launch a libel suit in the B.C. Supreme Court against the BCWF's Executive Director Doug Walker and President John Holdstock informing them of our intent. He got quick results. BCWF admitted its mistake. He negotiated and got BCWF to pull the fall issue of its magazine off newsstand shelves, place ads of apology in the major and local papers and pay for our entire lawyer's bill. (I never saw that bill but I'm sure it was a bundle!) Our success in getting this issue of BCWF's magazine banned was covered in a short article titled *Hunted down by the law* in the September 1996 issue of Canada's foremost national newsmagazine, *Macleans*. It was the only time I know of when WCWC's activities were reported in this influential periodical. The brief article began this way:

> *It was a case of ready, fire, aim for the BC Wildlife Federation, a group representing hunters in B.C. With 25,000 copies of the September/October issue of its magazine, Outdoor Edge, already delivered to their B.C. members, the B.C. group had to abruptly cancel the remainder of its distribution, pull 60 copies from store shelves, and print a public apology in Vancouver newspapers last week.*

But by the time the BCWF took corrective action, most of the damage had already been done because the magazine was already in the hands of its 20,000 BC members. The public apology did little to counter the utterly false "eco-terrorist" label the BCWF members had been fed. BCWF members kept harassing our canvassers. But I felt the BCWF's defamatory tactics contributed, more sinisterly, to the growing demonization of environmentalists by many rednecks, which culminated a few years down the road in Premier Glen Clark's infamous branding of environmentalists as *enemies of B.C.* This vilification snowballed into loggers' brutal physical attacks against peaceful protesters in the Elaho Valley and elsewhere in B.C.

After two months it became painfully obvious to me that we were not going to get the needed signatures. Worse, we were not even going to come close. At the two-thirds point in our signature-gathering period, we only had 20,000 signatures turned in to our office. We needed 220,000. Two keen canvassers had gathered over one thousand signatures each and quite a few others had gathered several hundred. But most of the canvassers had gathered only a handful and had not sent them in yet. In the Peace River North Electoral District we had only one canvasser trying to get signatures. This brave person said he would only sign up his sympathetic friends and would not go public because of possible reprisals from outraged redneck hunters.

We had to do better or we would become a laughing stock and do more harm than good for the cause. At another of our WCWC brainstorming sessions at Greek Characters, we came up with two brilliant ideas. We would send a signature-gathering expedition into the north and we would organize a canvassing blitz outside polling stations on Municipal Election Day.

On that day in mid November all across B.C. voters would be self sorted into electoral districts, waiting in lines and be thinking about democracy. That would make it easy for our canvassers to gather the signatures on the right petition sheet.

Some voiced concerns about using the municipal election as a springboard. "Aren't people prohibited from campaigning near the polling stations?" queried Andrea.

"It's got to be legal. Our bear petition has nothing to do with the local election. But just to make sure, I'll ask Greg McDade at Sierra Legal to check it out." I replied. Greg verified that since our bear petition had nothing to do with the local elections its circulation inside the exclusion zone was legal. The Municipal Act only forbids people from soliciting votes or doing things designed to influence the outcome of a municipal election (wearing signs, carrying party flags, leafleting regarding a candidate, etc.) within 100 metres of a polling station.

Bear caravan in front of our store taken at the launch of the trip to Prince George to gather signatures. Photo: WCWC files.

The bold expedition into the "redneck" north to collect signatures on the Initiative petitions

"Are there any canvassers who will to go to Prince George?" I asked Bonita. "Can we rent a van cheaply? Who will lead the group? Will we get media coverage?" I got a quick yes answer to the first three. Besides Anthony, who understood the importance of doing well on the Initiative and actually seemed to thrive on confrontation, Barney Kern, a plucky young environmental radical with a quick-witted mouth, volunteered to drive the van. Four other Initiative canvassers also volunteered to pay their own way and go along. Evelyn Kirkaldy, a talented artist who loved bears and who had already personally collected over a thousand signatures around Vancouver, made a giant banner to drape over the side of the van everywhere they stopped. Everyone heading north was more afraid of the cold than they were of potential violence from the pro bear hunters. This being Canada, we were certain that the ugliest of confrontations would only entail verbal abuse.

The TV media covered the departure of our ban-bear hunting caravan from our 20 Water Street office. Getting TV news coverage was important. We held hundreds of protest pickets and rallies over the years, but if the event didn't make the TV news (and most of them didn't), hardly anyone knew it happened. At our bear caravan launch, thank goodness, the TV and print media both came out and no one asked us questions that would have exposed how Don Quixotic this effort really was.

Neither the northern trip, nor the polling stations blitz worked out quite as well as we had envisioned. On their way up north, our valiant crew ran into miserable mid-November weather—freezing rain and heavy snowfall. They finally reached Prince George at 3 a.m. Exhausted, Barney pulled into a parking lot to spend the rest of the night. There was an arch above the entryway he judged he could easily clear. He didn't. There was a tremendous crunch and the nice warm comfortable van instantly turned into a huge uninhabitable deep freeze. Barney had forgotten that an air conditioner stuck up about a half-a-metre above the van's roof. The unit tore completely off, leaving a big gaping hole. Luckily, we had supporters living in Prince George who put our crew up in their homes.

When Barney called to tell me of the accident, I resisted saying, "What a stupid move," and instead said, "No problem. We'll handle it when we return the van." I was happy that they arrived safely and no one was hurt. In the end, it only cost us about a thousand dollars—insurance covered the rest—to get the van fixed.

Chilly temperatures and a posse of hunters who dogged our canvassers made signature gathering in Prince George difficult. But despite freezing pens and open hostility, the canvassers did a remarkably good job. Gordon Hoekstra summed up their effort in his article titled *Bear 'ban-wagon' gets cool reception* (Prince George Citizen November 16, 1996):

> Battling the wind and -10 C temperatures, WCWC canvassers from Vancouver set up tables Friday at the intersection of Victoria Street and 7th Avenue to gather signatures.... Hunter Brad Davis stopped to protest the bearskin propped on top of the 24-foot, banner-decorated motor home, which he thought was in bad taste.

On Municipal Election Day the crew set up shop outside one of the main polling stations in Prince George. Unlike many other places, the local returning officer ruled that there was no problem with them being outside the community centre's main door right next to this polling place. After a short while local polling officials took pity on our shivering canvassers and invited them to come inside the community centre and set up in the hallway right beside the entrance to the gym where the polling booths were stationed. This friendly reception was the exception. In most other municipalities our canvassers got a colder welcome from election officials. John Marshall, Victoria's chief electoral officer, after wavering a bit, decided that the 100-metre exclusion zone would have to be respected by our canvassers. When we complained to the provincial government's Municipal Affairs Department, a spokesperson there said that it was up to local electoral officers to interpret the provincial legislation that regulates permitted activity outside polling stations.

At about twenty percent of the polling stations our canvassers were allowed to stand on public land by the entrances and at the other eighty percent, they were forbidden. One young canvasser in the Okanagan demanded to exercise his democratic right and collect signatures by the entrance to the polling station. He was arrested, put in jail and released without being charged that evening after the polls had closed. Adriane and I stood in the school parking lot by our polling station (closer than 100 metres) on the Sunshine Coast and got several sheets full of signatures. We respectfully dialogued with several opponents who disagreed with a ban on black bear hunting but did agree with one on grizzlies. Needless-to-say, canvassers close to the polling stations did very well and those forced to remain far away did very poorly. Instead of the 100,000 signatures we planned to gather that day, canvassers got less than 30,000.

I swore to take this matter forward in a complaint to the B.C. Ombudsman. But after the campaign was over and it was obvious that, even if we had been allowed to be near every polling station we still would have fallen far short of the required number of signatures, I never pursued it.

As the campaign culminated in an intensive effort during the last few weeks, incidences of harassment, including hunters standing beside canvassers arguing with anyone contemplating signing the petition, were reported to me on a daily basis. There was really nothing I could do about it. I commiserated with and praised the canvassers for sticking it out in spite of the heavy-handed tactics on the part of some of our opponents.

A *Prince George Free Press* reporter, making no effort to hide his bias, wrote a column on November 29th titled *Bear hunting ban signers should be proud*. In it he said:

> I was not surprised by the intimidating, dirty behaviour of some wildlife killers during Western Canada Wilderness Committee's visit... Some wildlife killers and their supporters carried over their violent actions from defenceless animals to non-violent animal supporters, going so far as to tear up a petition sheet with signatures.

Concluding the campaign on a growl not a whimper

We decided to wrap up this campaign with a growl, not a whimper. At noon on the 90th day we held a rally on the steps of the B.C. Legislature. After several speeches imploring the government to acknowledge the will of the people who signed the petition and at least introduce the proposed Initiative legislation for debate in the Legislature, we paraded off to Elections BC headquarters on Pendergass Street eight blocks away. Four people in bear costumes, including a Kermode bear, led the way. Several people pushing wheelbarrows piled high with big boxes of petitions followed, with another 30 or so volunteers carrying placards behind. Bringing up the rear were a couple of tigers including one that looked like Tigger of Pooh Bear fame. It took about a half hour for us to parade on the sidewalk to Election BC's office on that rare nice, warm, sunny December day. We got there at about 1:30 p.m. It took us right up to the deadline of 4:30 p.m. to get all the petition sheets sorted and counted and the

Above: Gathering signatures on the ban bear hunting Initiative petition on the return trip from Prince George. Photo: WCWC files.
Below: Bears helping deliver Initiative petitions pose for a portrait in front of the Legislature buildings before parading to Elections BC's office. Photo: WCWC files.

final report filled out as required by law. Although the whole exercise was academic—everyone knew we fell far short of what was needed—the Elections BC officials were very accommodating and helpful. An article by *The Province* reporter Charlie Anderson published the day after titled *Drive for bear-hunting referendum falls short*, said that we were bloody but unbowed... 'The law itself is an ass,' said George [I was fed up with the impossible hurdles the Recall and Initiative Act posed], *who favours referendums based on the U.S. [Washington State] model. 'No issue, no group could ever get that amount of signatures all sorted by electoral district...'*

The Wildlife Federation saw it a different way, concluding that 96 percent of B.C.'s Voters rejected the Bear Referendum because we only got four percent of them to sign our petition. Elections BC had the final word. To this day you can find the results on Elections BC's website. All the thousands of hours of volunteers' efforts are distilled into two terse sentences: *The petition was returned to the Chief Electoral Officer on December 9, 1996. The petition contained 88,357 signatures, and therefore failed, as 222,272 valid signatures were required.* To this day B.C.'s Recall and Initiative Act stands as an unworkable piece of legislation because the "bar" is set too high. No government has been willing to amend it to give citizens a fighting chance to succeed in initiating laws via direct democratic referenda.

"Hey, 88,000 is a heck of a lot signatures. Don't be discouraged or feel bad, you did a great job," I said, praising Bonita and all our staff and volunteers that were in the office on the day after handing in the petition sheets. "How many other environmental causes get this many signatures?" I asked. We could only think of one: our petition asking the federal government to protect Clayoquot Sound in a Biosphere Reserve. We circulated that one in '93 and '94 across the whole country with no bureaucratic rules to slow us down and we still only gathered 120,000 signatures.

Although we had failed, we'd raised awareness and inched closer to our goal. I continued to push the issue, writing to the six MLAs in whose electoral districts we had gathered the required 10 percent of local voters' signatures. One of them was then opposition leader Gordon Campbell of Point Grey. I asked each MLA to take the Initiative's proposed ban bear hunting legislation, modify as they saw fit, and present it as their own Private Member's Bill. I knew full well that such a Private Member's Bill would never pass. Nevertheless, it would acknowledge the fact that so many people wanted to ban bear hunting, strengthen democratic debate and ultimately led to more protection for bears. I got no takers.

The next time Joe, Adriane and I had a meeting with Environment Minister Joan Sawicki, we asked her to increase poaching penalties. It was a bare minimum conservation measure that both WCWC and BCWF agreed upon and it would be something everyone could point to as a positive outcome of our Initiative effort.

We were rebuffed. Her advisors contended that judges wouldn't impose higher fines or sentences, so consequently such a move would not act as a greater deterrent. These advisors also thought that higher penalties might even be counterproductive, because they would be out of sync with those for other wildlife offenses. I failed to understand the logic of these arguments.

The campaign to protect bears continues

This was not the end of bear conservation campaigning. Our focus returned to grizzly bears, with the Raincoast Conservation Society (RCS) taking the lead on a campaign to protect their dwindling numbers along the mid coast of B.C. RCS pointed out that the methods and information the government used to estimate grizzly populations

Bears in front of the Legislature with the two wheelbarrows full of Initiative Petition sheets soon to be paraded. Photo: WCWC files.

were flawed. It urged the government to impose a moratorium and hire independent scientists to collect the best data possible to more accurately estimate the number of grizzlies in B.C. We held a rally at one of the opening day ceremonies of the spring session of the Legislature with large cut-out cardboard grizzly bears, each with a target on it. Each time a cannon went off in the 15-gun salute, a person tumbled over with his or her bear. Each bear symbolized 25 grizzlies killed that year. It was very effective theatre but it, too, failed to change the government's grizzly bear hunting policies.

WCWC hired bear biologist Wayne McCrory to check out the grizzly bear habitat in our Stoltmann Wilderness Proposal area. All down the coast below this wilderness area grizzlies had become extirpated (locally extinct). He found moderate to low habitat potential there. We also hired Wayne to analyze the proposed protected areas in the Lillooet region. There, he concluded, the habitat potential

Anti grizzly bear hunting demonstration on the Legislature lawn on opening day. Ken Wu in the background. Photo: WCWC files.

The cover of the WCWC anti-grizzly bear trophy hunting opinion poll mailer postcard that went to Premier Glen Clark. Document: WCWC archives.

was good but the survival of grizzlies there, and in southern B.C. in general and in the adjacent U.S. states, was threatened by the fragmentation of their habitat. Without protecting and connecting more roadless wilderness across our border and into the north, the population of grizzlies in the States would be permanently cut off from grizzlies in southern B.C. thus leaving the populations of B.C.'s southernmost grizzlies and the grizzlies in the lower 48 states isolated with gene pools too small to survive over the long term.

WCWC also continued to push for an Act to protect endangered species, which both Canada and B.C. lacked.

In the fall '98 a senior habitat protection biologist, A. Dionys de Leeuw, who worked for the Environment Ministry, circulated a paper that analyzed the method the B.C. Wildlife Branch was using to estimate grizzly bear populations and condemned the way bear hunting licences were issued. It totally contradicted the government's line that everything was OK with grizzly bear populations. But the NDP resisted our calls for B.C. endangered species legislation and banning grizzly bear hunting despite the fact we had strong scientific arguments and political will on our side.

For the first time in all the years I had been campaigning for wilderness preservation, I became discouraged. The sure-fire formula of getting the facts and the people on your side to achieve positive government action wasn't working. We had both in this case. The latest Angus Reid Poll had found that 75 percent of British Columbians supported a ban on grizzly bear hunting. What more could we do? I suspected that soon it was going to be like this with other conservation issues.

Former Premier Glen Clark, now backbencher, pushes private member's Bill to ban Grizzly bear hunting

Several years later in 2000, Andrea shouted out to me, "Glen Clark is on the phone for you." "Is it the Glen Clark?" I asked before picking up the phone.

"Sounds like him," she replied.

About eight months earlier Clark had stepped down from being Premier due to an investigation involving possible criminal corruption by him. He had been replaced by Ujjal Dosanjh. The demoted Clark now occupied a seat on the government's backbench.

"Hello," I said, not entirely convinced by Andrea's assertion that the caller was our ex-Premier. Immediately, I recognized his voice. He explained that he was putting forward a Private Member's Bill to ban grizzly bear hunting.

"That's great!" I exclaimed. "But why didn't you do this when you were the Premier and had the power to actually do it?" I asked.

"I wanted to but I couldn't," he replied, explaining that he could not get the cabinet consensus he needed to act when he was Premier because there was too much opposition from his northern MLAs. Now he could act freely on his own. He asked me if he could use my name as someone who backed his Bill and would speak positively about it to the media. I said yes, I'd be happy to. Some thought that Clark was just being mischievous. Maybe he was. But that didn't matter. Anything that pushed the issue forward was good. His announcement made a small publicity splash. Of course, the bill never was debated and quietly died like almost all such bills do.

Concern for the survival of grizzly bears remained strong, however. In February 2001, just three months before calling an election, Premier Ujjal Dosanjh, who had replaced Clark, established a three-year moratorium on grizzly bear hunting and struck an independent panel of bear biologists to determine the real facts. Were there less than 6,000 grizzlies left in B.C. as some claimed or were there in excess of 13,000 as claimed by B.C.'s Fish and Wildlife Branch? The studies to prove whose guesstimate was right were never completed. After the provincial election of May 2001, one of the first things the new Liberal Premier Gordon Campbell did was to end the moratorium on grizzly bear hunting.

Anthony Marr is violently attacked

The anger generated amongst some segments of B.C. society by our campaign to end the killing of bears was linked, we think, to an act of unconscionable violence. In late January '98, Anthony Marr was attacked and beaten by an unknown assailant in Vancouver. Anthony had parked his distinctive car in the lane by his parents' apartment for one of his regular visits. He documented what happened when he returned to drive home: *There was this guy waiting for me by my car. He advanced towards me a few steps and said, 'Are you Anthony Marr?' I said yes and he immediately attacked me.* The attacker, who Anthony described as being over six feet tall and around 200 pounds, punched him in the head and face several times, fracturing facial bones and damaging his eye socket.

As the attacker walked away, he said to Anthony, '*Let this be a lesson to you.*' Luckily, Anthony was not badly injured. After treatment in a hospital emergency ward, he was released.

The only enemies that Anthony knew he had were ones he had made while working for us. They came from two sources; certain hunters who hated him for his anti-bear-hunting campaigning and certain Asian merchants who he had hounded for selling medicines with endangered animal parts. Anthony had received anonymous phone call threats before, but they had not deterred him. And neither did this beating. It only strengthened his resolve to continue campaigning more vigorously to protect grizzly bears and tigers.

Anthony's attack was part of the increasing trend of violence towards environmentalists. I told Stephen Hume, a reporter for the *Vancouver Sun* who did a story about this assault, that it was time for

Famous bears lined up to sign the ban bear hunting Initiative Petition. First published in the *Times Colonist*. Artwork: Adrian Raeside.

police and government to take seriously the *...threats of violence and all the rhetoric that our people are subjected to. I think this [violent rhetoric] unleashes hate against environmentalists just as much as it does against Jews or people of a different sexual persuasion.*

Anthony Marr leaves us to start his own organization

In September of 1999, Anthony Marr who had been with WCWC since November 1995, left WCWC to start his own organization, Help Our Planet Earth (HOPE). His keenest interest was in animal rights issues and he had chaffed a bit working with WCWC because saving wilderness was our primary mandate. We hired another campaigner, Jacqueline Pruner, to take over supervising our CIDA-funded *Saving India's Endangered Tiger* program. Our partnership with Tiger Trust India lasted one more year. After it ended, Jacqueline left to work for the American-based group ForestEthics, which played a key role in the negotiations to protect B.C.'s Great Bear Rainforest.

Demonstrating at the opening of the Legislature in March 1998 to save the coastal ancient temperate rainforests. Photo: Joe Foy.

Most popular WCWC-Victoria poster promoting the Sooke Hills Wilderness. Coastal Douglas firs on Sugarloaf Mountain. Photo: Garth Lenz.

Chapter 54

Creative campaigning pays off
Clayoquot hit with landslides and another blockade

WCWC-Victoria crew writes and performs a "Broadway" musical to help save the Sooke Hills

Our Victoria staff and volunteers were very creative in devising ways to keep the fate of the Sooke Hills in the news and as a top-of-mind local issue in Victoria during the mid '90s. In February 1996, they created a play titled *Sooke Hills: The Musical*. Misty and her brother Cam wrote songs and a script to accompany a unique multimedia/musical slideshow. This was similar to the one they created (*The Stump Show*) for our second Stumpy tour of North America. Cam also produced another show for our WCWC-Victoria titled *Juvenile Delinquent Wood* about industrial-forestry.

Poster for the Juvenile Delinquent Wood show. Artwork: WCWC files.

The hilarious and satirical *Sooke Hills: The Musical* premiered in the McPherson Playhouse, Victoria's most prestigious venue, which seats over 800 people. An anonymous WCWC benefactor donated the rental fee so admission to the show was free. The audience was, of course, encouraged to donate to the cause, and our chapter made some money for their Sooke Hills campaign along with educating a great number of people not fully familiar with the cause.

Adriane and I attended the premier. An almost capacity crowd came out despite the foul, snowy weather that night. It was a genuinely professional and witty "Broadway styled" musical. It had villains named *Les Forest*, *Seymour Cash*, and *Vic Waters*. The songs had lyrics like "*Hooray, hooray for urban sprawl! Let's cut down the trees and put up a mall.*" The lines provided serious educational information. It was unmistakably based on the reality of WCWC Victoria's campaign to preserve the Sooke Hills as a sea-to-sea green-blue belt. By popular demand, our troupe put on *Sooke Hills: The Musical* several more times over the next few months to overflowing audiences in other smaller theatres in Victoria. The use of humour proved powerfully effective in exposing the backroom political resistance to this popular proposal to make the Sooke Hills a park.

About a month after the show's premier in April '96, with a provincial election looming on the horizon, Environment Minister Moe Sihota appointed a Special Commission on the Protection of the Greater Victoria Water Supply. He stated that it was *...in response to public criticism of the Greater Victoria Water District for many of its policies, particularly controversies about pricing of water for areas underrepresented in the present structure, about proposed uses of GVWD land in the Sooke Hills, and about the adequacy of public participation in key water supply decisions.* He appointed David Perry, a young Victoria lawyer, as the sole commissioner.

The NDP government then promptly called a provincial election. WCWC-Victoria's campaigners worked hard to make protecting the Sooke Hills wilderness a major election issue. They approached all candidates in the Greater Victoria region and got them to state their position on what should happen to the publicly designed non-catchment watershed lands in the Sooke Hills. Then, right before the election, in big ads in the local papers our branch published the names of each candidate along with their position on protecting Sooke Hills. This was not lobbying. It was educating: providing voters with vital information needed to make an informed choice in the voting booth!

In June, a month after the election, the Perry Commission held open houses followed by two months of stakeholder meetings. He concluded with a two-day public hearing in September of '96 before submitting his final recommendations.

In Vancouver we scrambled to print, collate and cerlox bind copies of the recently completed 37-page report on the Sooke Hills preliminary biological inventory research project that Alison had initiated two years earlier in time to present to the Perry Commission. Through their fieldwork in the Sooke Hills, the scientists had found 244 vascular plants, many of them threatened and some endangered. They also found 49 mosses, liverworts and lichens, 37 bird species and 11 mammals including Roosevelt elk and all of Vancouver Island's large predators—black bears, cougars and wolves. This research doubled the number of species known to exist in the Sooke Hills.

A major victory in the Sooke Hills campaign

At a press conference in Victoria on January 28, 1997, the provincial government announced that it was protecting three of the five major watersheds in the Sooke Hills—the Niagara, Waugh, and Veitch—in a Sooke Hills Wilderness Regional Provincial Park. Our Victoria Branch campaigners greeted the announcement with cheers and champagne. It was a giant step forward in protecting a contiguous Sea-to-Sea Green-Blue Belt for Victoria. At the media conference Andrew Petter, one of Victoria's MLAs and the government's Finance Minister, also stated that it was the Province's intention to fully implement all of Perry's 19 recommendations including the one to disband the current GVWD and create a new Regional Water Commission affiliated with the Capital Regional District.

Getting this Sooke Hills Park was great, but the campaign wasn't over. As they celebrated, our Victoria Branch campaigners simultaneously launched the second phase of the Green-Blue-Belt campaign—a drive to acquire the remaining two major unprotected valleys in the Sooke Hills—the Ayum and Charters—and some other small parcels of Crown and private lands needed to complete the dream.

Shortly after Alison Spriggs left WCWC to continue her Sooke Hills and other wilderness preservation work at The Land Conservancy (TLC), an organization focused on acquiring private lands for conservation. We hired Ken Wu as the new Executive Director for our WCWC-Victoria Branch and Kristin Lindell as the Sea-to-Sea-Green-Blue-Belt campaigner. Campaigning in tandem with Alison at TLC and other Victoria-based environmental groups, they achieved a major victory in January of 2000 when the Capital Regional District voters approved an average $10/ year household levy for ten years to raise an estimated $16 million for a parks acquisition fund that would help to complete the Sea-to-Sea-Green-Blue-Belt.

Huge storm triggers landslides in Clayoquot, most occurring in clearcut areas

On January 17, 1996, heavy rainfalls triggered a slew of landslides in Clayoquot Sound. As the storm raged, Adriane was driving the road to Tofino alone that night on her way to attend meetings the next day with First Nations about our Youth Services Canada project to fund the Flores Island trailbuilding. When she arrived she phoned to tell me that it was the scariest drive of her life. In places the road had turned into a river. Coming down the steep stretch to Kennedy Lake the car started to lose traction and move with the flowing water. At one point she had to pull over and wait for a particularly heavy deluge to stop; the wipers couldn't keep the windshield clear. According to weather record books, it wasn't an exceptional storm event. But nonetheless, perhaps because of its extreme intensity over a very short period, it caused more damage than any other rainstorm local residents could remember.

The disastrous impact was due, in part, to the fact that there were many more clearcuts in Clayoquot than when the last big storm hit a few decades earlier. Besides damaging the forest, the silt from the slides muddied many salmon streams.

Numerous landslides in clearcuts in Clayoquot Sound. The lake below is deep brown from the silt-laden runoff. Photo: WCWC files.

Strange giant rock pinnacle on Flores Island. Photo: Ian Mackenzie.

We immediately talked about documenting this storm event and the damage it triggered. Our aim was to prove that logging roads and clearcutting exacerbated erosion.

We jumped at the generous offer of WCWC director Ian Mackenzie to donate the money to hire a plane for a fly-over to assess the damage. On February 9, 1996, our aerial survey of Clayoquot Sound documented over 100 new landslides. The lakes and streams were still brown from the eroded silt. The overwhelming majority of slides started in clearcuts or along logging roads even though clearcuts covered only a tiny portion of the landscape.

Within a few weeks we published a 16-page report, titled *An Analysis of the Landslides that Occurred During Heavy Rainstorms in Clayoquot Sound in Mid-January 1996*. We illustrated our report with photocopies of the colour photos taken during the aerial survey, especially of the worst slides, muddiest streams and brown coloured lakes laden with silt. Our key conclusion was that the landslides occurred up to 20 times more frequently in the clearcut areas. As usual, we produced lots of copies of our report and sent one to every B.C. MLA.

During that flight over Clayoquot, Ian made a startling discovery—a narrow rock pinnacle poking up above the trees on the southwest side of Flores Mountain like an obelisk. It had to be at least 60 metres tall. Although we always wanted to bushwhack in and find this strange and mysterious anomaly, we never did. In 2004 at a meeting in Clayoquot Sound, Joe Foy met a native from Ahousaht who had visited this sacred, highly secret site.

Clayoquot logging spurs another big confrontation

In June of '96, while we were still immersed in our joint project with the Ahousahts training youth and building the Wild Side trail on Flores Island, events in Clayoquot erupted. Greenpeace and the Friends of Clayoquot Sound launched a surprise blockade of logging in the Bulson Valley. They figured the Upper Bulson was a pristine valley and thus, under the Clayoquot Scientific Panel requirements accepted by the B.C. government, it had to have full inventories of all forest values prior to any logging there. The B.C. government had contrarily decided that the Upper Bulson didn't qualify as a pristine valley because the lower valley was logged. Everyone was frustrated. The Nuu-chah-nulth were angry.

The blockade came as a complete surprise to the First Nations. Neither Greenpeace nor the Friends had respected their aboriginal title and rights by consulting with them before "the action" took place. The Nuu-chah-nulth immediately asked the two groups to suspend their blockade and called an emergency "summit" of all the groups and stakeholders involved in Clayoquot. It was held only a couple of weeks later, in July '96, in the old Tin-Wis residential school gym in Tofino. Adriane and I attended on behalf of WCWC.

It started like most meetings in Clayoquot did, with a series of speeches by First Nations' leaders. Each Chief made it clear that they were the rightful owners of the land and resources in Clayoquot Sound. They repeatedly went over the principles of protocol and respect for their rights. They stated that banning all logging in Clayoquot was not an option. They explained how tired they were of being caught in the middle of the war between the forest companies and environmentalists. They wanted peace in the Sound. WCWC was singled out and praised as a group that followed protocol in the right way. Greenpeace and the Friends of Clayoquot Sound were lambasted for their Bulson blockade.

Nelson Keitlah, then Chairman of the Nuu-chah-nulth Tribal Council, instructed representatives of Greenpeace, FOCS, MB and Interfor to go, right then, into a room at the back of the hall and work things out. They were given only half an hour! Adriane, I and the others left in the big room were all a bit stunned. We small-talked and watched the clock, wondering what was happening in the back room. After half an hour someone emerged to say that they couldn't come to an agreement. The chiefs gave them another half hour. At the end of the hour everyone emerged to say that, although they were still at an impasse, they had at least decided to set up a longer meeting in a couple of days time. No one held out much hope. Too many of the lead people in the back room were warriors, not negotiators.

As we were all getting ready to go, Adriane decided to approach Linda Coady, MB's newly appointed vice president of Environmental Affairs. Ever since Harcourt's infamous CLUD (Clayoquot Land Use Decision) of 1993, Adriane had been talking to all the environmental groups about using the NDP's promise to pursue a United Nations Biosphere Reserve in Clayoquot as a possible avenue to negotiate for more forest protection. The NDP had abandoned the Biosphere Reserve idea, having been told that the UN wouldn't touch an area so fraught with conflict.

But Adriane continued to believe in its potential and kept waiting for the right time to move on it. That moment was now. She told Linda very briefly about how the process to pursue a Biosphere Reserve could be used to bring people together to negotiate a solution to the Clayoquot controversy. Linda saw the potential right away and suggested they meet in her office later that week.

A few days later Adriane found herself sitting in Linda's big office in MB's executive suite high up in a downtown Vancouver office tower. The first question Adriane asked her was, "What does your company really want in terms of resolving Clayoquot?" Linda answered that all MB wanted was their customers to stop yelling at them and canceling contracts. This was a good start! Her answer was consistent with the message that Adriane had heard from another MB employee. A few weeks earlier after a meeting of the federal government's Biodiversity Convention Committee at a retreat centre near Ottawa, Adriane had approached Glen Dunsworth, MB's representative on that committee with the same question.

When he said, "All we want is to be liked again by our customers," she thought to herself, "There's hope! We can work out a deal based on that." One reason she wanted to meet with Linda, an MB VP, was to check out whether or not Dunsworth's thinking prevailed amongst the company's top decision-makers. Now assured that it did, Adriane launched into her explanation of how the NDP's statement about a UN Biosphere Reserve for Clayoquot (enshrined in a newspaper publication they sent to every household in B.C. after their April 13, 1993 decision) opened a window for further negotiations. "It's the only chance to get people to come to the table to talk and work out a new deal," said Adriane. Linda agreed. And thus, a new phase in the Clayoquot campaign was launched: the phase that would take the campaign to at least a temporary resolution.

Adriane and Linda had a lot in common. They were exactly the same age, with teenage daughters of the same age, too. They were both bright and likeable workaholics. And they were both consummate negotiators. Each went back to garner the support of their respective organizations to work together towards a solution within the biosphere reserve context. Next they arranged a meeting to explain their idea and seek the support of the Nuu-chah-nulth leaders.

That caused the first big Clayoquot campaign row in our office. Linda offered to fly Adriane to the meeting in Port Alberni in MB's helicopter. I hated the idea and thought that it would put WCWC in MB's debt. We agreed, in the end, that Adriane would drive herself to any meeting where advance notice made it possible. If the only reasonable way to get to an important meeting was via the MB helicopter, we would let her take it.

Adriane and Linda arrived in Port Alberni during one of the Nuu-chah-nulth's treaty negotiating sessions. All eyes were on them as they walked into the meeting together. It was a bit embarrassing, said Adriane, to have their arrival announced by the meeting's chair. At lunch, Cliff Atleo, met with the two of them and, on behalf of the Nuu-chah-nulth, authorized them to pursue working together on the Biosphere Reserve idea. He asked them to regularly report back to the First Nations.

After this authorization, Adriane and Linda were off and running. Within a few days they found several sympathetic and bright co-negotiators: Jim Walker, the Assistant Deputy Minister responsible for Clayoquot in BC's Environment Ministry and Ross McMillan, who worked for the Clayoquot Sound Central Region Board. The four of them talked and met regularly, working out the strategy to pursue UN Biosphere Reserve designation for Clayoquot. According to Adriane, it was the liveliest, most innovative and fun group of people she'd ever worked with.

From then on, everywhere she went, Adriane promoted the idea of a Clayoquot Biosphere Reserve. In October of 1996 the World Conservation Union (formerly called the IUCN-International Union for the Conservation of Nature), a global alliance of 880 agencies and governments from 133 countries, held their conference in Montreal. Paul Ramsey, B.C.'s Environment Minister, attended this conference. We knew that a resolution would be presented to designate Clayoquot Sound as a UN Biosphere Reserve, but that Ramsey was given direction to only support namby-pamby wording that endorsed the Clayoquot compromise announced by Premier Harcourt in 1993. Adriane was our envoy there and her mission was to get Ramsey on board for a cleverly-worded resolution that would open the door for more protection and possible resolution to the heated Clayoquot conflict. She and a number of other environmentalists, including Liz Barrett-Brown of NRDC and Vicky Husband of the Sierra Club, met with Ramsey. He listened and eventually supported the re-worded Clayoquot Sound Biosphere Reserve resolution, which urged all Clayoquot stakeholders to consider the importance of increased protection of pristine areas as well as creation of new opportunities for sustainable community development. It was a real victory and a big step forward in our Clayoquot campaign.

Just after the IUCN meeting in Montreal, the Clayoquot Central Region Board held its first community meeting about designating Clayoquot Sound as a Biosphere Reserve. There was suspicion and even anger about the IUCN resolution and resistance to the idea of re-opening the land use decision that the loggers, at least, thought

Staff outside the hotel where the federal government is holding a cabinet meeting attempting to get all our concerns addressed. Photo: Kerry Dawson.

"people should just get on with accepting." But generally, people calmed down when they were reminded that the B.C. government had committed in its household mailer to everyone in B.C. in 1993 to pursuing this designation and they were informed that a Biosphere Reserve had to have complete community buy-in to go ahead. And when they found out that it could garner some financial help from senior governments, some people were even downright enthusiastic about the idea, and follow-up meetings were scheduled.

Ongoing logging continued, however, to fuel conflict in Clayoquot. On January 1, 1997 high windstorms blew down a substantial number of trees on Vancouver Island, including some of the experimental "variable retention" patches of trees left standing under the new Clayoquot Scientific Panel rules. It demonstrated that even under "enlightened" logging rules, the logging was not very eco-friendly.

Meanwhile MB's customers were under continual harassment and the company was becoming increasingly frustrated by the slowdown in logging under the Clayoquot Central Region Board's authority. Finally, MB decided enough was enough. In January of 1997 the company laid off its Kennedy Lake Division workers—80 men—and announced that it was not going to log in Clayoquot Sound for at least 18 months as it worked out how to reconfigure its logging operations. MB publicly stated that the move was motivated by economic considerations. It had lost $7 million on its logging activities in Clayoquot Sound in 1996 due to the high operating costs, not because of the environmentalists' campaigns that had targeted markets.

That same January, Adriane achieved a real coup. On dozens of trips to Ottawa she had made the rounds on Parliament Hill and had met with, among others, staff in the office of Sheila Copps, Federal Minister of Canadian Heritage. Copps understood the importance of offering hope of federal support to spur on the Clayoquot negotiations, despite the Prime Minister's Office's reluctance to do so. While visiting Victoria to make a Gulf Islands National Park announcement, Copps was asked by the media what her government's position was on the idea of a UN Biosphere Reserve in Clayoquot Sound. Adriane had lined up the media question. She'd also met with Copps to line up the answer. Copps answered publicly that the federal Liberal government would support a Biosphere Reserve if all the stakeholders wanted it. It made headline news.

Meanwhile, Linda Coady was working diligently within MB to "reconfigure" its Clayoquot operations. In March of 1997 the Nuu-chah-nulth Central Region Chiefs and MacMillan Bloedel announced a new *Joint Venture Corporation*, 51 percent owned by First Nations, that would take over all logging operations in MB's Estevan Division, the northern part of its TFL—all of MB's holdings in Clayoquot Sound. The Joint Venture, based on a maximum harvest of 40,000 m^3 per year, would not log for three years while it investigated value-added forest product opportunities. Whether or not it would log in Clayoquot's pristine areas remained undecided.

Chapter 55

Environmentalists labeled "Enemies of BC"
Violence erupts in Elaho

Premier Clark rashly calls environmentalists "Enemies of B.C."

In April of 1997 Greenpeace, confident that the Clayoquot campaign was moving slowly to some sort of resolution, ramped up its new campaign front in the central coast of B.C. On April 21, 1997 it released a damning report titled *Broken Promises: The Truth About What's Happening to B.C.'s Forests*. Then Greenpeace campaigners sailed their ship, MV Moby Dick, into the heart of the Great Bear Rainforest to begin a summer of blockades against logging and road building in the pristine watersheds along the mainland coast. This was too much for NDP Premier Glen Clark. He went on a media tirade, declaring Greenpeace and other environmental groups campaigning internationally to save B.C.'s ancient temperate rainforest "Enemies of B.C."

Clark claimed that the environmentalists were engaged in a "misinformation campaign." But Greenpeace's *Broken Promises* report presented the photographic evidence to prove that B.C. was not doing "world class forestry." It was the same old-fashioned eco-destructive clearcutting.

Only in retrospect can I see what a big impact Clark's inflammatory rhetoric had on subsequent events. As far as I knew at the time, not one MLA or MP spoke out against the Premier's defamatory statement. Not one of them called it for what it was: hate mongering. At first I took it lightly. We'd been called tree huggers, eco-lunatics, cappuccino-sucking urbanites and other names in the past. But the loggers took Clark's statement as a signal from the highest authority in the province that anything goes when dealing with environmentalists. After all, we were declared enemies of the state.

Joe Foy experienced the repercussions the next day on his way back to Vancouver from the Upper Elaho. He stopped in at a small grocery store just outside of Squamish to purchase a coke when a man beside him said, "You're not going to serve him are you?" in a threatening voice to the storekeeper. "He's Joe Foy of the Wilderness Committee." Joe didn't want to see the extreme hostility erupt and handed back the coke. The store owner gave him back his money. As he drove away, Joe turned on the radio news and heard for the first time Clark's declaration that environmentalists were now the enemies of B.C. People were taking the Premier's word literally.

Disparagingly called "tree huggers" by some, we printed up bumper stickers for those who were proud to be tree huggers that wanted wilderness preservation. Artwork: WCWC files.

Upper Elaho research tent cabin. Photo: WCWC files.

WCWC sets up a research camp with a research tent deep in the Elaho's old-growth forest

All winter we planned to launch our spring '97 field season in the Stoltmann Wilderness with fanfare. We had located a great spot for a new old-growth research camp in the Upper Elaho Valley. With the Squamish First Nation's permission, in May '97 we transported via helicopter (and especially with the help of many hardy volunteers) hundreds of kilos of building and camp supplies to the research site. Our plans were to build a six-metre by eight-metre internal framed research tent big enough to house a team of rainforest researchers. We located it over six kilometres in on our Elaho trail route, far enough away from the logging roads and cutblocks to ensure that the research would not be disturbed.

By June, when the snow in the alpine had melted down a bit, hikers began using WCWC's surveyed Stoltmann Wilderness trail. Some proclaimed it to be the Lower Mainland's best multi-day backpacking experience.

Aggressive road construction in the Upper Elaho in 1997. Photo: WCWC files.

Interfor loggers blockade road to the Elaho illegally keeping WCWC researchers out

On June 19, 1997, less than two months after Premier Clark publicly declared that environmentalists were the "Enemies of B.C.," a bunch of loggers launched one of most bizarre battles in B.C.'s "war in the woods." On that day a couple of WCWC volunteers returned to our office saying they couldn't get into the Upper Elaho because they had been turned back by Interfor loggers who had set up a blockade across the public road. It was right beside Interfor's maintenance yard at mile 23 on the Squamish Valley Road, the only road access to the Elaho Valley, Sims Valley and all other points north in the Upper Squamish Watershed.

The blockade soon earned the nickname "Interfort," aptly, because of the company's maintenance yard's tall metal fencing behind which the loggers congregated. The loggers put considerable effort into constructing their blockade. It stretched for about 20 metres and consisted mostly of curled up strands of heavy, old logging cables strung across the road from the edge of the Interfor compound's chain link fence to the very edge of the thickly aldered road allowance on the other side. It was like the barbed wire barriers used in real wars, but much heavier and without the barbs.

The only access to the Upper Elaho was to drive through Interfor's gated maintenance yard and out on the upper side of the road above the barrier. Interfor loggers, acting like border guards, manned the two-metre-high gate into the yard day and night. When a car approached, one of the loggers came out to ask the driver and passengers to sign a petition verifying that they fully supported Interfor's logging of the Stoltmann Wilderness. Only those who signed their petition were allowed to pass.

We had only experienced one other such blockade—the one denying us access to the Carmanah Valley in October 1990. It lasted only one weekend and had obviously been set up to shield a gang of vandals (who were never apprehended) as they smashed our boardwalk, burnt down our research cabin and destroyed our footbridge across Carmanah Creek.

There were no scientists yet stationed at our newly established camp in the Upper Elaho but we were anxious to get the summer's field season started. Because it was located so far from the closest logging road, I didn't really worry about it being destroyed as our Carmanah camp had been. I was confident that this blockade, being on a well-used public road, not a remote logging spur like the road to Carmanah, would be short lived. Surely the RCMP, charged with protecting public interest and safety, would quickly see to it that this illegal blockade was removed. How wrong I was.

We tried for several more days to access our Elaho hiking route and research camp using the public road. Every attempt failed. The RCMP refused to uphold the law and arrest the guys blocking the public road arguing that if we wanted the blockaders removed we

Selena Blais, WCWC store manager, adds the word "illegal" to logger's sign warning motorists of the loggers' blockade a short ways down the Squamish Valley Road. Photo: WCWC files.

should seek a court injunction just like a logging company would have to do to remove anti-logging protesters if they were blockading.

We rejected that idea. If someone stood on a major public road and blocked traffic, the police would immediately arrest him for breaking the law, not wait until an inconvenienced car driver went to court, obtained an injunction, had the police read the injunction to the traffic-blocker and then, finally, arrested him if he failed to obey the injunction. We wanted the RCMP to uphold the criminal law just as they would in the case of a traffic-blocker.

This became the essence of a legal challenge we pursued with the help of the Sierra Legal Defence Fund. Our SLDF lawyers (*Supreme Court of British Columbia in Vancouver Registry No. A971763 - Between Western Canada Wilderness Committee, Petitioner, and The Royal Canadian Mounted Police and the Attorney General for the Province of British Columbia, Respondents*) asked the court for several orders.

They all focused on getting the RCMP and the Attorney General to *fulfill their statutory and common law duty to ensure that members of the public have free use of, and passage through, the public highway known as the Squamish Valley Road without obstruction or hindrance by illegal blockades.*

Although it was a good case, we had to do more to put pressure on the illegal blockaders and keep the issue in the media than simply

At the logger's blockade site. "Interfort" on the right; WCWC's counter protest camp across the road on the left. Photo: WCWC files.

going to court. We came up with a plan. To counter the Interfor loggers' road blockade and their pro-logging petition, we established our own counter-protest camp across the road from them. We staffed an informational picket and approached all the travelers either before or after they talked with the loggers (and sometimes both) giving people the opportunity to learn about the issue and sign our petition to save the Stoltmann Wilderness. Kerry Dawson, our volunteer coordinator and many volunteers and other staff including Chris Player our mapping expert spent lots of time on the line there. The whole show was quite a zoo.

Here's how one outdoor recreation enthusiast, Robin Tivy, described his experience (excerpted from an article titled *Interfort And Beyond [Notes on Illegal Interfor-IWA Blockade]* posted on his internet site biouvac.com):

> We rolled to a stop (there wasn't much choice), and it was as if a director of a movie said 'action!' First out stepped a young man for the Wilderness Committee to explain that there was a blockade ahead, and asked if we wanted to sign their petition. It was all very relaxed, as if by script. Since the petition seemed reasonable to us, we signed it. About three car lengths ahead was the main ramparts of the fort. Behind the chain link fence were a half dozen or so official looking people in orange vests and several underlings. It looked a bit like a nuclear power plant or something.
>
> After signing the WCWC petition, it was obvious that the next part in the drama was that we were to roll forward to the main gate. We did so without comment. The main gate of the fort swung open a crack, and we were asked by the official in charge if we wanted to sign their petition. I asked to see the petition, and a herald called out: 'Bring forth...the PETITION!' The petition was evidently some sort of very important document, similar to the Magna Carta that King John was forced to sign way back when the trees were young.
>
> However, the script wasn't entirely clear to some of the actors. There were some murmurs of protest at this point from the rank and file behind the wall. 'But he signed THEIR Petition! Whose side is he on?' a dutiful underling asked the man in charge. Ah yes, I remembered, 'whose side are you on?' the classic union song. However, it was obvious that there was a distinct pecking order in the troops beyond and the management quickly put the dissenter in his place, doing a bit of thinking for him: 'That's his prerogative to sign both, but he must sign ours.' And with that, the petition was offered to me with a flourish, as expectant faces looked on. After all it couldn't be certain I was one of the enemies of B.C., with two small boys in the back of the old Toyota.
>
> Anyway, I read the petition. It had been greatly reduced in size from their earlier Manifesto. Since my role in the drama at this point was to test what happens to dissenters, I politely said I didn't agree with the second paragraph, but still wished to continue up the public road to go hiking with my kids. The paragraph stated something about 'all foreign nations involved in blockades should be deported immediately.'
>
> The petition was immediately whisked away, and the gate swung shut! 'Then you go nowhere!' was the cry and with that, the keeper of the gate turned his back on me and walked away. However a few others from behind the steel walls come forth and conducted a brief interview, which is reproduced below.
>
> 'Where do you work?' was the first question.
>
> 'BIOUVAC.COM', I said, pleased to get in a plug for the website.
>
> 'Is your office locked today?'
>
> 'Well, yes,' I said. (I was a little bit apprehensive at this question having read too many stories of my ancestors' printing presses being wrecked.)
>
> 'Well, so is ours,' he said triumphantly, and I guess that was supposed to be the end of that.
>
> 'But the road is not your office,' I said. Well, that seemed like a new idea to the original speaker, so he fell silent. However, he was then rescued by a second speaker.
>
> 'If we let you go up there, you will be up there Monday morning and preventing us from going to work.'
>
> 'No, I won't,' I said. 'I've obviously got two children in the back, and I'll be back tonight.'
>
> At that point the questioners completely ran out of steam, and once again resorted to the technique they had apparently been instructed to do; turn their backs and walk away. I had read about this strategy in various interviews with business leaders: avoid discussion, keep the burly loggers off the front lines, and hopefully 'position' the environmentalists as 'law breakers'.
>
> So it was time for phase two—ask some questions. Their tactic, of course, was to stay far back from the wall with their backs turned, as if I was some kind of a nut case. So I announced that I was from the press, specifically Bivouac Mountaineering Directory, and wanted to report their side of the case. And I pulled out my mountaineering camera, and took a snapshot of the main ramparts of the fort. Their case was fairly obvious from the signs. As nearly as I can understand, here's their argument: Cutting down trees was good for them and their families, and since families are good, anyone against them was anti-family, and anti-worker. I guess my family doesn't count.
>
> And they came forth again to the wall, and I was told to move my car or the police would take it away. I asked where the police were, looking forward to perhaps seeing the representatives of law and order. But none seemed to appear, and it seemed that the speaker came to his senses and realized this fact as well. The police were hiding.
>
> All that appeared were a couple of official looking people behind the wall who stepped forward with walkie talkies, and said officially: 'Control. Control. Looks like we've got a troublemaker down here. Over.' Then a woman came forth with a very official looking camera, and concentrated on taking some mug photographs of my face, while others wrote down the licence plate number of my car. I guess they were trying to simulate an arrest or something. I recognized the woman's face from the paper as being the mayor of Squamish, and so I asked her if she was, and she confirmed and turned away.
>
> I felt I was fast officially becoming a criminal of some sort to be pursued by the forces of law and order at some future date. After all, these are the people in charge of the system. The masterminds of the 'grassroots' blockade obviously had a good design: a chain link fence to keep their mob from doing anything really stupid, and at the same time insulate them from any form of dialogue.
>
> However, there seemed to be one big public relations weakness to their Fort: once repulsed at the main wall, the victim falls back into the welcoming camp of the Western Canada Wilderness Committee, who are only too willing to discuss the issues.
>
> So I retreated back to the Western Canada Wilderness Committee tents, and was urged to turn in a report to the RCMP. They had at least two video cameras filming the whole thing. One woman came forth and told me she had just lost her job in Squamish for participation [in WCWC's informational campaign] at the barricade. The organizers told me that the RCMP National Security Investigation Section had been up to the blockade and photographed all the WCWC members, but didn't question the lawbreakers at all. And the kids went into the tents and were given nice posters of the big trees they wanted to see. Towering giants that take a thousand years to grow and once covered the entire valley, but now existed only behind the barricade.

I was told that WCWC was going to have roast salmon at some point, and I was invited to stay. They had a variety of entertainment, including some native boys who had danced for them the previous night. Apparently they were interrupted the previous night by a song from the fort of 'one little, two little, three fu**ing Indians...' The WCWC people let me know they thought this song was disgusting. Well, I thought to myself, I bet that didn't cause any great joy to the highly paid spin doctors in charge of the fort either....

Coyote native youth dancers providing entertainment for WCWC campers at the blockade. Photo: WCWC files.

We save the Elaho Giant

After the blockade had been up for several weeks, obviously to counter the negative publicity, the Squamish District Forest Manager announced that they'd designated the Elaho Giant as a *"resource feature"* under the Forest Practices Code. This meant it "deserved" special protection.

The Forest Service created a small reserve that included not just the giant tree but also 50 hectares of old-growth forest around it to make sure that the Giant wouldn't become a windfall when Interfor clearcut the surrounding forest. Then they threw a bone to the loggers, saying that they would reevaluate and determine later if such a large buffer was really needed.

"We've saved a tiny tree museum!" I crowed. "All our effort to save the Stoltmann so far, way more than it took us to save the whole Boise Valley, has at least got us something," I kiddingly said to Joe. There's no doubt that the Elaho Giant was a special tree...but it wasn't a representative ecosystem like the whole Stoltmann Wilderness was. Over the next few years, as they pushed their roads and clearcuts past the Giant, Interfor did nothing to promote the area. Some tape flagging which marked where the road bisected our trail to this tree was all that heralded this site.

For me it was a benchmark. When we first started the Stoltmann campaign it was a beautiful 20-minute hike from the nearest clearcut through the valley's old growth to get to the Elaho Giant. Within three years the clearcuts had advanced beyond the Giant and Interfor had clearcut the biggest and best groves of trees around the one big tree and the 50-hectare clump of old-growth trees the B.C. Forest Service had spared.

IWA sets up another blockade in its escalating war on environmentalists

Empowered by Clark's hate rhetoric and their success so far in blockading us from accessing our Elaho trail and research camp, IWA workers set up another blockade around the Greenpeace icebreaker Arctic Sunrise which had just tied up at the Main Street dock in Vancouver's harbour. It started with IWA workers mounting an information picket to discourage Greenpeace supporters from attending "open houses" and tours of the boat.

The ship was actually playing a key role in Greenpeace's international boycott campaign against B.C. forest products. Within a couple of days (July 2, 1997) the IWA escalated their action, surrounding the Arctic Sunrise and a smaller Greenpeace vessel (the "Moby Dick") with boom sticks, prohibiting their departure from Vancouver Harbour. Pilots refused to cross the union picket line, which was also supported by members of the International Longshoremen's and Warehousemen's Union, the United Fisherman and Allied Workers Union and other affiliates of the B.C. Federation of Labour.

We urged our friends at Greenpeace to use the blockade as an opportunity to stay longer and gather more public support for all of our forest campaigns, telling them that we would help bring people out. Their blockaded ship would have made a great rallying place. It may even have given us the chance to talk with some IWA members and persuade them that it was the big corporations, not environmentalists, who should be blamed for their recent job losses and insecure future. But we never got the chance to have the big showdown. Greenpeace's ships could not stay. They had a tight schedule to meet that included an important confrontation with some Russian whalers. Two days later, assisted by Vancouver Ports Police, both Greenpeace ships slipped away in the middle of the night. The IWA made a big thing of Greenpeace turning tail and running, including the fact that Greenpeace broke the law because the Arctic

Cartoon. Artwork: Geoff Olson.

Marion Parker dating a tree. Photo: WCWC files.

Sunrise did not have a ship's pilot as legally required. I felt that the withdrawal emboldened the loggers bent on bullying.

An implied threat

A couple of days after the Greenpeace ships left I got a call from one of my friends on the Sunshine Coast. "You've got to see what's being sent out to all the forest workers. I'll fax you a copy." Within a couple of minutes I had in my hand an "urgent" broadcast fax sent out by the Sunshine Coast Forest Coalition. It urged people to join with other workers to attend a "family picnic" the following Sunday (July 13th) at the loggers' blockade on the Squamish road to show solidarity with their brothers and sisters. It noted that WCWC's court case against the blockade was scheduled to begin on July 16th and after that the loggers could be ordered to clear the road and if they didn't, face arrest. The fax went on to say, *One of the Executive Directors of WCWC, Paul George, lives on Gower Point Road* [not the correct location of where I live—but the road on which another prominent environmentalist lived]. *Let's send a message to him and his kind that we will not tolerate any further assault on our employment. It is long past due that we turn the tables on these anti-loggers. Now is the time to show your support and pride in our industry.... Let's show these people that we are capable of pulling the same stunts they are!*

I immediately forwarded the fax to all the media. It was also a good opportunity to get more publicity on our campaign to save the Stoltmann Wilderness and our court effort to get the RCMP to enforce the law and pull down the loggers' blockade.

CBC radio's afternoon show invited me to come on air with the woman who wrote the memo. She said she did not intend people to come to my house and do any harm. She denied it was some sort of veiled threat and apologized if it seemed like that. "We are just afraid for our jobs," she said. I accepted her apology. Nonetheless, it was a hate-mongering memo and I believe strongly that exposing such statements for what they are to the public reduces the threat of real violence.

Pushed out of the "Interfort"

Adriane and I believed in standing up to bullies and so we decided we'd go up to the loggers' blockade on the day of the big picnic. We especially wanted to support WCWC's volunteers and staff at our counter-information picket there. Although we had to attend another meeting in Clayoquot on the day before the picnic, we drove back early on Sunday and headed up to Squamish. We got there just a few minutes before IWA union boss Jack Munroe arrived on his Harley Davidson motorcycle. When he recognized me he stopped to exchange some pleasantries. The pickup trucks of picnickers streamed in past him, disappearing behind the gate of the chain link fence.

After he zoomed in too, I went over to the gate and got into a discussion with some of the loggers standing behind it. All of a sudden a voice boomed over a loudspeaker summoning everyone to hear Jack speak. The gate was cracked open and I just walked in and strode on towards the centre of the compound where the loggers were gathered. "Where do you think you're going?" asked one of the gatekeepers.

"Rattling their cage" after being pushed out and denied entry into Interfor's maintenance yard to hear Jack Munroe's speech to the blockading loggers. Photo: WCWC files.

"I'm going in to hear Jack's speech," I responded.

"You're going nowhere!" he shouted, giving me a big shove with both his hands, catching me off guard and almost pushing me over backwards. He kept shoving me right outside the gate and then slammed it shut.

I yelled at him as he was pushing me. "Get your damn hands off of me." Locked out and shaken, I grabbed hold of the chain link fence with both my hands and shook it hard; to sort of symbolically "rattle their cage" and vent my frustration. This incident was caught on video by BCTV. The short clip they edited afterwards (not showing the guy pushing me) made it look like I was the violent one. The station played this clip repeatedly, using it every time there was a newscast about the conflict in the Elaho. You'd think I'd have learned from the clip of me just prior to the "Riot at the Legislature" urging people to get more aggressive in the fight to save Clayoquot Sound spliced into footage of the riot. But I didn't. I played right into the hands of the media who wanted to portray environmentalists negatively.

I have the greatest respect and admiration for all the WCWC volunteers and staff who took turns staying at our information camp alongside the Squamish Valley Road for the eight weeks that the logger's blockade was up. It was a "hard gig." The loggers must have been coached in brainwashing techniques, especially on how to break down resistance using sleep deprivation. They set up bright lights to shine on our camp and speakers to blare country western music at an extremely high volume all night long to keep our counter-protest campers awake. There were also the occasional frightening "incidents." One night when Joe was visiting there with his family, someone inside the "Interfort" compound started pitching big boulders at our camp. Joey, Joe's two-year-old son was playing beside the road.

One of the rocks landed right beside him and, if it had hit him on the head, could have easily killed him. Joe, who never looses his cool, completely lost it that time. He ran over to the fence yelling obscenities at the "blankity, blankity idiot," who nearly killed his son. The rock throwing immediately ceased and never occurred again.

WCWC Elaho research tent stolen and dumped at the Squamish RCMP station

A day after the loggers' picnic, a pickup truck dumped off the research tent that we'd set up deep in the Elaho wilderness before the Squamish road blockade started. It was all neatly disassembled, bundled up and stacked in the Squamish RCMP station's parking lot. The RCMP officers on duty got neither the name of the person or persons depositing it, nor the licence plate number of the truck that had carried it.

It was obvious from the way our tent was bundled up that a number of people were involved in the crime and they'd taken down our tent and packaged it for a helicopter to airlift to the logging road. That made sense given that we'd set up our research tent over six kilometres from the nearest road and it would have taken a huge effort to hike it out piece by piece. After being dropped on the nearest logging road, the bundles must have been loaded on the pickup and then brought to the RCMP station. We suggested to the RCMP that they check the flight logs of helicopters that were in the vicinity during the weekend of the blockade picnic. I doubt that they ever did. In fact, I don't remember them investigating anything except documenting our tent bundles left in their lot.

In the media, the loggers claimed that since our structure was a "trespass cabin" on crown land, there was no crime involved in someone removing it. We figured it was criminal theft, but because we got our tent back, it was a hard case to pursue. No one was ever ar-

All the pieces of WCWC's Upper Elaho research station, stolen when our access to it was blocked by the loggers' blockade are bundled up and left in the Squamish RCMP parking lot in July 1997. We demanded that the thieves be caught and charged. In December 1997 Crown Council ordered the RCMP to abandon the case without laying charges. Photo: WCWC files.

rested or charged. WCWC was out a lot of time, energy and money in setting up our tent! Of course, just as we had when our Carmanah camp was destroyed, we vowed to put our Elaho research tent back up as soon as we succeeded in getting the loggers' road blockade pulled down. The loggers were definitely getting bolder in their opposition to our efforts to save the ancient rainforest in the Upper Elaho Valley.

Beaten by the blockaders' action that makes our court case moot

Our court date was coming up. Of all the court cases we had ever taken on, I was sure we were going to win this one. We were on track to strike down the B.C. Attorney General's and RCMP's policy of not applying the criminal code and relying instead on civil actions and contempt of court punishment to deal with blockades and similar types of protests.

B.C.'s current method of dealing with civil disobedience was different from other provinces. The policy came out of the Haida's Lyell Island blockades in the mid '80s. It entailed a lengthy process of removing protesters using an injunction issued by the court. The crime that the protestors were charged, convicted and sentenced for committing was, in fact, not their action to stop logging but their disobeying a judge's orders. In legal language it was called contempt of court, or criminal contempt of court if the justice felt it was a really serious matter. Contempt of court is a unique crime in that there is no set schedule of what fine or jail time is appropriate. It's completely up to the justice trying the case to decide. In theory, a justice could sentence a person to life in jail. Of course, if a justice imposed such a harsh sentence, a person would have the right to appeal it.

When Interfor loggers blockaded the Squamish Valley Road to keep us out, our Sierra Legal Defence Fund lawyers asked the RCMP to simply apply the existing criminal code and charge those involved with public mischief. The RCMP refused, saying they had a policy in place that said we had to proceed with getting an injunction against Interfor and the loggers blockading the public exactly the same way that Interfor had to proceed by injunction to remove people blockading their logging roads.

We refused to go this route. Instead we sought a court order forcing the RCMP to uphold the regular Criminal Code in this matter. If we won, the court ruling would set precedent and, we believed, would lead to a more just and swifter way of handling these kinds of disputes.

During the week leading up to our day in court, we got a lot of publicity about our case. Joe was passionately steamed up about the RCMP's hands-off policy that was allowing "goons" to stop law abiding citizens and law abiding groups like WCWC from hiking into the Stoltmann Wilderness. In an article published in the July 5, 1997, *Vancouver Sun* newspaper, Joe was quoted as saying:

What is particularly galling to me is Greenpeace and the Western Canada Wilderness Committee can afford the time and resources to go to court [and try and seek redress for the loggers' blockades] but if you are mom and dad and the kids coming to see the country and you're turned back by these goons and you ask for help from the men and women in [RCMP] uniform they just stand there silent."

However, we never did get our day in court to settle this matter. On the morning of July 16, 1997, just as we were walking into the courtroom, the lawyers for the Attorney General and the RCMP informed us that the loggers had pulled down their blockade in the middle of the night and everyone could again freely travel on the Squamish Valley public road. There was no blockade; hence no need for the justice to rule on this matter. It was now all "moot."

"But," said our lawyers to the justice hearing our case, "our client is worried about another loggers' blockade and the long time it would take us to get back into court again." (It took just over a month to get this court date.) The justice understood our concern and put in his court order that, with notice, we could be back in court in 24 hours if the loggers' blockade went up again. We were happy with that, but disappointed as heck that we didn't get to set a new legal precedent. We figured that Interfor didn't really want fairer laws for blockaders and made sure that their workers cleared the road.

WCWC goes on a positive offensive with a door-to-door canvass in Squamish

While I was briefing our door-to-door canvass crew about our campaigns right after the loggers' blockade came down, I jokingly suggested that they try canvassing in the town of Squamish to see how many people there supported our Stoltmann Wilderness park proposal. "It will undoubtedly be tough," I said, knowing that the Squamish Chamber of Commerce supported the recent loggers' blockade and a Squamish town councilor who refused to sign the loggers' petition received a death threat on the phone.

I knew that our canvassers wouldn't make any money going door-to-door in Squamish so I said that WCWC would pay them their average nightly income for just going there, handing out our latest Stoltmann campaign newspaper and asking people a polling question as to where they stood on our Stoltmann Wilderness proposal. To my great surprise more than half of our canvass crew said that they would do it. So, in the middle of August, a handful of brave souls drove to Squamish in our old canvass van and canvassed door-to-door for two nights.

The results were amazing. Very few people they met at the door were outright hostile. About half of the people who revealed where they stood supported us. A sizable number refused to answer, perhaps being afraid of being harassed if their neighbours found out. Our canvassers even managed to sign up a few new members and get some donations. We put out a press release about the positive results of this door-to-door canvass in "enemy" territory.

Above and Below: Repairing our aluminum ladder suspension bridge over Lava Creek on our Elaho to Meager Trail in the late fall 1997. Photos: WCWC files.

Taking care of business

At WCWC we always kept focused on increasing our membership. In the summer of 1997 we took a hard look at our membership renewal rate and decided to put a big push on increasing it. The theory is that it is easier to keep a member than to get a new one. Beginning in July '97 we selected a test group of approximately 1,600 ex-WCWC members to identify the reasons why they did not renew. We found out that invalid addresses were the single biggest problem. About a quarter of this test group's mail had been returned to us with a Canada Post stamp saying, "invalid address." We'd automatically been taking these people off our list.

We ended up deciding that wasn't too smart a move. Instead, we established a new procedure to make sure that all of our bulk mailings had "return requested" marked on the envelope so that we would find out right away when someone had moved. We also hired someone at an hourly contract rate to specifically focus on renewing members, tracking down the "invalid addresses" as quickly as possible and get people back on our rolls. This new effort began in May of '98.

Newspapers stacked up in our main office awaiting distribution. Mailouts and other modes of mass circulation of our Wilderness Committee's educational newspaper like in our "eco-boxes" in places where free publications are available help bring about a paradigm shift in public consciousness. Photo: WCWC files.

We kept good track of our new efforts and soon discovered something quite unexpected. A sizable number of people we called to get new addresses from told us that they hadn't moved–the address on the envelope returned to us was still valid! One person said, "It happens all the time. We live at the top of a long flight of stairs and I figure some posties are just too lazy to climb them!" We diligently kept checking for the next five months and found that 17 percent of the mail that was returned to us during that period was actually deliverable! We requested a meeting with Canada Post and got it in November. By this time we had a record of eight months and the 17 percent rate of valid "invalid" returns remained constant.

Canada Post reimbursed us for our direct costs so far in conducting our study (about $1,000). The representatives said that they would look into it at their end. It took a strongly worded letter to get a follow-up meeting with Canada Post. At that meeting we gave them 53 returned envelopes with addresses that were deliverable that we had received during the last three months. Canada Post checked them out and found out that, indeed, the problem was theirs. At our next meeting they admitted that this problem was "systemic throughout Canada Post and therefore quite difficult to fix" but they were working on it. They expressed doubts of ever being able to reduce it to two percent, the level we thought to be acceptable. This situation never did get fully resolved. Our only recourse was to become more diligent ourselves.

From then on we no longer automatically took people off our mailing list if a letter was returned to us marked "invalid address." We always checked up on them to make sure it really was "invalid." We also made sure to get new members' phone numbers (and later email addresses) when they joined. Keeping track of a large membership is time consuming and costly, but it is the only way to keep an organization healthy and independent.

WCWC "attacks" the "propaganda forest" and wins

On August 24, 1997 the GVRD held a big day of celebration to mark the tenth anniversary of its "Demonstration Forest" in the Seymour watershed. Although access into the Upper Seymour was restricted because it supplied drinking water for Lower Mainland residents, when the Seymour Dam was constructed and the water intake moved up the reservoir it created, the lower third of the watershed became "off-catchment." There was no plan to build another dam further down the Seymour for at least 50 years, so these lands were even considered "surplus" to the future needs of the system. The area–about 5,200 hectares between Lynn Headwaters Regional Park and Mt. Seymour Provincial Park–had been heavily logged in the past. But it still contained some groves of spectacular giant old-growth trees and some nice second growth forest. In 1987 the Greater Vancouver Water Board opened up this area to the public.

It wasn't just opened up for recreation, which would have made it a really useful public asset. Instead, without any real public involvement, the area was developed as the Seymour Demonstration Forest where the forest industry was to showcase its best logging practices.

Fortunately, not much logging had taken place in the ten years since the demo forest was established. The few times that plans surfaced to log the big ancient trees there, Ralf Kelman, a volunteer big tree researcher for WCWC, SPEC and other organizations made a 'big stink' and the plans were scrapped. Even without much logging the forest functioned as a fabulous P.R. showcase for the industry.

When we heard about the August 24th celebration, we decided that this was the perfect time and place to launch a campaign to wrest this propaganda forest from the forest industry and make the area a park.

We rushed into print 50,000 copies of a full-colour four-page tabloid-sized newspaper titled OUR CHOICE: *Seymour Demonstration Forest OR Seymour Ancient Groves Park* and handed them out to everyone attending the celebration. On Friday, two days before the big event, we put out a press release (with a copy of our hot-off-the-press newspaper) announcing the launch of our campaign. In it I said:

> It's really not in the public interest to log this forest with its record-sized trees. Some are amongst the largest trees in the world. It's even more incredible that these trees grow just a short half-hour bus ride and a day hike away from downtown Vancouver.... The current Demonstration Forest designation acts as a great public relations tool for the forest industry giants that are clearcutting our old-growth coastal rainforest. It masks the negative impacts that their aggressive logging activities are having on old-growth dependent species, wild salmon and natural biodiversity.

A couple of days after the celebration the main editorial titled *Demo Dilemma* in the *North Shore News*, the big daily paper serving North and West Vancouver, supported our park idea. It said in part:

> Much backslapping and good will typified the 10th anniversary of the Seymour Demonstration Forest on Sunday.
>
> But a menacing cloud hung over the proceedings like so much Lynn Valley mist.
>
> Despite the 'demonstration' prefix, little logging has occurred in the area lately.
>
> That's good.
>
> Better, however, is the Western Canada Wilderness Committee (WCWC) proposal to rename the area the Seymour Ancient Groves Region Park–and change the land use to protected regional park.
>
> Through extensive legwork performed by Ralf Kelman, who is part amateur naturalist and part trailblazer, we now know there's lots of big old trees left in the Demonstration forest.... With much of the North Shore's unprotected wilderness vanishing to housing and strip malls, the Seymour Demonstration Forest land should be set aside as protected park so that future generations can enjoy and revel in the magnificent forests found in their own backyard.

Needless to say, the forestry types who were sponsoring the anniversary event were unhappy with our publication and campaign. But our park idea was very popular with the public. We continuously built support by publishing and selling a map showing the trails in the Lower Seymour, including ones built by Ralf Kelman. The trails took people to a number of remarkably big, ancient trees that Kelman had discovered and given names to, like the "Temple of Time." Initially, the GVRD asked us not to distribute our map because the trails were not official. We ignored their request and they did not pursue the matter further.

What really pushed this campaign to a successful conclusion was Will Koop's *Seymourgate*, a 100-page research document that chronicled the undemocratic way that the Lower Seymour area had become the Seymour Demonstration Forest. He widely circulated this document including copies to all the Greater Vancouver Regional District (GVRD) members.

The campaign had some bumps along the way. In February '99, Johnny Carline, Chief Administrative Officer of the GVRD, forwarded a report to the district's Water and Park Committees recommending that no action be taken on changing the designation of the Lower Seymour Demonstration Forest to park status. He gave three main reasons. The first: it would raise the public's expectation that protection would be permanent, which it couldn't be because eventually it would have to become a reservoir. The second: the forest industry was supporting the area financially to the tune of about $400,000 per year and that amount could grow to a lot more. Park status would threaten that money. The third: under the current status it could operate like a theme park with all the activities normally associated with a park and continue to be enjoyed "in an integrated way" by the annual visitors that currently numbered 300,000.

I immediately fired off a hard-hitting letter to the GVRD Parks Committee. In it I stated,

> Carline's argument that the Seymour Demonstration forest already is a 'theme park' based on logging and needn't be made a real park for nature protection is goofy. Vast areas of B.C. have already been converted into such 'theme parks'...they're not the kind of parks people want. Carline's other argument that the Lower Seymour Valley will be needed in the future as a reservoir also doesn't hold water. There is ten times more water already impounded in the Coquitlam reservoir that in all other existing GVRD reservoirs combined, including the upper Seymour. It would be an insane management decision to spend hundreds of millions of dollars to create a new reservoir of far smaller capacity in the Lower Seymour than to open up a larger tap on the Coquitlam reservoir.... What is really at stake in the Lower Seymour is the forest industry's prize propaganda show. Industry is desperate to keep its sham 'Demonstration Forest' in the heart of Greater Vancouver so it can continue touring school children and bolster its push to log our drinking watersheds, despite evidence that logging compromises drinking water quality....

Eventually, reason prevailed and in the spring of '99, the GVRD changed the name of the area to the *Seymour Conservation Reserve* and quietly dropped logging as one of the uses of the area. It was not quite what we wanted–a regional park–but it was the next best thing. Together with the name change came much increased public involvement in planning and developing a new management plan with the mandate to conserve and expand its recreational potential rather than log the area.

Elaho to Meager trail becomes legal

In mid September 1997, the Squamish Forest District unexpectedly approved our application to build a trail from the Upper Elaho

Chris Player, WCWC mapper, points out the Lower Seymour trail guide map for sale in our Gastown store. Selena Blais, Store Manager stands beside him.
Photo: WCWC files.

Volunteers lug in the plywood and all the other materials piece-by-piece back into the Upper Elaho and reassemble the Research Station by July 1997. Photo: WCWC files.

to Meager Creek. We had already surveyed the route and people were already using it, so I guess they thought they had better make it official. Upgrading the status of our hiking route to an officially recognized hiking trail by the B.C. Forest Service didn't mean very much as far as affording protection. It certainly didn't save any of the forest around it. In our application we had checked the box that said that we accepted the fact that logging rights came before any rights this trail had. If we had checked the other box, which said the trail must be sacrosanct and not logged over, we never would have obtained approval. There were some provisos that the Forest Service made in their letter of approval. We had to upgrade our route to B.C. Forest Service trail standards. We had to re-route some parts of the trail closer to the Elaho to be within a theoretical riparian protective zone. We also had to upgrade a couple of our river crossings from Burma bridges constructed out of aluminum ladders and ropes that we were currently using to more substantial cable car crossings. These new facilities had to be engineer-approved and were costly.

The Forest Service also stipulated that the project had to be done by the end of 1998. We had our work cut out for us when the snow melted in spring.

In all the media reports about our trail approval, the District Forest Manager repeated the fact that the area would continue to be managed for *"integrated resource use"* and *"forest harvesting activities."* He said, *"This is not a protected area and it's not going to be. It is a mixed use of the forest."* In spite of the fact that having this status did not mean we had saved any part of the Stoltmann Wilderness, we still viewed the trail approval as a tiny step towards eventually getting protection for the area.

Massive volunteer effort re-establishes our Elaho Research Station

Determined to rebuild our Upper Elaho research facility but unable to afford another helicopter to help, Joe and Kerry Dawson set up one of WCWC's largest ever volunteer work parties. The group had to pack in our research tent-cabin six kilometres piece-by-piece and then set it up again. By mid September they'd hauled all the pieces back in, including the full-sized sheets of plywood for the floor, all the framing two-by-fours, the large vinyl outside shell and the airtight stove. They worked in coordinated teams, passing the awkward pieces, especially the plywood sheets up the steepest sections. By the end of the month they'd re-assembled the tent and facility and put the word out that research could begin.

No sooner was their work done than we got an official notice from the B.C. Forest Service ordering us to remove our camp. It had ruled that our research tent cabin was a "permanent structure" not a temporary one, despite the fact that it had already been set-up, taken down, removed, brought back and set up again this year! The letter suggested that we do our research in the nearby, newly created Clendenning Park. We appealed this arbitrary decision, noting that our tent cabin was similar to those used by forestry workers in tree planting camps, which did not need permits and were not considered to be permanent.

In the appeal hearing we found out the Forest Service's rationale: our tent had a wooden frame and plywood sheets laid down for the floor, therefore it was different from the tree planter's tent cabins that had aluminum poles and dirt floors. The conclusion: all we had to do was to remove the wooden frame and replace it with aluminum poles and get rid of the plywood to make it legal. The only reason we had used a wooden frame was to save money. We'd put in the plywood floor to protect the forest floor.

We immediately ordered the aluminum poles. In June '98 our volunteers re-fitted our research tent-cabin with these poles and removed the plywood floor, thus making it a "temporary structure" in the eyes of the Forest Service. Some suggested that we burn the surplus pieces, but we felt that would disturb this wilderness site. Again our volunteers rose to the call, this time packing out all the wooden pieces some of them had worked so hard to hike in less than a year before.

I think the fear that the facility might not last was a key reason why, unlike our Carmanah Research Station, we were never able to attract any Ph.D. students to conduct long-term research using this station as a base camp.

Interfor logs two key sites "Magic Grove" and "Grizzly Grove" in the Stoltmann Wilderness

It's easy to understand why a few of WCWC's most passionate Stoltmann Wilderness supporters, especially those who volunteered on our trails and putting in our research camp, got discouraged. In the spring of '97 a few of them, along with some more radical environmentalists who felt that public education campaigns took too long in the fight to save wilderness, decided to form a new group called PATH (People's Action for Threatened Habitat). Barney Kern, who had led our Bear Initiative caravan to Prince George, was one of the main spokespersons of this group. They vowed to use non-violent direct action to save the Stoltmann. Their first acts of civil disobedience probably had more to do with instigating the loggers' blockade on the Squamish Valley road than did WCWC's trailbuilding and research activities. Because some of the PATH people were our ex-volunteers, Interfor loggers contended that WCWC started PATH and

Above: Before—In September 1997 volunteer Tony Hilton examines a giant red cedar in Grizzly Grove that is marked and about to be logged. **Right:** Joe Foy posing in 1997 by giant three sisters cedars in Grizzly Grove which is soon to be cut down. **Below: After—**Stump of the same tree shown above after it was cut down in November 1997. Photos: WCWC files.

that we were running this new group, which we didn't and weren't.

Truthfully, neither PATH's nor WCWC's efforts slowed down Interfor's logging of the Stoltmann Wilderness. This company was a lot meaner and more confrontational than MB had been in Carmanah and Clayoquot.

The new Forest Practices Code aided Interfor's aggressive logging regime by dictating that cutblocks had to be small. Interfor consequently was able to target the best groves of trees much faster than before the Code. They were also able to implement a deceptive practice called "grade setting" where they cheated the government out of fair stumpage payments by getting the valuable wood for junk wood prices. They would include an area with poor quality trees in each of their cutblocks and log them first, claiming that the original estimated "scale" (valuing for stumpage purposes) by the Forest Service for the whole block was too high. Then the Forest Service, without doing any checking up, would give them all the wood in the entire cutblock for a low stumpage rate assuming that there was a mistake in the original timber cruise that set the stumpage price for the timber in the block in the first place.

If there was ever any question about the B.C. Forest Service doing the bidding of the big forest companies, its turning a blind eye to the "grade setting" scam proved that this was truly beyond any shadow of a doubt. This corporate cheating put hundreds of millions of dollars that should have gone into the public purse into company pockets instead. Interfor was not the only company participating in this game of fraud. It became general practice by nearly all of the major companies during that era.

Along with the extra profits made from raw log exports (which the NDP government had also allowed to increase, much to the detriment of B.C. workers), "grade setting" took Interfor from a troubled financial position into a healthy one in a few short years. We actually didn't learn about the whole sordid story of "grade setting" until it was exposed by the Sierra Legal Defence Fund in a brilliant report in 2001. Meanwhile, all that we knew was that our campaign wasn't hurting Interfor in the pocketbook for they reported increasing profits during this time.

While we were concentrating mainly on the Elaho Valley area of the Stoltmann Wilderness, another group called the *Witness Project* was focusing on nearby Sims Creek, another remarkable wilderness valley within the proposed Stoltmann area. In 1995, when Interfor started clearcutting in the Sims, legendary B.C. mountaineer John Clarke and a young artist-photographer named Nancy Bleck who had helped us in our Carmanah campaign started car-pooling groups of people from downtown Vancouver to the Sims for a weekend of camping, exploration of the ancient rainforest and learning about the Squamish First Nation's culture. People witnessed both the magnificent old-growth forest as well as Interfor's destructive clearcutting.

Interfor was targeting the Sims' big cedars, all hundreds of years old and some more than a thousand, "mining" them out at a rate far faster than they could grow back. Interfor made a great deal of money logging the old-growth cedar trees and that's why it targeted this species when laying out its cutblocks. Far more importantly, cedars were intrinsic to the culture of the coastal First Nations, traditionally providing everything from houses and boxes to canoes and totem poles.

Witness Project participants pose for a portrait on the stump of a big red cedar recently cut down in "Magic Grove" in the Sims Valley. Photo: WCWC files.

Cedar bark was woven into blankets, clothing (including raingear), baskets and rope.

Big cedars were once plentiful on B.C.'s coast. But after over a century of logging, and especially with the more recent highgrading of them by companies like Interfor, they were becoming rare.

In 1996 Ta-lall-semkin, siem, Hereditary Chief Bill Williams of the Squamish Nation, co-founder of the *Witness Project*, made it truly a cross-cultural collaboration that increased peoples' respect for nature and enhanced dialogue between native and non-native communities. He introduced participants to the "Witness Ceremony" practiced by his culture continuously for over 8,000 years. In the following ten years the *Witness Project* raised the consciousness of hundreds and hundreds of citizens about what was happening to the wild forests of the Upper Squamish watershed of which the Sims and Elaho were a part.

One of peoples' favourite witnessing places in the Sims Valley was a spectacular stand of 800-year-old ancient red cedars people called "Magic Grove." Within Magic Grove was a massive red cedar with a small opening at its base leading into a large hollow centre. The hollow was large enough to hold 10 people. It had bear hair inside indicating that it was an active denning tree.

While the *Witness Project* was underway, Interfor applied for and received a cutting permit to clearcut Magic Grove. Seeing the cutblock ribbons, a concerned hiker told a B.C. Forest Service official in Squamish about the grove's bear-denning tree. The official then phoned Interfor, informing the company that the Forest Service would be sending out a biologist to verify the finding. If they verified that hibernating bears used the tree, it would have to be spared.

The very next day, to eliminate any possibility of the tree and perhaps some of the forest around it being protected, Interfor sent a faller in to find it and cut it down (which he did) before the government biologist had a chance to check it out. What Interfor did was entirely legal because it held a valid cutting permit, which gave them the right to cut the trees. But it revealed the company's greed and disregard for nature. Soon thereafter, Interfor clearcut the entire cutblock and Magic Grove was no more.

That fall Interfor also cut a road through another spectacular stand of cedars, this time in the Upper Elaho, that we had named "Grizzly Grove." It, too, was an important bear denning site. In our campaign to save the Stoltmann Wilderness we had made this area famous. People cried when they discovered it had been logged. It was hard on all of us at WCWC to lose, grove-by-grove, the ancient forest in the Stoltmann Wilderness as Interfor relentlessly cut its way up the valley.

The extremist of slideshows

On November 25, 1997, Joe Foy put on a Stoltmann Wilderness slideshow in the Howe Sound Secondary School in Squamish. I wasn't there, but from all accounts, it was one of the most intense, emotionally-charged events we'd ever held.

Viewpoints clashed, verbal insults flew and pamphlets opposing clearcut logging were ripped to pieces during a recent information meeting hosted by the Western Canada Wilderness Committee (WCWC) in Squamish. That's how the story began in *The Chief*, Squamish's weekly newspaper, titled *Boisterous loggers disrupt WCWC slideshow*.

According to Joe, about a hundred people attended. Most of them were local forest workers. They tried hard to shout down Joe, but Joe persisted, never getting angry. Finally, most of them left and the people who remained carried on a reasonable dialogue. The boorish behaviour of some of the loggers actually helped our cause more than harmed it. They ended by spurring a backlash that included an editorial in the local paper urging decent people to stand up and tell the hotheads to cool it.

All winter long Joe and his Stoltmann team put on *Save the Stoltmann Wilderness* slideshows in Lower Mainland communities to educate the public and get more people on board. We got warm welcomes at all the other events including the one in Whistler held just a couple of weeks after the unruly scene in Squamish.

Slabby — The Stoltmann Sequel to Stumpy

Behind the July '97 loggers' blockade on the Squamish Valley road, Interfor was busy logging. Despicably, during that time the company cut down one of the biggest Douglas firs we had found in the Upper Elaho Valley. That fall we posed many people on the huge stump. It was a photo that brilliantly portrayed how big the tree was. We used it over and over to fuel the outrage that such magnificent trees—the last of these marvelous giants—were still being logged.

Ken Wu standing on the stump of a giant Douglas fir determined by ring count to be 1,158 years old when it was cut by Interfor in 1997 in the Upper Elaho. With permission WCWC cut a round and eventually made two "Slabby" displays. Photo: WCWC files.

We asked one of our diligent volunteers to take up a high-powered magnifying glass and carefully count the stump's rings. To our amazement she counted over a thousand. We then applied to the B.C. Forest Service District Office in Squamish for permission to take a slab off this stump in order to get it officially counted by a UBC dendrochronologist and have it studied further. We couldn't lose. If the Forest Service refused our reasonable request we'd make a big stink and get a lot of publicity.

The Forest Service must have figured that out, too, and finally in late November gave us permission to slice off a slab of the stump. We didn't have much time. We had to get in, cut the slab and bring it

Teri Dawe chain sawing a round from the stump of a ancient Douglas fir recently cut in the Upper Elaho Valley, December 1997. This round was eventually turned into "Slabby"—a portable display with historical dates assigned to the tree's corresponding growth rings. Photo: WCWC files.

out before the snows closed off access. Otherwise we'd have to wait until the late spring.

We contacted Teri Dawe, the logger who had given our Stumpy from Clayoquot its "root canal," to lighten it up for its road tour. He agreed to meet us in the Elaho and saw off the slab right away. He was one of the few sympathetic loggers we knew. He was the only one we knew who had a chainsaw bar big enough to do the job. Best of all, he would do it for cost if we supplied the volunteers to help pack it out.

It took Teri a full day to do the job. The volunteers carefully numbered each piece as the ten-centimetre-thick slab naturally fragmented into chunks. The numbering was key; otherwise we wouldn't be able to reassemble it back together at our office in Vancouver. The volunteers also helped hike the labeled pieces out to the waiting pick-up truck. The following day it snowed hard. We got our slab out just in time.

The top floor at 55 Water (two floors above where our canvass and volunteers worked) was un-renovated at the time and vacant; that is if you discount the pigeons that lived there. Bird droppings were everywhere. We got permission from our landlord to store the slab pieces there so they could dry slowly in the unheated room. Joe reassembled the pieces on top of some pallets so that air could circulate underneath them. Every few months or so we would go up and check up on "Slabby" and dream about what it would look like as a finished display.

Joe Foy counts approximately 1,060 annual growth rings on the round cut from a Douglas fir stump in the Upper Elaho Valley of the Stoltmann Wilderness drying out in the vacant top floor at 55 Water Street. He dreams about touring this artifact of a once spectacular tree through North America to build support for saving the rest of the Stoltmann Wilderness from the chainsaw. (A dendrochronologist who examined the round after it was sanded smooth counted 1,158 rings.) Photo: WCWC files.

Slabby sat there drying until September 1998. But even after it had dried, it was still much too heavy for any kind of movable display. One of WCWC's longtime volunteers, Tony Briggs, came to the rescue. He had the carpentry skills and the vision of how to make a display from the pieces. He disassembled the round and transported all the pieces to a shop where he had each one carefully sliced in half with a large band saw and correctly numbered. He then brought them back and reassembled them. Now we had two future display slabs, each one light enough to transport around and easily display.

Each round was too big for one sheet of plywood so Tony glued the pieces onto two side-by-side sheets that were bolted together and could be easily taken apart. This made it possible for two people to transport the display via elevators and vans and to set it up and take it down. With a rented floor sander Tony spent a full weekend going over the glued down slabs time and time again with ever-finer-grit sand paper until the uneven surfaces became ultra-smooth and the growth rings highly visible.

After sanding them to a mirror finish, there was the varnishing. But before that, there was one more thing that Tony, the perfection-

Volunteer Tony Briggs sands the two Slabbies smooth working all day using a rented floor sander. Photo: WCWC files.

ist, felt he had to do. During the transporting and processing much of the natural bark had fallen off. One of the signature features of Douglas fir is its thick fire-resistant bark. It greatly contributes to this species' longevity, given that it grows in areas that are dry in the summer and prone to lightning-caused fires. The bark on a big old-growth Douglas fir, like one our slab came from, can be up to 30-centimetres thick and you can put a blowtorch to it and it will only smolder.

So Tony drove all by himself back up to the stump in the Upper Elaho to chip off some bark to glue onto spots where it was missing on our slabs. He got to the Elaho quite late in the day and discovered that, while he had remembered to bring a wedge, he had forgotten his sledge. All he had with him was a single-headed axe. He told me that he was surprised at how tightly the Douglas fir bark adhered to the tree's stump and how hard it was to chip off.

He thought that he had split enough off and transported it out to his truck. But when he had finished, just to make sure, he decided to slice off one last large piece. By now it was dusk. Holding the wedge with one hand, he absentmindedly picked up the axe with the other and swung it down with all his might. He had the wrong end pointing down. The razor sharp blade hit the plastic wedge right in the centre and sliced it neatly in two, just grazing Tony's fingertips in the process. It was a very near tragedy. Another one of our lucky stars was shining down on Tony and WCWC that evening.

With the bark glued on where needed, Slabby was finished, except for its dates and interpretative information. We hired a UBC Forestry School graduate student who was specializing in dendrochronology to date the slab. She counted 1,158 growth rings. I worked up a series of historical events going back to 840 A.D. It was easy finding interesting historical benchmarks for the outer 20 centimetres (the last 250 years). The earlier dates were sparser, vividly illustrating the accelerating pace of human-wrought change on our tiny planet Earth. Besides pasting onto the slab beside the corresponding tree ring clear plastic labels with dated historic information, we also pasted some information about our Stoltmann campaign on the plywood display board next to the slab.

Slabby on display outside building where Interfor is holding its annual shareholder's meeting in Vancouver. Photo: WCWC files.

The display was a great hit everywhere we took it. Slabby appeared at many demonstrations and as the feature attraction at WCWC booths. It went to Ottawa and stood outside the House of Commons and to Victoria several times to be displayed in front of the Legislature buildings during rallies. It almost got defiled by an angry logger at a rally in support of Greenpeace at the New Westminster Quay where a few years later the IWA had put a boom chain around a Greenpeace boat in protest of its support for protecting the Great Bear Rainforest.

Greenpeace offered to buy Slabby's identical twin to take to Europe to strengthen its call for a boycott of old-growth wood products from Canada's threatened ancient temperate rainforests. We couldn't do that because of the terms of our permit from the B.C. Forest Service. But we could, and did, lease Slabby to them for, amongst other things, study by scientists there. People in Europe loved Slabby. Many had never imagined a tree to be so big and old. Nor could they imagine a government allowing such trees to be cut down.

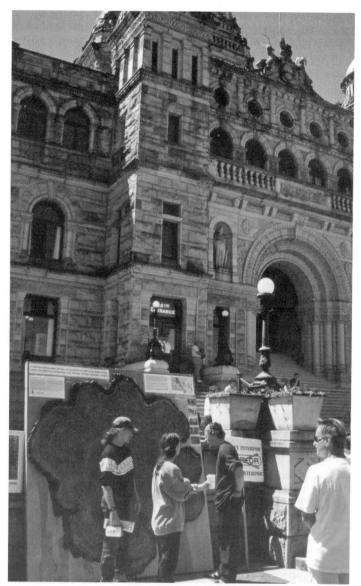

Above: Slabby on display outside the Legislature building at a WCWC rally. Photo: WCWC files.
Below: Slabby II at Greenpeace's Market Initiative launch at Mountain Equipment Co-op in Toronto. Slabby II was leased to Greenpeace for a European tour to gain support for saving B.C.'s ancient temperate rainforest. Photo: Greenpeace Canada.

Regaining the offensive in our Stoltmann campaign; going for National Park status

It was obvious that we were still getting nowhere with the B.C. government regarding more park protection in the Lower Mainland region. The NDP refused to budge even a millimetre from its October 1996 decision to only save 20 percent of the Stoltmann Wilderness. So we decided to repackage our Stoltmann campaign and pursue National Park Reserve status.

Ric Careless, head of BC Spaces for Nature, presents me with the first Wild Earth Award (which has been presented annually ever since to other environmentalists) at a BC Environmental Network (BCEN) annual meeting held on the Sunshine Coast in October 1997. Photo: WCWC files.

In the fall of '98, Joe redrew the boundary of our proposed Stoltmann Park, expanding its size from 260,000 hectares to 500,000 hectares. In the north our proposed park now bordered on Ts'yl os Provincial Park and took in the Upper Bridge River on the Chilcotin Plateau. Our expanded eastern boundary took in the Pemberton Ice Cap and the Upper Soo Creek Valley and made the proposed park much closer to the world-famous resort town of Whistler. We called on the federal government to designate the area a national park—the first in B.C.'s Coast Mountains. We reasoned that it would be good from both an ecological and economic point of view to have at least one big national park in the Coast Range to augment the great national parks in the Rocky Mountains.

In the winter of 1998 we published 130,000 copies of a WC^2 newspaper titled *Big Dreams Can Come True – Stoltmann National Park* (Educational Report Vol. 18 No. 6. Print run 130,000 copies.) The paper explained the change of direction in WCWC's Stoltmann campaign and called for a national park reserve (subject to co-management agreements and a treaty settlement with the Squamish First Nation). The paper's cover featured a spectacular 3-D relief map of our proposed park by Z-Point Graphics of Whistler, owned by Eckhart Zeigler who a few years later was elected to the resort village of Whistler's town council.

We got no support for our idea from the local MLA or from John Reynolds, the federal MP who represented the area. Nor did we get support from the surrounding municipalities nor from the First Nations who had aboriginal title there. Several First Nations representatives, besides giving the usual reason for non-support including that parks excluded native use, also asked, "Why would we support a park named after a white guy? What's so special about him?" There was some support, however, from a few councilors in Whistler and support from the Squamish Nation to protect key areas in their traditional lands.

As far as the federal parks planners were concerned, this region of Canada was already way over- represented having two parks, Pacific Rim National Park Reserve and Gwaii Haanas (South Moresby) National Park Reserve and Haida Heritage Site. No amount of reasoning with these federal bureaucrats could convince them that their huge region was ill-defined and the outer coast of Vancouver Island and Haida Gwaii are radically different from each other and both radically different from the Coast Mountains area encompassed by the Stoltmann Wilderness. As one Parks Canada employee told me, "We have regions in Canada with no park representation and we must get parks in place there before even considering an area like the Stoltmann." It was the same argument foisted on us 15 years earlier regarding South Moresby.

All of this did not deter us. A park like the Stoltmann comes from the people not from the bureaucrats. It was up to us to promote it. In the summer of '99 we published 80,000 poll cards, one going to Prime Minister Chrétien and the other going to Premier Clark. We asked people to "vote" either for or against the proposed Stoltmann National Park by filling out the cards and sending them in. We mailed our poll cards to every household and community bordering the proposed park including Pemberton, Mt. Currie and Whistler. We household mailed them to all of Squamish, too, knowing the town was the epicenter of opposition to the park. We never found out exactly how many cards were mailed to the Premier and Prime Minister, but were told that the pro-park cards greatly outnumbered the anti-park cards.

Chapter 56

A new head office and store
The new fisheries and pro-Great Bear Rainforest campaigns

Looking for a new home office

In the spring of '96, the store two doors west of our store in Gastown burned down. The store between us suffered serious damage but we were lucky. We briefly lost our electrical power and, with that, some data on our computer which we were able to recover. The damage to WCWC, however, had just begun. Because the burnt out store was a heritage building, city by-laws required that its front façade had to be saved. Workmen installed an iron reinforcing structure on the sidewalk to hold up the brick shell. Then they built a totally user-unfriendly walkway underneath the big iron bars. This ugly structure was a huge deterrent to tourists. Our store revenues dropped almost immediately by 30 percent and kept on dropping.

The tourists not only faced this burnt-out monstrosity but also more and more panhandlers who liked the "no-man's zone" created by the empty building. The tourists hated being accosted. Most of them turned around and headed back towards the Steam Clock and "safer" territory. In the summer of '98, as the degenerate drug scene in the lower part of Gastown got worse, we seriously started to think about moving.

After being at 20 Water for over eight years, it was a hard decision to make. It was made easier by the fact that our lease was up for renewal and the new property manager started negotiations at $6,000 a month—nearly double what we'd been paying.

Everyone at WCWC wanted to stay in the same general area of town. We'd just hired a new comptroller, Matt Jong, to replace the affable and competent Brian Conners who left us for a higher paying job. Matt had retired young and generously decided to "give back" to society by working full-time for us at an incredibly low wage. Luckily for us, Matt was not only an expert in accounting but also in real estate. He started diligently checking out every storefront on the market. After four months of searching he found nothing that would do. Just when we were about to despair, he happened upon the top floor of 341 Water, two blocks west of 20 Water Street and only a block from the waterfront terminus of the Sky Train.

It was a one-hundred-year-old building with an antique freight elevator with weird rolling doors that worked almost all of the time. Because it had once been a light manufacturing warehouse, the building's roof was covered in skylights, giving the top floor a great airy feeling. The floors were genuine hardwood. The space was larger than both our current premises—20 and 55 Water—combined and, the monthly rent was considerably less than both of them together. Everyone on staff could see the place's potential and loved its ambience. We leased it right away.

It took many volunteers working with a skilled carpenter and electrician, both of whom were sympathetic to our cause, several months to fix up our new place. Then came the big job of moving everything. We had accumulated hundreds of files, an inventory of posters and books, trailbuilding equipment, and junk over the years.

In our attempt to bring up the first big load of educational newspapers we learned that our antique freight elevator was overload intolerant. After fully loading it, a volunteer pushed the up button.

Joe Foy at work in his new space in our open office at 341 Water Street. His desk is just as cluttered as it was in the old office. Photo: WCWC files.

The elevator slowly sank and came to rest at the bottom of the shaft about a metre below the street-level loading dock. It took a day to fix. From then on we babied that elevator. No one wanted to walk bundles of papers up three long flights of stairs.

Now all we had to do was find another home for our store. About a block away from our old store at 20 Water was an old mission building (appropriately located on Abbott Street) that had recently been upgraded and converted into condominiums. Several ground-level strata-title stores were for sale. They had been on the market for over a year and had come down in price. Although they didn't meet the crucial "location, location, location" test so essential for retail success, we were convinced that they soon would. At that time, Vancouver's big new convention centre was proposed to be built on the waterfront next to Crab Park. As part of this redevelopment, Abbott Street was supposed to become the convention centre's gateway to Gastown and Chinatown. We figured our second source of customers would

Tony Briggs, volunteer carpenter, and our architect, Marianne Enhorning, clown for photo while working on our new store in September 1998. Photo: Adriane Carr.

367

come from the derelict heritage landmark Woodward's building, located half a block away, when it was renovated which was scheduled to begin soon.

In August of '98, with the help of our long-term supporters who collectively donated over $80,000 to a capital store fund, we purchased the smallest strata-title store located at 227 Abbott. It was a bare-bones cement-floored unit that needed a lot of creative vision and hard work to become the beautiful Wilderness Committee Store and Outreach Centre that it is today. We hired architect Marianne Enhorning who came up with a magnificent but expensive-to-build set of drawings that included a river of small rocks flowing in the centre of a slate floor from the back of the store down two stepped levels to the front entrance. We put out the call to our members asking them to send in their favourite naturally polished rocks from wild rivers and beaches around B.C. for the store's "stream."

Beautiful rocks came in by the package loads. Alice Eaton, our longtime full time volunteer coordinated a big work party to place the rocks artistically in the floor. We found a talented old-school

Director and longtime volunteer Alice Eaton, foreground, and architect Marianne Enhorning assemble the river of rocks on the new store's floor. Photo: WCWC files.

cabinetmaker to construct custom shelving and a special storage unit with dozens of sliding shelves for all our posters so everything would seamlessly fit in. On the main wall we prominently featured the best display from our old store—a round slice from the stump of Canada's oldest known tree, a 1,835-year-old yellow cedar. The tree had lived in the Caren Range on the Sunshine Coast before loggers cut it down in the early '90s. Ken Lay had originally prepared the slice and Marion Parker, our dendrochronologist, counting the rings. We'd pasted the dates of key historic events—like 1,000 AD when Leif Erickson "discovered" America—beside the appropriate growth ring to help people comprehend how ancient many of the trees growing in B.C.'s coastal temperate rainforests are.

The very last task of our move was to take down our beautiful carved and painted wooden Wilderness Committee Store sign from outside 20 Water Street and hang it up outside our new store at 227 Abbott Street. Just as we were about to finish this last job, the "sign police" told us we had to stop. Our sign was "non-conforming." According to the sign bylaws for this heritage area, our sign was too big relative to the square footage of our new store. I hit the roof. Our sign was not huge. Besides, it was part of Gastown's heritage, having hung there for nearly a decade. Officials at City Hall told us that the

A workman welds up our sign over the new store at 227 Abbott Street after getting it approved by the city's variance committee. Photo: WCWC files.

only way we could use it was to successfully plead our case at a meeting of the Board of Variance. Several weeks later, as we left for our hearing in City Hall, I picked up our sign and said, "I'm going to take it along so the decision makers can see how small and nice it is." Our sign sold itself. We got the variance.

We held our new store's grand opening in the late fall of '98. Several hundred members and reporters came to the all-afternoon and evening affair. Our store was beautiful and our hopes were high that it would bring in big revenues. But to date our store has yet

Store interior. Photo: WCWC files.

We publish our first newspaper explaining why open ocean net cage fish farms are bad

We were always a strong advocate for the protection of B.C.'s wild fisheries and critical of the damage done to salmon spawning habitat by clearcut logging on B.C.'s steep-sloped coast. We had collected many photos of logging-caused damage: mudslides flowing from logging roads and clearcuts into streams and creeks; pools filled in with gravel and silt. It appalled me that both federal and provincial fisheries ministries allowed the clearcutting and road-building on steep slopes above fish streams and their tributaries.

Another insidious problem for B.C.'s wild fisheries was the ever increasing concentration of fishing licences into fewer and larger boats. Interception fisheries using huge purse seines could and did gobble up whole salmon runs to smaller streams, causing those runs to go extinct.

It was hard to campaign for better fish management against our federal Department of Fisheries and Oceans (DFO), which was using its huge budget and vast resources to defend the increasingly corporate fishing industry. I found it particularly aggravating to see the growth in DFO of armchair theorists using computer modeling to rationalize and determine which areas should be open for fishing, the timing of openings, what type of fishing gear should be allowed, and how many fish should be caught.

No amount of computer modeling can replace the field monitoring of marine and freshwater habitats and the collection of escapement data stream by stream every fall to accurately determine what is happening with the runs. Yet in 1992 virtually all the stream fisheries officers on the Pacific coast were fired or phased out with the exception of a few charter boat people who were contracted to provide data. This lack of field monitoring gave the corporate lobbyists, armed with their computer modeled statistics and persuasive arguments, the advantage. There was insufficient ground-truthed data to counter industry's computer-based projections of stocks.

The result was industry's success in pressuring the DFO to allow more short-term over fishing on the Pacific coast, just as they had done on the east coast. The switch to computer modeling management also opened the way for the development of fish farming, as the effects on the wild stocks were not being directly monitored and reported annually.

Luckily, I was informed about all of this by an old friend of mine from the Queen Charlotte Islands, David Ellis. Ellis was a commercial fisherman for 12 years and a consultant for 12 years afterwards to commercial and sports fishermen including the Strait of Georgia Salish First Nations. He had a Masters of Science honours degree in fisheries management from UBC. While working for various Coast Salish First Nations, he saw and documented the decline of their food fisheries for herring and herring eggs and co-authored a scientific paper on it. Alarmed by this inside knowledge of the decline overall of the west coast fisheries, he started his own non-profit organization, Fish for Life, to do something about it. Every season Ellis came by our new office at 321 Water Street to brief me on the state of B.C.'s fisheries. **When you are a campaigner, you rely on people like Ellis who are experts in one field or another to supply you with facts, figures, and insights into what is happening and ideas about how to change things for the better.** Ellis was the first person to inform me that the NDP government, while putting a moratorium on the creation of new salmon fish farms, at the same time allowed existing ones to expand as fast as the market could handle more farmed fish. During the years of the NDP's so-called moratorium, the industry's production of B.C. farmed fish nearly doubled.

From left to right, Adriane, Joey, Joe and Christina, Joe's wife, pose beside our new store on Abbott Street. Photo: WCWC files.

to achieve the success we envisioned. The problem is beyond us. Gastown has been plagued with panhandlers and its renewal has proceeded at a snail's pace. It has taken eight years for the Woodward's building redevelopment project to get going. Vancouver's new convention centre is now under construction. However, it is located six blocks away on the non-Gastown side of Canada Place. It's hard to make money when you only get a few customers a day during the long winter rainy season.

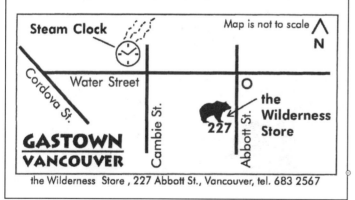

Our first flyer showing the location of our new store. Map: WCWC files.

Several years ago we prudently decided to keep our store open only during the high-traffic summer tourist season and Christmas time. The rest of the time we have a poster in the window instructing people to come to our nearby office. Gradually, our website store has gained popularity and all our posters, books, cards and other products are available to online shoppers. When the area around our Abbott Street store finally does redevelop, business is sure to thrive like it did in the early '90s at 20 Water Street. Ever the optimist, I still have hope that WCWC's income from sales, which has declined to less than half of what it was in the early '90s, will rebound.

I promised Ellis that our Wilderness Committee would help popularize the available information about the environmental problems associated with salmon farms. Pen-raising this carnivorous fish pollutes the local marine environment and transfers diseases to the wild salmon that swim by. We needed to build an overwhelming opposition to the proliferation of fish farms on the B.C. coast in order to save our wild salmon stocks.

We hired retired outdoors columnist Tony Eberts to write the text for a four-page, two-colour, tabloid-sized newspaper titled DON'T LET FISH FARMS DESTROY BRITISH COLUMBIA'S WILD SALMON MIRACLE (Educational Report Vol. 17 No. 2) which we published in the spring of 1998. The paper was primarily based on information presented in Net Loss: The Salmon Netcage Industry in British Columbia, a report by David W. Ellis for the David Suzuki Foundation, and an article in The Georgia Straight by Alexandra Morton titled Fish Farming Prompts Despair Over Fate of Salmon.

Calling on Fisheries Minister David Anderson to halt a fishery opening to save coho runs near extinction

While we were distributing the 50,000 copies of our *Fish Farms* paper, Dave Ellis came to me with an immediate problem. He had heard from inside contacts that the DFO was about hold a commercial sockeye salmon fishing opening in Juan de Fuca Strait that would likely drive some threatened coho salmon runs into extinction. It was a desperate situation due not only to past overfishing and competitive international "fish wars," but also to a recent downturn in marine survival of the stocks. No one really knew why, but fewer coho were surviving in the wild. The commercial and sports fisheries that threatened the coho had to be stopped.

Ellis knew all this because he had been involved in fisheries al-

Sea lice on a juvenile wild pink salmon in the Broughton Archipelago. Fish farms are suspected as the source. Photo: WCWC files.

location disputes between troll, seine and gillnet fleets, and because he had often attended the Canada/U.S. salmon treaty talks as an observer. He appealed to me to have WCWC put out a joint press release with his Fish for Life organization to try to stop what was obviously going to be a massively destructive fishery.

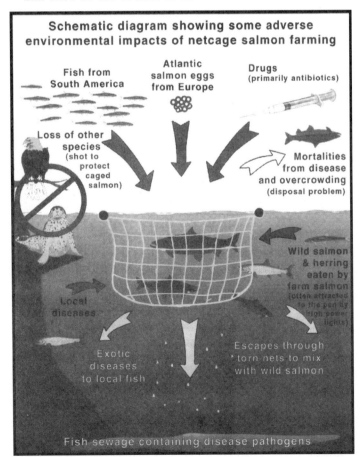

Diagram showing adverse affects of netcage salmon farming used in our *Educational Report* Vol. 17 No. 2, 1998. Artwork: WCWC files.

WCWC-Victoria carried on an active, successful campaign to get restaurants to serve only wild salmon on their menus and advertise to that effect. Photo: WCWC-Victoria files.

He pointed out how important the mouth of Juan de Fuca Strait was as nursery ground for chinook salmon and a major interception area for coho runs spawning both in the Fraser River and in Washington and Oregon States' rivers. It had long been a very "dirty" fishery with a high bycatch of immature chinook and coho. It had been allowed by DFO partly because a large percentage of the coho and chinook were returning not to Canada, but to the U.S., and therefore not managed by DFO for conservation. In fact, they were "managed" for the very opposite of conservation. They were being exploited as a crude 'fish war' lever in U.S.-Canada negotiations over the valuable Fraser River sockeye allocations under the international salmon treaty.

Ellis explained to me how interception fisheries in Canadian waters off Juan de Fuca Strait were opened in retaliation to U.S. having openings in Alaskan waters that intercepted fish bound for the Skeena and other B.C. rivers' spawning grounds. Another component of this "fish war" was DFO's setting of the coho and chinook quotas for the west coast Vancouver Island troll fisheries too high in a deliberate move to drive the U.S. coho stocks (a significant part of that catch as coded wire tags always showed) to extinction.

The DFO reasoned that "this is the only thing the Americans understand" in the negotiations for a bigger Canadian share of the Fraser sockeye catch between Canada and the U.S.—a big prize for the powerful corporations that dominated the Canadian commercial fishing industry. He also explained to me how wasteful and ecologically wrong the Juan de Fuca openings were, because so many juvenile salmon were by-caught, killed and wasted in the process. He told me he had once participated as a fisherman in a "fish war" opening called at Swiftsure Bank for trollers.

Here, for every adult salmon taken, five to 10 immature fish were caught and then shaken off the lines and dropped back bleeding into these waters which teemed with massive schools of dogfish sharks targeting the wounded salmon.

To stop the pending Juan de Fuca opening, we put out a press release stating that our organizations were:

...calling on the Canadian and U.S. governments to close all 'mixed stock' or 'test' coho salmon fisheries as well as completely closing all roe herring fisheries for 1999. The Western Canada Wilderness Committee and the newly created Fish for Life Foundation say the coho is on the verge of extinction and the roe herring fisheries take food away from salmon. 'The coho have been mismanaged and over-harvested for years and now El Nino is also reducing ocean survival,' said WCWC official Paul George. 'The situation is so critical for many runs that there is no room for any human kill of coho this year.'

We presented the biological information in our media release so that everyone reading it, including people unfamiliar with fisheries jargon, could understand the situation. We faxed it out to a long list of contacts including all the regular media as well as other environmental groups, native organizations and individual MPs from B.C.

When I personally asked DFO officials in follow-up on this release to explain exactly how they determined when and where to have commercial fisheries openings, I was told that it was "an art and a science." In a subsequent press release I called it "Voodoo science," saying that our government fisheries scientists and bureaucrat decision-makers were using the same kind of "science" that led to the total collapse of the east coast cod fishery. I've always found it effective to put out a series of hard hitting press releases as new information is revealed and the issue develops.

Sometime during this same time period, a retired U.S. fisheries scientist who had spent his whole working lifetime trying to bring back decimated salmon runs on the west coast of Washington and Oregon States called our office. I dropped everything I was doing to take advantage of this opportunity to candidly tap into someone knowledgeable and experienced in the field. I talked with him for nearly two hours over the phone asking him every conceivable question I could think of. I had to find out what had happened to their wild salmon, what they had done to try to recover them and what things he thought they should have done but didn't.

His story of dammed rivers, completely clearcut watersheds and overfishing were familiar. But what was shockingly new to me was

Bumper sticker. Artwork: WCWC files.

his realization after a lifetime of work in the field of salmon enhancement that once the genetic diversity of a particular salmon run is gone there is no way science or applied technology can get that run to recover. He explained it this way:

We've put millions and millions into trying to bring back the wild salmon with little positive results. It's not for lack of effort or that we should have put more money in or funded different projects that would have worked. What we've found out is that it's simply impossible to bring a run back once it's been reduced to only a few fish.

He went on to explain that once the specialized, diverse genetic information that a stock of salmon has developed to survive various local stream conditions is gone, the only way it can be recreated again is through hundreds of generations of natural selection. Having specialized in genetics in graduate school, this made sense to me. It made me realize that the salmon situation in B.C. was desperate.

B.C.'s coho crisis eventually attracted the involvement of most of the other environmental groups in B.C. In addition, the Fraser River First Nations, particularly the Shuswap, exerted their legal force to protect the last of the Thompson River coho, which were so low in numbers that the Shuswap First Nations had to forego community harvesting. The sportfishing community, led by Craig Orr and the Steelhead Society, demanded a coho moratorium too.

As a result of the pressure from all these players, the press picked up the coho story and began to follow it closely. For the first time, it was politically bad, not good, for the Minister of Fisheries to carry on a 'fish war' and continue to allow uncontrolled sportfishing. Responding to public concern, on May 21, 1998, Federal Fisheries Minister David Anderson made shockingly good policy by announcing that he would put conservation of B.C.'s endangered coho salmon ahead of commerce. The 1998 fisheries openings would be based on *Zero fishing mortality for upper Skeena and Thompson* [a Fraser River tributary] *stocks* [where the coho were most endangered] *and fishing elsewhere only where coho bycatch mortality will be minimal.*

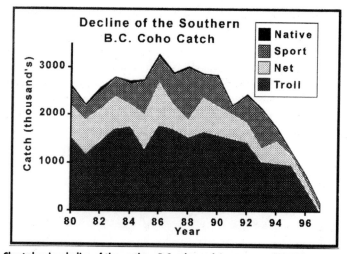

Chart showing decline of the southern B.C. coho catch in our paper BRITISH COLUMBIA'S WILD SALMON MIRACLE (Educational Report Vol. 17 No. 4). Note the Native fisheries' tiny proportion of the catch appears as a thin black line at the top of the graph. Graphical artwork: WCWC files.

We celebrated and reinforced this through the publication that spring of a four-page, two-colour, tabloid-sized newspaper titled *EVERYONE'S HELP NEEDED TO SAVE WILD COHO AND PRESERVE BRITISH COLUMBIA'S WILD SALMON MIRACLE* (Education Report Vol. 17 No. 4 with a press run of 50,000). We outlined a bold plan to save B.C.'s declining coho salmon stocks which included the elimination of non-selective interception fisheries. This would stop the by-catch of endangered stocks by both sports and commercial fishers. We also called for Canada to be a *conscientious objector* in the salmon war with the U.S. noting that each country catching the other's endangered salmon as a negotiating pressure tactic would wipe out wild stocks. Feeling the pressure, Anderson was eventually able to successfully re-negotiate a new Canada/U.S. salmon treaty that included a clause that protects coho. It was a proud achievement.

David Ellis went on to be appointed to sit on the Committee on the Status of Endangered Wildlife in Canada (COSEWIC) to represent the marine component of Canada's Pacific coast. He worked long and hard to get the very best experts to prepare accurate status reports on the endangered salmon stocks. In the face of blisteringly strong opposition by the Ottawa division of the DFO, which was under strong pressure from industry to oppose all fish and marine mammal listings, after two years he resigned from COSEWIC.

Before he left he appointed to the fisheries COSEWIC subcommittee Ken Wilson, a former DFO biologist who was well connected within the DFO, had all the data and had integrity. Wilson worked within COSEWIC for another very productive four years during which time COSEWIC reports were generated using DFO's own data to show the endangered nature of the coho as well many sockeye salmon stocks. But to no avail. The DFO, which is supposed to protect the salmon fisheries, was heavily lobbied by the corporate commercial fishing industry and also by some powerful elements of the sportfishing charter industry to block all attempts to get salmon runs designated as endangered.

Species a risk listing of near-extinct salmon stocks would go a long way towards halting, in particular, the decline of the Fraser River's salmon biodiversity. Such listing would effectively end fishing of co-migrating stocks. It would cost the fishing corporations lost revenues in the short term. But in the long term, commercial fishing would prosper due to abundantly rebuilt stocks, as was largely proven to be the case in Alaska.

Campaign to stop herring fisheries to allow the "biomass backbone" of our Pacific Ocean ecosystem to rebuild

We continued to collaborate with Dave Ellis and the Fish for Life Foundation, producing one more paper and engaging in our first ocean-going protest. Ellis taught us that to make the whole marine ecosystem healthier we had to allow more herring to survive and spawn. Old time fishermen and First Nations' elders remembered the days when herring was plentiful. Today there is about 10 percent of the original annual herring biomass left. Past decades of 'reduction' fisheries (catching herring to be rendered down into fish meal and oil) destroyed most of the herring biomass. A four-year near total closure in the late '60s and early '70s resulted in only limited rebuilding of herring stocks. Then the DFO, eager to accommodate the new, highly lucrative roe-herring corporate fishery, ignored the high abundance levels of the pre-reduction-fishery days. Instead they were seduced by the new, corporate-biased fisheries modelers who had the gall to say that herring were now at "historic highs."

There was also huge pressure to allow roe-herring fishing before B.C.'s stocks had fully rebuilt because the complete collapse of this fishery in Japanese waters had left demand and prices sky-high.

As plankton feeders, herring can rebuild their numbers in less time than needed by larger predatory fish. Dave Ellis was confident that, if there were a total ban on herring fishing for a few years, the stocks would rebound to historic levels of abundance. That's what happened following herring closures in Norway and on the North Sea in England. However the DFO, influenced by corporate-biased science, contended that fishing had little to do with the herring's current lack of abundance. Without any scientific proof, the DFO bureaucrats argued that the decline in herring was due to environmental factors.

Flush with our success in getting Fisheries Minister David Anderson to put in place strict measures to protect coho, Ellis pushed for an emergency meeting with the minister. In November 1998, Anderson met with WCWC and representatives of Islands Trust, other conservation groups and several Gulf Island communities where herring fisheries had taken place that year. At that meeting DFO's "industrially-biased" biologists tried to play down the herring crisis asserting that the herring fisheries were taking only 20 percent of the available biomass, leaving plenty of fish for spawning. We argued that this was not only taking from an already heavily depleted stock, but that the fisheries ended up targeting and destroying some stocks that, unlike the more common migratory stocks, were 'resident' and a key ecological component of the Georgia Strait all year round. The resident herring needed all of its remaining biomass to spawn and bloom again into the massive schools that had historically provided abundant food for coho and chinook salmon, Pacific halibut and cod, humpback whales, harlequin ducks and other seabirds to thrive.

It was difficult to keep Minister Anderson on topic. He completely dominated the conversation, telling us how difficult it was for him to close fisheries down to protect the coho. We kept constantly bringing the conversation back to herring and how it was just as important to give this species a chance to fully recover. Being a sports fisher himself, he understood our arguments. But he told us it was all he could do to take the one courageous stance he did to stop the coho bycatch fisheries. He said it was politically impossible for him to take on any more than one such battle at a time. We would need a much louder hue and cry from the public calling for such extreme measures to get

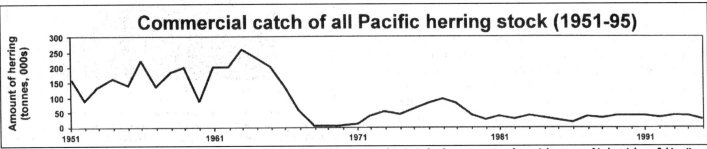

Chart showing decline in Pacific herring stocks (1951-95) used in our paper "B.C.'s Herring must be given the chance to recover from eighty years of industrial overfishing." Graphical artwork: WCWC files.

Stop the roe herring fisheries protest vessel leaving the Granville Island dock in the spring of 1999 for Hornby Island where the fishery was about to begin. From left to right unknown, Dave Ellis, Craig Delahunt and me. Photo: WCWC files.

him to stop the herring fishery. I had difficulty sympathizing with his position. How could he simply rubber stamp the plans of DFO bureaucrats, who were staunch advocates of the current corporate fisheries, to have a roe herring fishery on the depleted stocks when he himself knew that it was fundamentally wrong?

To try to build the loud public outcry that Anderson said was needed, we produced and published the third paper in our fisheries series in late February 1999: a four-page, two-colour, tabloid-sized newspaper titled *B.C.'s Herring must be given the chance to recover from fifty years of industrial overfishing* (Educational Report Vol. 18 No. 2 with a print run 30,000 copies). It called on the Canadian government to establish a four-year coast-wide moratorium on commercial roe and bait herring fisheries to give the depleted herring stocks a chance to fully recover. Our paper was full of alarming facts. It included a graph of the commercial catch of all Pacific herring stock from 1951-95 showing that the stock had never even come close to recovering from the near total collapse of the herring due to decades of over fishing that peaked in the early 1960s.

To support our fight to get a moratorium on the roe herring fishery, we launched our first and only sea-going protest. Carrying out the campaign on a shoestring budget of under $5,000, we chartered a boat bought by Craig Delahunt just for this purpose, although he used it afterwards as one of the first charter boats working in the mid coast's Great Bear Rainforest. Our expedition took us to Hornby Island in Georgia Strait to observe and report on the herring roe "Gold Rush." It took a great deal of courage to go among these very aggressive fishermen aboard a small protest boat.

Hornby was the last place in the Strait where herring still spawned in great numbers. From all reports about the fishery held there the previous year, it was an ugly circus. Seiners first scooped up much of the biomass including many immature fish. Gill-netters who, because there were not enough mature fish, then made set after set, straining through tiny juvenile herring in a desperate effort to get their quota. Even many old-timer fishermen privately admitted that this fishery had gone too far.

That spring of '98, our WCWC crew of volunteers on the protest boat documented a very sad spectacle of greed and unsustainable fishing. The media did not cover our voyage to Hornby like they cover Greenpeace and Sea Shepherd's sea-going campaigns to save whales. We felt pretty puny up against the hugely profitable roe herring industry and corporate-influenced DFO with its 'spin-doctor' bureaucrats and huge federal budget.

Rose Bay underway to protest the 1999 roe herring fishery in the Straits of Georgia. Photo: WCWC files.

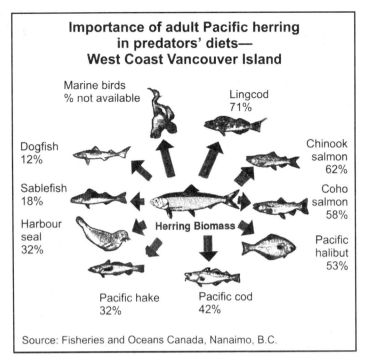

Diagram showing the percentage of herring in the diets of predators on the west coast of Vancouver Island. Artwork: WCWC files.

The depletion of herring and other small 'forage fish' is an ongoing tragic worldwide phenomena, much of it done to provide feed for the salmon farming industry. In fact, heavy fishing for the forage fishes leads to whole ecosystem collapse, forcing birds, mammals and other fishes to go into steep decline. Documenting this decline, and stopping these incredibly destructive fisheries worldwide, will be one of the most important environmental campaigns of the next few decades. Huge financial resources and very tough campaigners are needed for this cause. Worldwide, almost all fisheries have been maxed out. Catches are well beyond sustainable levels and they're crashing. What is needed in B.C. are much better funded eco-campaigns, including at WCWC, to fight for ocean wilderness protected areas and biodiversity-based management of all marine ecosystems.

A protected Great Bear Rainforest: not a clearcut mid coast timber supply area

In 1990, when all the focus was on Carmanah Valley, Walbran Valley and Clayoquot Sound on Vancouver Island, Peter McAllister, an outspoken advocate for wilderness preservation, founded the Raincoast Conservation Society and set sail on the first of many expeditions exploring, inventorying and documenting B.C.'s 20,000 kilometre-long mainland coast. Here, in the magnificent fiords, rainforests and mountains, he found Canada's final bastion of intact and threatened coastal wilderness. What he discovered was about 100 still intact valleys, many of them over 5,000 hectares in size. He reported that ...*each one is a jewel, a globally scarce and priceless masterpiece of nature*. In the estuaries and back along the rivers he discovered the largest concentration of grizzly bears in Canada. His son, Ian, a talented wildlife photographer, and Ian's wife Karen, a trained wildlife biologist, went along on many of these expeditions. Later the two of them got their own boat and continued to explore and photograph, gathering materials and keeping journals that were eventually used to produce a coffee table book that would help inspire public interest in this fabulous area.

Knowing that WCWC had produced successful coffee table books on Carmanah and Clayoquot, Howard White, owner of Harbour Publishing, invited me to his office on the Sunshine Coast to talk about the McAllisters' book. The title of the book was still not settled. He was not sure that a blatantly conservation-oriented book would appeal to a large enough audience. I assured him that it would. In fact, I said it would have a greater appeal than a book that was more of a travelogue. In the strongest way possible I urged him to keep it a conservation book as the McAllisters had intended and to use the catchy title Great Bear Rainforest with a photo of a grizzly bear on the cover. Then I put WCWC's money where my mouth was. I told Howard White that we would buy a thousand copies of the book and help distribute it.

Staying in regular contact with Karen and Ian, I decided to up the public profile of the area by co-producing a newspaper with the Raincoast Conservation Society. We titled it LET'S PROTECT CANADA'S GREAT BEAR RAINFOREST with the subtitle: *We have a choice – Clearcuts, tree farms and big stumps - Or grizzlies, wild salmon and big trees* (Educational Report Vol. 16 No. 4. Press run 120,000 copies). This paper's success was due to the fabulous images and the information provided by Ian and Karen and the detailed centrespread map produced by Baden Cross with the help of our cartographer Chris Player. The map revealed four large complexes of proposed protected areas: the Greater Ecstall Region (Northern Extension of the Great Bear Rainforest), the Central Great Bear Rainforest including the Spirit Bear area around Princess Royal Island that our good friends Colleen and Wayne McCrory of the Valhalla Wilderness Society were campaigning to protect, the Knight Inlet Region (Southern Extension of the Great Bear Rainforest) and the Stoltmann Wilderness which was our major campaign.

We stuck to our tried-and-true formula of having 50 percent of the paper covered in photos and illustrations and lots of big catchy headlines. We devoted a whole page to photos illustrating recent devastating logging under the triple banner headlines: *Government's claim that B.C. logging is now 'world class' is a whopping big lie! The truth: it's some of the worst in the world...rapacious clearcutting that's destroying ancient rainforests, wild salmon rivers and grizzly bears – Government and Industry spend millions telling you everything's OK in the woods BUT the Land SPEAKS for ITSELF.*

On the back page of this eight-page, full-colour, full-sized newspaper we presented an offer for the beautiful new book *Great Bear Rainforest: Canada's forgotten Coast* by Ian and Karen McAllister that people couldn't refuse. For only $35 (covering all taxes and shipping) we would send them a copy of this spectacular book that regularly retailed for $39.95. We also marketed Raincoast Conservation Society's powerful award-winning 10-minute video *Legacy* for only $15. This shocking video contained aerial footage showing the extent of the clearcut destruction on B.C.'s mainland coast and did a

Experiencing a Great Bear Rainforest wetland first hand. Photo: WCWC files.

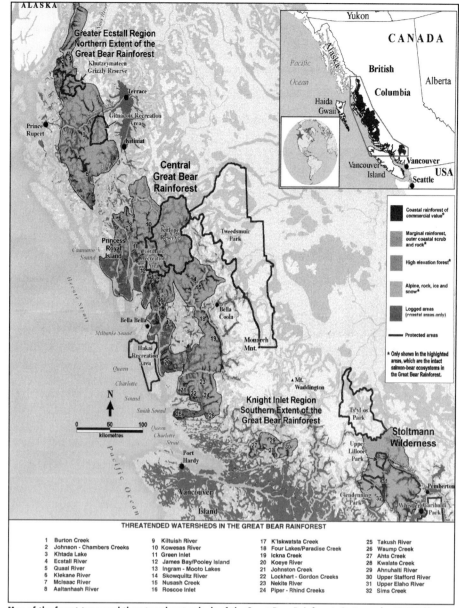

Map of the forest types and threatened watersheds of the Great Bear Rainforest. Cartography: WCWC files.

lot to get government and industry serious about negotiating with conservationists for more protection.

We householder mailed a copy of this newspaper to everyone with a postal address living along B.C.'s mid-coast—the area that nearly everyone was now calling the "Great Bear Rainforest." There are fewer than 5,000 households in this region (including Prince Rupert), so this mailout cost us less than $600 in total.

Raincoast and WCWC investigating new logging road being punched into the wild Johnston River, largest coho producer in Rivers Inlet 1998. Photo: Ian McAllister.

Local education, especially completely saturating every affected community with information, is vital to the success of a wilderness preservation campaign. It dispels myths and puts all the cards on the table for both friends and foes to see.

Ian and Karen kept pressuring us to come up and see some of the threatened areas we were campaigning to protect. We finally did. Adriane, Joe and I went on one of the trips, taking along a videographer, Daniel Gautreau, to record what we saw. Both visits were brief and the area was so magnificent and so vast that we only got a tiny glimpse of what was really at stake. I particularly remember the Johnston River, one of the best salmon rivers left that was about to be logged and wading through "the swamp" to get to it. Our expedition did not succeed in saving this watershed.

When Greenpeace, the Sierra Club and a new group, "ForestEthics," formed the Rainforest Solutions project and started negotiating with the B.C. government, logging companies and native and non-native communities to work out a Great Bear deal, WCWC's role in the campaign became minor. Raincoast Conservation Society and the Valhalla Wilderness Society stayed active as the two conservation groups on the ground in the Great Bear but in many ways they were marginalized by the big groups that took over.

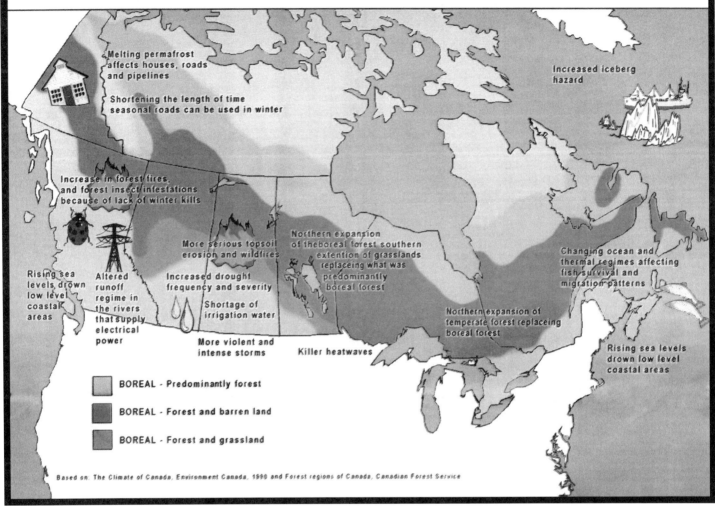

Above: Centerfold illustration showing the effects of global warming on Canada's boreal forest in WCWC's *Who Will Take Global Warming Seriously?* newspaper. Graphical artwork: WCWC files.

Below: Since we first put these two measured variables on the same graph and presented them in our newspaper *Who Will Take Global Warming Seriously?* we have come to realize that the content of ^{18}O (the stable oxygen isotope that makes the water molecule heavier and less volatile) in glacial ice is not simply dependent on the water temperature of the oceans. It is also dependent on the concentration of ^{18}O in the oceans which increases as ocean levels drop and more water is tied up in great ice sheets on the continents. Graphical artwork: WCWC files.

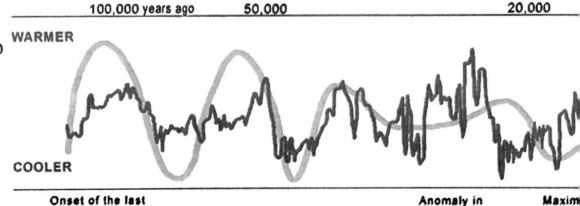

Chapter 57

Global warming looms
Lillooet LRMP table flounders

A visit to Gray Jones in Edmonton results in a WCWC newspaper on global warming

In the summer of 1999, on a visit to Edmonton to see how Gray Jones and our WCWC-Alberta Branch were doing, I found Gray working on a newspaper about global warming. He had just visited Fort McMurray with a team of Greenpeacers who were on an anti-global warming campaign. They focused on Alberta's Athabasca tar sands industrial development because it is one of the highest single-site CO_2 pollution sources in the world. Its CO_2 emissions were rising rapidly given that production of oil from the tar sands was increasing and industry had to "burn" on average the equivalent of more than a third of a barrel of oil for the energy needed to extract one barrel of oil from the vast but thin bitumen layer that is covered by deep overburden, capped by a thin veneer of muskeg soil and boreal forest.

The tale Gray told about what was happening around Fort McMurray shocked me. Few Canadians knew anything about the extent of environmental damage already done there, let alone the projected impact of the massive development plans on the books.

Greenpeace had focused its campaign on Suncor Energy Inc., the biggest tar sands oil producer at the time. According to Gray (he had a tendency to exaggerate) this effort by Greenpeace resulted in negotiations with Suncor that resulted in Suncor making a commitment to invest more than a $100 million into renewable energy developments world-wide. As a quid pro quo, Greenpeace backed down on its Suncor campaign and did not finish the video it was producing about the tar sands.

Gray wanted WCWC to take on the campaign where Greenpeace left off and produce the video to blast the Athabasca tar sands into public consciousness. Gray went on and on about how climate change was happening faster than people realized and that people just didn't get the seriousness of the situation. He asked me if I would work on his global warming newspaper. I said sure. Back in Vancouver I began immediately. It gave me a chance to study the science and produce a popular paper that would convince skeptics and motivate believers into doing something about global warming, not be overwhelmed into apathy by the enormity of the problem.

Most experts agreed then—and more do today—that the world's boreal forests will be one of the global ecosystems most affected by climate change. Through raging wildfires and insect outbreaks, scientists predict that anywhere from 50 to 75 percent of the boreal forests will become grasslands during the next fifty years. Already there were some pretty colossal events—1.7 million hectares of boreal forest burned during a two-week period in the summer of 1995, massive floods in the spring of 1997 in Manitoba and a giant ice storm in Quebec and Ontario in January of 1998. The mountain pine beetle epidemic was just beginning to take off.

Boreal forests represent 82 percent of Canada's forested land. As an ecosystem it is perhaps the most important in the world in terms of climate stabilization. Yet it was, and still is today, the least researched and least understood ecosystem in Canada.

As soon as I returned from Edmonton I read all the recent popular scientific magazines to get the latest publicly available information about climate change. It took me over a month of full-time effort to complete our eight-page full-colour global warming paper. We titled it, *Who Will Take Global Warming Seriously...and help protect the Earth's vast and vulnerable boreal forests?* (Educational Report Vol. 17 No. 3 with a press run of 80,000 copies.)

With the help of our in-house cartographer Chris Player, we combined and reconfigured the most interesting graphs found in other scientific publications, putting them on the same timeline to prove to skeptics that global warming was real. As far as I knew, all the graphs we assembled for the paper had never been juxtaposed as we presented them and made available in one publication before. They included a graph of earth's mean annual temperature showing its upward trend during the last 100 years and a graph showing the relative ice volume on earth compared to the average summer sunshine energy in watts per square metre over the last 750,000 years. The amount of solar energy reaching earth cycles up and down over long periods of time as dictated by the combination of our earth's constantly changing tilt, wobble and orbit. Putting these graphs together revealed the general and obvious relationship over time between less sunshine energy and more ice; warmer periods when solar energy increased and ice ages when solar energy declined. But during the last 100 years this relationship has fallen apart. The energy from the sun shining down on earth has been decreasing; which should be putting earth on the path towards another ice age. But instead the earth is heating up and the ice is melting. The only feasible reason is the build up of greenhouse gases that are trapping the heat at the earth's surface, causing non-solar-driven global warming (take a look at these graphs, up on WCWC's website www.wildernesscommittee.org).

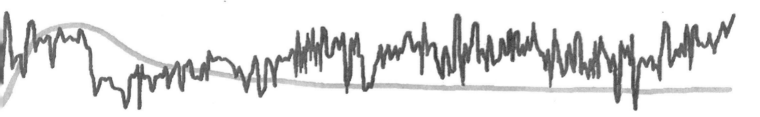

Another graph showed the annual cycle in the amount of CO_2 in the atmosphere, as measured at Mauna Loa on the Big Island in Hawaii, from the 1960s to present. The graph shows the expected seasonal dip in CO_2 during the northern hemisphere's summer months when increased photosynthetic activity in plants (especially trees and forests) are sequestering it and a rise in the amount of CO_2 in the winter when photosynthetic activity is at a low ebb and more fossil fuels are being consumed by people (primarily for heating).

Except for the seasonal fluctuations, the amount of CO_2 should stay fairly static year to year. Instead, over the last forty years the concentration of CO_2 in the atmosphere has risen from below 320 parts per million to over 360 parts per million. It's another scary confirmation of the human-caused build-up of greenhouse gasses that are causing global warming.

The centre of our *Global Warming* newspaper features a map of Canada titled *Impacts of Rapid Global Climate Change – Catastrophic and Astronomically Costly*. Done in a *National Geographic* style, it shows the extent of the boreal forest and regions affected by climate change including catastrophic storms, fires, forest diseases and droughts. Our lead article explains how cutting the boreal forest accelerates global warming through the drying out of boreal bogs releasing methane gas and the loss of their function as a carbon sink where a net fixation of CO_2 into cellulose is greater than its contribution of CO_2 to the atmosphere through respiration, decay and combustion.

Gray convinced Dr. D.W. Schindler, a professor of Ecology at the University of Alberta and one of the world's experts on the impact of climate change on water resources, to write a brilliant one-page article for the paper (titled *Boreal Mayhem: the effects of recent human activities on boreal landscapes*). He also got his friends Tooker Gomberg and Angela Bischoff, environmental activists, to write an article about positive things people could do to help curb global warming titled CLIMATE CRISIS DEMANDS SWIFT ACTION...*some are already rising to meet the challenge*.

Among the things we urged people to do was: *Order a portable family solar cooking oven for $50 from us and give solar energy a try*. In our efforts to save the tigers in India, we had experimented with solar cookers made in Saskatchewan by a solar activist. At our home on the Sunshine Coast we used one occasionally and it worked really well, at least on sunny days. We baked potatoes and cooked pots of rice with it. But I must admit it was more of a novelty than a practical kitchen appliance in our climate. We only got a few orders for the solar cookers and we lost money on the deal because it cost us more to get them built and shipped in and out than we got back in sales. But there is nothing like a solar cooker to demonstrate the potential of the sun's renewable energy to fulfill human energy needs.

Gray later told me he received a great deal of positive compliments about our *Global Warming* paper and many orders for classroom sets. But I knew we faced an almost overwhelming challenge to actually shift our civilization from its fixation on using fossil fuels. It also made me want to get WCWC more active in making this a top priority issue in the minds of people and politicians.

The opportunity to create a world-class protected area system in the Lillooet region is squandered

At the end of 1995, in conjunction with establishing the Stein Valley Provincial Park, the B.C. government announced that it would initiate a Lillooet Land and Resource Management Plan (LRMP) process. This LRMP finally got underway in 1997. Its task was to determine which lands in the Lillooet region would be preserved and which lands would be "developed." There were plenty of representatives for the environmental and recreational sectors, however most of them were from the Lower Mainland. Local representation at the LRMP table was predominately pro-logging and mining and anti any further protection in the area. The "rednecks" in the region didn't want mining and logging fettered by having to consider "wilderness values." The environmentalists living there were, for the most part, silent. It could be because of the undercurrent of threats and fear that permeated the area.

Of all the places in B.C., the Lillooet region has the most diverse and spectacular wilderness. Sadly, while the LRMP negotiations were underway, logging continued in many important wilderness areas that were candidates for protection. In the summer of 1998, with the help of conservation organizations at the table, WCWC published an eight-page full-colour full-sized newspaper titled *To be preserved? To be pillaged? Act now to save BC's Rainshadow Wilderness* (Educational Report Vol. 17 No. 5). We called for a bioregional approach, including preservation of the fourteen large wilderness areas in the Lillooet region that were being considered at the LRMP table. We appealed to the public to support the protection of all of them.

The 80,000 papers we printed up went out right away to all our members and to every household in the LRMP study area. The household mailings cost only nine cents each in postage—less than a thousand dollars. More than half of the people living in the Lillooet region are native and many live in villages on reserves. We believed that many of them would be natural allies of ours, in favour of more protection for the natural environment and their traditional hunting and trapping areas.

This LRMP table was scheduled to hand down its recommendations in 1999. But, not surprisingly, the process got bogged down. Those opposed to more parks realized that if they could hold out until after the next election, which at the very latest would be two years away, the increasingly unpopular NDP government would undoubtedly be rejected and replaced by a pro-development Liberal government. The Liberals would side with them instead of with the urban preservationists. So the LRMP process dragged on and on.

South Chilcotin Mountains Park support substantial

Believing in the need for public involvement, we asked the LRMP table to put on at least one public meeting in the Lower Mainland, because so many city people recreationally used this area on weekends to present its different land use options and gather public input. But to no avail. The table planned their final open house for Lillooet. We had no choice but to put on our own public meeting in Vancouver, mobilize supporters and bus them to Lillooet to this LRMP meeting.

An overflow crowd of four hundred packed into the event we sponsored at the Vancouver Planetarium in early February 2001. Everyone listened attentively to speakers from some of B.C.'s leading conservation groups, each calling for the B.C. government to immediately grant park protection to the Rainshadow Wilderness areas, the dry Lillooet region behind the Coast Mountains that squeeze out most of the moisture of the westward moving clouds. One of them, the South Chilcotin Mountains, has been a park candidate since 1937, making it the longest standing park proposal in the province. At the end of the meeting we called on people to join us in attending the LRMP's final open house. The response was overwhelming. We ended up organizing a convoy of three busses.

Everyone in Lillooet knew we were coming. We hadn't kept it a secret. In fact, we put out a press release announcing it. We didn't charge people to join our "tour" bus, knowing that many of the stu-

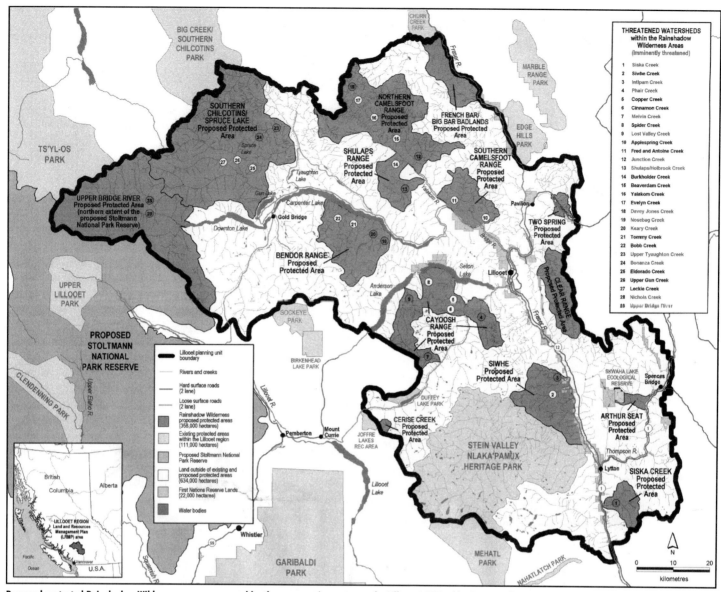

Proposed protected Rainshadow Wilderness areas preserved by the conservation sector at the Lillooet LRMP table. Cartography: WCWC files.

dents and seniors who wanted to come couldn't have afforded it.

In an effort to diffuse the hostility and resentment that local people had towards the outsider environmentalists "who were trying to tell them what to do," Chris O'Connor, then Mayor of Lytton, initiated a *Hug a [Tree] Hugger* campaign. He had some stickers printed

Typical pattern of clearcutting on the Chilcotin plateau in the rainshadow cast by the coast mountains. Photo: WCWC files.

up with this slogan on them. Quite a few locals were wearing them when our contingency of preservationists arrived.

When a particularly rabid guy started yelling at Joe Foy, O'Connor, who is a big heavyset man, rushed over and gave Joe a big hug. Then he stuck one of his little *Hug a Hugger* stickers on Joe's chest and one on the other guy's chest and pushed the two of them together saying "John, John, hug him. It's hug a hugger today!"

O'Connor's hugging campaign had limited success. In one encounter, Andy Miller, the naturalist heading up WCWC's spotted owl campaign, got into a conversation with an elderly woman. She adamantly opposed our stand to protect wilderness. Completely out of the blue, in the middle of their conversation, she punched Andy right between the eyes! Andy had no idea her fist was coming. He was stunned, but he just backed away and shook it off.

Not having been there, I asked Joe what he thought bussing in preservationists to Lillooet had accomplished. Joe said:

In the long run it helped prevent a worse rollback in the Southern Chilcotin Mountains Park than we got. It showed that people really were passionate about this and conveyed the idea that a large number of people really cared about protecting the area. Some of the good things happened later. Gary John, chief of the Seton Band saw that the Wilderness Committee was willing to stand its ground. In conver-

sations with him after we started working on Lost Valley, I learned that this encounter made him more comfortable with working with us because he knew we would be with them through thick and thin.

Even after that infamous public open house meeting in Lillooet, the LRMP negotiations continued to stall. Representatives of the conservation groups sitting at the LRMP table asked to meet in Vancouver with other environmental groups and leaders, especially about the Southern Chilcotin Mountains area. Both Joe and I attended this "summit" meeting of about 25 conservationists, which included former MLA Tom Perry and behind-the-scenes wilderness negotiator Ric Careless. Because the table was hopelessly deadlocked and the province was headed into an election, Ric had a plan that he had worked out with the NDP government to get the Lillooet region parks established before the government's mandate ended. But, he explained, it would take the cooperation of everyone. If all the conservation groups walked away from the table, informing the government that there was no hope of reaching a consensus recommendation and asked the government to make a decision, Ric assured us that there was just enough time for the government bureaucracy to work out the legislation to create the "rainshadow parks" and get it passed.

After considerable discussion, everyone attending the meeting agreed to this plan of action including all the representatives sitting at the LRMP table. I don't know what happened next or why the deal fell apart, but it did. The conservationists at the LRMP table opted to continue to sit at the table, work up their position as one of several "land use options," submit the options to government and abide by the government's decision whichever way it went. When they eventually did submit their packages, it was way too late for the government to act. It was only through the greatest amount of pressure by the conservation community that on the day before calling a provincial election, Premier Dosanjh established, by an Order in Council, a 71,400-hectare South Chilcotin Mountains–Spruce Lake protected area with equivalent status to a Class "A" provincial park in the Lillooet Region near Gold Bridge. We immediately put out a press release praising the government for creating this park. *What a wonderful present for all generations to come*, commented Joe Foy, *it's time to celebrate!!!!!*

As part of our celebration, waiting to roll off the press, was a thank you poster for the last minute creation of this park by the outgoing NDP. It featured a stunning view of Spruce Lake with snow-capped Coast Mountains in the background.

They were printed within a few hours after the announcement and we passed them out free at Dosanjh's campaign kick-off rally, from our store, and at other political meetings during the election. We hoped it would help make this good park decision stick. The move to protect this area was so last minute, we were afraid that it would unravel if the NDP failed to win the election—which was a sure bet given the NDP's unpopularity.

Our celebration was indeed short lived. The May 2001 election reduced the NDP from a majority government to holding only two out of 79 seats in the Legislature. Immediately after the election, the new Liberal government with its whopping majority reopened the Lillooet LRMP for reconsideration. The intention was clear—to roll back the boundaries of the new South Chilcotin Mountains Provincial Park. Thus began a whole new campaign of defending the park. By sheer luck, we had designed the poster so that the NDP thank you statement at the bottom could be cut off to make the leftover posters just fine for our new campaign!

Keeping this issue hot in the Lower Mainland succeeded in getting

South Chilcotin Mountains Park Celebration poster. Poster image: Garth Lenz.

the Liberal government to delay making a decision for a long time. We were convinced that the longer we could delay the decision to downsize the South Chilcotin Park, the greater the chance it would survive unscathed. In January 2002, we published a four-colour, four page newspaper titled *B.C.'s Fabulous South Chilcotin Park* (Vol. 21 No. 1- print run 50,000 copies) in defence of the NDP park boundary.

Over the next few years WCWC held rallies, information pickets in front of cabinet ministers' meetings, put on slideshows and held a Canada Day gathering in the South Chilcotin Park. We also picketed in front of the offices of Teck Cominco, a mining company that was openly backing the Mining Association of B.C.'s campaign to downsize the park by over 95 percent—to just 3,000 hectares around Spruce Lake. At one protest Joe, who was picketing in a Santa suit just before Christmas in front of Teck Cominco in downtown Vancouver, was kicked out of Teck's offices when he walked in and tried to get a meeting with company officials.

On April 21, 2004, we held a big third year birthday party for the park. I really feel that our unrelenting pressure forced the government to delay its decision several times. But finally, on July 22, 2004, the B.C. Liberal government announced its roll back of the park's boundaries. The heavy lobbying by B.C.'s powerful mining industry won out. The B.C. Liberals sliced off 14,600 hectares—20 percent—of the NDP's popular South Chilcotin Provincial Park, home to grizzly bears, mountain goats, bighorn sheep and a vibrant tourism industry. The deleted areas, it decreed, couldn't be logged; only roaded and mined. The government also stated that negotiations with the St'át'imc First Nations would have to be completed before the final roll-back of the park boundaries could be made law.

A small demonstration celebrating Park's Birthday gets lots of media. Photo: WCWC files.

Chapter 58

Logging in Vancouver watersheds halted

Canadian Senate challenged to fix a bad law

Environmentalists' persistent efforts end logging threat in Vancouver Drinking Watershed

Despite a decade of campaigning against logging in the three watersheds that provide drinking water for the Lower Mainland—almost half the population of B.C.—the Greater Vancouver Water District (GVWD) Board continued to advocate logging there. One of our less successful campaign tactics was a petition drive that featured my good friend Nik Cuff as "Mr. Drip." Nik duct taped a water faucet to the middle of his forehead and stood on a busy street corner trying to gather signatures. It didn't work. He scared the heck out of most people he approached. They quickly walked away without speaking to him, obviously concluding that he was a raving loony!

The campaign, which various groups participated in, did achieve partial success. In 1993 the GVWD stopped logging the drinking watersheds for commercial purposes. But it didn't stop all logging. In fact it began heavily promoting logging as a sound watershed management tool and asserted that "tree removal" in the future would only be done to enhance water quality, prevent future bug infestations and reduce the chance of fire. To this end, the GVWD Board hired various consultants to do a series of studies that were dubbed the "ecological inventory." The research took place over a series of years and seemed bent on understating the negative effects of logging and logging related activities on water quality. The GVWD spent $6.7 million dollars on this effort, which we all believed was nothing more than a colossal 'snow job' to rationalize continued logging.

In early 1997, the GVWD contracted a scientific review panel consisting of three members, Peter Pearse, Everett Peterson, and Rolf Kellerhals, to collectively determine the future management of the watersheds in relation to comments from public forums and comments from the scientific community. We decided to increase public pressure. Nik Cuff suggested that WCWC picket outside the GVRD office on Kingsway Avenue early in the morning once a month before their regularly scheduled meetings. For years SPEC had been observing and documenting these meetings which were open to the public. I thought the picket was a great idea.

So I agreed to sleep in the office the night before the meeting and grab an early Skytrain to be there about an hour before the meeting started. I think it was because of this protest activity that WCWC was invited onto the GVWD's citizens' advisory meeting. Watch out for being co-opted by such a consultation process. Government will "meet you to death" if you let them. I only attended a couple of these advisory meetings, mostly to glean what I could about the ecological inventory that the GVWD was close to completing.

Alternative forum outside the Greater Vancouver Water Board's eco-inventory presentations

At the last meeting I attended, the GVWD Board informed us that they had picked May 15, 1999 for the public viewing of their

From left to right - Joe Foy, Paul George, and Will Koop picketing in front of the Greater Vancouver Regional District Offices before a Water Board meeting. Photo: SPEC files.

Ecological Inventory. They opted to rent the large low ceiling room in Robson Square Media Centre and present their findings at dozens of information stations with photos, displays and the individual consultants who'd done the research beside each one. In this way, there wouldn't be a focused opposition or embarrassing questions and answers that everyone attending could hear. It was an effective way of defusing and blunting the opposition – and it became common practice from then on for all government public input "hearings."

As we left that last advisory meeting, Will Koop and I brainstormed the idea that WCWC should rent the Robson Square's Judge White Theatre, the big room right next to where the GVWD show would be, and hold an "alternative forum" on the watersheds at the same time. It was a brilliant idea. But I was sure the GVWD Board had rented the whole complex for the whole day. When I got back to the office I called to see if the theatre was available. I couldn't believe my ears when they told me it was! I booked it on the spot. It only cost a little over $500 for the day. I immediately called Koop and everyone else to tell them about our good fortune.

Things came together fast for our "alternative forum." For years Koop had been trying to get permission to enter the three GVWD watersheds to inspect the impacts of logging. By sheer perseverance he got permission and by sheer luck, there was a really big storm during the time he was allowed to enter the watersheds that provided lots of documentable evidence of massive erosion of the logging roads and massive landslides caused by the clearcutting and logging activities. Koop captured it all on 35-mm slides. Using our giant map plotter, we blew the slides up into three feet by four feet images and mounted them for display. It didn't matter that the GVWB had spent millions of dollars on consultants to conclude that logging was OK. Koop's series of images blew their biased studies out of the water. They proved beyond any reasonable doubt that logging roads and clearcuts on steep slopes were the problem.

We displayed Koop's images in the foyer outside the theatre with an explanation of where in the watersheds each one was taken and how it demonstrated that logging was harming the drinking water quality. Of course, everyone going to the GVWD's public input session couldn't miss our big display right next to them. We plastered posters all over the foyer, too, advertising the time of our alternative presentation later that day.

Realizing we had scooped them, GVWD officials asked if they could join us in the Judge White auditorium to present their findings and answer any questions from the public. Delighted because it was exactly what we wanted—a chance to publicly question their experts in a forum where everyone could hear the answers—I said sure, we'd share our space. As a bonus, they paid half the rental cost.

About 300 people came to the viewing and attended the show. We had our own set of experts who addressed the public. One was Dr. Michael Feller, a leading forest fire expert and professor in the University of B.C. Forestry School. He said that fire was not a threat in the Greater Vancouver watersheds because of the damp nature of the rainforest. He proved his point with historical data. There had been only two tiny fires covering only a couple of hectares since records had been kept. He also asserted that *not a single study demonstrates that logging of any type in old growth forests reduces fire risks*. Dr. Elaine Golds, a former McGill University biology professor said, *It's arrogant to say that we know enough about insect infestations to predict what might happen in the watersheds 120 years from now. We can't assume the only solution to possible insect infestation* [in the future] *is cutting down trees now.*

SPEC president David Cadman wrapped it up saying, *Prior to the early 1960s and the start of modern logging in the North Shore watersheds, Vancouver had the best drinking water of any city in the world.* He could not believe a report that came to the conclusion that ongoing logging is the preferred option for managing a watershed that supplies drinking water for 1.9 million people.

After our presentations concluded, we set up several big tables on stage for the GVWD. The expert representatives of the dozen consulting firms that had prepared the GVWD's "ecological inventory" sat stiffly behind the tables on stage. They were joined by the three members of the GVWD's Scientific Review Panel. Remarkably, there was not one woman among them!

Darth Vader warns the crowd

Just as they were about to begin, an actor hired by SPEC dressed in a Darth Vader suit of Star Wars fame walked down the aisle from the back of the theatre. There were two television cameras behind him. I was as stunned as everyone else, having no idea that this stunt was in the works. When he reached the front he said in his raspy Darth Vader voice, "Science fiction! You'll be hearing science fiction from these learned gentlemen." He brandished his glowing lightsaber and then strode back up the aisle and out of the theatre muttering over and over "Science fiction."

The GVWD's presentation took several hours to complete but, after Darth Vader's appearance, the credibility of the consultants and their "Ecological Inventory" was as un-salvageable as Humpty Dumpty's smashed egg.

The Water Board listens and does the right thing

Our victory came on November 10, 1999, just less than five months after the public viewing of the "ecological inventory." At the end of a two-hour meeting of the mayors on the Greater Vancouver Water District Administration Board and the Board, this five-point resolution was passed:

1. *The primary purpose of watersheds is to provide clean, safe water.*
2. *The watersheds will be managed to reflect and advance the commitment to the environmental stewardship and protection of those lands and their biological diversity.*
3. *The management plan will be based upon the minimum intervention absolutely necessary to achieve objectives.*
4. *The management plan will contain policies to return areas disturbed by human activities as close as possible to the pre-disturbance state consistent with the primary goal of protecting water quality.*
5. *The decision-making process will be transparent and open to the public.*

This resolution ended more than 30 years of commercial logging in the Greater Vancouver drinking watersheds.

The door was firmly closed to future logging on February 8, 2002. Members of the GVRD's Water Committee unanimously passed a recommendation that the GVRD Board cancel the water district's *Amending Indenture*, the 1967 agreement with the provincial government that had made the Greater Vancouver watersheds into a mini Tree Farm Licence. This was the final step needed to return the Greater Vancouver Water District to its original protected status under the powerful Land Act lease legislation granted in August 1927 through the recommendations of former Water Commissioner E.A. Cleveland. The lands and forests of the GVWD's three drinking watersheds are once again provincially protected watersheds.

Canadian Senate cannot or will not help strengthen a law to protect people from toxic substances

In July 1999, Adriane got a call from Elizabeth May, head of the Sierra Club of Canada, whom she knew well from working closely with her on the Clayoquot Sound and Endangered Species campaigns. Elizabeth asked for the help of WCWC in the fight to strengthen Bill C-32, *An Act Respecting Pollution Prevention and the Protection of the Environment and Human Health in order to Contribute to Sustainable Development*. Its short name was the Canadian Environmental Protection Act (CEPA). CEPA was first passed in 1988 as one of the Brian Mulroney Conservative government's most important pieces of environmental legislation. It dealt with everything from toxic chemicals, products of biotechnology and the international movement of hazardous wastes, to the environmental soundness of federal government operations.

The Chrétien Liberal government now claimed that its reform of CEPA (Bill C-32) made it the *most progressive environmental legislation in the world*. It was an outright lie. In truth, Bill C-32, as passed by the House of Commons, skirted around every tough issue and offered only the illusion of protection. It established a series of almost insurmountable hurdles to the Ministers of Environment and Health that inhibited the actions they could take to protect the health and environment of Canadians. Bill C-32 was so profoundly flawed that the members of the House of Commons Standing Committee on the Environment, who were charged with completing a clause by clause review of the proposed legislation, had presented hundreds of amendments in an attempt to deal with its weaknesses. The Prime Minister's Office, lobbied heavily by the chemical and biotech industry, didn't like all the changes that toughened up the Act. By the time the Act was presented for final vote in the House of Commons, the Environment Committee's key amendments were reversed, particularly in relation to toxic substances and biotechnology products. These changes made Bill C-32 even weaker than the law it was supposed to reform and improve.

Three Liberal members of the Environment Committee, two of them highly respected former Ministers of the Environment, ultimately voted against Bill C-32 including the Chair of the Environment Committee, Charles Caccia, and the Parliamentary Secretary to the Minister of the Environment, Clifford Lincoln. Of course it passed anyway. That's the way it works in Canada's parliamentary democracy. Adriane, who was in Ottawa for another one of her meetings, saw both Caccia and Lincoln the day after the vote. "I'd never seen them so depressed," Adriane said to me over the phone that night. They both had told her that it was a very sad day for democracy.

Almost every environment and health group protested Bill C-32. Among the gravest concerns were the watered down requirements for the virtual elimination of the most toxic chemicals, the disregard for the protection of children from harmful contaminants, the lack of regulation of products of biotechnology, and the shifting of power to determine the toxicity of chemicals from the Ministers of the Environment and Health to cabinet as a whole. The Bill required consultation with the provinces before the Minister of the Environment could take any substantive action, which included even the establishment of non-binding guidelines or codes of practice. Any steps that actually required action were now subject to cabinet approval, where the Ministers of Industry and Natural Resources held a veto on behalf of their industrial "clients."

The Bill also weakened the existing Environmental Protection Act by providing a mechanism for the cabinet to waive, at its discretion, the rule that all new substances, including products of biotechnology such as genetically engineered plants and animals, foods and drugs, be assessed for their potential impacts on human health and the environment <u>before</u> they are introduced into Canada.

This Bill is being rammed down the throats of Canadians by polluting industries and corporate interests, said Maude Barlow, head of the Council of Canadians, in a joint press release with the Sierra Club of Canada. In the release the two organizations appealed to the Senate to give the Bill *true sober second thought* to protect the people of Canada.

Adriane and I traveled to Ottawa in the last week of September 1999 to testify before the Senate Committee studying the Bill. In our brief to the Senate we outlined specific shortcomings we saw in the bill, especially its failure to ban the most toxic substances that bio-accumulate and cause birth defects and cancers. Frustrated by feeling powerless on an issue of such importance and noticing that many of the Senators sitting in Chambers were disinterested and not really paying attention to what I was saying during my oral testimony, I decided that I had nothing to lose by being forthright. Here is an excerpt of the last part of our testimony as reported by the Senate's Hansard:

Proceedings of the Standing Senate Committee on Energy, the Environment and Natural Resources

Issue 21 - Evidence, August 26, 1999 (afternoon meeting)

Heading on Senate Hansard website. Credit: Government of Canada.

Mr. George:
...I will take one more step to get my point across, and then I will conclude. I came from the U.S. when I was 28. I was taught when I first came here that the Senate was really useless; hence, we never sent a thing to the Senate. Then one day, about three or four years ago, someone said, "The Senate really is powerful. They can do things. You are really ignoring a great chance if you do not send your educational material to them." Hence, for the last three years, we have been religiously sending these things to you. However, I still do not know personally how powerful the Senate is. Could it stop this bill? Listen to these statistics about people in Canada. Can it stop it? Can it use its force to make a better and more solid bill? That is my question to you. If you can, then for God's sake, do it.

Senator Buchanan: The answer is absolutely, yes, we can.

Senator Spivak: We can.

Mr. George: My next question is: Has the chemical lobby got to you?

Senator Spivak: Right.

Senator Buchanan: You mean it has?

Senator Spivak: Sure.

Senator Chalifoux: They have not talked to me.

The Chairman: Mr. George, we have every opportunity to amend bills or to veto bills. That is within the power of the Senate if the Senate so determines to do so. Whether it will or not is, of course, up to the individual senators and the way they wish to deal with the subject matter.

Senator Spivak: The Senate treats this power in a very restrained way. We are not elected and to compete with the will of the house is a very rare occurrence.

It has happened a few times since I have been here; for example, the abortion bill. In a matter of extreme importance and a matter that touches the conscience of people, the Senate has indeed voted to defeat bills.

Senator Taylor: The Senate can also kill bills. If you had a choice between a dead bill, going back to the way things were a few years ago, or this bill not being as good as what you want, which would you choose?

Mr. George: I would like to see you go through it point by point, clause by clause, listen to the experts, improve it, and send it back. You can say, "These are the amendments that would make this into a decent, strong environment bill, one that we can be proud of in Canada." Are you allowed to do that?

Senator Buchanan: Yes.

Senator Taylor: I was trying to get a between-this-and-that answer out of you. If that does not get done, is the bill as it is still better than nothing?

Ms. Carr: No. The bill in its current form is not worth allowing to be passed.

Mr. George: It makes things worse, and it sets a bad trend. I would say you should kill it.

Ms. Carr: It is important. When you work as a public advocacy group, as an environmental group, frequently you work hard to achieve new legislation. If that legislation gets watered down to the point where it really is ineffective yet permits government and industry to promulgate the myth that there has been improvement it really sets back the democratic process. It is difficult for citizens to understand what has happened. If you do not choose to improve this bill, then I believe you ought to let it die and let process take over to create something. I believe you need to pass legislation that you can feel good about standing behind.

In the weeks that followed, Environment Minister David Anderson threatened to drop the Bill rather than take Senate changes back to the House. Despite a controversial debate, the Senate made no changes to the legislation passed by the House of Commons. Bill C-32, the new Canadian Environmental Protection Act, received Royal Assent on September 14, 1999 and was proclaimed into law on March 31, 2000.

Chapter 59

Stoltmann campaign grinds on CMT research conducted MOU "saves" Clayoquot's pristine valleys

Stoltmann Campaign dominates our campaign plate

During the last few years of the 1990s, WCWC's efforts focused more and more on the spectacular big-treed valleys of the Stoltmann Wilderness. On June 20-21, 1998 we hosted a Summer Solstice Gathering —a "Tree Huggers' Weekend"— in the Stoltmann. About 200 people showed up. We put on an outdoor slideshow, celebrated the grand re-opening of our Stoltmann Research Station (after 'persons unknown' had dismantled it and dumped it in the Squamish RCMP yard during the logger's blockade the year before) and launched our Elaho to Meager Creek Trail Project. Events like this are key to building campaign camaraderie.

During the following weeks, our volunteers began to upgrade our "surveyed hiking route" from the Upper Elaho to Meager Creek Valley to the standards of a Forest Service recognized trail. We replaced the suspension bridge over Lava Creek with an engineer-approved cable car to meet those requirements. It was a costly, time-consuming project.

Then, just when we thought things were improving, Interfor cut down most of the magnificent Grizzly Grove of big trees. It was terribly hard on the volunteers who had first discovered and explored this grove. To add insult to injury, Interfor did this just before it closed its sawmill in Squamish, laying off 165 workers. We photographed the stumps in Grizzly Grove, to forever remind people of the demise of the ancient cedars there. It was a poignant testimonial to the fact that we were losing the fight to save the Stoltmann Wilderness. But we were not going to give up.

All summer, hundreds of people hiked our Elaho Valley trail each month. In August we printed 80,000 Stoltmann Wilderness National Park opinion poll postcards and distributed them to our members throughout Canada. We also sent one to every household in Whistler, Pemberton, Mt. Currie and Squamish. Like all our opinion poll cards, those who were against what we wanted could check the "no" box and send their card straight to government. I'm sure that the loggers in Squamish sent in a lot of "no" cards, but we heard that the cards supporting the creation of a national park in the Stoltmann greatly outnumbered them. Doing something new and positive for the campaign kept everyone busy. It left our volunteers and staff with no time to brood about the continuous losses.

A search for new arthropods in the treetops of the ancient firs in the Upper Elaho

Searching for new scientific evidence that the forest should be saved, we contacted Dr. Neville Winchester, who had collected so many new-to-science species during his research in our canopy platforms in the Upper Carmanah valley's big Sitka spruces. He was now a professor at the University of Victoria. We asked him if he would undertake a similar research project in an ancient Douglas fir grove near WCWC's Elaho Research station, offering to build a similar platform station. He told us that he was now collecting samples from many different forest types in B.C. and couldn't do a long-term study in the Elaho. But he would come for a weekend with a few graduate students to gather samples there. On a weekend in late August, Dr. Winchester and his team arrived. They climbed 15 stories up into the tree-tops to gather insects. Before he left with his samples, Dr. Winchester publicly called for an immediate halt to logging to preserve the area's unique ecology.

Did Winchester's team find new species there?

Above: Dr. Neville Winchester ascends a giant ancient Douglas fir in the Upper Elaho Valley to collect arthropods in the canopy. Photo: Joe Foy.
Facing page: CMT expert David Garrick, centre, takes notes and instructs Heiltsuk students from Bella Bella taking a CMT survey workshop sponsored by WCWC in September 1998. Photo: Adriane Carr.

I asked Neville this question seven years later. He immediately said "yes." In fact, he found new species in every valley he looked in. He was uncovering them faster than taxonomists could classify and name them. What had emerged through his research was an extraordinary richness in arthropod species diversity in the rainforests of B.C. What astounded Winchester as much as these findings was the fact that the politicians and governments didn't care. They continued to authorize the liquidation of B.C.'s old-growth forests knowing that it was causing massive speciocide, thinking that the extinction of these unnamed creatures bore no consequences.

Book and CMT workshop focus attention on the importance of documenting aboriginal use of B.C.'s coastal forests before the evidence is destroyed

As fast as species were being eliminated in the coastal rainforest, so were the ephemeral traces of the aboriginal cultural use of the forest. One person who had extensively studied culturally modified trees (CMTs) was David Garrick. His academic training was in anthropology and he was one of B.C.'s original environmentalists, having been a cook and chronicler on the *Phyllis Cormack* on Greenpeace's first sea-going anti-whaling campaign in 1975 and active in SPEC during its early days. I got to know him when he was working as MP Jim Fulton's research assistant.

Garrick had a cabin on Hansen Island near Orcalab, the research facility from which Dr. Paul Spong studied the killer whales in Johnston Strait. Marion Parker had visited Hanson Island, studied some of the CMTs there and passed on his knowledge and expertise to Garrick. What made Garrick's work unique was his years of phenomenally thorough in-the-forest research and his powers of detailed observation. Garrick had combed the entire island and contended that there were a lot more culturally modified trees (shaped cedars as he called them) in the coastal forests of B.C. than those found by the professional archeological consultants who were hired primarily by the big forest companies to clear this impediment to logging. He also came to the conclusion that much more information was contained in the trees' scars, especially in the patterns of bark stripping, than was generally thought by others specializing in this field.

When I saw his manuscript, with the many detailed illustrations he'd hand-drawn and the fabulous photos, most of them taken by Greg McIntyre who had visited the island dozens of times, I knew it was information that had to "get out" and be more widely distributed. It was the kind of book that a commercial publisher wouldn't touch. Its audience would be way too small. But it was easy to convince the rest of WCWC's E-team that this should be our next publication. We had long recognized the key role that this proof of aboriginal use and occupancy of the land played in preservation of the coast's ancient rainforests. Garrick's text would serve as a guidebook to enable others to collect this profoundly important information.

In early September, six weeks prior to the scheduled publication date, David Garrick came and stayed for a couple of weeks at our house on the Sunshine Coast while he, Adriane and I went over the final edits of the book. After Nik Cuff printed it, a team of volunteers, including Garrick and McIntyre, worked long hours for a whole week going around and around a couple of tables collating 1,000 copies of this 390-page book. Greg McIntyre estimated that he traveled a total of 27 kilometres walking around the tables gathering up the pages of the book while doing his share of the task. On October 20, 1998, we publicly released Garrick's book *Shaped Cedars and Cedar Shaping – a guidebook to identifying and documenting, appreciating and learning from Culturally Modified Trees.*

In the forward to the book Jack Woodward, who specialized in native law, wrote:

> ...until the 1980s most of us had not realized the vast scale of archaeological destruction that was occurring in British Columbia with the logging of old-growth cedar forests. The 1985 Meares Island injunction brought to wide public attention the fact that the forests were full of these fascinating and valuable artifacts. The late Mr. Justice Seaton, in granting the injunction said: 'It appears that the area to be logged will be wholly logged. The forest that the Indians know and use will be permanently destroyed. The tree from which the bark was partially stripped in 1642 may be cut down.' We now know that he could have been describing almost any place on the Coast; and for most of these places of beauty, history and nature, it is now too late.

A year later, when we learned that Hanson Island was in imminent danger of being logged, which of course would destroy the CMT evidence of aboriginal use that Garrick had so painstakingly documented, we published a four-page full-colour tabloid-sized newspaper titled *Yukusan-Hanson Island – an urgent Conservation Plea* (Educational Report Vol. 18 No. 5 – Summer 1999 with a press run of 50,000 copies). This paper highlighted some of the information found in Garrick's book, explaining the uniqueness of this island. I believe that Garrick's book and our paper helped get Hanson (Yukusan) Island protected. But it was the Kwakwaka'wakw First Nation's resolve to protect this island of theirs that truly won the day.

Heiltsuk students on field research trip during the CMT workshop put on by David Garrick sponsored by WCWC. A board was removed from this cedar many years ago. Photo: WCWC files.

We also contracted Garrick to put on a CMT workshop in Bella Bella for some keen Heiltsuk First Nation members, including several youth. We coordinated the workshop through the Heiltsuk Band Council. Joe, Adriane and I tagged along, keen to meet with the Heiltsuk and learn more about CMTs. In all we spent an extraordinarily enlightening week talking with people, including several Heiltsuk elders, and being guided through their territory. I clearly remember one of the days we joined Garrick's workshop group for its practical fieldwork. We boarded a local fish boat that WCWC had hired to take Garrick's students to a survey site, traveling up a cliff-

Adriane on the radio phone on the boat in Bella Bella on the CMT Survey course for the Heiltsuk youth September 1998. Photo: Joe Foy.

faced inlet during low tide. Ancient pictographs adorned a vertical rock face under an overhanging rock. Then we passed some small stone rock walls in the intertidal zone around several creek mouths. The native guide explained that these were fish traps used historically by aboriginals. Now there were no salmon in these small streams. Commercial interception fisheries and illegal "creek robbing" had finished them off.

When we got to the forest fieldwork sites, Garrick carefully instructed the participants on how to identify, classify, measure and take field notes on the CMTs that they found. Everyone was excited. They were documenting archeological artifacts that may help secure their aboriginal title and control over the land. But I kept thinking about how much work it took to build those fish traps and how sad it was that the salmon runs were no more.

Clayoquot First Nations logging company signs MOU with environmental groups to protect pristine areas

At the same time that we increased our efforts in campaigns to protect the Stoltmann Wilderness and Great Bear Rainforest, Adriane kept diligently working to secure "peace in the woods" in Clayoquot Sound. The critical piece in the puzzle was transferring the Clayoquot part of MB's tree farm licence (TFL) to a new First Nations forest company, Iisaak Forest Resources, and developing a Memorandum of Understanding (MOU) between Iisaak and the environmental groups to ensure that Clayoquot's pristine areas wouldn't get logged.

It took several years of negotiations between the First Nations, MB and the environmental groups, starting in late 1997, to work out the innovative deal that respected the First Nations' rights and aspirations as well as the environmentalists' campaigns and concerns. Adriane, as the lead negotiator for the environmental groups, spent hours every day working on Clayoquot, talking with MB's Vice-President of Environmental Affairs, Linda Coady, and representatives of the key environmental groups, especially Valerie Langer from Friends of Clayoquot Sound, Karen Mahon and Tamara Stark of Greenpeace, Vicky Husband and Lisa Matthaus of the Sierra Club of B.C. and Liz Barratt-Brown and Matt Price of Natural Resources Defense Council. She told me there would be no deal if all the key groups weren't on board so it was her job to consult, clarify the bottom lines and check for consensus.

There were some very tense moments, especially at the final negotiating meeting held in Ahousaht on January 22, 1999, between the environmental groups and the Nuu-chah-nulth Central Region chiefs. Adriane recalls the meeting this way:

I remember every detail of that meeting. I was charged with making the lead presentation on the draft Memorandum of Understanding to all the Chiefs. The room in Ahousaht was packed with important stakeholders, including hereditary chiefs. There were forestry workers as well as environmentalists who didn't want one more tree to fall in Clayoquot there. We passed out copies of the draft MOU and I started my speech. It was hard to concentrate because at one end of the table some of the key Nuu-chah-nulth spokespersons, including Cliff Atleo and Francis Frank, were whispering amongst themselves. When I finished, Cliff rose to speak and stated, calmly and clearly, that there was a problem with our MOU. The chiefs did not like the naming of the 'pristine' areas identified as places not to be logged on the map attached to our MOU. My heart was pounding. I really feared everything would fall apart. Then Cliff continued. The chiefs, he said, wanted the areas to be renamed in the Nuu-chah-nulth language as "eehmiis" areas – places that are very, very precious. Other than that, they approved of our MOU. You can't imagine the relief that I felt!

The formal signing ceremony for this pivotal Memorandum of Understanding was held in Tofino on June 16, 1999. It established the environmentalists' support for First Nations' rights and their control over local forestry and committed Iisaak both to eco-forestry in the already logged parts of Clayoquot and to designating the pristine "eehmiis" areas in Clayoquot for no logging.

The only major environmental group involved in the Clayoquot conflict that did not sign onto the agreement was the Friends of Clayoquot Sound (FOCS). But they had been consulted in detail every step of the way. FOCS's position had always been that there would be no more logging of old growth in Clayoquot. Period. Not even in the watersheds where logging had already occurred in past years. The Friends did not oppose the MOU (Adriane wouldn't have moved it forward if they had). They simply decided to position their organization to be the watchdog to make sure it worked as planned.

Joseph Campbell of Ahousaht, Chairman of Iisaak's Board, signing the Memorandum of Understanding with Environmental groups which commits Iisaak to maintaining "eehmiis" (undeveloped areas that are, in Nuu-chah-nulth, "very, very precious") for activities listed below that maintain their pristine nature and spiritual and sacred values and that generate non-timber benefits for the region. Photo: WCWC files.

Here is the full text of this historic MOU:

MEMORANDUM OF UNDERSTANDING
Between Iisaak Forest Resources Ltd. and Environmental Groups

Environmental groups from within Canada and outside have long recognized Clayoquot Sound as a place of exceptional natural beauty and bio-diversity. The big-treed ancient forests of Clayoquot have been the focus of a land use conflict that has lasted for decades and that has reached global proportions.

Iisaak Forest Resources Limited (Iisaak) is a forestry company formed through a partnership between Ma-Mook Natural Resources Limited, owned by the Central Region First Nations of the Nuu-chah-nulth Tribal Council, and MacMillan Bloedel Limited (MB). Majority control of Iisaak is in the hands of Ma-Mook Natural Resources Limited. Iisaak operates in the Clayoquot Sound portion of Tree Farm Licence #44 held by MacMillan Bloedel Ltd.

Through this agreement, the Central Region First Nations of the Nuu-chah-nulth Tribal Council, whose traditional territories include Clayoquot Sound, MacMillan Bloedel Limited and the undersigned Environmental Groups have come together to promote the resolution of the historic land use conflict in a way which respects First Nations' traditional ownership of their territories, enhances local sustainable economic development opportunities, provides stability for local communities by reconciling parties that have been involved in social conflict, and protects the natural beauty and bio-diversity of Clayoquot Sound.

On the condition that IISAAK, in consultation with the First Nations in whose traditional territory the proposed activities would occur, will:

1. Gain control of the forest tenure in Clayoquot Sound.
2. Operate within its tenure:
    a) according to the spirit, principles and recommendations of the Clayoquot Sound Scientific Panel;
    b) giving respect to the traditional values of First Nations, and
    c) in a holistic way so as to sustain:
        i) bio-diversity and all timber and non-timber resources;
        ii) water systems and water-related resources, including salmon;
        iii) eco-tourism, recreational and scientific research opportunities;
        iv) spiritual and sacred values
        v) traditional cultural uses; and,
        vi) economic development which is culturally, socially and ecologically sustainable.
3. Designate already fragmented areas for ecologically sustainable commercial forestry and other ecologically sustainable commercial and non-commercial uses, as determined and prioritized through ongoing research.
4. Prioritize production of ecologically sustainable volumes of high-value-added, second-growth forest products so as to minimize the production of forest products from old-growth areas, ensure that all old-growth forest characteristics be well-maintained within harvest areas on the stand as well as on the landscape level, and nurture the long-term transformation of second growth forests into old-growth forests.
5. Designate "eehmiis" (undeveloped areas that are, in Nuu-chah-nulth, "very, very precious") (see attached map) for activities listed below that maintain their pristine nature and spiritual and sacred values and that generate non-timber benefits for the region (see Section 13), as determined by ongoing research:
    a) ecologically sustainable harvest of non-timber forest products;
    b) eco-tourism;
    c) traditional cultural uses;
    d) scientific research; and,
    e) other activities agreed to by the signatories to this agreement.
6. Operate under an independent, internationally recognized certifying agreed to by both the Central Region First Nations and the undersigned Environmental Groups. The Undersigned Environmental Groups then agree to:
7. Endorse and actively promote Iisaak as a globally significant model of ecologically sustainable forestry.
8. Actively support the consolidation of existing Crown forest tenures in Clayoquot Sound into the control of Central Region First Nations of the Nuu-chah-nulth Tribal Council.
9. Actively assist Iisaak, within the capacity of each Environment Group, in Iisaak's institutional development, including research, financing and capacity-building endeavours.
10. Actively assist Iisaak, within the capacity of each Environment Group and as enabled by FSC-accredited certification, in the marketing of Iisaak's forestry and associated value-added forest products.
11. Actively assist Iisaak and the Central Region First Nations, within the capacity of each Environment Group, in the development and marketing of specified non-timber forest products.
12. Actively assist Iisaak and the Central Region First Nations, within the capacity of each Environment Group, in the development and marketing of specified eco-tourism products.

Together, IISAAK and the undersigned Environment Groups agree to:

13. Establish, participate in and fund a small and effective joint working group that will:
    a) identify options for and assist in the establishment of a non-timber tenure and/or similar mechanisms as an overlay to existing forest tenure that will enable Iisaak and the Nuu-chah-nulth Central Region First Nations to manage and derive benefits from non-timber and eco-tourism resources in Clayoquot Sound;
    b) draft a corporate Code of Ethics for Iisaak;
    c) identify and recommend specific opportunities for strengthening the relationship between Iisaak and the undersigned environment groups:
    d) identify and recommend ongoing mechanisms for sustaining cooperation between Iisaak and the undersigned environment groups; and,
    e) table recommendations on sections 13 to the signatories to this agreement by November 1999.

Singed in Tofino, B.C. on this 16th day of June, 1999
FOR: Iisaak Forest Resources Limited
Joseph Campbell, Chairman of the Board of Directors, Iisaak
Linda Coady, Secretary of the Board of Directors, Iisaak
FOR: Environmental Groups
Tzeporah Berman for Greenpeace International
Peter Tabuns for Greenpeace Canada
Elizabeth Barratt-Brown for Natural Resources Defense Council
Vicky Husband for Sierra Club of B.C.
Adriane Carr for Western Canada Wilderness Committee

B.C. government transfers MB's TFL lands in Clayoquot to the Nuu-chah-nulth controlled Iisaak

With the MOU between the environmental groups and the Nuu-chah-nulth's Iisaak Forest Resources signed, the next task was to get the TFL transferred from MB into Iisaak's hands. One of the biggest hurdles was to get the B.C. Forest Service to back down from its insistence that Iisaak log a large volume of wood annually. Normally B.C.'s Chief Forester assigns an AAC (Annual Allowable Cut) to every TFL based on the theoretical maximum volume of wood that the forest would grow annually to eventually replace it. But to do the kind of eco-forestry mandated by the Clayoquot Science Panel and stipulated in the MOU and, especially, to not log in any of the pristine "eehmiis" areas, the annual rate of cut had to be greatly reduced. Linda Coady and Adriane worked hard to get the government to agree to this. But the best they could achieve was in the wording of the letter of October 1, 1999 from Forest Minister David Zirnhelt to Iisaak and MB giving approval in principle to the subdivision of MB's TFL 44 and the transfer of MB's holding in Clayoquot Sound to Iisaak Forest Resources Limited.

In the letter Zirnhelt noted the *unique operating environment in Clayoquot Sound* and the condition that Iisaak *meet the spirit and intent of the Scientific Panel Recommendations for Clayoquot Sound*. But there was no written commitment to a reduced AAC, nor special consideration of the higher costs of operating this unique TFL. As I complete this book, this failure has come back to haunt the people who worked so hard to create a Clayoquot solution.

The joint environmental group press release applauding the new forestry tenure sent out when the tenure was transferred forewarned of this problem.

CLAYOQUOT TENURE TRANSFER APPLAUDED BY ENVIRONMENTAL GROUPS
For immediate release October 4, 1999

(Tofino, BC) Western Canada Wilderness Committee, the Sierra Club of BC, Greenpeace and the Natural Resources Defense Council applauded the government today on its transfer of the Clayoquot Sound portion of Tree Farm Licence 44 from MacMillan Bloedel to Iisaak Forest Resources. Iisaak is a joint-venture company, 51% owned by Clayoquot First Nations, and 49% owned by MacMillan Bloedel.

'This is a great step forwards for a sustainable future in Clayoquot Sound,' said Adriane Carr of the Western Canada Wilderness Committee. 'The government should be commended for recognizing the golden opportunity that Iisaak represents to develop a new model of conservation-based forest management in BC. MacMillan Bloedel should be commended, too, for initiating this transfer of its tenure to a First Nations' controlled company. Although there are still some details to be worked out, it's towards the kind of tenure reform that we've been wanting, and we're happy to thank the B.C. government, MB and First Nations for working hard to break through traditional forest industry thinking and establish a type of forestry we can be proud of,' stated Carr.

A cautionary note was also struck, however, as the government attached conditions to the transfer according to BC's volume-based forest policy and laws. Despite the government's adoption in 1995 of the Clayoquot Sound Scientific Panel recommendations that specified an end to volume-based forestry in Clayoquot (forest management that requires a pre-determined volume of wood be cut every year), the government is proposing to take back a certain volume of cut. The conservation groups believe these volume-driven conditions are not appropriate in Clayoquot, and are hopeful that the Ministry of Forests will work with Iisaak over the next two months to ensure that the solution building in Clayoquot is not compromised by volume-driven requirements.

'There are clearly some things to work out,' said Lisa Matthaus of the Sierra Club of BC. 'We are building a relationship of trust with Iisaak based on a value-based approach to forestry, and this trust does not extend to volume-based operations in Clayoquot. We're looking forward to discussing strategies with the Ministry of Forests to build a lasting trust for the region as a whole.'

Iisaak is expected to begin operations either later this year or early next year. A memorandum of understanding signed in June of this year between Iisaak and the four environmental groups commits the parties to cooperate on a conservation-based approach to forest management, including managing undeveloped old growth watersheds for eco-tourism and other non-timber values, and marketing timber and non-timber products. Iisaak is intended to be a model of sustainable forestry, which will require continued flexibility and a commitment to innovation from all parties.

Adriane Carr, Western Canada Wilderness Committee
Lisa Matthaus, Sierra Club of B.C.
Tamara Stark, Greenpeace Canada
Matt Price, Natural Resources Defense Council

PEACE WILL COME TO ALL HUMANKIND

Friendly grey whale in Clayoquot Sound allows whale lover to touch it. Photo: Rod Palm.

WHEN WE MAKE PEACE WITH THE WHALES AND HEAR THEIR SONG.

an ancient prophesy brought in 1998 to the West Coast by Geronimo's Great-granddaughter

Published by the Wilderness Committee, 227 Abbott Street, Vancouver BC V6B 2K7.   Copyright WCWC 1998.   All rights reserved.   Printed in Canada.
All proceeds raised by this poster will be used for the Western Canada Wilderness Committee's educational campaign to protect whales.

Chapter 60

No to Makah grey whaling
No to land trade for parks
No to foreign multinational takeover of MB

WCWC protests the Makah's grey whale hunt in front of the U.S. Embassy in Vancouver

Scientific evidence points to human hunting as one of the main causes of the post-glacial extinctions of big land animals all around the world: mammoths, moa birds and the giant beaver, to name just a few. But gigantic ocean mammals—the whales—survived in huge numbers into the early 20th century. Then human technology caught up with them and they were almost wiped out. The plummeting whale populations forced an end to commercial whaling in most countries, although Japan, Norway and Iceland continue to hunt whales despite international objections.

Since commercial whaling on the west coast of North America ended about a half-century ago, grey whales have been making the best comeback of all the decimated whale species along the Pacific coast. In fact, they have become objects of a lucrative new industry—commercial whale watching. In 1999 the greys were under a renewed threat. That spring the Makah First Nation in Washington State declared that it was going to resume its traditional whale hunt, an aboriginal right firmly entrenched in a treaty it signed long ago. Their declaration put many environmental groups, including ours, in a quandary. On the one hand, WCWC had always strongly supported aboriginal rights. On the other hand, there are good and moral reasons to stop hunting whales. Scientific studies reveal that these sea mammals are highly intelligent sentient beings. On humanitarian grounds alone many believe, like I do, that whales should not be hunted and killed.

We also feared that the resumption of aboriginal whale hunting on the west coast would set a precedent for other nations that traditionally hunted whales in generations past to begin commercial whaling again. So WCWC, along with some other more radical environmental groups, decided to actively oppose the Makah whale hunt.

A few years earlier Paul Watson, founder of the Sea Shepherd Society, explained to me why he thought it was all right to destroy property in his efforts to protect whales. He defended his action of sinking rogue whaling ships by saying that doing this was no different than destroying the ovens in Auschwitz. Such an act would be violent civil disobedience but, nonetheless, it would be justified because it would save people's lives. Whales, in Paul's eyes, were like people and had the same right to life. He was just destroying the hardware that brought about their demise.

The Makah proposed to hunt the grey whales as they passed by their reserve during this whale's annual spring migration from their winter calving grounds in Baja California northward. I was first alerted to their plans by my old friend from Clayoquot Sound, Steve Lawson, who strongly opposed it. He told me that Geronimo's great-granddaughter had recently come to the west coast and brought with

Facing page: Poster of grey whale in Clayoquot Sound that appeared to enjoy having people touch it. Poster image: Rod Palm.

her this ancient prophesy: *Peace will come to all humankind when we make peace with the whales and hear their song.* Immediately I thought about the friendly grey that had hung around Flores Island all summer long a few years earlier. It would swim right up to the whale watching boats and "spy-hop" (come out of the water vertically to look at the whale watchers). It also seemed to like being touched until one day, according to local legend, a dog jumped out of a whale watching boat onto the whale's back, barking and snarling. Freaked out, the whale dove and no one ever saw it again.

"Someone must have a picture of that whale's friendliness to people," I thought to myself. "Such an image, along with the native prophecy, would make a great poster." I immediately contacted Clayoquot-based photographer Adrian Dorst. He didn't have the image I was looking for but he knew someone who did, Rod Palm, a local wilderness guide. In exchange for some posters, Rod let WCWC use his image.

When I got the slide in the mail and held it up to the light, I was delighted. There was the whale, two people in a zodiac, and Lone Cone on Meares Island in the distant background. One person in a bright red floater jacket was leaning over the back of the boat with one hand touching the head of the spy-hopping whale. Neither person was close enough to identify so we didn't need their release to use it.

But then, upon looking at the photo more closely with an eye loupe, I realized there was a big flaw. A bluish exhaust plume hovered above the water behind the outboard. The boat's engine had been idling. It was against the law to pet a whale and the running engine made it even worse. It could pose a danger for the whale. Worse, the blue cloud of pollution broke the mood.

But, hey, I'm a practical guy. I reasoned that we were not making an award-winning poster, just a campaign poster to make the point that these whales were friendly, sentient beings. We decided to go ahead with the un-retouched photo, flaws and all. Nik printed up a couple of thousand at cost. We still have 500 left today.

We gave a lot of the posters away. We gave some to Anthony Marr, who went to the Makah Reserve in Washington State on his own behalf (not as a representative of WCWC) and handed them out. Not all of the Makah natives supported the whale hunt. In Vancouver we decided to set up a picket in front of the U.S. Embassy to pressure the U.S. government to get the hunt stopped. As a result, we got a meeting with the Consular General who explained to us that Makah's right to whale was clearly spelled out in the Makah Treaty of 1855. He asserted that the hunt had to go ahead because there was no way that this clause could be changed outside of passing a constitutional amendment.

The Makah were intent on exercising their right to whale. It took them many attempts over several weeks to succeed in their mission. Shortly after dawn on May 17, 1999 a seven-man Makah hunting party in a traditional cedar canoe stalked a grey whale about 250 metres offshore and harpooned it. The whale tried to swim away and ended up dragging the canoe behind it. Within minutes of the initial hit, Makah hunters in powerboats surrounded the whale and shot it several times with 50-calibre rifles. It was the first whale killed by the tribe in 70 years. They then towed in the whale and beached it in front of the Makah village. Several thousand Makah First Nations and members of other coastal tribes (some from B.C.) feasted for two days, stripping the meat and blubber from the carcass. One of the Makah band members explained their actions this way in a newspaper interview. *We view this as having cultural significance and is, in a way, part of religious freedom. People who don't understand us call us savages. I call them extreme missionary zealots.*

Several other WCWC staff and I protest the Makah gray whale hunt in front of the U.S. consulate office in Vancouver because our organization believed that killing these sentient beings is unnecessary and morally wrong. We also feared that this hunt could also set a precedent leading to the opening up of aboriginal commercial whaling elsewhere. Photo: WCWC files.

Anthony Marr was extremely upset by this, believing that whale hunting was a tradition that should die like slavery, which also was traditionally practiced long ago by many west coast First Nations. In an article in the *Victoria Times Colonists*, Marr railed on that this whale was sacrificed *in a vain and vainglorious attempt to revive an obsolete tradition*. Marr went on to say

> ... to many in the environmental movement, it is a day that will go down in infamy.... Should native cultures with whaling traditions have special rights to whale? In my opinion, no, just as I say no to the Chinese culture having special rights to use bear gall bladders, tiger bone and rhino horn in traditional medicine, nor European cultures having special rights to practice their bloody trophy hunting tradition...

After the whale-killing debacle in 1999, the U.S. Consul General's assertion to us that the Makah's treaty rights trump everything has not panned out. A series of court challenges have prevented the Makah from carrying out another whale hunt since then.

Fee simple land for parkland? B.C. citizens say no!

In March of 1999, in a move that some say was a desperate attempt to stem the tidal wave of public support shifting from the NDP to the corporate-friendly Liberals, the B.C. NDP government announced a tentative deal to turn over up to 30,000 hectares of public land to the logging giant MacMillan Bloedel (MB). This privatization of public lands was to compensate MB for the NDP government's designation of about 7,700 hectares of timber-licenced lands in TFLs 39 and 44 to create the Carmanah, Tsitika and other parks.

It was one of two options reached in an out-of-court settlement of MB's compensation claim for loss of logging rights arising from the Vancouver Island CORE land use allocation process. The other option was to pay MB $83.75 million in cash.

The government hired Victoria lawyer David Perry to conduct a series of public consultations on the settlement agreement. Was it to be land or cash? WCWC fervently opposed giving away publicly owned lands to settle this debt. It was the thin edge of the wedge opening the gate to privatization of all of B.C.'s publicly held forest, an idea constantly being pushed by the big forest companies.

Perry's consultation process was held during a three-week period in June 1999—the beginning of summer when people were thinking about enjoying the warm weather, not about participating in a political process. But the issue was a hot one and many groups in addition to WCWC were strongly opposed to this proposed privatization, including most First Nations whose unresolved land claimed lands would be negatively impacted.

Perry scheduled meetings in a number of affected communities as well as solicited written public input. As soon as we got the schedule of public meetings, we sent out an emergency action alert to all our B.C. members. We especially concentrated on getting our members out to the Vancouver and Victoria meetings or, if they were unable to attend, to comment by mail. Both the Victoria and Vancouver meetings were held in large hotel venues that held over 400 people each. Both meetings were filled to capacity. I attended both of them and recognized many of the speakers as long-term WCWC members. The public's verdict was clear. It was an overwhelming resounding "NO" to the giveaway of public lands!

Many who spoke at the meetings not only did not want to give land to MB, they did not want to give MB money either! They felt that MB did not deserve anything at all for the park "claw back." Some went even farther suggesting that MB owed us—the citizens, communities and First Nations of B.C.—compensation for environmental damage caused by their bad logging practices in past years. At the grassroots level, resisting the privatization of our publicly owned forests and promoting community control of these forests and respect for the rights and title of First Nations ran deep.

In total, about 1,400 people attended the consultation meetings. Perry received over 1,000 written submissions, almost all of them opposed to privatization of TFL lands. It wasn't only environmentalists and First Nations that spoke out. IWA locals, Council of Canadians chapters, and various local governments were opposed to the compensation agreement, too, especially to providing land in lieu of cash. I don't know how many of the written submissions came from WCWC members, but I'd bet it was well over half. WCWC had, and continues to have, a very active membership that participates when needed to protect public lands, defend the environment and establish new parks.

The government released Mr. Perry's report, *MacMillan Bloedel Parks Settlement Agreement Decision*, a few months later. Perry recommended that none of the proposed land transfers take place. In September 1999 the B.C. government accepted Perry's recommendations and announced that it would compensate MB with cash not with land.

WCWC opposes U.S. forestry giant Weyerhaeuser's bid to takeover MB

In June of 1999, before the government had even announced its compensation decision, MB informed B.C. that it was moving towards merging with the U.S. forestry giant Weyerhaeuser Inc. Headquartered in the town of Federal Way, Washington State, Weyerhaeuser was the world's largest softwood lumber company. Although the deal was announced as a "merger," it really was a takeover by Weyerhaeuser. In order for the $3.6 billion deal to go through, B.C. Forests Minister David Zirnhelt had to approve the transfers of MB's TFLs and other Crown forest tenures to Weyerhaeuser. The government again contracted lawyer David Perry to conduct a public input process. We again roused our members to participate and oppose this foreign corporate takeover and increased concentration of ownership in our forest industry.

It is ironic that the founder of MB, H.R. MacMillan, had said to the 1955 Sloan Commission on Forestry *It will be a sorry day for British Columbia when the forest industry here consists chiefly of a very few big companies holding most of the good timber– or pretty near all of it.* A little over 20 years later, in 1976, after a few large corporations had bought up most of the smaller tenures, the Pearse's Royal Commission on Forestry came to a similar conclusion regarding the increasing concentration of corporate control of B.C.'s publicly owned forest and advocated greater diversification of tenures. The B.C. Social Credit government ignored this commission's advice too, although in 1981 the Socreds did scuttle the purchase of MB by Noranda, supposedly to protect our B.C.-grown company from takeover by this eastern Canadian corporate giant.

Now the B.C. NDP, despite an even more recent call by the Forest Resources Commission in 1991 to turn 50 percent of the forest tenures in B.C. over to smaller-scaled local forest licences, was looking at allowing a merger that would put B.C.'s most successful forest company in foreign hands.

The public understood why monopolistic foreign control of B.C.'s forests was bad for B.C. citizens, forestry workers and the forests themselves. There was as much public opposition, if not more, to Weyerhaeuser's takeover of MB as there was to the move to privatizing public forest lands and giving them to MB in compensation for new parks. Jim Fulton, executive director of the David Suzuki Foundation, explained why people opposed the deal this way:

> *The last thing we need in this province is more corporate concentration in the forest industry. What we need is tenure redistribution... that will give us more options, diversity and innovation in our forests. This means investing in the value-added sector and moving away from the volume-based focus of large companies like Weyerhaeuser.... MacBlo is a B.C. company and it has taken years of tremendous pressure to get its executives to address environmental concerns. It's hard to imagine that a U.S. company will respond to that kind of pressure when only 5,900 of their 35,000 employees are based in Canada.*

Some thought that the takeover was a set-up deal

A year before the Weyerhaeuser takeover of MB hearings started, MB began to position itself as a "good and progressive" logging company. On June 10, 1998, its American imported CEO, Tom Stephens, announced with a great deal of fanfare that MB was phasing out clearcut logging and replacing it with "variable retention" logging. The phase out of clearcutting was to be completed within five years. Some of the big environmental groups, including Greenpeace, heaped compliments on the company saying it was the "beginning of a new era." WCWC was a little more circumspect, lauding the move as a "good first step." Privately, we were highly skeptical. Then in August 1998, MB quit the B.C. Forest Alliance, the pro-industry advocacy group fronted by Patrick Moore, the eco-movement traitor. The environmental community took this as another good sign that MB was going to be doing things differently.

After the announcement that the "merger" between MB and Weyerhaeuser was underway, and before the public hearings commenced, Adriane and I attended a "new forestry practices evaluation session" sponsored by MB on Vancouver Island. The point of it was to take representatives of B.C.'s major environmental groups and eco-forestry experts (including Herb Hammond) on a tour of some of MB's recent logging sites to see what the new variable retention logging looked like and get our approval. MB still cringed in memory of Greenpeace's international boycott campaigns that had resulted in customers canceling contracts because of its clearcutting of Clayoquot Sound. The company badly wanted to become the new forestry "good boy."

We quickly discovered that "variable retention logging" was mostly just verbally disguised clearcutting. Certainly the areas with minimal retention of trees (10 percent or less) wouldn't preserve the ecological integrity of the naturally self-regenerating old-growth coastal rainforest. Only the cutblocks with the highest retention, about 30 percent of

"Variable Retention" logging in the Walbran Valley. Photo: WCWC files 2004.

the original stand, looked like they might be more ecologically sound. Thirty percent retention, however, was the rare exception not the rule in MB's new "non-clearcut" logging operations. The vast majority of MB sites retained 10 percent or less of the forest. Moreover, the trees retained appeared to be mostly the scrubby un-merchantable ones that the company was only too happy to leave standing. To cut them down and haul them away would have cost the company more money than it would have made back from milling and chipping them.

I quickly concluded that MB's new "variable retention" logging was plain old-fashioned "high grading," a forest practice that had been condemned and abandoned more than a generation earlier. It wasn't the real change to sustainable eco-forestry that B.C. needed. But even I had to agree that the company managers who toured us around seemed genuinely committed to doing things differently.

I was far more skeptical as to the intentions of those at the very top, like Tom Stephens. In hindsight, I believe the quick image makeover pushed forward by Stephens and the other structural changes he made, like spinning off some of the company's pulp mills into a new company, made MB more attractive for a takeover.

The takeover worried us. We were especially worried about whether or not Weyerhaeuser would follow through on MB's commitments to the First Nations in Clayoquot Sound. Would the new company transfer its tenures to Iisaak and participate in the new rules for logging in the Sound established by the Science Panel? Weyerhaeuser did not say yes it would right away. After a month of hesitating, it finally announced that it would honour these commitments. Karen Mahon, Greenpeace's forest campaigner, succinctly expressed our apprehensions regarding this belated announcement when she commented: *Is this real and substantive...or are they just trying to smooth things over so they can get the deal?* This was a very different tune than her comment the year before: *It's the beginning of a new era!* when MB announced it was phasing out clearcutting and adopting variable retention logging.

This time it really was the beginning of a new era... of big corporate takeovers in B.C.'s forest industry.

Perry's public input process concerning the takeover turned out to be a sham. The vast majority of participants spoke eloquently against the NDP government allowing the transfers of the logging tenures. Only a few spoke in favour and most of those were lower level MB employees obviously trying to curry favour with their soon to be new corporate "boss."

Our input made no difference. The die was cast.

At the meetings we learned that MB was a perfectly viable company at the size that it was. It wasn't in any financial trouble. All of us at WCWC were convinced that if the MB management staff on stage could have spoken freely, they would have said that it was a hostile takeover and not in the best interest of MB's current employees or that of all British Columbians to have their company foreign owned.

During the hearings a number of citizens and First Nations raised concerns about the compensation that would be due under NAFTA to Weyerhaeuser as a foreign investor in Canada if the B.C. government did anything that affected the profitability of its operations. West Coast Environmental Law Society presented a legal opinion stating that any changes to B.C. forest or land use policy in the future that result in a reduction in this U.S. company's logging rights or in other ways affected the company profitability of operations on public or private lands (like establishing tougher eco-forestry rules or taking away tenured lands to provide land claim settlements) could be considered "tantamount to expropriation." Thus under NAFTA's Article 1110, Weyerhaeuser could claim, and the B.C. government would have to pay, multi-million-dollar compensation.

The MB representatives sitting on the stage up front, including Linda Coady who had worked closely with Adriane and the others intimately involved in resolving the Clayoquot conflict, were noticeably uncomfortable. We were tipped off beforehand as to why. They had to dodge questions from the public. They could not speak out honestly about the negative aspects of the deal because of a clever tongue-tying clause that Weyerhaeuser had put into its takeover bid. In short, if anyone in MB's management spoke out against the

Cartoon used to characterize Premier Glen Clark anti-environmental forest policies. It was a very popular T-shirt image. Artwork: Geoff Olson.

deal and this led to the takeover collapsing, MB would have to pay Weyerhaeuser $80 million dollars in compensation!

I felt that this was an outrageous bullying tactic that, if not illegal, was morally wrong. I asked for and got a special private meeting with Perry to brief him on this penalty clause and how it prevented the MB representatives from truthfully answering questions from the public at the meetings he was holding. Perry told me, in so many words, that he really didn't have much power over what was going to happen.

Basically, the Clark government was playing the trade-off game. We (the environmentalists) had won the compensation round. Now it was the industry's turn to win. I don't think Perry even wrote up a report. It was already a done deal before public input started. Premier Glen Clark had decided that giving provincial approval would demonstrate to the world that B.C. was "open for business" and welcomed foreign investment. Liberal opposition Leader Gordon Campbell did not oppose the takeover either. The only opposition came from the people of B.C. and it was loud and clear, and totally dismissed.

In early October of 1999, with the B.C. government's approval in principle the transfer of the tenures, the last hurdle left to clear was MB shareholder's formal approval—a mere formality. MB held its last shareholder's meeting in late November in a hotel near Canada Place, only a few blocks away from our Gastown office. A group of us from WCWC went down to protest on the sidewalk outside the meeting with picket signs asking the shareholders to turn down the takeover bid. We knew it was a lost cause, but we had to do it! There were tears in some of the shareholders eyes. Someone pointed out to me several relatives of H.R. MacMillan himself. For B.C., this was the death of a B.C. business icon. The "sorry day" that the founder of MB decried in 1955 had finally arrived!

Chapter 61

Stoltmann wilderness campaign peaks in conflict

An MP supports the proposed Stoltmann Park

Every time either Adriane or I went to Ottawa to educate our parliamentarians about issues we always visited MP Charles Caccia's office on Parliament Hill. After Jim Fulton, an NDP MP for 14 years, retired in 1993, Caccia's office replaced Fulton's as our temporary "headquarters" on the Hill. In early 1999 we asked Caccia if he was interested in forwarding a private member's bill to create a Stoltmann National Park Reserve. He was.

In June 1999 Caccia forwarded a request to the House of Commons Legal Council to draft a bill that would establish a big 500,000-hectare park that extended westward from the boundary of the resort village of Whistler and included Meager Creek Hot Springs—the same boundaries we had established for our proposed Stoltmann National Park Reserve. On the same day that Caccia announced he was preparing a private member's bill, Interfor stated falling giant 1,000-year-old Douglas firs along a roadway towards Lava Creek in the Upper Elaho part of the Stoltmann. The company loggers were headed towards the popular Douglas Fir Loop Trail that John Clarke had surveyed the previous fall through the finest grove of ancient firs left in the valley.

Caccia had never visited the Stoltmann Wilderness. When opposition MPs and the press made a big thing of it, he told *Vancouver Sun* reporter Larry Pynn, *I have a notion of what it's all about. I know we are running out of sites of this kind. I don't think one should wait for a physical impression. I can put two and two together.* He said he would be visiting the area on his next trip to B.C., which would most likely be in September after his bill was drafted.

Reform MP John Reynolds, whose electoral district encompassed most of the Stoltmann Wilderness area, totally panned the idea. He told a reporter for the Squamish paper, *The Chief*, that it was just a publicity stunt. *Basically it's a waste of taxpayers' money even drafting a private member's bill.... I can guarantee it would never get through the House of Commons,* declared Reynolds, who went on to assert that not only were his Reform party colleagues going to vote against it, but so would Caccia's own Liberal party.

The Squamish City Council also opposed to the creation of the park and wrote a letter to Prime Minister Jean Chrétien urging him not to support the bill. On the local political scene, only Whistler councilor Ken Melamed expressed support.

Despite the substantial opposition that Caccia's bill announcement generated before it had even been drafted, it had already accomplished a prime purpose...fuel the debate and give us faint hope. When drafted, I was surprised that the bill was so short—it simply stated that the National Park Act be amended to add the Stoltmann Wilderness area to the existing list of park reserves. However, it was not a sure thing that Caccia's bill would be introduced in the House of Commons. Lady luck had to approve of it first. Because of time limitations, only the first 30 private members' bills selected through a random draw lottery process out of all the proposed bills (50 that round) would get heard in the House. In this particular lottery we were too lucky. We got picked in the ninth draw. This meant that we would be getting our second reading in just a couple of weeks. This early date did not give us time to get our members to send in

Tagged giant Douglas fir that volunteers found and measured as part of the Wilderness Committee's big tree survey in the Stoltmann wilderness. Photo: WCWC files.

letters of support and draw the maximum media attention to it that we could have if it had come up later. Number 30 would have been perfect, but not being drawn at all would have been a lot worse. We couldn't complain.

The discovery of a 1300 year old Douglas fir fans the fire of conflict in the Upper Elaho

The summer of '99 was a time of intense activity on both sides of the Stoltmann Wilderness conflict. In July, Interfor's road builders reached the Lava Creek canyon and began constructing a bridge across it. This put Lava Grove, a magnificent stand of ancient Douglas fir trees, at immediate risk. The area between Lava and Cessna Creeks was truly the heart of the remaining wilderness. It contained the last big concentration of ancient Douglas firs in the entire Stoltmann Wilderness area.

Kitchen tent at our millennium tree research camp. Photo: WCWC files.

To draw public and political attention to the heritage trees at stake, we launched a Millennium Tree Research project. Our goal was to measure and catalogue every big old Douglas fir in the grove and beyond. We hired James Jamieson to run the camp. Chris Player joined him in the field to provide GPS support. The previous year John Clarke had surveyed a Doug Fir Loop Trail, which branched off our trail to Meager Creek and meandered through about 200 hectares of incredibly beautiful ancient rainforest. Volunteers cleared the trail and it instantly became a popular day hike. Interfor subsequently got approval for a cutblock that would wipe out about a third of it.

All along Interfor had targeted the best and biggest trees, driven as much by increasing their bottom line profits as thwarting our

Stump of a giant cedar in Cut Block 72-20 in Sims Creek. Photo: Joe Foy.

904-year-old Douglas fir cut down in the Upper Elaho Valley. Photo: WCWC files.

dream of saving the area. For the last four years we had watched one spectacular grove after another destroyed by this aggressive company. Every time we found a grove of remarkable trees, named it, and worked as hard as we could to bring it to public attention, Interfor would build a road to it and cut the big trees down. In 1997 Interfor clearcut Magic Grove in the Sims Creek Valley leaving a grove of huge red cedar stumps behind. Tree ring counts revealed that many of the giants were over 800 years old. In 1998 Interfor completed its construction of a bridge across to the west side of the Upper Elaho Valley and cut down most of the huge old red cedar trees in a grove we had discovered and named Grizzly Grove. This time Interfor left a few big cedars as a tiny island of old growth surrounded by clearcuts. In the press Interfor claimed that it "had preserved the grove." Just before the big snows closed things down in November, 1998 Interfor cut down the thousand-year-old Douglas fir trees in another special stand we named Serenity Grove, just south of Lava Creek.

If Interfor proceeded with its 1999 logging season unchecked, it would destroy much of the biological heartland of our proposed National Park. I knew that at the rate Interfor was going, there would be little left to save in two more years.

In July, Andy Miller cored one of the biggest, oldest looking Douglas firs along the Doug Fir Loop Trail and came up with an estimated age of at least 1300 years. The 24 inch boring tool was not long enough to reach the centre of the tree but he found over 1100 rings in the core that he extracted and the un-cored distance to the centre was so long that even a conservative estimate put it at another 200 years worth of growth rings.

We announced in a media release that we had found the oldest grove of Douglas firs in B.C. and called for a moratorium on logging the area. No one questioned Andy's finding. What the Forest Service did question, however, was the uniqueness of this grove of higher elevation ancient Douglas firs. Interfor and Forest Service foresters all asserted that it was a nice stand, but that there were others like it elsewhere, implying that it didn't matter if this one was cut down. To prove that they were right, the Forest Service commissioned a comparative study of our "millennium grove" with other Douglas fir sites in the region. It took a long time for that study to be completed and a lot occurred between this announcement and the release of the results. Meanwhile, logging plans proceeded as usual and confrontations ensued.

A forest fire gets out of control and almost destroys the Upper Elaho Valley

In the early afternoon of August 5th, a fire started in a clearcut in the Upper Elaho just south of Lava Creek. An Interfor crew was in the area yarding out the dry logs (the block had been logged the year before), despite the fact that the fire hazard was extreme. Sheldon Neufeld, a photographer who contributes images to WCWC, was hiking high on the mountain above the valley at the time. He saw the smoke billowing up at 2 p.m. In his journal he wrote:

> For the next 10 hours we watched as flames first headed north towards Lava Creek and Canada's oldest known fir trees... Only massive water and fires retardant bombs dropped from an entourage of helicopters and planes prevented the fire from being out of control beyond the edges of the clearcut.

James Jamieson happened to be monitoring the radiophone at the time listening to the chat of the Interfor loggers. From the flurry of messages going back and forth he got the impression that Interfor's loggers tried to put the fire out by themselves and called for help only after they had failed, and the fire was raging totally out of control. This lost precious time. From then on, it was touch and go for hours

Giant old-growth western hemlock log — Upper Elaho Valley. Photo: WCWC files.

as to whether or not the Forest Service's Mars water bombers could control the fire. It was very close to being a major disaster.

Bar talk in Squamish pinned the blame on environmentalist protesters. The facts, however, cast suspicion on Interfor loggers who were working at the time in the clearcut where the fire started. They were there during this time of fire closure when logging activities elsewhere were curtailed doing what a company representative called "low risk yarding." It is a good bet that a spark from a choker chain scraping across a rock as it dragged in one of the big logs ignited the tinder dry wood waste.

About 40 acres of felled and standing timber burnt to a crisp. The company estimated that it "lost" about 5,000 cubic metres of wood. At $80 per metre that amounted to $400,000 worth of timber incinerated. This figure did not include the cost of fighting the fire, which had to be over $50,000. The B.C. Forest Service investigated to try to determine the cause of the fire, questioning all of our witnesses. Joe called the Forest Service investigators literally dozens of times to find out the status of this investigation. He always got the same answer. "Soon" they would be coming out with a report. If Interfor were found guilty of starting the fire, it would have to pay the fire fighting costs and a fine. As far as I know there never was a report. Interfor somehow wiggled off the hook, a sure testimonial to the "pull" that this company had inside the B.C. government.

Out of control forest fire that originated in an Interfor clearcut as viewed from a mountaintop in the Upper Elaho Valley. Photo: Sheldon Neufeld – August 5, 1999.

Whistler expressed interest in protecting the Stoltmann Wilderness in a park

On the political front that August, undoubtedly responding to Caccia's Private Members Bill, the Resort Municipality of Whistler unanimously passed a resolution to commission a study of the national park idea. Their study was to be completed by mid-September. Our hopes were dampened by the opposition that came forward. Besides the Squamish and Pemberton municipal councils, both of which we'd expected to pass resolutions opposing the national park, Chief Bill Williams stated to the press that the Squamish Nation opposed removing land from the negotiation table to create a park while treaty talks continued. Chief Williams also stated, however, that members of the Squamish Nation opposed Interfor's logging in the Upper Elaho Valley.

Caccia wasn't getting any support from the Parks Canada bureaucracy either. MP John Reynolds released a letter he had received from the Ministry of Canadian Heritage, which now included federal parks in its portfolio. It said that Parks Canada had no interest in creating the Stoltmann Park because the region (No. 1 – Pacific Coast Mountains) was already more than fully represented by the South Moresby and Pacific Rim National Park Reserves (Parks Canada's goal is to have one park in each of its 39 regions). It didn't matter that these two areas represented completely different ecosystems or that the Coast Mountains with Canada's biggest and tallest mountains, Parks Canada's namesake for the region, had no federal park.

The first violent loggers' attack

PATH (People's Action for Threatened Habitat) was a small group of fewer than 50 activists who came together to protect the Stoltmann Wilderness from the chainsaws. The core group, dedicated to trying to stop the logging through personally engaging in acts of non-violent civil disobedience, was a lot smaller. They thought it was the only way to bring the issue to a head and give it the added media attention they believed would end the painful grove-by-grove loss of this wild rainforest. They set up a series of "tree-sits" beginning in July 1999.

Interfor continually claimed that WCWC was orchestrating the civil disobedience, which we weren't. Although we'd employed PATH's spokesperson, Barney Kern, four years earlier to work on the Stop Bear Hunting petition, Barney and the rest of his group were acting entirely on their own. Interfor kept repeating this myth, stating in a local paper: *WCWC says it doesn't condone illegal protests,*

Aerial photo showing the extent of the damage (estimated at over a half a million dollars worth of logs and tens of thousands of dollar in fighting the fire) caused by Interfor loggers working during "fire closure" due to high forest fire risk. Photo: Jeremy Williams.

but all they do is use wingnut groups like PATH to do their dirty work for them. Joe Foy responded with complete honesty in the same article: *It's important we have a good relationship with both the protesters and the RCMP. That's the sort of tightrope we walk.*

In truth PATH was quite anarchistic. There wasn't any mastermind behind the scenes running the group.

In late July of 1999 Interfor went to the Supreme Court of B.C. seeking an injunction to stop PATH from interfering with its road building and logging activities in the Upper Elaho Valley. A Sierra Legal Defence Fund lawyer represented WCWC's interests as an intervenor to make sure that the injunction would not prevent us from carrying on our lawful research and trailbuilding activities.

The small PATH protest camp constantly complained that Interfor loggers were using intimidation tactics. At the same time Interfor counter-claimed that the protesters were vandalizing their equipment. I personally did not think the protesters were doing the damage, but that didn't matter. The "war of words" was part of the "war in the woods" which was being fought as much in the news media as on the ground.

All summer long every newspaper in Whistler and Squamish featured articles on the various incidents. It amounted to a stack of clippings three centimetres high. For example, in an article in early August in the *Sea to Sky Voice*, Dave Miller, operations manager for Interfor was quoted as saying:

> We are going to leave them (the protesters) alone. The worst thing in the world we could do is go in there and lay a whipping on these guys. We'll leave the loonies alone in the trees and the bears and the blackflies will chase them out of there.

Barney Kern disputed Miller's claim. PATH had proof that Interfor was not "leaving them alone." He recounted an August 11th incident when several loggers wearing balaclava face masks arrived in Interfor trucks, got out their axes and shovels and trashed PATH's camp. PATH captured part of the assault on video. At first Interfor denied that company loggers were involved. Confronted with the visual evidence, Interfor had to admit that they were. The company later claimed it had "disciplined" the four loggers who were involved in that incident. This video footage was later used in Daniel Gautreau's 25-minute video *Hoods in the Woods* that documented the escalating conflict.

On August 12th the Supreme Court granted Interfor its injunction. But the PATH protesters did not heed the injunction. Individuals continued to climb up into trees near the leading edge of the road building. Each tree-sit created work stoppages and prevented road blasting until RCMP officers were able to remove the tree-sitter.

Interfor gets a "bubble zone" injunction around its Upper Elaho logging sites

PATH protesters had just established their 4th tree-sit on September 1st when Interfor went back to the Supreme Court and sought a large 500-metre "bubble zone" injunction around its active logging and road-building sites in the Upper Elaho. Karen Wristen, our SLDF lawyer, argued against it, contending that such an injunction zone would prevent WCWC from having access to the forest we were studying. The Justice granted the injunction, but he made an arrangement for us to issue identification cards and a letter of permission to our law-abiding volunteers that they could carry with them. This would make it legal for them to go through the area under injunction when there were no active road building or logging activities going on.

It was a sort of a win for WCWC. It allowed us to host an already planned tour of European parliamentarians and MP Charles Caccia through the area.

Joe Foy is the dignitaries' tour guide

On Saturday, September 11, the Honourable Charles Caccia toured the Upper Elaho with seven European Parliamentarians and Joe Foy as their guide. We rented a helicopter so they could see the big picture, too. They were shocked at the clearcuts and impressed by the massive ancient cedars and Douglas firs, the likes of which Europe never had. Regarding the opposition to the idea that the area should become a park, Caccia said to the reporter for the *Sea to Sky Voice* covering the tour:

> The world is in big trouble if we depend on the likes of John Reynolds and Jack Munro. Reynolds does not see past the next election. Munro is a real calamity to our forests. [Caccia was a forester by trade and a forestry graduate of the University of Vienna]. The older generations have always looked at natural resources as something to be exploited.

Honourable Charles Caccia September 12, 1999 on his visit to the Stoltmann Wilderness. Caccia is the only MP to visit this area. Photo: Jeremy Williams.

> After the helicopter tour Cassia criticized the logging this way. You can't trust those multinational companies. It's all cut and run. Look what they have done!

By September 12th, people had been arrested for breaching the court orders not to impede Interfor's lawful construction of its logging road. Four days after the dignitaries' tour, things turned really ugly. On Wednesday, September 15th, Interfor was back in court to get its injunction bubble zone extended even further. We, of course, had our SLDF legal representation with intervenor status in the court to protect our rights of lawful access and use of the area. PATH had one protester up in a tree.

Our man James calls Joe for help

Early in the morning on that same September 15th, James Jamieson, our millennium Tree Research Camp Coordinator, left his four WCWC volunteers surveying for large ancient trees near the Cessna Creek area, and made the one-hour hike to PATH's camp. Their camp was located along the logging road just south of the Lava Creek Bridge, outside the then court-ordered no-go 500 metre injunction zone. He was headed into town to get fresh supplies for the camp and told them that he would return that evening.

James' normal practice was to monitor our VHS radio, which was set to the frequency used by the Interfor workers. When he got to his van he turned it on and overheard Interfor loggers arranging "truck pooling" rides and plans to bring buckets of cold water to douse the environmentalists in the PATH camp. Shortly before noon, Interfor foreman Derek Sayles and an Interfor contractor arrived and parked inside the "no-go" zone. They called out for James to leave the PATH camp and meet them at a halfway point. At this meeting point Sayles told James that there soon would be 50 loggers on the north side of the Lava Creek Bridge (inside the injunction zone) and that he, Sayles, couldn't control them. He told James that people in the PATH camp had 30 minutes to leave and that James would be "beat up" if he remained.

James went back and told the PATH people what Sayles had said. James decided not to leave. With angry loggers rolling in, he reasoned that he should stay with the group. It was obvious that the company was involved in the planning of the attack, for James saw about 10 more pick-up trucks arrive, some with Interfor's blue and white colours and logos, and foreman Sayles smiled and waved at the 30 or so men in them as they got out. Interfor's operations manager,

Dave Miller, was also there.

James set-up our satellite phone in his van and phoned Joe Foy. Getting no answer, he left a message about his conversation with Sayles. Joe's cell phone is nearly always on, but at this time it was turned off because he was in the Supreme Court of B.C. with our SLDF lawyer, Angela McCue, fighting Interfor's application for an extended injunction zone that would cover the very area James was calling from. Interfor's lawyer was arguing in court that a larger zone was needed in order to separate the loggers from the environmentalists to prevent violence!

James called back about 20 minutes later and this time reached Joe who had just turned on his cell phone because the court was on lunch recess. He told Joe and Angela about the brewing mob. He estimated that there were at least 70 or 80 men gathered who were about to attack the camp. During this conversation, James saw through the rear window of his van two men speaking with Sayles and then walking towards him. They were followed by 20 or 30 men who spread out into the forest and tore down PATH camp's banners while yelling threats of physical harm to the environmentalists. James asked Joe to phone the police and then there was only static. The mob of angry loggers had disabled James' satellite phone. That was the last communication we had with James for many hours.

Joe immediately phoned the Squamish RCMP. He was informed that police officers had already been dispatched to the Upper Elaho and would be arriving there in an hour and a half—at about 2 p.m.

After phoning the RCMP in Squamish, Joe called me (I was in our Gastown office) and relayed the message that the loggers were in the process of smashing the PATH camp. I immediately phoned Lyn Krailing, an RCMP officer who had been visiting our office on a regular basis since the Elaho confrontations began and whose job it was to cool down potential conflicts. He was in his car headed who knows where. I told him that we had a real emergency and that he should call and get the Squamish RCMP out to the Elaho fast. He said he would call. I waited with apprehension all day long not knowing what had happened.

In the late afternoon, James finally called and told Joe what had happened. When the loggers struck there were seven protesters in PATH's camp (outside the current injunction zone) plus one more PATH protester tree-sitting high up in one of the Douglas firs within the injunction zone. There were also four WCWC volunteers in our research camp an hour's hike into the Elaho Valley.

After his satellite phone conversation with Joe was abruptly terminated, the loggers systematically stole and destroyed all the communication and photography equipment they could see in the camp, including James' VHS radio. They assaulted all the environmentalists. They dragged James out of his van and kicked him around.

As a result of the assault, two protesters and James sustained minor injuries. They went to the hospital for treatment and were released. James sustained a sprained hand when the loggers pulled him from his van and forcibly spread his fingers apart. He also had bruises from the kicks he was given when he was down on the ground.

Excerpts from James' account of the incident as told to *The Radical* newspaper

"It was terrifying," James says in an interview published in the October 1999 issue of *The Radical*, an independent newspaper published out of Quesnel, B.C. in its lead article titled ATTACK OF THE REDNECKS!!! In its extensive interview *The Radical* asked James whether or not the attackers were talking with them while they attacked them. James replied:

> Yah, they were saying things like 'we're going to kill you, there's no evidence' We were saying to them that they'd never get away with it and they said, 'don't worry the police are never going to get here on time.' It just went on and on…Towards the end, this one guy… forced me into a ditch. He had me down and he picked up a big rock weighing maybe 10 to 15 pounds, about 12 inches across. He held it up and was going to smash in my head. It took me a moment to realize, 'Hey, he really means it.' Luckily, a couple of his friends talked him out of it.

Finally all the environmentalists in the PATH camp (except for the one guy up the tree) managed to crawl into James' van.

> …we split about 2:30. We just had to go…. There was just nothing left to do. We finally got everyone back into the vehicle but they started to rock the vehicle and threatening to roll it into the ditch.
>
> We ran into the RCMP about fifteen minutes after that [driving towards the confrontation site]… and explained to them what had just happened to us and they said 'Okay, okay, okay.' We said we will go back there with you and will all sit down and get this story all down. And they said 'Oh no. No. You guys go into Squamish and file your report there. And we want to go up there and do our investigation and check out the situation.' So we headed off towards Squamish.
>
> Then, within about ten minutes of leaving the conversation with the police officers, we were starting to get run off the road by a logger's truck. It went by, passing me on the left within an inch or so at super high speed, skidded across the road in front of me and came to a stop. I had to slam on the brakes to avoid hitting them. Then two guys got out of the truck and picked up rocks and started smashing the van. So I had to throw it into reverse and back up around the corner.
>
> And so these were the exact same guys that had assaulted me earlier in the day and the RCMP just let them go. Not only did they not get names or licence plates or question the people, but they just let them go to get me again!

Joe drives to the site of the crime arriving at dawn on the morning after the attack

Joe left at 3 a.m. the next morning to go up and rescue the WCWC volunteers that James had left in our research camp the day before. He wanted to drive the logging road and get there before the loggers did. He got to the site of the confrontation by the Lava Creek Bridge at around 6 a.m. and looked around. Everything was cleaned up. The only trace of the PATH camp was a big pile of ash and charred rubble in the remains of the fire beside the road that the loggers used to burn up all of PATH's equipment. Joe took a photo.

He then hiked into our camp and found everyone there safe and sound. Some wanted to stay, but he made everyone pack up and pull out. We couldn't leave them there without a vehicle and with no staff person in charge.

An Elphinstone campaigner's chance trip to the Upper Elaho; photos tell the truth about the attack

On the weekend after the attack, Rick O'Neill, a retired photographer living on the Sunshine Coast, came to my house to bring me some images for possible WCWC cards. Just as he was leaving, I apologized to him for not seeming too interested, explaining to him that I was pre-occupied with the violence up in the Elaho where our trail crew boss was assaulted a couple of days earlier. He said, "Ya, I know. I was there."

After about 20-30 trucks passed, my anger turned to fear for the lives of my friends. I knelt at the side of the road with my hands together... in prayer. Excerpted from John Vessey's statement to RCMP regarding the attack on the Elaho protesters' camp by Interfor's loggers on September 15, 1999. Note loggers making their way towards protesters' camp from down the road. Photo: Rick O'Neil.

"What?"

Rick went on to explain that he had decided to get away to recuperate after being on a blockade on Mount Elphinstone for a month straight. He had decided to take a relaxing hike along WCWC's trail in the Elaho. He saw a lot of trucks heading up the Elaho road and a lot of parked logging trucks, some loaded and some unloaded, on the side of the road on the way up there that morning, which he thought was sort of unusual, but then thought nothing more about it. By sheer chance or serendipity, he arrived at the protesters camp just moments before the loggers attacked. He pulled out his camera and photographed what happened. I was dumbfounded and asked him for the photos. He gave us the originals and we made copies and also scanned them and made big blow ups of his 35 mm slides with our mapping inkjet printer, the same one we used to make big posters of clearcuts for our rallies.

Here is Rick's personal recount of what he saw happen on September 15, 1999:

It was quite a traumatic experience for me. I had never felt that threatened before or seen that kind of mob violence. I got there around noon about 10 minutes before it started. I stopped at the camp where the protesters were and asked them how to get to the Wilderness Committee research camp. They were kind of jittery and said, 'Well it's not a very good day to be here. We think they are going to attack our camp.'

I said, 'Oh, really?'

I drove down the road to turn around and saw all these loggers gathered—at least 50 or 60 of them. They tore down the protest banner strung up high across the road. I drove a little farther on, turned around and drove back out and parked beside the road on the far side of the protest camp headed back towards Squamish. I walked back with my camera and by this time the whole gang of loggers had finished tearing down the banner.

I tried to stay out of the way and be as inconspicuous as possible and still get some pictures. I think the only reason that I got away with it was because I wasn't supposed to be there and nobody knew who I was. The young people in the camp thought I was one of the loggers and the loggers didn't know who the hell I was.

They [the loggers] knew who were the ones who had cameras and videos and went for them right away and smashed them. So I was able to stand there with this mêlée going on around me and stay there quite a while before they started to go after me. They were just like mad bulls. I tried to talk to one or two of them, but couldn't talk to them. At one point there was this huge guy coming straight towards me just glaring at me. I started to walk backward. It's funny. I kind of think my experience two years earlier photographing grizzly bears in the Khutzeymateen helped me. I did a lot of thinking then about what I would do if a grizzly charged me and I had all that in the back of my head. He was coming at me faster than I could walk backwards. So when he got just a few feet away from me I just stepped off to the side and he just kept going like a mad charging Rhino. He just kept going. I don't know if he was coming for me or not.

Then after a while, they realized that I was taking pictures and was up to no good. There was one young guy there—you usually think of the young guys being the hotheads and the old guys being calmer,

This image and a few others was the only proof that the loggers attacked the protesters at the PATH camp on September 15, 1999 because the first thing the Interfor loggers did in the attack to drive the protesters out was to systematically destroyed all the cameras and video in the camp and before these images came to light they denied that the attack had occurred. Photo: Rick O'Neil.

but in this case it seemed like this young guy was the only one holding his head together. He came over to me and said, 'You should just get out of here. Get out while you can. It's not a good place to be.' I managed to hang around a little while longer and then a couple of them started to threaten me and a couple of them picked up rocks and were going to start throwing rocks at me. So I said, 'OK, I'm going.'

I guess the reason I managed to get away with the pictures is the way I carry my camera. I normally carry it over my shoulder so it's actually kind of behind my back except when I bring it up and take a picture. I'm not trying to hide it; it's just a convenient way to carry it. So it wasn't too obvious that I was taking pictures. They were so busy destroying the camp they just didn't pay any attention.

I walked to my car and got in it. But I still didn't want to leave. I knew those guys [the protesters] needed help. So I didn't pull away right away. Then a couple of guys picked up big rocks and started coming. Then I left. There was no point in trying to stay there. I saw one police car coming in when I was about half way along the road driving out.

It was a very shocking experience. I saw one person thrown to the ground very hard. I saw them tearing down tents and cutting ropes with mattocks. But I didn't see them burn the tents. That happened after I left. I knew the pictures were valuable, but when I went to tell the RCMP I didn't tell them that I had them. I was afraid that they would take the undeveloped film from me and.... They said they knew about [what happened]. They seemed more interested in getting the protester up in the tree out of the tree than protecting anybody.

I stayed in Vancouver that night and got the film developed at Custom Colour and then came home [to the Sunshine Coast] and saw you on the weekend. I heard an IWA representative on the radio that weekend talking about it, saying that nothing happened. They were just trying to erase history, the lying bastards.

In all Rick took 14 pictures clearly showing the beginning of the attack. One of the most dramatic photos caught a protester praying on his knees in the middle of the road as the hoard of loggers marching toward him. Rick also got a clear picture of Mr. Haffey, who worked for Interfor, videoing the whole scene. It is extremely fortunate that Rick succeeded in documenting this well-orchestrated act of violence. For without his hard irrefutable evidence, Interfor and the loggers would have almost certainly gotten away with it, scott free. There was a bias against the young kids who were protesting in the Elaho. There is little doubt that the RCMP would have believed the loggers and not the protesters when it came to one's word against the other. This incident was just the culmination of a series of lesser incidents. The other ones were just ignored by the authorities.

It shocked all of us when not one MLA spoke out against this violence. Adriane, in particular, was incensed. In fact, it profoundly changed the direction of her life. When asked about the attack, Premier Glen Clark sympathized with the loggers saying he could understand the frustration they felt that led them to do it. When Adriane, who was elected the leader of the B.C. Green Party in 2000, was asked why she left WCWC to enter politics she says that the final straw was the cowardice of the MLAs refusing to denounce Interfor

The remains of the PATH protesters tents, sleeping bags, equipment and other personal belongings in the ashes of the fire on the road were the attacking loggers burned them up. Joe removed any food garbage so as not to attract bears. By noon, some person or persons unknown had completely cleaned up this mess and all this evidence was gone. Photo: taken at 6 a.m. on September 16, 1999, the day after the attack by Joe Foy.

and the violent loggers for beating up young people trying to protect the Stoltmann's ancient forests.

WCWC and other organizations, including the B.C. Civil Liberties Association, asked Ujjal Dosanjh, then B.C.'s Attorney General, to assign a special prosecutor to investigate why the RCMP took so long to respond to the call for help from the protesters and to recommend what charges should be laid. He refused to do it, saying that such a move was only warranted when some sort of conflict with an elected official was in question.

Interfor gets its injunction zone expanded, which triggers more blockades

It was incredibly fortunate that we were in the Supreme Court of B.C. fighting Interfor's request for a larger "bubble zone" to their injunction at the same time as the loggers' attacked PATH's camp in the Elaho. Otherwise, the testimony of James would never have got into a court record so fast nor acted as a catalyst to get the RCMP to take the assault seriously.

I watched the court proceedings on the 16th, the day following the loggers' attack. I was heartened when Justice Parrett accepted Joe Foy's affidavit chronicling what had happened in the Elaho, as told to him by James Jamieson. The Justice reacted swiftly to the information that David Haffey, a company employee, had videotaped the assaults, snickering as he taped. Justice Parrett ordered that Haffey's tape be kept and not destroyed. He also disapproved of the loggers mustering within the existing injunction's bubble zone. Only workers involved in the logging were supposed to be there.

However, when RCMP officers went to Interfor's office in Squamish, they did not find the tape. The fate of this tape became quite a soap opera in the months to come. In an attempt to act even handedly, in his new injunction order Parrett also specifically barred Interfor's employees and contractors from the injunction zone except to carry out their job—a backhanded acknowledgment that they had illegally used the injunction zone as a shield when they attacked the protesters on September 15th.

On September 17th B.C. Supreme Court Justice Glen Parrett came down with his decision. He expanded the size of the bubble zone injunction area. Now it included a stretch of road one kilometre long from the end of the road that Interfor was punching into the Elaho Valley and a 200-metre-wide "no go" zone on either side of the road. WCWC staff and volunteers again were excluded from the order, but we had to carry identification and register with Interfor every time we went into the Upper Elaho. The injunction effectively banned the media from seeing the magnificent forest that was being destroyed unless we escorted them through the injunction zone.

I was shocked. It seemed to me that the loggers were being rewarded for their violent behaviour.

By the end of the weekend following the attack, a couple of PATH protesters had re-established another camp down the road outside the one kilometre injunction zone.

Betty Krawczyk joins the fight to save the Elaho

Betty Krawczyk, a grandmother already famous for civil disobedient acts to save what was left of the ancient rainforests in Clayoquot Sound and for her articulate and passionate explanations as to why she felt it was so vital to do so, was incensed at the violent attack on the young people in the protest camp. So she decided to help them save the Stoltmann. She had already served jail time for blockading several times in Clayoquot.

Less than two weeks after the attack, in the middle of the night with the help of her "Women of the Woods" friends, she set up camp in the middle of the Squamish River road at mile 21, many miles to the south of the bubble injunction zone. It was right near a large banner, signed by all the loggers, that Interfor had strung up across the road which read: *Welcome to our Tree Farm. We pledge to work safely, peacefully and responsibly in your forest.* Her choice was a strategic one, for it was on the main road into the Squamish Valley before it branched out. There was no way around her. Every logger going to work in TFL 38 had to drive up this road.

The RCMP were present at Betty's blockade on the first day when Interfor's crews arrived. The loggers (about 15 trucks and 40 workers) turned back without any hint of violence. Betty held them off for three days, the time it took for Interfor to get an injunction and a court order to have her removed. The RCMP kept a watchful eye night and day. But who would beat up a 71-year-old grandmother? She did civil disobedience in the classy classic way, accepting her jail time punishment as the price she had to pay. The longer and more unjust the sentence, the stronger she felt she embarrassed them and made her point that logging these precious forests was not right.

During the next six weeks Interfor completed its road building

Betty Krawczyk blockading at mile 21 on the Squamish Road October 5, 1999. Photo: Jeremy Williams.

north of Lava Creek. They shut down work for the winter when the snows started to stick. We closed down our Upper Elaho Millennial Tree Camp. The snow was too deep for travel. Logging and protesting in the Stoltmann Wilderness was over, until the following spring.

MP Charles Caccia's Stoltmann National Park Reserve Bill debated in the House of Commons

On October 18, 1999, Honourable Charles Caccia, MP for Davenport and Chair of the House of Commons Standing Committee on Environment and Sustainable Development, introduced his Private Member's Bill (Bill C-236) in the House of Commons. Caccia's bill amended Canada's National Park Act to include a 500,000-hectare Stoltmann National Park Reserve. This Bill would get its second reading and one hour of debate within the month. We did not know exactly when that would be until the government put it on the order papers the week before.

When we got the news from Caccia as to the day of his Bill's second reading, Adriane and Joe flew to Ottawa to witness the debate. They were sitting on the edge of their seats in the public gallery at 5:30 p.m. on November 24, 1999, when it began. Joe described what went on in a press release put out right after it was over:

> ...a heated debate between people with very different visions. One vision championed by Charles Caccia is of creating a National Park that would protect the environment including a threatened population of grizzly bears and Canada's oldest known Douglas fir trees and that would greatly expand employment opportunities in the vibrant and growing tourism industry. The other vision forwarded by John Reynolds, Reform MP for West Vancouver-Sunshine Coast, is a tired old vision of more clearcuts, environmental degradation and unreasonable protection of an industry that is producing fewer and fewer jobs at the same time as it forecloses options for tourism.
>
> 'Not only is Mr. Reynolds against the National Park, he clearly has no idea of what is going on in Interfor's logging shows,' fumed WCWC's Joe Foy. 'Hasn't he seen the destruction and massive clearcuts in Interfor's operations in the Elaho Valley? It astounds me that Mr. Reynolds accused "protesters" of spreading "mistruths... about Interfor wanting to cut 1,300 year old trees"' He ought to know that Interfor's current logging plans include clearcuts in the exact same areas in which the Wilderness Committee has identified 1,300-year-old Douglas firs,' explained Foy.

No other MP besides Caccia spoke out in favour of his bill. Several other Liberals, who had expressed support to us privately, failed to show for the debate. NDP MP Svend Robinson, who had expressed his support for the park in a letter to Heritage Minister Sheila Copps, would not publicly support it because we did not have the support of the Squamish First Nation. During the debate Caccia informed the House of Commons that this national park reserve would include the recognition of native land rights. He stated that, *An option could be a co-managed national park reserve such as the one that was set up and established quite successfully in South Moresby in the Queen Charlotte Islands* [with the Haida].

The Stoltmann Park Bill, as per parliamentary rules, officially died without a vote after this one-hour-long debate. However, the public debate and our efforts did not subside. Bill-C236, like Betty's stand of civil non-violent disobedience, was a principled piece of the building campaign.

Chapter 62

WCWC-Victoria takes off
WCWC-Manitoba gets started

A fully committed, creative campaigner joins our Victoria WCWC Team

In June 1999, in between campaign crises in the Stoltmann and celebratory events in Clayoquot, our E-team hired a dynamic young activist, Ken Wu, to head up our Victoria office and campaigns on Vancouver Island. Impressed by Ken's ability to motivate people, his broad grassroots experience in the environment movement and his work with us, I had been pushing to get him to devote his full-time energies to WCWC for a long time.

Ken grew up on the prairies. After completing his high school in Calgary in 1991, he moved to Vancouver to attend the University of British Columbia. He graduated from UBC's Ecological Sciences program in 1995 with an emphasis on evolution and conservation biology. While in university, and for a few years afterwards, Ken canvassed for WCWC and Greenpeace, earning a pretty decent living at it because he believed in the cause and was so knowledgeable. An activist by nature, Ken got heavily involved as a volunteer in many smaller environmental groups including the UBC Student Environment Centre, FAN (Forest Action Network), PATH (People's Action for Threatened Habitat), and Earth First!

Many people see Ken's activist side, especially his emails urging you to do something. Few see his analytical side; constantly trying to figure out the best way to achieve his campaign goals. Ken once told me that my answer to a question he asked during the 1993 Clayoquot Sound blockades really stuck in his mind. It was something he later incorporated into his endeavours to change the world for the better.

His question to me was: "What's the point of WCWC printing hundreds of thousands of campaign newspapers?" He had been thinking to himself, "So what if people read the articles? The trees still get cut." Ken was also questioning the campaign tactics of other organizations, "What's the point of the Clayoquot Sound blockades where hundreds of activists stand on the road for 10 minutes, are hauled off to jail and the logging trucks still pass through for another day's work?"

During the first few weeks while at the Clayoquot protests, Ken had thought that the blockaders should get more militant—perhaps lock their necks to the machinery or tree sit to completely halt the logging for hours or days. But the Friends of Clayoquot Sound had specifically asked everyone not to undertake such actions at their blockade.

As the protest unfolded, Ken came to the realization that if the Friends of Clayoquot had used 'harder' civil disobedient tactics, they wouldn't have attracted 12,000 people onto the logging road at the Kennedy Lake Bridge to protest. Using more passive and peaceful blockade tactics that kids, grannies, businessmen and other average Canadians could readily participate in ultimately resulted in over 900 arrests and massive political pressure.

It was my answer that, Ken says, changed him in the end. When he asked me about WCWC's mass production and distribution of newspapers, I replied, "An educated citizenry exerts the ultimate lobby pressure on government. It's the electorate, once informed about issues and mobilized en masse, that pressures the government to legislate new protected areas and stronger environmental laws. It takes outreach to new people on a massive scale over time."

Ken realized that people cannot lock themselves to grapple yarders forever (they have other things to take care of in their lives). Saving trees requires governments to enact legislation. Just as I had figured out while campaigning for South Moresby, he concluded, too, that the secret to success is building up massive citizen support. It forces the government, given its worry about public opinion and getting re-elected, to act. Environmentalism is ultimately a political battle. This is how Ken puts it:

> The blockades in large part were a catalyst to draw attention to the issue so that environmental groups like the Wilderness Committee could educate a greater segment of the general public. The giant stacks of Wilderness Committee newspapers were a good way to effectively educate and mobilize people to write and call the elected decision makers.

Ken focused right away on creating a core of dozens, then hundreds, of actively involved citizens who would readily show up at rallies, write letters, call their MLAs, circulate petitions and volunteer when called upon to protect ancient forests. His goal was to have "an army of thousands of such activists." It was a tough job, because the populist environmental activism of the early 1990s had somehow fizzled away by '99. He had to rebuild it.

He was in the right place. If wilderness activism was going to revive, Victoria was the place that it would happen first. It had the right demographics—lots of middle-class youth, students, alternative folks, educated professionals, civil servants and elderly people who loved gardens, nature and wildlife—just the right kinds of caring and progressive people needed to rebuild a popular grassroots environmental movement.

When Ken Wu arrived, our WCWC-Victoria office was campaigning to get the Capital Regional District (CRD) to hold a referendum in the upcoming municipal election, asking voters if they would agree to an average household levy of $10 per year for 10 years (2000 until 2009) to help expand regional parks. The money—about $16 million over time—would go directly into a "parks acquisition fund" for the CRD to use to buy up private land to add to its regional parks system.

Alison Spriggs, who had headed up WCWC-Victoria's operations for many years before Ken, was now employed by The Land Conservancy (TLC) and could not help on the campaign during working hours. However, she worked furiously after hours and on weekends directing the campaign to convince people to pass the park levy. Ken and Kristin Lindel, who campaigned for us in the summer and fall of 1999, followed Alison's lead.

They printed 10,000 mail-in cards for our members and supporters to send in to the CRD indicating whether or not they supported the referendum. Several thousand cards expressing support flooded the municipal politicians' offices in September and October. The CRD, seeing the tangible support, decided to hold the referendum.

On municipal election day, an overwhelming majority of those who voted—69 percent—voted in favour of the parks acquisition fund. In January of 2000, the Capital Regional District approved the levy. That set the stage for government to start purchasing private lands to complete the Sooke Hills Sea-to-Sea Green-Blue Belt. The ultimate goal was to create a contiguous large protected area stretching from the Sooke Basin in the south, to the Saanich Inlet by Goldstream Provincial Park, and then north along the Saanich Peninsula all the way to Saltspring Island.

The Sooke Hills campaign that Alison launched with WCWC had already won protection for the Niagara, Veitch, and Waugh Valleys

Facing page: Ken Wu poses beside Castle Giant, the biggest cedar in the unprotected part of the Walbran Valley. Photo: WCWC files.

(Victoria Water Board "off-catchment" lands) in 1997. Still unprotected, were the private lands in Ayum and Charters Valleys in the Sooke Hills. In February of 2000, the federal government through Liberal MP David Anderson, with the provincial government headed by Ujjal Dosanjh and the CRD announced that they were together contributing over $5 million to buy the 3,400 acres of private lands in the Sooke Hills. The Land Conservancy is still raising the remaining several hundred thousand dollars needed to complete the purchase.

Ken's first big rally

In August 1999, two months after he started working for the Wilderness Committee, Ken Wu organized his first rally at the legislature. He aimed to bring attention to our efforts to recruit thousands of volunteers to gather signatures on our Stoltmann Wilderness petitions. Ten people showed up (including Ken who was in a bear costume). Four were his friends. Disappointed, Ken had the urge to joke, "OK folks, don't riot now! Keep cool," but he didn't.

Instead he just worked harder. A year later at the end of September 2000, on the opening day of the Legislature, Ken organized an "Ancient Forest Rally." This time over 500 people showed up to urge the government to save the Great Bear Rainforest, Stoltmann Wilderness, Mt. Maxwell on Saltspring Island, Slocan Valley, and Walbran Valley. This was the first of many large rallies organized by Ken and his Victoria Wilderness Committee volunteers.

Ancient Forest Rally at the Legislature, September 2000. Photo: WCWC-Victoria files.

Ken Wu's formula for a successful campaign

Ken's explains how his multi-pronged approach to campaigning on a wilderness issue works this way:

> *Since 1999 I've placed a particular emphasis on organizing events like slideshows, rallies, public camping trips, and conferences, and directly involving as many people in the campaigns as possible, combined of course with garnering as much media coverage as possible. I believe the 'double-whammy' of a few thousand regulars writing and phoning government and rallying, along with substantial media coverage to reach the broader public, creates a tremendous amount of pressure on government. Governments know that while there is a popular movement, and at the same time a large segment of the public is following the issue and the campaign through the media, they had better pay attention because the potential exists for the movement to explode and become a passionate pervasive household concern like Clayoquot Sound became in the early '90s.*

WCWC-Victoria starts to use e-mail to build a renewed wilderness preservation movement

As an individual, not as a representative of WCWC, Ken protested at the World Trade Organization (WTO) meeting in Seattle in November of 1999. His experience there taught him the power of electronic communications. Many of the people who came to this rally did so in response to receiving emails informing them about it. "There must have been close to a hundred thousand people there. It was by far the most impressive demonstration I had ever been to," Ken told me on his return. The movement against the WTO was growing—especially amongst youth who deplored the WTO's modus operandi. The WTO promoted "forced trade" not "free trade" and advanced trade rules favouring the supremacy of multinational corporations' rights over nation states' rights, including a country's or province's rights to protect its own environment, resources and jobs.

Back home in Victoria, Ken Wu began to develop a similar e-mail action alert and information network for the Victoria Wilderness Committee—a campaign tool that we had never used before. Prior to attending the Seattle WTO protest, Ken had always built phone trees where volunteers called another set of volunteers. Even with a well-organized phone tree and good volunteers doing the phoning (not people who flaked out or avoided this time-consuming chore), he managed to reach only a few hundred people. And that often took several days to accomplish. Another problem was the number of people who didn't answer their phones or who used answering machines and voice mail that allowed only 30-second-long messages. WCWC's message had to be brief. Volunteers often could not convey enough information about the issue to get people well informed and motivated. They could only let them know the time and place of an upcoming event.

Ken began collecting the email addresses of everyone who came into the office, attended rallies, attended slideshows, or, when contacted by one of our door-to-door canvassers, expressed interest. By the spring of 2000, he had almost 300 people on his Victoria Wilderness Committee email list. It was an exciting beginning.

Over the years, the e-mail alerts have become the bread and butter of the educational and organizing efforts in Victoria. Today, there are almost 9,000 people on this email list, largely consisting of people living on Vancouver Island. Of course, not everyone regularly reads his or her emails, but a substantial number do and, of those that do, a substantial number heed the message.

Our Victoria Wilderness Committee is currently able to generate a thousand letters (via email) to government within a couple of weeks using the e-alert system. Within a day, it can get out dozens of people to an impromptu emergency rally. With longer notice it can get out hundreds, simply by notifying them with "a click of the mouse." Via the email newsletters Ken is able to transmit significant amounts of detailed information about the issues, too.

Ken explains the e-alerts mode of communication this way:

> *It has really been a godsend for building up a core of several thousand regulars who follow the issues. Of course, we've also been collecting people's phone numbers at the same time as their emails, and we now have over 2,000 phone numbers of people who have shown up at one of our events or shopped at our store over the past few years who do not correspond by email. Our volunteers call people on the phone list only for major events and mobilizations.*
>
> *The people on the phone and the e-mail lists are far more active than our average annual dues-paying members are, as they have specifically signed-up requesting to be kept informed by this method. Many of the people on our email and phone lists, in fact, are quite*

young—largely university and college students—and are not dues-paying members of WCWC. Many of WCWC's dues-paying members don't come out to our events, but rather support us by sending in donations every year as their vital contribution to the movement. We need both kinds of support to win a campaign.

I never put on a slideshow or rally without having volunteers work through the entire crowd collecting the e-mail and phone numbers of everyone who will give them to us. That has really been the secret to how we have been able to swim against the current culture of apathy and continued to actively involve several thousand people in the environmental movement while most of the active popular wilderness movements in North America have temporarily fizzled away over the past decade.

The "Save the Upper Walbran Valley" campaign

When Ken took the job in Victoria I urged him to revive the floundering campaign to save the Upper Walbran Valley. The lower half of the Walbran valley—5,500 hectares—was protected in 1994 along with the Upper Carmanah Valley as recommended by the Commission on Resources and Environment (CORE). However, the 7,500-hectare Upper Walbran Valley, with thousands of massive red

Another view of the Castle Giant cedar in the Walbran Valley. Photo: WCWC files.

cedars including "Castle Grove," the most spectacular dense stand of giant ancient cedars remaining in B.C. was left open to logging.

I knew Ken had been part of the 1991 direct action campaigns in the Walbran. He told me he watched a tree sitter elude arrest by taking off all his clothes and smearing his naked body with human feces.

Walbran research camp in the Castle Grove. Photo: Joe Foy.

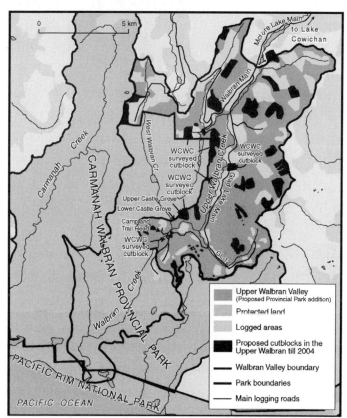

2003 map showing the unprotected Upper Walbran Valley in relation to Carmanah Walbran Provincial Park and Pacific Rim National Park Reserve's West Coast Trail. Cartography: WCWC files.

The man then descended the tree and ran away through the forest. The RCMP took chase, but never quite caught up to him, allowing him to escape. This, and other acts of civil disobedience, however, culminated in saving the Lower Walbran Valley. However, since then, in the Upper Walbran Valley year after year, cutblock by cutblock of the most magnificent unprotected old-growth rainforest remaining on the lower half of Vancouver Island was being destroyed.

Ken set to work in the summer of 1999 to build up our WCWC Walbran campaign, organizing public camping trips and guided hikes to take people to see the Walbran's spectacular trees. In the summer of 2000, Ken and Camosun College Environmental Technology student Geoff Huber launched a Forest Guardians Research Project. Ken hired Geoff for three months to coordinate and lead some biological survey expeditions. Most of the participants were University of Victoria biology and geography undergraduate and graduate students. Every weekend the students drove to the Upper Walbran

Victoria Wilderness Committee Walbran weekend campout. Photo: WCWC files.

In the summer of 2001, our Victoria Wilderness Committee published a full-colour, four-page, tabloid-sized newspaper titled *Park the Upper Walbran Valley* (Vol. 20, No. 1 with a print run of 100,000 copies). It summarized the research results from the scientific surveys the summer before and made the point that half valleys do not make ecologically viable parks. Although nearly 50 percent of the old growth in the upper valley had by now been logged, the rest was still worth saving, including "Castle Grove," named for the fortress-like "candelabra" tops of the giant cedars growing there.

This paper made all the right compelling arguments. But none of them appeared to have touched the hearts of those in government. The Walbran campaign is one of the most frustrating, disappointing—tragic—campaigns the Wilderness Committee has ever waged. Joe Foy has observed that once the network of logging roads is established in a watershed, as it has been in the Upper Walbran, no single focus of the protest or line to hold remains. The logging cutblocks are dispersed every year and it's impossible to build up a campaign to save each one. I think that as the number of holes in the fabric of the ancient rainforest increase and get bigger, the losing battle gets more depressing and many people give up hope.

"Grandma Betty," one of Canada's largest (in circumference) Douglas firs growing in the unprotected Upper Walbran Valley. Posing beside it are U.S. and Chilean Forest Campaigners touring B.C.'s old growth in April, 2003. Photo: WCWC-Victoria files.

Measuring and counting the growth rings in 2002 on a gigantic cedar stump, over four metres in diameter, in the Upper Walbran. Photo: WCWC-Victoria files.

Valley and surveyed the vascular plants and vertebrates in the proposed cutblocks there. At the time, it was Weyerhaeuser and Timber West who were logging in the Upper Walbran valley—extending a web of roads and punching large clearcuts into the formerly contiguous ancient forest.

Ken figured if we created an "identity" of animals and plants that would be destroyed in each cutblock, it would help build public sympathy for protecting the area. The researchers found many nesting marbled murrelets (they could hear the birds perched in the trees above and flying overhead with their distinctive "keir keir" calls in the early June dawn), a pair of Queen Charlotte goshawks hunting through the woods (a red-listed subspecies of goshawk), red legged frogs (blue-listed), cougar scats, Roosevelt elk tracks, and several relatively uncommon plant species.

After a few weeks of surveys, Ken brought *Times Colonist* writer Malcolm Curtis and photographer Ray Smith to the Upper Walbran along with several of our volunteer biologists (UBC ecologist Richard Feldman and University of Victoria ornithologist Katie Christie). Later that week, the *Times Colonist* ran a Walbran story on its front page and a large feature piece in its Capital Region section with many photos and a map. It was a major leap forward for the campaign, but somehow not enough to stop the logging.

Last project with CIDA falls through

After Adriane had left, embarking on a political career as the newly elected leader of the British Columbia Green Party, Andrea Reimer took her place on the E-team. However, there was no one employed at the Committee who could fully fill Adriane's shoes and carry on with the same drive in directing the international WILD campaign that she had launched more than a decade before.

Just before Adriane left, she initiated discussions with a tribe in Irian Jaya who wanted to map and demarcate their traditional lands in an effort to stop the logging that was poised to destroy their wild tropical rainforest. Larry Commodore, a former Soowahlie elected chief who had visited Irian Jaya and met with the tribe on a First Nations exchange program, was hired by WCWC to explore the partnership and project. But the contacts petered out. Throughout 2002, various WCWC staff, including Andrea Reimer, tried to re-establish communications and were frustrated in every attempt.

The E-team logically decided it was time to reassess WCWC's capacity and interest in undertaking another international project and continuing the WILD campaign. It took a huge commitment of time and money to work at the global level, including raising matching funds to qualify for CIDA funding. No one at WCWC had the experience, passion and energy to overcome the many hurdles in the way of getting international projects off the ground. I think it was a relief to many WCWC staff (although sad for some like Sue Fox who had diligently kept communicating with WILD contacts) when the E-team finally determined we'd simply drop the international campaigns and focus solely on protecting Canadian wilderness.

Manitoba campaigner comes to us for help

In November 1999, someone none of us at WCWC knew, named Ron Thiessen, called our office several times asking to speak with me. By happenstance, every time he called I was not in. As a rule, I always take calls, even if I don't know the person on the chance that it might be a fantastic new contact. I'm not great, however, at returning calls. Finally in this case, Alice Eaton, our full-time volunteer receptionist, told me that I had better call him back. The fellow was determined to speak with me and would just keep on calling if I didn't.

Heeding Alice's advice, I picked up the phone right that moment and called. Ron and I hit it off right away. He told me about his experiences canvassing for Greenpeace (which had recently shut down its canvassing operations in Winnipeg) and his idea of opening a WCWC office and canvass there.

After checking out Ron's references, confirming that he had eight years experience with Greenpeace and that people there liked him, I talked to Joe, Adriane and Matt and we decided to fly him out to Vancouver at our expense to discuss his proposal in person. I called Ron back and booked his flight. He arrived at our Vancouver office about a week after our initial phone conversation.

We all liked Ron's attitude and energy and agreed to try out a test canvass in Winnipeg right away. The idea was to send canvassers door-to-door seeking support for a campaign to ban logging in Manitoba's provincial parks. It had to start within a few weeks, as bitterly cold weather would soon arrive making door-to-door canvassing impossible.

Ron returned to Winnipeg. Within a week he obtained a temporary permit to canvass and produced a fact sheet explaining how logging was damaging Manitoba's parks. Logging was legally permitted in over half the areas within three of Manitoba's largest and most popular provincial parks (Whiteshell, Nopiming, and Duck Mountain).

Bumper Sticker. Artwork: WCWC-Manitoba files.

Ron launched a 10-day pilot canvass in Winnipeg in early December with several experienced canvassers and found that Winnipeggers were "highly responsive." When they were made aware of the park logging they were opposed and financially supportive of our fledgling campaign.

With the proof that Ron could make a success of it in Winnipeg, we hired him to open a Wilderness Committee office there. No other conservation organization in Manitoba was using anything similar to our massive public educational campaigning style to promote park protection.

Starting as a door-to-door canvass operation concentrating on getting full protection for existing parks, the Manitoba Wilderness Committee quickly grew to be a full-fledged Branch campaigning on a variety of issues. In their first 18 months, our door-to-door canvassers visited over 100,000 homes in the Winnipeg area. Canvassers continued to find that the vast majority of people were unaware that commercial logging was going on in Manitoba's big parks. The canvassers' efforts built up a large opposition to continued industrial activities in the park.

In all of Canada, Manitoba had one of the stingiest park systems. It had permanently protected less than seven percent of its land base—about half the amount of land that Ontario, B.C., and Alberta had set aside in protected areas. Yet there still existed the chance to protect a lot more wilderness because a lot of wilderness remained. Manitobans just needed more political will.

Manitoba has one of the largest areas of roadless intact wild boreal forests left on Earth. Bounded by the eastern shores of Lake Winnipeg on the west and the Ontario border on the east, the region called the East Shore Wilderness is over 1.5 million hectares in size. It had the largest herds of endangered woodland caribou in all of Manitoba. Helping achieve protection for this area soon became one of the biggest Manitoba Wilderness Committee's campaigns.

Like almost all the remaining unprotected wilderness areas in Canada, the nearly pristine state of the East Shore Wilderness was under threat. Proposals for hundreds of kilometres of all-weather roads, a hydroelectric power line corridor and several big forestry operations were on the drawing boards. However, the Manitoba government had a moral and legal responsibility to seek consent from the many First Nations communities that would be affected before proceeding with any plans. We knew it also had to have public and First Nations' support to protect wilderness there.

In July 2001, our fledgling Branch published its first campaign newspaper—a three-colour, four-page, tabloid sized paper titled *Manitoba Parks at Risk* (Educational Report Vol. 20 No. 3 - press run 50,000 copies). The paper gave full details about the impact of logging in Manitoba's three big parks and the reasons why it should be stopped. It also highlighted proposals to protect the East Shore Wilderness and to establish new provincial parks to protect high-value areas like Sturgeon Bay and the Manigotagan River corridor, a well-loved canoe route.

Vindicated—the Lava Creek firs are unique and merit protection

In May 2000 Betty Krawczyk blockaded for a second time in the Elaho to save the Stoltmann Wilderness. Once again she was arrested and jailed for defying the court order not to interfere with Interfor's logging operation.

Shortly before Betty's arrest, on April 14, 2000, the B.C. Ministry of Forests issued an Information Bulletin titled *Report Complete on Elaho Douglas Firs* which stated that: *A Ministry of Forests Study concludes that a stand of Douglas fir in the Upper Elaho Valley is unique and should be managed for research and conservation.*

We *were* right! There were no other big ancient Douglas fir groves like the one above Lava Creek growing elsewhere, including nearby Clendenning Provincial Park where big Douglas firs were few and scattered. Interfor was wrong when it cavalierly claimed that there were many groves like those we'd found in the Elaho.

This Forest Service information bulletin went on to state that:
Interfor has agreed to extend a wildlife tree patch in an adjacent cutblock to include a portion of the Lava Creek stand of Douglas fir,

sparing them from cutting. Interfor has also voluntarily withdrawn two proposed cutblocks from its 2000-2004 forest development plan. The wildlife tree patch and the proposed cutblocks contain the main concentration of old, large Douglas fir in the Lava Creek area.

WCWC-Victoria gets a storefront office

In March 2000, after a two-year search, our Victoria office manager Selena Blais finally found a suitable place for a storefront office, 651 Johnson Street. She called to give me the rundown. It was about six blocks away from the Legislature Buildings and a half-block off Douglas Street, the city's main commercial drag. It wasn't ideally located in terms of tourist traffic (it was several blocks away from the main tourist haunts), but it was the right size—three times larger than their current office space—and the right price. All the stores she had looked at in the prime tourist zone were far too expensive.

Our comptroller Matt Jong and I went to Victoria to check it out. We both liked the place at first sight and agreed that it had great potential. Because it was not located in the best part of downtown, Matt was able to negotiate a reasonable long-term lease with reasonable monthly payments with its sympathetic owner.

Heidi Sherwood, store manager outside the Wilderness Committee's store at 651 Johnston Street in Victoria. Photo: WCWC-Victoria files.

Rainforest Store managers Heidi Sherwood (left), and Selena Blais (back right) in our Victoria storefront office. Photo: WCWC-Victoria files.

It only took minor renovations to ready the premises for us to move in. Our store took up the front half of the space leaving enough room behind for Ken Wu, the staff and our volunteers to work. There was even enough space for a couch for our door-to-door canvass crew to sit on while getting briefings. It was great to be on the street level after years on the 3rd floor in the "bleeding hearts building" on View Street (nicknamed so because so many non-profit organizations had offices there). We were up and running at 651 Johnson by March 2000.

Getting this storefront was an incredibly smart move. While sales were slow at first, by 2004 this store under store manager Heidi Sherwood's leadership was bringing in about $60,000 annually in gross sales and donations. Not only that, the new location tremendously helped increase the energy and profile of the Wilderness Committee in Victoria. Since it opened, just about every environmentally-minded person living in Victoria has stopped into our Rainforest Store. Working out of this visible location helped Ken Wu build up a large volunteer base. The media liked its easy accessibility, too!

Chapter 63

Cathedral Grove under attack
Elaho campaign rolls toward resolution

WCWC launches campaign to expand and protect B.C.'s most popular park—Cathedral Grove

U.S.-based Weyerhaeuser Inc. took over MacMillan Bloedel's operations in September '99. It didn't take long for the new company to show its ugly side. This American company understood that the real gold in MB's holdings was its undervalued private lands—over 300,000 hectares worth—much of it located on Vancouver Island and the Sunshine Coast. It began logging these private lands much more aggressively than MB ever had.

For years, conservationists had been pressuring the provincial government to expand the world-famous 157-hectare MacMillan Provincial Park in Cathedral Grove. Located half way between Parksville and Port Alberni on the highway to Clayoquot Sound, this tiny park was being visited by over half-a-million tourists per year.

The first complaint letter that WCWC wrote to government after forming in 1980 was to decry the clearcutting on the western edge of Cathedral Grove that opened up the big trees in the park to blowdown. The government did nothing to stop the logging. The government's reply to our letter was that the logging was being done on MB's private land. During every major windstorm the big trees on the leading edge of Cathedral Grove blew over. MB removed the wind-felled trees and did some selective logging to try to protect the park. It knew its clearcutting had put the park at risk. From then on MB left most of the big trees on its private lands on both sides of the road west of the park alone, actually giving the illusion that MacMillan Park was really much larger than it really was. I believe that MB, being a Canadian company, had some pride in leaving these grand trees for people to enjoy.

Clearcut behind Cathedral grove. After the blowdown tree is cut, the stumps often spring back to near upright position. Photo: WCWC files.

This company ethos did not carry over to the new foreign owner. In late October 2000, an old time logger, sick of what was happening, tipped off Annette Tanner, Chairperson of WCWC Mid-Island, that Weyerhaeuser was logging in Cathedral Grove near MacMillan Park. The company had been very sneaky about it. Usually road building activity starts at the same time as falling. But this would have drawn attention to the logging. Instead, Weyerhaeuser sent in fallers ahead to do "a drop show" i.e. fall all of the trees, one on top of the other and worry about yarding them out later. They did this along the entire one-and a half kilometre road allowance that paralleled the existing Port Alberni highway, only 50 metres from its north edge. The new roadway was more than twice the width of the existing highway's cleared right-of-way. It ended up being a long linear clearcut. The 50-metre screen of trees shielded the logging from public view. The people driving by had no inkling that Cathedral Grove was being pillaged.

Annette, of course, followed up on the logger's tip, discovered the fallen giants, and in November made front page *Victoria Times Colonist* news. The article featured a picture of her beside one of the giant logs. People were furious, especially when they learned that Weyerhaeuser was planning to do a lot more logging in Cathedral Grove in the near future. To stop the rest of the unprotected grove from being logged, WCWC Mid-Island began asking local Municipal Councils and Regional Districts to urge the provincial government to acquire the land adjacent to the existing MacMillan Park boundary while it still had its big trees intact.

Weyerhaeuser, meanwhile, finished its road, yarded out the logs and then voluntarily put a temporary moratorium on further logging there, installing a locked gate across the new road's entrance onto the highway. The "harvesting" of the magnificent trees in this long, linear clearcut should have satiated a great deal of this giant forest company's greed. Rumour had it that the big beautiful old-growth firs that Weyerhaeuser cut down garnered over seven million dollars in profit. Because these trees came from the company's private lands, they were undoubtedly exported as raw logs to Weyerhaeuser's plywood mills in the U.S.

The new logging road stopped about half-a-kilometre short of MacMillan Park right at a parcel of land that was in the Park's Master Plan to be eventually purchased and added to the park. All along the road the trees were magnificent, just like the old growth protected in the park. In August of 2001 Weyerhaeuser ribboned three new cutblocks next to its road, targeting some of the finest and largest stands of trees. It planned to start falling the cutblocks in the fall of 2001. We argued publicly that if the big trees there were cut, the new openings could have a disastrous effect on the tiny park. The logging would also destroy the option to increase the park's size.

In the next few months the Alberni-Clayoquot Regional District, the City of Port Alberni, the Town of Qualicum Beach City of Parksville, the Regional District of Nanaimo, the Tourism Association of Vancouver Island, the Parksville Chamber of Commerce and other groups passed motions expressing their support for expanding the size of MacMillan Park. They all informed the Honourable Joyce Murray, Minister of Water, Air and Land Protection, of their support. Wilderness Committee volunteers with clipboards took shifts in the parking area along the Alberni Highway beside MacMillan Park gathering signatures from people on a petition in support of enlarging the Park to include the whole of Cathedral Grove. Visitors from all around the world signed it. Annette presented the completed petition sheets to the Minister. In the accompanying letter Annette wrote to Murray that *MacMillan Park is one of the most frequently visited Provincial Parks in the Province. The economic benefits of this World Famous Tourist Treasure to the central Island, are enormous in that this Park is unique because it is an old-growth forest accessible to everyone in the world.*

In response to all the pressure, the B.C. government added 21 hectares to MacMillan Park. The Weyerhaeuser land trade that accomplished this was said to be worth 1.7 million dollars. Excluded

Carmanah Zimmermann, a WCWC volunteer, hugs a record-sized Pacific yew tree in Cameron Canyon, part of the proposed expansion to MacMillan Park. Photo: Phil Carson.

from the deal, however, were 20 hectares of huge trees on the west side of the highway, which the Park's 1992 Master Plan also said should be acquired. To put this deal in an even larger context, WCWC was advocating that approximately 500 hectares of Cathedral Grove old growth be protected. The province was thinking way too small!

Ostensibly worried about the traffic hazard that park visitors created during peak summer tourist months, the Liberal B.C. government then fast-tracked plans to build a huge $1.3 million two-hectare, 200 car and 20 bus parking lot on the east side of the park inside the newly acquired parcel of land. The government was poised to start constructing this lot to have it ready for the 2001 summer tourist season.

The location—on the Cameron River's flood plain close to the river—was picked without any public consultation and without applying common sense. The scheme involved using concrete barriers to line the highway on both sides to prevent people from parking next to the road and walking directly into the big tree groves as they had done for the past 50 years. Instead, people would be forced to park in the new lot and walk for 800 metres along a wide new footpath that BC Parks planned to build parallel to the highway. After visiting the big trees in the park, they'd have to walk 800 metres back to their car or bus.

Photo that ran in the *Times Colonist* on March 8, 2000, Annette Tanner kneels by swath of old-growth trees cut to make room for a super-wide new logging road in Cathedral Grove. Photo: Ray Smith/*Times Colonist*.

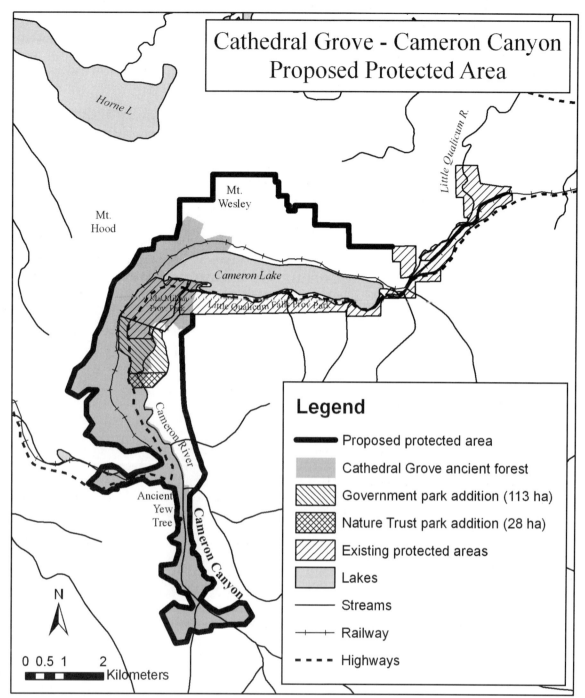

Map of proposed expanded Cathedral Grove (MacMillan Park). Cartography: WCWC files, 2006.

On a visit to Victoria, Scott and Annette Tanner clocked with their car's odometer the distance from Victoria City Hall on Douglas Street to the Legislative building—a popular route often used for provincial protest marches. It turned out to be exactly the same distance as a one-way hike from the proposed parking lot to the big trees in MacMillan Park. Annette used this fact in her presentations thereafter to put the 1,600 metre round-trip into a mind-frame that people on Vancouver Island could relate to.

The parking lot and long hike into the park was a crazy idea. But, like Weyerhaeuser who wanted to profit from logging as many of these huge trees as it could, the Liberal government had its own greedy plan to make money off this living tree museum. Parking fees! Someone actually estimated that the take on parking meters in the proposed parking lot would amount to over half a million dollars per year! We realized the true motivation for the parking lot plan only after the Liberals installed parking meters in 40 other B.C. parks the following year, one of the most expensive financial failures and unpopular moves that a B.C. government has ever made in terms of our parks!

WCWC Mid-Island made a 'big stink' in the media about the parking lot plan. Everything was wrong about it. It was located in a floodplain that was subject to periodic flooding. It was in prime Roosevelt elk winter habitat and blocked the elks' migration route. It would accelerate blowdown in the park.

On short notice, Minister Joyce Murray invited Annette and Scott Tanner and a few other concerned people to meet with her in Victoria about the issue. They all agreed to visit Cathedral Grove together on the following weekend. Joe and Annette were joined by several others including Gillian Trumper, the Mayor of Port Alberni. The goal was to inspect the site with Minister Murray and her staff and try to resolve the conflict. Joe pointed out evidence that everything Annette and other conservationists had said was wrong with the plan was true. He and Annette suggested alternatives to Minister Murray, including known traffic calming methods like a lower speed limit through the grove and flashing yellow lights by the pedestrian highway crossing. She seemed to agree but was non-committal. As far as WCWC was concerned, the parking lot plan was a total no go.

In the late fall of 2001 we rushed into print a two-page, three-colour, tabloid-sized newspaper titled *Expand MacMillan Park – Protect all of Cathedral Grove* (Educational Report Vol. 20 No. 4 - press run 100,000). It made a strong plea to expand the 157-hectare park to include all the lands around Cameron Lake and the entire valley bottom at the head of the Cameron Lake where many giant unprotected Douglas firs still stand. The paper expressed WCWC's strong opposition to the B.C. government's plan to put a parking lot right beside the tiny park on the upwind side of the existing park that was already suffering from blow down.

Our paper's lead article started out this way:

One of BC's most beloved parks is about to be wrecked by stupidity and greed. Its official name is MacMillan Park, but it's commonly called Cathedral Grove. Millions of people have come to know and love this park. It is so accessible–Highway 4, the only road to Port Alberni and Clayoquot Sound–passes right through it. The star attraction is, of course, the park's magnificent trees that befit the name cathedral.

Who hasn't stopped beside the road and taken a 20-minute stroll among the park's Douglas fir giants? Most are nearly 400 years old and some old veterans are over 800 years old.

Just after we published our paper the government put its parking lot plans on temporary hold. But they didn't totally abandon them. The fight to permanently axe the parking lot and expand MacMillan Park to encompass all of Cathedral Grove raged on. Over the next five years it became a full blown environmental conflict with tree-sits, court cases, and a Friends of Cathedral Grove defender's camp that occupied the proposed parking lot site.

A little win came in the spring of 2005 when the B.C. government announced an expansion of the park from 157 to 280 hectares through a $5.5 million "groundbreaking" partnership with Weyerhaeuser. Then, in the spring of 2006, the government publicly stated it would shelve the parking lot idea.

Although the expanded park includes some fantastic old-growth on the west side of the highway heading toward Port Alberni, a big part of Cathedral Grove is still unprotected. Island Timberlands, the company that bought Weyerhaeuser's private forestland in 2005, still plans to log the incredibly valuable trees on its privately owned lands next to the park. Given the incredible rarity of these old-growth Douglas firs and their importance in maintaining the integrity of the whole grove, the Wilderness Committee intends to keep fighting to expand MacMillan Park and protect what's left.

Elaho fight brings surprises in 2000

Early in 2000, six months after the loggers attacked the PATH protesters and destroyed the Lava Creek protest camp, the Squamish Crown Council laid criminal charges against five Squamish area men involved in the incident. There were many more men that took part in the assault and there's no doubt, at least in our minds, that their employer–Interfor–had "aided and abetted" their violence. We believed that Interfor and others should have been charged for their part in the crime but, nonetheless, having five people brought before the court was a clear sign that the wheels of justice were at least turning in the right direction.

After the attack in the Elaho, Interfor showed signs that it may be changing its ways. It initiated an effort to come to an agreement with the Squamish First Nation, hiring John McCandless, who had been so instrumental in working with First Nations to save the Stein Valley, as a consultant to facilitate communications. John came to meet with Joe and me several times to get our viewpoint. It seemed strange to see him working for Interfor. But he certainly didn't push Interfor's industrial vision. He just listened and asked questions. In April 2000, Interfor held its Annual General Meeting in the Squamish Nation's Totem Hall just outside of Squamish, B.C. in what was obviously a gesture of good will. It was the first time the company had ever held an AGM outside Vancouver.

Adriane and I and other staff members and volunteers picketed along the highway near the meeting hall, holding up signs asking Interfor to "Save the Elaho" and "Halt Raw Log Exports." Inside the hall, Interfor's CEO Duncan Davies announced that his company wanted to talk to all groups in an effort to resolve the Elaho conflict. He also promised a temporary hold on logging north of Lava Creek to allow for talks to occur. In May, a month later, before we even had a chance to formally meet with the company, Interfor did an about-face. The company announced that it would immediately begin logging the ancient forest north of Lava Creek.

We reacted immediately, calling on conservation groups worldwide to mount a "Boycott Interfor" campaign. We printed up a "Boycott Interfor" poster and sent it everywhere. We reopened our Millennium Tree Camp at Lava Creek (which we'd closed for the winter) and volunteers move back in to continue inventorying the ancient Douglas firs. It looked like this campaign would never end.

Meanwhile, the 13 environmental protesters arrested in the Elaho in 1999 were in court. Their trial started in March 2000. Four of them pleaded guilty to civil contempt of court charges and were summarily sentenced to 21 to 51-day jail terms—the longest sentences ever handed out in B.C. for this crime. The trial of the remaining eight went on for 43 days and included testimony about the Interfor attack, including lies by an Interfor employee about what happened to the videotape the Interfor employee had made of the assault. Ultimately all the accused protestors were convicted.

In mid-September protesters Betty Krawczyk, the 72-year-old grandmother and heroine of many blockades to save B.C.'s ancient forests, and Barney Kern, the young activist who had driven our "Anti Bear Hunting Initiative" caravan to Prince George nearly four years earlier, were given one-year-long jail terms for criminal contempt of court for their efforts to protect the Elaho Valley. Justice Parrett also sentenced the other non-violent civil disobedient Elaho defenders to unprecedentedly long jail terms.

A big psychological victory

In June 2000, Justice Vickers of the B.C. Supreme Court, based on evidence presented by Sierra Legal Defense Fund Lawyer Angela McCue who was acting on WCWC's behalf, struck down the bubble injunction zone set out in Justice Parrett's order dated September 17, 1999. It had already resulted in the arrest of a dozen environmentalists. In a strongly worded written judgment, Justice Vickers, referring to the beating up of environmentalists by loggers and the lying and destruction of evidence by Interfor employees, stated: *The court must denounce vigilante actions, misrepresentations and lies in the strongest possible way.* Noting that the plaintiffs (International Forest Products Limited) were *not before the court with clean hands*, Justice Vickers concluded his ruling with this explanation: *The injunction must be set aside in order to preserve the integrity of the court.*

Our poster calling on people to boycott Interfor. Document: WCWC archives.

Interfor began logging north of Lava Creek in cutblock 102-51. Friends of Elaho (FOE), a new organization of activists, peacefully stood in the way of the loggers and halted logging. Interfor had to start all over again in court to get another injunction, which they did. It seemed like the summer of 2000 would be a repeat of the summer of 1999.

Wedge to the Ledge Trek

"Sooner or later we are going to run out of big trees to inventory. We have to do something new and upbeat to catch people's imaginations," I ranted at a staff meeting in early June where we were discussing how to invigorate our campaign in the Elaho Valley that summer. As we brainstormed, someone came up with a unique idea that really could spark the campaign. It was bold, had never been done before, yet it did not involve spending a lot of money that we didn't have.

We would trek to Victoria, carrying with us all the way to the Legislature buildings, two large wedge-shaped pieces of wood cut by

Volunteers with one of the wedges that they trekked to the Legislature. Photo: WCWC files.

a faller out of the trunks of two big thousand-year-old Douglas fir trees when he felled them. The wedges were part of the wood waste left behind in the clearcuts to rot. We'd give the wedges to Premier Dosanjh, who had refused our offer to come on a guided tour with us to see the old growth in the Upper Elaho. We hoped that it would help him grasp the size and antiquity of the Douglas fir trees being cut down in the Stoltmann Wilderness. Perhaps after seeing them he would use his power to stop this eco-carnage.

It was not legal for us to take wood wedges left behind in the clearcuts because Interfor had the salvage rights to all the wood left behind in the clearcuts in its TFL. We knew that the company would never give us a salvage permit like the one we'd got from MB to salvage 'Stumpy' to use the wedges "for our purpose." However, we were sure that Interfor would not try to stop us because of the negative media it would get if it did.

These wedges were big and heavy chunks of wood. It took two people to carry each one from the clearcut to the logging road. Each one filled a full-sized wheelbarrow. The logistics involved in wheelbarrowing the wedges the 260-kilometre-long trek to Victoria were a bit of a nightmare. One of the biggest problems was the stretch of Sea-to-Sky Highway from Squamish to Horseshoe Bay. We all agreed that it was too dangerous to trundle our bright yellow wedge-filled wheelbarrows down that route. The road was too narrow, too curvy and in many places had no shoulder. Moreover, too many people drove that road like maniacs. We decided to bypass this section by paddling down Howe Sound and into Vancouver. The whole thing became a bit like an iron man biathlon. At first we wanted to use a large cedar dugout canoe from the Squamish First Nation, but none of their canoes were available. We settled on a 32-foot-long voyageur canoe that we could rent, which actually was a much more culturally appropriate boat, given the cargo.

On June 19, 2000, six volunteers launched our twelve-day-long "Wedge to the Ledge" trek. They loaded up the wedges into the wheelbarrows and starting to roll them down the toughest stretch, the long dusty 40 miles of gravel logging road to the pavement which started at Mile 21 (21 miles from Squamish). I thought there would be a lot of media interest in this event. But despite sending out a media advisory, there wasn't any... except our own. Anticipating inadequate media coverage, we had contracted Jeremy Williams to video the entire journey and make a documentary of this epic journey and Chris Jeschelnik to produce it.

The six volunteers took turns pushing the wheelbarrows as logging trucks whizzed by. A few pickup trucks sped by extra fast, spinning up rocks. But there was no real harassment, nor nasty surprises. The loggers knew by radiophone our whereabouts at all times. At the outskirts of Squamish, we arranged to have breakfast in a friendly cafe where we were sure we wouldn't be refused service. At Squamish City Hall, Mayor Corinne Lonsdale and some loggers with their own decorated wheelbarrows greeted our trekkers. It's all documented on our 45 minute *Wedge to the Ledge* video. Lonsdale, who championed the loggers' point of view, gave her speech and left, not even giving Joe the courtesy of listening to him speak. The trekkers also met with Squamish Forest Service officials who politely listened to our side of the issue.

The canoe paddle from Squamish to Bowen Island, where our

Rolling down the logging road out of the Elaho Valley with the wedges on the first leg of the trek to the "Ledge." Photo: WCWC files.

Protest at Squamish City Hall on the Wedge to the Ledge trek. Photo: WCWC files.

trekkers stayed overnight, took just a day and was uneventful. The next day—on Sunday—they paddled to Vancouver. That afternoon Joe, Adriane, about 30 Wilderness Committee supporters and I gathered on English Bay to welcome their arrival. Again, despite our best efforts, the media turnout was pathetic. No one came except for

Paddling down Howe Sound in a cargo canoe loaded with the wheelbarrows and wedges towards Vancouver on the second leg of the trek to the "Ledge." Photo: WCWC files.

one lone City TV cameraman who showed up after it was all over. We wheeled the wedges along the sidewalk from the beach to our Gastown store where the wheelbarrows were parked for the night. One of them had worn out its wheel bearings and was squeaking and getting hard to push. The building supply store where we had bought it 10-days earlier replaced it free of charge with no questions asked. No one imagined that a wheelbarrow could be worked so hard over

Rolling down the causeway to catch the ferry to Swartz Bay with the wedges on the second to the last leg of the trek to the "Ledge." Photo: WCWC files.

so short a period of time!

The next day our band of wedge trekkers headed out to the Tsawwassen Ferry Terminal. The ranks had thinned to only a few by then. They spent the night in Delta. In the morning they really had to rush to make the ferry. In fact, they ran the last bit to get on board, not wanting to be late to meet the team waiting for them in Swartz Bay. They were looking forward to the extra help pushing the wheelbarrows the final leg of the journey to Victoria. We celebrated their arrival in Victoria at a big party celebrated that night. The next morning the trekkers and their wedge laden wheelbarrows arrived on time for the big welcoming rally that Ken Wu and volunteer Dr. Keith Ferguson had organized at the legislature. At the rally, which the media did cover, David Ball, a young University of Victoria student who had trekked the entire distance, gave a passionate speech. Premier Dosanjh refused to come out and meet us to discuss the fate of the Stoltmann Wilderness. No one from the government would accept the gift of the two large wedges. So we wheeled them to our Rainforest Store on Johnston Street where the bigger of the two wedges is still on display to this day.

A little less than a year later, in April 2001, we held our *Wedge to the Ledge* video premier in the Vancouver Museum and Planetarium. It captured both the spirit of this epic journey and the urgency of the issue.

Endangered bird discoveries do little to slow Interfor's logging

With all the research and exploring going on in the Elaho forest in the spring of 2000, the discovery of an endangered species was inevitable. One hiker got a good photo of an unusual owl. Looking at the photo, government biologists verified that it was a spotted owl. Andy Miller, WCWC's owl expert, investigated and heard the owl's calls, but couldn't find a nesting site. He thought it was probably a young male owl searching in vain for a mate. Hearing this endangered species' call and having a picture of it taken within a cutblock that was soon to be felled meant nothing to the authorities.

No laws in B.C. or Canada require that the habitat of endangered species be protected. We have laws that protect only some nesting birds. As logging progressed in the cutblock, the spotted owl ceased its nightly calls.

Researchers found a pair of northern goshawks, the largest of the three North American forest hawks, nesting in a tree in one of the cutblocks. These birds fared better. In a court order granting an injunction against environmental protesters interfering with the logging of Elaho cutblocks 102-51 and 102-52, Justice Pitfield also granted a temporary injunction to the goshawks! He ordered Interfor not to log near their nesting site—but only until the first of September. After the young fledged and the nest was no longer occupied, the company had the right to cut down all the trees, including the one with the now empty nest, even though northern goshawks are known to use the same nests in following years. Protection of occupied nests is all that these endangered birds get in law.

In the year 2000 there was no repeat in the Stoltmann of the violent episodes of 1999. A large contingent of RCMP moved into the Upper Elaho Valley and maintained a permanent encampment at Lava Creek. Twice a day a helicopter flew in to bring them supplies and change personnel. Friends of the Elaho, a new group that had taken over from PATH, continued its actions to block the logging. The RCMP arrested four "tree sitting" protesters. Interfor continued aggressively logging north of Lava Creek.

Final stage of Wedge to the Ledge Trek and Rally, the culmination of the long 240 kilometre trek from a clearcut in the Upper Elaho Valley to bring evidence of the destruction of the thousand-year-old Douglas firs, in the form of wedges (undercuts from two felled giant Douglas firs) to the Premier in an effort to get him to save the area.
Left top: Ken Wu gives dedicated volunteer David Ball (in front) a break and wheels one of the wedges on an overpass on the way to the Legislature on the last leg of the journey. Left bottom: Ken Wu talks in front of TV cameras at the Legislature.
Right top: David Ball, the only volunteer to walk the entire Wedge to the Ledge Trek, speaks at the final rally in front of the Legislature in Victoria on June 21, 2000.
Right bottom: Wedges being delivered to the Legislature. Adriane Carr, WCWC campaigner, speaking to those attending the rally. Photos: WCWC files.

Independent study reveals National Park would generate more jobs than logging the ancient forest would

In August 2000, One Whistler, a local tourism-oriented Whistler organization made up of business and government representatives, released an economic study it had commissioned to assess our proposed Stoltmann National Park Reserve. Grant Thornton International, a large multinational consulting firm, conducted the study. It concluded that the economic benefits of a Stoltmann National Park included up to 730 jobs that could be sustained through the "preservation and tourism" use of the Stoltmann—far more than the 230 logging jobs that the Stoltmann Wilderness area could sustain over the long run. This was grist for yet another one of our Stoltmann Wilderness education newspapers.

Production of another Stoltmann newspaper caps the year's campaigning efforts

In the fall of 2000, we produced our 6th newspaper about our Stoltmann campaign. It was an eight-page, full-colour, tabloid-sized newspaper titled: *Save the Stoltmann Wilderness and its 1,000-year-old trees –FIVE YEAR FIGHT TO SAVE THE WILD ELAHO VALLEY* (Educational Report Vol. 19 No. 3 with a press run of 150,000 copies). In it we told the story of our "Wedge to the Ledge" trek. Like all of our issue papers we wrote it so that anyone reading it could get a full grasp of the issue, even if they had never heard of our campaign

Cover of our 6th newspaper on the Stoltmann Wilderness. Document: WCWC archives.

before. We particularly liked to contrast "pro and con" arguments. In this paper our *ANTI-PARK PROPAGANDA VS. PRO-PRESERVATION TRUTH* article stated:

| **Anti-park propaganda:** | **Pro-preservation truths:** |

Anti-park propaganda: The B.C. government has already protected 14% of the Lower Mainland Region in provincial parks. In the Squamish Forest District 27% of the land area is protected. This is higher than the provincial 12% goal and therefore is more than enough.

Pro-preservation truth: The current Lower Mainland park system does not adequately protect biodiversity. This system, including portions within the Squamish Forest District, is mostly high-elevation scrub forest, alpine meadows, rock, and ice. While beautiful, these parks encompass only about 8 percent of the Lower Mainland's low and mid elevation forests. To conserve biodiversity and thousand-year-old trees for future generations we must preserve now-rare forested wilderness including the Upper Elaho Valley and other remnants of original forest in the Stoltmann Wilderness.

Anti-park propaganda: In 1996 the B.C. Government chaired a land use planning table in the Lower Mainland Region. It unanimously decided not to protect all of the Stoltmann Wilderness. The Wilderness Committee chose not to participate so they shouldn't complain now.

Pro-preservation truth: This NDP government "public" land use planning process was a sham. The planning table was hand-picked by the government and WCWC was not invited to participate. Interfor and other logging companies were allowed to continue logging in proposed park areas during the process. The public was not allowed to observe the planning team's meetings. No open house public meetings were ever held to gather the public's input. Economic and biological studies were not done to determine the best options and amount of land base that should be protected to sustain the region's economies and biodiversity. The amount of protection was predetermined based on politics not ecology or economics.

Anti-park propaganda: Two wonderful parks, Clendenning Valley and the Upper Lillooet Valley, were created as a result of the Lower Mainland Planning Process. These two areas do an adequate job of protecting wilderness and wildlife.

Pro-preservation truth: While the Clendenning and Upper Lillooet Valleys are important additions to B.C.'s Provincial Park system, they encompass only 10 percent of the Stoltmann Wilderness and are made up of mostly high elevation or steep-slope non-forested areas. These two parks alone will not sustain the area's threatened grizzly bears or the region's complex ecosystems.

Anti-park propaganda: Careful logging under B.C.'s Forest Practices Code, which is the toughest in the world, will succeed in protecting wildlife, tourism and recreation values in the Stoltmann Wilderness.

Pro-preservation truth: The B.C. Forest Practices Code is extremely weak; much weaker than logging rules for National Forests in the U.S. The vast majority of logging since the B.C. code's inception five years ago is clearcut logging of old growth. Nearly identical to earlier logging, it cannot be termed "careful logging." A growing list of endangered species, damaged salmon rivers, landslides, and the increasing scarcity of low-elevation ancient forests attest to its failure. We must save the few high biodiversity places that remain – including the 500,000-hectare Stoltmann Wilderness.

Anti-park propaganda: The loggers, their families and the economy will suffer if a park is created. We can not afford to save any more wilderness in the region.

Pro-preservation truth: According to a recent Tourism Whistler economic study, continued logging of the Stoltmann Wilderness could sustain up to 220 timber industry jobs. As a protected area it would sustain up to 730 tourism industry jobs. Halting Interfor's raw log exports to foreign mills and establishing an ecoforestry-based/labour intensive Squamish Community Forest on lands currently in Interfor's TFL 38 lands would more than make up the shortfall of timber jobs caused by preservation. Preserving the Stoltmann will result in more employment for local families...not less.

Anti-park propaganda: The B.C. government can not re-visit its land use decisions or there would be chaos. We must all move on and compromise.

Pro-preservation response: There is chaos in the Elaho now because of the bad land use decision. People are being jailed for peacefully trying to protect the Stoltmann's ancient forests. Endangered species and their habitat are being destroyed. New information since the decision was made—a grizzly bear habitat study, a B.C. Forest Service Report on the uniqueness of the area's groves of thousand year old trees, and the One Whistler Tourism Study—all demonstrate the shortcomings of the original decision. The B.C. government must reconsider its Lower Mainland land use decision now and take action to protect wildlife, preserve thousand year-old-trees and create tourism jobs. The Stoltmann Wilderness, including the entire Upper Elaho Valley, must be granted park protection now.

We also made sure that this paper inspired the already faithful followers of the issue and motivated them to continue to support us and write letters and get their friends involved. We did this in the lead article in the paper. This is how it began:

Will Interfor's aggressive anti-conservation tactics covered-up by a clever public relations machine succeed in destroying this gem? NO WAY! say thousands of Canadians - we've just begun to fight! Read on to find out about the amazing Elaho Valley in the Stoltmann Wilderness and what you can do to help preserve it for future generations.

The big turning point in our Elaho campaign

Over the summer of 2000, on a sandbar beside Sims Creek, Squamish Nation artist Aaron Nelson-Moody carved a seven-foot-tall statue he called Cedar Woman. Hundreds of people who participated in the Sims Valley Witness Program watched her shape emerge from the cedar log over the months it took him to carve her.

When it came time in September for the Squamish Nation people to carry Cedar Woman to the top of Bear Bluffs in the Elaho Valley, the place they'd chosen for her to stand, they invited everyone, including eco-activists and the RCMP, to lend a helping hand. Interfor loggers even cleared the trail through their clearcut up to the top of the bluff. Everyone carrying Cedar Woman up the steep mountainside helped the healing begin.

Joe Foy who was there described the day this way:

> The statue arrived at the end of the logging road on a pickup truck from the Sims. It was wrapped in blankets. The day was a mixture of emotions. On the one hand, it was a real chapter mark. On the other, it was like a picnic with kids running around with sandwiches and juice boxes having a good time. Everyone was getting along, including RCMP and protesters who had clashed only a few months before.
>
> It took less than an hour with a couple of rest breaks to get it to the top. There were plenty of helping hands. When a person got tired they just stepped aside and someone rushed in to take their place. There were drums and ceremonies at the top. During this same time the Squamish Nation was undertaking its land use planning for their territory and Cedar Woman was a physical reminder of Squamish Nation's increasing presence on the land.
>
> And as history has revealed, once Cedar Woman went up, there was no more logging!

Cedar Woman will stand on Bear Bluffs in the Elaho Valley for a long time. She overlooks the massive Interfor clearcuts of the Lower Elaho Valley. At her back is the pristine wilderness of the Upper Elaho Valley.

Since she was placed on Bear Bluffs, no additional logging roads or clearcuts have been pushed further up the Elaho. Year 2000 was the final big year for civil disobedience and arrests in the Elaho Valley.

Celebrating Clayoquot Biosphere Reserve after Twenty Years of Campaigning

Year 2000 was also a very big year for Clayoquot Sound. Adriane was jumping with joy when she opened the envelope with her special government invitation to the May 5, 2000 official announcement of the Clayoquot Sound Biosphere Reserve at Long Beach. It was such a good news story that Prime Minister Jean Chrétien, who rarely visited the west coast, would attend in person. He apparently had a soft spot for Long Beach having been the Minister in charge of National Parks in 1971 when Pacific Rim National Park Reserve was established. Knowing that the Wilderness Committee was always cash strapped, NRDC's Liz Barratt-Brown offered to host Adriane and me at one of the beautiful Tofino resorts for the event; a generous offer that we happily accepted. This was the culmination of years of efforts by Adriane and many, many others.

The idea to designate Clayoquot Sound as a UN Biosphere Reserve was actually a decade old. It was first promoted by the U.S. group Ecotrust, who set up a biosphere research office in Clayoquot, and also by the Wilderness Committee after we first learned about Biosphere Reserves at WILD's inaugural conference in Hawaii in 1990. When the NDP made its awful 1993 decision to log two-thirds of Clayoquot, it simultaneously promised to *vigorously pursue United*

Above and facing page: Working together Natives, protesters and RCMP easily carry Cedar Woman to her outlook over the Elaho Valley. Photos: Joe Foy.

Above top: Aaron Nelson-Moody, Squamish artist who carved Cedar Woman speaks at ceremony. Photo: Joe Foy.
Above bottom: Squamish Chief Bill Williams and Witness leader Nancy Bleck beside Cedar Woman on Eagle Bluffs. Photo: Joe Foy.

Nations Biosphere Reserve status for Clayoquot. Adriane saw this as the only opening to pursue more protection. After an attempt failed in July 1996 by the Nuu-chah-nulth to get Greenpeace and the logging companies talking, Adriane took her idea of a Biosphere Reserve as a framework around which to resolve the Clayoquot conflict to Linda Coady at MB. Linda and Adriane jointly presented the idea to the Nuu-chah-nulth and they directed the two women to explore it further.

A small victory came in October 1996 at a meeting of the World Conservation Union in Montreal, just after the re-election of the NDP in B.C., when Adriane, Liz Barratt-Brown, Vicky Husband and others (including Jim Walker, an assistant deputy minister in the Ministry of Environment in Victoria) talked B.C.'s new Environment Minister Paul Ramsey into supporting a resolution that opened the

Local government official signing the letter asking the federal government to seek Biosphere Reserve designation for Clayoquot Sound. Photo: WCWC files.

door to more protection as part of a Clayoquot Biosphere Reserve. Then, in January 1997, Adriane met one-on-one with federal Heritage Minister Sheila Copps who was on the west coast to announce new additions to the Gulf Islands National Park.

Adriane loved to tell me the story: "I had to prime someone in the media to ask Sheila a question about Clayoquot, which he did. Copps' answer put on public record that the federal government would be receptive to backing a UN Biosphere Reserve with the proviso that *all the stakeholders have to be on board*. Her response made media headlines and prompted the various sides in Clayoquot to work together. Adriane heard later from Sheila's staff that the Prime Minister's office was furious with her. But without her gutsiness, things would have stalled.

That statement by Sheila motivated Linda Coady and I to convince all the different stakeholders to sign a letter to Prime Minister Jean Chrétien asking for federal support for a Clayoquot Biosphere Reserve. Eventually everyone's signature was on it: all the First Nations chiefs, the environmental groups, local mayors, the logging companies, even the IWA. I think we totally surprised people on the Hill! The only problem was that the letter arrived in Ottawa just days before the June, 1996 election, too late to exact an election promise. But it set the groundwork for the next government to hire the former secretary of the Central Region Board, Ross McMillan, to coordinate a multi-stakeholder process to prepare a Biosphere Reserve application. And it led to the final celebration in May 2000.

After arriving and checking into our hotel room overlooking the open Pacific Ocean the night before the official Biosphere ceremonies, we headed into Tofino to celebrate with Liz Barratt-Brown, Valerie Langer, Vicky Husband, Tzeporah Berman, Tamara Stark, Susan Jones, Ross McMillan and all the others who had worked so hard to make this event happen. Vicky was so happy she bought all of us dinner! Such celebrations are too infrequent in the circles of environmentalists.

Later that night Adriane talked to me about the people behind the scenes who might never get the credit they deserve, including Nuu-chah-nulth leaders like Nelson Keitlah, Francis Frank, Cliff Atleo, Larry Baird, Louie Frank, Stanley Sam and Moses Martin; and the political staff in Sheila Copps office who set up meetings for her in Ottawa and helped with the 'quiet diplomacy' on the Hill. Not too many people knew about the constant meetings and phone calls that went on between Adriane, Linda Coady, Ross McMillan and Jim Walker, each of whom worked in their own arena to move forward a Clayoquot solution. "It was the brightest and best brainstorming group I've ever worked with," Adriane told me that night.

The next morning we all left early to drive to Wickaninnish at the end of Long Beach in Pacific Rim National Park Reserve where the Biosphere Reserve announcement was going to take place. We wanted to be sure we got there early so we could meet and talk to people beforehand.

Blockading Prime Minister Chrétien's Celebration

Just a short distance past Schooner Cove we ran into a long line of cars stopped on the road. There were at least 50 vehicles already stopped and the line-up grew longer by the minute. We got out of our car and joined the people milling around. They told us what was up. About 100 people were blocking the road ahead. Treaty-related negotiations had just gone sour over the Tofino airport and the Tla-o-quiaht First Nation were protesting the federal government's decision to sell off the airport lands. The Tla-o-quiaht, in whose territory the airport was located, wanted these lands granted to them as part of their land claims settlement package.

There was a great deal of controversy over this blockade. Everyone understood the anger over the airport deal. But not even all of the Nuu-chah-nulth agreed that this was the time to protest. Of course, when else would the Tla-o-quiaht have the chance to impress the Prime Minister with their concerns? Unfortunately, besides us, many of the Nuu-chah-nulth representatives, including dancers from Ahousaht, who were invited to be part of the celebration, were also prevented from attending. The blockaders were letting no one through.

All of a sudden a couple of RCMP escort cars and several limousines passed by on the left hand side of the road and drove right up to the blockading natives. "That's the Prime Minister and his entourage," someone said explaining the obvious. Some of the security guys got out and started talking to the people in charge of the blockade. We were quite far back and couldn't hear what was being said. Liz Barratt-Brown kept asking us, "Why doesn't your Prime Minister get out and talk with them? There's no way [President] Clinton would just sit inside his limousine. He'd get out and ask them 'What's the problem?'" Chatting with people, even upset people, was this American president's style, Liz claimed. She went on and on about how Bill Clinton was unafraid to meet with protestors to find out first hand what was bugging them.

After holding up the Prime Minister's motorcade for less than ten minutes, the protesters let it through. But they continued to stop everyone else. I personally thought this was not very smart, but I held my tongue. At this point Adriane and I walked up to talk with the blockaders who included several chiefs whom we knew. As we learned more about the issue it sure seemed to us like the federal government had not negotiated in good faith.

By now it was raining pretty hard—a typical May day on the wet coast of Vancouver Island. Finally, more than an hour later—about a half an hour after the ceremonies were scheduled to be over—the protesters stood aside and let everyone pass. The official ceremonies were taking place about 10 kilometres further down the road. There were so many people still milling around there that I thought for a minute that the government had delayed the proceedings. But, alas that was not true. It was all over.

Later we read the speech made by Prime Minister Jean Chrétien at Long Beach that May 5, 2000, officially designating Clayoquot Sound as a UNESCO Biosphere Reserve. He got it right:

> Clayoquot Sound is a place of wonder, one whose beauty takes the breath away. It fills you with a sense of our sacred responsibility as stewards of this very special place. Small wonder that its preservation has prompted such passion here and around the world.

Although the ceremonies were over, the Prime Minister was still there. I shook his hand and thanked him. Adriane did the same. Obviously told by one of his aides who she was, he thanked her for her work. During their brief chat the government's official photographer took a picture of the two of them talking. A copy of the photo, signed by the Primer Minister, came in the mail several weeks later.

Adriane Carr shaking hands with Prime Minister Jean Chrétien at the Clayoquot Biosphere Reserve ceremony. Photo sent to her by the Prime Minister with his hand written greeting at the bottom. Photo: Government of Canada.

After the hand-shaking at the official ceremonies ended, all the environmentalists regrouped at Café Pamplona in the Tofino Botanical Garden. George Patterson, who owned it, was a long-time Clayoquot supporter. We toasted each other with champagne and sang a rousing version of Bob Bossin's Clayoquot anthem "Sulphur Passage": *Come you bold men of Clayoquot, come you bold women. There is a fire burning on the mountain, the sting of smoke blowin' in the wind...no pasaran Megin River, no pasaran Clayoquot River...no pasaran Sulphur Passage... .*

That night Adriane and I talked over our 20 years of involvement with the campaign and tabulated the monumental efforts of our staff and volunteers: publishing and distributing 1.4 million copies of 12 different Clayoquot newspapers, three books, five research reports, two hiking maps, seven posters, dozens of T-shirts, bumper stickers, buttons and videos and too many media releases to even count. We wrote grants, conducted scientific research and documented dozens of expeditions into Clayoquot's wilderness. We made hundreds of trips to Tofino and Victoria, organized dozens of public meetings and rallies, put on slideshows, set up information kiosks, organized conferences, worked with First Nations to build three trails and hauled the stump of an ancient cedar left in a Clayoquot clearcut from coast to coast across North America and even to Europe. In one year we gathered 150,000 signatures on a petition. Adriane made 52 trips to Ottawa to talk to the federal Liberals. The mailouts were endless. Our members and supporters were phenomenally generous. In the end we probably raised and spent over 3 million dollars to help save Clayoquot. But the statistic I like the best was that in 1996, when WCWC ran a trailbuilding project in Ahousaht on Flores Island, we employed more people in Clayoquot Sound than logging giant MacMillan Bloedel did at the time.

During the May 5 ceremonies Chrétien announced a $12.5 million federal grant to endow a Clayoquot Biosphere Trust (CBT) to support the new Clayoquot Biosphere Reserve. It was great to get a line item in the federal budget, but unfortunately it was too little to develop the Biosphere Reserve and a new economy. In fact, it was a lot less than the $32.8 million that the federal government granted to the Gwaii Trust. The Gwaii Trust was set up as a locally controlled, interest-generating endowment fund when the South Moresby National Park Reserve/Haida Heritage Area was established in 1987. Its original goal was to have the money spent within the following eight years to help transition the local economy from forestry to tourism. It took until 1995 for the natives and non-natives to agree on the makeup of the committee, how it would operate and transform the trust into a permanent fund to help create "a sustainable Islands community." In the interval, the fund grew to $75 million. Today, managed by a board of both Haida and non-natives living on Haida Gwaii, the annual interest earned is parceled out each year in grants to local applicants for various community projects, like the Haida Gwaii Rediscovery program. By 2005 (during its first ten years in operation) the Gwaii Trust had granted $25 million in total to worthwhile local projects.

The lack of sufficient funding to fully support the economic transition in Clayoquot continues to be a major problem. Iisaak Forest Resources, trying to manage forest resources under a very different set of rules than the big companies, struggles to make money. Yet the protection of Clayoquot's pristine watersheds and the hopes for a lasting Clayoquot solution are pinned on Iisaak and its MOU with the environmental groups. As I finish this book in 2006, Adriane Carr, Valerie Langer, Tzeporah Berman, Liz Barratt-Brown, Vicky Husband, Francis Frank and many others who worked to protect Clayoquot in the 1990s are back in touch to prevent the Clayoquot solution from unraveling.

Adriane Carr leaves the Wilderness Committee to lead the Green Party of B.C.

During her spare time in 1999, Adriane began to get active again in B.C. Green Party politics. She was one of the Green Party's founders in 1983 and served for two years as its first spokesperson. What she had realized back in '83 (that neither the NDP nor the Liberals would ever make the big changes needed to protect the environment) was even truer now, 16 years later. Both parties were traveling down the same unsustainable development road, just traveling in different cars. Ecological concepts were not part of either of these parties' core worldview.

It began with the final straw that snapped and sent Adriane back into politics when <u>not one</u> MLA stood up and condemned the Interfor loggers for beating up the peaceful protesters in the Elaho Valley on September 15, 1999. Non-violence is a fundamental principle of all Green Parties. "The cowardice of all our politicians appalled me. How could they not condemn the use of violence, especially against youth who were peacefully trying to protect ancient forests that shouldn't be logged in the first place?" stated Adriane.

Adriane had become particularly frustrated by the Clark NDP government's anti-environment agenda. They'd jailed people for protesting logging in drinking watersheds and Clayoquot, re-opened the door for off-shore oil and gas development, fast-tracked drilling for oil and gas in the north, doubled the capacity of fish farms and increased raw log exports. Premier Glen Clark had even called protesters in the Great Bear Rainforest "enemies of B.C." She concluded that the most effective way to influence the lawmakers was to play in the same league. They'd take environmental issues seriously if they lost votes to another political party that truly championed these issues.

At the very least, Adriane believed, the stronger the B.C. Green Party grew, the more the other parties would change in positive ways to incorporate green ideas and principals to try to garner the votes of green leaning people.

When Adriane was elected the leader of the Green Party in the fall of 2000, she immediately resigned from WCWC and Andrea Reimer replaced her on the E-team. For me, who had been working together with her on Wilderness Committee projects both at home and in the office for over 17 years, it was the beginning of a hard time.

Other campaigns need attention

I didn't lack things to do. The Wilderness Committee always had many campaigns on its plate. I was already involved in the campaign to protect the "back yard" wilderness near where Adriane and I live on the Sunshine Coast—a 1,500-hectare park on Mount Elphinstone behind Roberts Creek. Despite a thriving tourism industry, less than two percent of the Sunshine Coast region's forestland was protected. The Elphinstone forest, with its beautiful trails and bike paths, was a tourism gem. It also had the highest diversities of fungi found anywhere in Canada and Adriane and I loved to harvest chanterelle mushrooms there each fall. In July, WCWC published 50,000 copies of a four-page full-colour tabloid-sized newspaper titled *Help Make Mt. Elphinstone a Provincial Park now!* (Educational Report Vol. 19 No. 2). This was our second newspaper on the issue. The first one was published in 1995. We mailed both of them to all the residents on the lower Sunshine Coast.

Blue Mountain second growth with a residual cedar tree. Photo: Jeremy Williams.

The vast majority of local residents and the Sunshine Coast Regional District government supported protecting Mt. Elphinstone. Yet the area was slowly being destroyed, cutblock by cutblock, through B.C. Timber Sales, an independent organization within the B.C. Ministry of Forests that auctions blocks of Crown timber to logging companies. This was one issue that really discouraged me. It would take so little to save this special place. Yet the things people did to try to save it, including petitions, letters and local government resolutions—things that are supposed to work in a democracy—were unsuccessful.

Joe had greater success on a local campaign in the Fraser Valley. He was working with Danny Garek, the 1995 recipient of our Eugene Rogers award, to fight a proposed gravel mine in the Upper Pitt River watershed. The mine site was located only a few kilometres from Boise Creek which was now protected in Pinecone Burke Provincial Park. Nearly every local politician of every political stripe was against the mine because of the huge threat it posed to Pitt system fisheries. Joe spent a great deal of time on this campaign. Finally in June 2000 the provincial government pulled the plug on the mine and we celebrated.

Joe immediately moved into another local campaign: to protect Blue Mountain. This second growth forested area abutted the south shore of Alouette Lake (on the north side is Golden Ears Provincial Park). Most of Blue Mountain was logged in the early 20th century. Its beautifully maturing forest had become a de facto park with many bike and hiking trails following the old logging roads.

Working with local conservationists, we called for a total ban on logging in the Blue Mountain area and campaigned to have the entire 7,500-hectare south slope and ridge rolled into Golden Ears Provincial Park. In November 2001 we published a two-page three-colour tabloid-sized newspaper titled *Blue Mountain – Time to Protect Maple Ridge's BACKYARD WILDERNESS* (Educational Report Vol. 20 No. 5 – press run of 100,000 copies). Joe Foy went to countless meetings and even sat for a while on a Forest Service Planning Team to promote this sensible park expansion. Over the summer in 2002 our volunteers helped staff an information picket near the entrance to Golden Ears Provincial Park. They were there every weekend right up until Labour Day. The support for this park addition was overwhelming. We collected 2,117 signatures on our Save Blue Mountain petition, which we then sent to B.C. Parks Minister, Joyce Murray.

Final Chapter in the September 15th 1999 Loggers' Attack on the Elaho Protesters

In January 2001, justice was finally dealt to some of the loggers who attacked the PATH protesters at Lava Camp in the Elaho Valley on September 15, 1999. B.C. Provincial Court Justice Ellen Burdett sentenced five International Forest Products (Interfor) employees and contractors for their roles in the attack. They were part of a mob of approximately 70 loggers who viciously attacked, terrorized and destroyed the property of the peaceful protestors. Also attacked was WCWC's millennium tree research camp boss James Jamieson who happened to be passing through PATH's camp the morning of the attack. Justice Burdett noted in her sentencing that *at the very least, Interfor gave tacit corporate approval* to the violence. Justice Burdett also concluded that the attacks were pre-meditated and directly involved at least ten to fifteen of the seventy men at the scene at the time of the attack.

While eight citizens including women and a 15-year-old boy were attacked and three of them required hospital treatment, only one man, Richard James, was convicted for these assaults. The others were convicted only for uttering threats.

Despite these strong words from Justice Burdett regarding the vigilante nature of the crimes and the damage to individuals, property and the community as a result of the attacks, none of the men were sentenced to serve any time in jail. Instead, they were given suspended sentences with one-year probation. The men were also required to take anger management classes, send letters of apology to the victims and undertake 40 hours of community service. Four men were required to pay $1,250 each in restitution. The only man sentenced for criminal assault was not required to pay restitution to the victims. Interfor was not charged with any offence despite its role in the affair (written into the September 15th square on the desk calendar of one of Interfor's supervisors were the words "Ethnic Cleansing Day"). But, nonetheless, Interfor paid a high price for this despicable day in the court of public opinion.

Justice hardly seems the right word to use when you contrast the light sentencing of the violent loggers with the heavy sentencing of peaceful environmentalists four months earlier. On September 15, 2000 (ironically, the one-year anniversary of the mob attack by Interfor loggers) Betty Krawczyk and Barney Kern each got one-year prison terms for peacefully blockading logging in the Elaho, with no time off for good behaviour. Supreme Court Justice Glen Parrett imposed these harsh sentences on Betty and Barney for violating the court injunction that Parrett himself had ordered.

It's hard to understand why contempt of court is a worse crime than intimidating people, beating them up and destroying their property. But from a purely tactical viewpoint, the longer and harsher the sentences against environmentalists the better they serve our cause. The harsh sentencing of Betty and Barney incensed people across the nation. People believed that cutting down the last of the giant thousand-year-old trees and destroying the last large tract of old-growth rainforest in the Greater Vancouver area were the real crimes.

Sea to Sky LRMP offers hope of more protection for more of the Stoltmann park proposal area

In January 2001 the provincial government kicked off the Sea to Sky Land and Resources Management Process (LRMP), as a follow-up to the Lower Mainland Protected Area Strategy that left local land use plans unfinished. Included in the LRMP were the Elaho, Sims, Callaghan, Soo, Meager, Salal and Boulder valleys—all within our Stoltmann National Park proposal. Finally, during this LRMP, Whistler made its voice for wilderness preservation heard! Knowledgeable people from Whistler sat at the LRMP table and introduced land use zoning that would protect more of the Stoltmann Wilderness including the Upper Soo watershed. By the end of 2005 the LRMP planning phase was complete except for negotiations with First Nations such as the Squamish and Lil'wat.

Squamish Nation designates Wild Spirit Places to be protected in its territory

In the spring of 2001 the Squamish Nation, spurred on by the rapid loss of the natural forest in their traditional territory in the Upper Elaho and Sims Valleys, conducted an extensive community-based land use planning process for the forests and wilderness areas of their territory. On June 19, 2001 the Squamish Nation released its draft plan entitled *Sacred Land* (*Xay Temixw* in the Squamish language).

The Squamish's draft land use plan called for the establishment of four "Wild Spirit Places" covering 8.5 percent of the Squamish Nation's traditional territory: the Upper Elaho Valley, Sims Valley, the middle west side of the Squamish River, and the Upper Callaghan Valley. These areas would be off-limits to industrial logging and other commercial developments and would be managed by the Squamish Nation for cultural purposes protection.

In addition, the well-thought-out Squamish Nation *Sacred Land Plan* included three other zones, the largest of which was its "Forest Stewardship Zone" where logging would be carried out in an ecologically sound way under some form of co-management with the current logging industry.

Within the Forest Stewardship Zone were "Restoration Zones" where a lot of logging and industrial development had already occurred and where the natural and cultural values needed to be restored. Also within the Forest Stewardship Zone were "Sensitive Zones" where relatively large amounts of old-growth forest still remained. Here, care had to be taken to ensure that the wildlife and cultural values were not compromised as further development in that area occurs.

WCWC put out a press release praising the Squamish's Sacred Land Plan titled: *The Wilderness Committee Applauds Squamish Nation's Land Use Plan. Implementation of the Squamish Nation's vision could end Elaho Valley "war in the woods."*

'It's the most comprehensive land use plan we've seen for the forests and mountains of this region. We need this kind of First Nations' land use planning throughout all of BC,' said Joe in the release.

Interfor stops logging the Upper Elaho

When logging season began again in the Stoltmann Wilderness in the spring of 2001, Interfor left the area above Lava Creek and moved operations across the Elaho River to concentrate instead on pushing a road up the west side of the Elaho Valley. The year before, Joe and several volunteers had hiked up the west side of the Elaho to a point where they were finally able to cross the river on a log to the east side and come back down our Elaho to Meager Trail. They found no deep canyoned tributaries on the west side, which meant that Interfor could easily build a road right to the top of the watershed, turning the Upper Elaho into just another roaded, logged valley. It would effectively end our campaign to save this wilderness.

Interfor had just begun to bulldoze that road when the Squamish Nation "put their foot down" and demanded that Interfor cease logging in the Upper Elaho and Sims Valleys. The Squamish Nation was in negotiations with Interfor, so the company complied, stopping its road building in the Elaho immediately and several months later in the Sims too.

The Squamish Nation's success in getting Interfor to stop logging brought a quiet de-escalation of WCWC's Stoltmann Campaign. We'd been pouring a large portion of our time and resources into this campaign since 1995. We carried out expeditions and produced publications about the Elaho and Sims, but at a much reduced level. The freeing up of time and money allowed us to pursue other campaigns, including a photo expedition to East Creek Valley in the Bugaboo Rainforest—part of the world's only inland temperate rainforest.

Looking back, the year 2001 marked the end of the very big campaigns like the Stoltmann and Clayoquot and the beginning of a new period when the Wilderness Committee became engaged in more new issues than we had been since 1995. In December 2001 we produced our last Stoltmann Educational Report, a four-colour, four-page, tabloid-sized newspaper titled *Saving the Stoltmann Wilderness Valley by Valley - No logging in the Upper Elaho and Sims for the past year!* (Vol. 20 No. 7 with a print run of 120,000 copies).

Another big win

On July 24, 2001, WCWC celebrated another win. The Alberta government established a 591,008-hectare Caribou Mountains

Wildlands Provincial Park. It abuts the southwest border of Wood Buffalo National Park and is the last stronghold of woodland caribou in Alberta. Gray Jones, our Alberta Wilderness Committee director, championed the protection of this special remote area for years. He mounted expeditions there, consulted with the Red Creek Cree, brought it to media attention and made sure the issue was in our annual wilderness calendar.

The park was exactly what we had asked for in our fall 1996 four-page, full-colour tabloid-sized newspaper titled *Caribou Mountain Wilderness* (Educational Report Vol. 15 No. 14 - press run 50,000 copies).

Gray was away when the government made the announcement. He never issued a press release praising the government for the park. And we got no credit in the media for this win. But there is no doubt in my mind that without the Alberta Wilderness Committee and Gray's efforts over the years, this spectacular plateau of tundra, lichens and woodland caribou would never have been protected.

A new campaigner brings fresh energy to WCWC

In March of 2001, eight months before I took an extended leave of absence from WCWC, our E-team hired a new campaigner, Gwen Barlee. She had worked on filming the Outer Limits TV show in Vancouver for the previous two years but wanted to do something more meaningful. A friend of hers saw our tiny ad in the *Vancouver Sun* for a campaigner. Gwen, who had a background in political science, was an easy choice for us. Her diverse educational and employment experience included social work, film production and web development. She was a past executive team member of the Victoria Chapter of the Council of Canadians. But most importantly, she had a passion for environmental and social advocacy. Although I don't know by how much, I do know she took a big pay cut to work for us. That's commitment.

We assigned her to be our Interior Rainforest campaigner, a new campaign we were just starting to get involved in with the Valhalla Wilderness Society in the Columbia Mountains that was being logged to death. Then, after just her first 10 days at work getting up to speed on that campaign, we abruptly changed her assignment. Jacqueline Pruner, our Endangered Species campaigner abruptly quit WCWC for another job. She and Joe were scheduled to go to Ottawa in a couple of days to push for a much stronger federal Endangered Species protection law than was being proposed.

Highly adaptable, Gwen got up to speed on this issue quickly. At 2:30 p.m. on May 15, 2001, with our giant inflated grizzly bear standing on Parliament Hill as a backdrop, she, Joe and Eve Adams (formally the Executive Assistant to the Honourable Charles Caccia on Parliament Hill) made public our Wilderness Committee's *Bill C-5 Species at Risk Act Comments Brief*, which they had presented earlier that day to the House of Commons Standing Committee on Environment and Sustainable Development. In the brief we advocated a strong Species at Risk legislation that would force action to save endangered species and their critical habitats across Canada whenever provinces, like B.C., failed to do so.

Our seven-page brief bluntly called for the federal government's proposed Species at Risk legislation (Bill C-5), to be substantially strengthened or tossed out all together. Our press release stated:

'We explained to the Standing Committee that the Province of BC is pushing two threatened populations of animals to their demise in the Chilliwack Valley and the surrounding North Cascade Mountain Range;' said Joe Foy, WCWC Campaign Director.

'Without some sort of powerful intervention the North Cascade

Gwen Barlee at David Anderson's office attempting in vain to call him via cell phone after a funeral parade for extinct species (like where the spotted owl is headed) in an effort to convince David Anderson to intervene using his powers under Canada's Species at Risk Act (SARA). Photo: WCWC files.

Mountains populations of spotted owl and grizzly bear on the Canadian side of the international boundary, will be finished off because of aggressive logging practices, which are permitted by the government of B.C.,' explained Gwen Barlee, WCWC endangered species campaigner...

'We told the Standing Committee that - it's sad to say - but in its present form Bill-C5 is dishonest legislation and really is worse than having no endangered species legislation at all,' said Foy.

Our Wilderness Committee delegation in Ottawa also contended in its brief that to enact Bill C-5 in its present toothless form would give timber companies a public relations tool which they could use to fend off international market pressure. The weak Bill C-5 would do nothing to protect species like the spotted owl and grizzly bear in Canada.

In fact, it would actually speed up their demise by calming the concerns of domestic and foreign buyers of Canadian timber products by making them think that we have laws to protect Canada's species at risk when we don't. Bill C-5 wouldn't stop the logging of the spotted owl and grizzly bear's critical habitats.

On the road across Canada to promote strong federal endangered species protection legislation

In early fall 2001, Gwen planned to travel across Canada with Paula Neuman, another WCWC staffer, to raise public awareness about endangered species legislation. Their departure date was delayed by the terrorists' attack on the World Trade Center in New

May 15, 2001 demonstration in front of the Parliament Buildings in Ottawa to get strong federal endangered species protection (SARA, Species at Risk Act) legislation. Eve Adams, who was working on WCWC's Endangered Species campaign, is standing on the far right. Photo: WCWC files.

York on September 11th. The whole trip ended up being a bit strange under the pale of this shattering event. They took with them our giant inflatable grizzly bear whose nose wasn't quite right (many thought it looked like a pig). "It really got people's attention though, drawing crowds of people asking 'what is it?'" explained Gwen.

They stopped at three or four places in each province, putting on slideshows, gathering signatures on our Endangered Species petition, handing out Endangered Species poll cards (we had printed up 120,000 of these cards) and talking to literally thousands of people all the way to Ottawa. "Nine out of ten people thought Canada already had endangered species legislation," said Gwen. They arrived in Ottawa with 15,000 signatures on our petition. Someone in government told them that 30,000 poll cards had been sent in. What a successful trip!

Subsequent public opinion polling showed that 85 percent of the people in Canada wanted strong endangered species legislation. An obvious question was: 'With that level of support why didn't the federal government deliver on strong legislation?' Unlike two previous attempts (Bills C-33 and C-65) that died on the order papers, the House of Commons passed Bill C-5, the weak and ineffectual Species at Risk Act, on July 11, 2002.

"Canada is a paper tiger. We sign international agreements (like the Biodiversity Convention we signed in December 1992) and take the accolades and clapping. But when it comes to actually implementing something on the ground, we are not there. It's catching up to us. We are becoming an international embarrassment," says Gwen Barlee, now the policy coordinator at the Wilderness Committee and member of the E-team.

Successive Canadian governments have refused to do anything that curtails industry. Effective legislation that truly protects endangered species would mean stopping certain developments and activities in critical habitats in order to save them from being forced into extinction. No government to date has had the guts to pass such legislation.

Chapter 64

Defeating the NDP's "Working Forest" Legislation

Protection prospects look good for the Great Bear Rainforest

Continuing the fight to save endangered species

Extremely unpopular NDP tries to curry favour with big forest companies by proposing "working forest" legislation

The Clark government's press release of February 12, 2001 took us totally by surprise. It read in part:

The B.C. government will introduce legislation this spring to protect families and forest communities by defining and securing working forests in British Columbia, Forests Minister Gordon Wilson announced today....

Working forests of B.C. were defined as *the lands where timber management and harvesting is a priority use or where timber production will be secured over future multiple rotations. After local communities have made their land-use plans, working forests would have a status in law like the status given to provincial parks.*

Most disturbing was the short time allocated for public input. It was limited to 15 days and the government encouraged people to submit their comments via an online web-based comment form.

Our reaction was swift and harsh. In our press release Joe Foy said:

The people of this province have spent literally millions of person hours in land use planning meetings over the past ten years. The result of this phenomenal investment of time is the NDP protecting just 3% more of the province's forests for a grand total of 5% of provincial forests set aside in provincial parks. Now the NDP is trying to give away 95% of the public's forest lands to private corporations with just 15 days of public input...

This is the biggest corporate land grab in B.C. history – involving almost half the province. Yet the B.C. public is not being given a fair chance to have their say. It's a disgrace!

We pulled out all stops to get as many comments in as possible during the short public input time. It was hard to believe that the NDP would think such a law would be popular. The citizens of B.C. cherish their publicly owned forests and this "working forest" legislation was tantamount to treason. It was one of the biggest blitzes we ever pulled off. Ken Wu's email list came into good use. We also scheduled a big rally on March 14, 2001 against this sell-out idea. On the same day we held our rally the B.C. government issued this press release:

The Ministry of Forests has concluded consultations on the proposed working forest legislation, Forests Minister Gordon Wilson announced today.

'We have received a great deal of input and comments, and the legislation as it is currently proposed is clearly not ready for consideration by the legislature,' said Wilson. 'I believe British Columbians want legislation that provides greater certainty for forest jobs and forest communities, and that will only be achieved through dialogue and consensus-building. I will recommend to the premier and cabinet that government do further work with all parties to set this important issue right.'

Well, Wilson and the NDP really messed up on this! I read their press release to the crowd at our rally (we received it just minutes before I spoke) to rousing cheers. But the damage had been done. This ploy set the stage for the soon-to-be-elected Liberal government to "consult" with the same big companies and introduce virtually the same legislation, although much more cleverly written. During the next three years, Ken Wu passionately and tenaciously fought the Liberal's "Working Forest" initiative.

Facing page: Waterfalls in the Roscoe Inlet of the Great Bear Rainforest, an area that was eventually protected. Photo: WCWC files.
Below: Protest against the Working Forest legislation proposed by NDP Forest Minister Gordon Wilson. Photo: WCWC files.

Rally at the Legislature against Working Forest legislation held in May 2001 right before the writ is dropped indicating a provincial election. In my hand is the press release WCWC had just received from the outgoing B.C. government saying it was not going to pursue Working Forest legislation. Photo: WCWC files.

The NDP protects some of the Great Bear Rainforest and sets up a process to protect some more

On April 4, 2001, just days before the drop of the provincial election writ, the government announced with a lot of fanfare an Interim Land Use Plan for B.C.'s 4.8 million-hectare Central Coast—the area known as the Great Bear Rainforest. The big ceremonies, held in Vancouver and attended by all the major environmental groups including us, also featured the signing of a General Protocol Agreement with the First Nations of the region. The Great Bear had been the focus of a Land and Resource Management Plan (LRMP) process since June of 1997. The planning area stretched from northern Vancouver Island through the mainland mid coast to Princess Royal Island, home of the white-coated "Spirit Bear." Although the area is large, only about 4,500 people live in the region. More than half of the population is aboriginal including members of the Nuxalk, Heiltsuk, Kitasoo and Oweekeno Nations. The major population centres, all unincorporated, include Hagensborg, Bella Coola, Firvale, Bella Bella, Ocean Falls, Shearwater, Klemtu, Namu, and Oweekeno.

The Interim Land Use Plan for the Central Coast identified 20 new areas for protection encompassing 441,000 hectares where industrial development would be strictly prohibited. These new areas brought the total protection of the Central Coast to 20 percent of the land base. The main driver of this conservation move, ironically, was the blockades that included European activists and the market campaign boycotts that Greenpeace spearheaded.

The Plan also designated another 17 areas encompassing 534,000 hectares as "Option Areas." In these 17 areas, development was put on hold pending completion of land use planning and development of a new framework for ecosystem-based management.

Our press release heaped praise on the government. Joe Foy said, *It's a fantastic achievement! By protecting these ancient forests for future generations, Premier Dosanjh has done something everyone in B.C. can be proud of.*

In that same press release we stated that we would also be launching a *Cedar is Sacred* campaign to raise public awareness about the diminishing supply of old-growth red cedar trees on B.C.'s mainland coast and in B.C.'s interior temperate rainforest. We highlighted the need to conserve B.C.'s remaining old-growth red cedar for native cultural and heritage purposes. We then contracted a native researcher who wrote us a report confirming that there was growing interest in the aboriginal communities on the coast for such conservation. We followed this up with a report co-published with the David Suzuki Foundation titled *A Vanishing Heritage: the Loss of Ancient Red Cedar from Canada's Rainforest* that statistically documented the demise of this species. We also published the *Last Voyage of the Black Ship* by Michael Nicoll Yahgulanass, a graphic novel about the destruction of the ancient red cedar ecosystem on Haida Gwaii. This is a campaign that still needs much attention. One of the greatest tragedies of B.C. forestry is that nearly all of the giant ancient red cedars, the backbone of the traditional west coast aboriginal culture have been "liquidated."

Grassroots conservation organizations marginalized in the new mid coast planning process

The Valhalla Wilderness Society, the group that first proposed and campaigned to protect the Kermode ("Spirit") bears on B.C.'s mid-coast, and the Raincoast Conservation Society, the group that originally inventoried and launched the whole mid-coast wilderness campaign, were two major groups sidelined in the big Central Coast LRMP announcement. The reason was: they, like WCWC and the David Suzuki Foundation, didn't sit at the well-funded LRMP negotiating table. Who should sit at the table and how thoroughly they should consult with the grassroots groups had been bones of contention right from the beginning. Most of the infighting between the groups "inside" and "outside" of the process focused on whether any environmental group had the right to negotiate away key wilderness areas that a fellow group was fighting on the ground to protect.

Remembering the frustration of environmental groups at the Lower Mainland Protected Area Strategy table negotiating away our Stoltmann Wilderness campaign, we tended to sympathize with the Valhallas and Raincoast. We kept working with both of these on-the-ground groups, although we kept up good relations with the big enviro-groups sitting at the LRMP table, too. We knew that without market campaign pressure generated by Greenpeace there certainly would be less wilderness protected.

In November 2001, Gwen Barlee conducted a fly-over photo expedition of the Ecstall River in the Great Bear Rainforest near Prince Rupert. This was a big contentious area that Ian and Karen McAllister of Raincoast suggested we focus on. We had tentatively planned to produce an educational campaign newspaper about this big, beautiful watershed. But difficulty in gaining First Nations' permission for our campaign, and increasing tensions between the conservation groups working in the area made it too difficult for WCWC to operate effectively.

With so many environmental groups working on the mainland coast, we decided to put our time and money into other areas that were being sidelined and ignored in the big focus on the Great Bear area. Our contribution to the Great Bear campaign over the years had been two newspapers, a half dozen posters, three expeditions, a workshop, feature spots in several calendars and lots of phone calls.

Our work all culminated in a draft proposal for a new type of conservation designation—a salmon sanctuary. This new designation was aimed at engaging the federal government, through the Department of Fisheries and Oceans and the Department of Indian Affairs, in protecting the "salmon forests" of B.C.'s coast. The Ecstall would be an ideal candidate. The trouble was that by the time we had it together the Ecstall was deeply entrenched in the provincial North Coast LRMP process, and the opportunity to engage the federal government had diminished. After that, we concentrated on alerting people to the remarkable record of aboriginal use of Hansen Island off the north end of Vancouver Island which was ironically part of the same LRMP for reasons that only make sense bureaucratically. We were very pleased when Hansen Island was ultimately protected.

Prolonged pitched battle to save the spotted owls

In spring 2001 the Wilderness Committee intensified a smouldering campaign that is more aptly described as a "spotted owl survival battle." This battle continues on in the courts and in the media today. The battle scars are the "scorched earth" clearcuts that were once-verdant old-growth spotted owl habitat. Logging of their habitat has reduced the owls from an estimated 100 pairs when we first started our campaign in the early '90s to six pairs today. In 2006, the nearly extinct owls produced, as far as we know, just one chick in B.C.

In the spring of 2001 a young provincial government biologist, Carla Lenihan, was so upset with a Ministry of Forest's decision to log the remaining old-growth forest in an area called Siwash Creek where a female spotted owl was residing, that she blew the whistle on government's ongoing disregard for its own scientific experts who had strongly recommending against logging the area. In her affidavit to the court, Carla said: *"The last thing I wish to say is that I give this affidavit because I believe that it is my ethical and moral obligation as a professional biologist..."* Carla paid a heavy price for her courageous stand. In the years to come she was forced out of her job by a bureaucracy that appeared to be more concerned with creating the illusion of protecting this endangered species than it was in actually dealing concretely with the reality of the spotted owls' logging induced decline.

Saving the owl meant getting our Sierra Legal Defence Fund lawyers, armed with Carla's affidavit and a video-tape evidence of the owl in a tree right on the edge of a freshly built logging road, to seek an injunction to halt Cattermole Timber from logging. The only hitch was that the Wilderness Committee could be responsible, in the event that we lost the court case, for any costs to Cattermole resulting from the delay of logging. It came out in court that because the logs were being cut under a raw log export permit that expired in a month, the costs award to the Wilderness Committee could be as high as $175,000! Imagine our anger to find that this last ancient forest stand for spotted owls was not only scheduled to be logged against the advice of the government's own biologists, but that the raw logs would be exported to enrich and employ workers in another country. Despite our anger, the reality of a potential huge hit to our always 'wing-and-a-prayer' finances was terrifying. If we lost—always a possibility in B.C. courts as there are very few laws that protect the environment—we would lose not only the spotted owl but probably WCWC as well. In the end there really wasn't a choice. We had our lawyers let the court know we were prepared to put up the deed to our strata title store in Gastown as collateral.

Justice Anderson of the Supreme Court of British Columbia granted WCWC a temporary injunction to halt the logging from proceeding for a few days, in order to give our SLDF lawyers the opportunity to argue the case in a three-day-long judicial review court hearing to quash the logging permit.

To gather more information about the owl and video evidence, Joe Foy, Andy Miller (our owl expert), Andy's five-year-old daughter Roan, and Daniel Gautreau, a videographer, got the key to the gate on the logging road from Cattermole's head office in Chilliwack and explored the area. They could not find the owl. Cattermole had already logged about one third of the block before we got the injunction, including the area where the owl was last seen perched.

When they tried to open the gate on their way back, the key wouldn't work. Some smart aleck employee had changed the lock since they entered earlier that day! Joe had some shovels and a mattock in the truck and the three of them worked for about six hours to make a narrow roadway around the gate. It was cold and Andy wanted to get Roan home. Finally, at about 3 a.m. he was able to slip by with his smaller vehicle. Joe's truck couldn't make it. He and Daniel stayed the night. When the loggers opened the gate in the morning on their way to work, they were laughing and joking and generally gave Joe a hard time.

Our $175,000 gamble paid off! A few weeks later we won our court case. The Justice struck down the District Manager's approval for the three cutblocks and sent the matter back for another Forest Manager's decision regarding logging in those proposed cutblocks requiring that the new decision maker consider the information that we had presented in court. The Forest Service gave the job of reassessment to Cindy Stern, a Manager in another Forest District. She came up with a compromise. Two blocks were saved, one lost. She claimed that the logging in the "lost" block would be done in such a way as to enhance the owl's habitat! The block Stern allowed to be logged was still critical spotted owl habitat and the only way to enhance it was to allow no logging at all. We challenged that second decision in court and lost. This time the justice assigned court costs against us—over $10,000—that we had to pay to Cattermole.

That was a lot of money for us. But looking at the bright side, the court case got tons of publicity and brought public attention to the issue. It ended up being the first of a long series of court cases and other efforts to save the spotted owl. What became patently clear over time is that the B.C. government publicly said that it would do anything to save the spotted owl, except the one thing that would actually work—stop the logging of their old-growth habitat.

A visit to the world's largest single source CO_2 pollution site

In the early fall of 2000, Gray Jones called to invite me to join him on a weeklong visit to Fort McMurray, the bitumen-rush boom city located in the boreal forest 435 kilometres northeast of Edmonton. He wanted me to help shoot a video about the tar sands and global warming that he had finally decided to produce for WCWC-Alberta. He also requested a couple thousand dollars from "head office" to hire a helicopter, because the only way to capture the truly colossal size of the fossil fuel extraction developments was from the air. He had lined up several professional videographers who were prepared to volunteer their time on the project. This was to be the video that Gray had lambasted Greenpeace for never producing when it gave up its campaign against Suncor. I leaped at the chance.

We spent a week in Fort McMurray interviewing people and videoing the industrial sites, the slag ponds and the so-called reclamation areas, staying with some of Gray's eccentric friends. I found the people in Fort McMurray very guarded and afraid to say anything negative about the tar sands development. No one would even comment on the obvious local air pollution problems. Gray's friends told me that several years earlier a marriage councilor in private practice had spoken out on the issue and consequently was blackballed. He went from being extremely busy with referrals from the major compa-

Panoramic view of spotted owl habitat intensively logged in 2001. A female spotted owl was last sighted on June 14, 2001 (at "A") at the edge of the old-growth corridor

nies to no one using his services and he eventually had to leave.

One day, while we were near the giant Syncrude plant, it "vented." Something went wrong and every valve opened up and spewed out toxic fumes and steam—whatever was backed up in the pipes—in order to prevent the plant from literally blowing up. The air stank. People living nearby told us that similar events occurred fairly frequently. How they could stand it, I do not know.

The sheer size of the tar sands operations was mind-boggling. The power shovel machines that dug out the sands were five stories high. Workers were transported to and from them by helicopters that landed on the top of the machines! They lived in those huge machines for shifts that lasted several days. One ex-employee, whom we promised not to identify, told us he had to conduct fake sampling of the Athabasca River, which received Suncor's wastewater. It made me sick to think of this, especially when new medical reports were released in the spring of 2006 about the extremely high rates of unusual cancers amongst the native people who live downstream of the tar sands.

We documented the phony land reclamation project where a company put buffalo to graze to indicate that everything was all right. We documented Syncrude's vast settling ponds. But then there was a colossal problem that wasn't easy to document that dwarfed the others: the discharge of a tasteless and odourless pollutant that is changing the face of the whole planet—CO_2. The tar sands produce more CO_2 greenhouse gas per barrel of refined oil than any other fossil fuel. This is mainly due to the energy needed to extract it from the ground.

Gray went up in the helicopter with our videographer. It was a low overcast day with great clarity below the thick low-lying blanket of clouds. The clouds forced the helicopter to travel in restricted airspace and, although warning bells went off in the cockpit, there was nothing he could do about it. It was a godsend for us. We got incredible close up shots of the plants, a plume of brown coloured water being discharged into the Athabasca River and the colossal machines that looked like something out of a science fiction movie.

I came away from there with a profound feeling of distress and powerlessness. There was no way we could stop this. As I flew out from Edmonton, Gray assured me that although the people doing the editing could not get to our video before the New Year, it would be done by spring.

But the video was never completed. Gray got interested in helping out victims of gas well flaring and left the employ of WCWC in early 2003 before this project was finished.

Taking a leave to work on Adriane's Pro Rep Initiative

Starting in April 2001, with the drop of the provincial election writ, I took a month off without pay to help Adriane with her election campaign. The Green Party did fantastically well getting 12.4 percent of the vote across B.C. Adriane did the best of any Green in B.C., getting over 27 percent of the vote in her electoral district and tying for second place with incumbent Gordon Wilson, the NDP forest minister who introduced the idea of "Working Forest" legislation that WCWC helped squash in the dying days of government. Although no Green candidate won a seat in the legislature, it was one of the highest popular votes that a Green Party had achieved anywhere in the world. Greens have been elected to state and national governments in 30 countries where there are fairer, more democratic proportional voting systems.

After that election, Adriane and the Green Party decided to push for a mixed member proportional voting system, like the systems in New Zealand and Germany, where a party gets the same percentage of seats in the legislature as its share of the popular vote in the election. The best way to achieve that goal was to undertake an initiative under B.C.'s Recall and Initiative Act, like the anti bear hunting initiative, backed by WCWC, for which I was the proponent six years earlier. Adriane would be the proponent for a "Proportional Representative Initiative" and I really wanted to help her. I decided to ask for a six-month unpaid leave of absence from the Wilderness Committee starting in January 2002. The board of directors granted it. The six months ended up being eight months due to bureaucratic delays in the initiative process. That ended up being long enough to rediscover that I truly loved working with Adriane. I started thinking about a different future than continuing to hands-on manage the Wilderness Committee.

Last Christmas appeal letter and calendar write-ups

In the fall before I left to work on the "Pro Rep Initiative," I wrote my final set of inspirational descriptions of the issues featured in our 2002 Endangered Wilderness and Endangered Species annual calendars. I'd written these synopses for 22 years. It's true that the Internet made it much easier than before. But, after doing these write-ups for

on the left of Siwash Creek Valley near the South Anderson River. "B" shows cutblock 36-10A, logged in June 2001. Photo: WCWC files.

so many years, I had become weary of the task. It was one job I was happy to pass on to Joe, Gwen and other staff members to do in future years!

The other task that I was pleased to pass on was the annual Christmas-time appeal letter. Each year we had to come up with a new idea. Every year I start with the sure belief that somewhere in my heart and mind is the ultimate letter that will deeply touch all of our members, especially those who faithfully pay their Wilderness Committee annual dues but never give more than that (over half of our members).

This time I procrastinated longer than usual, because I just couldn't think of a good theme. The night before the absolute final deadline to print it I had a vivid dream. In the dream I was sitting at the reception desk in the Wilderness Committee's office a few days before Christmas when the mail arrived. There were bags and bags of postpaid return envelopes containing cheques for $50, $100 and even some for $1,000, totaling hundreds of thousands of dollars.

But instead of being overjoyed, I was very sad. Why? Because it was too late!

There is nothing we could effectively do to save wilderness with all that money. All the big unprotected areas were already roaded, logged and/or mined. Global warming was spiraling out of control, long past the point where we could do something about it. There was no way to halt the catastrophic loss of species. Their critical habitat had been destroyed and there was no way to recreate it.

I woke up with a start in a cold sweat. Thank goodness it was just a dream!

That morning I reiterated this dream in my appeal letter to our members and supporters. It worked well, one of the Wilderness Committee's most effective appeals of all time. Although people gave more generously to this appeal than to others, the great piles of money envisioned in the dream did not come pouring in. I fervently hope that my dream was not a prophetic nightmare!

Making the big leap to greener pastures

When the time rolled around for me to return to the Wilderness Committee in September of 2002, I realized that WCWC was doing quite well without me! Andrea Reimer had taken on the job of Executive Director. She was doing an excellent job, excelling at it as she had in fundraising and writing the grant proposals after Adriane left. Campaigns were rolling along as best as could be expected under the anti-environmental Liberal government (the new government didn't even have an environment ministry!). Finally the Committee was reducing its debt. Everything seemed to be running smoothly. Sooner or later I'd have to leave anyway, so I figured now was as good a time as any to take the leap. I liked not having to commute regularly to Vancouver from the Sunshine Coast. And I especially liked working with Adriane again, helping out as a volunteer on Green Party campaigns.

But the Wilderness Committee had one last job that everyone said that only I could do. WCWC really needed a chronicle of its 25-year history of activities and antics that illustrated our campaign style and philosophy. If it wasn't published in a book, our history was sure to be lost. Those who write it make history, they say.

The issue of "the book" had been brought up before and I had flatly turned down the job every time. I knew that it would be a lot of work and couldn't be done while managing operations and running active campaigns. Now the Committee offered to pay me $500 a month to work on the book part-time beginning in late 2002. I accepted. It turned out to be way more difficult and time consuming than I, ever in my wildest estimates, imagined. I learned fast that nobody remembers exactly when anything happened and everybody's version of history is different. This book is my version.

Joe Foy writing personal messages on Christmas appeal letters. Photo: WCWC files.

Above: 400 people took to the streets in Victoria on May 7, 2005 at a WCWC pre-election rally before the May 17th 2005 election. UVic WCWC club members Cayce Foster (left) and Ziya He (right) hold the banner. Ken Wu on far left holds the megaphone.

Below: Later, as part of the rally the protesters below lie forming an aerial art "Tree of Life" image telling voters to "Wake Up" and think of the environment when they cast their ballots. Photos: WCWC-Victoria.

Chapter 65

Digging in to protect what we have gained

A tsunami of environmental rollbacks in B.C.

B.C.'s new Liberal government was elected by a landslide on May 17, 2001, winning with 58 percent of the popular vote an unbelievable 77 out of 79 seats. Even before they were elected everyone knew they would give environmental protection a low priority. Premier Gordon Campbell had made only one firm environmental promise—to protect Burns Bog—during his election campaign. In answer to the Wilderness Committee's 13-item questionnaire sent out to all the candidates prior to the election, the Liberal Party provided blanket answers for all of its candidates. Instead of a yes or no answer to our yes or no questions, there were lots of run-around paragraph-long answers that boiled down to lots of "No's."

No, the Liberals were not going to continue the blanket moratorium on grizzly bear hunting. No, they were not going to support full protection of community watersheds by banning of pesticides, herbicides, logging and other industrial disturbance in them. Instead, they were going to use *results-based* environmental protection (a newspeak for lower or no standards of protection). No, they were not going to enact endangered species protection legislation. Instead, they simply promised to *meet the government's duty to protect all wildlife species*. No, they would not ban open net cage salmon farming. They supported the fish farming industry.

Amongst the "Yes" answers was a "Yes" that we didn't want. Like the NDP before them, the Liberals would pursue "working forest" legislation to give companies more secure tenure over public lands.

One "Yes" made us happy. A Liberal government would ban raw log exports. But, once the Liberals got into power, instead of banning raw log exports they increased them.

Few of us forecast how swiftly and massively the Liberals would launch their assault on environmental legislation. Within a few days of assuming power, the Liberals scrapped the three-year moratorium on the trophy hunting of grizzly bears imposed by the NDP a few months before the NDP dropped the election writ. The Liberals wouldn't wait to give scientists a chance to comprehensively inventory the population and assess this species at risk.

The anti-environment bent of the government was clear when Premier Gordon Campbell re-organized the ministries and formed his first cabinet. He eliminated the word "environment" from a ministry title.

Another bad sign was the new "Ministry of Deregulation." Its job was to ensure that every ministry cut bureaucratic red tape and the "unnecessary regulations" that were a "barrier to investment." During a publicly telecast cabinet meeting, the new Liberal ministers mindlessly discussed whether they should cut one-third of the number of regulations or one-third of the pages of regulations. They never once questioned what the regulations were intended to protect. Under the B.C. Liberals, more environmental protection laws have been rolled back than at any time in the history of B.C.

The new government cut budgets for parks, forest management and environmental protection that were already cut to the bones by the former NDP government. With fewer forest service staff, environmental protection officers and park rangers, vital environment oversight and protection, even over water quality, diminished. It reorganized the Agricultural Land Reserve and made it easier for developers to take prime agricultural lands out of protected status.

The Liberals were particularly excited about scrapping the loathed Forest Practices Code. They replaced it with weakened regulations that no longer required government foresters to approve cutblocks in tenured public forests or even keep forest inventory maps of the publicly owned forest lands. The Liberals' new "results-based forest practices code" actually stipulated that "timber supply" could not be "unduly restricted" by environmental constraints, even in critically endangered species habitat. Getting out volumes of timber became, as it was in the 1950s, the supreme goal of forest management. They also watered down pulp mill pollution standards designed to limit the discharge of cancer-causing dioxins.

Gordon Campbell's B.C. Liberals, like Ralph Klein's Alberta conservatives, opposed Canada's ratification of the Kyoto Accord. The federal Liberals ratified it anyway, committing Canada to reducing its greenhouse gas emissions. B.C. completely ignored the commitment and raced ahead with plans to greatly increase fossil fuel developments. The new Liberal government opened the door to coal-fired power production, provided incentives for coal bed methane exploration and pressured the federal government to lift the moratorium on offshore oil and gas development along B.C.'s coast. They gave huge subsidies to northeast gas developments to speed up production for export and turned a blind eye to massive pollution problems in the area.

Intent on getting B.C.'s old-style resources-based economy rolling again, the new government made mineral claim staking easier and weakened local government controls over environmentally damaging "provincially significant" development projects. They lifted the moratorium on new salmon farms and embarked on an aggressive expansion of salmon farms that endangered wild salmon.

Fish Farm in the Broughton Archipelago. Photo: WCWC files.

Despite saying it wanted to increase tourism revenues, the Liberals treated B.C. parks as if they were a net government liability. Over their first four years they cut the B.C. parks' budget by 30 percent. They also cut park ranger staff by 30 percent, discontinued park interpretative programs, allowed logging inside park boundaries, shut down many campgrounds and took out pit toilets. Even more troubling, the Park Act was weakened in November of 2003 to allow, at the discretion of the Minister, private commercial development inside parks. The government even started marketing sites in parks to developers. It also tried to make money off the parks by installing parking meters and charging for firewood. The result has been a dramatic 25 percent province wide decline in the number of people visiting our parks. B.C.'s park system, once world class, is now in decline.

Leaked documents tip the government's hand and mobilize opposition

One of the big things that changed under the Liberal government was how many leaked documents we received from alarmed government staff. One such document leaked to us in 2002 revealed that the B.C. government aimed to get out of managing parks altogether! The rationale behind this radical plan was that provincial parks don't in and of themselves make money. The government's strategy was to incrementally privatize our provincial parks with the 20-year goal of having the B.C. government totally out of the park business! We forwarded the leaked document to the *Vancouver Sun* and it made front-page news. Because of some brave soul, these dastardly plans were at least temporarily derailed!

In 2004 the Valhalla Wilderness Society got and gave to the *Vancouver Sun* a leaked document that revealed which of B.C.'s most treasured provincial parks were being eyed for development. The "B.C. Parks Lodge Strategy" described how certain parks should have lodges ("containing up to 80 beds") to provide tourist accommodations in wilderness areas that, until now, had been accessible only to those willing to hike in, carrying tents. Among the parks short listed for development were Mount Robson, protecting the highest peak in the Rocky Mountains, and Tweedsmuir Park, known as one of the greatest wilderness parks in North America.

The government started selling the park lodge idea to the public claiming that these accommodations would make it easier for average B.C. families who did not want to camp to be able to enjoy our parks. This was not really true. Gwen Barlee, our policy adviser, found out that the government Park Lodge Strategy included a range of roofed accommodations from Yurts/Cabins to "major resorts."

The government documents talked about "high-end" lodges for "well-heeled" tourists. They were going to be high-end luxury lodges for wealthy international tourists. The B.C. government was clandestinely, but aggressively, marketing this "investment opportunity" to big developers.

This whole lodge scheme completely disregarded previous public consultations confirming that British Columbians do not want commercial operations like this in our parks. Our group and other conservation groups like the Valhalla Wilderness Society blasted the idea of privatizing B.C.'s parks. Public response on radio talk shows and in newspapers' letters to the editor confirmed peoples' outrage at the idea of selling private concessions deep within our wilderness parks. Our exposure of this scheme, while not achieving a total abandonment of the plan, forced government to slow down its plans. Keeping our big parks pristine is still a battle that has to be fought and won.

Parking charged to visit paradise

In 2003 the B.C. government took its first step in privatizing parks by installing parking meters in 41 of B.C.'s most popular parks. It didn't even bother consulting first with the public. It turned out to be a very unpopular and expensive move, made even more costly by the obvious haste of the whole fiasco.

Installing the meters cost $1.2 million. People quickly discovered that some of the machines only accepted U.S. quarters. Canadian quarters jammed them. It would take another $360,000 to replace these "defective" meters but the government opted to not fix them. During the Wilderness Committee's tour of parks to galvanise opposition to this new "tax," volunteers noticed that parking lots were

Elaborate and expensive parking meter in Golden Ears Provincial Park. Photo: WCWC files.

only about half full. Yet the year before they were jammed full of vehicles. One plausible explanation was that people were so angry about the meters that many of them refused to visit the metered parks. Our volunteers also noticed that the vandalism of meters was extremely high. We learned that someone had winched out a meter from one of the provincial parks along the Sea to Sky highway to Squamish, put it in the back of their pickup and dropped it off on the lawn of the Squamish City Hall in the mistaken belief that the city was responsible for its placement.

Gwen, who had become an expert in using the Freedom of Information (FOI) Act to access information that the B.C. government should readily make available but doesn't, put in a FOI request to find out how much money the meters were bringing in and how many park visitors there were before and after the meters were installed. The government resisted giving out this specific information. "You can always tell if you have a good one [FOI request] when the government fights giving you the information tooth and nail," commented Gwen.

Eventually she extracted the fact that visitors to the 14 parks in the Lower Mainland area had dropped by 20 percent—one million visitors—from the year prior to the meters being installed. Overall the attendance in B.C. parks across the province had declined 25 percent and some of that is attributable to the meters. It was at its lowest level since 1986.

Other documents accessed through FOI revealed that public compliance with parking meters averaged less than 25 percent. Most people parked and didn't pay. Despite issuing 15,000 parking violation tickets, zero revenue was collected in parking fines. Why? People discovered by word of mouth that the parks officials did not have the legal authority to ticket and collect the money.

Gwen also learned that the meters brought in $700,000 and the $1.2 million "cost" didn't include the cost of vandalism, maintenance, staff time and contracts to service the meters. When all those expenses were taken into account, the government had lost lots of money. Even more important, local communities and tourism-related businesses lost considerable revenues from the decline in park visitors and the government lost even more money that would have come as spin off tax revenues from those businesses.

Since the installation of the meters there have been more than 40 articles in the print media that cast negative publicity on the provincial government for their park parking tax. Despite proof that the meters were a costly mistake, however, the government has not reneged. Gwen explains the situation this way: *The current Liberal government digs in its heels when it gets bad publicity for something it does. On the parking meter issue it has got nothing but grief from the beginning. The government thought that the outcry would calm down after a year. But it hasn't. It still isn't making money from them and park visits are still down.*

"Park Lovers' Tour" in Manning Provincial Park in 2004. Photo: WCWC files.

Park Lovers' Roving Kiosk Tour

To build public support for our campaign to stop government from further park privatization and to get funding restored to BC Parks, we launched a B.C. Park Lovers' Roving Kiosk Tour in the summer of 2004. Starting at Manning Park on June 18th the tour visited 10 popular parks across British Columbia. The tour lasted through the Labour Day long weekend. The focus was on educating park visitors about how the government's funding cuts, new user fees and new permits allowing industrial and commercial developments in parks damage B.C.'s provincial park system. In conjunction with this campaign we published *B.C. Parks - A WORLD FAMOUS LEGACY* newspaper (Educational Report Vol. 23 No 5 with a print run of 90,000 copies).

New environment minister takes a second look at park parking meters

Our campaign has had impact. After the Liberals were re-elected in 2005 they re-established an Environment Ministry. The new minister was Barry Penner, a former park ranger, who had earned the respect of many environmentalists because of his strong opposition to the Sumas II natural gas burning electrical generation mega-project proposed just south of the border in the Fraser Valley airshed. One of the first things Minister Penner did was to commission a study on the parking meters in parks. Although the study was completed in early 2006, it had not been released as of publication date of this book. Gwen contends that what recently happened in Washington State gives us hope. Washington put in new user fees in their State Parks about the same time as B.C. did, provoking the same negative response. In the spring of 2006, the state cancelled the fees. Gwen believes B.C. is sure to follow.

Manitoba Wilderness Committee amplifies its efforts

In June 2002, our Manitoba Wilderness Committee responded to the Manitoba government's East Side Planning Initiative, a land-use planning process for this big boreal forest wilderness region east of Lake Winnipeg. The government's document suggested that some developments proceed before the planning process was completed. We presented a different vision for the area in a newspaper titled *Manitoba's East Shore Wilderness* (Educational Report Vol. 21 No. 3 with a print run of 70,000 copies). The paper lauded the government's planning process as an incredible opportunity to create a large, interconnected network of protected areas in this largest piece of intact boreal forest on the planet. It also made the point that there was only one chance to "get it right" with this area. It urged the government not to permit any industrial developments in the area before the planning process was completed.

That same summer the Manitoba Wilderness Committee held a *Save our Parks Festival* at The Forks, a National Historic Site at the junction of the Red and Assiniboine Rivers in Winnipeg. The Forks is <u>the</u> "meeting place" where major festivals and other special events are held. At the *Save our Parks Festival* many speakers, poets, local musical acts and other performers helped raise awareness and mobilize

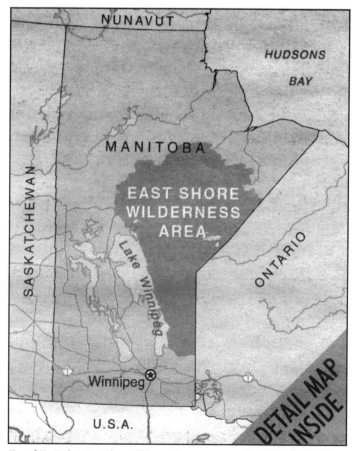

Map of Manitoba's East Shore Wilderness from newspaper. Document: WCWC files.

the public in support of our Wilderness Committee's campaign to stop logging in Manitoba's existing provincial parks and make sure every park is a fully protected area.

As part of our Stop-Logging-In-Parks campaign, Ron Thiessen also worked on a video exposing the damaging effects of park logging. During a videotaping session of active clearcut logging in Whiteshell Provincial Park, angry loggers confronted Ron and his video crew. Ron truthfully informed them that his crew was videotaping for a Red River Community College student project. Ron didn't, however, mention the other component of the project, which was the Wilderness Committee's use of the footage to protest logging in provincial parks. Accepting his half explanation, the loggers became quite friendly, asked how they could help and provided some very useful information.

A victory for a few nesting marbled murrelets

Throughout our history we have always tried to lend a hand to local grassroots environmental groups in various ways such as co-publishing newspapers with them and becoming a co-petitioner in court cases. When the Sunshine Coast Conservation Association (SCCA) came to us in 2002 and asked if we would support its attempt to stop the logging of an old-growth forest heavily populated with nesting murrelets we said yes. The SCCA was seeking a judicial review of the B.C. Forest Service District Manager's approval of a cutblock in Jervis Inlet north of the Caren Range. Our name on the petition would perhaps give it more clout. The outcome: a rare court victory.

In the Wilderness Committee's celebratory press release Joe concluded: *This is a significant victory for Marbled Murrelets and all threatened and endangered species. There is very little habitat left for this old-growth forest dependant seabird on the Sunshine Coast. We thought the decision to allow logging was flawed, and we were correct.*

This fascinating, now endangered, bird has become one of the keys to saving the old-growth forest left along the B.C. coast. In the Sunshine Coast forest district less than 10 percent of the original old growth remains. A great deal of it is not suitable marbled murrelet nesting habitat (no large limbs with heavy moss pads). Scientists studying marbled murrelets conclude that this species' entire remaining nesting habitat must be saved in order for it to survive. But the B.C. government's politically-motivated policy allows the net "loss" of only one percent of the annual allowable cut for the protection of wildlife and forest dependent species, including endangered species. It's an arbitrary barrier that's based on maintaining logging revenues not sustaining biodiversity. No one believes that the "one percent solution" is enough to save the marbled murrelets and the other old-growth dependent species. In practice, it ends up being an excuse to log where logging should not occur. Forest Service officials continue to approve cutting permits in endangered species' habitat with the excuse that the "one percent habitat protection" will surely be achieved somewhere else.

Book cover. WCWC publications file.

One of the last projects I completed before going on leave in January 2001 was to help produce and publish Paul Jones' book *The Marbled Murrelets of the Caren Range and Middlepoint Bight*. It recounts his detailed observations of adult marbled murrelets feeding their chicks in the first active nest found along the west coast of Canada. Without scientific accounts like Paul Jones', people would have little understanding and appreciation of the fascinating lives of this species that now faces extinction.

Chapter 66

Callous Canadian governments refuse to protect Canada's endangered species

Embarrassing Canada in the Netherlands

In April of 2002, Gwen Barlee flew to the Netherlands to attend the sixth meeting of the Conference of the Parties to the Convention on Biological Diversity. One of her chief goals was to hold the Canadian government accountable for its failure to enact a law to protect endangered species in Canada. Such a law was long overdue. Ever since Canada ratified the International Biodiversity Convention in December 1992, the Wilderness Committee had argued that our governments should move quickly to pass strong Endangered Species Legislation as part of our international obligations.

The Wilderness Committee's press release sent out before the

Gwen Barlee promoting biodiversity in the Netherlands. Photo: WCWC files.

April 7-19 conference began with this statement:

'I know it sounds strange that I have to go to the Netherlands, but I've already been to Ottawa and Victoria asking for strong endangered species legislation. It's a sad comment on the lack of environmental leadership shown by our federal and provincial governments that we need to go abroad to hold them accountable,' remarked an angry Gwen Barlee, endangered species coordinator with the Western Canada Wilderness Committee.

Gwen told me later that she went to this conference to counter the false "goody-two-shoes" impression that the Canadian government presented to the world. What they said publicly and what was actually happening in Canada were worlds apart. She produced a full-colour handout detailing Canada's dismal record on endangered species and handed the flyers out to as many of the delegates as she could. She also left some in prominent places for other delegates to pick up. After a while she noticed a guy running around behind her, picking up her flyers as soon as she left an area. "We caught him red-handed. He worked for the Canadian government. They were really desperate to keep their false image. Our country was the 'Great Pretender' when it came to environmental protection. Brown to the core," explained Gwen who said that quite a few delegates came up to her and asked, "What's going on with Canada?"

At this conference, environmentalists presented Canada with a "stump of the day" award for "its unwillingness to agree on international priorities," greenwashing its own forest industry, and despite being the wealthiest forested nation in the world, blocking prioritization of forests within the agreement. Canada, a former leader in environmental protection, had become a follower of the U.S. in seeking a weakening of international standards of environmental protection.

Gwen found that for Canada this conference was all about trade. Canada was even pressing some countries to come forward with resolutions that would undermine the integrity of the Biodiversity Convention. The bywords were 'keep your options open' and 'don't let it [the biodiversity conservation] impinge on any profits'.

Canadian government passes an endangered species Act knowing it doesn't protect endangered species

Canada passed SARA (the Species At Risk Act) in June 2002. It came after two earlier attempts at legislation failed. Sadly, the final product was as weak and toothless as the first ones put forward.

SARA is of virtually no help to almost all of Canada's endangered species because it applies only to federal lands and doesn't require protection of the critical habitats of the species at risk. In B.C., federal lands encompass less than one percent of the province: Primarily Indian Reserve lands, Department of Defense lands, ports, Canada Post office properties and federal prisons.

Why did Canada write and pass an act that obviously would not work? Good question. It was not from lack of environmental groups, including the Wilderness Committee, clearly spelling out what the act should say to be effective. Nor was it due to lack of public will. Polling showed that a large majority of Canadians support strong, effective endangered species legislation.

People get that we're in a biodiversity crisis. In fact, we're in the midst of Earth's sixth major mass extinction of species. The sixth in all of geological time! But for the first time, one species is the prime culprit—human beings, who are destroying, altering, and fragmenting other species' habitats and introducing exotic species that crowd out native ones.

The consequence of this mass species' extinction is not known. But virtually every scientist agrees that there will be irreversible negative consequences for future generations and every species we hold on to is part of our safety net. Unfortunately, governments don't yet get this. Or maybe they are just too fearful of alienating the corporate interests that profit from habitat and species destruction.

Calling for strong B.C. habitat protection legislation that will save endangered species

Disgusted with the federal government's lackluster laws, many environmentalists refocused their efforts on fighting for provincial species-at-risk legislation. It was one of the top priorities highlighted by environmental groups in the lead-up to the 2005 B.C. election.

Then, in an Ancient Forests Conference on March 18, 2006 that Cassbreea Savage and Ken Wu of WCWC-Victoria helped organize, Neville Winchester turned the concept of Endangered Species legislation on its head. "We should be calling for endangered habitat legislation," he said. It made total sense. It's only by protecting habitats that we protect species.

I've thought a lot about this idea ever since Neville mentioned it. The point is to stop the fragmentation and loss of the natural habitats that species need to survive. Such legislation would have to include a science-based listing of threatened ecosystems at risk and a legislated mandate to halt further damage to these ecosystems. It must also require mandatory ecosystem recovery plans and sufficient funding to do the job. If it had teeth and money, such legislation would save and recover all the endangered species in a threatened ecosystem both large and small.

The big question: will those who wield economic power and make money off the current development path be willing to reduce profits and change their business plans to do this? Will political parties have the courage to take on the powerful interests that are blocking this rational approach?

Maintaining B.C.'s biodiversity is especially important because we are the most biologically diverse province in Canada. We are also the province where habitat degradation has impacted the greatest number of species. As of 2006, BC Conservation Data Centre lists 1,367 species "at risk."

Only a few species get some theoretical "protection." And that is only if their protection is negligible in terms of impacting resource extraction. It is no exaggeration to call B.C.'s record on protecting species abysmal.

Provincial government's claim that other Acts protect endangered species is bogus

What appalls me more than anything is the B.C. government's flagrant lies about its strong legislation and efforts to protect endangered species. For years the government could have used its powers under the Wildlife Act to list species as endangered. But from 1980 to June 2006 the provincial cabinet listed only four species as endangered: the burrowing owl, white pelican, sea otter and Vancouver Island marmot. Prior to 2006, 34 vertebrates were listed "at risk" under the Forest and Range Practices Act. Then with great fanfare, in June of 2006, the government added another 45. When the additions were made to the list, the Wilderness Committee issued a media release panning this move *as a cynical attempt to avoid federal requirements to protect BC's growing list of species at risk. This listing process is simply meant to side step federal obligations to protect endangered species in British Columbia.*

BC is doing everything it can to give the appearance it is taking action when in actuality they are letting species at risk fall between the cracks.

Simply listing species is only a first step. In and of itself it doesn't do anything to help. Listing must be coupled with the design and implementation of recovery plans, including plans to protect and restore critical habitat. B.C. has no such plans for any species!

On March 3, 2005, the B.C. Forest Practices Board in a press release titled *Provincial Systems Fail to Protect Threatened Species* found a systemic failure in government policy to protect threatened species such as marbled murrelets on crown forestlands. Its conclusion was based on its investigation of a 1999 complaint concerning logging approvals on southwestern Vancouver Island in areas that are prime marbled murrelet nesting habitat. The Board actually released its findings back in November of 2001, right after it had completed its investigation of the 1999 complaint. At that time the Board recommended that the provincial government act promptly to complete detailed land-use planning and establish wildlife habitat areas for threatened and endangered species. The B.C. government did nothing to correct the situation! Hence the Board tried again.

In the 2005 release on the matter, Forest Practices Board Chair Bruce Fraser said:

This is just one example of a recurring breakdown in the habitat conservation system, which begins with a lack of clear targets set by government for population levels of threatened species such as marbled murrelets. Furthermore, ministries have limited resources for timely completion of land-use plans on the landscape level, which is needed to provide guidance to resource managers and forest companies for conservation objectives in their operating areas. Many companies are doing valuable work on habitat assessments but these efforts require a specific government objective for murrelet conservation to be truly effective.

Government has imposed a one per cent timber supply constraint on the creation of wildlife habitat areas, which are key tools for preserving murrelet habitat. This leads agencies and forest companies to propose habitat areas, which are driven by arbitrary policy targets instead of relying on science-based assessments of wildlife habitat needs. It would be more logical to use science to determine the habitat needs of species first, and then government can decide what impact on other sectors it is willing to accept to protect wildlife habitat.

The B.C. government's failure to adequately protect species contravenes Canada's international and national commitments, risks disastrous economic repercussions and ignores the moral obligation to pass on to our children an undiminished biosphere. This irresponsibility of current politicians will haunt future generations.

Near the tipping point

Joe Foy thinks that people are at the point where they understand it's time to stop all old-growth logging in the Lower Mainland-Fraser Valley area. He explains why he has come to this conclusion:

When I was a kid you could fish in the upper Chilliwack River. Then at some point they found declining fish populations. There weren't that many wild fish left in the river. So sports fishing in the upper river was banned.

We are reaching the point now where we have fished-out the old-growth forest. There is not much old-growth forest left in these valleys. It's reasonable now to say 'you can't fish old-growth forests in these valleys.' It is a conservation decision we apply to things when they get scarce. We make that choice as a society.

Less than 10 percent of the Lower Mainland's low elevation old-growth forests are left. Old-growth dependent species like the spotted owl are on the brink of extinction. If there ever is a Land and Resource Management Plan (LRMP) done for the Fraser Canyon Harrison region—and there hasn't been one yet—such a plan would reasonably call for an end to old-growth logging. It is the only hope of saving the spotted owl and other old-growth species. It is also our only hope for, one day, figuring out how old-growth forest ecosystems really work.

Manning Provincial Park picket to save critical spotted owl habitat from being logged

Operating our information kiosk at Sutton Pass during the hot Clayoquot summers of '93 to '94, we learned that a highly visible roadside booth is a very effective way to alert the motoring public about a wilderness issue. When Joe learned that Interfor planned

Protest to stop Interfor's proposed logging road through Manning Park. Photo: WCWC files.

to build a road through Manning Park to log spotted owl habitat in several valleys adjoining the park, he came up with the obvious campaign response. He set up our old kiosk beside the Hope-Princeton Highway, put up big banners and collected signatures on a petition against the road and the logging. The valleys (18-Mile and 20-Mile Valleys) were virtually unknown. But Manning Park is well-loved by British Columbians.

Over the Canada Day long weekend in July 2002, the Wilderness Committee volunteers staffing our kiosk got a very warm reception from people. Many waved and honked their horns. People who stopped to talk with our volunteers were shocked to hear that the provincial government was even entertaining Interfor's request to build a road through the park. All but two people who stopped signed our petition. After a few more weekends of our Manning Park kiosk campaign, Interfor withdrew its application to build the road and log the valleys. It was a clear victory for the Wilderness Committee's campaign style.

Trying to stop logging, cutblock-by-cutblock, to protect specific spotted owls

In 2002 we also succeeded in a campaign to get two big companies (Interfor and Canfor) to stop logging spotted owl habitat. Our victory came right after we co-published a report with the Sierra Legal Defence Fund (titled *Logging to Extinction*) that listed these companies as the biggest culprits in the demise of spotted owls. We distributed it widely. But our victory was short-lived. Almost immediately afterward the B.C. government embarked on a tenure reform program requiring the province's big forest tenure-holders to give back 20 percent of their cut to be redistributed to smaller companies and local tenures. Both Interfor and Canfor sold back to the government their cut in spotted owl habitat. Smart move. Then the B.C. government issued the cutting permits and, through its small business and timber sales programs, became the largest logger of spotted owl habitat. Our win against Interfor and Canfor meant nothing.

In August 2002 the Supreme Court of B.C. handed down another big loss for us. We were petitioning the court to strike down a Forest Service decision to approve Cattermole Timber's plans to log in spotted owl habitat where there was an active owl site. The owl was there when the trees fell!. The court turned down our request. While the

Old-growth logs originating from prime spotted owl territory in the lower Fraser Valley. Photo: Jeremy Williams – August 2002.

Justice found that the spotted owls in the specified area were indeed in grave risk of extirpation, he basically came to the conclusion that B.C.'s forestry laws did not protect a species from extinction. If the provincial government had intended to prevent extinctions, it would have passed a specific law such as an endangered species protection act, which of course no provincial government has done.

We 'upped the ante'. In October of 2002, faced with the B.C. government's active involvement in logging spotted owl habitat and the B.C. court's failure to offer any hope of redress, the Sierra Legal Defence Fund on behalf of Western Canada Wilderness Committee and Washington-based Northwest Ecosystem Alliance filed a petition with the United States Secretary of the Interior. Our petition sought to have the U.S. place a ban on the importation of B.C. wood products originating from critical spotted owl areas. It requested that the President of the United States certify Canada as a country in breach of international treaties to protect endangered species. The petition asked for a ban on importing B.C. wood products unless B.C. logging companies certified that the wood did not come from spotted owl habitat.

This was a great campaign initiative and got a lot of publicity at the time, but the petition went nowhere. Under no obligation to respond, the U.S. Secretary of the Interior didn't. And nothing ever came of it.

In November, 2002, the *New York Times* (NYT) wrote a feature article about the spotted owl situation in B.C. that was not flattering to the B.C. government. In response, Premier Gordon Campbell wrote a letter claiming that B.C. had more old-growth forests today than it had 100 years ago! WCWC along with other environmental organizations bought an advertisement in the NYT pointing out the error in Campbell's infamous claim.

B.C. District Forest Service office in Chilliwack subtly resists our efforts to save the spotted owls

Andy Miller, the Wilderness Committee's wildlife biologist in charge of our spotted owl campaign, described the hurdles he had to jump to get the cutblock information we needed for one of our spotted owl court cases this way:

We'd phone in advance to get a map and they would say yes and then you'd drive all the way out there and they'd say no... Just to yank our chain. One time I phoned the Chilliwack Forest District Office and asked for a particular Forest Development Plan, a public document that shows where they were cutting. They were non-committal as to whether or not they would release it. I drove out there with Roan, my four-year-old daughter. When I got there and started to talk to the secretary about what I wanted, all of a sudden I heard doors slamming in the back of the office. All the staff were leaving. It was 2:30 in the afternoon!

'What's going on?' I asked.

'Oh, I guess everyone finished their meeting and all the foresters have left. There is no one here to give you the information you want. Sorry.'

'I am going to stay in this office until you give it to me or call the police to drag me out of here,' I replied.

I called a reporter at the Chilliwack Progress and told her what was happening and invited her to come down to the Forest Service office to cover the story. I also called Devon Page, our SLDF lawyer, and explained the situation to him. We were in court at the time and Devon was in close touch with Crown Council–the Attorney General's lawyers.

Meanwhile the secretary called the police. Soon the phone rang and the secretary answered. A government lawyer was on the line telling her that what I wanted was a public document and demanded that she find someone to give it to me. She explained that there was no one there to do it. After discussing the matter over for a short time, they came to a solution. The District Office would give me the Forest Development Plan that I wanted the first thing the next morning. I said that was OK and called the reporter who was on his way and told him not to come, explaining how the matter had been resolved. The secretary called off the police. What a fiasco in terms of government responsibility.

Our losing fight to save the spotted owl continues

Andy Miller has a firm grasp on the Wilderness Committee's strategy to save the spotted owl.

We know we might lose the court cases. A lawsuit also gets a lot of public attention. It's a fairly cost effective thing for the Wilderness Committee to do. We got new attention including even international attention. It reaches a different audience than the educational newspapers do. The legal cases are only one component of a campaign. Over the last couple of years we have been going out, taking pictures and issuing press release, owl site by owl site, of habitats that are being logged, vilifying one company after another responsible for the logging. We've taken tons of reporters with us.

During the one and a half years that Andy was on the government's Spotted Owl Recovery Team (SORT) he got all the documents about the few remaining owl locations. Andy resigned from SORT in February 2004 with a flare. He released his alternative recovery strategy on CBC Almanac, inviting other Recovery Team members to support his strategy. This is what Andy had to say about his experience on the Recovery Team and what he believes the future has in store for spotted owls in B.C.:

My Alternative Strategy contained everything that the Recovery Team wanted to do, which was all kinds of habitat improvements like thinning young forests, and population augmentation like helping single owls find mates. The few remaining single and baby owls were so surrounded by clearcuts that they could not travel to locate mates. So you would catch these birds and move them to where another single owl lived. The new recovery team on which I sat was specifically directed by our ministry bosses not to recommend new habitat protections but to leave that up to them, despite clear direction from the federal government to do just that. I wanted to do everything that they wanted to do...plus protect the bird's habitat.

They were told that they could not advocate for habitat protection. Sitting in the back room, members of the team from the Canadian Wildlife Service and the Ministry of the Environment who constituted about half of the Team membership were openly supportive of me. But they were political realists.

When you become a bureaucrat you become an incremental change person. You don't look for big change. You've had it beaten into you that that doesn't happen. They acknowledged that I was right, but they couldn't go there.

What I learned while sitting on the recovery team was that the extirpation of the spotted owl had been exquisitely managed, perhaps even planned; funding for spotted owl management had been severely restricted, scientific opinion ignored, and government scientists were not allowed to search for new spotted owls. I did let myself get too optimistic. The previous recovery team scientists, having made recommendations to protect owl habitat, were disbanded by the B.C. government and replaced by bureaucrats who were directed to minimize habitat protection. The result was a 3 percent reduction in logging. The few owls they knew about were kept secret while their estimates, which were greatly inflated, were public. At the time of the initial recovery efforts in the early 90s there were 50 sites. Half were pairs and half were singles in 1993. They never allocated much money to surveys. They only went back to the sites they knew about.

Then it came to a crisis with the timber industry, which was saying that there were a whole lot more out there. In 2004 –2005 the industry put up a whole lot of money and hired a consultant to do a thorough job. They genuinely thought that they were going to find a lot more. They did find about 20 more owl sites—about what the scientists had predicted.

2003 was another bad year for the spotted owl. In a desperate attempt to save five baby spotted owls born in British Columbia that year, the provincial government's biologists captured and radio-tagged the birds. Andy Miller, who was still on the Spotted Owl Recovery Team at the time, explained to me that it was done in response to the dramatic decline in the spotted owl population, to determine exactly where and when the baby owls were dying in B.C. None of these birds survived.

About the same time, we had to courier a $11,855 cheque to Cattermole Timber Co., the flagrant logger of spotted owl nesting habitat. It was full payment for their court costs which was awarded to them by the Appeals Court of B.C. at the conclusion of our unsuccessful appeal of its logging a cutblock in spotted owl habitat located in Anderson Valley, just beyond Hope, B.C.

Joe summed up the situation this way: *We lost the appeals case in July, Cattermole timber began to log the spotted owl habitat in late September and now we've lost our money in October. It hasn't been a very good time for us… or the owl.*

Realizing that the B.C. government wasn't going to do anything to save the owls, on February 27, 2004, the Sierra Club of Canada, Wilderness Committee, ForestEthics and the David Suzuki Foundation, petitioned federal Environment Minister David Anderson to invoke the emergency provisions of Canada's Species At Risk Act. Our joint petition cited the sharp decline in spotted owl populations: there were only 14 spotted owls recorded in B.C. in 2003. We detailed not only the B.C. government's failure to act to prevent the owl's extinction in Canada but also its continued approval of logging in the owls' critical habitat despite the advice of government scientists who called for a halt to the logging.

Our lawyer, Devon Page of Sierra Legal Defence Fund (SLDF), represented the four conservation organizations on the petition. He said in a press release issued at the time: *We are appealing to Minister Anderson to immediately ask his cabinet colleagues to issue an emergency order under SARA to protect the habitat of Canada's spotted owl. Otherwise, this owl will definitely go extinct in Canada within this decade.*

On May 5, 2004, Anderson replied to our petition stating that while he agreed that the spotted owl faces a high risk of extirpation (extinction in Canada), he was not yet prepared to agree to our request because protection for the spotted owl is the responsibility of the B.C. government.

Since then there has been no move on the part of the federal government to step in and make B.C. protect the spotted owl. The federal Species At Risk Act is a complete sham.

In 2006, the SLDF, on behalf of the Wilderness Committee, launched the first court case under the federal Species At Risk Act to test the power of the Act to force the federal government to intervene

Cover of Andy Miller's Alternative Recovery Strategy report on the spotted owl. Document: WCWC archives.

when a provincial government fails to protect endangered species. This case in the federal courts was to review the Federal Minister of the Environment's de facto decision to let the spotted owl go extinct in B.C. Our point is that, like Liberal Minister David Anderson before her, the new Conservative Minster, Rona Ambrose, is not exercising the ministerial powers that she has under SARA to prevent the extinction from happening. Our case is winding its way through the courts.

After talking at length with Andy Miller, it's my belief that the government's Spotted Owl Recovery Team was a con job right from the start. The forest industry and Ministry of Forests representatives always held the balance of power. Written right into the terms of reference was 'no net loss' in annual cut for the timber industry. A supposed conservation-oriented "compromise" was to allow logging in the active spotted owl sites as long as 67 percent of the old growth remained. But many of the sites didn't even have that amount of old growth left! The lower elevation good owl habitat was logged and the higher elevation poorer habitat was spared. The spotted owls' population continues to decline.

The B.C. government's spotted owl management strategy has been a dismal failure. It is a case of too little too late. Spotted owl conservation is still constrained by being limited to only having a one percent diminishment of timber supply. The prime problem, of course, has been the continuous logging in the old-growth forests that the owl depends on to survive. How can anyone expect to save the owl and not impact logging? Government itself has caused the spotted owls' demise.

Chapter 67

"Working Forest" monster held off

Brand new campaigns

Leading the charge against the "Second Coming" of Working Forest Legislation

On January 22, 2003, the B.C. government released its proposed *Working Forest Initiative for British Columbia*. The new plan was even worse than the one tested by the NDP in 2001. It proposed legally designating B.C.'s public forest lands (the 45 million hectares outside of existing parks) as a guaranteed logging zone. This would effectively put an end to the establishment of new protected areas. But it also suggested streamlining the privatization of Crown forest lands by eliminating any ministry other than the Ministry of Sustainable Resource Management from the approval/review process in Crown land sales. It would be simple, one-stop shopping for those who wanted to buy, own and privately develop Crown forest lands.

Although the sale of public forest lands is currently just for "higher" purposes like suburban sprawl, agriculture and golf courses, it looked suspiciously like the "Working Forest Initiative" would open the door to privatizing lands for logging. Many big corporations have pushed this for decades.

The clandestine route to privatizing B.C.'s forest lands would simply be to rescind Provincial Forest designation and replace it with a "Working Forest" designation. The protections under the Forest Act that forbid the sale of the Provincial Forest for forestry purposes would thus be gone. The Wilderness Committee raised this issue repeatedly, much to the ire of the government. Although they always retorted that "the laws don't allow the sale of Crown lands for forestry" I never heard them promise to include the Forest Act's prohibitions against privatization in their "Working Forest" law.

At any rate, we had already determined that the proposed Working Forest's Crown land zonation system was a form of privatization, as it would increase the compensable rights of private logging companies on public lands. Working Forest legislation would force taxpayers to pay greater compensation to private logging companies for lost logging rights—say, to create a new park or settle a native land claim. The idea of compensation was not new. In the 1990s many people criticized the idea of paying money for corporate "loss of logging opportunity" when the new Vancouver Island parks were created. After all, they said, the people of B.C. own both the lands and the trees growing on them and have the right to decide that they want a park instead of logging. Because a Working Forest designation would entrench the corporate rights to the trees even more, the cost of "buying the trees back" would get bigger.

Relentlessly campaigning, Ken Wu stirred up a public groundswell against the proposed Working Forest legislation. He and the campaigners in our Vancouver office hammered the government particularly hard on the "streamlining" of Crown land privatization. By May of 2003, fully 97 percent of the 2,700 respondents to the government's public input process stated their opposition to the Working Forest Initiative. The results of the public input process became a giant stumbling block making it impossible for the government to easily proceed with implementing their proposal.

The campaign continued to grow from there. Ken Wu spent nearly all of his waking hours campaigning to defeat the Working Forest. More and more people responded to his constant appeal for their help as they realized how bad this legislation was for British Columbians. Ken gave dozens of slideshows to more than a thousand people over the next year and a half. He worked relentlessly to educate people, organise rallies, produce papers and press releases, send out emails and newspapers, conduct petition and letter-writing drives and garner media attention on the issue.

Heidi Sherwood, office manager during this time, worked hard, too, helping Ken organize this campaign. During this time Ken helped establish a University of Victoria (UVic) Wilderness Committee Club. Led by English student Glenys Verhulst, it became an extremely effective outreach arm of the Committee that got students out to rallies and circulated petitions. The UVic campus probably had, and continues to have, a higher concentration of environmentally aware and active students than any other Canadian campus.

In April and again in June of 2003, great grandmother Betty Krawczyk protested the Working Forest Initiative by blockading Weyerhaeuser's logging operations in the Walbran Valley. She, too, helped draw attention to the proposed legislation and its ultimately bad consequences.

Ken Wu and Betty Krawczyk in the unprotected Upper Walbran Valley, April 2003. Krawczyk is arrested later for blockading Weyerhaeuser in protest of the "Working Forest" legislation. Photo: WCWC-Victoria files.
Facing page: "Candelabra" cedar in the unprotected Castle Grove in the Upper Walbran Valley. Photo: Jeremy Williams.

The Wilderness Committee held over a dozen protest marches and rallies between 2002 and 2004 opposing the Working Forest Initiative. With each rally the public could see through media reports that this issue was building up a lot of energy behind it. Many of the demonstrations incorporated the use of street theatre. One particularly good act had people dressed up like "Ents," the giant tree creatures that crush the evil forest destroyers in the *Lord of the Rings*.

The movie rendition of *Two Towers*, Tolkien's second book in his epic fantasy trilogy, had just premiered and become a blockbuster. Ken realized that the movie was an incredibly popular expression of the need to protect ancient forests. On April 1, 2003, at the Wilderness Committee's rally at the legislature about 300 protesters showed up. Many wore the papier-mâché "Ent" costumes that volunteers had made. NDP leader Joy MacPhail spoke out publicly against the Working Forest Initiative for the first time at this rally. A contingent of Ents stood behind her.

On June 11, 2003, Ken showed up at the Legislature with a group of about 30 protesters. He brought with him a taped-together printout

printout of more than a thousand pages of public statements to the Working Forest public input process (he printed it from the government web site). His group of protesters zigzagged this ream of public input—97 percent against the Working Forest Initiative—across the Legislature lawn in a stunning visual display that the media reporters photographed and filmed. There was no way that Ken was going to let the B.C. government get away with quietly dismissing the results of their own public input process.

Again, Ken held another fun protest on June 29, 2004. He pulled out a hundred people to the Legislature to witness the unveiling of about 40,000 signatures on thousands of anti-Working Forest petitions. He had tacked all the pages of signatures onto a giant "petition" board for display to the media. It celebrated the results of over 1,500 volunteers who had circulated and gathered names on the petition over a six-month period.

During the campaign, the Wilderness Committee sent out thousands of backgrounders and "calls to action" via fax and email. Our mailing list included not just environmentalists but tourism businesses, naturalist clubs, rod and gun clubs, churches, municipal councils and First Nations groups. By the fall of 2003, opposition to the Working Forest started to mount from these sectors. Opposition statements came from the Tofino Chamber of Commerce, the municipalities of Tofino and New Denver, the Islands Trust (Gulf Islands), tourism businesses across the province, various church ministers and many First Nations affiliated with the Union of B.C. Indian Chiefs.

In October of 2003, Ken undertook a "Ken-squared tour" (a play on our popular nickname from the early days—WC²). Ken Wu joined up with Ken James of the Youbou TimberLess Society, a group of out of work mill workers who had lost their jobs because the NDP government let TimberWest export the logs rather than mill them locally. They toured around the southern mainland and Vancouver Island giving presentations to mixed groups of environmentalists and forestry workers on the threats associated with the Working Forest Initiative and the Forestry Revitalization Act. Many forestry workers, thanks to the work of Ken James and Roger Wiles of the Youbou TimberLess Society, as well as some of the leaders in the Pulp, Paper, and Woodworkers of Canada Union (PPWC) and the Communication, Energy, and Paperworkers Union (CEP), voiced their opposition to the Working Forest Initiative. They all understood that increased privatization of Crown land would be extremely bad for workers in B.C.

In October of 2003, NDP leader Joy MacPhail put up a spirited, days-long debate in the Legislature against Bill 46, the Working Forest's "enabling legislation." The legislation eventually passed on October 29, 2003. (The two lone NDP members in a legislature of 77 Liberals could only hold it off for so long.) It was a coup to get the NDP on board as a major ally in the campaign. Of course the B.C. Green Party, under the leadership of Adriane Carr, right from the start had lambasted the Working Forest Initiative too.

However, Bill 46 wasn't the Working Forest itself. It only empowered the B.C. Cabinet to establish various land-use designations, such as the Working Forest Initiative, through Orders in Council. It permitted the government to more easily establish legally binding land use designations in closed door cabinet meetings without having to take the decisions through legislative debate.

This is how Ken describes the fight after Bill 46 passed:

> I was pretty choked about the passing of Bill 46, but I knew the battle continued. I emphasized to people that the fight was far from over, as the enabling legislation wasn't the Working Forest itself. For it to be established, the government would have to rescind the Provincial Forest designation, then issue a Cabinet Order in Council to establish an overall Working Forest designation in its place. After that, it could issue a series of Orders in Council to designate legally-binding Timber Targets (guaranteed logging zones in hectares or as a percentage of the landbase) in the commercially productive forests within the Working Forest. So there were still several major moves the government had to make to get their Working Forest up and running. The fight was still on!
>
> I wrote two versions of our Working Forest educational tabloids, which I originally worked on with Paul George. Paul was such a grumpy guy and we debated so much, that eventually we decided I should finish the newspaper by myself. I always thought that Paul was a grumpy old guy who is hard to work with, but his politics are pretty much bang-on with my own.
>
> In October 2003 we printed a first run of 100 000 newspapers [Proposed Working Forest Denies BC Citizens Their Public Land – Vol. 22 No. 6] which we paid Canada Post to drop off in strategic neighbourhoods across the province in October of 2003. When the papers ran out in February of 2004, we printed and distributed another 50,000 copies.

Largest forest rally at the legislature since Clayoquot

On March 26, 2004, WCWC-Victoria organized the largest environmental protest at the B.C. Legislature since the 1993 Clayoquot Sound rally. Over 600 people came out. At the rally Ken had a street theatre troupe act out a hilarious satirical skit called "Gordoman versus our Public Forests," or "Lord of the Right-Wing." In the skit, a character with a long white beard and hair named "Gordoman" (a contraction of the names of B.C.'s Premier Gordon Campbell and the Lord of the Rings' evil wizard Saruman) unleashes his two "Orks" Abbott and De Jong (B.C.'s Minister of Sustainable Resource Management and B.C.'s Minister of Forests) to destroy public forest lands and eat spotted owls. They were beaten back by the "Ents," the good giant ancient tree creatures, just like Ents beat back the Orks in the *Lord of the Rings* movie. The costumes were incredible, particularly the professional set of Ent costumes that Greenpeace lent to us. (They had made them for their own visually stunning anti-Working Forest protest at the Legislature in 2003, which drew national attention to the B.C. government's plans.) With stilt walkers inside, the Ents towered above the rallying crowd.

Forest defenders with a papier-mâché "Ent" in tow in April 2003 on their way to the anti-Working Forest rally. Photo: WCWC-Victoria.

"Ents," ancient forest protectors, stand behind speakers at an anti-Working Forest rally at the Legislature. Photo: Victoria Wilderness Committee.

There was also a powerful line-up of speakers that came out strongly against the Working Forest at this protest rally. They included Chief Stewart Phillip, President of the Union of B.C. Indian Chiefs; Arnold Bercov, Forestry Officer with the Pulp, Paper, and Woodworkers of Canada; Ken James, Director of the Youbou TimberLess Society; Saul Arbess, President of the Council of Canadians Victoria chapter; Jessica Clogg, West Coast Environmental Law Association's forestry lawyer; Colin Campbell, Chairperson of the Sierra Club of B.C.; Dr. Neville Winchester, Adjunct Associate Professor of Biology at the University of Victoria; Adriane Carr, Leader of the Green Party of B.C.; Paul Nettleton, a former Liberal MLA now sitting as an Independent; and, of course, Ken Wu, who had organized the whole thing with his large crew of Wilderness Committee volunteers.

It finally culminated

In campaigner Ken Wu's own words, this is what happened:

As a result of all of the slideshows, the newspapers, the protests, the outreach to non-traditional allies, the public input process, the 40,000 petition signers, the constant flood of letters and phone calls to B.C. Liberal MLAs, and more media coverage than we had garnered on any issue previously, on July 29, 2004, the B.C. Liberal government announced that there would be no legal implementation of the Working Forest Initiative, that is, it would not issue any Orders in Council or pass new legislation. Victory!!

Instead, it announced a vague 'Working Forest Policy', which had no legal teeth, which essentially said nothing new, only rehashed other pre-existing initiatives. George Abbott, Minister of Sustainable Resource Management, was quick to point out to Ken over the phone on the day of their announcement that there would be no legal implementation of the Working Forest, but that the government still supported the "concept" or "principle" of the Working Forest. It was the same as it was under the NDP government in 2001, when it backed down from the legal implementation of its Working Forest proposal, while it still supported the "principle" of it, before becoming opponents to the initiative when the Liberals came to power.

Whether or not this beast is revived in the future, is to be seen. But we'll be ready to battle it again if so. For now, the Working Forest Cave Troll has retired back to his cave, and it's still possible to establish new parks.

In many ways, however, it was a bittersweet victory. In the months to come the B.C. government stripped the B.C. Forest Service of its oversight role and gave forestry tenure holders on Crown lands near total freedom to log their tenures however they wanted to, as if the land belonged to them. The government weakened all environmental prescriptions to meaninglessness, tacking on the phrase "whenever possible without undue constraint on logging" to every environmental target. It's now a virtual laissez-faire for corporate logging rights on our "so called" public forest land.

The Stanley Park of Chilliwack?

It's hard to write about a losing campaign. On the face of it, the proposed 2,000-hectare Elk Creek forest (five times the size of Stanley Park) in the Fraser Valley just to the south of Chilliwack seemed like a perfect park candidate—an easy win. It was the last large unprotected stand of old growth left in the Fraser Valley. Its ancient Douglas firs were taller than the ones in Cathedral Grove on Vancouver Island. There were more than two dozen of these photogenic "mega flora" behemoths, towering taller than 22-storey office towers (over 60 metres high). There were many more trees between two and three metres in diameter and lots of not-quite-so-large, but nonetheless magnificent, beautiful ones. Though the resident elk were wiped out of this rainforest by over-hunting in the early part of the twentieth century, populations of marbled murrelet, Pacific giant salamander, mountain beaver and spotted owl were among the rare and endangered wild animals that still lived there. It was a spectacular forest.

Elk Creek Falls. Photo: Jeremy Williams.

Bumper sticker. Artwork: WCWC-Vancouver files.

Nik Cuff took on the job of trying to save Elk Creek from the chainsaws, working part time on contract for the Wilderness Committee. One of the first things he did was to produce and print (at cost) a large full-colour flyer for the recently formed local group, the Elk Creek Conservation Coalition. The flyer's title captured the essence of the campaign: *A Community Vision for Elk Creek – the Green Heart of Chilliwack*. One of the major points highlighted in this flyer was that Elk Creek was important as a drinking water source for Chilliwack in the past and as a backup today. The big problem was that this wasn't true! The local opposition made a heyday of this misinformation gaff. I asked Nik about it and he said he relied on local supporters for the facts and never even thought about checking them out himself. **Herein lies a good lesson. Always double check your facts before you put them into print, especially if you are using them as a key argument in your campaign! Even a small error, let alone a big one, can discredit everything you do.**

But this wasn't a fatal flaw in the campaign. The campaign was also plagued by the fact that the area was not identified as a potential protected area during the Lower Mainland's RPAC process, the same closed-door land use process that the Wilderness Committee had bitterly protested in 1994-95 that had quashed the Stoltmann Wilderness. At the time, we were so tied up with campaigning for the Stoltmann Wilderness that we did not even look at potential areas in the Fraser Valley. The local Chilliwack people who participated in RPAC defended their decision to let Elk Mountain be logged in favour of protecting other areas, as did the government. Both viewed the RPAC decision as final.

Although several hundred people turned out to an Elk Creek rally held at the Cheam Band Hall to protest the first cutblock, Nik never did get a big cohesive group of supporters in Chilliwack to fight for the Elk Creek Park. Strong local support is crucial in any campaign. But the single biggest problem, according to Nik, was that people could not easily access the area. People could not just go there on their own to see this magnificent forest. You had to cross private property to get there. Access was by guided tour only.

In spring of 2003 the Wilderness Committee published a four-page full-colour tabloid-sized newspaper titled *Fraser Valley Rain Forest ELK CREEK – Chilliwack's Wild Wonder Worthy of Protection* (Educational Report Vol. 22 No. 3 - press run 45,000).

Unmoved, the B.C. Forest Service proceeded to issue a permit to log in the heart of the proposed protected area, including in the relatively flat big-treed area below the spectacular Elk Creek falls (which are big enough to be visible from Highway 1). Our efforts forced the Forest Service to make it a helicopter logging show rather than a truck logging show. The proposed road construction would have gone through an extremely environmentally sensitive wetlands. But this was no real victory.

We found out later through a Freedom of Information request that Cattermole Timber, the company that did the logging, paid only 25 cents per cubic metre for the magnificent wood they took out. The people of B.C. got less than $10 per logging truckload for this prime 100-year-old Douglas fir wood. About 10 percent of the area was eventually logged. Cattermole left the biggest trees and selectively cut smaller ones, doing a clean job of heli-logging the site. But such "good" logging should take place in regular forests, not in ideal park candidate areas like Elk Creek. Some direct action attempts to stop the logging failed. Although there were protesters in the woods, the fallers just went ahead and felled the trees anyway.

Joe Foy put out a press release describing the situation one morning this way:

> Monday morning at about 10 a.m. eight protesters at Elk Creek had to run for their lives again – this time from an out-of-control logger who charged them while revving his chainsaw. A supervisor who was looking after the logging crew observed the man's chainsaw charge, but did nothing to stop it according to some of those who witnessed the attack. And, once again the RCMP were nowhere to be seen.

The logging continued despite the fact that wildlife biologist Andy Miller located a spotted owl in the vicinity and informed Ministry of Environment officials. But that made no difference, either. Eventually Cattermole Timber Co. was able to log everything it wanted to at Elk Creek. The Wilderness Committee's campaign to save the area had failed.

I asked Nik, "Is it still worth saving what's left up there?"

"Absolutely! But until you get public access, I don't think that

Ralf Kelman, big tree researcher, clowning around with "goat's beard lichen" on an expedition in spring 2002 to Elk Creek looking for big trees. Photo: WCWC files.

Searching for marbled murrelets in Elk Creek. Andy Miller, WCWC's spotted owl expert on far left. Photo: Anton von Walraven.

there is a hope in hell. You need some trails in there, especially to get up the steep mountainside to the plateau above the falls. We'd need to protect the fragile soils on the steep mountainside. It's a beautiful area and it shouldn't be logged," replied Nik. Joe Foy, National Campaign Director said the fight has just begun. "We've lost cutblocks elsewhere but have ultimately won!"

Paul Morgan, on WCWC board of directors, inspects a Douglas fir stump in the helicopter logged area in our proposed Elk Creek Provincial Park. Photo: WCWC files.

Burns Bog is finally saved

In February 3, 1999, the B.C. NDP government thought it had found a creative solution to the Burns Bog preservation controversy. Without formally consulting the Burns Bog Conservation Society, the lead group that had been fighting for over 12 years to save the whole bog, the NDP government announced that it has loaned the owners of Burns Bog $25 million of taxpayers' money in exchange for the company's commitment to donate 1,200 hectares (a little more than half of the undeveloped bog lands) for a park. This tentative deal also included the provincial government's commitment to back the company's plan to develop the remaining 800 hectares of bog as the relocation site for Vancouver's big Pacific National Exposition (PNE) and as an *integrated, themed retail-leisure-entertainment centre*, whatever that meant.

The Wilderness Committee was not the only group adamantly opposed to this deal. It generated a huge public outcry. The Delta Municipal Council unanimously rejected the company's application to rezone the 800 hectares for the PNE. The B.C. government back-

Nik Cuff beside one of the giant Douglas firs. Photo: WCWC files.

tracked and decided, again, to try to purchase the whole bog and designate it a protected area as originally planned.

In spring of 1999, we published a four-page, full colour tab titled BUY BACK BURNS BOG NOW! – *Protect Forever Vancouver's Biggest Remaining Urban Wilderness*, printing and distributing 300,000 copies. Our paper presented the already familiar arguments for preservation and urged the provincial government to expropriate the land if the owners refused to sell it at a reasonable price. Shortly thereafter, the B.C. government appointed Radcliff and Co. lawyer Greg McDade to chair an Ecological Review of the bog. This was good news! We'd worked a lot with Greg when he was the head of the Sierra Legal Defence Fund and I knew he would do a fair and thorough job. I was also confident that "the science" would ultimately say "save it all" because common sense suggested that protecting only a portion of this rare domed bog would mean that it would not survive over time. The Ecological Review did indeed verify this and we all took heart. The campaign seemed on the road to sure victory.

In follow-up to the Ecological Review, the B.C. government appointed McDade to negotiate with the landowners and other levels of government to purchase the bog. At the Wilderness Committee we used every possible opportunity to pressure different levels of government to cough up the cash to buy Burns Bog. In November, right after the Ecological Review report came out and before the federal election of 2000 the Wilderness Committee, Burns Bog Conservation Society and SPEC put out a media release urging Prime Minister Chrétien (who was visiting Vancouver at the time) to save the bog. In this release I bluntly stated that, *If Chrétien is unable to commit, during*

The last big demonstration at Robson Square to get the governments moving on purchasing Burns Bog. Photo: WCWC files.

this election, to investing about $30 million federal dollars to help buy back Burns Bog, I believe he's saying the Liberals aren't interested in protecting BC's priceless natural heritage.

After a lot more negotiation, rallies and citizen pressure, on February 5, 2004, all levels of government signed an agreement in principal with the owners of Burns Bog to purchase and preserve in a park over 2,100 hectares of the Bog (over five times larger than Vancouver's Stanley Park) for $73 million. It was about the cost of creating the South Moresby National Park Reserve fourteen years earlier, but considerably less than the cost of a FastCat, the B.C. NDP ferry fiasco.

On the day of the announcement, the Wilderness Committee put out a press release heaping praise on government leaders, including B.C. Premier Gordon Campbell and Delta Mayor Lois Jackson. Under the deal to purchase the bog, the B.C. government kicked in the biggest share of money—$28.6 million followed by the federal government at $28 million, then the Greater Vancouver Regional District at $14.2 million. The city of Delta provided $7.9 million, a huge amount for a municipal government. The deal was not an easy one to broker. Western Delta Lands (WDL), the private landowner of the bog, was not the most willing of sellers. In the end it was ordered by the courts to sell its bog lands by the end of January 2004 in order to pay back mortgage holders $50 million, including the primary mortgage holder, the provincial government.

This successful deal was the culmination of a 15-year campaign spearheaded by Eliza Olson and the Burns Bog Conservation Society. It was a huge success but not a complete victory. The Wilderness Committee and Burns Bog Society are still fighting to protect several hundred hectares of critical bog habitat left out of the park acquisition. We also know that in the years to come we'll have to fight incursions into the bog for roads and other developments, including the proposed Gateway Project, a multi-billion dollar road, bridge and port expansion to make the shipment of goods easier at the expense of the region's livability. And we will have to get the Vancouver dump (landfill) that was unthinkingly put into the bog years ago closed down for good and the leachates from it monitored and controlled. It would be nice to think every campaign could have a perfect ending, but that's rarely the case in our complicated geo-political world.

Five hundred people pose in a Walbran clearcut

Another incredible event that Ken Wu and WCWC-Victoria organized was a "Tree of Life Celebration" on October 9, 2004. It was held in, of all places, a clearcut. We partnered with the San Francisco-based Rainforest Action Network (RAN) and the Gettin' Higher Choir, a 300-voice no-audition Victoria choir, and LA artist/activist John Quigley to take 500 people in 10 school buses to a recent Weyerhaeuser clearcut in the Upper Walbran Valley. There, while the choir got everyone singing "eco-songs," marshals got everyone to take positions along a ribbon that had been strung out earlier in the clearcut. From the air, they formed a poignant picture of a felled tree, teary faces of a mother and her child, and the words "Wake Up."

When the human outline was complete, a helicopter flew overhead to document it. Incredibly, even though it had been raining heavily and steadily since the participants boarded the buses in Victoria, Nanaimo and Duncan that morning, the clouds parted for exactly 10 minutes. This provided just enough time for the photographer in the helicopter to take the photos. Then the clouds closed in again and the rains returned. There were 500 witnesses to this mini miracle that day.

Unfortunately, due to the uneven terrain in the clearcut, the image didn't quite work out perfectly. The tree and the words were clear, but what were supposed to be the mother and child's faces looked more like a pointy-headed ogre with an earring, or perhaps a small beaver gnawing on the tree. This greatly upset John Quigley, the artist-creator who felt the project was a failure.

But it wasn't. We garnered great media coverage and the hundreds of people who participated were very moved by the experience. It was, in fact, the biggest forest protection event in B.C.'s woods since the Clayoquot Sound blockades over 10 years earlier. A good deal of the organizing of the event was done by RAN's Rharon Smith and Victoria's office manager, Cassbreea Savage, who was good at both administrative work and campaign work, a rare combination of abilities.

How much such stunts help save wilderness is anyone's guess. But they do get publicity, something that campaigners always hunger for and campaigns thrive on. This one, besides garnering media, certainly opened the eyes of a lot of people to the clearcutting going on and gave them a unique opportunity to experience a bit of the majesty of B.C.'s old-growth temperate rainforest. Before they all returned, the participants got to see one of Walbran's unprotected, yet still standing, giant trees.

Fending off offshore oil and gas developments on the B.C. coast—a never ending battle

After the May 2001 provincial election, the B.C. government really put pressure on the federal government to lift its 1972 moratorium on off-shore exploration and development of oil and gas reserves on the west coast. Federal Environment Minister David Anderson was adamantly opposed to lifting the moratorium and he managed to hold off other federal Liberals who were more development minded. In the winter of 2001-2002 the Wilderness Committee published a four-page three-colour tabloid-sized newspaper titled *OIL SPILL? – A grim reality accompanies oil and gas development on B.C.'s stormy coast* (Educational Report Vol. 20 No. 6 - press run of 50,000 copies). It gave the reasons why the moratorium on offshore oil and gas exploration should be maintained and promoted viable alternative sources of power, the development of clean, renewable solar, wind and tidal energy.

In November 2004, the Wilderness Committee launched a new national campaign to build Canada-wide support to protect B.C.'s wild Pacific coast from oil and gas development. We sent out thousands of information packages to potential allies that included a new petition calling on the federal government to maintain its morato-

Top left: Busloads of "Tree of Life" participants arrive in the rain to the Walbran clearcut site, October 2004.
Middle left: Constructing the human "Tree of Life" in the Walbran clearcut.
Bottom right: "Tree of Life participants get to experience a bit of the threatened ancient forest in the Walbran before they bus home.
Top right: "Tree of Life" human formation figure in an Walbran clearcut.
Bottom right: Aerial view of "Tree of Life" with Inset showing schematic drawing of the "Tree of Life" performance art shape.
Photos: Shel Neufeld and Jeremy Williams.

rium on off-shore oil and gas. Following on Ken Wu's success in building a broad coalition against the Working Forest Initiative, we reached out not just to other environmental organizations but also to churches, university activist clubs, and green businesses to speak up in defence of Canada's Pacific coast.

The heightened campaign thrust coincided with the federal government's release of the *Report of the Public Review Panel on the Government of Canada Moratorium on Offshore Oil and Gas Activities in the Queen Charlotte Region, British Columbia* (Priddle Report) in November 2004 that detailed the results of the spring '04 public input process on offshore oil and gas. Of 3,700 respondents, 75 percent opposed lifting the moratorium. Most importantly, the Haida and most other coast First Nations expressed their strong opposition to off-shore oil and gas developments in their marine territories.

A provincial election was looming in May 2005. The Wilderness Committee kept up the pressure. In March 2005, a small group of Wilderness Committee picketers unfurled two large banners—"Ban Oil Drilling and Seismic Testing Off BC's Wild Coast" and "Keep our Coast Oil Free"—and handed out leaflets outside a B.C. Oil and Gas Opportunities Summit meeting in Victoria. The event was spon-

UVic Wilderness Committee club makes the media aware of the offshore oil threat with this message composed of 130 students in 2005. Photo: WCWC-Victoria files.

sored by the B.C. Chamber of Commerce.

The keynote speaker at this conference was Richard Neufeld, B.C.'s Minister of Energy and Mines, a big promoter of coastal oil development. At the time Minister Neufeld was promoting seismic testing (oil exploration) off B.C.'s coast under the guise of "filling in science gaps." The B.C. government in its recent Throne Speech proclaimed that it would be spending a total of $11.55 million over four years to fund an Offshore Oil and Gas Team, a public relations lobby group to push the federal government into lifting the moratorium. That's a lot of taxpayer's money to spend on selling something the majority of people don't want!

At the protest Ken Wu said:

The BC Liberals are slick - like oil on water. They're promoting seismic testing under the guise of 'doing science,' when it's simply a sneaky, foot-in-the-door strategy to eventually get the moratorium lifted. Once companies spend hundreds of millions of dollars to undertake oil exploration—seriously harming fish, whales, and crabs with sonic shock waves through the ocean—and if major deposits are found, then it'll be extremely difficult to keep oil drilling away by that point.

The campaign continued in November 2005 with a catchy event organized by the Wilderness Committee's University of Victoria (UVic) club. A large group of students stood side by side on the UVic campus, forming the words "Oil Free Coast." Media photographers and cameramen were able to access the McPherson Library roof to capture the image. Ziya He, one of the co-ordinators of the UVic Wilderness Committee club stated:

We believe the federal government should be obligated to adhere to the results of its own public input process and publicly announce that it will maintain the moratorium, or better yet, simply ban oil and gas development from Canada's Pacific waters. With a federal election coming up, it's time the federal government acted in the public's interest.

It's a strong collaborative effort that's holding off offshore oil and gas development in B.C., including the Haida and other coastal First Nations, fishermen, tourism operators, environmental groups and people in general who want to keep our coastal waters oil-free and who understand the pressing need to cap fossil fuel production and reduce fossil fuel consumption to slow global warming. I'm proud of the role the Wilderness Committee has played to help keep the moratorium in place despite all the big moneyed efforts to lift it.

Successful conclusion to a major Manitoba Wilderness Committee campaign

The establishment of Manitoba's 79th provincial park in December 2004 was a sweet victory for our Wilderness Committee Branch in Manitoba. Ron Thiessen had campaigned long and hard to get a park with real protection along the Manigotagan River. In 2002, our Committee distributed a *Manigotagan River Park Reserve* postcard mailer with a better vision of the park than the government's. We proposed a protected area twice the size with full protection. The government's proposal allowed for logging in every part of the park.

Ron launched our Manigotagan River park campaign without speaking first with the local communities that would be affected by it. This ruffled a few feathers! When Ron made a surprise visit to Manigotagan's community office, one displeased community leader commented, "I'm surprised that you came alone." Locally Ron was being called the "enemy of Manigotagan."

Ron acknowledged his error and apologized. From then on he worked with the provincial government and the affected communities. Several meetings later, a common vision emerged. The Manigotagan River Provincial Park was established and consists of a 1,500-metre-wide protected backcountry area on both sides of this challenging white water canoeing river. It stretches 45 kilometres from the northwestern tip of Nopiming Provincial Park and ends downstream just outside the community of Manigotagan on the east side of Lake Winnipeg. The government designated over 99 percent of the park's 7,432 hectares "backcountry" (protected from mining, logging and hydroelectric development). In order to accommodate an existing mine, less than one per cent of the corridor (16 hectares) was classified as "recreational development," that provides weaker protection.

Ron Thiessen lauded the decision in a Wilderness Committee media release:

"Protecting the Manigotagan River and surrounding area is a momentous victory for conservation and communities," said Thiessen. "I commend Premier Doer and Minister Struthers for sharing the vision to preserve this area for future generations and wildlife. The leadership of the Manigotagan community and Hollow Water First Nation deserve a great deal of respect for their efforts in preserving their traditional lands and water."

Our campaign to get industrial development out of Manitoba's parks was working well, at least in keeping logging out of new parks! Every new park created since our Manitoba Wilderness Committee Branch began in 1999 prohibits industrial logging. To get logging out of existing parks still remains a big challenge and a dream.

The Manitoba Wilderness Committee Branch continued to grow and take on more campaigns. In 2004 it organized *Embrace: A Celebration of Wilderness and Culture*. Held at the Winnipeg Art Gallery and featuring First Nations' musicians and speakers, our event promoted the incredible conservation opportunities in Canada's boreal forests. It highlighted, too, the need for sustainable communities.

That same year our Manitoba Branch held its 5th Anniversary Party at the Pyramid Cabaret. Speakers and musical guests entertained the 300 Wilderness Committee supporters who attended and celebrated the branch's achievements.

By the end of 2005, Manitoba Wilderness Committee members numbered nearly 3,500—nearly one eighth of all of our total membership. In January 2006, Ron Thiessen, who had worked for six years to build our organization from scratch, left the Wilderness Committee to become the Executive Director of CPAWS Manitoba (undoubtedly a better paying job!). Since then Ron and CPAWS Manitoba have worked closely with the Wilderness Committee on campaigns. In the Spring/Summer of 2006 we co-published with CPAWS Manitoba a four-page full-colour tabloid-sized newspaper titled *Protect Manitoba's Big Wild East Side Lake Winnipeg* (Educational. Report Vol. 25 No.3).

The Web provides another avenue for wilderness education

Today, besides publishing "hard copies," the Wilderness Committees newspapers are published online. In early 2006, after working nearly two years on the project with volunteers, volunteer coordinator Anton van Walraven succeeded in getting all of the Wilderness Committee's 136 educational newspapers up on our website www.wildernesscommittee.org. Today you can hardly Google any major wilderness campaign in Western Canada that occurred in the last 25 years and not get Wilderness Committee references to come up on the first page.

Anton came from the Netherlands in 1996 and spent that summer in B.C. as a volunteer helping build our trail in the Stoltmann Wilderness. Over the next few years, he became more involved with environmental advocacy both in the Netherlands and with the Wilderness Committee. His activism resulted in several longer trips to B.C. In 1999 Anton finally moved to B.C. and, in 2001, volunteered to co-ordinate our volunteers. In 2002 the E-team hired him to fill that position as a paid staff person. Today Anton figures out the various volunteer and "intern" jobs, recruits people to fill them, and ensures that the experience is rewarding for both the volunteer/intern and the Wilderness Committee. Among his ongoing duties is to maintain the Wilderness Committee's campaign library archives including video footage. He designed and helped assemble the material on the DVD that is in the pocket on the inside back cover of this book.

A warning sign to heed - local killer whales going extinct sparks a toxics campaign

In 2003 the Wilderness Committee, along with the Sierra Legal Defence Fund and the Georgia Strait Alliance, joined U.S. environmental groups in a court action to try to force the U.S. government to list the trans-boundary Southern Resident population of orca (killer) whales under the U.S. Endangered Species Act (ESA). During the 1990s this subspecies of killer whales experienced a 20 percent decline in their population, raising concerns about their future. Only about 85 individuals of Southern Resident orcas exist today. The U.S.'s ESA has the teeth to help save this whale, unlike Canada's useless Species at Risk Act (SARA).

Our collective efforts were successful. In December 2004 the U.S. federal government listed the Southern Resident Orcas as "threatened." According to ESA criteria, a species is listed as threatened if it is at risk of becoming endangered and a species is listed as "endangered" if it is at risk of becoming extinct.

Less than a year later, the U.S. agency in charge of administering the ESA re-evaluated the status of this orca population and "upgraded" it to the endangered list, citing that vessel traffic, toxic chemicals and limits on availability of food, especially salmon, continue to put these whales at risk. Their population has always been small, making them particularly susceptible to catastrophic events, such as disease outbreaks or oil spills. The endangered listing requires the U.S. federal agencies to make sure their actions are not likely to harm the whales. But saving these border-crossing whales will take a massive effort that has to include action on the part of Canadians to clean up our sources of pollution and rebuild our wild salmon stocks.

Window sticker. WCWC product.

There are major gaps in our knowledge about all orcas, including this southern resident population, such as where they go when they're not in local waters. Given the fact that killer whales can live up to 90 years in the wild, existing data collected over the last 30 years doesn't cover even half the life span of the older animals.

Living at the top of the food chain in polluted seas, these orcas are being poisoned. In fact they are amongst the most polluted mammals on the planet. Tests have revealed that they are currently four to five times more contaminated than the highly toxic beluga whales of the St. Lawrence Seaway in Central Canada. Persistent pollutants like PCBs and DDT, although banned for years in North America, still abound in the environment and have bioaccumulated in the whales' blubber, leading to endocrine disruption, reproductive disorders and lowered immunity. This toxic load is augmented by a slew of other dangerous chemicals that bioaccumulate which are still in use, including dioxins and PBDEs. PBDEs, used as flame retardants in the manufacturing of electronics, foam cushions, computers and even toys, are very similar in their chemical structure to PCBs. They are perhaps one of the most poisonous and widely distributed chemicals in North America and have been banned in Europe.

To rouse people out of their complacency about the growing threats to wilderness and wildlife, the Wilderness Committee has started a campaign that directly connects the bad things happening in the "non-human" world to the bad things happening to humans, especially toxic contaminants' threat to our health.

Not even the dire consequences of rapid climate change hits peoples' consciousness as hard as the threat of man-made toxic chemicals that currently permeate our biosphere. The body of every human being is affected. Every species on our planet is affected. Impacts are seen in the rising rates of cancers and reproductive disruptions. Man-made toxins attack the very basis of the biosphere's overall health. Wilderness Committee Executive Director Andrea Reimer reports that people are responding with as much enthusiasm and financial support to this campaign as they did to our Carmanah campaign 17 years earlier.

Even more importantly, it has allowed us to make good alliances we never imagined before with teachers, church groups and unions. Also it enabled us to engage people in new ways.

In the era when more people camped and tromped in the woods, it used to be enough to give them beautiful images and trail access to motivate them to help protect a threatened wilderness area. Now, with most people stuck in their electronic worlds, it may make sense to appeal to their concerns about their own health.

It is not enough to just get people to be aware of an issue and its impact. It still takes a great deal of work and skill to capture peoples' interest enough to get them actively engaged in supporting a campaign (with money) or working on a campaign (writing a letter, signing a petition or coming to a public meeting).

It does not take much deep thought to connect the dramatic increase in cancers found in the population of Beluga whales in the chemically contaminated Saint Lawrence estuary, which number only 650 individuals, to our own health. Over the past decade scientists have found cancers in 27 percent of the dead adult Belugas that they examined—a cancer rate similar to that found in humans in industrialized countries. Elsewhere amongst wild whales, cancers are rare.

The plight of these whales is telling us that we live in an interconnected world where chemicals that poison our waters and pollute our air have no boundaries. The whales living in polluted seas are like the canaries that coal miners used to take into their coalmines. If the air was "foul" the canary, with its small body mass, quickly died. It was a warning to the big guys carrying the caged bird that they had better get the hell out of there quickly or they would die, too!

The warning bells are ringing as loudly as nature can ring them. A newspaper co-published by the Wilderness Committee and the Labour Alliance Society titled *Turning the Tide, Protecting our health & the marine environment from toxic pollution* (Educational Report Vol. 24 No. 2 Winter 2005 with a press run of 85,000 copies) details both the warning signs and the practical things that people can do to reduce their exposure to pollutants.

Recently the Wilderness Committee followed up with an updated version of this popular publication. Our toxics campaign continues to "build steam." And as of the publication of this book, public pressure has generated a listing under Schedule 1 (most toxic substances) of the Canadian Environmental Protection Act for PBDEs (flame retardants). Two school districts have signed on to our Toxic Free School campaign.

A "secret" wilderness conserving campaign

In the summer of 2002 the Wilderness Committee embarked with no fanfare on a campaign to save a precious pocket of old-growth forest in the Cayoosh Mountain Range. The area is called Lost Creek. Over 10,000 hectares in size, and 20 kilometres long, it is the largest remaining wild valley in the Lillooet region and the campaign to protect it is still ongoing. Andy Miller, who has spent weeks in the Lost Creek Wilderness for WCWC, describes how we got involved:

> Some forest ecologist had been studying forest succession in the Lillooet region and noticed that there were these stands of ancient forest, pockets of coastal rainforest in the mouths of north and east facing hanging valleys throughout the Lillooet country. They were little refugia where cold air pockets and wet air pockets exist all year round. When it is 40 degrees in the surrounding areas in summer, inside these forests it's still cool and wet. They range in size from 100 to 500 to 1,000 hectares.
>
> That was the attraction of Lost Valley. To check it out, Will Koop, who knew people in leadership positions, including chiefs, in the St'at'imc Nation, lined a trip up to take a look, getting permission to go in by helicopter with a local native archaeologist. We helped pay for the rental of the helicopter.
>
> The pilot let us off in the big snow in the back of the Valley. The first thing we saw were two grizzly bear goat kill sites. They were really interesting.
>
> On the way down the valley our native archaeologist guide pointed out all kinds of stuff he had never seen before, including pits that were used for temporary food storage. Then we got into the glacial refugia which is, about 1,000 hectares in size, with enormous red cedars and all the coastal rainforest plants right in the middle of the Chilcotins.
>
> There was a tremendous downpour. We had poor raingear and were not prepared to deal with it. The archaeologist called out that he had found an ancient rock shelter. Long ago a giant bolder had rolled down from above and got hung up on some other boulders making this beautiful shelter. We packed in all our gear, lit a big fire and dried out all of our stuff. There was even some dried wood in there from the last person who used it...
>
> We explored around the forest and found evidence that some foresters had recently been there. They had laid out 30 cut blocks. One of them had the audacity to write in permanent marker pen a note and put it in a plastic Ziploc bag and nailed it on a tree saying that the camping site was brought to us by the B.C. Forest Service and we should send our $10 camping fee to a certain address. I'm sure it was a joke, but it really irritated the natives.
>
> There were all kinds of cedar CMTs everywhere. They obviously didn't have to trade with the coastal people. They just came here and got the cedar bark themselves. We picked up pieces of a really old trail throughout the valley. It had to be hundreds of years old. We reported our findings to the chiefs and got support to work together to find and reopen the trail. The Wilderness Committee is now four years into the project with the Seton Band and we will finish the trail down Melvin Creek this year.
>
> Jeremy Williams has made three short videos and we have hired B.J. Williams to set up a tour of all the native villages in St'át'imc

Rock shelter in Lost Valley. Photo: WCWC files.

Overlooking Melvin Creek, home of proposed ski resort. Photo: BJ Alexander – August 2005.

[pronounced Stat le um] country and are working on setting up other joint projects where we protect a place for its spiritual and cultural values.

Despite the fact that the St'at'imc have developed and adopted their own land use plan which designates Lost Valley as a protected area, it is still on the chopping block. The B.C. Forest Service recently tried to give the cutting rights to the nearby D'Arcy Band, but the Band refused. The government is still trying to find someone else to log Lost Valley.

The Seton Band is working on setting up trapping and other cultural development workshops. They have identified those who traditionally owned the trapping rights there. They have also expressed interest in working towards being the first toxin-free community in Canada where no toxic chemicals like chlorine-based cleaning products would be used in schools and in public buildings. There are about 600 natives living in Seton Portage. It's an amazing place.

In January 2005, after three years of quietly clearing the traditional trail in Lost Valley in partnership with the Seton Lake Indian Band, its traditional St'át'imc owners, the Wilderness Committee co-published with this band a four-page, full-colour tabloid-sized newspaper titled *Save Lost Valley* (Educational Report Vol. 24 No. 1 - press run of 50,000 copies). It told the story of their efforts to protect this incredible wild valley and our joint trail clearing expeditions. After it came out, B.C. Forest Service officials told us verbally that the Forest Service had put logging plans on hold there because of "First Nations concerns."

Pre-election "Vote Wild" environmental education campaign launched in the spring of 2005

With the date of the B.C. provincial elections now set in law, it was only natural to use the pre-election time to get political parties and candidates to clearly state their positions on key environmental issues. Elections are times when politicians are more likely to listen to citizens' concerns; and a time when citizens are more interested in hearing about what government has or has not done.

In the spring leading up to the May 17th 2005 election, the Wilderness Committee coordinated its activities with other B.C. environmental groups and launched with other environmental groups our first joint "Pre-Election Environmental Education Campaign." One component of our campaign was to mobilize hundreds of volunteers to go door-to-door in as many neighbourhoods as possible and hand out information. People really wanted to learn about the B.C. government's environmental track record. The changes had been so sweeping that most people did not know the full scope of them. We trained the door-to-door "eco-election" educators about a multiplicity of issues, from fish farms to forest protection to parks to endangered species. In total, volunteers visited over 60,000 households in B.C. alone between February and the May 17, 2005 provincial election.

In designing the campaign literature, the Wilderness Committee was extremely careful to keep well on the safe side of the line that divides "education" from "lobbying." Charitable societies are not allowed to lobby. The educational four–page tabloid newspaper we produced, called *Vote Wild*, stuck to the facts. It outlined the track record of the B.C. government on key environmental issues and listed the websites of all three major political parties (Liberals, NDP, and Green Party) so people could learn more about their positions if they wanted to. It did not tell people how to vote or express support for any candidate or political party. It only urged the readers to exercise their democratic franchise and vote.

Despite this, on April 18th, Elections BC informed Joe Foy by e-mail that it deemed our *Vote Wild* publication (Educational Report Vol. 24 No. 3 - press run of 125,000 copies) to be "partisan advertising." It stated that the Wilderness Committee had to register as a partisan advertiser with them—which we couldn't, because it would mean losing our charitable status—or cease and desist from distributing this paper starting the next day (April 19th) until after the citizens went to the polls on May 17th.

The reasons Elections BC gave for ruling that our paper was "partisan advertising" were statements such as, "More environmental laws

Cover of newspaper that Elections BC tried to ban. Document: WCWC archives.

have been rolled back than by any previous government in B.C.'s history." They concluded that we were criticizing the B.C. government in a partisan way. We disagreed. We were just stating a fact! We also contended that B.C. Liberal candidates, staff, and members could very well take different policy positions and stances while they campaigned for the upcoming election than the current government. In our criticisms, we didn't name any political parties.

In effect, Elections BC's ruling said that the Wilderness Committee couldn't criticise the government six weeks before an election—a very dangerous position to take in a democratic nation—and that it intended to suppress the distribution of our newspaper and freedom of speech.

That same day we issued a press release titled, *Wilderness Committee to keep distributing Vote Wild newspaper despite Elections BC warning.*

The Western Canada Wilderness Committee (WCWC) announced today that it would continue to distribute their Vote Wild! newspaper, despite a warning from Elections BC that they regard the educational newspaper as partisan election advertising. The newspaper highlights five key environmental issues, the government's record on those issues and possible solutions.

'We've been publishing our little four page educational report newspapers regularly for 24 years now. Over those two and a half decades, we have suggested ways to safeguard BC's environment and sometimes that means being critical of the environmental record of the various governments who've been in power,' said Foy...

A couple of days later a sympathetic reporter notified Ken Wu that Elections BC had scheduled a press conference about our paper, letting Ken know where and when the conference would be held. Ken and his assistant Pearl Gottschalk showed up, of course, and started to debate with Elections BC officials in front of all the news cameras. Ken argued that Elections BC was wrong in trying to suppress our newspaper. The media were very sympathetic, naturally being

Ken Wu defies Elections BC's order to cease and desist from distributing our *Vote Wild* newspaper. No one tried to stop him as he distributed copies from the steps of the Legislature. Photo: *Victoria News* newspaper.

inclined to defend freedom of the press. After the news conference, Ken and Pearl went to the B.C. Legislature with 8,000 of the banned newspapers and, in defiance of the ruling, started to distribute copies of them to all the passers-by in front of the media and the guards. You can't be timid in defending your rights.

No one tried to stop them or anyone else from continuing to distribute this paper. Considering the anti-democratic nature of the ruling, we were sure that Elections BC would relent. Or, if Elections BC pressed charges, the courts would inevitably rule in our favour and uphold the rights of freedom of the press and freedom of speech.

On May 5, 2005, Elections BC "threw in the towel" and "unbanned" our newspaper. In the letter to Joe Foy telling him that the Wilderness Committee was free to distribute its *Vote Wild* newspaper, Elections BC explained its change of heart this way: because *Vote Wild* was one of the Wilderness Committee's regularly produced Educational Report publications, it therefore could not be classified as elections advertising.

Webcam to highlight the increasing raw log exports is of limited success

One constant barrier to wilderness protection is the cry that the cost is too high due to the number of jobs that would be lost in the forest industry. Meanwhile, way more jobs, thousands more, are lost every year as the volume of raw unprocessed logs keeps climbing annually. Statistics from the Ministry of Management Services show that B.C. exported 14 million cubic metres of logs during the first four years of Liberal rule compared to five million cubic metres during the previous four years under the NDP who themselves increased raw log exports significantly during their ten years in power.

Nik Cuff lives high up in a condominium overlooking the Fraser River in New Westminster. Every time I visited him, we sat for a bit on his balcony and watched raw logs being loaded onto barges and ships for export. The raw log terminal is right across the river

Log export terminal across from Nik Cuff's condominium. Banner erected on breakwater in WCWC's ongoing protest against raw log exports. Photo: Joe Foy.

from his place. Nik says that the loading goes on from dawn till dusk most every day, often with two ships being loaded at once. As a part time campaigner for the Wilderness Committee, Nik suggested at one campaign team meeting that the Wilderness Committee should take advantage of his place to get some video footage showing the unbelievably high volume of raw log exports.

After brainstorming the idea, the campaign team decided to launch the world's first live "raw log webcam" to let people see for themselves what was happening. After considerable technical hassles in which it was decided that the tripod with the video camera be placed inside Nik's sliding glass balcony doors to protect it from the cold and rain when it blew up a storm, the webcam was ready to go on air. Every thirty seconds it would automatically post a new "snapshot" of the loading activities onto the Wilderness Committee's website. In the media release announcing our live raw log broadcast, policy director Gwen Barlee said:

It's our version of reality TV. We thought it was so shocking that people really needed to see it with their own eyes.

That evening when the webcam began broadcasting the only thing people could see was a guy walking around and cooking supper in his kitchen. What the technician who set it up had failed to realize was that, at night, when it was dark outside and the lights were on inside, the glass on the balcony door acted like a mirror. Thank goodness the majority of people watching the first episode were largely Wilderness Committee staff and volunteers!

After that got fixed, our "raw log webcam" broadcast faithfully over the Wilderness Committee's website, except for the time when Nik stumbled over the tripod and the lens shifted for half a day to an area on the river with no logs. Our webcam ran 24 hours a day 7 days a week taking a continuous series of photographs of ships and barges loading raw logs at the Fraser River Port in Surrey, one of British Columbia's busiest raw log shipment points.

At noon on June 11th after five weeks and half of broadcasting (a dozen ships had been loaded to export B.C. logs for manufacture in foreign mills) the Wilderness Committee retired its raw log webcam. Our media release put a very positive spin on this effort. *"The purpose of the webcam was to raise public awareness about the export of jobs and logs from British Columbia - and I think we helped achieve that goal,"* quipped Gwen Barlee.

Truthfully, it had not been the most exciting video entertainment. "It was a lot like watching paint dry," explained Nik. "The most interesting thing that happened was seeing the tugs go by."

Despite the efforts of the Wilderness Committee and other groups including the Youbou Timberless Society and millworker unions, raw log exports kept increasing and mills kept closing. In 2006 Port Alberni loggers invited the Wilderness Committee to join with them on the protest lines and speak from the same podium to stop the greedy corporations from over-cutting, raw logs from being exported and Vancouver Island's mills from closing. Together we are calling for a halt to raw log exports and the adoption of sustainable forestry practices so that there will be a work-filled future for those who live in B.C.'s forestry dependent towns.

Parks Day 2005 – A day to protest, not celebrate

On Parks Day, Saturday, July 16, 2005, the B.C. government hosted events in 27 provincial parks to celebrate the more than 800 protected areas in the province. We could not join in the celebrations because for over a decade the B.C. parks system has been run into the ground and new park creation had virtually come to a standstill. On that day, the Wilderness Committee called on the B.C. government to halt its assault on parks including logging within them, downsizing them, decommissioning services and privatization existing facilities

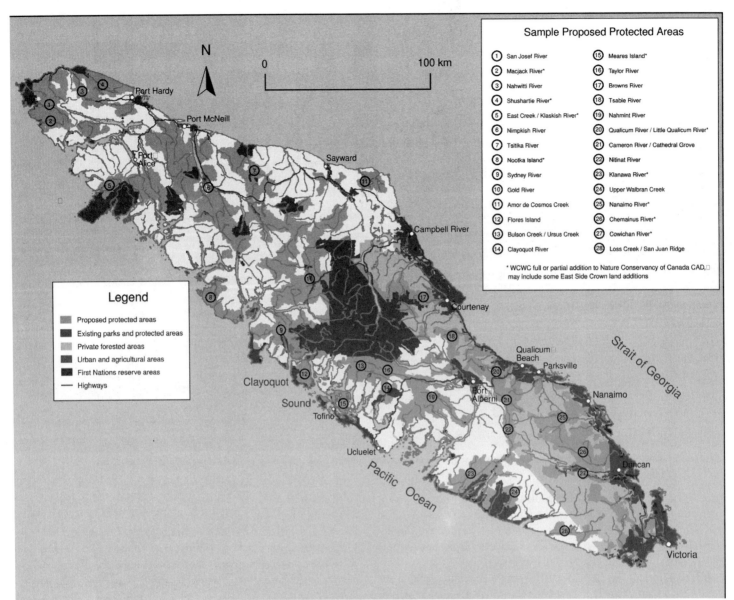

Updated *Vancouver Island Vision* map proposing that 41 percent of the Island be wilderness. Cartography: WCWC files.

within them, permitting private enterprises to build lodges in them, and instituting new user fees. We also called on the government to expand B.C.'s protected areas system through initiating a new Protected Areas Strategy to create a second wave of new B.C. parks.

At noon that day a team of Wilderness Committee volunteers and staff including Ken Wu leafleted visitors in Goldstream Provincial Park near Victoria. The educational papers they handed out informed people about the anti-parks policies of the B.C. government and the Wilderness Committee's vision to protect 41 percent of Vancouver Island from industrial developments.

This new eight-page, four-coloured, tabloid-sized *Vancouver Island Conservation Vision* paper published in June (Vol. 24 No. 50) updated the Vancouver Island Vision paper we put out 11 years earlier in response to the CORE park process for Vancouver Island. The liquidation of the old-growth forest on Vancouver Island had proceeded relentlessly over the past 11 years. Now there was less than 25 percent of the Island's old growth left. At this rate almost all would be gone in the near future. If government didn't act quickly, we'd lose the ancient forests and land base needed for parks, wildlife, clean water, wild salmon and high-value-added forest products.

In the spring of 2006, as a natural next step to make this conservation vision a reality, Ken launched a campaign to phase out all old-growth logging on Vancouver Island just as the governments of New Zealand and Southwestern Australia have done. Ken explains that old-growth logging will stop anyway when the ancient forests run out, so choosing to stop it before the last remnants of old growth are gone is a smart idea. Squandering these last pieces of priceless, irreplaceable, species-rich and yet-to-be-fully-understood wilderness is a tragedy that must be averted.

Chapter 68

An organization that grows, keeps on creatively campaigning and never gives up on its goals

A roast to ensure that I would actually write this history book

Although I had vehemently said that I did not want to be the object of a roast when I "retired early" to write this book, Andrea and others at the office continued to secretly organize one. When I found out about it in early December 2002, about a week before my roast was to take place, it was too late to stop it. As the guest of honour, I had to go.

Unbeknownst to me, they invited Richard Krieger who co-founded the Wilderness Committee with me and he decided to come. Since Richard and I had our "falling out" 19 years earlier, I'd only exchanged a few words with him. My roast was a chance for us to bury hatchets and we did. He promised to help me reconstruct the history of the Wilderness Committee's early years for this book and provide images, and he did. It's been great re-establishing our friendship.

Wilderness Committee holds roast to put my feet to the fire to ensure this history book is written. Photo: Greg McIntyre.

The event ended up being a lot of fun. Many people, some of whom I hadn't seen in years, came forward with funny stories and reminiscences—great grist for this book. About 200 people attended and Costa, who used to own Greek Characters in Gastown (our after-work hangout for years) provided the food. Ken Lay who had quit the Committee 10 years earlier took on the job of MC. They even had a papier mâché bust of me because a few months earlier I had jokingly said after visiting the David Suzuki Foundation office, "Why don't you have a bust of me in the entrance to our office like they do of David in the Suzuki Foundation's office?" It really was a great evening. The Wilderness Committee has grown into a large and wonderful extended family.

Several people came up to me inquiring as to my health, thinking that I might be bowing out of "running the Committee" because I had some horrible health problems. I don't. Questioning my sanity for undertaking this huge book-writing project would have been more appropriate.

In retrospect I'm sure the new E-team held this roast to ensure that I would actually focus and finish this book. They undoubtedly guessed what a big project it was. After the big roast, they figured that I could not disappoint everyone...I had to follow through.

The promotional mail-out inviting people to the event had this pitch: 'About two hundred folks who have known Paul throughout three decades of campaigning on behalf of conservation are coming out to get even,' said Joe Foy, a cohort of Paul's and one of the roasters. Earlier this year Paul George stepped back from the front lines to author a book on the history of B.C.'s environmental movement from the perspective of B.C.'s largest conservation organization.

Incredibly, writing this book took three and a half more years to finish!

Impossible to tell the whole story

There are too many campaigns, too many people and not enough room to give them all their due—even in a book of this length!

Often we forget to recognize the many people who work behind the scenes of campaigns and organizations who keep everything on track. The Wilderness Committee has a great volunteer Board of Directors that currently include the Chair, Mike Gildersleeve, as well as Bob Broughton, Jennifer Campagnolo, Alice Eaton, Ian Mackenzie, Paul Morgan, Devon Page, Tom Perry and Annette Tanner. Some of these folks, like Bob, Alice and Ian, have been on the Board for more than a decade; others are new.

The day-to-day management of the Wilderness Committee is headed up by Executive Director Andrea Reimer who started working for WCWC in 1992 and the other Executive Team members: Joe Foy, National Campaign Director (1988), Matt Jong, Comptroller (1998) and Gwen Barlee, Policy Director (2001). They deserve huge accolades for putting the Wilderness Committee on a financially sound path and ensuring it keeps maturing as an effective force for wilderness protection. Only a tiny portion of their contributions are covered in this book.

There are dozens of Wilderness Committee employees who work behind the scenes doing the essential work needed to keep our large membership-based organization growing and running smoothly. These stellar employees currently include Monalisa Amirsetia, Finance Director since 1991; Nick Chamchuk, Canvass Director since 1999 in Alberta; Katherine Colby, Volunteer Nights Coordinator since 2001; Nik Cuff, Lower Mainland Campaign

Canvassers getting their briefing in the canvasser's area at our Vancouver office on Water Street. Photo: WCWC files.

Director since 2003; Deborah Drouin, Membership Services since 1997; Sue Fox, Communications Director since 1989; Bill Granger, Office Manager in Manitoba since 2006; Tammy Lea Meyer, a canvasser since 2002 and Website Manager since 2004; JP LeFrank, Printer and Distribution Coordinator since 2004; Shaine Macleod, Toronto Outreach Director since 2003; Andy Miller, Staff Scientist since 1998; Bri Drouin, Membership Services since 2002; Andrew Radzik, Public Outreach Director since 2005; Ian Russell, Canvass Director since 2005; Matthew Sasaki, Executive Assistant since 2005; Geoff Senichenko, former canvasser since 2003 and now Research and Mapping Director since 2004; Diana Vander Veen, part time Sales and Marketing since 1992; Anton van Walraven, Volunteer Coordinator since 2002; and Ken Wu, Victoria Campaign Director since 1999.

Finally there are the canvassers who take the Wilderness Committee's messages and campaigns out to the people. The Wilderness Committee would not have over 30,000 members if it were not for the diligence of the Wilderness Committee canvassers on their doorsteps: Mark, Jules, Tammy, Steve, Adam, Ryan, Steph, Micah, Jenn B., Nikita, Aimee, John. This list is a lot, lot longer.

There are ongoing campaigns and new publications that should have been included in this book but, for lack of space, are not. They include, among other, efforts to protect Canada's woodland and mountain caribou, both of which are on the verge of extinction, and our sporadic campaigns to save the East Creek and the Nahmint Valleys on Vancouver Island, some of the last remaining old-growth valleys that are being logged.

There is also a big new cooperative campaign to protect the world's only inland rainforest. It's located in the Kootenays and includes another East Creek with fabulous, giant cedars. We've been working on the Inland Rainforest Campaign for several years with the Valhalla Wilderness Society, publishing two newspapers so far: *Bugaboo Rainforest, Canada's Inland Rainforest* and *Rarest of the Rare – World's only Inland Rainforest Under Threat*.

We also co-published with the Union of B.C. Indian Chiefs an educational newspaper titled *Wild Fish need Wild Rivers and Oceans...* explaining the dangers of salmon fish farms to wild fish and destruction of spawning habitat by bad logging practices.

Above: Skit performed at the Ministry of Sustainable Resource Management's office in Victoria in December 2002 to try to save East Creek on northern Vancouver Island. Top: Ken Wu. Bottom: Volunteers Magali Ringoot (left) and Sarah Pawliuk (right). Photos: Mark Unrau.

Top: A giant western red cedar growing in East Creek in the Inland Rainforest. Photo: Gary Diers.
Bottom: View of the Bugaboo Spires from East Creek in the Inland Rainforest. The Valhalla Wilderness Society and the Wilderness Committee are calling for park protection for this unique and threatened wilderness. Photo: Joe Foy.

During the last five years the Wilderness Committee has published about eight educational reports and several research reports per year. The problems keep escalating and increasing in number because so few of them get solved. Yet for most of the campaigns we tackle, feasible and reasonable solutions exist. Governments simply refuse to implement them.

There are also campaigns that have almost come to a successful conclusion. Chief Bill Williams of the Squamish Nation informed us that negotiations with the provincial government over the legislative protection of the Wild Spirit Places—including the Elaho and Sims—are going well. In late December 2005, after Interfor closed and dismantled their lumber mill in Squamish, the Squamish First Nation purchased Interfor's TFL 38. They are now in the driver's seat to both protect their Wild Spirit Places including the Upper Elaho and the Sims and institute true sustained yield eco-forestry. By June 2006 we had great hopes that the announcement of legislated protection for the Wild Spirit Places was "imminent." In the summer of '06, upon the invitation of the Squamish Nation, Wilderness Committee volunteers are back again clearing trail in the Elaho Valley, something we hadn't done for several years.

One very major celebration occurred on February 7, 2006: the B.C. government announced its Central Coast and North Coast LRMP decision to protect one-third of the Great Bear Rainforest. It was the conclusion to an unprecedented collaboration between First Nations, industry, environmentalists and governments. The combined protected areas total approximately 1.8 million hectares. Included for protection are 200,000 hectares of the white Kermode (Spirit Bear) habitat—about 80 percent of the habitat that bear biologist Wayne McCrory, who co-authored the first Spirit Bear proposal, says is critical. Not everything happening out there is negative!

How times change – Gwaii Haanas rated North America's top park

A few weeks before the Wilderness Committee's 25th anniversary (August 7, 2005) the July-August issue of the National Geographic Traveler magazine hit the newsstands with a ranked list of the top 55 national parks in North America. The National Geographic had some 300 experts in sustainable tourism, destination quality, and park management evaluate each park. On a scale of 1 to 100, not one scored 90 or above, designating the park *as unspoiled and likely to remain so.* In other words, all of the great national parks in North America are under threat. But the park that topped the list, with a score of 88, was Gwaii Haanas, the park proposal that had sparked the Wilderness Committee into existence.

This high score was not simply because the place is spectacular and relatively less threatened because its remoteness brings fewer than 3,000 visitors a year. According to the National Geographic, it was due to the fact that the residents of Haida Gwaii display a real stewardship ethic and there is a unique partnership between Parks Canada and the native Haida people in the management of the park. As one of the experts involved in the rating said, *Beautiful and intact. A great model for other regions.*

When the threat of offshore oil development near Gwaii Haanas is eliminated, its score will surely rise.

An editorial titled *Haven from logging honoured as top park* published in the July 10, 2005 *Province* newspaper stated:

> What, because of its remoteness and natural beauty, was termed 'Galapagos North' during the heady days of confrontation has proven to be just that to the destination park enthusiasts. Not the sort of park you'd pop into to walk the dog, but a worthwhile save from the chainsaws.

Richard and I and the others who started the Wilderness Committee and the many who joined its ranks as staff, volunteers, members and supporters were right all along.

We were right in the beginning about the need to protect and safeguard Gwaii Haanas. And we are right about the need to protect all the wilderness park candidates that are still threatened today. We are right about lifting the 12 percent cap on wilderness protection and letting science be our guide as to how much wilderness needs to be protected. We are right about the need to reduce our emissions of global warming gasses and about the unprecedented havoc that rapid climate change will cause. In fact, I'm sure we are on the right side of all the conservation issues that our Wilderness Committee continues to champion!

It's not about balance, trade-offs or expedience. It's about survival!

It took a colossal effort in both public education and political willpower to divert a place once called South Moresby on the Queen Charlotte Islands from becoming just another remote logged-out area and ensure it remained one of the most precious of wild places on the planet: Gwaii Haanas of Haida Gwaii.

I have a dream. It's not about bags of cheques arriving at the Wilderness Committee every Christmas (although that would be nice). My dream is that the Wilderness Committee's 50th anniversary in 2030 will celebrate the fact that Canadian governments listened to us and did what was right; that they left enough wilderness for mother nature to conduct its business, free from the outwash of the heavy crushing footprint of humankind.

My dream is that the Wilderness Committee is honoured for its decades of work in the best way possible: that sufficient wilderness is safeguarded to keep Earth healthy, wild species thriving, and generation after generation of Canadians to come thrilled by the experience of the wilds.

Calling all tree huggers to celebrate

In early July of 2005 the Wilderness Committee sent out a notice—*Calling all Tree Huggers*—urging our members to "grab their gear" and come to a campout in the Elaho Valley to celebrate the Committee's 25th anniversary on the August 1-3 long weekend. What better way to celebrate B.C. Day than to have a campout in an endangered wilderness area?

Already the event had grabbed a bit of local media attention. An article in the Squamish Chief reported that:

> While some Squamish residents celebrate logger sports at one end of the corridor, others will make merry alongside some of the oldest living trees in Canada. The Western Wilderness Committee is holding a gathering in the Elaho Valley to commemorate the conservation group's 25th anniversary. The party coincides with Squamish Logger Days, but the committee didn't give the event a thought when they chose the B.C. Day weekend for their celebration, said Wilderness Committee national campaign director Joe Foy.
>
> 'The reason is it rains a lot in this province,' laughed Foy. 'And there's one – and I mean one – long weekend in the year when you can pretty well count on good weather.'

I hitched a ride from Horseshoe Bay to the campsite with Joe Foy who was also taking up a couple of young volunteers on Friday morning in his truck. It was cloudy and the forecast was for rain. But this didn't dampen our spirits. There hadn't been any logging in the Upper Elaho since 2001. Recently Interfor had shut down their sawmill in Squamish and announced that they were going to sell their Tree Farm License. Rumour had it that the Squamish First Nation

Sitting around the fire at the 25th anniversary campout in the Upper Elaho. Photo: WCWC files.

was interested in buying it (which they did several months later). Four years earlier the Squamish Nation had declared the Upper Elaho area where we planned to camp a protected area—a Wild Spirit Place.

A couple of days earlier Andy Miller and JP LeFrank had taken up several volunteers to prepare the site for our big campout celebration. It was located at the end of an abandoned Lava Creek spur logging road on the north side of Lava Creek. Andy and his volunteer crew built several outhouses, strung up banners and some big tarps so there would be a dry area in case it rained, set up a kitchen area, built a big fire pit far from forest combustibles and marked out some campsites along the road.

I had not been to the Upper Elaho for many years. In fact, the last time I'd been there was in July 1997, when I was stopped at "Interfort" (the loggers' blockade) at the height of the Stoltmann conflict. Joe and I reminisced on the way in and talked about current campaigns. It was a great chance to get back in touch. I'd been immersed in writing this book for two years.

The road into the Upper Elaho was long and dusty; about 60 miles of deteriorating gravel. Joe explained that the upper part of the road was seldom used since logging had ceased four years earlier.

We were all feeling confident that the *Wild Spirit Place* we were going to camp in would soon be protected. Joe had brought along copies of the Wilderness Committee's latest newspaper: Kwa Kwayexwelh-Aynexws: *Wild Spirit Places* (Educational Report Vol. 24 No. 6 with a press run of 75,000 copies) which had been published in co-operation with the Squamish Nation a month earlier. It told the story of the land use planning process undertaken by the Squamish Nation in early 2000 and the seven Wild Spirit Places, including the Upper Elaho, that they had designated in their traditional territory.

Just as we were about to enter one I asked:

"Hey Joe, isn't this the clearcut where we launched our Stoltmann Campaign ten years ago?"

"Yes, it is."

"Where's the Elaho Giant that we all hiked to that weekend?"

"We'll be driving close by it in a few moments."

We drove along in silence, for what seemed like a long time, but perhaps it was only a little over a kilometre.

"There it is. See that flagging tape. It's only a short walk in to it."

We continued to drive along in silence past recent clearcuts and further into what was part of the wild forest when we first began the Stoltmann campaign... the hardest campaign with the most heart-breaking moments that we had ever undertaken.

It seemed like forever, but it was only about another two kilometres when Joe broke the silence and pointed out that we were approaching Lava Creek Bridge, the site of the big "ethnic cleansing day" attack by the Interfor loggers in September 1999, six years earlier. We parked beside the road only a few hundred metres beyond the bridge. Our camp was already set up, banners flying. All I had to do was put up my own tent and join Joe for a 'cuppa joe' that Joe brewed for us.

All afternoon people kept pouring in and the rains held off (although it poured heavily every night). I only knew about a quarter of the approximately 100 or so people who arrived. Many of those I knew, I hadn't seen for years. It was a mixed group including a lot of kids who ran rambunctiously around, obviously enjoying themselves in the outdoors and around the fire like tens of thousands of generations of kids had done before them.

One family there had received a gift membership from a relative last Christmas and on impulse decided to come and find out what the Wilderness Committee was about. They got a flat tire coming in and during the time that they were there, two more tires on their car went flat. Joe and others helped get the tires fixed. Their flat tires didn't dampen their enthusiasm for the campout's berry picking contest, chili cook-off, fireside talks and music.

Around the fire that night the kids did what all kids love to do: put a stick in the fire until it catches fire, take it out, twirl it around and revel in the glowing patterns it makes in the dark. The parents let them do it (like I had let mine when they were young and my parents had let me), making sure that they did not hit anyone nearby or stumble into the fire.

In a society that's become so obsessed with "safety," it was heartening to see this wild side of child-nature expressed and enjoyed. While people sat around the fire I talked for hours relating some of the stories of the "good old days" which are in this book. Others joined in with their stories and Joe added details that I'd never heard before. As night settled in and the sprinkles turned into showers and then into a steady downpour, people retired to their tents until there were only a few of us diehards left by the fire, sharing good humour and good tactics. Joe got everyone that night to participate in a letter-writing campaign aimed at convincing the provincial government to officially recognize and protect the Squamish *Wild Spirit Places* in law. The Committee still lived by the rule of never missing an opportunity to further the cause.

The next morning we split up into three groups. One was a work party to remove the remnants of the no-longer-needed cable crossing and ladder bridge across Lava Creek because a logging bridge now spanned the creek. Some of the young, fit and energetic campers hiked with Joe Foy a long ways along our trail to Cesna Creek and back. Those who were out of shape, had young children or simply wanted to take it easy hiked up to the top of Eagle Bluffs to the lookout to see the ten-foot-tall Cedar Woman, the Squamish carving that stood guard over the Upper Elaho Valley.

I took the easier hike with the kids whose joyous curiosity abounded and who were keen to learn about nature as I pointed out things along the way that they would have otherwise overlooked. As I climbed up to the viewpoint, I thought about Joe helping hoist the statue up the steep trail four years earlier. PATH protesters, the RCMP, natives, had all hiked up together on the same trail that we were on, taking their turns lifting up the statue. I had seen pictures of it, but it's not the same as being there or visiting the place yourself.

There is a spiritual side to nature that we will never understand. I stood beside Cedar Woman resolutely overlooking the string of clearcut scars that had proceeded up the Elaho Valley and then

Looking out over the clearcuts down the Elaho Valley beside Cedar Woman. Photo: Paul Morgan.

stopped, as if heeding her aura that said stronger than words ever could: "Thou shalt go no further." Behind her, up the rest of the valley, was wilderness as far as the eye could see.

Wilderness, the handmaiden of nature, that's what we have stood on guard to protect since the Wilderness Committee began less than 10,000 days ago. And with the help of our large extended family of members, supporters and colleague organizations wilderness is what we'll continue to save, protect and restore for 10,000 years into the future on our tiny blue green planet.

Appendix I

Protected Area Accomplishments

Since forming in 1980, Western Canada Wilderness Committee working with other environmental organizations, First Nations and concerned citizens has helped successfully establish the following protected areas:

1983 - Valhalla Provincial Park, BC

1987 - South Moresby (Gwaii Haanas National Park Reserve), BC

1990 - Lower Carmanah Valley (Carmanah Pacific Provincial Park), BC

1992 - Khutzeymateen Grizzly Bear Sanctuary, BC
Maplewood Mud Flats Conservation Area, BC

1993 - Megin Valley addition to Strathcona Provincial Park, BC

1994 - Tatshenshini-Alsek Provincial Park, BC
Chilko Lake (Ts'yl-os Provincial Park, BC
Upper Carmanah Valley (Carmanah Walbran Provincial Park), BC
Lower Carmanah Valley (Carmanah Walbran Provincial Park), BC
Nasparti Valley addition to Books Peninsula Provincial Park, BC
Lower Tsitika Valley Provincial Park, BC
Tashish-Kwois Provincial Park, BC
Kitlope Valley Protected Area, BC
Niagara Valley (Cariboo Mountains Provincial Park), BC

1995 - Power Valley addition to Brooks Peninsula Provincial Park, BC
West Arm Provincial Park, BC
Akamina-Kishinena Provincial Park, BC
Indian Arm Provincial Park, BC
Pinecone Burke Provincial Park, BC
Surrey Bend Regional Park, BC
Stein Valley Nlaka'pamux Heritage Provincial Park, BC
Tetrahedron Provincial Park, BC
Jedidiah Island Provincial Park, BC
Boundary Bay Regional Park, BC

1996 - Clendenning Valley Provincial Park, BC
Upper Lillooet Valley Provincial Park, BC
Caren Range (Spipiyus Provincial Park), BC

1997 - Greendrop Lake addition to Chilliwack Lake Provincial Park, BC
Skagit Valley Provincial Park, BC
Cummins Valley addition to Cummins Lakes Provincial Park, BC
Sooke Hills Wilderness Regional Park Reserve, BC
Northern Rocky Mountains Provincial Park, BC

2000 - Clayoquot Sound Biosphere Reserve, BC
Grand Canyon of the Stikine (Stikine River Provincial Park, BC

2001 - Greystokes Provincial Park, BC
Snowy Protected Area, BC
Spirit Bear Protected Area, BC
Koeye River Protected Area, BC
Southern Chilcotin Mountains (Spruce Lake Protected Area), BC
Caribou Mountains Wildland Provincial Park, AB
Sturgeon Bay Park Reserve (protection extended), MB

2002 - Manigotagan River Park Reserve (protection extended), MB

2003 - Addition to Manigotagan River Park Reserve, MB
South Atikaki Provincial Park, MB

2004 - Poplar/Nanowin Rivers Park Reserve (protection extended), MB
Burns Bog Protected Area, BC
Manigotagan River Provincial Park, MB

2005 - Cathedral Grove/MacMillan Provincial Park, BC (protected area expanded)

Appendix II

A chronology of the key events pertaining to the environmental movement primarily in Canada and especially in B.C. with an emphasis on Western Canada Wilderness Committee's involvement

Circa 52,000-13,000 BP (before present) Waves of Paleo-Siberians arrive in the Americas from Asia, across the Bering Strait Land Bridge called Beringia and disperse throughout the Americas.

13,000 BP-200 BP First Nations developed many different cultures in North America as the glaciers receded and habitats change, some of them were sustainable over long periods of time.

6,000 BP Climate change along the northwest coast of North America brings wetter conditions and western red cedar's range expands and this species becomes one of the dominant tree species along the coast.

October 7, 1763 A Royal Proclamation by the King of England sets the boundaries of lands in North America. It recognizes Aboriginal Title to these lands and sets out the rules of extinguishment of this Title for land acquired by the British Empire.

1778 Captain Cook, an English explorer, visits Nootka Sound on the west coast of what is now called Vancouver Island, British Columbia. He is the first European to trade goods with the natives.

June 12, 1792 Captain George Vancouver arrives, charts and names Point Grey, the Strait of Georgia, Point Atkinson, English Bay and Burrard Channel.

1804 The worldwide human population reaches one billion.

1844 The last two Great Auks clubbed to death on an island in the North Atlantic. It was the first species of birds to be driven to extinction by humans in North America. The five-kilogram 70-centimetre-tall Great Auk, the only flightless bird in the Northern Hemisphere, occupied a niche similar to that which the penguins do in the Southern Hemisphere.

1846 U.S. government under terms of the Treaty of Oregon ceded the territory, which is now British Columbia to England without extinguishing aboriginal title. This Treaty established the border between Canada (Vancouver Island) and the United States, at the 49th parallel. The United States wanted the border to be at the 54th parallel. It provides evidence that aboriginal title and rights existed prior to 1846 and have not been extinguished in B.C.

1849 Great Britain made Vancouver Island a Crown Colony and leased it to the Hudson's Bay Company for seven shillings per year.

1851 Henry David Thoreau delivers an address to the Concord (Massachusetts) Lyceum declaring that, *"in Wildness is the preservation of the World."* In 1863, this address is published posthumously as the essay titled *"Walking"* in Thoreau's *Excursions*.

1858 Colony of British Columbia is established.

1858 James Douglas, governor of Vancouver Island officially sets aside a 75-hectare (200 acre) plot of land in the town of Victoria. It was named Beacon Hill Park after a pair of masts strategically placed on a hill in the park to act as a beacon and navigational aid to mariners approaching Victoria's inner harbour. It was designated as a park in 1882.

1859 Charles Darwin's Origin of Species published. It presents the theory of evolution of life on earth and the role of natural selection.

1866 The word "ecology" is coined by the German biologist Ernst Haeckel to embody the concept of the study of living things within the context of their surround living and non-living environment.

1866 Gregor Mendel publishes his paper on his experiments on heritable traits in peas pioneering the science of genetics.

1867 The Federation of Canada is established through the passage of the British North America (BNA) Act.

March 1, 1872 Yellowstone National Park comprising 2.2 million acres (890,000 hectares) of wilderness is set aside for "the benefit and enjoyment of the people" in the U.S. It becomes the world's first National Park. Preserved within Yellowstone are Old Faithful Geyser and the majority of the planet's springs and geysers—about 10,000 in all.

1884 The province of B.C. deeded almost one million hectares of land on southeastern Vancouver Island (almost 20 percent of the entire Island) to the Esquimalt & Nanaimo Railroad Company as compensation for building and operating a railroad on the Island.

1885 A 2,600-hectare Banff National Park nearby the new Trans Canada rail line—Canada's first National Park—is created as a tourist attraction around a cave with hot springs in the Rocky Mountains. Over the next 45 years, more and more area is added to this small parcel of land until 1930, when it reached its present size of 664,100 hectares.

1886 Robert Dunsmuir receives 1.9 million acres of land on Vancouver Island including Cathedral Grove by Cameron Lake in return for building part of a Vancouver Island railway.

June 8, 1887 By resolution, the Canadian government establishes the Government Reserve lands in Vancouver—1,000 acres (400 hectares) of a largely logged peninsula with some veteran old-growth trees for park and recreation purposes, known today as Stanley Park.

June 4, 1892 John Muir and a group of associates meet in San Francisco, California to found the Sierra Club, devoted predominantly to outings and mountain climbing expeditions. It eventually evolved and became explicitly dedicated to the preservation of wilderness.

1901 Dr. James Fletcher, Dominion botanist and entomologist, urges the B.C. government to preserve the forests around Cameron Lake.

1905 A Vancouver entrepreneur Mr. Burns buys the big bog beside the Fraser River near the city and names it after himself.

1905 William Bateson coined the word "genetics."

1905-1907 To raise money the B.C. government sells off the timber on one square mile parcels of prime forestland as Timber Licences (TLs) that "timber prospectors" staked at $10 each. The purchaser owned the timber from that time on and when the old growth was cut the TL was to expire and the land revert back to the "Crown."

1907 The Jasper area in the Canadian Rockies is assigned protection as a Forest Park. It underwent various excisions and extensions, until it was established as 1,295,000-hectare Jasper National Park in the National Parks Act passed in 1930.

1908 U.S. President Theodore Roosevelt held the first National Conference on Conservation at the White House.

November 1, 1908 The last Dawson's caribou is shot on the Queen Charlotte Island as a museum specimen for the Royal Museum in Victoria B.C.

1909 The Canadian government establishes a Commission of Conservation.

1910 Canada passes a National Parks Act.

1911 Canada passes the Dominion Forest Reserves and Parks Act which places the federal parks under the administration of the world's first national parks branch, known variously over the years as the Dominion Parks Branch, the National Parks Branch, Parks Canada, Canadian Parks Service and now the Parks Canada Agency.

March 1, 1911 The B.C. government establishes B.C.'s first provincial park, a 250,000-hectares triangle of mountainous virgin land in the centre of Vancouver Island. It is named Strathcona, after the man who drove the last spike on the Trans-Canada Canadian Pacific Railroad, Donald Alexander Smith, the Baron of Strathcona.

1912 The British Columbia government establishes the B.C. Forest Service.

1913 A 218,795-hectare Mount Robson Provincial Park, located around the highest peak in the Canadian Rockies, is the second park to be established in B.C. In 1967, 739 hectares are added to this park.

1913 A major rockslide at Hell's Gate (resulting from railway construction) narrowed an already narrow channel and created a five-metre-high waterfall almost completely blocking salmon access to the upper Fraser River and Thompson River spawning beds. By 1915, the removal of 45,000 cubic metres of rock from the channel had eased the problem, but the 33-metre-wide Hell's Gate gorge still remained one of the toughest obstacles salmon have to negotiate on their way to their spawning grounds.

September 1, 1914 The passenger pigeon becomes extinct. The last one of its species dies in a U.S. zoo.

1917 The first commercial logging operation starts in the South Moresby region of the Queen Charlotte Islands (Haida Gwaii).

1918 Professor John Davidson founds the Vancouver Natural History Society (VNHS). Its objectives are to promote the enjoyment of nature, foster public interest and education in the appreciation and study of nature, encourage the wise use and conservation of natural resources, work for the complete protection of endangered species and ecosystems and promote access to, and maintenance of, natural areas around Vancouver B.C.

1918 The Strathcona Park Act is amended to allow for mining in the park. From this date to 1965, more than 400 claims are staked inside the park in addition to the very few that existed when the park was first created.

July 2, 1918 Canada and the U.S. sign a Migratory Bird Treaty. It implements the 1916 Convention between the U.S. and Great Britain (for Canada) for the protection of migratory birds including provisions to protect nesting birds.

1922 Canada establishes the 4,480,700-hectare Wood Buffalo National Park in northern Alberta and the NWT consisting of northern boreal interior plains to protect the habitat of a small herd of wood bison whose population had dropped from an estimated 40 million in 1830 to less than 1000 by 1900.

1922 The B.C. government establishes the Greater Victoria Watershed District by statute protecting the area for the sole purpose of supplying water for the area's inhabitants.

1925 A 1,300-hectare forest fire started by the Capilano Timber Co. in the Capilano watershed where people in Vancouver, B.C. get their drinking water is the catalyst for the formation of the Greater Vancouver Water District to protect from logging the watersheds from which Vancouver draws it drinking water.

February 3, 1926 The Greater Vancouver Water District is formed as mandated by the Greater Vancouver Water District Act passed by the B.C. government in late 1924.

1927 The B.C. government issues a 999-year lease for Crown lands in the Capilano and Seymour watersheds to the Greater Vancouver Water District to give it control of these lands and guarantee there will be no future logging to protect Vancouver's drinking water.

1927 The Earth's human population reaches two billion.

1927 The B.C. government establishes the 194,650-hectare Garibaldi Provincial Park, consisting mainly of mountaintops and glaciers, near Whistler, B.C.

May 9, 1927 The 38,750-hectare Prince Albert National Park is established in Saskatchewan on the southern edge of Canada's great boreal forest. Its most famous resident was Grey Owl who used the park as a home base between 1931 and 1938.

1929 Green Timbers, the last remaining low elevation old-growth forest of giant Douglas firs in the Fraser Valley forest (approximately 2,000 hectares in Surrey), is clear-cut despite proposals to have the grove protected in a park. Replanted in the '30s as an urban forest, a 260-hectare parcel of the original Green Timbers lands is now home to an impressive second growth stand protected as the Green Timbers Nature Reserve.

December 28, 1929 The 297,3000-hectare Riding Mountain National Park is established in Manitoba. This park is now an "island" completely surrounded by developed lands.

1930 The Canadian government passes Canada's first National Parks Act. It ensures that no new parks can be established or any change made in the boundaries of existing parks except by an Act of Parliament. Mineral exploration and development is prohibited and only limited use of green timber essential for park management purposes is allowed.

January 1935 The Wilderness Society is formed in the U.S. to protect American wilderness and to develop a nation-wide network of wild lands through public education, scientific analysis and advocacy.

November 30, 1936 Greater Vancouver Water District Commissioner Ernest Cleveland states: *"The district's (Greater Vancouver Water District) policy is to preserve all the timber–both commercially loggable and otherwise–in the watershed...so that neither now, nor in the future, will filtration or sterilization of the water be required."*

1937 The Vancouver Natural History Society made a presentation to the provincial government to make the Southern Chilcotin Mountains region a provincial park.

1937 The International Pacific Salmon Fisheries Commission (IPSFC) is established for the protection, preservation, and extension of the Fraser River sockeye and pink salmon, and to rehabilitate stocks depleted by passage problems and over-fishing.

1938 The B.C. government establishes B.C.'s largest park, the 981,000-hectare Tweedsmuir Provincial Park near Bella Coola in the Chilcotins.

1938 Ducks Unlimited (DU) is founded by a group of conservation minded U.S. hunters. Noting that waterfowl numbers were plummeting and wetlands were steadily disappearing from drought and agricultural and urban expansion, DU raises funds in the U.S. to initiate habitat conservation projects in Canada, where 70 percent of North America's waterfowl nest. DU has been working on habitat protection in B.C. since 1969.

1939 The B.C. government establishes a 500,000 hectare Wells Gray Provincial Park in South Eastern B.C. to protect a small part of the Cariboo Mountain Range.

1941 The B.C. government establishes the 1,009,112-hectare Hamber Provincial Park in the Southern Rocky mountains as the then largest provincial park in the province. Twenty years later in 1961 and 1962 the B.C. government reduces Hamber Park to its present size of 24,500 hectares. Hamber Provincial Park is a historic, dramatic example of the fact that parks are never completely secure. Governments can always create parks, alter existing park boundaries and abolish parks.

1942 After the Greater Vancouver Water District discovers that the B.C. Ministry of Forests permitted a timber sale in the Coquitlam Watershed (where some Greater Vancouverites were drawing some of their drinking water) without approval from the Water District, the Water District obtains a 999-year lease for the Coquitlam watershed from the provincial government.

1944 B.C. Premier Hart offers H.R. MacMillan, a new manager of Victoria Lumber & Manufacturing Company (later to become MacMillan Bloedel Ltd. logging company) an exchange of government owned land for the big giants in a grove of large Douglas firs called Cathedral Grove at the head of Cameron Lake on Vancouver Island. MacMillan accepted the exchange and "donates" 136 hectares of land–part of Cathedral Grove for a park.

1945 The Sloan Royal Commission on forestry in B.C. takes a look at the unregulated forest practices in B.C. and recommends a policy of sustained yield forestry, the establishment of a rate of logging that is equal to the overall annual rate of re-growth of the forest. The result is the awarding of Tree Farm Licences (TFLs), area-based tenures that gave the owner exclusive rights to cut trees in the area.

1945 The International Pacific Salmon Fisheries Commission (IPSFC) completes first fishway at Hell's Gate, on the Fraser River to aid upstream migrating salmon impeded by a slide precipitated by railroad construction in 1913.

1945 Forester Merv Wilkinson begins to practice selectively logging on an even-flow sustained yield basis in the woodlot he purchased in 1938 near Ladysmith on Vancouver Island. In 2002 he was on his 13th cut on his 137-acre mixed-species forest and he had managed this woodlot so that it had more standing volume on it then than it had when he started.

November 16, 1945 UNESCO, the United Nations Educational, Scientific and Cultural Organization is formed.

1947 The B.C. government establishes MacMillan Provincial Park to protect part of Cathedral Grove at the head of Cameron Lake on Vancouver Island.

1948 The Audubon Society of Canada is established–a bird preservation organization named after the American naturalist and wildlife painter John James Audubon, who lived from 1785 to 1851. The first Audubon Society formed in the U.S. in 1886.

1948 The World Conservation Union (IUCN) is founded. The IUCN's mission is to influence, encourage and assist societies throughout the world to conserve the integrity and diversity of nature and to ensure that any use of natural resources is equitable and ecologically sustainable. IUCN is the world's largest environmental knowledge network and has helped over 75 countries to prepare and implement national conservation and biodiversity strategies. Today it brings together 75 States, 108 government agencies, 750 plus NGOs, and some 10,000 scientists and experts from 181 countries in a unique worldwide partnership. IUCN is a multi-cultural, multilingual organization with 1000 staff located in 62 countries. Its headquarters are in Gland, Switzerland.

1949 *Sand County Almanac* by Aldo Leopold, a professor of game management who had come around to believing that wolves had as much right to live as people, is published posthumously. Aldo Leopold urges his readers to adopt a new land ethic that rests on a community defined to include all living things.

1950 B.C. and Alcan enter into the "1950 Agreement" which provides Alcan a number of rights in perpetuity to the water in the Nechako and Nanika Rivers, in exchange for building an aluminum smelter plant in Kitimat. It was the first environmental issue in B.C. studied by the U.S. Sierra club because of its widespread ecological consequences.

1951 The B.C. Wildlife Federation is incorporated under the B.C. Societies Act. Its purposes are to ensure the sound, long-term management of British Columbia's fish, wildlife, park and outdoor recreational resources in the best interests of all British Columbians, and to coordinate voluntary agencies, societies, clubs and individuals interested in that objective.

1952 American financiers proposed building the biggest dam in the world on the Fraser River at Moran Canyon 30 kilometres north of Lillooet, B.C. The plan called for a $500 million, 270-metre-high dam, backing the Fraser River up into a 260-kilometre-long lake, stretching upstream to Quesnel. This project is permanently shelved in 1972 due to concerns over the survival of Fraser River salmon, then the largest salmon run in the world.

August 1, 1952 William Andrew Cecil (WAC) Bennett, Social Credit, becomes Premier of B.C. and holds that position until September 15, 1972.

January 24, 1955 Tofino Tree Farm Licence (TFL) 20 is awarded to MacMillan Bloedel Ltd. (MB). This perpetually renewable licence, awarded without any consultation with the First Nations affected, grants the company exclusive rights to log in more than half of Clayoquot Sound. Under the Forest Act at the time (1947), TFLs are called Forest Management Licences (FML).

March 1, 1955 The B.C. government grants Alberni Tree Farm Licence (TFL 21) to MB. In exchange, MB agrees to provide jobs and a mill and to log on a sustained yield basis.

May 18, 1955 Maquinna Tree Farm Licence (TFL 22) is awarded to British Columbia Forest Products Ltd. (BCFP) granting this company the exclusive logging rights to almost all the rest of Clayoquot Sound also without any consultation with the First Nations affected. Granted under a cloud of corruption, the Forest Minister of the day is eventually convicted and jailed for accepting bribes in connection with this matter. This licence was later sold to Fletcher Challenge and subsequently, in 1992, sold to International Forest Products (Interfor). During the 1970s, clearcutting in Clayoquot more than tripled over the initial rate of cut established when the licence was granted.

1956 The third Royal Commission of Inquiry into Forest Resources, the 1956 Sloan Commission, assesses the government's implementation of the sustained yield policy. It recommends management of individual, self-contained watersheds on a sustained yield basis—a recommendation not heeded by government.

1956 A timber cruiser reports seeing a mammoth Sitka spruce measuring 7.9 metres in diameter and 94.2 metres tall growing in Carmanah Valley on the west coast of Vancouver Island.

1957 The B.C. government creates a Department of Recreation and Conservation that included a Parks Branch. For the first time in the history of B.C., parks are independent of the B.C. Forest Service.

October 4, 1957 The Russians launch the Earth's first artificial satellite—Sputnik 1.

October 20, 1958 Nitinat Tree Farm Licence (TFL 27) is awarded to Moore-Whittington Lumber Company Ltd.

1958 The B.C. government awards TFL 24 to Alaska Pine & Cellulose Ltd. a 21-year renewable licence totaling 115,521 hectares on Moresby, Lyell and Burnaby Island. The company is authorized to harvest 212,376 cubic metres annually.

1959 The world wide human population exceeds three billion.

1960s The B.C. government allows huge tracts of prime old-growth forest to be cut within Strathcona Park to compensate companies for logging rights taken away to get smaller areas for new parks, including Cape Scott, Long Beach, Rathtrevor Beach and others on Vancouver Island.

1961 Bowron Lake Provincial Park is established.

1961 Alaska Pine bought out by Rayonier Canada Ltd., who consequently now owns TFL 24.

1961 The B.C. government grants Tree Farm Licence (TFL) 38, which includes most of the Squamish River System, to Empire Mills Ltd., without public consultation and without the permission of the Squamish First Nation.

1961 Western Mines, later called Westmin, gets government permission for open pit mining and builds a 35-kilometre-long access road into Strathcona Park.

April 12, 1961 Yuri Gagarin becomes the first human in space by completing one full orbit around the Earth.

September 27, 1962 *Silent Spring,* a book by Rachel Carson that blew the whistle on the devastating effects of the widespread use of pesticides, is published in the U.S.

1963 The National and Provincial Parks Association of Canada (NPPAC) later renamed Canadian Parks and Wilderness Society (CPAWS) is founded. It focuses on establishing new parks and making sure the needs of nature come first in park management. A B.C. chapter is founded in 1979.

1964 The U.S. government passes the Wilderness Act, establishing a process for permanently protecting special federal lands from development.

1965 Backcountry enthusiasts, ranchers and outfitters in rural southwest Alberta form the Alberta Wilderness Association (AWA) to protect wilderness values in Alberta.

1967 World Wildlife Fund Canada (WWF) founded as an affiliate of World Wildlife Fund International. Its mission is to stop the degradation of the planet's natural environment and to build a future in which humans live in harmony with nature by conserving the world's biological diversity, ensuring that the use of renewable resources is sustainable and promoting the reduction of pollution and wasteful consumption.

1967 The B.C. government grants Cream Silver Mines permission to explore and drill in Strathcona Park.

March 7, 1967 The B.C. government places an Amending Indentures to the three preexisting 999 year leases held by the Greater Vancouver Water District which open the way up to industrial logging in the three watersheds (the Capilano, Seymour and Coquitlam) which supplies the drinking water for Greater Vancouver citizens.

April 20, 1968 Pierre Elliott Trudeau, Liberal, becomes Prime Minister of Canada.

1969 Sierra Club of B.C. is founded. It is dedicated to protecting B.C.'s environment with a broad and comprehensive mandate including the conservation of wildlands and wildlife.

1969 The "granddad" of B.C.'s environmental movement, the Society Promoting Environmental Conservation (SPEC) began in the home of Gwen and Derrick Mallard. With activists and academics from Simon Fraser University as the original nucleus, it began with four campaigns: offshore oil drilling, strip mining, Burrard Inlet pollution and pesticides. SPEC remains an urban environmental organization concerned primarily with water quality, air quality, pesticides, and energy conservation.

1969 A small group of University of Toronto students launch *Pollution Probe* to protect Canada's environment.

1969 B.C. government increases Moresby TFL 24's Annual Allowable Cut from 212,376 cubic metres annually to 305,822 cubic metres annually.

1969 The Nisga'a launch first native land claims court case in the Supreme Court of B.C. (Caulder case).

July 20, 1969 The first human, Neil Armstrong, steps onto the moon—Apollo 11 mission.

October 1969 The Don't Make A Wave Committee is formed in Vancouver to protest the proposed underground atomic testing in Amchitka, Alaska. It was the precursor organization that transformed into Greenpeace.

October 30, 1969 The Seattle City Council votes to construct High Ross Dam despite the environmental impact in the United States and Canada.

1969 The R.O.S.S. (Run Out Skagit Spoilers) Committee is formed as a coalition of numerous interest groups, which organized protests against raising the Ross dam. Plans to flood the Skagit Valley were delayed and eventually shelved in light of the public outcry.

1970 The Natural Resources Defense Council (NRDC) is founded in the U.S. to protect the environment and human health through advocacy, litigation, scientific research and education, not to mention publication.

1970 Citizens who are alarmed at the decline of B.C.'s wild salmon and steelhead stocks establish the Steelhead Society of British Columbia, a registered, non-profit, watershed conservation organization that fights to protect, enhance and restore British Columbia's wild salmon and wild salmon habitat.

April 22, 1970 The first Earth Day is celebrated. Twenty million people across North America hold events marking the birth of the modern environmental movement. Founded by Senator Gaylord Nelson of Wisconsin and first organized by Denis Hayes, it put environmental issues on the map in the U.S. Earth Day has grown to be an annual worldwide affair.

January 26, 1971 A 49,996-hectare Pacific Rim National Park Reserve is established on the west coast of Vancouver Island. It comprises three units—Long Beach, the Broken Island Group consisting of approximately 100 islands and rocks located in Barkley Sound and West Coast Life Saving Trail from Port Renfrew to Pachena.

1971 The B.C. government passes the Ecological Reserves Act. Twenty-nine Ecological Reserves (ERs) are created, including Cleland Island, a seabird colony west of Tofino, and the smallest reserve, Canoe Islet, a 0.61-hectare island off the southeast tip of Valdes Island.

1971 The B.C. government increases TFL 24's AAC again. It goes from 305,822 cubic metres annually to 410,594 cubic metres annually.

1971 Don't Make a Waver Committee adopts the name Greenpeace and makes headline news around the world when a handful of people set sail from Vancouver in a hired fishing boat for the U.S. atomic test zone of Amchitka, Alaska to protest nuclear weapons testing in an attempt to pressure the U.S. to abandon this largest underground nuclear test explosion to date.

1971 The Canadian Audubon Society expands its mandate and becomes the Canadian Nature Federation (CNF).

May 1971 A group of Dalhousie University students with the help of a government grant found the Ecology Action Centre in Halifax. It's two goals: are to convince Haligonians to recycle and to be a source of environmental information for the public.

1972 Federal government moratorium issued to forbid crude oil tanker traffic and exploratory drilling in Dixon Entrance, Hecate Strait and Queen Charlotte Sound due to concerns about negative impacts on B.C. fisheries.

1972 The Fish and Wildlife Branch of the B.C. government proposes that the entire Tsitika Watershed be made into an ecological reserve. The proposal stated, "This is the last untouched watershed on the east coast of Vancouver Island, and as such is an invaluable study area."

February 1972 Greenpeace Foundation is established and registered under the B.C. Societies Act. Today, Greenpeace is a global environmental organization operating in more than 27 countries with over 2.5 million members around the world.

February 22, 1972 The Canadian government establishes the 2,201,500-hectare Kluane National Park Reserve in the Yukon and the 476,560-hectare Nahanni National Park Reserve in the North West Territories.

July 26, 1972 The first satellite dedicated to studying the Earth and its resources is launched—Landsat 1.

September 15, 1972 Dave Barrett, NDP, becomes Premier of B.C. and holds this position until December 22, 1975.

1973 The B.C. government institutes a logging moratorium covering 120,00 hectares in the Tsitika Valley-Schoen Lake area, including the entire 39,000-hectare Tsitika watershed.

1973 Roy Mason of the B.C. Mountaineering Club submits the first Stein Valley park proposal to the B.C. government.

1973 The B.C. government establishes 10 Ecological Reserves (ERs) including the Krajina Reserve No. 45. Located around Port Chanal and including Hippa Island off Graham Island on the Queen Charlotte Islands (Haida Gwaii), the Krajina Reserve includes 9,834 hectares of old-growth forest with larger trees than ones protected in any other Ecological Reserve in the province.

December 28, 1973 The U.S. passes an Endangered Species Act. It requires the U.S. Fish and Wildlife Service to maintain a list of endangered and threatened species in the U.S. and to implement a plan for the recovery of each one of them.

October 1973 Office of Ombudsman is established in B.C. under the Ombudsman Act 1973 to enquire into or investigate complaints against Victorian government departments, public statutory authorities and the officers of local councils.

1974 The Earth's human population reaches four billion.

1974 Island Protection Society is formed by local citizens concerned about the rate of logging on the Queen Charlotte Islands (Haida Gwaii) and specifically dedicated to saving the South Moresby area from being logged. It submits to the provincial government the Southern Moresby Wilderness Proposal (SMWP), first proposed by Gary Edenshaw (Guujaaw) and Thom Henley, advocating protection of all the land below Talunkwan Island and the Tangil Peninsula.

1974 The federal government accepts the Nisga'a Tribal Council's Land Claim for negotiation.

1974 West Coast Environmental Law (WCEL) Society is established to empower citizens to participate in forming policy for, and make decisions about, protecting our environment. WCEL provides free legal advice, advocacy, research and law reform services. Through its Environmental Dispute Resolution Fund since its inception up until 2004 it had provided a total of over $2,000,000 in assistance to hundreds of citizens' groups across B.C. to help them solve environmental problems in their own communities.

1974 The B.C. government forms the North Island Study Group (NISG), an interdepartmental civil service study team, to assess management options for the Tsitika/Schoen Lake logging moratorium area.

1974 Moresby TFL 24 AAC is raised yet again. It increases from 410,594 cubic metres annually to 436,079 cubic metres annually because, according to the B.C. Forest Service, the new analysis includes lower than previous estimates of physical inaccessibility and smaller estimated losses due to fire, insects and disease.

1974 Rowland and Molina publish an article in *Nature* magazine reporting the danger posed to the ozone layer by the use of chlorofluorocarbons (CFCs) in aerosol cans, refrigerators and elsewhere. The two eventually win the Nobel Prize in 1995 for their work in this area.

February 1974 The B.C. government announces a two-year moratorium on any development in the Stein watershed in order to conduct a "Stein River Basin Study" to evaluate different management plans.

July 1974 *THE VALHALLA PROPOSAL –A Brief Concerning a Proposal for a Nature Conservancy Area in the West Kootenays, British Columbia* is submitted to the B.C. government by A. Eweson. This is the first formal proposal that Valhalla Mountains be protected.

October 1974 The plan by Rayonier Logging to move its contract (Frank Beban's) logging operations from Talunkwan Island to Burnaby Island located deep in the South Moresby region of Haida Gwaii is met with strong opposition from the Skidegate Band Council who fear traditional food sources will be threatened.

November 1974 Five hundred residents on the Queen Charlotte Islands (Haida Gwaii) sign a petition requesting the B.C. government impose a moratorium on development in South Moresby so that the South Moresby Wilderness Proposal can be fully studied by the government, industry and the public.

December 1974 Bob Williams, B.C. Minister of Lands, Forests and Water Resources in the NDP B.C. government, announces that a five-year moratorium will be placed on logging on Burnaby Island within the South Moresby Wilderness Proposal area on the Queen Charlotte Islands (Haida Gwaii).

1975 The Valhalla Wilderness Society is founded by a group of local citizens dedicated to protecting the Valhalla Mountain Range in southeastern B.C.

1975 The North Island Science Group recommends that the Tsitika Watershed be either permanently preserved as a "benchmark" study area, or left untouched for at least a further five-year research period followed by strictly controlled logging only in limited areas of the watershed to be determined by the research.

1975 The novel *Ecotopia* by Ernest Callenbacin is published.

1975 The Federation of Mountain Clubs of B.C. completes its two-volume study that recommends that the Stein Valley in southwestern B.C. be preserved.

February 1975 A petition with over 500 Queen Charlotte Islands (Haida Gwaii) residents' signatures is delivered to government calling for an immediate moratorium on all logging in the South Moresby Wilderness Proposal (SMWP) area until environmental impact studies are complete. The provincial government promises to have its Environment and Land Use Committee Secretariat (ELUCS) that vetted all land use decisions place the SMWP on its agenda.

April 1975 The B.C. Forest Service issues road building and cutting permits on Lyell Island in TFL 24 at the north end of the SMWP area. Beban Logging, the contract company working for Rayonier, moves its logging operations from Talunkwan Island to Lyell Island instead of to Burnaby Island and starts logging with large clearcuts on steep slopes Lyell.

1975 France ends atmospheric nuclear tests after Greenpeace protests at its South Pacific test site.

December 22, 1975 William Richards Bennett (Bill Bennett), Social Credit, becomes Premier of B.C.

1976 Canada signs World Heritage Convention; Nahanni National Park is designated the world's first natural World Heritage Site by UNESCO.

1976 The Australian Wilderness Society a national, community-based, environmental advocacy organization with a mission to protect, promote and secure the future of wilderness and other high conservation areas is founded. From its startup through 2004 it had achieved secure protection for more than seven million hectares of wilderness in Australia.

January 1976 *Nature Canada*, the Canadian Nature Federation's magazine, publishes an article about Southern Moresby Wilderness Proposal titled *Queen Charlotte Wilderness–Unique and Threatened*. It brings to a wide audience the beauty of this unique area and the efforts of local citizens to protect it.

1976 The fourth Royal Commission of Inquiry into Forest Resources, the Pearse Report, makes significant recommendations to reduce corporate concentration in the B.C. forest industry. This recommendation was not incorporated into to the new Forest Act of 1979.

1976 The B.C. government holds public hearing only in northern Vancouver Island logging towns regarding the fate of the Tsitika watershed while conservation groups in Victoria, Nanaimo and Vancouver continue to campaign for the protection of the entire watershed.

May 12, 1976 The two-year government moratorium on logging in the Stein watershed ends with an announcement by the B.C. Environment Minister that the area would be developed for "multiple use" (the code words for logged). The study document upon which this decision is based is not released to the public and no public hearings are held.

June 1976 After a one-year delay, ELUCS begins to study the South Moresby Wilderness Proposal.

1976 B.C. Forest Products Ltd. announced plans to begin logging in the Stein Valley.

1977 From an idea that began in the Umfolozi Wilderness, the First World Wilderness Congress (WWC) is convened in Johannesburg, South Africa. Sponsored by the WILD Foundation, 2,500 delegates from 27 countries attend. The Congress establishes wilderness as an international issue of importance to developing countries, not just to Western developed cultures. WWC is the oldest continuing international public forum on wilderness and related conservation concerns.

1977 The B.C. government passes the Heritage Conservation Act. This Act establishes a comprehensive scheme for the protection and conservation of "heritage" properties imposing a restriction on all persons in the province from interfering with sites containing physical evidence of human habitation or use before 1846.

1977 The B.C. government decides to open the Tsitika Valley to logging and sets up a Tsitika Planning Committee (TPC) consisting of logging industry, union and government representatives to prepare a "multiple use" logging plan for the watershed. In 1978 the TPC comes up with a plan to log over 95 percent of the productive forest lands in the watershed and establish seven small ecological reserves totaling five percent of the watershed area, recommending that those reserves be in non-productive forest areas.

1977 Paul Watson founds the Sea Shepherd Conservation Society (SSCS), an organization dedicated to protecting the marine environment with a particular focus on halting illegal fishing activities. It accomplishes its mission through support and assistance in the upholding of international treaties, laws and conventions through the investigation, documentation and, where appropriate and where legal authority exists under international law or under agreement with national governments, enforcement of violations of these treaties, laws and conventions.

1977 The Save the Stein Coalition formed. It includes all of B.C.'s major environmental and recreation organizations.

March 1977 Paul George, zoologist, and Richard Krieger, photographer, future founders of Western Canada Wilderness Committee, begin field studies in South Moresby in the QCIs (Haida Gwaii), which go on all summer and into the fall, calling themselves the Northern Galapagos Book Collaborators.

March 1977 In response to the growing criticism of logging practices, the B.C. Forest Service holds a public meeting on the QCIs to set up the Queen Charlotte Public Advisory Committee (QCPAC) to discuss forest practices and TFL renewals on the islands.

September 1977 Islands Protection Committee and the Northern Galapagos Book Collaborators present slideshows at the logging camps on Lyell Island in South Moresby and in Sewell Inlet in Central Moresby showing the beauty of the area and the destruction caused by logging and engage in dialogue with the loggers working there.

1977 The Committee on the Status of Endangered Wildlife in Canada (COSEWIC) is founded. Its job is to maintain a list of vulnerable, threatened and endangered species in Canada.

November 1977 The QCPAC passes a motion calling for full public hearing on the renewal of all Tree Farm Licences (TFLs) on the Queen Charlotte Islands (Haida Gwaii).

November 26-27, 1977 A symposium is held at the Skidegate Museum on the Southern Moresby Wilderness.

December 28, 1977 Paul George, on behalf of the Northern Galapagos Book Collaborators, formally proposes to the Ecological Reserves Unit of the B.C. government that Windy Bay, located on Lyell Island in the proposed Southern Moresby Wilderness Area, be considered for ecological reserve status. Ecological Reserves Unit subsequently accepts this proposal and begins to investigate the merits of it.

1978 In response to 50 B.C. organizations calling for a public judicial inquiry to determine whether uranium should be mined in B.C., the B.C. government sets up a Royal Commission on how uranium should be mined in B.C. Dr. David Bates heads up this commission.

January 1978 The Pacific Seabird Group, an international body representing biologists from 39 countries, unanimously passes a resolution at their convention in Victoria, B.C., calling for the SWMP area to be granted protected status to prevent further destruction of critical seabird nesting habitat found there.

May 1978 Tom Waterland, B.C. Forest Minister, introduces Bill 14, a new Forest Act, that would automatically renew all TFLs for 25 years with "evergreen" automatic renewals thereafter, strengthening corporate hold over forest lands in B.C. The Act also removes the mandate to log the public forest on an even-flow sustained yield basis and mandates maximum fibre production from B.C.'s publicly owned forests. A few citizens form the Committee for Responsible Forest Legislation (CRFL) and try to slow the pace of passage of this Bill and inform the public about its shortcomings and get the legislation amended, modified and improved.

Summer 1978 The first Rediscovery Camp for native and non-native youth to learn about nature and indigenous culture is held at Lepas Bay on the West Coast of Graham Island on Haida Gwaii with Thom Henley as camp director.

June 29, 1978 The Social Credit government passes B.C.'s new Forest Act (Bill 14), changing the terms of tenure for companies operating in provincial forests, requiring increased use of contractors in TFLs, and releasing the Forest Service and companies from the requirement to manage the logging on a sustained yield basis. Social needs and the need to supply mills supplant consideration of biological limits governing the rate of forest regeneration in setting the rate B.C.'s forests are to be cut. All members of the opposition parties in the Legislature (Liberal, Conservative, and NDP) vote against it.

December 1978 The Islands Protection Society (IPS) is incorporated as a B.C. Society to legitimize the organization and enable it to pursue court action against the B.C. Minister of Forests decision to renew Moresby TFL 24 without holding public hearings.

1979 *Gaia: A New Look at Life on Earth*, written by National Aeronautics and Space Administration (NASA) scientist James Lovelock, is published. In it Lovelock proposes that the Earth and all life on it form one organism, which he called Gaia, after a Greek goddess who drew the world out of chaos.

1979 Earth First!, an environmental group that advocates "direct action" including ecotage, is founded in the U.S. by Dave Foreman *"in response to a lethargic, compromising, and increasingly corporate environmental community."* Earth First! takes *"a decidedly different tack towards environmental issues."* Earth First! believes in *"using all the tools in the tool box."* Members' activities range from grassroots organizing and involvement in the legal process to civil disobedience and criminal monkey wrenching.

1979 The B.C. government approves logging plan for the Tsitika Valley. MB begins building roads and logging in the Upper Tsitika Valley.

1979 LightHawk, a voluntary organization of pilots form to provide aerial reconnaissance for environmental groups gather scientific data and photographic images in support of conservation.

January 1979 The Environment and Land Use Committee Secretariat (ELUCS) completes its *SOUTH MORESBY ISLAND WILDERNESS PROPOSAL, AN OVERVIEW STUDY*. It is a very brief and flawed work. It is not released to the public.

January 26, 1979 After being totally frustrated in its efforts to be involved in the B.C. government's decision making process regarding the replacement of TFL 24 (that encompassed South Moresby wilderness) for 25 more years, Garth Evans, lawyer for Islands Protection Society and others files a Petition seeking a Judicial Review of the decision.

February 21-23, 1979 IPS, Nathan Young, Haida Chief of Tanu, and Guujaaw, Haida carver and hunter gather and others' Petition is heard by Mr. Justice Murray in the Supreme Court of B.C. It asks for a declaration that the Minister of Forests, Tom Waterland had a duty to act fairly and consider their interests in the government's renewal of TFL 24. It is an attempt to get a public hearing regarding this tenure, located on South Moresby on the Queen Charlotte Islands (Haida Gwaii), before it is replaced for another 25 years under the New Forest Act.

1979 Roger Freeman and David Thompson's book, *Exploring the Stein Valley*, is published. It provides detailed information about trails and hiking in the Stein watershed near Lytton, B.C.

March 6, 1979 Justice Murray hands down a 29-page written decision on petition to get the Minister of Forest to consider Haida interests regarding the renewal of Moresby TFL 24. Murray grants standing to all petitioners except Islands Protection Society but dismisses the petition, ruling that the Minister of Forest still had time before the May 1, 1979 renewal date of TFL 24 to act fairly and consider the interests of the petitioners.

March 1979 A few days after Justice Murray's ruling, the B.C. government releases ELUCS' *SOUTH MORESBY ISLAND WILDERNESS PROPOSAL, AN OVERVIEW STUDY* to the public. ELUCS recommends setting up a South Moresby Resource Planning Team. The B.C. Forest Service immediately sets up this Team with representatives of the government ministries, forest and mining industries, Skidegate Band Council, Islands Protection Society and the public given seats. This Team meets regularly until June of 1983, handing in its "Options Report" to the B.C. government in January of 1984.

March 28, 1979 The Three Mile Island nuclear power plant in Pennsylvania, U.S. almost has a catastrophic meltdown, giving the nuclear power industry a permanent black eye.

April 1979 The B.C. Forest Service ignores registered letters from Nathan Young, Chief of Tanu, and Guujaaw, a Haida hunter and gatherer, asking for their right to input on the terms and conditions of the replacement of TFL 24.

April 19, 1979 Garth Evans, lawyer for the petitioners Young and Guujaaw, files a new petition with the Supreme Court of B.C. The B.C. Forest Service immediately responds by telegram inviting the petitioners to a meeting in Victoria with the top brass of the B.C. Forest Service including the Chief Forester and several deputy Ministers. During over five hours of negotiations, the Ministry agrees to make several changes to the TFL licence agreement including giving the B.C. Forest Service the right to ask TFL holders to provide information and requiring companies to gather that information if it was not available.

April 30, 1979 Not satisfied with the minor changes, the petitioners proceed with their court case. The Forest Ministry swears in affidavits that it has listened to the petitioners and made appropriate changes based on the consultation to the TFL document. The judge orally dismisses the case in light of the Forest Service affidavit saying that only after the final decision is made can the court review the decision.

May 1, 1979 TFL 24 owned by Rayonier expires (is due for renewal) and the B.C. government offers a replacement. The new TFL 24 agreement document that is offered does not contain any of the changes that the Deputy Minister of Forests had sworn in court that the B.C. Forest Service had made. These omitted changes were terms and conditions of the TFL agreement that accommodated the Haida petitioners' interests. These clauses were thrown out, according to inside sources, because the cartel of big multinational companies, the Council of Forest Industry (COFI), refused to accept them.

June 4, 1979 Joe Clark, Progressive Conservative, becomes Prime Minister of Canada.

1979 A B.C. chapter of the Canadian Parks and Wilderness Society (CPAWS) is established.

1979 The Friends of Clayoquot Sound (FOCS) is founded as a non-profit B.C. society to stop old-growth logging in Clayoquot Sound on the west coast of Vancouver Island, B.C. Clayoquot Sound stretches from Ucluelet in the south to the height of mountains along the spine of the Island to the east and extends to the Hesquiat Peninsula in the north and is roughly 362,000 hectares in size. The FOCS immediately begins to campaign to save the old-growth rainforest on Meares, the second largest island in the Sound, from logging by MacMillan Bloedel.

September 1979 The Queen Charlotte Island PAC members vote to disband right after it passed a resolution to automatically accept all Forest Service decisions without question. This was the culmination of the local B.C. Forest Service office's clamp down on providing information about logging plans and more logging industry representatives joining the PAC, giving the industry a voting majority at the meetings.

1980 Canadian Environmental Advisory Council (CEAC), the academic community and industry representative organization that provides advice to the federal minister of the environment, recommends that Environment Non-Government Organizations (ENGOs) be funded directly by the Department of the Environment. This would build a "viable National ENGO structure" to provide input to the government. Federal Environment Minister John Roberts notes in discussions leading up to this federal funding of the environment movement that, *"The challenge is to find forms of government assistance to voluntary organizations which do not become self-defeating by undermining the independence on which vitality depends."*

1980 The U.S. Federal Alaska National Interest Lands Conservation Act designates over 100 million acres (405,000 square kilometres) of parks, wildlife refuges, and wilderness areas in Alaska.

1980 Peter Hamilton founds Lifeforce Foundation, a Vancouver based ecology organization, to provide a public service to help people, animals and the environment. Among other things, it rescues marine mammals and educates whale watchers so their activities do not harm the whales.

1980 The B.C. government established the Tsitika Follow-up Committee (TFC) to monitor the adherence of the logging in the Tsitika to the approved plan. The United Fisherman and Allied Workers Union refuses to participate in the Committee stating that *"...the cutting plan is totally incompatible with the preservation of the Tsitika as a wildlife, salmon, and recreational habitat and we have no desire to preside over the desecration of the last major watershed area* [on the east coast of Vancouver Island] *that fleetingly had some hope of retaining its natural state."*

1980 The first B.C. Rivers Day organized by Mark Angelo under the umbrella of the Outdoor Recreation Council of B. C., is held. From then B.C. Rivers Day is held on the last Sunday every September. It has become the largest river celebration of its kind in Canada with a series of locally planned events taking place across B.C. to highlight the recreational, environmental, economic and aesthetic importance of B.C.'s rivers.

1980 Doman Industries joins with two other British Columbia forest products companies in forming Western Forest Products (WFP), which then purchases the British Columbia timber resources and manufacturing facilities of ITT, including Moresby TFL 24, from ITT Rayonier.

1980 The B.C. Forest Service forms the Meares Island Planning Team. Representatives of the Nuu-chah-nulth Tribal Council and the Friends of Clayoquot Sound are members. Environmentalists are hopeful that scientific studies initiated on archeology, fish, migratory birds, mariculture, and eagle nests will win the case for preservation of Meares Island.

1980 Clearcut logging expands rapidly from the southeast of the Clayoquot Sound area along Kennedy Lake.

February 27, 1980 The B.C. government announces a moratorium on all uranium exploration and mining in B.C., calling an end to the Bates Commission before it had completed its hearings. Premier Bill Bennett states: *"What uranium resources British Columbia has will be left in the ground until the people are prepared to do otherwise."*

March 3, 1980 Pierre Elliott Trudeau, Liberal, elected as Prime Minister of Canada.

Spring 1980 Islands Protection Society publishes *All Alone Stone IV* an 80-page perfect bound magazine about conservation issues including information about the South Moresby Wilderness Proposal area on the Queen Charlotte Islands (Haida Gwaii).

August 7, 1980 Paul George, Richard Krieger and three other B.C. citizens found the Western Canada Wilderness Committee (WCWC) (WC2) as a B.C. society. Its constitutional objects are: *1) to educate the public concerning Canada's wilderness heritage and reasons for preserving representative areas for future generations; 2) to conduct research concerning wilderness values; and, 3) to obtain and distribute information on areas in Western Canada which have potential for protective wilderness status.*

October 16, 1980 The fifteen Nuu-chah-nulth First Nations of the west coast of Vancouver Island sign and present their Declaration and Claim to their traditional lands and waters on the west coast of Vancouver Island to the Government of Canada. The Canadian government accepted this land claim for negotiation in 1983.

November 1980 WCWC publishes its first *Western Canada Wilderness Calendar*—a full-colour wall calendar for 1981. Print run 10,000. Each month features a different threatened wilderness area proposed for protection with map and "thumb-nail" explanation of the issue and with a tear-off form to join and contribute to the conservation groups most concerned with that preservation issue. This calendar has been published annually with out interruption ever since.

January 12, 1981 Finally, WCWC gets a letter from Revenue Canada granting the society charitable tax status. This comes after a strongly worded letter from Ian Horne, QC, who was the acting Clerk of the B.C. Legislature at the time, championing WCWC's application. The approval is backdated to the founding of the Committee on August 7, 1980.

1981 The Sierra Club of B.C. proposes that a provincial park be established covering the lower reaches of the Tsitika watershed. In a document made public four years later, that same year (1981) the B.C. Ministry of Lands Parks and Housing completes a park feasibility study for the Lower Tsitika. This document recommends that a Class A Provincial Park be established covering 3,400 hectares in the Lower Tsitika Valley, as well as 1200 hectares of Robson Bight marine waters and foreshore.

1981 The B.C. provincial government reaffirms its moratorium on offshore drilling for oil and gas along the whole coast of B.C.

March 26, 1981 Appeals Court orally dismisses petition asking to reverse a lower court decision which effectively denied Petitioners Nathan Young, Chief of Tanu, and Guujaaw, a Haida hunter and gatherer, the right to have their interests fairly considered in the renewal of TFL 24 (part of which encompassed most of the South Moresby Wilderness Proposal area on the Queen Charlotte Islands) when the Minister of Forests renewed this TFL. TFL 24 is the first TFL to be replaced under the new terms and conditions conforming to the new Forest Act of 1979.

1981 With the help of federal government money, the British Columbia Environmental Network (BCEN) is established to exchange and share information among member environment and conservation groups and to consult directly with provincial and federal governments.

1981 The 25-year replaceable Moresby TFL 24 (which encompassed all the merchantable forest within the South Moresby Wilderness Proposal) is finally signed.

1981 The Friends of the Stikine, a conservation group dedicated to the "preservation of the Stikine River in its natural state," is established.

July 1981 WCWC makes a presentation to the *Pearse Commission on Pacific Fisheries Policy*. Among other things WCWC recommends that all fisheries be managed with planned under-harvesting to provide a needed biological reserve to get long-term stable harvests and that there be more on-the-ground creek-watchers and technical personnel to observe and collect data during salmon runs to help stop creek-robbing.

October 1981 The Haida Nation presents its *Declaration and Claim to the Haida Gwaii* (Queen Charlotte Islands and the surrounding seas) to the Government of Canada. The Canadian government accepted this land claim for negotiation in 1983.

October 1981 The Haida Nation declares the establishment of the Duu Guusd Tribal Park on the northwest coast of Graham Island on Haida Gwaii.

October 1981 WCWC experiences a growth spurt and now has 35 members.

November 1981 After intensive negotiations at a First Ministers' conference, the federal government and all the provincial governments except the Parti Québécois government of Quebec, agreed on a package of constitutional amendments. The agreement does not alter the fundamental distribution of powers between the provincial and federal levels but includes a Canadian Charter of Rights and Freedoms, entrenchment of the principle of equalization payments to the poorer provinces, a strengthening of the provinces' control over natural resources, an affirmation of existing aboriginal rights and a comprehensive amending formula.

1981 After over three years of intense effort by environmentalists, TFL 24 Annual Allowable Cut (AAC) reduced slightly from 436,079 cubic metres annually to 432,375 cubic metres per year by the Chief Forester of B.C. to accommodate concerns over *"terrain stability, streamside protection, wildlife protection, archaeological site protection and local aesthetics."*

1981 Federal Private Members Bill C-454 to protect South Moresby is introduced into the House of Commons by MP Ian Waddell. WCWC polls all the MPs regarding where they stand on this bill that calls for National Park protection for the South Moresby region on Haida Gwaii, ascertaining that well over 50 percent of them support it "in principle."

1981 WCWC publishes its first poster titled *Canada's Tallest Trees...must not be logged* featuring a black and white image of towering Douglas fir on Nimpkish Island in the Nimpkish River on Vancouver Island, an area proposed as an Ecological Reserve.

1982 The B.C. government decides against protecting the Lower Tsitika valley in a park and instead establishes a 1,248-hectare water only Ecological Reserve in Robson Bight. Meanwhile, blowdown occurs in the leave strips between clearcut blocks in the Upper Tsitika. Salvage logging there results in massive clearcuts over 500 hectares (850 football fields) in size.

1982 Last residential school for Nuu-chah-nulth youth close down (Tin-Wis near Tofino). The buildings there are torn down and replaced by a Best Western Hotel owned and managed by the Tla-o quiaht First Nation.

1982 Canadian EarthCare Society, a non-profit environmental organization with a mandate to protect Earth's environment through advocacy and public education, is founded. EarthCare encourages the use of dialogue and a non-confrontational approach to resolving environmental issues.

1982 The Friends of Clayoquot Sound led by Adrian Dorst begin building the "Big Cedar Trail" on Meares Island to give people an opportunity to see the incredible rainforest there. Dorst discovers the "Hanging Garden Tree," an ancient red cedar 18 foot in diameter, the second largest of its species in B.C.

1982 The Tourism Industry Development Committee releases a joint federal-provincial study advising that the Stein should be preserved as a Class A provincial park because of its strategic proximity to Vancouver and its unique status as an untouched watershed.

March 20, 1982 WCWC presents a brief titled *IWA vs. WCWC?* to the NDP caucus's resources committee at a public meeting it holds in Victoria on "The Future of the Forest Industry in B.C." The brief contains a request that the NDP bring together environmentalist and union representatives to a table to discuss wilderness and environmental issues.

April 17, 1982 The Canadian Constitution is patriated, without the consent of the Quebec Legislature. The Supreme Court of Canada subsequently rules despite this, the patriation process had respected Canada's laws and conventions, and that the Constitution, including the 1982 Constitution Act, is in force throughout Canada.

May 31, 1982 Dynamite explodes, destroying four transformers on Vancouver Island that are part of B.C. Hydro's Cheekeye-Dunsmuir transmission line project, a project which is opposed by many law abiding environmentalists. "Direct Action" (eventually found out to be five misguided individuals not connected to any other group) sends out communiqués claiming responsibility for this act of eco-terrorism. Later called "Squamish 5," this group continues to commit criminal acts of vandalism and terrorism until they are arrested on January 20, 1983. They are eventually convicted and sent to jail released and reintegrated back into society.

June 2, 1982 Randy Stoltmann accompanies Dr. Bristol Foster of the B.C. Ecological Reserve Unit on a tour of the Carmanah Valley on the west coast of Vancouver Island. Arriving by helicopter and landing on a sandbar in mid-valley, they spend eight hours exploring Carmanah's old-growth forests on the ground. While not finding the legendary record-breaking-sized mammoth Sitka spruce, they found, in Stoltmann's words, "...*a forest too magnificent to cut down*."

Spring 1982 Adrian Dorst presents his *Eagle Nest Survey* report to the Meares Island Planning Team. On contract to the Nuu-chah-nulth Tribal Council to provide input in support of the no logging option, Adrian found and mapped 24 Eagle nests on Meares Island.

1982 The B.C. government establishes the 1,248 hectare, totally marine Robson Bight ER in Johnston Strait on the east side of Vancouver Island by the Tsitika River to protect killer whale (orca) rubbing beaches.

Summer 1982 WCWC Founder Paul George goes on 27 day Kayak trip in South Moresby with Adriane Carr. They fall in love with each other and strengthen their commitment to work tirelessly to protect the area.

1982 With a grant from Environment Canada, WCWC publishes its first coloured poster, *Canadian Wilderness – Environments Worth Protecting*, printing 10,000 copies, and sends one to every federal MP and every MLA in B.C.

1982 Ahousaht historian John Jacobson charges MB with violating Canada's Fisheries Act with its log booming activities in Steamer Cove, Flores Island, claiming and proving that bark and debris from the logs has driven out the aboriginal roe herring fishery and polluted and buried shellfish beds. MB warns judge that ruling in favour of the Ahousahts would cost the industry millions of dollars and shut down all booming grounds in the waters along the coast. Judge says the Ahousahts are right, but damage is minimal and inconsequential, ruling in favour of MB.

1982 WCWC co-publishes with the Valhalla Wilderness Society a full colour poster titled *Visit the Valhallas–Canada's Shangri-La*. It features beautiful alpine flowers in the foreground with Slocan Lake and the proposed Valhalla Mountains Park in the background. 10,000 copies printed.

September 1982 Nuu-chah-nulth Tribal Council (NTC) produces and publishes the *"Save Meares Island"* poster featuring children holding hands around a giant ancient red cedar poster. Print run 10,000. WCWC Director Richard Krieger took the photo and WCWC helps distribute the poster.

October 16, 1982 WCWC holds its first art auction in Newcombe Auditorium sponsored by Robert Bateman. It raises over $19,000. Haida artist Bill Reid contributes his first ever scrimshaw work, a frog pendant to the fundraiser.

October 28, 1982 Adriane Carr purchases a house on Trafalgar Street in the Kitsilano area of Vancouver, B.C. It becomes the home of WCWC's office and later the temporary home of the Green Party of B.C.

December 10, 1982 The United Nations Convention on the Law of the Sea is launched in Montego Bay, Jamaica.

1983 The International Tropical Timber Organization (ITTO) is created by treaty and its headquarters are established in Yokohama, Japan. It is the cartel of companies active in logging the tropical rainforest of the world.

1983 The German Green Party wins its first seats in the Bundestag with 5.6 percent of the countrywide popular vote. It is the first ecologically based party to achieve electoral success.

February 4, 1983 Green Party Political Association of B.C. is established. It is the first Green Party to be established in North America. The Party's major objects are to work towards: *achieving the goal of nuclear and general disarmament and world peace; achieving the goal of a conserver society; building an economic system based upon sound environmental and ecological principles; and, establishment and respect for basic human rights.*

February 23, 1983 The B.C. government establishes 49,600-hectare Valhalla Class A Provincial Park, bringing to a successful conclusion the local Valhalla Wilderness Society's thirteen-year-long relentless effort to save this magical place in the West Kootenays.

March 16, 1983 MB withdraws from (quits) the Meares Island Planning Team to develop and submit its own plan for logging Meares Island to the B.C. provincial government.

Spring 1983 WCWC produces and Environment Canada designs and prints a large brochure titled *British Columbia/Yukon Territories Environmental HOTSPOT Directory*. Print run is 10,000 copies. Environment Canada temporally halts distribution of this *HOTSPOT Directory* because the subject matter is too hot, but it quickly relents after banning it becomes a big news story.

May 5, 1983 The B.C. voters elect Bill Bennett's Social Credit government to a second term.

1983 The federal government accepted the Haida claim to all the Queen Charlotte Islands and surrounding waters, the Nuu-chah-nulth Land and Sea Claim to territories on the west coast of Vancouver Islands, the Heiltsuk Nation, Nuxalk Nation, and four other First Nations' claims for negotiation.

1983 Trevor Jones publishes his study titled *Wilderness or Logging: Case Studies of Two Conflicts in B.C.* putting forth the argument that it is completely uneconomic to log in the Stein Valley. He concludes that the government and company estimates of merchantable wood volumes in the Stein are greatly inflated.

November 10, 1983 The B.C. government cabinet's Environment and Land Use Committee announces that logging could go ahead on Meares Island TFLs with a 20-year delay on logging those parts visible from Tofino. Harry Teileman, a Tofino resident and one of the leaders opposing logging on Meares, calls the decision, *"The death of an Island...a complete sell-out."*

December 12, 1983 Paul George settles libel suit against Doug Macleod, forester for the company that owns Moresby TFL 24, on the steps of the B.C. Supreme Court in Vancouver. It concerned Macleod calling George "deceitful and dishonest" regarding wording of a petition seeking public support for making Windy Bay on Lyell Island in the Proposed South Moresby Wilderness Area an Ecological Reserve and George calling Macleod "unprofessional."

1984 The B.C. Forest Service's Stein Public Advisory Committee submits its final report to the government on how to best log the Stein. The chair of this Committee points out the futility of the exercise because it is uneconomical to log the Stein without taxpayer subsidies.

1984 Ivvavik National Park of Canada (1,016,800-hectare Northern Yukon National Park Reserve) is established to protect the porcupine caribou herd's calving grounds in the Yukon Territory. It is the first national park in Canada established under a land claim agreement.

January 1984 After four years of deliberation, the South Moresby Planning Team hands its recommendations to the B.C. government's Environment and Land Use Committee in the form of four land use options. All the while during the Team's deliberations and afterwards logging in the South Moresby area progresses without any slowdown. Haida begin to hold meetings with Tony Brummet, B.C. Minister of Parks and other

Ministers during the next two years while public pressure grows to protect the place.

April 1984 The Skagit Valley Treaty is signed which kills the High Dam Project that would have flooded over 11 kilometres of the Skagit Valley in B.C. Portions of Big Beaver Creek and Little Beaver Creek now within the North Cascades National Park are also saved. In exchange, B.C. agrees to provide Seattle, Washington with equivalent power. The Skagit River Treaty also establishes an innovative and unusual Skagit Environmental Endowment Commission that grants money every year to conservation projects in the watershed area. This is a clear win for the ROSS (Run Out Skagit Spoilers) Committee, the citizen group that championed the protection of the area.

April 20-22, 1984 (Easter weekend) The Tla-o-quiaht Band Council and Hereditary Chiefs of the Tla-o-quiaht First Nation declare Meares Island a Tribal Park and erect a "Meares Island Tribal Park" sign on Meares Island near Mosquito Harbour. Hundreds attend a Meares Island celebratory event in Tofino.

April 21, 1984 The Tla-o-quiaht First Nation issues a declaration designating Meares Island a Tribal Park.

May 29, 1984 Federal Environment Minister Charles Caccia makes public his letter to B.C.'s Minister of Lands, Parks and Housing, Anthony Brummet, in which he proposes exploring the possibility of creating a national park and a national marine park in the South Moresby area of the Queen Charlotte Islands (Haida Gwaii).

June 1984 The raising of a giant carved cedar welcoming figure by the Tla-o-quiaht First Nation in front of the Legislature buildings in Victoria marks the beginning of natives and non-natives working together to protect Meares Island and later all of Clayoquot Sound.

June 30, 1984 John Napier Turner, Liberal, becomes Prime Minister of Canada.

Summer 1984 WCWC produces its very first newspaper, a four-page, black and white, tabloid-sized newspaper titled *South Moresby–A Special Issue – Summer 1994* about protecting South Moresby region on the Queen Charlotte Islands (Haida Gwaii) from logging. Print run 30,000. The cover story features the letter from Federal Environment Minister Caccia to the B.C. Minister of Lands Parks and housing suggesting that the whole area become a park and proposing that the two levels of government talk.

Summer 1984 Arcas Associates, funded by MB, undertakes the first big study of aboriginal forest use on Meares Island with Marion Parker, dendrochronologist, analyzing samples of scar re-growth to obtain dates of modification. Culturally Modified Trees (CMTs) dating from 1642 to 1948 are found. Arcas's research findings prove to be crucial evidence of continues aboriginal use of Meares Island's cedar forests which greatly strengthens the Nuu-chah-nulth claim to Meares and their aboriginal right to continue to use its forest.

August 1, 1984 MB's TFLs 20 and 21 are amalgamated to form TFL 44. The newly created TFL 44 agreement is granted to MB for a term of 20 years and 5 months. TFL 44 covers much of area surrounding Barkley Sound, North of Nitinat Lake and South of Strathcona Park including much of Clayoquot Sound. It has a total area of about 453,000 hectares.

September 17, 1984 Martin Brian Mulroney, Progressive Conservative, becomes the Prime Minister of Canada.

Fall 1984 Islands Protection Society's *Islands at the Edge: Preserving the Queen Charlotte Islands* (Vancouver: Douglas & McIntyre) a 144-page hardcover coffee table book is published giving the campaign to save the South Moresby Wilderness Area a big boost.

Fall 1984 WCWC co-publishes its second newspaper with IPS and NPPAC (CPAWS) printing 80,000 copies of a four-page black and white tabloid-sized newspaper titled *South Moresby Fall 1984 - New Federal Government continues push for National Parks in South Moresby; B.C. Gov't must make first move*. By the end of the year almost all copies are distributed. The co-publishers agree to share the donation returns on the appeal clip-out coupon on the paper. An offer to order IPS's book *Islands at the Edge* is also included on this coupon asking for campaign donations.

October 20, 1984 1200 people rally on the Legislature lawns in Victoria in support of the Meares Island Tribal Park raising Joe David's huge carved Cedar Man. It now stands at the entrance of the Anthropology Museum at the University of B.C. in Vancouver.

November 10, 1984 The B.C. government's Environment and Land Use Committee (ELUC) decides to accept most of MB's plan to allow logging on 90 percent of Meares Island—all but 800 hectares of the 8,000-hectare Island.

November 17, 1984 The B.C. Forest Service issues permits for road building and clearcut blocks on Meares Island.

November 18, 1984 Local citizens clear a site for a place to build a protectors' cabin at C'is-a-quis (Heeloom Bay) on Meares near the site where logging is scheduled to begin.

November 21, 1984 MB loggers arrive at Heeloom Bay on Meares Island ready to begin work. Local natives and non-natives who have set up the first logging blockade in Canadian history meet them. Tla-o-quiaht Chief Moses Martin says Meares is "their garden," and the loggers are welcome to visit the island but they had to leave their chain saws in the boat.

November 1984-March 1985 Logging protesters maintain five camps on Meares Island with hundreds of people ready for mass arrest, if need be, to stop the logging. The Friends of Clayoquot Sound (FOCS), the local environmental group coordinating the protest, spend $10,000 on diesel fuel alone for boats servicing the camps.

November 23, 1984 MB seeks injunction against Tla-o-quiaht Chief Moses Martin and the Friends of Clayoquot Sound to restrain them from obstructing MB beginning logging operations on Meares Island.

December 3, 1984 B.C. Supreme Court Justice Reginald Gibbs grants MB a temporary injunction to remove the blockaders on Meares, which is only served on the non-native protectors.

December 1984 The Tla-o-quiaht and Ahousaht apply for a counter-injunction to prevent MB from logging Meares.

December 1984 WCWC co-publishes with the Friends of Clayoquot Sound a poster titled *For the Children – Make MEARES ISLAND a PARK* featuring a giant 19-foot (5.9-metre)-diameter red cedar growing within the proposed logging zone on the island. Complimentary copies of this poster are given to every MLA.

December 1984 The Nlaka'pamux First Nations submit a comprehensive Land Claim, which includes the Stein Valley, to the federal government for negotiation.

1984 MB presents for public viewing its TFL 44 1985-89 Management and Working Plan. The plan shows no development occurring in the Carmanah/Nitinat area during that period. Other MB documents show no logging in Carmanah Valley until 2003.

1985 The Rainforest Action Network (RAN) is founded in San Francisco to protect rainforests and the human rights of those living in and around those forests.

1985 Terry Jacks, famous pop singer of the Poppy Family, founds Environmental Watch, an environmental group focused mainly on stopping pulp mill pollution on the B.C. coast.

1985 The Canadian government passes a Historic Sites and Monuments Act to protect Canada's rare and exceptional natural features.

1985 The Seymour Advisory Committee, comprised mostly of professional foresters for logging interests, is formed to counteract a proposal to establish a public park in the Lower Seymour, an off catchment area in the Greater Vancouver Water District lands. This Advisory Committee eventually names the area the *Seymour Demonstration Forest* and endorses proposals for continued logging of the Lower Seymour.

1985 British scientists discovered a hole in the high atmospheric protective ozone layer over Antarctica.

1985 MB shifts its logging quota from areas in Clayoquot Sound into the Carmanah/Nitinat area, revising its five-year logging plan to include Carmanah Valley without public knowledge or review.

1985 *Deep Ecology* written by sociologist Bill Duvall and philosopher George Sessions is published. It promotes an eco-centric viewpoint. Some attacked it as "ecological fascism" for its willingness to sacrifice people's individual interests for the greater good of a supposed ecological community.

1985 Dr. William Newmark, an American conservation biologist, releases a study on species extinctions in protected areas in North America showing that the smaller the park, the higher the rate of extinction due to natural catastrophes, genetic inbreeding and other adverse phenomena. He found that in all but the very largest parks, large mammals gradually are lost.

January 1985 Colleen McCrory, a WCWC Director, begins organizing a National Save South Moresby Committee. By the spring of 1985, 20 famous Canadians have joined including MP Jim Fulton, Honourable Charles Caccia, federal environment critic, and artist Robert Bateman.

January 1985 WCWC publishes its third newspaper; a four-page two-colour tabloid-sized one on Meares Island titled *Meares Island - Peaceful Protest Halts Logging*. Print run 50,000 copies. It features the text of the Clayoquot Band's Tribal Park declaration and a pictorial history of the confrontation between loggers and preservationists on Meares.

January 25, 1985 Supreme Court of B.C. Justice Gibbs grants injunction to MB to stop protesters from halting logging activities on Meares Island. He refuses an application by the Clayoquot and Ahousaht First Nations to halt logging until the completion of a full trial on their claim to aboriginal title of the Island. The First Nations appeal this decision.

February 1985 Forest Ministry announces that road building up the Stein Valley will commence as soon as possible in preparation for logging.

February 11, 1985 WCWC has the only demonstration on opening day of the Legislature. "One lone picketer," reported the *Globe and Mail*, "is on lawns (WCWC) asking government decision to log Meares Island be reversed."

February 27, 1985 WCWC publishes 20,000 copies of a four-page tabloid-sized newspaper written and illustrated by Randy Stoltmann Titled "*a proposal to create...CANADA LANDMARKS – PROTECTION FOR OUR LARGEST, TALLEST AND OLDEST TREES.*" The first proposal to protect Carmanah Valley is found in this publication.

March 1985 WCWC moves its office from Adriane Carr's home on Trafalgar Street in Vancouver to her new apartment on Alberta Street in Vancouver.

March 13, 1985 B.C.'s new Minister of Environment Austin Pelton meets with four WCWC Directors including Colleen McCrory. He promises to discuss possible park options for South Moresby with his federal counterpart Honourable Suzenna Blais-Grenier. Pelton personally favours the park, but says he *"...needs demonstration of widespread public support in order to win over more skeptical cabinet members."*

March 27, 1985 The B.C. Appeals Court overturns Justice Gibbs' decision to allow logging on Meares. This court rules in a written Reasons for Judgment by Justice Seaton that MacMillan Blooded Ltd. cannot log Meares Island before the issue of Indian land claims is settled in court. This case is delayed for several years and then adjourned to allow for negotiation to take place between the First Nations involved and the provincial government.

June 1985 WCWC produces and publishes its second Educational Report on Meares Island (an eight-page tabloid-sized newspaper) entitled *Meares Island NEWS – Court Saves Meares –Until November- Hopes Soar for Tribal Park – Summer 1985*. Press run 50,000. It provides an update on the campaign to save Meares Island.

Summer 1985 WCWC publishes *South Moresby, Queen Charlotte Islands–Misty Wilderness Gem of the Canadian Pacific 1985* tabloid-sized newspaper updating people on the progress being made to gain National Park protection for the area. Print run 50,000.

Summer 1985 WCWC publishes *Wild Watershed: The Stein–Summer 1985*. This four-page two-colour tabloid-sized newspaper promotes the three-day Voices for the Stein festival being held in the Stein alpine on Labour Day weekend. Print run 50,000.

July 1985 WCWC with the help of FOCS publishes 4,000 copies of a 64-page, full-colour, soft cover book titled *Meares Island: Protecting a Natural Paradise*. It is WCWC's first book publication. Retail price is $12. It tells the story of the efforts to save Meares Island complete with beautiful images and ugly ones from clearcuts nearby to illustrate the threat to the island. It has been reprinted several times.

July 10, 1985 French Secret Service agents bomb and sink the Rainbow Warrior, Greenpeace's flagship, in Auckland harbour. One Greenpeacer dies and two French agents ultimately plead guilty to the crime. Greenpeace's effective protest against French nuclear testing in the South Pacific had upset the French government.

Summer 1985 B.C. Parks Minister Brummet tells the Haida that no more cutting permits will be handed out on Lyell Island while the talks with the Haida are going on, or before the government makes a formal decision regarding the whole of South Moresby which would come following the conclusion of the talks with the Haida Nation. Before talks with the Haida resume, the B.C. Minister of Forests announces that logging will continue and more cutting permits are issued for Lyell Island.

August 15, 1985 Austin Pelton, B.C. Minister of the Environment, meets with the Haida Nation and claims to have no knowledge of Brummet's broken promise regarding no further issuance of cutting permits on Lyell Island. The Haida Nation informs Minister Pelton that any logging after 30 days would be seen by the Haida as an act of aggression against the Haida Nation. At a subsequent meeting, Pelton gives renewed assurances that no more logging permits would be issued, yet three more cutting permits are subsequently issued by the B.C. Forest Service in a critical area on the east coast side of Lyell (Athlii) Island.

August 31-September 3, 1985 The Lillooet Tribal Council sponsors a "Voices for the Wilderness" Festival in alpine meadow beside Brimful Lake, located two kilometres above sea level in the Stein watershed. Its purpose is to bring public awareness to the Stein Valley wilderness and better understanding between Natives and non-natives working to protect the Stein. Over 500 people hike in over a not-so-easy, five-kilometre-long trail to attend this unprecedented event, celebrate and dedicate themselves to preserving the Stein watershed.

September 1985 The federal government holds Parks Canada's centennial celebration of the establishment of the first National Park—Banff—in Banff National Park. At this "Heritage for Tomorrow" celebration, the B.C. environment movement launches a big drive to get the federal government to create a South Moresby National Park Reserve. Leaders of WCWC, the Valhalla Wilderness Society, the Sierra Club, Island Protection Society, CPAWS and other park supporters attend. At the closing session, Thom Henley of IPS gives a passionate speech and the newly appointed Federal Environment Minister, Tom McMillan, upon hearing Thom's speech makes an impromptu commitment in his keynote address to working towards making South Moresby a National Park.

September 1985 The Lytton natives of the Nl'akapxm Nation formally join with the Lillooet Tribal Council to oppose Logging of the Stein.

October 1985 Lyell Island logging resumes with the RCMP standing by to enforce injunctions. Haida begin blockading the logging road.

October 1985 WCWC publishes *"The STEIN WILDERNESS is in Danger of Immediate Destruction"* a four-page tabloid-sized newspaper reporting on Voices for the Wilderness Stein Festival held on the past Labour Day weekend and rumours that Forest Minster Tom Waterland was about to approve funds to build the bridge across the Fraser River so the Stein could be logged. Print run 50,000.

October 18, 1985 The B.C. government establishes a Wilderness Advisory Committee (WAC) chaired by Vancouver lawyer and future B.C. Supreme Court Chief Justice Bryan Williams to evaluate the validity of protecting Gwaii Haanas (South Moresby), as well as 16 other proposed protection areas including the Stein Valley and Tatshenshini-Alsek. An insider tells WCWC that WCWC's annual endangered wilderness calendars were used as source for the list of 16 threatened wilderness areas to be considered by the WAC. The WAC has only one environmentalist sitting on it, Ken Farquharson, who returned from living in Scotland to serve on it. The WAC has only four months to bring forward its recommendations to the provincial government.

October 20, 1985 The Nuu-chah-nulth hold a big rally to protect Meares Island in front of the B.C. Legislature and erect Joe David's giant Cedar Man Welcome Figure. This figure now stands outside the U.B.C. Anthropology Museum in Vancouver.

November 4, 1985 Meares Island Land Claims trial set to be heard on this date is postponed and later is adjourned.

November 7, 1985 Western Forest Products seeks an injunction in B.C. Supreme Court to prevent Haida from interfering with logging on Lyell Island. Haida appear without council arguing that they had more at stake than the logging company.

November 9, 1985 Court rejects Haida position and grants Western Forest Products an injunction.

November 15, 1985 Haida elders stand on the road, vowing to protect South Moresby from further logging, and form a human blockade across a road on Lyell Island denying access to Frank Beban's loggers—72 Haida and one non-Haida (MP Svend Robinson)—are eventually arrested. This chain of events is widely reported in the media and it brings the South Moresby wilderness preservation issue to national and international attention. Subsequently, after much hesitation and backtracking on promises, the Bennett government places a logging moratorium on Lyell Island, leading eventually to a permanent halt on Lyell Island and the entire South Moresby Wilderness Area being protected as a National Park Reserve.

December 1985 WCWC surpasses the 200-member mark.

December 3, 1985 B.C. Appeal Court Justice Alan MacFarlane grants counter injunction to Nuu-chah-nulth delaying MB's logging on Meares Island giving time for the Nuu-chah-nulth to launch its case in Supreme Court of B.C. Meanwhile, clearcutting continues elsewhere in Clayoquot Sound.

1986 The 10,915-hectare Akamina-Kishinena Recreation Area, the 28,780-hectare Brooks Peninsula Recreation Area, and the 58,000-hectare Gitnadoix River Recreation Area are established. (All later become Provincial Parks)

1986 Chlorine linked to the formation of an "ozone hole" in the upper atmosphere appearing in the springtime over Antarctica, which was discovered by scientists a year earlier.

1986 The proposed Windy Bay Ecological Reserve on Lyell Island in South Moresby is protected.

1986 Entomologist E. O. Wilson coins the word "biodiversity" in a report for the first American Forum on biological diversity, organized by the National Research Council.

1986 The Committee on the Status of Endangered Wildlife in Canada (COSEWIC) adds the northern spotted owl to Canada's endangered species list.

1986 The B.C. Forestry Association establishes the *British Columbia Register of Big Trees*, which records the biggest individuals for each of B.C.'s native tree species. It is now housed at the B.C. Conservation Data Centre (CDC) in Victoria and available online (www.elp.gov.bc.ca/rib/wis/cdc/trees.htm).

February 1986 Haida Nation publishes special edition of *Haada Lass (good people) Journal of the Haida Nation* (VOL II NO 2). It has an article about Duu Guusd Tribal Park and a lengthy article presenting the history of the conflict over logging in South Moresby and the reasons the Haida were compelled to stand on the road to prevent further logging on Lyell Island in their efforts to protect Haida Gwaii and the South Moresby Wilderness Area.

March 5, 1986 The Save South Moresby Caravan trail departs from St. John's Newfoundland and travels 7,500 kilometres by rail across Canada, arriving in Vancouver on the afternoon of March 15th to be greeted by over 5,000 people expressing their support for protecting South Moresby in a National Park Reserve.

March 7, 1986 The B.C. Wilderness Advisory Committee releases its report *The Wilderness Mosaic*, after four months of consideration, recommending resource development in nearly all of the 16 study areas under consideration including the Stein Valley and the Tatshenshini. From a conservationists point of view it was a disaster. For South Moresby it recommends the logging company's Option "B" Ecological Reserve proposal—a leave-strip along the ocean and along the lower banks of Windy Bay Creek—only ten percent to be saved and recommends that only the area south of Lyell Island and the scrubby forestland on Moresby Island become a park. It did, however, recommend that no road be constructed up the Stein Canyon without a formal agreement between the Lytton Indian Band and the provincial government. Subsequently, as a result of their recommendations, the Forest Act was amended in August 1987 to recognize wilderness as a distinct resource and a legitimate land use.

Spring 1986 WCWC published eight-page 8 1/2 X 11" newsprint two-colour booklet titled *Special Issue – Save South Moresby National Caravan*. It chronicles the epic 10-day 7,500-kilometre-long train journey across Canada in early March rallying the support of Native, church, conservation, student and human rights groups at every stop along the way to make South Moresby a National Park Reserve. The caravan culminating with a welcoming party of Haidas and a huge rally of thousands at the train station in Vancouver in the afternoon of March 15. A petition to the B.C. government to Save South Moresby is featured on the back cover. Press run 25,000 copies.

April 1986 100,000 protesters march through downtown Vancouver, across Burrard Street bridge, as part of an annual End The Arms Race event. Founded in 1982 by Vancouver and District Labour Council executive Frank Kennedy, the event brought together union, church and labour groups and many citizens concerned about escalating nuclear tensions, especially between the U.S. and the former Soviet Union. Attendance at peace walks eventually dwindles and 1991 marks the last year of the event.

April 26, 1986 A massive radioactive explosion at a nuclear power plant at Chernobyl in the Ukraine, caused by human error, led to 31 rapid deaths, the evacuation of 130,000 and the likelihood that many thousands, including children, will die prematurely from exposure to radiation. The nuclear power industry has never recovered from the effects of the publicity given to this, the worst nuclear accident to date.

April 29, 1986 Nuu-chah-nulth's *Ha-Shilth-Sa* newspaper publishes a 16-page special edition (Vol. 13 No. 2) on their efforts to protect Meares Island.

May 1-9 ,1986 Under contract to WCWC dendrochronologist Marion L. Parker, studies tree rings on the Queen Charlotte Islands and writes two reports *Preliminary Investigations of the age and dendrochronological Quality of four Coniferous Species on Graham Island* and *Notes on Culturally Modified Trees on the Queen Charlotte Islands*. He found the trees he studied very old and slow growing—the oldest being a 1294-year-old yellow cedar. Regarding the Culturally Modified Trees (CMTs) he studied he stated "The CMTs contain much information about cultural practices and occupation dates and need to be extensively studied and preserved."

May 26, 1986 Federal Environment Minister Tom McMillan announces the creation of a task force to examine co-operative means of establishing and funding new national parks. This is an important step towards making South Moresby a national park reserve.

August 6, 1986 William Nick (Bill) Vander Zalm, Social Credit, becomes Premier of B.C. and holds that position until April 2, 1991.

Summer 1986 Under the auspices of the Lytton and Mt. Currie Bands, WCWC recruits volunteers and upgrades the Stein Heritage Trail from Cottonwood Creek to Stein Lake.

Summer 1986 The Friends of Strathcona Park is founded to protect Strathcona Provincial Park from further mining encroachment. The Friends decry to the Wilderness Advisory Committee recommended that central area of the park be downgraded to "B" status to allow a new silver mine to open within its borders. Citizens win a protracted battle to stop mining exploration in Strathcona Park, but not before 64 people are arrested for disobeying a court injunction and blockading the road.

1986 WCWC publishes *The Last of the Giants – Canada's Largest Tree*, a four-page two-colour tabloid-sized newspaper written and illustrated by Randy Stoltmann It is WCWC's second newspaper calling for the creation of a Canadian Landmark to protect the Red Creek Fir, growing near Port Renfrew, B.C., Canada's largest known living single organism, and for other Canadian Landmarks to protect other record trees in B.C. Print run 50,000.

September 1986 Provincial and federal governments agree to "fast track" negotiations for a South Moresby National Park with September 1987 set as the deadline.

October 1986 The Lytton and Mt. Currie First Nations host the second annual Stein Gathering. It is held at the mouth of the Stein River. Over 700 supporters attend.

October 1986 The B.C. government establishes the Upper Stein and Lower Stein Recreation Areas where no logging can take place. However the Middle Stein, where all the commercial timber grows, is left open for logging.

November 1986 Haida build a longhouse in Windy Bay on Lyell Island. It reinforces their claim of traditional use and occupancy of the South Moresby region and is available for use if possible Haida blockades are needed to stop logging in Windy Bay.

November 26, 1986 The federal government accepts the Nlaka'pxm Nation's land claim, which includes the Stein Valley, for negotiation.

December 23, 1986 The B.C. Forest Service approves a five-year cutting plan for Lyell Island, which includes logging in Windy Bay in 1987 and no slow down in the liquidation of the old-growth forest left on Lyell Island.

1987 Human population worldwide reaches five billion.

1987 The B.C. government establishes the 123,000-hectare Hakai Recreation Area, the largest Marine Park on the on the central coast of British Columbia; and the 91,000-hectare Fiordland Recreation Area (6,645 hectares of marine area and 84,355 hectares of land area) encompassing Kynoch and Mussel Inlets, located about 100 kilometres northwest of Bella Coola on the mid coast of B.C.

1987 The B.C. government establishes the 127,690-hectare Kakwa Recreation Area. Located on the Continental Divide, its main physical features include Mount Sir Alexander (3270 m), Mount Ida (3189 m) and Kakwa Lake. The government also established the 68,000-hectare Height-of-the-Rockies Wilderness stretching northwest along the Continental Divide between Elk Lakes Provincial Park in southeastern British Columbia and Banff National Park and Peter Lougheed Provincial Park in southwestern Alberta. (Height-of-the-Rockies becomes a Provincial Park in 1995.)

1987 The B.C. government establishes the 16,680-hectare Cascade Recreation Area lying within the Hozameen Range of the Cascade Mountains, adjacent to the northwest portion of Manning Provincial Park. This recreation area is downsized later to 11,858 hectares in order to allow for logging development.

1987 Conservation International is founded in the U.S. to work towards conserving biodiversity worldwide.

February 1987 The first "Raging Grannies" group forms in Victoria, B.C., when elderly women peace activists who had been doing street theatre begin dressing up in outrageous hats and singing satirical songs to protest against nuclear submarines, uranium mining, nuclear power, militarism, racism, clear-cut logging, and corporate greed. Since their beginning, raging grannies have sometimes been arrested, but never taken to court. Today raging grannies groups are found throughout Canada and the U.S.

Winter/Spring 1987 WCWC publishes 100,000 copies of a four-page two-colour tabloid-sized newspaper titled *Battle for Windy Bay – Last Chance to save WINDY BAY* and immediately sends copies to all federal MPs and all B.C. MLAs. It features a recent Landsat satellite photo showing clearcuts completely surrounding Windy Bay on Lyell Island and photos of the Haida longhouse being built in Windy Bay.

April 21, 1987 WCWC moves its office from Adriane Carr's apartment on Alberta Street to 1520 West 6th Ave., across the alley from the Vancouver Press Club.

April 27, 1987 The Brundtland World Commission on Environment and Development Overview Report *Our Common Future* is released. It champions the concept of "sustainable development" to solve global environmental problems.

Summer 1987 WCWC publishes a six-page three-colour tabloid-sized newspaper titled *Lower Mainland Pocket Wilderness Coalition – VANCOUVER AREA WILDERNESS –It's time to save the few pockets left* (Educational Report Vol. 7 No. 3) written by then volunteer Joe Foy. Press run 50,000. It has a centrefold map showing 20 areas around Vancouver B.C. that merit protection. Six of these areas are featured, each on a separate page, in the newspaper.

Summer 1987 WCWC volunteers continue to work on the Stein trail system completing the trail from Blowdown Pass to Cottonwood Creek.

June 1987 WCWC and Islands Protection Society petition the B.C. Supreme Court to judicially review and declare that "Wildlife Sundry Permit CO 17993" to allow the capture of 10 Peales peregrine falcon chicks on the Queen Charlotte Islands (Haida Gwaii) be struck down. The B.C. government withdraws the permits before this case goes to court citing threat of violence by people, not the extreme low numbers of birds remaining, as to their reason.

June 1987 WCWC produces a 16-page black and white tabloid-sized newspaper *B.C. Environmental Hotspots* with a full-colour centrefold map with help from many people including Adriane Carr's Vancouver Community College Langara Campus Environmental Geography students who, as a special class project, did the background research and drafted the articles. Published by the New Catalyst, 20,000 copies are printed as an insert to the Summer 1987 (Issue Number 8) New Catalyst Quarterly newspaper and 20,000 are printed as stand alone copies for WCWC to distribute.

Summer 1987 WCWC volunteers continue to work on the Stein Heritage Trail system, completing the trail from Blowdown Pass to Cottonwood Creek. The Stein Heritage Hiking Trail system is now 90 kilometres long.

Summer 1987 WCWC co-publishes with the Friends of Strathcona Park an eight-page two-colour tabloid-sized newspaper titled *SOLD OUT TO MINING INTERESTS 1978 – Strathcona Provincial Park*. Press run 50,000. In the ensuing months local citizens blockade the logging road in a protracted battle with 64 people arrested for disobeying a court injunction and blockading the new mining road to Cream Lake. The provincial government finally relents, bringing to an end the threat of establishing new mines within existing parks.

July 11, 1987 Government of Canada signs a memorandum of agreement with government of B.C. to establish a 147,000 hectares South Moresby National Park Reserve now known as the Gwaii Haanas National Park Reserve/Haida Heritage Site. At the same time, the Canadian government commits to establishing a 305,000-hectare Marine National Park Reserve to protect the surrounding waters of this new terrestrial park in the near future.

Summer 1987 WCWC publishes an eight-page two-colour tabloid-sized newspaper titled *Special wildlife centenary issue – GRIZZLY BEARS Western Canada – A last stronghold of the monarch of North American wildlife*. Written by bear biologists Wayne McCrory and Erica Mallam, it makes a strong case for creating the Khutzeymateen Grizzly Bear sanctuary. Print run 50,000. WCWC also premiers and sells its new poster *HER CHILD DESERVES A FUTURE – KHUTZEYMATEEN – SAVE THIS GRIZZLY SANCTUARY* featuring a mother grizzly bear and her cub.

August 1-3, 1987 Third annual Stein Voices for Wilderness Festival, sponsored by the Lytton and Mt. Currie Indian Bands, is held in the Stein alpine where the first festival was held. David Suzuki, Farley Mowat, Thomas Berger and Miles Richardson speak and John Denver, Gordon Lightfoot and others perform, donating their music talent to help the cause. Over 2,000 people gather to hear the inspiring speeches, music and express their support for keeping the Stein wild.

September 14-17, 1987 The 4th WORLD WILDERNESS CONGRESS is held in Denver, Colorado. 2,000 delegates from 64 countries attend, including Adriane Carr and Paul George of WCWC who had a booth with WCWC publications at the convention.

September 30, 1987 At 11 a.m., WCWC gets an injunction to stop Whonnock Industries Ltd. from logging above Greendrop Lake, a key "Pocket Wilderness Area" WCWC is seeking to protect, near Vancouver. It is WCWC's first injunction, but destined to be short lived. At 4 p.m. the same day Whonnock gets an ex parte hearing before another Supreme Court Justice and gets WCWC's injunction quashed. The Greendrop cutblock is immediately logged.

1987 WCWC publishes its second book titled *Hiking Guide to the Big Trees of Southwest British Columbia* by Randy Stoltmann. It contributes to eventually getting protected nearly all the places featured.

October 5, 1987 The Lytton and Mr. Currie Indian Bands issue their "Stein Declaration" proclaiming *"we will maintain the Stein Valley as a wilderness in perpetuity for the enjoyment and enlightenment of all peoples and the enhancement of the slender life thread on this planet."*

October 1987 WCWC publishes an eight-page full-colour tabloid-sized newspaper titled *STEIN VALLEY – The choice is ours. Where do you stand?* Print run 50,000. The centrefold features the *Stein Declaration* signed by Lytton Chief Ruby Dunstan and Mt. Currie Chief Leonard Andrew.

November 1987 WCWC publishes an eight-page two-colour tabloid-sized newspaper titled *NO SAFE WAY TO MINE URANIUM! –Government in 'hot' water for lifting moratorium*. Print run 50,000 copies. Copies of this paper are mailed to every household in the Kootenay-Boundary Regional District. Voters there overwhelming vote "yes" in a non-binding referendum question on the November 21 ballot to continuing the ban on uranium mining in B.C.

1987 The Canadian Nature Federation premiers its video *Save South Moresby Caravan Documentary*.

December 1987 WCWC publishes 25,000 copies of a full-colour poster titled *Stein Valley – Joy to the World*. Over 20,000 rolled up gift copies are hand out by 20 volunteer Santa Clauses in downtown Vancouver in one day right before Christmas to raise awareness and support for protecting the entire Stein Valley.

1987 The Montreal Protocol is adopted. It sets up a schedule for the phase-out of compounds that deplete ozone in the stratosphere, including chlorofluorocarbons (CFCs), halons, carbon tetrachloride and methyl chloroform.

1988 The human population reaches 5 billion.

1988 The 18-hectare Nimpkish Island Ecological Reserve established in the Nimpkish River on Vancouver Island to protect a sample of some of the tallest Douglas firs left standing in B.C.

1988 The Boundary Bay Conservation Committee is formed to promote public awareness and appreciation of, and protection for the Boundary Bay Ecosystem which is used by 1.5 million migratory birds along the Pacific Flyway between Arctic nesting grounds and Central and South America wintering habitat.

1988 Burns Bog Conservation Society forms to *"conserve and preserve Burns Bog for all life in perpetuity."* With this new society, WCWC co-published its first newspaper calling for full preservation of the entire remaining Burns Bog located near Boundary Bay in the Fraser River's delta.

1988 A group of Robson Valley residents form the area's first conservation organization, the Canoe-Robson Environmental Coalition (CREC). It takes on a variety of conservation issues including pesticide use and the protection of old-growth forests.

1988 The B.C. government adds 412 hectares of land to the 1,248-hectare all water Robson Bight Ecological Reserve. With another 100 hectares of land added later in 1989, a narrow ocean-side land buffer averaging 200-400 metres wide is established to protect the killer whale rubbing beaches in Robson Bight.

February 4, 1988 B.C. Forest Minister Dave Parker meets with the Lytton and Mt. Currie bands and promises that no road will be built into the Stein Valley without a formal agreement with these bands.

February 29, 1988 A court hearing begins on WCWC's petition to stop the helicopter wolf kill in the Muskwa Valley. On March 7, 1988, Madam Justice Huddard of the Supreme Court of B.C. grants WCWC standing in wolf case and rules that the B.C. government issued the wolf kill permit in question illegally, thus quashing the proposed helicopter hunt to eradicate wolves in the Muskwa Valley.

February 1988 WCWC publishes an eight-page full-colour tabloid-sized newspaper titled *Save the Stein – S.W. BRITISH COLUMBIA'S LAST MAJOR WILDERNESS VALLEY*. Press Run 50,000. It features the *Stein Valley – Joy to the World* poster as a centre spread. Written by Joe Foy, Ken Lay, John McCandless and others, and the design help of Emily Carr College students. It is WCWC's best paper yet in the growing campaign to save the Stein.

February 1988 WCWC publishes an eight-page two-colour tabloid-sized newspaper titled *British Columbia Professional Foresters FOR SHAME*. Two press runs totaling 75,000 copies are distributed. It features photos of "unethical" logging practices permitted by B.C.'s foresters. It asks B.C.'s professional foresters to live up to the professional code of ethics they have all sworn to uphold.

March 15, 1988 WCWC presents a brief to Cyril Shelford's *Hearing on Peales's peregrine falcons*. It makes strong biological arguments for why the B.C. government should not issue any more permits to falconers to take Peales's Peregrine Falcons chicks from nests in the wild. At the conclusion of the hearing, Shelford recommends that no permits be issued.

April 1, 1988 WCWC members Randy Stoltmann and Clinton Webb discover a newly constructed logging road that extends to the edge of "remote" Carmanah Valley on the west coast of Vancouver Island. They find road survey tape extending to the magnificent Sitka spruce grove in the heart of the Carmanah Valley, far beyond anything shown on the five-year planning documents in the B.C. Forest Service office presented to the public only a few years before.

April-May 1988 WCWC publishes a four-page full-colour tabloid-sized newspaper titled *Lower Mainland Pocket Wilderness – SAVE WHISTLER'S BIG TREES*. Press run 50,000. Inserted as a supplement to the Whistler Question paper, which is distributed free to the residents of the world-famous ski resort town of Whistler, it calls for the protection of the huge cedars on nearby Rainbow Mountain.

May 13, 1988 The Heritage Forest Society and the Sierra Club of B.C. present to the B.C. and federal governments, and to MB and the media a brief titled *A proposal to Add the Carmanah Creek Drainage with Its Exceptional Sitka Spruce Forests to Pacific Rim National Park*.

May 15, 1988 1,500 International Woodworkers of America (IWA) union members from Port Alberni travel to Victoria to demonstrate in favour of logging the Carmanah Valley in front of the B.C. Legislature. The 7,000 strong PPWC (Pulp and Paper Woodworkers of Canada disagree with the IWA's position and send a letter to Premier Vander Zalm calling for a halt to logging plans.

May 19, 1988 MB voluntarily halts all its road construction into Carmanah Valley for one month to allow for a study of the valley.

May 30, 1988 Three determined WCWC volunteers led by WCWC Director Joe Foy set up "Hell Camp" on the side of MB's muddy logging road on the rim of the Carmanah Watershed. They begin building a trail down to "Heaven Grove" in Carmanah's mid-valley alluvial flood plain where giant towering Sitka spruces grow. The next day MB officials ask them to stop, but they refuse.

June 10, 1988 During a helicopter survey of Carmanah Valley, MB engineers discover a 95-metre-tall Sitka spruce, the "Carmanah Giant" growing very near the Pacific Rim National Park Reserve boundary and the west coast lifesaving trail. It proves to be Canada's tallest known tree and the world's tallest recorded Sitka spruce.

June 20, 1988 Joe Martin, Tla-o-quiaht canoe builder, stands with 20 others on the point of land at the head of Sulphur pass that leads to Shelter Inlet and the Megan River Valley (the largest remaining pristine watershed left on Vancouver Island). They conduct a ceremony confirming their commitment to stop the logging road extending into this area. Fletcher Challenge, the company building the road, obtains an injunction to stop the blockades. Those defying it are charged with contempt of court. Eventually 35 people are arrested including Ahousaht Hereditary Chief Earl Maquinna George. Charges against Chief George are eventually dropped.

June 29, 1988 MB serves a Notice of Motion to WCWC that MB will be in court at 9:45 a.m. on July 7, 1988 seeking a court order to force WCWC and others to halt all Meares Island trailbuilding activities and remove all traces of the Meares Island trail WCWC is constructing.

June 29, 1988 MB proposes that two reserves be created in Carmanah Valley; a nine-hectare area around the "Carmanah Giant" to protect it and a 90-hectare area, encompassing some of the largest Sitka spruces, in the mid-valley.

July 1-3, 1988 WCWC, having completed its first phase of trailbuilding—a trail down to the mid valley bottom, hosts the Carmanah Caravan, a Canada celebratory event for hikers and campers in Carmanah's "Heaven Grove." Over 150 people attend.

Early July 1988 MB seeks and gets, by consent of WCWC, a court order varying Justice Gibbs' injunction of January 1985 to specifically include trailbuilding, thus stopping WCWC and FOCS from continuing to build and complete their Meares Island circuit hiking trail.

July 15, 1988 Federal MP Bob Wenman introduces in the House of Commons an amendment to the National Parks Act to include the entire Carmanah Valley within the boundaries of Pacific Rim National Park. This amendment fails to get enacted.

July 22, 1988 MB seeks a B.C. Supreme Count injunction to stop WCWC's trailbuilding activities in Carmanah Valley.

July 26, 1988 The B.C. Supreme Court rejects MB petition to stop WCWC's trailbuilding activities in Carmanah Valley. The ruling establishes for the first time in B.C. history the public's right of access to tree-farm-licensed lands, as long as it does not directly interfere with the licensee's exclusive right to harvest the area's wood. The B.C. Forest Service requests that MB prepare a revised logging plan for Carmanah Valley to present to the government by the end of September 1988.

July/August 1988 WCWC publishes a four-page full-colour tabloid-sized newspaper titled *Carmanah, the world's tallest Sitka Spruce deserve protection*. Press run 50,000. This is WCWC's first educational report newspaper calling for the preservation of the Carmanah Valley.

August/September 1988 WCWC publishes a four-page two-colour tabloid-sized newspaper titled *Time to sustain rather than destroy*. Press run 50,000. It tells the story about the fight by local native and non-native citizens of Clayoquot Sound to stop the construction of the Sulphur Pass logging road in the Sound.

August 1988 WCWC premiers its first video *Carmanah Forever*. It was filmed and produced for WCWC by Susan Underwood.

August 28, 1988 WCWC trail crew reached the "Carmanah Giant." This trail from the mid valley to edge of Pacific National Park Reserve is eventually closed for safety reasons and to limit access to Park's West Coast Trail.

Summer 1988 WCWC co-publishes with Burns Bog Conservation Society a four-page full-colour tabloid-sized newspaper titled *Burns Bog mega-project would put farmland and Fraser fisheries at risk*. Press run 50,000. This is WCWC's first publication in support of protecting all that remained of this rare domed bog in Fraser River delta.

August 30, 1988 WCWC publishes *Maplewood Flats – Reasons for Protection and Enhancement* an 18-page report about the importance of preserving North Vancouver's last estuarine wetland and foreshore wildlife area. The area is eventually protected.

Labour Day 1988 Construction of the Sulphur Pass road is finally halted.

September 1988 *STEIN–The way of the River*, a coffee table conservation book written by Michael M'Gonigle and Wendy Wickwire, is published.

1988 Fletcher Challenge takes over British Columbia Forest Products Ltd. (BCFP) and announces a moratorium on logging in the Stein.

October 1988 The Valhalla Wilderness Society publishes a large B.C. endangered wilderness map titled *British Columbia's Endangered Wilderness– A Proposal for an Adequate System of Totally Protected Lands*. It has on it all the protected areas environmental groups had proposed up to that time with explanatory information about them printed on the back. The document recommends that the total of the protected areas cover 13 percent of B.C. including all the new areas on the map. Western Canada Wilderness Committee, Canadian Parks and Wilderness Society (CPAWS), Sierra Club of Western Canada and the Friends of Ecological Reserves endorse this publication.

October 6, 1988 MB releases revised plans for Carmanah, which include an increase from 99 hectares to 275 hectares that the company is willing to preserve out of Carmanah's total area of 6,730 hectares.

October 27, 1988 WCWC publishes a four-page full-colour tabloid-sized newspaper titled *Carmanah – forest forever*, its second newspaper designed to rally support for saving the entire Carmanah Valley. Press run 75,000.

December 23, 1988 An oil barge is punctured off the coast of Washington State. In all 875,000 litres of Bunker C oil escapes in this Nestucca spill.

1989 World's human population exceeds 5 billion.

1989 WCWC launches a Forest Watch community outreach program. Mark Wareing, WCWC forester, conducts six workshops in locations across the province.

1989 The Canadian Nature Federation (CNF) and its affiliates, in partnership with World Wildlife Fund Canada (WWF), launch the Endangered Spaces campaign seeking as a goal only 12 percent of Canada's land base to be placed in protected status, safeguarded from resource developments.

1989 West Coast Environmental Law Society establishes the Environmental Dispute Resolution Fund (EDRF) which provides funding to individuals and citizen groups needing support for environmental litigation or alternative dispute resolution.

1989 Some residents of the Sunshine Coast form the Tetrahedron Alliance to protect the remaining 6,000 hectares of some of Canada's oldest trees in the Tetrahedron wilderness. WCWC supports the proposal and sends a letter to Ministry of Forests requesting that the BC government preserve the area.

January 1989 *Time* magazine named "The Endangered Earth" as its "Person of the Year."

January 1989 The Nestucca oil spill off Washington State hits Clayoquot Sound, and the Tofino community rallies together to clean it up. WCWC sends two directors and volunteers to help with the cleanup and publishes an educational report newspaper warning of the dangers of oil spills. This spill impacts every sea bird Ecological Reserve on the west coast of Vancouver Island. An estimated 30,000 - 60,000 birds are killed. It is the worst spill that ever impacted the B.C. coast.

February 20-25, 1989 The First National and International Gathering of the Indigenous Peoples of the Xingu, funded in large part by Canadians through efforts of David Suzuki, is held in Altimera, a frontier town in the Brazilian Amazon. 800 people including representatives of Amazonian Indians from 20 tribes attend to fight the proposed Kararao hydroelectric dam and 79 others planned for the Amazon Basin.

Spring 1989 WCWC Director Adriane Carr and three other women establish WILD (Wilderness Is the Last Dream) as a special campaign of WCWC to work on international wilderness preservation. It is based on the conviction that wilderness is the wellspring of life and that we must act now on a global scale to do something concrete to stem the tide of its destruction. WILD's first ambitious project is to map all the remaining unprotected wilderness areas left on the Earth and denote the threat(s) to them.

Spring 1989 WCWC publishes a four-page two-colour tabloid-sized newspaper titled *OIL SPILL*. Press run 50.000. It tells about the Nestucca oil spill clean-up efforts in Clayoquot Sound and makes a plea to "never let it happen again" and for the B.C. and Canadian governments to continue their moratoriums on offshore oil exploration and development on the west coast.

Spring 1989 *Adbusters – the Journal of the Mental Environment* is launched in Vancouver by former advertiser Kalle Lasn and environmental filmmaker Bill Schmalz. The first issue of the magazine was devoted to calling attention to the sophisticated techniques used by corporate forestry companies to manipulate public opinion and distort reality in its *Forests Forever* television campaign. *Adbusters* skillfully produced its own TV "anti-commercials" to address this issue and many other aspects of excessive consumption, and it has gone on to build international recognition for its demonstrating the links between advertising, conspicuous consumption and environmental destruction.

Spring 1989 WCWC helps David Suzuki put on an Amazon Benefit to help Chief Paiakan of the Kayapo protect his tribe's lands from being flooded by a hydroelectric project. Over 1,000 show up to attend it in the Vancouver high school auditorium, which only held 750 people. Those attending donated nearly $16,000 to the cause.

Spring 1989 WCWC's WILD team publishes an eight-page full-colour tabloid-sized newspaper titled *Canadians vow to help save the Amazon rainforest*. Press run 50,000. It calls on Canadians to help the Kayapo protect their Amazonian homeland.

Spring 1989 WCWC publishes an eight-page full-colour tabloid-sized newspaper titled *FORESTRY MALPRACTICE ON RISE IN British Columbia* (Educational Report Vol. 8 No. 4). Press run 50,000. It documents bad logging practices and calls for the enactment of a "Forest Practices Act" to set eco-forestry standards.

March 24, 1989 Exxon Valdez oil spill disaster occurs in Prince William Sound, Alaska.

April 1989 WCWC publishes Thom Henley's *REDISCOVER Ancient Pathways – New Directions* book featuring forwards by Bill Reid and David Suzuki. This innovative 288 page full-colour paperback guidebook becomes a landmark work in outdoor education. Press run 5,000 copies.

April 1989 WCWC initiates in Vancouver its first door-to-door canvass to raise membership and funds.

Spring 1989 WCWC publishes and distributes 75,000 copies of a four-page two-colour tabloid-sized newspaper titled *DELTA UNDER ATTACK! – Urban development threatens to destroy the beautiful Fraser Valley Forever!* (Educational Report Vol. 8 No. 5). It is householder mailed to local citizens to get people out to a public meeting in opposition to removing land from B.C. Agriculture Land Reserve for development.

June 27, 1989 The UN's General Conference of the International Labour Organization (ILO), at its seventy-sixth session, adopts Convention (No. 169) concerning Indigenous and Tribal Peoples in Independent Countries. This convention guarantees the rights of indigenous peoples, including the right to ownership of their traditional lands and to maintain and develop their own unique cultures. It enters into force September 5, 1991.

July 21, 1989 WCWC holds the grand opening of its 20 Water Street store and office in Vancouver's historic Gastown. The Raging Grannies sing a song of protest against clearcuts and steep slope logging while the red ribbon is being cut. Over the 10 years the store was operating at that site, it sold over a hundred thousand WCWC posters and made money to help fund WCWC's campaigns.

July 25, 1989 B.C. Premier Van der Zalm, after seeing the huge clearcut beside the road to Ucluelet and Tofino called the "Black Hole," declares logging in Clayoquot Sound to be a disgrace. He announces he will form an appropriate planning committee to resolve the Clayoquot Sound conflicts.

August 1989 Marion Parker, a dendrochronologist under contract to WCWC, produces a 15-page report titled *Preliminary Dendrochronological Investigations in the Carmanah Creek Area*. Analyzing stumps in clearcuts next to Carmanah Valley, he found the trees there to be extremely old, including a 300-year-old western hemlock only 24 centimetres in diameter at breast height.

August 4, 1989 The B.C. government establishes the Clayoquot Sound Sustainable Development Task Force, sponsored by the Ministries of Environment and of Regional and Economic Development. This Task Force is charged with preparing a consensus-based sustainable development strategy for Clayoquot Sound, including determining which areas should be logged or protected during the one-year term of its planning process. Environmentalists call for preservation of all the remaining pristine valleys and islands in the Sound with eco-forestry (single tree selection logging) to be practiced in the fragmented areas. The local environmental group, Friends of Clayoquot Sound, boycott this "talk and log" process.

August 23, 1989: The Tsilhqot'in People of Xeni, known as the Nemiah Valley Indian Band, issue a declaration stating that the lands of their traditional territory be known as the Nemiah Aboriginal Wilderness Preserve where no industrial logging or industrial mining is permitted to take place.

Summer 1989 WCWC published a 12-page two-colour tabloid-sized newspaper titled *Taking the preservation message to the world – We decide forever...* (Educational Report Vol. 8 No. 6). Press run 50,000. It features the opening of the new storefront office in Gastown and reports on all the various current activities of WCWC.

May-August 1989 WCWC hosts four artists' expeditions into Carmanah Valley. Over 100 artists attend, most donating a work for inclusion in an Art Auction to be held later in the fall.

September 1989 More than 200 people (Ahousahts, representatives of other First Nations and non-natives), gather on a logging road in the Atleo River Watershed in Ahousaht Territory in an unsuccessful attempt to stop the further clearcut logging destruction of this once salmon-rich stream. The logging company gives its workers the day off to avoid confrontation.

1989 Doman Forest Products Ltd. buys out the other partners of Western Forest Products, but keep using the WPP name for the Moresby TFL 24 ownership.

1989 Two years after the signing of the agreement to establish a South Moresby National Park Reserve, 52 percent of the TFL 24 lands are taken to create it. Along with an overcut correction, this results in a 73 percent reduction in the Allowable Annual Cut in this TFL to 115,000 cubic metres annually.

1989 WCWC Victoria chapter is established with Derek Young acting as its executive director. It opens a storefront office in premises donated by former Victoria Mayor Peter Pollen at 1002 Wharf Street.

1989 SmartWood, the first certification process designed to improve the effectiveness of sustainable forestry in conserving biodiversity and providing equity for local communities, fair treatment to workers, and creating incentives for businesses so that they can benefit economically from responsible forestry practices, is initiated by the Rainforest Alliance, an international nonprofit environmental group based in New York City.

September 1989 MB plans to begin clearcutting near Robson Bight, which would be visible to whale-watching tourists and cruise ship passengers. Public outcry grows.

September 16, 1989 WCWC holds its first (and last) *Walk for the Environment*. Thousands of citizens walk from Kits Beach to Queen Elizabeth Park in Vancouver. The uphill route is chosen to illustrate the long road ahead to healing and protecting the environment.

September 20, 1989 Volunteers finish WCWC's Carmanah trail to the top of the valley, linking up trails in the middle and lower valley, and completing WCWC's Carmanah Valley trail system.

October 15, 1989 WCWC premiers its Carmanah Art Show featuring a silent auction of nearly a hundred works of art donated by artists to help raise funds to save Carmanah Valley. WCWC publishes 15,000 copies of a 168-page hard cover coffee table book title *CARMANAH – Artistic Visions of an Ancient Rainforest* by Ken Budd of Summerwild production and Raincoast Books featuring artworks and statement donated by 70 artists who had visited Carmanah Valley the previous summer. The show tours around to various galleries in B.C. for two months with people attending the show having an opportunity to put in a silent bid.

Fall 1989 WCWC publishes 300,000 copies of a four-page full-colour full-sized newspaper titled *CARMANAH – Canadian rainforest deserves protection* (Educational Report Vol. 8 No. 8). It tells of a recent scientific expedition, the completion of the WCWC trail system and the Artist project that culminated in the publication of *CARMANAH – Artistic Visions of an Ancient Rainforest*. It has a tear-off coupon to order the Carmanah book at a discount price.

November 17, 1989 Over 400 residents of the west coast communities of Ucluelet and Tofino, most of them associated with the logging industry centred in Ucluelet, attend the first public meeting of the *Share the Clayoquot Society* held in Ucluelet, and organization dedicated to opposing the preservation (advocating the logging) of the old-growth forests in Clayoquot Sound.

November 1989 WCWC staff and volunteers carry out a fact-finding mission in the logged areas of the Upper Tsitika, documenting environmental damage caused by MB's logging.

1989 WCWC's WILD (Wilderness is the Last Dream) publishes an eight-page two-colour tabloid-sized newspaper titled *Squalid resettlement camps no substitute for Penan's Jungle homeland. Help save Sarawak's remaining rainforest* – HALT MULTI-NATIONAL RAIN FOREST DESTRUCTION (Educational Report Vol. 8 No. 6). Press run 100,000. It is an attempt to rally international support to save the Penan and their hunter and gatherer cultural existence that depends on this rainforest.

1989 Ten protesters jailed for participating in the Sulphur Pass blockades. The IWA sues protesters for lost wages – the first SLAPP (Strategic Lawsuit Against Public Participation) suit in B.C.

December 1989 The B.C. Forest Service approves clearcut logging in Tofino Creek in Clayoquot Sound.

Winter 1989-1990 WCWC co-publishes with Environmental Watch a four-page full-colour full-sized newspaper titled *STOP the Killing of Howe Sound NOW!* (Educational Report Vol. 8 No. 9). Print run 150,000 copies. This paper "blows the whistle" on the excessive pollution coming from the two pulp mills in Howe Sound near Vancouver B.C. and reveals the fact that nearby Squamish has the highest lung cancer rate for both males and females of anywhere in B.C.

Late 1989 Ric Careless, Johnny Mikes and Dona Reel, recognizing the need due to the threat of a massive opening up of a massive copper mine to quickly mount a campaign to protect the nearly 1,000,000-hectare Tatshenshini wilderness area in the northwest corner of B.C., found Tatshenshini Wild.

1990 Peter McAllister founds the Raincoast Conservation Society to work in partnership with scientists, First Nations, local communities and non-governmental organizations to build support for decisions to protect the marine and rainforest habitats on B.C.'s central and north coast.

1990 Conservation International releases a study which reveals that out of 89 primary watersheds (river systems that drain directly into the sea) over 5,000 hectares in size on Vancouver Island, only six have never been logged. The study also reveals that of the twenty-five primary watersheds in the coastal temperate rain forest in B.C. larger than 100,000 hectares, none remain in an entirely pristine condition. One of them, the Kitlope Watershed (275,100 hectares), was pristine except for a tiny 30-hectare island in the lower valley that had been logged. The study concludes that the Kitlope is the largest undeveloped watershed left in B.C.'s coastal temperate rainforest.

1990 The Indigenous Environmental Network (IEN) is formed in the U.S. at a national gathering of tribal grassroots youth and indigenous leaders to deal with environmental assaults on indigenous lands and chart a common course of action to restore their environmental health and harmony.

1990 WCWC publishes its first children's book titled *The Magical Earth Secrets*. Featuring a story about an "Eagle Child" written and illustrated by Della Burford. With multicultural paintings, it becomes an instant hit.

1990 The B.C. government appoints a Spotted Owl Recovery Team to prepare a recovery plan for this endangered species in B.C.

January 4, 1990 WCWC directors set up the World Wilderness Committee as a registered non-profit corporation in Washington State.

February 1990 The B.C. Ministry of Forests orders MB to release a copy of MB's Tsitika logging plan to WCWC in response to WCWC launching a legal action to get a copy of the plan.

February 1990 WCWC, along with the Sierra Club of Western Canada, commissions a preliminary hydrological study of Carmanah Creek Watershed. The ensuing report titled *Evaluation of a proposed Hydrology study for Carmanah Creek* by E. Kranka and Associates pinpoints the areas where future hydrological studies are needed.

March 2 1990 WCWC volunteers converge on the Upper Carmanah Valley and begin the construction of a scientific research camp, including boardwalk access, a large wooden floored internal framed research tent and interconnected platforms high in the canopy of several closely-growing large old-growth Sitka spruce trees.

March 10, 1990 WCWC Volunteers complete the first canopy research platform 38 metres up the trunk of an old-growth Sitka spruce in the Upper Carmanah. By June, there are five platforms in the complex of three Sitkas and scientists begin inventorying the insects in the canopy.

March 19-23, 1990 Globe '90 is held in Vancouver B.C. It is hailed as the first fully integrated conference and trade fair to address issues of economic development and environmental protection. The B.C. government, the federal government, as well as major national and international associations and financial institutions officially sponsor it. Gro Harlem Brundtland, author of *Our Common Future*, gives the keynote address.

Spring 1990 WCWC co-publishes with Tatshenshini Wild a four-page full-colour tabloid-sized newspaper titled *TATSHENSHINI - Ice Age Wilderness –Help Protect North America's Wildest River* (Educational Report Vol. 9 No. 1). Press Run 100,000. It clearly presents the threat of the proposed Geddes copper mine and the merits of keeping this big river and surrounding wilderness wild.

March 1990 WCWC Victoria publishes its first newspaper, a four-page full-colour full-sized newspaper titled *TSITIKA/ROBSON BIGHT - Nature's Mystery Deserves Protection* (Educational Report Vol. 9 No. 3). Print run 150,000. This paper calls for a permanent halt to all industrial development including road building in the Lower Tsitika/Robson Bight area, as well as stronger guidelines and enforcement for protection of the killer whales (Orcas) at Robson Bight.

April 10, 1990 B.C.'s Forest Minister Claude Richmond announces the creation of the 3,600-hectare Carmanah Pacific Provincial Park encompassing the Lower Carmanah Valley. (The entire watershed is only 6,700 hectares— 16 times larger than Vancouver's Stanley Park). He further announces that the Upper Carmanah will be opened to logging only if studies to be conducted by MB show that such logging would not hurt the new park downstream.

1990 Future New Democratic Party (NDP) Premier Michael Harcourt calls the Social Credit government's decision to only protect half of Carmanah Valley and allow logging in the other half "cowardly." He vows that an NDP government would end the "war in the woods" by protecting the remaining old-growth forests left on Vancouver Island.

April 24, 1990 The Hubble telescope is deployed in space.

April 26, 1990 B.C. Environment Minister John Reynolds presents Western Canada Wilderness Committee and the Orphaned Wildlife Rehabilitation Society "Citizens' Group of the Year" awards.

May 6, 1990 WCWC's *CARMANAH - Artistic Visions of an Ancient Rainforest* wins both the Roderick Haig-Brown Regional Book Prize and the Bill Duthie Booksellers' Choice award.

May 30, 1990 WCWC submits the *Lower Tsitika Valley –A case for Preservation*, a 52-page comprehensive report and critique of current logging in the Tsitika researched and written by forest technician Clinton Webb for WCWC, to the Tsitika Follow-up Committee.

June 17-23, 1990 WCWC's WILD hosts MAPPING THE VISION to Protect Earth's Natural Ecosystems, an International working conference of wilderness experts and activists from around the world. Held in the East-West Centre at the University of Hawaii in Honolulu, participants from 23 different countries attend and begin mapping the remaining wilderness in their countries with a view to protect as much of it as possible.

June 1990 WCWC co-publishes with the Tetrahedron Alliance a four-page two-colour tabloid-sized newspaper titled *Protect the TETRAHEDRON* (Educational Report Vol. 9 No. 4). Press run 75,000. It makes a strong case for park protection for this mountaintop watershed area on the Sunshine Coast of B.C.

July 1990 WCWC published a report titled *Carmanah Valley Old-Growth Research Project*. The big report details the research conducted from the WCWC camp in the Upper Carmanah. It has information about WCWC's newly constructed canopy research station in a grove of old-growth Sitka Spruce. It is widely distributed to governments, industry, universities and media. WCWC research camp staff reports that MB has not yet started its on-the-ground scientific studies.

July 1990 The Friends of Ecological Reserves publish a four-page, full-colour, tabloid-sized newspaper titled *Ecological Reserves – Natural Areas for Tomorrow*. Print run 80,000 copies. It explains the origin of the Ecological Reserve Act in B.C. and seeks more public support for the establishment of additional Ecological Reserves noting that the pace of establishing new reserves has slowed in recent years. WCWC provides support and helps distribute this publication.

Summer 1990 WCWC publishes a four-page full-colour full-sized newspaper titled *Carmanah Valley Campaign Phase II – Working Together to Save the Upper Carmanah Valley* (Educational Report Vol. 9 No. 5) Press Run 100,000. It features the canopy research station being built in large, tall, old-growth Sitka spruces in the Upper Carmanah Valley—the world's first in a temperate rainforest.

1990 The Georgia Strait Alliance (GSA), then named Save Georgia Strait Alliance, is formed to stem the mounting pollution and the degradation of the marine habitats in the inland waters of the Strait of Georgia.

August 2, 1990 The first marbled murrelet nest in Canada is found on the moss pad on a tree limb high up in a giant old-growth Sitka spruce growing in Walbran Valley on the west coast of Vancouver Island.

September 1990 WCWC-WILD publishes 5,000 copies of a full-colour paperback coffee table book authored by Thom Henley and Wade Davis titled *Penan-Voice for the Borneo Rainforest*. It features the verbal testimony of Dawat Lupung, a young Penan, about the devastation of his rainforest homeland through logging.

September 1990 The David Suzuki Foundation is founded to find ways for society to live in balance with the natural world that sustains us. Focusing on four program areas—oceans and sustainable fishing, forests and wild lands, climate change and clean energy, and the web of life—it uses science and education to promote solutions that help conserve nature.

September 22, 1990 At its AGM, WCWC adopts bylaw changes to bring in an "outside" Board of Directors (who are not WCWC employees) at the next AGM. However, at this AGM all 14 directors elected are employees of WCWC.

Fall 1990 WCWC publishes an eight-page full-sized two-coloured newspaper titled *A first for Canada – Members Paper* Educational Report Vol. 9 No. 6). Press run 50,000. It features a picture of the press conference about the discovery of the first marbled murrelet nest in B.C. and articles about WCWC's currently active campaigns.

September 28, 1990 The World Funding for Life Foundation puts on Rainforest Benefit II with proceeds going to the Amazon Fund and WCWC's WILD campaign. Dr. David Suzuki, Dr. Wade Davis, Miles Richardson, Thom Henley and Adriane Carr speak. Doug and the Slugs, Jim Byrnes, Long John Baldry, Ann Mortifee, Sarah McLachlan and others play. WILD uses its share of the money raised to pay for the printing of WCWC's *Penan-Voice for the Borneo Rainforest* book, which is premiered at this gala. Each person attending is given a copy of this book.

Fall 1990 WCWC publishes and distributes 150,000 copies of an eight-page full-colour tabloid-sized newspaper titled *Save the Wild Side of Vancouver Island* (Educational Report Vol. 9 No. 7). It features images and the wilderness map from WCWC's soon to be published *Clayoquot on the Wild Side* book.

October 1990 The Clayoquot Sound Sustainable Development Task Force set up by Premier Vander Zalm, failing in its task, disbands, coming to the conclusion that a re-shaping of its structure and process is required if a sustainable development strategy is to be achieved for Clayoquot Sound.

October 1990 Vandals burn down WCWC research tent cabin, smash the boardwalk and destroy the bridge into Upper Carmanah. Over $30,000 worth of material and over 8,000 hours of volunteer labour are lost. They spare, however, the research spruces and the canopy platforms in them. WCWC immediately launches "Phoenix Project" to rebuild everything.

October 10, 1990-November 25, 1990 WCWC sponsors and helps organize a tour by three indigenous people of Sarawak, Malaysia, including two Penan, to 13 countries and 25 cities around the world in an effort to raise public awareness and pressure to halt the aggressive, rapid logging that was destroying the Penan's tropical rainforest homeland and way of life.

October 29, 1990 WCWC launches its *Clayoquot on the Wild Side*. Written by Cameron Young with photographs by Adrian Dorst in the Vancouver Museum's planetarium dome. This hard cover full-colour 144-page coffee table book presents a vision of establishing a series of protected wilderness areas on the west coast of Vancouver Island. At the launch Dorst shows several 360-degree panorama views of Clayoquot Sound taken with a special camera set-up to give a feel for this spectacular area.

October 1990 According to WCWC comptroller Mike Rodgers, WCWC's debt now exceeds $700,000.

November 1990 WCWC publishes a four-page two-colour tabloid-sized newspaper titled *CRISIS IN THE WOODS* (Educational Report Vol. 9 No. 8). Written and researched by WCWC staff Forester Mark Wareing, it offers "new forestry alternatives" to the destructive clearcut forestry currently being practiced.

November-December 1990 WCWC-Victoria holds a Tsitika Art Show and silent auction titled "Wilderness Treasures by Land and Sea" to raise money for its campaign to save the Lower Tsitika Valley.

Fall/Winter 1990 WCWC-WILD publishes a four-page, full-colour, tabloid-sized newspaper titled *HELP THE PENAN SAVE THE BORNEO RAINFOREST* (Educational Report Vol. 9 No. 10). It is dedicated *"to the three courageous Sarawak natives who, during the fall of 1990, embarked on a world tour to reach out to all who believe in social justice and environmental protection for help in achieving their dream – the preservation of their forest homeland."*

November 1990 WCWC published a four-page two-colour tabloid-sized newspaper titled *West Vancouver's Ancient Forest* (Education Report Vol. 9 No. 11). Press run 75,000. It spearheads a successful campaign to save this Vancouver backdrop and prime recreational resource from development.

October-November 1990 Protests and blockades are held in the Lower Tsitika Valley to protect the orca (killer whale)-rubbing beaches of Robson Bight. Thirty one people are arrested in total for contempt of court for defying the injunction.

Winter 1990-1991 WCWC-WILD co-publishes with the Steelhead Society of B.C. a four-page two-colour tabloid-sized newspaper titled *HELP*

STOP JAMES II & Kemano II – Two proposed nature-destroying hydroelectric megaprojects (Educational Report Vol. 9 No. 12). Press run 50,000. It makes the argument that the potential megawatts of electricity that could be produced by these two proposed mega projects are not worth the ecological and cultural devastation the projects would cause.

Winter 1990 WCWC-Victoria publishes a four-page two-colour tabloid-sized newspaper titled *Tsitika - Your Voice needed now!* (Educational Report Vol. 9 No. 13). Press run 75,000.

December 11, 1990 The Canadian government announces the "Green Plan." Two of the plan's goals are to stabilize greenhouse gas emissions at the 1990 level by the year 2000 and stop the use of all chlorofluorocarbon compounds (CFCs) by 1997.

December 21, 1990 Despite widespread local opposition and government commissioned environmental reports recommending against it, Alberta Premier Don Getty announces his government's approval of the proposed ALPAC project. Heavily subsidized with taxpayers' money, this world's largest pulp mill is built near Athabaska, Alberta and a Forest Management Agreement (FMA) to log vast tracts of Alberta's boreal forest is granted.

December 1990 The Sierra Legal Defence Fund is founded in B.C. as a national non-profit organization. It is dedicated to enforcing and strengthening the laws that safeguard Canadian's environment, wildlife and public health.

1991 The Protocol on Environmental Protection to the Antarctic Treaty (Madrid Protocol) is agreed to. It comes into force in 1998 when all 26 (now 27) Antarctic Treaty Consultative Parties agreed to it.

1991 Ahousaht First Nations sign onto a revised *Meares Island Tribal Park Declaration* with the Tla-o-quiaht First Nation.

January 1991 The Clayoquot Sound Sustainable Development Strategy Steering Committee is established to replace the failed Clayoquot Sound Sustainable Development Task Force. It, too, flounders because logging continues in areas that are prime candidates for preservation including Bulson Creek, compromising the outcome of the process. In May 1991, environment and tourism representatives quit this committee in protest.

1991 Ecotrust is established in Portland Oregon by a small group of diverse people with the goal of fostering the development of a conservation economy. Ecotrust actively works to ensure that the conservation-based development it backs has positive benefits for all members of the communities, particularly those who have been economically marginalized. Ecotrust's efforts are concentrated on the "Salmon Nation" along the west coast of North America. Soon after Ecotrust Canada, an affiliate of Ecotrust, is founded.

1991 In the B.C. Supreme Court's written decision in Delgamuukw v. B.C., Chief Justice McEachern asserts that the province owes a fiduciary duty to aboriginal peoples.

January 2, 1991 WCWC Mid-Island Chapter opens up a drop-off recycling depot in Nanaimo.

February 1991 WCWC publishes an eight-page three-colour tabloid-sized newspaper titled *MAJOR OPPORTUNITY TO HELP CREATE B.C. WILDERNESS AND PARKS SYSTEM* (Educational Report Vol. 10 No. 2). Print run 50,000. It urges people to participate in the Ministry of Parks and Ministry of Forests joint *Provincial Parks & Wilderness for the '90s* process, giving people the places, dates and times of the public meetings that the government has scheduled.

Winter/Spring 1991 WCWC publishes a four-page full-colour tabloid-sized newspaper titled *HALT WATERSHED LOGGING* (Educational Report Vol. 10 No. 3). Press run 75,000. It presents the growing evidence that links the clearcut logging in Vancouver's Capilano, Seymour and Coquitlam watersheds to muddy, silty tap water and calls for the end to all logging in these watersheds.

Spring 1991 WCWC-WILD publishes, in association with the Hellenic Society for the Protection of Nature and the Hellenic Ornithological Society, a 16-page tabloid-sized three-spot-colour newspaper titled *SAVE WILD GREECE* (Educational Report Vol. 10 No. 6). Research and text is provided by Stamatis Zogaris, a Greek foreign student studying at Capilano College. The centre features – the first time ever in any publication – a map of Greece with important natural areas demarcated on it. The body of the paper outlines special features and the threat to each of the numbered area on the map.

March 1991 David Suzuki appeared on a German TV show titled *A Paradise Despoiled* in which Canada is called 'Brazil of the North,' condemning forestry practices in B.C.

March 16, 1991 At the WCWC AGM members elect an outside board of directors made up of only volunteers who are not employed by WCWC.

April 1991 WCWC and the Sierra Club of Canada present a 25-page joint submission titled *Koeye River Watershed Wilderness Reserve Proposal* to the B.C. government's Parks & Wilderness for the '90s public process, making the case that this 17,000-hectare wild salmon rich watershed on the mid coast of B.C. be protected.

April 1991 In response to opinion polls that show widespread public distrust of Canada's timber industry because of clear-cutting practices and pollution from mills, 13 of Canada's largest timber companies hire the infamous PR firm Burson-Marsteller to improve their image and found the B.C. Forest Alliance to blunt the environment movement. Its first-year budget is $1 million. Patrick Moore, former executive director of Greenpeace Canada, is placed on the Board of Directors and he becomes an apologist for the forest industry.

April 1991 The Canadian Parks and Wilderness Society initiates a campaign, joined by the East Kootenay Environmental Society, to upgrade the Akamina-Kishinena Recreation Area in the East Kootenay to full Class A Provincial Park status and to surround it with a Special Management Area.

April 2, 1991 Rita Margaret Johnston, Social Credit, becomes the Premier of B.C.

Spring 1991 With the help of a grant from Patagonia, WCWC publishes 100,000 copies of a full-sized full-colour newspaper titled *Save the Kitlope –Protect the Kitlope* (Education Report Vol. 10 No. 5). It features the Haisla Nation's *Kitlope Declaration* to protect their aboriginally titled lands, including the entire Kitlope watershed.

Spring 1991 WCWC WILD publishes an eight-page three-colour tabloid-sized newspaper titled *WILD GOES TO BRAZIL* (Educational Report Vol. 10 No. 5). Press run 50,000. It reports on WILD's projects underway to protect wilderness areas around the world and especially in South and Central America.

April 1991 The Forest Resources Commission (FRC) recommends, in a report on the state of forest management in B.C., the adoption of a *"single, all-encompassing code of forest practices."* This Code would consolidate the existing legislation, regulations and guidelines governing forest practices in B.C.

May 1991 US-based Earth First! activists conduct a civil disobedience training workshop in Tofino.

May 1991 Michael M'Gonigle of Greenpeace Canada tells *Zeit* magazine in Germany "*the last hope* [for Clayoquot] *is a European boycott of Canadian wood*"; Peter McAllister of the Sierra Club also calls for a boycott of Canadian forest products to create pressure to stop massive clearcutting in B.C.

May 2-3, 1991 WCWC's Forester Mark Wareing presents a detailed brief critiquing the Greater Vancouver Regional District's Watershed Management Plan. In the brief Wareing makes a strong case that all logging should stop in these watersheds that provide the drinking water for the greater Vancouver region because the logging is causing increased erosion and siltation of the reservoirs and muddy drinking water.

May 17-24, 1991 WCWC WILD holds an eco-wilderness mapping conference in Picinguaba National Park in Brazil hosted by the local environmental group SOS Mata Atlántica. Funded by the Canadian International Development Agency (CIDA), 50 invited participants attend this working conference.

Summer 1991 WCWC volunteers and staff member Joe Foy, following up on stories of a legendary cedar grove somewhere west of Pitt Lake, launch an expedition into the Boise Valley that is about to be logged. They come back and WCWC launches a campaign to save the area.

Summer/Fall 1991 WCWC co-publishes with Doug Radies of Silver Moon Educational Project, 100,000 copies of a four-page full-colour full-sized newspaper titled *Save the Cariboo Mountains Wilderness* (Educational Report Vol. 10 No. 8) advocating the preservation of a large wild area between Wells Gray and Bowron Lake Provincial Parks.

1991 The B.C. Forest Service approves logging in the contentious Bulson River Valley in Clayoquot Sound. Vandals burn down the bridge to the logging area. Six people are arrested during the Bulson blockade.

June 7, 1991 Federal Environment Minister Jean Charest awards Canada's Outstanding Environmental group of the Year to WCWC at a ceremony in Montreal, Quebec.

June 17, 1991 The Parliament of Canada unanimously passes motion M-330: *"That, in the opinion of this House, the government should consider the advisability of preserving and protecting in its natural state at least 12 percent of Canada by working co-operatively with the provincial and territorial governments and assisting them to complete the protected area networks by the year 2000."* This remarkable motion to double the protected lands in Canada comes about through the close cooperation of Jim Fulton NDP environment spokesperson, Conservative Environment Minister Jean Charest, Liberal Environment Critic Paul Martin, and the endorsement of the BLOC Québécois.

July 1991 The B.C. Ministry of Forests releases *A Forest Practices Code Discussion Paper* seeking input from industry and the public.

1991 *Reach for Unbleached!* forms as a national foundation and Canadian registered charity. Founded by Delores Broten in response to the dioxin contamination of shellfish beds by pulp mills bleaching with chlorine compounds, it works for a sustainable pulp and paper industry by making pulp mills clean up and through promoting the use of clean paper.

1991 The *Watershed Sentinel*, a bimonthly news magazine from Cortes Island, British Columbia, a ""think globally, act locally" publication is launched.

It provides in depth stories and behind-the-scenes information that is not found in other media about environmental issues in B.C. Editors Delores Broten and Don Malcolm are helped by many volunteers who love B.C.'s environment to put out this publication, which has continued without interruption until the present.

1991 Tatshenshini International is established as a network organization that links together the top 50 conservation organizations in North America, representing about 10 million people in total, to protect the Tatshenshini and Alsek Rivers in the north west corner of B.C.

September 1991 WCWC-WILD publishes and widely distributes its 81-page *Voices for the Borneo Rainforest – 1990 World Tour Report*. The report chronicles the fall 1990 six week Penan world tour to raise public awareness and pressure the Malaysian government to halt the aggressive, rapid logging that was destroying the Penan's tropical rainforest homeland and way of life.

September 24, 1991 First reading in the House of Commons of MP Robert Wenman's Private Members Bill *An Act Respecting Endangered Species and Biological Diversity*.

October 1991 WCWC publishes and widely distributes, right before the October 17, 1991 B.C. provincial election, a perfect bound book titled *The Faceless Ones –Environmental Issues – Candidate Questionnaire Results*. It features the answers to where the 80 candidates who responded to the questionnaire stood on environmental issues (out of 317 candidates running) and the NDP's position on the questions. The NDP gave one set of answers for all their candidates.

November 5, 1991 Michael Harcourt, New Democrat, sworn in as B.C.'s Premier.

November 1991 WCWC publishes as part of the UNCED process *GLOBAL BIODIVERSITY: PROTECTING THE HERITAGE ESSENTIAL TO THE SURVIVAL OF THE PLANET*, a six-page paper commissioned by the Canadian Council for International Cooperation. Written by Adriane Carr and Paul George, it calls for a UN Convention on Biological Diversity that enshrines the principle of the inalienable right of all species to exist; initiates a massive effort to catalogue species and ecosystems and monitor the state of protection of natural biodiversity in every nation; establishes measures and elaborates on obligations to save traditional and ancient varieties of cultivars; and, provides both incentives and disincentives that are significant enough so that all countries maximize the conservation of biological diversity.

November 16, 1991 WCWC holds *Celebrate Wilderness!*, a gala ("Pizza, Black Tie & Black Sneaker Event and Live Auction") event at the Harbourview Room in the Vancouver Trade and Convention Centre. A huge windstorm strikes and cuts off all power on Vancouver's North Shore. Attendance is poor and the event looses money. Hon. John Cashore, newly appointed Environment Minister of the NDP Harcourt government elected a few weeks earlier gives his first speech (impromptu) reaffirming the government's commitment to create new parks.

November 1991 WCWC presents a slideshow and brief entitled *A Proposal to Work More Closely Together to Find Ways to Protect and Preserve the Ancient Rainforest on the 'Wild Side' of Vancouver Island* to the Nuu-chah-nulth Tribal Council at their Annual Assembly held November 21-23, 1991. Nothing came of this proposal at that time.

Fall/Winter 1991 WCWC publishes a 16-page three-spot colour tabloid-sized newspaper titled *WILDERNESS REPORT – New Hope for Wilderness – New British Columbia Government* (Educational Report Vol. 10 No. 10). Press Run 50,000.

December 1991 Gray Jones founds WCWC's Alberta Branch going door to door in Edmonton canvassing for memberships and support for a campaign against the sell-out of Alberta's boreal forest to multinational forest companies.

December 7, 1991 WCWC holds its AGM and elects, under new bylaws adopted at the previous AGM, a system of alternating two-year-term-directorships with all board members being non-WCWC employees. Allan McDonnell, a Vancouver lawyer, becomes chairman of the board.

December 1991 WCWC publishes a 28-page brief titled *The West Coast Trail Rainforest – A Proposal for Completion of the West Coast Trail Unit of Pacific Rim National Park Reserve*. Researched and written by Randy Stoltmann, this paper recommends that 32,300 hectares of primary rainforest be added to the 22,896 hectares already protected along Pacific Rim National Park Reserve's West Coast Trail. This worthy proposal falls on deaf ears.

Winter/Spring 1991-1992 WCWC co-publishes with the Friends of Caren, a local B.C. conservation group on the Sunshine Coast, a four-page two-colour tabloid-sized newspaper titled: *Canada's oldest tree on Sunshine Coast...Chainsaw found it first* (Educational Report Vol. 10 No. 11). It features a photo of the stump of the then oldest known tree, a 1,717-year-old yellow cedar, on the cover and an impassioned plea to save the old growth of the Caren Range upland forest with similar and possibly older living trees nearby.

1992 The B.C. Ministry of Forests and Lands issues a special use permit for WCWC's Carmanah Canopy Research Station charging $200 per year. Subsequently, the Nanaimo/Cowichan office of the B.C. Assessment Authority assesses the value of WCWC's station at $32,000 and WCWC begins paying annual property taxes.

1992 Taiga Rescue Network (TRN) is established to give a voice to those wanting to see only sensitive development in the boreal region. The Boreal Forest Network (BFN) is established as the North American affiliate of the Taiga Rescue Network. More than 150 organizations belong to this network, which is the only international network of non-governmental organizations, indigenous peoples and individuals working to defend the world's boreal forests.

January 21, 1992 The B.C. government announces the formation of the Commission on Resources and Environment (CORE), to *"help resolve valley-by-valley conflicts throughout B.C."* It is given an 18-month mandate to solicit public input and make recommendations regarding park protection for all of Vancouver Island except Clayoquot Sound. Premier Harcourt adopts a plan to *"log around contentious areas"* of old-growth forest that are possible park candidates.

January 9-10, 1992 WCWC presents a report to the Chilko Lake Study Team titled *The Nemiah Aboriginal Wilderness Preserve*, recommending that this area on the Chilcotin Plateau be protected.

Winter 1992 WCWC-WILD publishes a four-page, full-colour, full-sized newspaper titled *British Columbia's Temperate Rainforest - A GLOBAL HERITAGE IN PERIL* (Educational Report Vol. 11 No. 1). It features a map and tables showing the 118 primary watersheds along the B.C. coast that are still pristine or only slightly modified by industrial activity. This information was complied by B.C. RPF Keith Moore for Conservation International and Ecotrust Canada.

February 10-24, 1992 WCWC-WILD Director Adriane Carr attends the International Union for the Conservation of Nature's (IUCN's) 4th World Parks Congress held in Caracas, Venezuela, handing out WCWC's Tatshenshini, Kitlope and WILD campaign educational newspapers to all 1,500 participants. The congress adopted a resolution stating that: *"The world community must adopt...sustainable use of the environment and the safeguarding of global life-support systems."*

1992 Silva Foundation Forester Herb Hammond's long awaited book titled *Seeing the Forest Among the Trees: The Case for Wholistic Forest Use* is published. It explains the principles and practice of eco-forestry.

1992 WCWC publishes the results of a research project conducted by Clinton Webb and funded by the Bullitt Foundation titled *The Status of Vancouver Island's Threatened Old Growth Forests*. This report summarizes the rapid liquidation of the remaining old growth forest pointing out that of 316 remaining unprotected wilderness sub-areas identified 215 were potentially threatened or definitely scheduled for logging between 1992-96. It predicted a loss of over 6,000 logging and milling jobs because the current allowable cut on Vancouver Island was 41 percent higher than the projected Long Run Sustainable yields (LRSYs).

Spring 1992 WCWC-Alberta publishes an eight-page three-colour tabloid-sized newspaper titled *YOU MUST HELP US STOP THE CLEARCUT DESTRUCTION OF WOOD BUFFALO NATIONAL PARK* (Educational Report Vol. 11 No. 3). Press run 60,000. Noting that already 60 percent of the Park's old growth has been logged and that only 400 hectares of prime white spruce, one of the rarest most distinctive and awe inspiring forest ecosystems in Alberta, remains in the park.

April 1992 Colleen McCrory of the Valhalla Society is one of six persons awarded the prestigious U.S. $60,000 Goldman Environmental Prize. It is given in recognition of her courageous wilderness preservation work in B.C. and Canada. In her acceptance speech, Colleen strongly condemns, yet again, the current bad forest practices in Canada, claiming Canada is the *"Brazil of the North."*

April 27, 1992 WCWC's Board of Directors establish an Executive Team (E-team) comprising Joe Foy, Adriane Carr, Paul George and WCWC's comptroller, giving it the mandate to run WCWC's day-to-day business and "downsize" the organization to bring its soaring debt under control.

April 30, 1992 The Nisga'a and the Province of B.C. sign a Memorandum of Understanding (MOU) setting up a joint Nisga'a/B.C. Parks Committee to manage the Nisga'a Memorial Lava Bed Provincial Park.

May 1992 WCWC publishes an *Environmental Activity Guide*, a 128-page soft cover book written and illustrated by artist Della Burford as a companion to her *Magical Earth Secrets* book published by WCWC two years earlier.

Summer 1992 WCWC publishes a two-page two-colour tabloid-sized newspaper titled *SAVE LASCA...an area of magnificent ancient–old growth–forest in the heart of the Southern Columbia Mountains of B.C.* (Educational Report Vol. 11 No. 5). Press run 50,000. Produced with the help of a grant from the WWF, it calls for increased support for the preservation of a 50,000-hectare wilderness area near Nelson, B.C. and chronicles the extensive efforts made so far to save it.

Summer 1992 WCWC-Victoria publishes a four-page three-colour tabloid–sized newspaper titled *Logging YOUR watershed lands! – YOU CAN STOP IT* (Educational report Vol. 11 No. 6). Press Run 75,000. It educates Victorians about the harm caused by logging in the Greater Victoria drinking watersheds and calls for an end to logging there.

1992 WCWC publishes *Wild and Beautiful Clayoquot Sound –TWO DIFFERENT FUTURES – WHICH ONE WILL YOU HELP CHOOSE?*, a four-page full-colour tabloid-sized newspaper (Educational Report Vol. 11 No. 9). Print run 100,000 copies. It advocates that Clayoquot Sound become the *"core protected area of a world-famous UN Biosphere Reserve."*

June 2, 1992 The 44,300-hectare Khutzeymateen "Valley of the Grizzlies," located 45 kilometres northeast of Prince Rupert, is protected by the government of B.C. as Canada's first grizzly bear sanctuary.

June 3, 1992 WCWC-WILD campaign premiers its large format Endangered Peoples Endangered Spaces 1993 wall calendar to celebrate the upcoming UN's International Year for the World's Indigenous People United Nations. Published in all the official languages used at the UN, WILD campaign staff members take copies of this calendar with them to the big UN Conference on Environment and Development (UNCED) in Brazil to distribute there.

June 3-14, 1992 Five WCWC staff members attend UNCED held in Rio de Janeiro, Brazil. This Earth Summit, bringing together the largest gathering of heads of state in the history of the world, captures the world's attention. WCWC has a booth there and presents the WILD mapping project at the Global Forum and the Earth Summit, the NGO parallel conferences to UNCED. WCWC reports that, in light of the failure of the forest convention, the watering down of the climate change convention by removing specific targets, U.S. President Bush's refusal to sign the biodiversity convention and the restricted discussions on population and military issues, *"the governments of the world are still hesitant to move. We can, and must, take the initiative."*

Summer 1992 WCWC volunteers clear the 12-kilometre-long Boise Valley section of the "Fool's Gold Trail" that transverses the wilderness between Coquitlam and Squamish.

1992 During a five month period in 1992, 65 citizens are arrested for blockading logging in the Clayoquot Arm region of Clayoquot Sound. Six are jailed.

Summer 1992 Burke Mountain Naturalists publish a four-page two-colour tabloid-sized newspaper titled *Lower Mainland's Backyard Wilderness - LET'S PROTECT IT NOW!* It presents the case for preserving Widgeon Valley and Burke Mountain.

July 1992 Meares Island court case recesses to allow for a possible negotiated settlement between the Nuu-chah-nulth and the B.C. governments.

August 7, 1992 The 1,227,400-hectare Aulavik National Park is established on Banks Island (nearly one-fifth of this most westerly island in the Canadian Arctic Archipelago) in the Northwest Territories to protect the Western Arctic Lowlands Natural Region. The Thomsen River, the park's principal feature, is reputed to be the most northerly navigable river in the world.

Fall 1992 WCWC Alberta publishes an eight-page full-colour tabloid-sized newspaper titled *Save our Boreal Forests – The Mystery & The Heritage* (Educational Report Vol. 11 No. 7). Press Run 60,000. It features an excellent centre-spread article by Dr. Jim Butler about how little is known and how much more research is needed on the boreal forest, the world's largest ecosystem.

Summer/Fall 1992 WCWC publishes a two-page two-colour tabloid-sized newspaper titled *FOOLS GOLD REGION – BRITISH COLUMBIA – Share in a Wilderness Adventure of a Lifetime – Help Rebuild the Fools Gold Heritage Trail* (Educational Report Vol. 11 No. 8). Print run 60,000 copies. This paper is part of WCWC's continuing campaign to save the Boise Valley.

Summer/Fall 1992 WCWC publishes a four-page full-colour tabloid-sized newspaper titled, *Wild + Beautiful Clayoquot Sound* (Educational Report Vol. 11 No. 9). Press run 100,000. In the centrefold is featured a 1991 Landsat satellite image showing the large clearcuts that have already occurred in Clayoquot Sound under the headline, "Make Clayoquot Sound a UN Biosphere Reserve Core Area."

October 28, 1992 The Clayoquot Sound Steering Committee is disbanded, having failed to reach consensus after three years of study and debate. The entire issue of logging in Clayoquot Sound is sent back to the B.C. government, along with volumes of factual data amassed in the process.

November 1992 WCWC publishes a 16 page full-colour tabloid-sized newspaper titled *WILDERNESS Committee Report* (Educational Report Vol. 11 No. 10). Press Run 50,000. It features information about all WCWC Branches' and Chapters' campaigns. In the centrefold is a detailed drawing of WCWC's dream to build a B.C. Ancient Forest Museum celebrating old-growth forests in Gastown. This dream is never fulfilled. Another dream outlined in the paper to conduct an "ancient log tour" eventually morphs into an ancient stump tour and comes true.

November 1992 New Star Publishing published *NEMIAH, The Unconquered Country* by Terry Glavin for WCWC. This book is the culmination of a two-year writing project sponsored by WCWC. It relates through stories by Nemiah natives the power of their landscape and why they are fighting so hard to protect their valley from being invaded by logging companies.

November 1992 Conservation director of the Canadian Parks and Wilderness Society, George Smith, and president of the Chetwynd Environmental Society, Wayne Sawchuk, jointly initiate a campaign to protect 4.4 million hectares of wilderness in the northern Rocky and Cassiar Mountains in northern B.C.

November 20-12, 1992 The pro-logging Share BC holds its 3rd annual conference. A speaker attacks preservationists and calls them, *"a bunch of fruits, nuts and flakes,"* and calls for an increase in the AAC. Share frames the "war in the woods" as a clash of urban environmentalists versus rural realists. Many speakers assure the audience that the forest industry is changing and doing better.

December 4, 1992 Canada becomes the first industrial nation to sign the International Biodiversity Convention, pledging to conserve biodiversity in Canada. It takes more than ten years for Canada to enact a Species at Risk Act that is extremely weak and ineffectual and that almost all environmental groups condemn.

1993 BC WILD (the Conservation Alliance of British Columbia), an environmental group funded mostly by U.S. private foundations, is established *"to provide research, mapping, skills, development opportunities and other resources to individuals and groups in the wilderness conservation sector."* This organization's primary objective is to ensure that the B.C. provincial government meets its commitment to protect at least 12 percent of each of the province's representative ecosystems by the year 2000.

1993 The Forest Stewardship Council (FSC), an international non-profit organization, is formed at a meeting in Toronto. Headquartered in Oaxaca, Mexico it supports environmentally appropriate, socially beneficial, and economically viable management of the world's forests through a certification process.

1993 Several Ahousaht women form "Walk the Wild Side," a First Nations Eco-tourism Initiative, to help secure jobs and steady income for their community. In the spring of 1993 they clear the traditional trail from the Village of Ahousaht to the first wild side beach on Flores Island.

1993 Orca Publishing publishes Randy Stoltmann book *Written by the Wind*. It recounts Randy's exploratory expeditions into five of his favourite wilderness areas in southwestern B.C.

1993 The B.C. government changes the terms of reference of its Spotted Owl Recovery Team from preparing a scientifically based recovery plan for this endangered species, to preparing an "options report." At the same time the government reduces the funding for this effort.

1993 The Manitoba government publishes its *Woodland Caribou in Manitoba* report that estimates Manitoba's population of woodland caribou has decreased by 50 percent since 1950.

1993 The B.C. government agrees to negotiate the Meares Island case, instead of continuing fighting in the courts. Clearcut logging continues elsewhere in Clayoquot Sound, but at a reduced rate.

1993 WCWC raises funds to support Tla-o-quiaht project to maintain and improve the *Big Cedar Trail* on Meares Island by building a boardwalk to the "Hanging Garden Cedar," B.C.'s second largest red cedar tree.

January 1993 Canada's Future Forest Alliance, a coalition of environmental groups across Canada founded by Colleen McCrory of the Valhalla Society to protect the boreal forest, publishes a 36-page tabloid-sized newspaper, with colour cover and centrefold, titled *Brazil of the North*. This paper features articles about clearcutting across all of Canada and makes a strong case in photos and words that Canada's boreal forests are more overcut and poorly managed than those in Brazil.

January 1993 A coalition of environmental groups (which does not include WCWC) places a full-page advertisement in the New York Times seeking international support for preservation of Clayoquot Sound. The ad reads: *"Will Canada do nothing to save Clayoquot Sound, one of the last great temperate rainforests in the world?"*

January 1993 WCWC accepts the B.C. government's offer of a seat on the Study Team considering the Pinecone/Boise/Burke area as a park. Director Joe Foy is WCWC's representative on the Team.

January 30, 1993 The Haida Nation and the Government of Canada sign The Gwaii Haanas Agreement, which establishes an Archipelago Management Board to manage the Gwaii Haanas (South Moresby) National Park Reserve. First of its kind in precedent setting.

February 9, 1993 The B.C. government purchases additional shares in MacMillan Bloedel as a retirement fund investment, bringing the government's interest in the company up from 1.5 percent to about 3.5 percent, making the province of B.C. one of the largest stakeholders in this company that owns the largest TFL in Clayoquot Sound.

February 1993 Friends of Clayoquot Sound spokespeople Garth Lenz and Valerie Langer tour Canada and Europe with a slideshow that documents the devastation of Clayoquot Sound by clearcut logging.

February 1993 Premier Harcourt travels to Europe with members of the industry-funded B.C. Forest Alliance to counter calls for a boycott of Canadian forest products. He asserts that B.C.'s logging practices are *"World Class."*

March 2, 1993 The B.C. Cabinet begins deliberations on Clayoquot Sound. WCWC directors Joe Foy, Paul George and Adriane Carr "campout" for three days on the lawns of the B.C. Legislature seeking a meeting with Premier Harcourt before his cabinet makes its Clayoquot decision. They finally get a meeting scheduled for March 16th.

March 16, 1993 WCWC Directors Foy, Carr and George, meet with Premier Harcourt and plead for a half hour with him not to make a decision to log Clayoquot Sound. (WCWC finds out later that the government's decision to log had already been made several weeks earlier, through testimony given later at the conflict of interest public inquiry regarding the B.C. government ownership of a big block of MB shares). Harcourt meets with IWA representatives right after meeting with WCWC and then announces to the media that "now that he has talked to everyone concerned" he can make the Clayoquot decision. Nuu-chah-nulth leaders watching Harcourt make this statement on TV are outraged. Having never been consulted, they launch an ombudsman complaint.

March 18, 1993 WCWC organizes and holds a Clayoquot rally on the B.C. Legislature lawns on the opening day of the spring legislative session. Nearly 1,000 demonstrators showed up. Some at the rally turn it into a riot as 400 protestors break into the Legislature, injuring a security guard and occupy the rotunda in front of the Legislature chambers. Some pound on drums; someone pounds on a stained glass window breaking several panes. Loud chanting delays the throne speech for almost two hours. WCWC's Joe Foy addresses protesters, calms them and helps persuade them to leave. Although WCWC did not instigate or orchestrate the sit-in, WCWC apologies for the incident and ultimately pays $3,500 for the damage done to the stained glass windowpanes in the ceremonial door to the Legislature chambers.

March 29, 1993 WCWC publishes a satirical alternative cover to the B.C. Forest Alliances paper about public participation titled "PUTTING YOUR TWO CENTS IN," titled *Choices – Either the multinationals own the forests or the people of B.C. do. Your opinion is worth more than two cents.* On the back of WCWC's mock cover is a cartoon "movie" poster titled *CUT-AND-RUN FILMS PRESENTS JAWS - A labour Insensitive Fantasy*, decrying the increased use of grapple yarders in clearcut logging operations on B.C.'s coast which cause greater environmental damage than older logging and yarding methods that involved less road building.

Spring 1993 WCWC publishes and distributes 75,000 copies of an eight-page full-colour tabloid-sized newspaper produced by Tatshenshini Wild with partial funding provided by WWF Endangered Spaces program titled *TATSHENSHINI –North America's Wildest River – B.C. government to decide immediately. Your help needed now.* (Educational Report Vol. 12 No. 4). It graphically shows the threat to this wilderness area posed by the proposed huge Windy Craggy "mega" acid rock copper mine.

April 8, 1993 WCWC holds an informational picket outside MB's shareholders' meeting in Vancouver. More riot squad police are in attendance than picketers.

April 13, 1993 On the top of Radar Hill near Tofino, the NDP government of British Columbia announces its "Clayoquot Sound Land Use Decision." Harcourt claims it balances the environmental, economic, and social needs of the area. The decision, claimed to be without prejudice to aboriginal treaty negotiations, allows the logging of 74 percent of Clayoquot Sound's ancient temperate rainforest. Only the 25,000-hectare Megin watershed and a few other smaller areas are preserved. Sierra Club research maps show most of the protected areas are bog and marginal scrub forest, while most of the big treed forested areas in logging zones. Local people, not giving up their efforts to stop the logging in Clayoquot, vow to blockade MacMillan Bloedel logging operations near Kennedy Lake. People from all over the nation come to support them. More than 900 peaceful protesters eventually are arrested.

April 28, 1993 The B.C. government appoints the Honourable Mr. Justice Seaton to hold a public inquiry into whether or not the B.C. government was in a conflict of interest in buying shares in MB prior to its April 13 decision to allow the logging of over 70 percent of Clayoquot Sound's commercially valuable wood.

May 1993 WCWC-WILD publishes a 47-page report titled *Mapping Natural Ecosystems in Latin America*. It includes full colour maps of remaining wilderness in 10 countries in Central and South America. This publication is the culmination of a two-year WILD project funded by CIDA.

May 16, 1993 Three people, including a Friends of Clayoquot Sound director and founder, are arrested after arsonists' attempt to burn MB's Clayoquot Arm Bridge (main access to the majority of logging sites in Clayoquot Sound). They eventually plead guilty.

May 31, 1993 With the permission of the Tla-o-quiaht First Nation, WCWC volunteers started work on building a 20-kilometre-long trail (the Clayoquot Witness Trail) through the threatened Clayoquot Valley to draw attention to the beauty and threatened clearcut devastation of the pristine old-growth rainforest there. The river that drains this watershed (Clayoquot River) empties into the Clayoquot Arm of Kennedy Lake. Clayoquot River is home to the best remaining sockeye run in Clayoquot Sound.

Summer/Fall 1993 WCWC publishes a four-page full-colour tabloid-sized newspaper titled *SAVE CLAYOQUOT VALLEY* (Educational Report Vol. 12 No. 5). Press Run 100,000. It features WCWC beginning efforts to build a Witness Trail through the pristine 7,679-hectare Clayoquot Valley in Clayoquot Sound. The back page headline reads in big type "*Refuse to accept the B.C. government's 'clearcut Clayoquot' decision.*"

June 15, 1993 The Friends of Clayoquot Sound, Greenpeace, WCWC and the Sierra Club of Western Canada representatives meet with leaders of the Nuu-chah-nulth's five Central Region Bands. The Chiefs do not want to employ confrontational tactics, but are also opposed to the government's decision to log most of Clayoquot Sound. Environmentalists request permission to open a protest camp on native lands. Meetings regarding Clayoquot between the environmental community and native leaders continue on a regular basis in the years to follow.

Summer/Fall 1993 WCWC publishes an eight-page full-colour tabloid-sized newspaper titled *SAVE UPPER CARMANAH VALLEY – Home of the world's first temperate rainforest canopy research station* (Educational Report Vol. 12 No. 6). It provides a detailed account of the discoveries of over a hundred new insect species by ecologist Dr. Neville Winchester who is working out of WCWC platforms high up in the canopy of several tall ancient Sitkas trapping arthropods there in the then still unprotected Upper Carmanah Valley.

June 23, 1993 Premier Harcourt announces the establishment of the 958,000-hectare Tatshenshini-Alsek as a Class A Provincial Park. Situated in the very northwest corner of British Columbia, it's nestled between Kluane National Park and Reserves in the Yukon and Glacier Bay and Wrangell-St. Elias National Parks and Preserves in Alaska. Combined, these parks create the largest protected area in the world, approximately 8.5 million hectares in size.

June 25, 1993 Kim A. Campbell, Progressive Conservative, becomes the Prime Minister of Canada.

June 26, 1993 Sir David Attenborough, star of countless BBC nature programs, along with a video crew from the Ted Turner satellite network, trek into the Upper Carmanah Valley and visit WCWC's research station there. They are hoisted into the canopy research station to do a story on the research findings' contribution to our knowledge about biodiversity—findings there have already doubled the estimate of insect species believed to exist in B.C.

July 1993 The B.C. government establishes the Protected Areas Strategy for British Columbia committing itself to developing and expanding a protected areas system that would protect 12% of the province by the year 2000.

July 1, 1993 With the permission of the Tla-o-quiaht First Nation, WCWC sets up an information kiosk at Sutton Pass, the entryway providing the only road access to Clayoquot Sound. Two huge banners state: *Welcome to Clayoquot Sound – Nuu-chah-nulth First Nations Territory* and *Save Clayoquot Sound – Protect B.C.'s Ancient Rainforest*. During the summer over 5,000 people stop at WCWC's kiosk to get information, buy posters and sign petitions.

July 1, 1993 The Friends of Clayoquot Sound and Greenpeace establish a "Peace Camp" in the "Black Hole," a huge clearcut beside Highway 4 just before the turn-off to Tofino and Ucluelet, as a base from which to conduct peaceful civil disobedient protests of the ongoing clearcut logging in Clayoquot Sound. Daily blockades and arrests soon follow at the nearby Kennedy River Bridge, the major logging road access route into the heart of Clayoquot Sound.

July 5, 1993 Clayoquot blockades begin with 15 protestors refusing to leave the Kennedy Lake logging road, including Svend Robinson, a Canadian MP.

On July 15, 1993 At about 8:00 a.m. the Australian rock band Midnight Oil, in support of the blockades that are taking place every working day morning at 6:00 a.m. at the nearby Kennedy River bridge, puts on a free concert at "The Black Hole," a big clearcut along the Port Alberni-Tofino highway on Vancouver Island near the Tofino-Ucluelet road junction where the Clayoquot logging protesters were camped. Over 1,500 attend.

July 15 1993 A MarkTrend poll of 500 British Columbians finds that 52 percent support the B.C. government's Clayoquot Compromise, while 39 percent oppose it. A total of 57 percent oppose the blockades, with 35 percent in favour and eight percent undecided.

July 19, 1993 The *San Francisco Chronicle* publishes a front-page article titled *Canada's Endangered Rainforests – Ancient Trees Disappearing in 'Brazil of North'.*

July 22, 1993 MB issues a special use permit to WCWC for it to use "waste" cedar wood from MB's clearcut in the Upper Kennedy watershed to "*further the Committee's objectives*" and build a trail through the Clayoquot River Valley, which WCWC is using as part of its public relations campaign to prevent logging in the area.

July 28, 1993 Commissioner Hon. Mr. Justice Seaton tables his report on the government conflict of interest in buying MB shares before making its pro-logging Clayoquot decision. It vindicates the government.

July 31, 1993 Chief Francis Frank of the Tla-o-quiaht First Nation brings attorney Robert Kennedy of the National Resources Defense Council to see the Clayoquot Witness Trail. Along with a host of Nuu-chah-nulth chiefs and elders and some media, Robert Kennedy hikes down the boardwalk trail in the pouring rain and stops by the tarp roofed kitchen camp for a good meal of hot soup and bread.

August 2, 1993 3,000 people march and rally in Vancouver to "Ban Clearcuts in Clayoquot." First Nations spokespersons are key participants in the event's opening and closing ceremonies.

August 1993 About 200 litres of human excrement is dumped at night where WCWC puts its information kiosk at Sutton Pass during the day. This is followed by fetid water with rotted fish guts being dumped time after time at night in the same spot.

August 17, 1993 A logging truck driver is arrested for stealing a WCWC banner at Sutton Pass.

September 1993 WCWC extracts a huge 4,000-kilogram red cedar stump, called "Stumpy," remnants of a 400-year-old tree left behind in a clearcut by MB loggers in the Upper Kennedy River near WCWC's Witness Trail in Clayoquot River Valley. The extraction is legally done under the terms of a salvage permit that MB issued earlier in the summer to WCWC. Stumpy is sent on a tour across Canada, Europe and through the U.S.

September 1993 Al-Pac—the largest single-line bleach Kraft pulp mill in the world—begins pulping Alberta's boreal forest in a vast region around Athabasca, Alberta.

September 7, 1993 The RCMP makes a mass arrest of 242 people who are attempting to stop logging in Clayoquot Sound by blockading a logging road at the Kennedy Bridge. Many are charged with criminal contempt of court for disobeying a court ordered injunction.

September 26, 1993 WCWC holds a rally in front of the Victoria Legislature buildings, launching the first leg of Stumpy Canada road tour to gain support for protecting all of Clayoquot Sound. The goal is to reach Ottawa and hold a rally in front of the House of Commons right before the upcoming federal Election Day.

October 22, 1993, The B.C. government establishes the Scientific Panel for Sustainable Forest Practices in Clayoquot Sound (Clayoquot Sound Science Panel). The government appoints a special panel of scientists and First Nations representatives with the mandate to review the existing forest management standards and to recommend changes to these standards appropriate to the ecological conditions of Clayoquot Sound. The Panel's mandate is to develop world-class standards for sustainable forest management in Clayoquot Sound by combining traditional and scientific knowledge. It is not allowed to look at preserving more forest wilderness.

October 21, 1993 Parksville NDP MLA Leonard Krog writes a letter to WCWC criticizing the Committee's attacks on the Clayoquot "compromise" decision. The letter warns that if WCWC keeps *"pissing in the government tent... somebody is going to close the flap and you can howl in the wilderness."* Premier Harcourt says in response to complaints about Krog's letter that he understands his colleague's frustration.

October 1993 "Stumpy" reaches Ottawa in time to hold a rally right before the federal election is held. Candidate Jean Chrétien, Leader of the Liberal Party, promises to help protect Clayoquot Sound if his party forms the government.

October 25, 1993 Jean Joseph Jacques Chrétien, Liberal, is elected to be the Prime Minister of Canada.

November 2 1993 The results of the latest Angus Reid poll show that a solid majority—59 percent of British Columbians—supports the B.C. government's decision to allow logging in Clayoquot Sound. The poll also suggests that the environmental movement may be discrediting itself in the public's eye because of its protest tactics. Many environmentalists attribute the results to public ignorance of the issues involved.

November 1993 The B.C. Ombudsman releases *Public Report No. 31 - Administrative Fairness of the Process Leading to the Clayoquot Sound Land Use Decision.* It concludes that the consultation with the First Nations in this case was less than satisfactory. It recommends, *"...that the provincial government continue to consult the Nuu-chah-nulth First Nations to ensure their present and future interest in the land and resources of Clayoquot Sound is meaningfully considered for incorporation into the Clayoquot Sound Land Use Decision."*

November 1993, The Friends of Clayoquot Sound close down their "Black Hole Peace Camp" and end the blockades of MB logging operations at the Kennedy River Bridge. During the four months the protest was ongoing, over 12,000 people came to the Friends' camp including two trainloads of activists from across Canada. Over 900 peaceful blockaders were arrested—the largest action of civil disobedience in Canadian history.

November 9, 1993 Greenpeace International declares Clayoquot Sound an international priority. Greenpeace activists are arrested for acts of peaceful civil disobedience at Canadian consulates in Germany and Austria.

November 1993 The B.C. government releases its much-anticipated new Forest Practices Code, a rules based system with many loopholes which incorporates many of the regulations already in effect in Clayoquot Sound. The code proposes tougher regulatory enforcement, independent audits, clearcutting restrictions and fines topping $1,000,000 and/or the loss of cutting privileges for serious contraventions. It also makes trailbuilding without prior B.C. Forest Service permission in provincial forests including Tree Farm Licence areas illegal.

November 15, 1993 Canadian environmentalists, including Colleen McCrory of the Valhalla Wilderness Society and Mark Wareing RPF of WCWC, meet with international journalists in Japan about how they can pressure to get better logging practices in B.C.

December 1993 Greenpeace Germany convinces four of the largest publishers in Germany to no longer purchase pulp or paper from B.C. suppliers because of B.C.'s bad forestry practices.

December 10, 1993 The provincial government and the Nuu-chah-nulth Tribal Council announce that they have come to an agreement on an *Interim Measures Agreement* (IMA), which sets out a process for joint decision-making during treaty negotiations. This agreement is heavily criticized by a few First Nations people and environmentalists as a buy-out of native interests by the government and MB, but the Nuu-chah-nulth leadership and most environmentalists defend the agreement, which gives the Nuu-chah-nulth virtual veto power over developments in Clayoquot Sound, as being "a historic achievement for First Nations."

Winter 1993-94 WCWC publishes an eight-page full-colour tabloid-sized newspaper titled *A Conservation Vision for Vancouver Island – More jobs and more wilderness – It's a realizable dream* (Educational Report Vol. 12 No. 7.) Print run 300,000 copies. The centrefold has a map showing proposed community forests and new protected areas. Copies of this paper are inserted into all the freely distributed community newspapers on Vancouver Island.

Winter 1993-94 Mass trials of protesters who were arrested for contempt of court (blockading the Kennedy Lake logging road in Clayoquot Sound) are held in Victoria B.C. Sentences are in the range of 45-60 days in jail, with fines from $1,500 to $3,000.

Winter 1993-94 WCWC publishes an eight-page two-colour tabloid-sized newspaper titled *HOW TO SAVE JOBS IN THE B.C. WOODS* (Educational Report Vol. 12 No. 8). Press run 50,000. Written by Ken Drushka (before he became an apologist for the big forestry corporations) it reports on research done by professional forester Ray Traverse for WCWC that calculates how many more jobs and how much more wood could be produced by selectively harvesting maturing second growth forests on Vancouver Island rather than clearcutting them.

1994 WCWC-Victoria launches a campaign to protect the Sooke Hills and the Sea-to-Sea Green Blue Belt, a network of interconnected terrestrial and marine parks stretching from the Sooke Basin to the Saanich Inlet, north to Salt Spring Island.

1994 The Haida Nation establishes the Haida Forest Watch program that extends Haida stewardship responsibilities to the forests of Haida Gwaii. The program initiates strategies to reform forest tenure and management systems, to identify and protect areas of high cultural and ecological significance, and to work in partnership with other communities of Haida Gwaii to develop a self-sustaining islands-based economy.

January 1994 An Angus Reid poll finds that 67 percent of Canadians oppose the practice of clearcut logging and only 14 percent of Canadians feel that the B.C. government's decision to allow clearcut logging on approximately two-thirds of the land area in Clayoquot Sound was a good one.

January 13, 1994 The B.C. government establishes the 233,000-hectare Ts'yl-os Provincial Park (nearly half the size of Prince Edward Island) around Chilko Lake, comprising rugged mountains, clear blue lakes, glaciers, alpine meadows, and waterfalls. It's bordered by the rugged peaks of the Coast Mountains to the west and the dry Interior Chilcotin Plateau to the east. The Tsilhqot'in People of Xeni (Nemiah Valley Indian Band) and the Province of B.C. sign a Memorandum of Understanding establishing the Ts'yl-os Management Board to manage it. The *Vancouver Sun*, B.C.'s largest daily newspaper, does not run even a small story about it.

February 1994 Greenpeace leases WCWC's Stumpy for a four-month tour of Europe to win support for saving Clayoquot Sound's ancient rainforest.

February 2, 1994 Sierra Club Books/Earth Island Press publishes *Clearcut – The Tragedy of Industrial Forestry,* a large folio-sized book produced by the Foundation for Deep Ecology. It features shocking images of clearcut forests across North America, along with articles explaining how forestry could be done in a more ecologically sustainable way. WCWC helps widely distribute complimentary copies to the media and political leaders in B.C. and the rest of Canada.

February 9, 1994 Stephen Owen's Commission on Resource and the Environment (CORE) tables its Vancouver Island final report. This report, according to one prominent provincial newspaper reporter, brings *"howls of discontent from both environmentalists and loggers."* Owen recommends that 13 percent of Vancouver Island be in protected areas, which would include, in total, 7.8 percent of the island's forest. Owen's compromise brings neither peace in the woods nor ecological integrity.

Winter/Spring 1994 WCWC publishes 100,000 copies of an eight-page full-colour tabloid-sized newspaper titled *Save Vancouver's Wilderness Backyard* (Educational Report Vol. 13 No. 1). The paper urges people to attend one of the B.C. government's upcoming public input meetings regarding the establishment of a Pinecone/Boise/Burke provincial park near Vancouver.

March 1994 Greenpeace International, in its efforts to get market pressure to bear on protecting Clayoquot Sound, successfully pressures Scott Paper (UK Division) to cancel a pulp contract with MacMillan Bloedel Ltd. worth $5.4 million.

March 19, 1994 The *Clayoquot Sound Interim Measures Agreement* (IMA) between Her Majesty the Queen In Right of the Province of British Columbia and Nuu-chah-nulth Central Region First Nations; the Hawiih of the Tla-o-quiaht First Nation, the Ahousaht First Nation, the Hesquiat First Nation, the Toquaht First Nation and the Ucluelet First Nation is formally signed. This IMA establishes the Clayoquot Sound Central Region Board (CRB). It gives First Nations the right to review all proposed resource developments in Clayoquot and power to stop those they believe harm the environment or their native cultural heritage. In the following years this IMA is modified and extended several times.

March 21, 1994 27,000 people attend a massive rally on the lawns of the B.C. Legislature in Victoria sponsored by the IWA to protest Owen's CORE recommendations to create new parks to preserve a total of 13 percent of Vancouver Island, which were released earlier in the year. A handful of WCWC staff and volunteers, who support much more protection than CORE recommends, stand in the crowd holding up pro-park signs including "Save Jobs–Stop Raw Log Exports" in the sea of forest industry supporters who are chanting "12 percent and no more."

March 1994 Sierra Legal Defence Fund lawyers, who are acting for the Western Canada Wilderness Committee, the Sierra Club and six local citizens who are concerned that tree cutting and road building in the watershed which supplies area's residents with drinking water seriously jeopardize water quality by increasing sedimentation, contamination and the need for chemical disinfection, obtain a B.C. Supreme Court order stopping all commercial logging in the Greater Victoria watershed. This marked the beginning of an intensive WCWC campaign to protect Victoria's Water District's off-catchment lands in a park.

Spring 1994 WCWC publishes an eight-page full-colour tabloid-sized newspaper titled *VANCOUVER ISLAND PARADISE – Lost or Saved* (Educational Report Vol. 13 No. 2). Print run 150,000. It is WCWC's response to the February CORE Commissions report that recommends protecting only 13 percent of Vancouver Island. This newspaper features in the centrefold WCWC's Vancouver Island Vision Map with CORE's proposed protected areas and lands CORE recommends be classified as "regionally significant" superimposed upon it.

April 1994 Sponsored by Mountain Equipment Co-op, Randy Stoltmann, noted B.C. writer, conservationist, and mountaineer, presents a 32-page written proposal titled *The Clendennning/Elaho/Upper Lillooet Wilderness "Stanley Smith Wilderness"* to the B.C. government's Protected Area Strategy team. It advocates the preservation of a 260,000-hectare wilderness area that encompasses the Upper Elaho, Sims, Clendennning Valleys (all within the Squamish Watershed and TFL #38) and Upper Lillooet Valley. WCWC reprints this brief shortly after Randy's tragic fatal accident and widely distributes it. Subsequently, WCWC renames the proposed Stanley Smith Wilderness the "Randy Stoltmann Wilderness" and vigorously campaigns to protect this area.

April 12, 1994 WCWC's Adriane Carr presents to Deputy Prime Minister Sheila Copps WCWC's official Save Clayoquot Sound petition containing over 120,000 signatures. The petitioners ask the federal government to initiate negotiations with the First Nations of Clayoquot Sound and the provincial government with the view to ending the big forest company logging rights and protecting Clayoquot's irreplaceable wild forests. Ms. Copps insinuates that this petition is not a strong enough show of support to get her government to act.

April 13, 1994 MB holds its AGM on the first anniversary of the Clayoquot logging decision. Tzeporah Berman, representing Greenpeace prepares an 'Erratum' to MB's Annual Report that lists all logging violations and toxic spills. The erratum is distributed to their shareholders at the meeting. Several shareholders give their proxy votes to Berman and representatives from Friends of Clayoquot Sound so they can attend the AGM. The meeting is disrupted by continuous questions about MB's operations in Clayoquot Sound and their environmental record. Earth First! activists arrive at the meeting and unfurl a banner that reads "No More BS MB." Other Earth Firsters disrupt the meeting by shoveling manure onto the sidewalk in front of the hotel. The Forest Action Network (FAN) hangs a banner from the top of the twenty-story building that reads, "As MB Lies Clayoquot Dies." Seven people are arrested. A few WCWC volunteers and staff meekly hold up a giant banner that says "Save Clayoquot Sound – Nuu-chah-nulth First Nations' Territory" on sidewalk alongside the hotel near the pile of manure.

April 15, 1994 WCWC forwards a written brief titled *WCWC Proposed Protected Areas for the Vancouver Region of B.C.* in response to the government's Protected Areas Strategy's Lower Mainland Regional Protected Areas Team's request for proposals. WCWC's submission details 13 areas needing park protection.

Spring 1994 WCWC co-publishes with the World Wilderness Committee a four-page full-colour tabloid-sized newspaper titled *Cascade International Park – Help complete the dream of a fully protected Ecosystem* (Educational Report Vol. 13 No. 4). Press run 80,000. It calls for five additional protected areas added to the already protected areas in the Cascade—four in Canada and one in the U.S. along the US-Canada boarder. U.S. partners do not like this paper because it shows firm lines demarcating the proposed protected areas, which was not part of their negotiating strategy.

May 10, 1994 The Scientific Panel for Sustainable Forest Practices in Clayoquot Sound releases its second progress report entitled *Review of Current Forest Practice Standards in Clayoquot Sound*. It calls for an end to logging and road building in undeveloped watersheds until there are adequate inventories and recommends that alternatives to clear-cutting be introduced and used elsewhere.

May 21, 1994 Randy Stoltmann, a young and accomplished wilderness activist and conservationist and former WCWC director and employee, loses his life in a tragic mountaineering accident on a ski expedition in the Kitlope area in the central Coast Mountains region of B.C.

1994 The 36,100-hectares Churn Creek Provincial Park, located approximately 60 kilometres southeast of Williams Lake at the confluence of the Chilcotin and Fraser rivers in the south central region of B.C., is protected as part of the Cariboo-Chilcotin Land Use Plan. This protected area grows to its present size with the government's acquisition of the Empire Valley Ranch in 1998. The Churn Creek Provincial Park protects one of the few remaining natural grassland habitats in B.C.

1994 The Clayoquot Rainforest Coalition (CRC) is established. CRC uses market forces in its campaigns focused on saving Clayoquot Sound, which has the largest remaining tract of intact temperate rainforest on Vancouver Island. This organization changes its name to ForestEthics and broadens its mission in 2000 to encompass protecting all endangered forests by redirecting markets toward ecologically sound alternatives.

June 17, 1994 WCWC reopens it kiosk at Sutton Pass, the only road entry point into Clayoquot Sound, for the summer to continue to build support for protecting the remaining primary forest there and serve as a base for WCWC continuing work on its Clayoquot Witness Trail through the Clayoquot River Valley.

June 22, 1994 The B.C. government establishes 24 new park areas on and around Vancouver Island (some of them additions to existing protected areas) that were recommended by the Vancouver Island CORE (Commission on Resources and the Economy). They include, among others, the 3,343-hectare Upper Carmanah; the 9,500-hectare Lower Walbran (additions to Carmanah Walbran Provincial Park); 10,829-hectare Tashish-Kwois; the 22,851-hectare Brooks-Nasparti-Power River (addition to Brooks Peninsula Provincial Park); the 5,514-hectare Lanz-Cox Islands; the 3,744-hectare Lower Tsitika (addition to Robson Bight Ecological Reserve); the 6,750-hectare Nahwitti-Shushartie (addition to Cape Scott Provincial Park).

Summer 1994 WCWC publishes a four-page full-colour tabloid-sized newspaper titled *CLAYOQUOT a heritage worth protecting* (Educational Report Vol. 13 No. 5). Press run 150,000. It provides an up-date of WCWC's continuing campaign efforts to save Clayoquot Sound and invites people to help.

Summer 1994 WCWC-Alberta's Boreal Forest Campaign publishes an eight-page full-colour tabloid-sized newspaper titled *Al-Pac: Mitsubishi's Attack on Alberta* (Educational Report Vol. 13 No. 7). Press run 60,000. It tells the story about the largest pulp mill in the world built with the greatest taxpayers' dollar subsidies ever given to a private company, which totaled well over seven billion dollars. It contains an incredible chronology of events and pleads for the Alberta government to give the control over Alberta's boreal forest to local communities.

Summer 1994 WCWC publishes a 16-page tabloid-sized newspaper with a full-colour cover and centrefold titled *WILDERNESS COMMITTEE REPORT 1994* (Educational Report Vol. 13 No. 6). Press run 50,000. It gives a report on all of WCWC's campaign activities including the cross Canada Stumpy tour.

July 1994 The B.C. Legislature passes Bill 40—the Forest Practices Code of British Columbia Act. It is touted as a tool that will improve logging practices in B.C.

July 1994 WCWC membership stands at 15,121 members.

July 1994 The Sea-to-Sea Greenbelt Society is formed in Victoria B.C. to preserve the Greater Victoria Water District surplus lands as parkland. It holds its first public meeting on August 4, 1994.

August 16, 1994 The B.C. government establishes the 317,000-hectare Kitlope Protected Area on the mid-coast of B.C. While it is true that it becomes the world's largest protected temperate rainforest watershed, only about three percent of the steep, mountainous Kitlope watershed has commercially valuable timber growing on it.

August 17, 1994 The Gitsi'is Tribe and the Province of B.C. sign a Memorandum of Understanding regarding the management of the K'tzim-a-deen Grizzly Sanctuary (also known as the Khutzeymateen Provincial Park).

August 23-27, 1994 WCWC-Alberta hosts the "People of the Snow Forest," a Taiga Rescue Network international conference bringing together boreal forest people from around the world in efforts to find ways to protect the boreal forest. Over 200 attended from more than 15 different countries.

August 1994 WCWC begins negotiations with the Ahousaht Band Council over opening a WCWC storefront office, on reserve, in the village of

Ahousaht on Flores Island in Clayoquot Sound to support native women in Ahousaht who had started an eco-tourism project in 1993 called "Walk the Wild Side." Walk the Wild Side offers hiking tours to the outer beaches of Flores with interpretive talks about traditional uses of the forests and the native history of the area.

September 1994 MB publishes correspondence between MB, the Nuu-chah-nulth and Greenpeace and claims in a public relations package that the Clayoquot conflict is solved. Greenpeace demands that clearcutting stop if MB is going to publicize the working group. MB refuses. Greenpeace cancels eco-forestry trips and joint meetings with MB and the First Nations. In anger, Nuu-chah-nulth Chiefs accuse Greenpeace of not respecting First Nations' rights and cancels projects with the WCWC because Greenpeace is not cooperating.

1994 Harbour Publishing Co. publishes *Forestopia – A Practical Guide to the New Forest Economy*, a soft cover 119-page book co-authored by Professor Michael M'Gonigle and Ben Parfitt. The authors make a strong case for community ownership of forests, a competitive log market

November 1994 WCWC begins a joint field research project in the Ursus Valley with the Ahousaht First Nations.

1994 Upon Greenpeace's return of WCWC's Stumpy from Europe, Alison Spriggs, Misty MacDuffy, Zane Parker, and Cheri Burda of Victoria WCWC Chapter organize a "Clayoquot Stump Tour" across Canada. They tow the trailer with the massive 400-year-old red cedar stump ("Stumpy") to shopping mall parking lots and university campuses all over eastern Canada and across the U.S. to educate the public about B.C.'s endangered ancient rainforests and the need to protect Clayoquot Sound.

November 24, 1994 The B.C. government protects part of the Cariboo Mountains Wilderness, the Mitchell and Niagara watersheds linking Wells Gray and Bowron Lakes parks. It is part of the Cariboo-Chilcotin Land Use Plan new protected areas, which total 462,000 hectares in all.

Winter 1994 WCWC-Victoria publishes a four-page full-colour tabloid-sized newspaper titled *Save the Sooke Hills...from this!* (Educational Report Vol. 13 No. 8). Press run 60,000. It launches WCWC's campaign to save a 9,300-hectare green belt from Sooke Harbour to the Saanich Inlet in greater Victoria.

December 15, 1994 UNESCO designates B.C.'s one-million-hectare Tatshenshini-Alsek Wilderness Park a World Heritage Site—the world's largest one.

1995 WCWC current membership is 22,000.

1995 The B.C. government places a moratorium on the creation of any new fish farms in B.C., but allow existing ones to expand capacity to meet demand for their product.

Winter 1995 WCWC publishes 80,000 copies of a four-page full-colour tabloid-sized newspaper titled *SAVE SURREY BEND* (Educational Report Vol. 14 No. 1). Written by well-known outdoorsman Tony Eberts, it champions the protection of 925 hectares of un-diked floodplain in lower Fraser River just east of the Port Mann Bridge. Shortly thereafter the B.C. government protects the entire area as a regional park.

January 1995 The Ahousaht First Nations and WCWC publish a 45-page report of their joint expeditions into the Ursus Valley made in September and December 1994. Titled *Preliminary Investigations of Culturally Modified Trees (CMTs) by Aboriginal Use of the Ursus Valley in Ahousaht Territory of Clayoquot Sound* it details extensive aboriginal forest use in the 6,567-hectare Ursus Watershed.

January 1995 WCWC re-names the wilderness north of Squamish the Stoltmann Wilderness Area in honour of Randy Stoltmann who originally proposed that the area be protected and had recently died in a tragic mountaineering accident. WCWC expands its campaign to get this wilderness area protected.

February 1995 MB refuses to show WCWC representatives its planned Ursus Valley logging roads and clearcut blocks at its public review of TFL 44 Management and Working Plans held in Vancouver.

February 1995 The B.C. government establishes a Land and Resource Management Plan (LRMP) process for the Okanagan-Shuswap region to "help design the kind of park system local people want." WCWC-Okanagan is actively involved during the initial phase of the process.

February 14, 1995 The federal government establishes Vuntut National Park (4,345 square km is located in northern Yukon, bordered by Ivvavik National Park to the north and the Arctic National Wildlife Refuge in Alaska to the west) in the Yukon Territory.

February 24, 1995 The Sierra Legal Defence Fund, on behalf of the Haida Nation files a Petition in the Supreme Court of B.C. seeking to prevent the Minister of Forests from replacing MB's Tree Farm Licence 39, part of which is on Haida Gwaii (the Queen Charlotte Islands) because the lands in question are encumbered by First Nations' title and rights to the lands.

March 6, 1995 The B.C. government establishes a 243-hectare Jedediah Island Provincial Park after a long campaign conducted by The Nature Conservancy to buy this piece of private property. Jedediah Island is located between Lasqueti and Texada Islands in the Sabine Channel of Georgia Strait.

March 13, 1995 The B.C. government establishes a 79,500-hectare Goat Range Provincial Park, Laska Creek Provincial Park and a 10,921-hectare Akamina-Kishinena Provincial Park in the Kootenay region of B.C.

1995 The B.C. Recall and Initiative Act (SBC 1994 c.56) comes into force providing a mechanism to recall sitting MLAs and to bring citizen initiated legislation before the Legislature or to province-wide referendum.

Spring 1995 WCWC publishes its first newspaper advocating park protection for the Upper Squamish and Lillooet Rivers titled *Randy Stoltmann Wilderness Area – Save it Now* (Educational Report Vol. 14 No. 7). Written by retired Province outdoor columnist Tony Eberts, it describes the outstanding features of a 260,000-hectare roadless area that Stoltmann had studied and reported on shortly before his death (which he called the Stanley Smith wilderness area).

March 28-April 7, 1995 The signatory parties to the 1992 Climate Convention hold their first meeting in Berlin, Germany to set goals to reduce greenhouse gasses. WCWC's Stumpy, held up in shipping, fails to make a scheduled appearance at it.

April 20, 1995 The Province of B.C. and the Greater Vancouver Regional District jointly announce the creation of the Lower Mainland Nature Legacy, a program of new parkland acquisition that will see the amount of park and conservation land in the GVRD more than triple.

1995 Ahousaht Band Council and WCWC jointly apply for a Youth Services Canada Grant to hire youth to fix up and boardwalk the tradition Wild Side Trail to the outside beaches on Flores Island.

1995 WCWC publishes and distributes 30,000 *Clayoquot Sound the Hawaii of British Columbia*. This "3-part-mailer" with a "tongue in cheek" comparison between the Hawaiian Islands and Clayoquot Sound consists of two big postcards held together by a perforated fold; one showing an aerial photo of pristine Clayoquot; the other showing the clearcut destruction of Side Bay north of Clayoquot Sound. The ugly postcard to be signed and returned to Prime Minister Jean Chrétien contains a polling question as to whether the sender thinks he should keep his promise made during the last election to save the ancient temperate rainforest in Clayoquot Sound or not.

Spring 1995 WCWC jointly publishes with the Ahousaht First Nations 130,000 copies of a four-page full-colour tabloid-sized newspaper titled *PROTECT URSUS VALLEY – AHOUSAHT TERRITORY*. It showcases the extensive evidence of aboriginal use of the old-growth forest in Ursus Valley discovered on a joint Ahousaht-WCWC expedition there a few months earlier. It urges people to support the Ahousaht First Nation in its effort to keep MB from logging this pristine, salmon rich watershed that is part of the Ahousaht's traditional territory.

Spring 1995 The provincial government sets-up a thirteen member Regional Public Advisory Committee (RPAC) to quietly find consensus on which of the Lower Mainland's few remaining wild places should remain wild and which ones should be open for development—especially logging. Invited to sit on this Committee was Interfor, and the IWA. The B.C. Wildlife Federation, BC Wild and the Canadian Parks and Wilderness Society, environmental groups known for their moderate stance are the invitees to represent preservationists. Left off this Committee was the B.C. Provincial Park Branch. Instead the Lands Branch, notorious for its pro-development, represents the B.C. Environment Ministry.

Spring 1995 WCWC publishes for the Canadian Coalition for Biodiversity with partial funding from Environment Canada, 120,000 copies of an eight-page full-colour tabloid-sized newspaper titled *Protect Canada's Biodiversity – A world of Differences* (Educational Report Vol. 14 No. 4) calling for strong and effective federal legislation to protect Canada's endangered species.

April 11, 1995 The provincial government establishes the Power River Provincial Park on Vancouver Island.

April 20, 1995 The B.C. government establishes the 6,826-hectare Indian Arm Provincial Park. Adjacent to Mount Seymour Provincial Park in North Vancouver, this park protects the eastern and western shorelines of the upper portion of the 18-kilometre-long fjord that extends northward from Vancouver's Burrard Inlet.

May 1995 Ahousaht First Nations and WCWC conduct a joint research expedition into Easter Watershed and Young Bay in the northern reaches of Clayoquot Sound and publish a 17-page report titled *Preliminary Investigations of Cultural and Recreational Features of Easter Watershed in Ahousaht Territory in Clayoquot Sound*. The report states that the area "contains a rich history of aboriginal use" and is "an outstanding candidate for nature-based recreation."

Spring 1995 WCWC publishes four-page full-colour tabloid-sized newspaper titled *STOP ALL LOGGING in GREATER VANCOUVER'S WATERSHEDS NOW...before it's too late!* (Educational Report Vol. 14 No. 6). Press run 100,000 copies. By showing the damage that logging has caused already in the drinking watersheds, it makes a strong case to stop all logging there and save the remaining ancient temperate rainforest that is "...natures unbeatable water purification system."

May 1995 The Clayoquot Scientific Panel releases its report with 128 recommendations regarding logging in Clayoquot Sound. Among them is one stating that no logging should proceed in currently unlogged old-growth watersheds until complete inventories of all forest values are completed. The provincial government responds to the report by saying it needs a month to evaluate it. Some environmentalists threaten to resume blockades unless the government implements all the recommendations.

May 1995 WCWC discovers that International Forest Products (Interfor), which recently purchased TFL 38 in the Squamish Valley, is constructing a new logging road up Sims Creek Valley in the Stoltmann Wilderness.

June 1, 1995 The B.C. government establishes the 6,164-hectare Tetrahedron Plateau Provincial Park on the Sunshine Coast.

June 8, 1995 The B.C. government announces the establishment of the 38,000-hectare Pinecone Lake-Burke Mountain Provincial Park. Situated south of Garibaldi Provincial Park and west of Pitt Lake, the largest fresh water tidal lake in North America. This new park is the largest of the Lower Mainland Nature Legacy Parks. It includes all of the Boise Valley that WCWC had been campaigning to protect. At the same press conference announcing the new park, NDP Premier Harcourt announces his government is opening up 175,000 hectares of Spotted Owl Conservation Areas (SOCA's) for logging to the "boos" of many of the environmentalists attending the media conference.

June 8, 1995 The Katzie First Nation and the B.C. government sign a Memorandum of Understanding regarding the management of C'elc'al-s (also known as Pinecone Lake - Burke Mountain Provincial Park).

June 15, 1995 The provincial government adds, as part of the Lower Mainland Nature Legacy program, 142 hectares of land adjoining the existing 16-kilometre-long dike to the Boundary Bay Regional Park in Tsawwassen, B.C. Boundary Bay, located in the Fraser River Estuary on the Pacific Flyway is one of the top ten migratory bird-staging sites in the world.

June 15, 1995 B.C.'s Forest Practices Code comes into force.

June 1995 (Summer Solstice) WCWC holds a gathering in a big clearcut on the edge of the remaining wilderness in the Upper Elaho Valley in the Stoltmann Wilderness. Over 100 attend, hear motivational talks by Joe Foy, John Clarke and others and vow to help save this wilderness area.

Summer 1995 WCWC-Alberta publishes a four-page full-colour tabloid-sized newspaper titled *Logging on private land: Not in my own back yard!* (Educational Report Vol. 14 No. 8). Press run 20,000. It decries the extremely rapid logging of private lands in Alberta with the raw logs being shipped to B.C. to be milled.

Summer 1995 WCWC publishes a four-page two-colour tabloid-sized newspaper written by the Elphinstone Forest Watch titled *MOUNT ELPHINSTONE FOREST – Eco-Forestry Proposal* (Educational Report Vol. 14 No. 9). Print run 20,000 copies. It advocates a community forest licence for the 8,000-hectare area where sustained yield eco-forestry be practiced and the establishment of a core 1,500-hectare Mt. Elphinstone Park protected area.

July 1995 The B.C. government accepts the 128 recommendations of the Clayoquot Science Panel including a moratorium on logging in pristine watersheds so inventories of all values can be conducted. The Panel details new logging rules, including an ecosystem-based "variable retention" logging, that environmental groups hope, if they are adhered to in spirit and practice, will "end clearcutting in Clayoquot."

Summer 1995 WCWC publishes a four-page full-colour tabloid-sized newspaper titled *PROTECT the SOOKE HILLS – Wilderness now or never!* (Educational Report Vol. 14 No. 10). Press run 80,000. It showcases scientific inventories done by volunteer researchers in the off-catchment lands less than 20 kilometres away from downtown Victoria and promotes the establishment of a sea-to-sea green-blue belt protected area that includes these off-catchment and other lands.

July 1995 WCWC, with permission from the Squamish Nation, begins surveying a trail to access the Stoltmann Wilderness's wild reaches in the Upper Elaho Valley. WCWC surveyors discover a huge Douglas fir–ten metres in circumference–in the Elaho Valley, which they name the "Elaho Giant." It is B.C.'s third largest (in girth) known living Douglas fir tree.

July 1995 The Natural Resources Defense Council (NRDC), in cooperation with WCWC and other environmental groups, mounts a two-day symposium in Ahousaht on Flores Island in response to First Nations' request for information about alternatives to industrial logging in Clayoquot Sound, including non-timber and value-added forest products and eco-tourism. The forum is well attended by First Nations and the local non-native community.

Summer 1995 The Uts'am Witness Project, a cross-cultural collaboration designed to promote respect for nature and enhance dialogue among native and non-native communities, is launched by legendary B.C. mountaineer, John Clarke, and photo artist Nancy Bleck, who car-pool groups of people from downtown Vancouver to a temperate rainforest about to be clearcut logged in Sims Creek in the proposed Stoltmann Wilderness for a weekend of camping. Ta-lall-semkin, siem, Hereditary Chief Bill Williams of the Squamish Nation, joins this project in 1996. He introduces the 'Witness' ceremony practiced by his culture for over 8,000 years.

August 31, 1995 The B.C. government calls the first meeting of the 13-member Regional Protected Areas Public Advisory Committee, (RPAC) whose task is to decide which Lower Mainland areas are to be preserved and which are to be logged. WCWC protests this "closed door" (not an open public) process that is mandated by the provincial government to abide by the government's arbitrary 12 percent cap on wilderness protection in the region.

Summer 1995 After spending $150,000 in donations from concerned citizens and utilizing 15,000 hours of volunteer labour, WCWC completes its Clayoquot Valley Witness Trail.

Fall 1995 WCWC launches a campaign to open up the government's RPAC (Regional Public Advisory Committee) to the public. Behind closed doors, RPAC is in the process of deciding which areas in Vancouver's Lower Mainland region will get park protection.

September 22, 1995 The B.C. Forest Service authorizes WCWC to construct and maintain the Elaho Giant Trail to the big Douglas fir in the Upper Elaho Valley in the proposed Stoltmann Wilderness, lifting the stop work order it had issued to WCWC on July 21, 1995.

September 29, 1995 The B.C. government renews MB's TFL 44 that encompasses, among other areas on the west central coast of Vancouver Island, more than half of Clayoquot Sound for a term of 25 years beginning August 21, 1994.

Fall 1995 WCWC-Alberta publishes a four-page full-colour tabloid-sized newspaper titled *Take a Stand with the Last of the Lubicon Cree* (Educational Report Vol. 14 No. 12). Press run 50,000. It calls for people to support the Lubicon efforts to get a just land claim settlement that gives them the right to control and stop the wilderness-destroying logging and oil exploitation of their traditional lands.

October 1995 WCWC volunteer trail surveyors cease work for the year, having surveyed a 10-kilometre hiking route up into the Upper Elaho Valley.

Fall 1995 WCWC publishes a four-page two-colour tabloid-sized newspaper titled *Forest Hiking Trails or Urban Expansion* (Educational Report Vol. 14 No. 13). Press run 40,000. It advocates protecting land adjacent to Mt. Seymour Park in North Vancouver.

November 15, 1995 Harcourt announces his resignation as Premier of B.C. saying he will stay on until the party chooses a new leader.

November 22, 1995 Premier Mike Harcourt and Chief Byron Spinks of Lytton First Nation announced the creation of the 107,000-hectare Stein-Nlaka'pamux Heritage Park to be jointly managed by the Lytton First Nation and B.C. Parks. It is the fruition of 25 years of conservationists' and natives' campaigning for wilderness preservation to prevail over industrial development of this special watershed.

November 23, 1995 The Lytton Community of the Nlaka'pamux Nation (Lytton First Nation) and the Province of B.C. sign a Stein Valley Co-operative Management Agreement; a Fish and Wildlife Sub-agreement and a Cultural Heritage Sub-agreement. It establishes a Management Board to manage the Stein Valley Nlaka'pamux Heritage Park, A Living Museum of Cultural and Natural History also known as the Stein Valley Nlaka'pamux Heritage Park.

December 1995 WCWC hosts the launch of Raincoast Conservation Society's award winning Video *Legacy* which documents the clearcut destruction of the mid coast region of B.C.

Winter 1995/96 WCWC-WILD co-publishes with CODEFF a four-page full-colour tabloid-sized newspaper titled *Eco-Forestry Needed Now!* (Educational Report Vol. 14 No. 14). Press run 30,000 (20,000 English and 10,000 Spanish) copies.

1995 By the end of 1995, there are 131 Ecological Reserves (ERs) that protect 160,000 hectares in total in B.C. of which two thirds of the area covered is terrestrial and one-third marine. After 25 years of effort, this is only about 17 percent of Dr. Krajina's (the man who originated the idea of having Ecological Reserves in B.C.) original conservation vision of having one percent of B.C. protected in ecological reserves when the Act creating these reserves was first passed in 1971.

Winter/Spring 1995/96 WCWC publishes a 12-page full-colour tabloid-sized newspaper titled *WCWC Members Wilderness Report* (Educational Report Vol. 14 No. 15). Press run 50,000. It informs WCWC's members about WCWC's current activities and campaigns.

1996 Athabasca Sand Dunes Provincial Park is established in Saskatchewan, just south of the Northwest Territory's border to protect some, but not all, of the sand dunes in the region. The Athabascan dune complex is the largest in Canada, and the most northerly dune complex in the world.

Winter/Spring 1996 WCWC publishes a four-page full-colour tabloid-sized newspaper titled *CREATE the CAREN RANGE PROVINCIAL PARK to persevere Canada's oldest known forest* (Educational Report Vol. 15 No. 1). Print run 30,000. It advocates creating an 8,500-hectare sea level to mountaintop provincial park to protect the ancient trees and nesting marbled murrelets found there.

Winter 1996 WCWC-Alberta publishes a four-page full-colour tabloid-sized newspaper titled *SAVE THE GRIZZLY BEAR – The Ultimate Symbol of Canadian Wilderness* (Educational Report Vol. 15 No. 2). Press run 50,000. It explains how the national parks in Alberta are killing not conserving grizzlies and calls for a total ban on grizzly bear hunting in Alberta because only very few are left.

Winter 1996 WCWC publishes a four-page two-colour tabloid-sized newspaper titled *Port Moody needs its forested greenspace NOT more urban sprawl* (Educational Report Vol. 15 No. 4). Print Run 15,000 copies. It urges people to attend a rezoning meeting where the local City Council is considering turning this heavily recreationally used city-owned undeveloped land into subdivisions rather than a park.

Winter 1996 WCWC-WILD co-publishes with the Chilean conservation organization CODEFF/ Friends of the Earth a four-page full-colour tabloid-sized newspaper titled *TEMPERATE FORESTS IN CRISIS!* (Educational Report Vol. 15 No. 6). English print run 20,000. Spanish print run 20,000. It advocates that eco-forestry be practiced in both countries and explains the concept of how eco-labeling wood products will help, through consumer choice, to bring about better forestry practices.

January 1996 Canada Youth Services funds a joint Ahousaht Band-WCWC *Ahousaht Wild Side Heritage Trail and Eco-Tourism Project* to train youth in eco-tourism and clear and boardwalk an ancient trail to the outside beaches on Flores Island.

January 17, 1996 Heavy rainfall triggers over 100 landslides in Clayoquot Sound, most of them in clearcut areas.

January 22, 1996 WCWC sets up a protest "camp-out" on the lawn of the B.C. Legislature in Victoria, demanding open public hearings on the fate of the Lower Mainland's remaining wild areas, including the Stoltmann Wilderness. Lasting for two months, and featuring a "flying tent," it becomes the longest running Legislature protest camp in B.C. history. The government does not listen and refuses to open up the RPAC process to the public.

February 9, 1996 WCWC conducts an aerial survey of Clayoquot Sound to assess the damage caused by the big rainstorm on January 17, 1996.

February 16, 1996 The Haisla Nation and the Province of B.C. sign an agreement to establish a Committee to manage Huchsduwachsdu Nuyem Jees, also known as the Kitlope Heritage Conservancy.

February 20, 1996 The Valhalla Wilderness Society publishes an eight-page full-colour tabloid-sized newspaper titled *B.C.'s Rare Spirit Bears and their Rainforest Home: A Lost Legacy or an Enduring Trust?* It seeks support for creating a large sanctuary for the Kermode or Spirit bears (white-coated black bear) on Princess Royal Island and the surrounding area.

February 22, 1996 Glen Clark, NDP, becomes the Premier of B.C. after winning the leadership race to replace Mike Harcourt who had resigned.

March 1996 20 youth—half native, half non-native—begin working on upgrading the Wild Side Heritage Trail on Flores Island. The project receives additional support from many sources including thousands of WCWC members, MB, Long Beach Model Forest and Forest Renewal B.C. The project is successfully completed seven months later.

March 1996 WCWC publishes a 16-page research report based on aerial reconnaissance and photography on February 9, 1996 following the big storm event in Clayoquot Sound. Titled *An analysis of the landslides which occurred during heavy rainstorms in Clayoquot Sound in mid-January 1996*, it concludes that landslides occurred up to 20 times more frequently in clearcut areas. In this report are numerous photos of the new erosion and of the muddied lakes and streams resulting from the logging-induced landslides that are obviously harming these salmon streams.

March 11, 1996 The Information and Privacy Commissioner of B.C. issues a 15-page report (Order No. 91-1996) on WCWC's request that he overturn a decision by the Ministry of Environment, Lands and Parks to withhold TRIM Digital Map Data from the Western Canada Wilderness Committee by requiring an excessively high price for this information already paid for by taxpayer dollars. The Commissioner rules against WCWC, but suggests the government review its policy to allow the public and interest groups access to information.

Spring 1996 WCWC publishes a four-page full-colour tabloid-sized newspaper titled *Canada's Spotted Owl – Who's to blame if it goes extinct?* (Educational Report Vol. 15 No. 7). Press Run 100,000 copies. It calls for a halt to all logging within SOCAs (Spotted Owl Conservation Areas) in B.C. to save this bird from being driven to extinction in Canada.

Spring 1996 WCWC publishes 50,000 copies of a four-page tabloid-sized newspaper titled *Protect CENTRAL OKANAGAN'S MOUNTAINS AND CANYONS* (Educational Report Vol. 15 No. 8) advocating certain areas to be protected in the LRMP process that is underway in the region.

Spring 1996 WCWC publishes a four-page full-colour tabloid-sized newspaper titled *Southern Chilcotin Mountains – B.C.'s "secret, gentle wilderness"* (Educational Report Vol. 15 No. 9). Press run 80,000. It calls on the B.C. government to make this outstanding, longstanding park candidate a Class A provincial park.

April 1996 Clayoquot Interim Measures Agreement with the Nuu-chah-nulth is renegotiated and extended for three more years.

April 1966 WCWC-Victoria publishes a four-page demi-tabloid-sized black and white newspaper titled *The Sooke Hill Wilderness–Your Urgent Help is Needed Now to Save it* (Educational Report Vol. 15 No. 5). It is part of a big push to head off plans by the Greater Victoria Water District Board to trade off-catchment lands to a logging company for private lands inside the watershed rather than protect those off-catchment lands in a park.

April 1996 WCWC's Joe Foy presents an *Ahousaht Wild Side Trail Survey Options Report* to the Ahousaht Band Council outlining various possible routes for the Wild Side Trail, a project being undertaken jointly by WCWC and the Ahousaht First Nations.

Spring 1996 WCWC-Alberta publishes a four-page full-colour tabloid-sized newspaper titled *Our Great Boreal Forest* (Educational Report Vol. 15 No. 10). Press run 80,000. This informative paper makes a passionate plea to protect change the way the boreal forest is being logged and to preserve a lot more of it.

April 16, 1996 Mark Wareing, the forester who blew the whistle on the deleterious effects that logging in the watersheds that supply Greater Vancouver's drinking water and who campaigned to stop this logging, passes away at the age of 49.

April 24, 1996 The 1,147,500-square-kilometre Wapusk (Polar Bear) National Park is established on Hudson's Bay in Manitoba. Canada's 37th National Park, it protects the largest polar bear denning area in North America.

April 29, 1996 The Champagne and Aishihik First Nations and the Province of B.C. sign the Tatshenshini-Alsek Park Management Agreement establishing a Park Management Board to jointly manage the Tatshenshini-Alsek Park.

May 1996 WCWC trail surveyors push deeper into the Elaho Valley, discovering new groves of big trees.

May 28, 1996 Glen Clark, NDP, elected Premier of B.C. after winning the majority of seats in the Legislature but not a majority of the popular vote in the provincial election.

Spring 1996 WCWC publishes 100,000 copies of a four-page full-colour full-sized newspaper titled *STOTLMANN WILDERNESS – Save the entire 260,000 hectares* (Educational Report Vol. 15 No. 11). Press run 100,000. This is another paper designed to inform existing supporters of our progress (or lack of progress) in this campaign and recruit new supporters.

June 1996 Interfor changes logging plans in an obvious effort to aggressively eliminate the big tree forested area where WCWC is trailbuilding in the Stoltmann Wilderness. It gains B.C. Forest Service approval to build a logging road into the Upper Elaho Valley, without giving public notice. WCWC demands a halt to the road building, goes to court charging that Interfor is breaking the Forest Practices Code because no public notice was given. The B.C. government amends the Act to make it legal and WCWC looses the court case because the court decision is moot. The road building goes ahead.

July 18, 1996 The provincial government launches the Central Coast Land and Resources Management Plan (LRMP). WCWC decides not to participate because key areas are not deferred and logging continues in proposed wilderness areas. Greenpeace and the Sierra Club of B.C. participate in this negotiation and trade-off process.

June 21, 1996 WCWC holds a Council of All Beings Solstice Celebration on the Victoria Legislature lawns. Bill Devall, one of the fathers of Deep Ecology, speaks. Participants dress up as animals and "re-affirm their rootedness in nature."

June 1996 Frustrated by government's decision to not conduct full inventories of forest values in the pristine Bulson Valley prior to logging as called for by the Clayoquot Scientific Panel, Greenpeace and FOCS blockade the Bulson Valley logging. First Nations persuade them to suspend their blockade (on protocol issues) and agree to meet with logging companies to begin negotiating a solution to the conflict.

Summer 1996 WCWC publishes 100,000 copies of its seventh Clayoquot paper, an eight-page full-colour tabloid-sized newspaper titled *Beautiful Clayoquot Sound–Ancient Rainforests-Pristine Salmon Streams the Fifteen-year fight to preserve this precious heritage continues* (Educational Report Vol. 15 No. 12). Press run 100,000.

July 6, 1996 The Nuu-chah-nulth chair an all stakeholders meeting to discuss future of Clayoquot's still wild watersheds. Unions, forest companies, the CRB, communities, First Nations and Greenpeace, FOCS, RAN, SCWC and WCWC attend. Environmentalists refuse to participate in Science Panel implementation that would lead to an outcome that may put pristine areas in logging plans. MB and Interfor agree to a proposition by environmentalists that they participate in an economic transition strategy based on the possibility that logging may never be allowed in the remaining pristine watersheds. FOCS and Greenpeace agree not to blockade until at least August 9th, pending further negotiations regarding the fate of the pristine areas.

July 1996 WCWC petitions the B.C. Supreme Court asking for a court injunction against the road building in the Upper Elaho Valley. The court refuses to halt Interfor's road building activities, but agrees to hear the case at a later date.

July 1996 Over one hundred B.C. artists hike into the Stoltmann Wilderness along WCWC's surveyed hiking route to paint, draw and sculpt works of art, which they then donate to WCWC to raise awareness and funds for WCWC's fight to preserve the Stoltmann.

August 1996 WCWC presents a report prepared by James MacGregor of

Ecoplanet Ltd. titled *Ahousaht Wild Side Heritage Trail and Ecotourism Development and Management Options* to the Ahousaht First Nations. It details significant economic opportunity presented by the just completed Ahousaht Wild Side Heritage Trail.

August 24, 1996 Ahousaht First Nations, WCWC and Youth Service Canada hold a celebration ceremony to congratulate the youth of the Ahousaht Wild Side Heritage Trail and Eco-tourism project for the construction and successful completion of *"the most remarkable 16-kilometre-long oceanside trail on the planet!"* WCWC subsequently publishes a 42-page report about this successful joint project.

September 9, 1996 The 90-day signature gathering begins on *"An Act to Prohibit the Hunting of Bears in B.C."* Initiative Petition forwarded by Paul George. WCWC sponsors the Initiative in an effort to end sport and trophy hunting of grizzly bears in B.C.

Fall 1996 WCWC prints 80,000 copies of a four-page full-colour tabloid-sized newspaper titled *BEARS Need More Protection Now* (Educational Report Vol. 15 No. 13). Initial press run 100,000. Reprinted 50,000. Written by Anthony Marr and Paul George, it features the Bear Initiative, currently being forwarded by Paul George under B.C.'s unique Recall and Initiative Act and supported by WCWC. This newspaper presents a strong case for why trophy hunting of bears should be banned in B.C.

Fall 1996 The American Fisheries Society releases a report that states 142 salmon stocks have gone extinct in B.C. and the Yukon and a further 642 Canadian salmon stocks are at high risk due primarily to habitat destruction.

September 25, 1996 WCWC-Victoria releases a 39-page research report titled *Biodiversity in the Sooke Hills Wilderness – An Inventory of Species and Ecological Values* the results of a research project that many local scientists from the University of Victoria had voluntarily participated in over the last two years.

September 11, 1996 The B.C. government establishes the 217,000-hectare Stikine River Provincial Park to protect the wild and dramatic Stikine River, and especially the portion that carves its way through the rocky surrounding plateau to form the stunning Grand Canyon of the Stikine. This 80-kilometres-long canyon is the largest and most spectacular in Canada. Its steep walls are at times as much as 300-metres-high and its width ranges from 200-metres-wide to an incredibly narrow and turbulent two-metre-wide gap that all the water of this great river must pass through.

Fall 1996 WCWC–Alberta publishes a four-page full-colour tabloid-sized newspaper titled *Caribou Mountain Wilderness* (Educational Report Vol. 15 No. 14). Press run 50,000. This paper advocates that the unique plateau adjacent to Wood Buffalo National Park be protected as an Alberta wildland park.

October 1996 RPAC makes its recommendations that 23 area comprising 136,000 hectares of mostly rock ice and scrub forest become the new parks in Vancouver's Lower Mainland. This recommendation includes only 20 percent of the proposed Stoltmann Wilderness Area (Clendenning and Upper Lillooet) be protected. The largest forested valley, the Elaho Valley, is given to Interfor to clearcut. Sims, Salal, Boulder and North Creek Valleys also within the proposed wilderness, area are not included RPAC's package of parks.

October 28, 1996 At a press conference in Vancouver, Premier Clark announces the creation of 22 new parks in the Lower Mainland. They contain very little old-growth forest. The CEO of B.C.'s second largest logging company, Interfor, stands beside Clark at the podium sporting a big smile. WCWC booed the decision because it failed to protect 80 percent of the Stoltmann wilderness and gave most of the best timbered, most biodiverse areas to the logging industry, including the Upper Elaho Valley to Interfor. WCWC vows to fight on until the government protects the entire Stoltmann Wilderness.

Fall/Winter 1996-97 WCWC publishes a 16-page full-colour tabloid-sized newspaper titled *Wilderness Committee Annual Members Report* (Educational Report Vol. 15 No. 15). Press run 70,000.

November 1996 In Europe, Greenpeace leaders launch an international boycott against B.C. forest products to put pressure on the B.C. government to save B.C.'s coastal ancient temperate rainforest.

November 1996 WCWC trail crew successfully completes surveying a 30-kilometre-long hiking route up the Elaho Valley, over the Hundred Lakes Plateau and into the Meager Creek Valley.

November 1996 The B.C. Supreme Court finally hears WCWC's case regarding Interfor's illegal road building in the Upper Elaho Valley. It contravenes the new Forest Practices Code that requires public notice of changed logging plans. However in the intervening time between when this case is launched and when it is heard by the court, the B.C. government weakens the Forest Practices Code, no longer requiring public notice and review of changed plans. The Court thus rules the case moot and dismisses it.

December 9, 1996 The Initiative petition to ban bear hunting in B.C. is returned to the Chief Electoral Officer (the end of the 90-day signature-gathering period) by proponent Paul George. It contains 88,357 signatures, and therefore fails, as 222,272 valid signatures are required for success.

1996 Interfor becomes the first company to be convicted under the new Forest Practices Code for violation regarding setbacks from a salmon stream along Rolling Stone Creek in Clayoquot Sound.

1996 The World Conservation Union endorses the call for UNESCO Biosphere status for Clayoquot Sound.

1997 The Burns Bog Conservation Society presents a 25,000-signature petition to the B.C. government calling for it to preserve Burns Bog as an ecological reserve.

January 1, 1997 High windstorms blow down substantial numbers of trees on Vancouver Island, including some of the experimental "leave strips" designed under the Science Panel's "variable retention" logging rules in Clayoquot Sound.

January 1997 MacMillan Bloedel Ltd. (MB) suspends logging in Clayoquot Sound. The company lays off its Kennedy Lake Division workers announcing that it will not log in Clayoquot Sound for at least 18 months as it works out how to reconfigure its logging operations. MB lost $7 million logging Clayoquot Sound in 1996.

January 1997 While in Victoria, B.C., the Honourable Sheila Copps, Federal Minister of Canadian Heritage, announces that her government would support a Biosphere Reserve solution in Clayoquot Sound if all stakeholders want it.

January 1997 A group of environmentalists form the People's Action for Threatened Habitat (PATH), a new organization to protect wilderness, wildlife, First Nations' rights and recreational opportunities in the Stoltmann Wilderness.

January 28, 1997 The B.C. government accepts the Special Commission on Greater Victoria's water supply report adding 4,900 hectares of non-catchment land to the Capital Regional District's regional park system. The protection of the three valleys, the Niagara, Waugh, and Veitch, in a Sooke Hills park is a stunning campaign victory for WCWC-Victoria and all the other groups and persons working towards that goal.

February 1997 Community and environmental organizations' representatives from Greater Victoria, the Sunshine Coast Regional District, Greater Vancouver, and the Slocan Valley in southeast B.C. form the B.C. Tap Water Alliance.

March 1997 WCWC hosts a two-month-long art show and silent auction, raising over $20,000 for its Stoltmann campaign with works of arts donated by the artists who visited the Stoltmann Wilderness as guests of WCWC.

March 1997 The Nuu-chah-nulth Central Region Chiefs and MacMillan Bloedel Ltd. announce a new *Joint Venture Corporation*, 51 percent owned by First Nations, that will take over all logging operations in MB's Estevan Division—the northern part of its TFL in Clayoquot Sound. The Joint Venture, based on a maximum harvest of 40,000 cubic metres per year, will not log for three years while it investigates value-added forest product opportunities. Logging in pristine areas remains undecided.

Spring 1997 WCWC publishes a four-page full-colour tabloid-sized newspaper titled *Join the fight and save the ANCIENT RAINFORESTED VALLEYS of the STOLTMANN WILDERNESS... Before it's too late!* (Educational Report Vol. 16 No. 1). Press run 100,000 copies. The map in the centre shows the proposed clearcut from 1997-2001 in the proposed park area that was not protected in the B.C. government's 1996 Lower Mainland parks decision.

Spring 1997 WCWC publishes a four-page full-colour tabloid-sized newspaper titled *It's time for Algonquin Park to become a genuine Wilderness Park* (Educational Report Vol. 16 No. 2). Press run 60,000. It advocates ending all logging activities within the boundaries of Algonquin Park, Ontario's largest and oldest provincial park, giving it a chance to "heal" from the damage caused by the 130 years of industrial logging there. WCWC door-to-door canvassers in Toronto are the main distributors of this paper.

April 21, 1997 Greenpeace releases its report titled *Broken Promises: The Truth About What's Happening to B.C.'s Forests*. The report provides proof that, contrary to the government assertions, B.C. is not doing *"world class forestry."* In reaction to this report, Premier Glen Clark calls Greenpeace and other B.C. environmentalists who criticized B.C. forest practices *"Enemies of B.C."* claiming they are engaging in a *"misinformation campaign."*

May 1997 With the Squamish Nation's permission, WCWC transports materials via a helicopter deep into the Upper Elaho Valley and builds a large temporary scientific research tent-cabin to house rainforest researchers.

June 1997 Hikers begin using WCWC's surveyed Stoltmann Wilderness trail, proclaiming it to be the Lower Mainland's best multi-day backpacking experience.

June 2, 1997 Jean Chrétien, Liberal, is re-elected to be the Prime Minister of Canada.

June 19, 1997 With the help of Interfor, loggers "block the blockaders" at the 24-1/2 mile mark on the public road leading into the Stoltmann Wilderness. For six weeks they prevent WCWC researchers and trail surveyors from entering these public lands and anyone else who will not sign a petition saying they supported Interfor's logging. The RCMP refuses to make the loggers remove the blockade. WCWC takes the RCMP to court in an effort to make this organization enforce the law that makes the blockade a criminal offense. WCWC sets up a camp alongside the road at the blockade and distributes information about the Stoltmann Wilderness to the public being stopped there.

June 28, 1997 IWA workers set up an information picket line on the Vancouver Harbour Authority Main Street Dock, near the Greenpeace icebreaker *Arctic Sunrise*, which is playing a role in Greenpeace's international campaign to boycott B.C. forest products.

June 21, 1997 The Siska Indian Band councilors and elders sign a Siska Band Heritage Park Declaration for the Siska watershed making this part of their traditional territory off limits to industrial developments.

July 2, 1997 IWA union members surround the Arctic Sunrise icebreaker and a smaller Greenpeace vessel, the "Moby Dick," with boom sticks, prohibiting their departure from Vancouver Harbour. The loggers are protesting Greenpeace's action a few weeks earlier that had briefly halted some logging on the mid coast of B.C. and had cost a few IWA workers some lost wages. Pilots refuse to cross the union picket line which is also supported by members of the International Longshoremen's and Warehousemen's Union, the United Fisherman and Allied Workers Union and other affiliates of the B.C. Federation of Labour.

Summer 1997 WCWC co-publishes with the Raincoast Conservation Society 100,000 copies of a full-sized eight-page full-colour newspaper titled *LET'S PROTECT CANADA'S GREAT BEAR RAINFOREST. We have a choice – Clearcuts, tree farms and big stumps - Or grizzlies, wild salmon and big trees* (Educational Report Vol. 16 No. 4). It features a centre spread map produced by WCWC's Chris Player showing four complexes of proposed protected areas. They are the Greater Ecstall Region - Northern Extension of the Great Bear Rainforest; the Central Great Bear Rainforest including the Kermode (Spirit Bear) protected area on and around Princess Royal Island; the Knight Inlet Region –Southern Extent of the Great Bear Rainforest and the Stoltmann Wilderness.

June 30, 1997 The Chilliwack Lake Provincial Park, 150 kilometres east of Vancouver in the upper Chilliwack River Valley characterized by a valley-bottom lake, old-growth forested slopes, and spectacular sub-alpine and alpine ridges, is expanded to 9,122 hectares.

Summer 1997 WCWC publishes an eight-page full-colour tabloid-sized newspaper titled *Protect Clayoquot Sound – A Biosphere Reserve Vision* (Educational Report Vol. 15 No. 5). Print Run 80,000 copies. The centrefold has a comprehensive map of Clayoquot Sound showing the remaining pristine watersheds and the approximate boundaries of the First Nations' territories in the Sound. A separate map shows the boundaries of the Tree Farm Licence areas.

July 4, 1997 Assisted by Vancouver Ports Police, both Greenpeace ships, the Arctic Sunrise, and the Moby Dick bolt from the IWA loggers log boom blockade around the ship in Vancouver's Burrard Inlet. The icebreaker departs the harbour without a ship's pilot, as required by law.

July 8, 1997 The B.C. government establishes the 21,000-hectare Cummins River Valley Provincial Park.

July 23, 1997 The B.C. government establishes the 32,577-hectares Skagit Valley Provincial Park.

August 1997 The loggers take down their blockade on the Squamish Valley road to the Upper Elaho Valley shortly before a B.C. Supreme Court justice is to rule on WCWC's petition asking the court to force the RCMP to act to remove this illegal roadblock. During the blockade, persons unknown remove WCWC's research tent-cabin from the wilderness in the Upper Elaho Valley and dump the pieces at the Squamish RCMP station. None of the culprits are ever apprehended.

August 7, 1997 WCWC signs a formal agreement with B.C.'s Squamish Forest District, which permits WCWC volunteers to upgrade the 30-kilometre-long Stoltmann "surveyed hiking route" from the Elaho to Meager Creek drainages to hiking trail standards. Thus the B.C. Forest Service officially recognizes this trail.

August 24, 1997 On the tenth anniversary of the establishment of the 5,200-hectare Seymour Demonstration Forest between Lynn Headwaters Regional Park and Mt. Seymour Provincial Park, WCWC launches its campaign to change the designation of this area to a regional park where no logging can occur. WCWC publishes 50,000 copies of a full-colour four-page tabloid-sized newspaper titled *OUR CHOICE: Seymour Demonstration Forest OR Seymour Ancient Groves Park* (Educational Report Vol. 16 No. 6), and passes a copy out to everyone attending the anniversary celebration at the entrance to the then demo-forest.

September 1997 WCWC volunteers pack in the research tent-cabin piece by piece and set it up again in the wilderness in the Upper Elaho where it had stood before. An increasing number of people come to hike in the Stoltmann Wilderness.

Fall 1997 WCWC publishes *Ahousaht Wild Side Heritage Trail Guidebook* by Stanley Sam Sr. of the Ahousaht First Nations. This book, besides providing information about the Wild Side trail on Flores Island in Clayoquot Sound also relates the recent aboriginal history of the area.

October 1997 Interfor loggers cut down "Magic Grove," a stand of 800-year-old red cedar trees in Sims Valley within the Stoltmann Wilderness Proposal area. The B.C. Forest Service rules that WCWC's research tent-cabin is a "permanent structure" and orders WCWC to remove it because it has a "permanent" (cheaper) wooden rather than an aluminum pole frame.

October 8, 1997 The B.C. government passes legislation adopting the recommendations of the Mackenzie Land and Resource Management Plan (LRMP). It establishes an internationally significant conservation complex of parks nearly 2,000,000 hectares in size in the northern part of the Mackenzie region in the northern Rocky Mountains. Among them are the Muskwa-Kechika parks and protected areas covering 1,170,000 hectares (approximately twice the size of Prince Edward Island) surrounded by 3,240,000 hectares of special management areas. This unroaded wilderness represents one of the greatest large mammal predator-prey systems in the world.

Fall 1997 WCWC publishes a four-page full-colour tabloid-sized newspaper titled *Tiger, tiger, burning dim...* (Educational Report Vol. 16 No. 8). Press run 50,000. It calls for removal of all outlawed tiger bone medicines from Canadian stores and reports on WCWC-WILD's CIDA sponsored project with Tiger Trust India to help protect tiger reserves in India.

October 1997 *The Great Bear Rainforest – Canada's Forgotten Coast*, a 144-page coffee table hard cover book by Raincoast conservation Society's campaigners Ian and Karen McAllister and published by Harbour publishing, is released in Canada. Released the following year in the U.S. by Sierra Club Books, it becomes a major force for conservation of this threatened area in B.C.

October 1997 The B.C. government announces that it will support the recommendations of the Fort Nelson and Fort St. John Land and Resource Management Plans (LRMPs) and establish a 4.4 million hectare Muskwa-Kechika Management Area in the Rocky and Cassiar Mountain area of northern B.C. This internationally significant conservation management area to be created is approximately the size of Nova Scotia, with 3.3 million hectares of special management zones buffering over 1.1 million hectares of Provincial Parks. This roadless wilderness area contains one of the greatest remaining large mammal predator-prey systems in the world.

November 1997 Interfor cuts a logging road through "Grizzly Grove," a beautiful stand of huge ancient red cedar trees in the Upper Elaho Valley that is an important grizzly bear denning site.

November 1997 A B.C. Appeals Court decision confirms that the Haida claim of aboriginal title to their traditional lands does indeed constitute an encumbrance to existing and future logging licences on the same landbase.

December 1997 With permission from the B.C. Forest Service, a WCWC volunteer cuts off a slab from a very large stump left behind when Interfor loggers cut the Douglas fir tree down in July 1997 within the proposed Stoltmann national park. The big slab was then cut into two slabs, mounted on plywood boards, sanded and finally dated by a dendrochronologist. It was found to contain 1,158 tree rings (the tree was at least 1,158 years old when cut down). WCWC tours across Canada to raise support for protecting the Stoltmann Wilderness with "Slabby," mounted on two plywood frames that bolt together, complete with historic dates beside the corresponding tree rings to illustrate its antiquity. The other, identically mounted and dated slab, it leases to Greenpeace for use in its campaign in Europe to help save B.C.'s ancient temperate rainforest.

December 10, 1997 Julia Butterfly Hill, a 23-year-old woman climbs into a 55-metre-tall California coast redwood tree and begins a "tree sit." Her aim is to prevent the destruction of the tree and the surrounding forest where it had lived for a millennium.

1998 Forestry company MacMillan Bloedel (MB) announces plans to end clearcutting (practice variable retention logging) and to seek certification for their operations from: the Canadian Standards Association (CSA), the International Standards Organization (ISO 14001), and the Forest Stewardship Council (FSC). The environmental community expresses cautious optimism.

1998 Labour Environmental Alliance Society (LEAS) is founded in B.C. as a unique environmental organization, based on an alliance model that brings together workers and environmentalists, unions and environmental groups to find solutions to environmental problems based on social justice. WCWC joins the Alliance in 2002.

January 1998 WCWC increases the size of its proposed Stoltmann Wilderness protected area from 260,000 hectares to 500,000 hectares by including the Pemberton Ice Cap, Soo Creek Valley and the upper Bridge River Valley, and linking the proposed park to nearby parks and the town site of world-famous Whistler. WCWC now calls on the federal government to make the area a national park—the first in B.C.'s Coast Mountains.

January 16, 1998 The Tsleil-waututh Nation (Burrard Indian Band) and the B.C. government sign the Indian Arm Provincial Park Management Agreement establishing a Park Management Board to jointly manage Say-nuth-khaw-yum Heritage Park also known as Indian Arm Provincial Park.

February 1998 Interfor begins accelerating the export of raw logs from B.C. to U.S. lumber mills for processing.

March 1998 The B.C. Wildlife Branch publishes a report titled *Wildlife in B.C. at Risk – the Northern Spotted Owl*.

March 1998 The Fraser Headwaters Alliance is formed to serve as an umbrella organization for groups concerned about conservation issues in the headwaters of B.C.'s mightiest river, the Fraser.

March 24, 1998 Several hundred people protest in Victoria on the Legislature lawns on the opening day of the B.C. Legislature to remind the government that key intact areas of the Great Bear Rainforest on the B.C. coast are being clearcut logged and they need immediate protection.

April 1998 WCWC's campaign coordinator, Joe Foy, meets with Premier Clark about the Stoltmann Wilderness. Foy presents Clark with a "Jurassic Clark" T-shirt, explaining to him that his poor record on environmental protection has taken B.C. back to the dark ages. Clark responds by saying he would be willing to revisit his Stoltmann decision if new information comes forward.

Spring/Summer 1998 WCWC publishes a four-page two-colour tabloid-sized newspaper titled *DON'T LET FISH FARMS DESTROY BRITISH COLUMBIA'S WILD SALMON MIRACLE* (Educational Report Vol. 17 No. 2). Press run 50,000 copies. Information presented in this paper is based on *Net Loss: The Salmon Netcage Industry in British Columbia*, a report by David W. Ellis for the David Suzuki Foundation and an article in *The Georgia Straight* by Alexandra Morton titled *Fish Farming Prompts Despair Over Fate of Salmon*.

April 2, 1998 The B.C. NDP government announces amendments to its Forest Practice Code that weaken it and make it easier for companies to operate in B.C. Among other things it "*Encourages industry to increase the average cutblock size.*"

May 21, 1998 Federal Fisheries Minister David Anderson announces that the 1998 salmon fisheries would be based on a "zero fishing mortality for upper Skeena and Thompson coho stocks (where the coho are most endangered). Other fishing would be allowed only where "coho bycatch mortality will be minimal."

June 1998 WCWC volunteers re-establish the committee's research tent-cabin in the Upper Elaho wilderness using aluminum poles instead of a wooden frame and leaving out the plywood floor, thus making it a legal "temporary structure." WCWC also begins upgrading the Elaho surveyed hiking route into a trail. In a Herculean effort the volunteers hike out the offending wooden components, the 2 x 4s and plywood sheets, to the logging road to be trucked back to town. WCWC also replaces its suspension bridge over Lava Creek with a cable car.

June 1998 The Muskwa-Kechika Management Area Act is passed by the B.C. government. It officially creates the 4.4 million hectare Muskwa-Kechika Management Area (MKMA) in northern B.C. protecting one of the world's most abundant wildlife areas using a leading edge conservation reserve management model. It also formally establishes and funds a MKMA Advisory Board, a multi-stakeholder body, including two wilderness campaigners, to oversee the management of the area.

June 15, 1998 WCWC prepares and presents a brief titled *Assessment of Wilderness Remaining and Preservation Options for the Lillooet Land and Resource Management Plan (LRMP)*. It presents detailed GIS maps and evaluations prepared by WCWC's Chris Player of the various options for wilderness preservation in the region.

Summer 1998 WCWC-Alberta publishes an eight-page full-colour tabloid-sized newspaper titled *Who Will Take Global Warming Seriously...and help protect the Earth's vast and vulnerable boreal forests?* (Educational Report Vol. 17 No. 3). It presents the evidence for global warming in a new graphic way and makes the case for ceasing the clearcutting of Canada's boreal forest so it can act as a carbon sink to absorb more atmospheric CO_2.

Spring/Summer 1998 WCWC publishes a four-page two-colour tabloid-sized newspaper titled *EVERYONE'S HELP NEEDED TO SAVE WILD COHO AND PRESERVE BRITISH COLUMBIA'S WILD SALMON MIRACLE* (Education Report Vol. 17 No. 4). Press run 50,000. It outlines in point form WCWC's bold plan to save B.C.'s declining coho salmon stocks by ending interception fisheries and thus by-catch of endangered stocks.

July 1998 Interfor cuts down most of Grizzly Grove. Interfor closes its sawmill in Squamish, laying off 165 workers. Interfor has already exported 1,000 logging truckloads of logs to U.S.A. mills since the beginning of 1998. During the summer several hundred people a month use WCWC's Elaho Valley to Meager Creek hiking trail.

Summer 1998 WCWC publishes an eight-page full-colour full-sized newspaper titled *To be preserved? To be pillaged? Act now to save B.C.'s Rainshadow Wilderness* (Educational Report Vol. 17 No. 5). Print run 80,000. It showcases 29 proposed wilderness areas in the Lillooet region being considered at the local Land and Resource Management Plan (LRMP) table and asks the public to support the protection of all of them.

August 1998 WCWC prints 80,000 Stoltmann Wilderness national park opinion poll postcards and begins distributing them to WCWC members throughout Canada and to every household in Whistler, Pemberton, Mt. Currie and Squamish.

August 1998 WCWC moves its Wilderness Committee Store from 20 Water Street, where it has operated for 10 years, to 227 Abbott Street a block away in Vancouver's Gastown. A few months earlier WCWC purchased this strata title store and outreach centre and completely remodeled its interior. The store features a display of a slice from Canada's oldest known tree a 1,835-year-old yellow cedar that grew in the Caren Range on the Sunshine Coast before loggers cut it down in the early 1990s. WCWC's main office moves at the same time from 20 Water to the top floor at 341 Water Street, two blocks away from WCWC's new store.

August 1998 MB quits the B.C. Forest Alliance. Tom Stephens, CEO of MB had announced earlier that his company was phasing out clearcutting, replacing it with "variable retention" logging to a lot of fanfare.

September 1998 Dr. Neville Winchester, University of Victoria professor, investigates the upper canopy in the Douglas fir grove near WCWC's research station. Dr. Winchester and his team climb 15 stories up into the treetops to gather insect samples to classify in a search for species new to science. After visiting the site, Dr. Winchester calls for an immediate halt to logging to preserve the area's unique ecology.

October 1998 A team of bear biologists, under contract to WCWC and using WCWC's Elaho Valley research station as a base camp, conduct a survey of the grizzly bear habitat in the Upper Elaho Valley.

November 26, 1998 Okanagan Nation Alliance and the Province of B.C. sign a Co-operative Working Agreement to establish a working group to manage the following provincial parks: Echo Lake, Ellison Fintry, Granby, Kalamalka Lake, Kekuli Bay, Mabel Lake, Monashee, Silver Star and Truman D. Locheed, and the following Ecological Reserves: Buck Hills Road, Campbell, Cougar Canyon Brown, Kingfisher Creek, Lily Pad Lake, Upper Shuswap River and Vance Creek.

Winter 1998 WCWC publishes 130,000 copies of a newspaper titled *Big Dreams Can Come True - Stoltmann National Park* (Educational Report Vol. 18 No. 6). Print run 130,000 copies. It premiers a change of direction in WCWC Stoltmann campaign, nearly doubling the size of this proposed protected area from 260,000 hectares to 500,000 hectares and calling for national park reserve status instead of provincial park status. On the cover it features a spectacular 3-D relief map by Z-Point Graphics of Whistler.

1998 The Friends of Clayoquot Sound join First Nations in Clayoquot Sound in demonstration against fish farming in Clayoquot Sound. The summer of '98 is the first one since WWII that sees no logging in Clayoquot Sound.

1999 The world's human population exceeds six billion.

January 1999 Wilderness Committee co-publishes with the Rainforest Conservation Society an eight-page full-colour tabloid-sized newspaper titled *Canada's Great Bear Rainforest - Under Attack* (Educational Report Vol. 18 No. 1). Press run 150,000 copies. It reports, "*Since 1990, 32 large intact watersheds in the Great Bear Rainforest have been lost to industrial logging. If conservationists fail to thwart logging company plans, in the next few years the companies will road and initiate clearcutting in virtually every remaining unprotected pristine valley on the Mainland Coast.*"

January 28, 1999 The Clayoquot Sound Biosphere Reserve application is signed by First Nations, local, provincial and federal government officials and sent on to United Nation offices in Paris.

February 3, 1999 The B.C. government announces that it has loaned the owners of Burns Bog $25 million in exchange for the company's donation of 1,200 hectares (about half of the bog) for a park and plans to allow the company to develop 800 hectares of the remaining bog as a site for the relocation of the PNE and an "*integrated, themed retail-leisure-entertainment centre.*" This plan generates a huge public outcry. The company's application to have the 800 hectares rezoned is unanimously rejected by the local municipal council. The B.C. government backtracks and holds out for purchasing the whole bog for a protected area.

Spring 1999 WCWC publishes a four-page two-colour tabloid-sized newspaper titled *B.C.'s Herring must be given the chance to recover from fifty years of industrial overfishing* (Educational Report Vol. 18 No. 2). Print run 30,000. It makes a strong argument for the federal government to institute a five-year moratorium on all roe herring fishing to give the depleted stocks a chance to recover.

Spring 1999 WCWC publishes *BUY BACK BURNS BOG NOW! – Protect Forever Vancouver's Biggest Remaining Urban Wilderness*, printing and distributing 300,000 of this four-page tabloid-sized newspaper (Educational Report Vol. 18 No. 3). It urges governments to purchase and preserve this domed bog.

April 1, 1999 The B.C. government announces plans to compensate MB for the approximately 7,700 hectares of Timber Licence (TL) lands the company has relinquished to create new parks on Vancouver Island including Carmanah and Tsitika Provincial Parks with 30,000 hectares of public lands to become MB's "fee simple" ownership lands, or instead, pay MB $84 million in cash. A public outcry ensues regarding privatizing public forestlands.

May 1999 Federal MPs David Anderson and Art Eggleton overturn a signed agreement and launch the first expropriation of provincial territory in Canadian history, permitting U.S. warships sailing up Georgia Strait to carry nuclear warheads and conduct trials in Nanoose Bay, north of Nanaimo in the Straits of Georgia.

May 1999 WCWC protests the Makah grey whale hunt in the Strait of Juan de Fuca at the American Consulate in Vancouver and publishes and distributes a poster decrying the hunt.

May 17, 1999 Members of the Makah tribe in Washington State hunt and kill a 50-foot-long gray whale off the tip of Washington State's Olympic Peninsula, about 60 miles west of Victoria. It was the first whale killed by this tribe in 70 years. The whale was dispatched with 50-calibre rifles after being symbolically harpooned from a cedar canoe.

June 1999 WCWC publishes 150,000 copies of a four-page tabloid-sized newspaper titled THE FIGHT TO OWN B.C.'s FORESTS –B.C.'s Big Timber Corporations Want it All (Educational Report Vol. 18 No. 4) and distributes them to raise public awareness about the threat to privatize public lands in exchange for cutting rights being withdrawn. In the ensuing months the government holds public hearings and decides, because of public pressure to keep public lands public, to pay cash compensation to MB instead of land for forest tenured lands "taken away" from this company and others to make new parks.

June 1999 Sponsored by WCWC, the world-renowned wilderness educator, John Seed, conducts a "Council of All Beings" in the Stoltmann Wilderness. Thirty people attend and experience "deep ecology."

June 16, 1999 The Memorandum of Understanding is signed between Iisaak Natural Resources Ltd. (the First Nations/MacMillan Bloedel joint venture logging company that replaced MB's TFL holdings in Clayoquot Sound), and Greenpeace Canada, Greenpeace International, Natural Resources Defense Council, Sierra Club of B.C. and Western Canada Wilderness Committee. Iisaak, the forest company with 51 percent ownership residing with the Nuu-chah-nulth, agrees not to log in the remaining pristine watersheds in Clayoquot Sound and log only according to the rules developed by the Clayoquot Science Panel, while the environmental organizations agree not to oppose Iisaak's logging and help the Iisaak market its wood and other wild forest related products. The Friends of Clayoquot Sound does not sign this MOU in order to maintain its independent watchdog position.

June 21, 1999 After a B.C. government commissioned public hearing process in which the vast majority of the people participating express opposition to government approval of the sale, the B.C. government goes ahead anyway and OK's U.S. softwood industry giant Weyerhaeuser's buyout of MB. During the process, Weyerhaeuser promises to honour MB's commitment to variable retention logging and to implement the Clayoquot Sound Memorandum Of Understanding (MOU) between the environmental groups and Iisaak Natural Resources Ltd.

Summer 1999 WCWC publishes a four-page full-colour tabloid-sized newspaper titled Yukusan-Hanson Island – an urgent Conservation Plea (Educational Report Vol. 18 No. 5). Press run 50,000. This paper tells the story of the extensive record of First Nation's forest use on Hanson Island told by culturally modified trees there documented by researcher David Garrick.

Summer 1999 WCWC-Alberta publishes a four-page full-colour tabloid-sized newspaper titled Halt the Clearcut destruction of Central Alberta's West Country (Educational Report Vol. 18 No. 6). Press run 100,000. The cover features an aerial photo showing the extensive patchwork of clearcut that already dominates the area.

Summer 1999 WCWC publishes a four-page full-colour tabloid-sized newspaper titled Save Whistler's 1300-Year-Old Douglas Firs (Educational Report Vol. 18 No. 7). Press run 150,000. It furthers WCWC's campaign to protect the Upper Elaho and the rest of the Stoltmann Wilderness in a national park by showing the most recent destruction of the ancient rainforest there by Interfor and ancient firs still standing that can be protected.

August 1999 The Supreme Court of B.C. awards Interfor an injunction that establishes a 500-metre-wide bubble exclusion zone effective 24 hours per day around all of Interfor's active work sites in the Upper Elaho Valley to keep out the public and protesters who might interfere with Interfor's logging activities.

August 20, 1999 Attorney General Ujjal Dosanjh announces that B.C. Premier Clark is the subject of a criminal investigation. The next day, August 21, 1999 Clark resigned as Premier.

August 25, 1999 Dan Miller, NDP, becomes Premier of B.C. and holds that position until February 24, 2000.

September 1, 1999 Interfor issues a permit to WCWC to allow Joe Foy, MP Charles Caccia and some Members of the Council of European Parliamentary Assembly's Committee on the Environment, Regional Planning and Local Authorities to enter (on September 11th) the 500 metre bubble exclusion injunction zone in the Upper Elaho Valley established by the court a short time earlier in order for the party of concerned elected officials to use WCWC's hiking trail to see the old-growth forest that WCWC is campaigning to save.

September 14, 1999 The Canadian Environmental Protection Act becomes law. This Act was opposed by environmentalists because of the weak protection it provided. In the course of its passage, all amendments to strengthen it were defeated.

September 15, 1999 At about 1 p.m. about 100 loggers attack the protesters' camp beside the logging road in the Upper Elaho valley, injuring three people and destroying all the protesters' belongings including almost all film documentation of this attack. WCWC Millennial Tree Research Camp coordinator, James Jamieson, is also assaulted and WCWC VHF radio is stolen. Not one B.C. MLA speaks out against this violence.

September 28, 1999 The Vancouver Sun publishes an eight-page full-colour full-sized insert to their paper titled THE MILLENNIUM/GEORGIA STRAIT: Special Report. It features an article titled Sentinels of Time about the ancient Douglas firs discovered by WCWC researchers in the Upper Elaho Valley.

October 1, 1999 B.C. Forest Minister David Zirnhelt gives approval in principle to the subdivision of MB's TFL 44 and the transfer of the MB's holding in Clayoquot Sound to Iisaak Forest Resources Limited, a company that has 51 percent First Nations' control.

October 18, 1999 Honourable Charles Caccia, MP for Davenport, introduced his Private Member's Bill (Bill C-236) in the House of Commons calling for an amendment of Canada's National Park Act to include a 500,000-hectare Stoltmann National Park Reserve, located near the world-famous resort community of Whistler, B.C.

November 10, 1999 The Greater Vancouver Water District Administration Board passed a five-point resolution to fully protect the three watersheds that provide one-half of B.C.'s population with its domestic water supply from industrial use, a successful conclusion to eleven years of research and consistent campaigning by local environmental activists.

November 24, 1999 MP Charles Caccia's Private Member's Bill C-236 to amend the National Park Act (Stoltmann National Park) is hotly debated in the House of Commons for an hour. It then "died without a vote" as is customary with Private Member's Bills.

Fall 1999 WCWC-Victoria publishes a four-page full-colour tabloid-sized newspaper titled Victoria Sea-to-Sea Green-Blue Belt (Educational Report Vol. 18. No. 8). Press run 50,000. It chronicles the successes so far in the conservation campaign to protect the Sooke Hill and outlines what remains to be protected.

December 1999 Film-maker Daniel Gautreau premiers his video documentary Hoods In the Woods. It tells the tale of the September 15, 1999 violent attack by employees of International Forest Products (Interfor) on environmentalists and the destruction of their camp, which was located alongside the logging road in the Upper Elaho Valley.

December 18, 1999 Julia Butterfly Hill descends from Luna (the name she had given the ancient redwood she had been "tree sitting" in continuously for over two years) after concluding a deal with Pacific Lumber/Maxxam Corporation to save "Luna" and a three-acre buffer zone around it.

December 1999 WCWC becomes registered as a non-profit society in Manitoba and WCWC canvassers begin going door-to-door in a campaign to stop logging in Manitoba's provincial parks.

January 1, 2000 Research scientists report that only 22 percent of the world's original forests remain intact. Seventy percent of that is found in three countries—Russia, Brazil and Canada.

January 10, 2000 UNESCO designates Clayoquot Sound on the west coast of Vancouver Island, B.C. a Biosphere Reserve.

February 10, 2000 Criminal charges are laid against three men for the September 15, 1999 assault on the environmentalists protesting in the Upper Elaho Valley.

February 24, 2000 Ujjal Dosanjh, New Democratic Party leader, becomes Premier of B.C. and governs until June 5, 2001.

March 2000 The B.C. NDP government proposes to pass "Working Forest" legislation to give forest corporations more secure control over B.C.'s publicly owned forests. It is killed by a huge outcry from the public. This Act would have locked up the existing public forestland, allowing no further withdrawals for parks or protected status without compensation or an equivalent amount of forestland withdrawn from existing parks. Forestland for land claim settlements would be further alienated and more difficult to acquire if this legislation had been enacted.

March 23, 2000 The Panel on the Ecological Integrity of Canada's National Parks (launched in 1998) releases its report. The panel concluded, "ecological integrity in Canada's national parks is under threat from many sources and for many reasons." Glacier National Park was one of Canada's big parks that had the least ecological integrity i.e. it does not encompass the full range of the animals found within it.

Spring 2000 WCWC publishes a 16-page full-colour tabloid-sized newspaper titled 1999-2000 Members Report – Entering a New Millennium of Wilderness Conservation (Educational Report Vol. 19 No. 1). Press Run 50,000. The centre spread highlights campaign activities from 1997-2000.

May 2000 The Manitoba government releases its Woodland Caribou Conservation Strategy for Manitoba. It classifies the Nopiming caribou herd as being at "High Risk," with the main threat to its survival being "timber harvesting operations," yet it does not recommend that logging in critical habitat be banned.

May 5, 2000 At a ceremony on Long Beach near Tofino, Clayoquot Sound is officially designated as a UNESCO Biosphere Reserve. Prime Minister Chrétien attends the celebration and announces a $12 million federal grant to endow the Clayoquot Biosphere Trust (CBT), the cornerstone of the new Clayoquot Sound UNESCO Biosphere Reserve.

May 30, 2000 The Squamish Nation and Interfor sign a Letter of Intent to develop a negotiated agreement that would see substantive forestry issues from both groups addressed to meet the needs of both parties in a mutually beneficial manner. Interfor agrees to limit logging in the Upper Elaho to existing approved cutblocks and not seek the approval of any new ones.

June 7, 2000 Interfor becomes the first logging company in B.C. history to have its injunction, which was in force for 11 months against logging protesters in the Upper Elaho Valley, struck down by the B.C. Supreme Court. Justice Vickers who heard the case put forward by the Sierra Legal Defence Fund on behalf of WCWC writes in his judgment *"I have reluctantly concluded that the plaintiffs [Interfor]...are not before the court with clean hands. [Referring to the September 15, 1999 assault on protesters that the company participated in.] The injunction must be set aside in order to reserve the integrity of the court's process."*

June 16, 2000 Representatives of WCWC, Valhalla Wilderness Society, Friends of Clayoquot Sound, Raincoast Conservation Society and the Forest Action Network call for a boycott of Interfor's wood products because of this company's destructive logging practices and its workers' violent behaviour towards conservationists trying to protect the old-growth forests the company is logging.

June 19-30, 2000 WCWC launches its "Wedge to the Ledge [Legislature]" campaign, an 11-day-long trek from the Upper Elaho to the B.C. Legislature Buildings in Victoria by about a dozen volunteers supporting the protection of the Stoltmann Wilderness. Traveling by mostly land and partly by sea, the volunteers transport by wheelbarrow a big wedge of wood from the undercut of a thousand-year old Douglas fir felled by loggers. WCWC holds a rally to try to save this wilderness on June 30, 2000 when the wedge and the trekkers arrive in front of the Legislature. Later WCWC releases a video documenting this trek titled *Wedge to the Ledge*. This wedge is now on display at WCWC-Victoria's storefront office on Johnson Street in Victoria.

July 2000 WCWC publishes 50,000 copies of a four-page full-colour tabloid-sized newspaper titled *Help Make Mt. Elphinstone a Provincial Park now!* (Educational Report Vol. 19 No. 2). It is part of the long-standing effort to save 1,500 hectares of forest on the lower slopes of Mt. Elphinstone near Roberts Creek on the Sunshine Coast. This forest has the highest diversities of fungi found anywhere in Canada. Less than one percent of the Sunshine Coast region's forestland is in any kind of protected status.

August 2000 *One Whistler*, a local Whistler business and government tourism organization releases an economic study it commissioned Grant Thornton International to undertake to look at the benefits of a Stoltmann park. The study concludes that the economic benefits of protecting the forest in a park far exceeds the benefits of logging it. It predicts that up to 730 jobs could be produced through the preservation and tourism use of this wilderness area.

September 15, 2000 Betty Krawczyk, a 70-year-old grandmother, and Barney Kern, a young activist, are given one full year in jail with no time off for good behaviour for contempt of court in their efforts to protect the Elaho Valley. Justice Parrett of the B.C. Supreme Court, after 43 days of testimony, besides sentencing Krawczyk and Kern, sentenced three other non-violent civil disobedient Elaho defenders who had blockaded the logging road in protest of the logging to unprecedentedly long jail terms.

November 14, 2000 The B.C. government announces that it will adopt the recommendations of the Mackenzie LRMP and add 1.9 million hectares to the Muskwa-Kechika Management Area in northern B.C., making the total conservation management area 6.3 million hectares. Park and protected areas are to comprise a quarter of the conservation management area; with special management areas comprising the rest. The government also increases the MKMA Advisory Board funding to $3 million annually. The government eventually establishes this "package" in legislation in March 2001, culminating over eight yeas of campaigning by the Canadian Parks and Wilderness Society and the Chetwynd Environmental Society.

Fall 2000 WCWC publishes an eight-page full-colour tabloid-sized newspaper titled *Save the Stoltmann Wilderness and its 1,000-year-old trees –FIVE YEAR FIGHT TO SAVE THE WILD ELAHO VALLEY* (Educational Report Vol. 19 No. 3). Press run 150,000. It tells the story of WCWC's 11-day "Wedge to the Ledge" trek made by WCWC volunteers taking a slice of a 1,000-year-old Douglas fir from a clearcut in the Stoltmann Wilderness to the Legislature in Victoria in June 2000.

2001 Canada signs on to the Kyoto Protocol, committing Canada to stabilizing its greenhouse gas emissions at six percent less than the country's 1990 level by the year 2012. That was over 30 percent lower than the nation's level at the time of signing.

January 2001 Betty Krawczyk is released from jail after serving over seven months in total of her one-year sentence for an act of non-violent civil disobedience in blocking logging trucks going into the Elaho Valley in the proposed Stoltmann Wilderness.

January 18, 2001 The 26,000-hectare Snowy Provincial Park along the U.S. Canadian Border and other parks totaling 97,000 hectares are established in the Okanagan region of B.C.

February 2001 The remaining two unprotected valleys in the Sooke Hills Sea-to-Sea Green Blue Belt (the Ayum and Charters Valleys) become protected. Only a few other blocks of private property remain to be acquired to complete the belt.

February 2001 A federal Species at Risk Act is introduced into the House of Commons. Environmental groups roundly criticize this legislation for being too weak and not applying to all of Canada. The proposed Act only protects endangered species on federal lands.

February 8, 2001 The NDP government announces a three-year moratorium on grizzly bear hunting while independent studies are undertaken by scientists to establish the population size and long-term viability of species.

Spring 2001 The Squamish Nation conducts an extensive community-based land use planning process for the forests and wilderness areas of their territory.

April 4, 2001 After intense negotiations, environmental groups, First Nations, logging companies, workers, communities and the provincial government agree to a new approach to conservation and sustainable management in the Great Bear Rainforest and Haida Gwaii (the Queen Charlotte Islands). If implemented in its entirety, this agreement could set British Columbia on the path to becoming a global leader in environmental stewardship. The B.C. provincial government also announces a "Spirit Bear Protection Area" and promises protection of approximately 135,000 hectares—more than half of the Valhalla Wilderness Society's Spirit Bear Conservancy Proposal (but not enough to protect this unique bear according to bear biologists who have studied its habitat needs).

May 2001 The Squamish Nation releases its land use plan, called Xay Temixw (Sacred Lands), that calls for the protection of the Upper Elaho, Sims Creek and other areas to be managed by the Squamish Nation as "Wild Spirit Places" where no industrial activities are allowed take place.

May 2001 The day before calling a provincial election, B.C. Premier Ujjal Dosanjh establishes, by Order in Council, a 71,400-hectare South Chilcotin Mountains–Spruce Lake Provincial Park in the Lillooet Region near Gold Bridge.

June 5, 2001 Gordon Campbell, Liberal, elected the Premier of B.C.

Summer 2001 WCWC-Victoria publishes 100,000 copies of a four-page tabloid-sized paper titled *Park the Upper Walbran Valley* (Educational Report Vol. 20 No. 1) advocating adding this 5,000 hectare region, which still has more than half of its original old-growth rainforest with huge trees still intact, to the Carmanah Walbran Provincial Park.

Summer 2001 WCWC publishes *The Marbled Murrelets of the Caren Range and Middlepoint Bight* by Paul Harris Jones. The book contains his detailed observations—the first ever—of active nesting Marbled Murrelets.

Summer 2001 WCWC produces a four-page two-colour demi-tabloid-sized newsletter titled *Bow Valley UNDER SIEGE* (EDUCATIONAL Report Vol. 20 No. 2). Press run 5,000. It calls for the full preservation of the narrow wildlife corridor in the fragile Bow Valley near Canmore, Alberta.

July 19, 2001 The new Liberal B.C. government reinstates grizzly bear hunting in B.C. by lifting the three-year-long moratorium that had been put in place by the previous government only six months before.

July 24, 2001 The Alberta government establishes the 591,008-hectare Caribou Mountains Wildlands Provincial Park. This area, which Gray Jones, WCWC-Alberta director, had championed for years to protect, abuts on the southwest boarder of Wood Buffalo National Park and is the last stronghold of woodland caribou in Alberta.

Summer/Fall 2001 WCWC publishes 100,000 copies of a four-page three-colour tabloid-sized newspaper titled *Manitoba Parks at Risk* (Educational Report Vol. 20 No. 3). Press run 50,000. It calls for the end to logging in Manitoba's Provincial Parks.

September 14, 2001 XwaYeN (Race Rocks) becomes Canada's first Marine Protected Area. A tiny area 17 kilometres southwest of Victoria, B.C., was originally protected in an Ecological Reserve established in 1980. XwaYeN is comprised of a 251.4-hectare marine area extending down to 38 metres below sea level, and a 1.5-hectare land area totaling up to 251.9 protected hectares.

November 2001 The new B.C. government decides to review the decision to create the Southern Chilcotin Mountains Park, stating that it will come up with a new decision by March 2002.

Fall 2001 WCWC publishes a two-page three-colour tabloid-sized newspaper titled *Expand MacMillan Park – Protect all of Cathedral Grove* (Educational Report Vol. 20 No. 4). Press Run 100,000. This paper makes a strong plea to expand the 157-hectare park along the road to Port Alberni and Clayoquot Sound to include all the lands around Cameron Lake and the entire valley bottom at the head of the Cameron Lake where many giant unprotected Douglas firs still stand. The paper expresses WCWC's strong opposition to the B.C. government's plan to put a parking lot right beside the tiny park on the upwind side of the existing park that is already suffering from blow down.

Winter 2001 WCWC publishes a four-page three-colour tabloid-sized newspaper titled *OIL SPILL? – A grim reality accompanies oil and gas development on B.C.'s stormy coast* (Educational Report Vol. 20 No. 6). Press run 50,000. It provides reasons for continuing the moratorium on offshore oil and gas exploration and to develop alternative wind and tidal energy clean renewable energy. It is WCWC's response to the B.C. government's vigorous efforts to get the moratorium lifted and exploratory drilling underway in the Hecate Strait and other areas off the coast of B.C.

Winter 2001-2002 WCWC publishes a two-page three-colour tabloid-sized newspaper titled *Blue Mountain – Time to Protect Maple Ridge's BACKYARD WILDERNESS* (Educational Report Vol. 20 No. 5). Press Run 100,000. This paper makes a strong plea for the B.C. government to protect the older second growth forest on the 7,500-hectare side hill and ridge on the south side of Alouette Lake across from Golden Ears Provincial Park in B.C., an area that local environmentalists are fighting to save.

Winter 2001-2002 WCWC publishes its 9th newspaper in its campaign to save the Stoltmann Wilderness, a four-page four-colour tabloid-sized newspaper titled *Saving the Stoltmann Wilderness Valley by Valley* (Educational Report Vol. 20 No. 7). Press run 120,000 copies. It reports on the *2001 Squamish Nation's Land Use Plan* that calls for the establishment of Wild Spirit Places including the 17,753-hectare *Nsiiwx-nitem tl a sutch* that encompasses all the wilderness forest north of Lava Creek (above the existing clearcuts) in the Upper Elaho Valley.

Winter 2002 WCWC publishes a four-page full-colour tabloid-sized newspaper titled *B.C.'s Fabulous SOUTH CHILCOTIN PARK – WILDERNESS IN PERIL* (Educational Report Vol. 21 No. 1.). Press run 50,000 with reprint of 50,000 in the summer of 2002. It tells the story of the recently elected government's move to dismantle the community land use planning process and lift the logging moratoriums in the Lillooet area and cancel the South Chilcotin Mountain Park created by the NDP. It implores the public to rally to save this precious rainshadow wilderness area.

February 2002 The B.C. Supreme Court quashes the decision of the B.C. Forest Service to allow logging in endangered marbled murrelet nesting habitat in Jervis Inlet on the Sunshine Coast. (Less than 10 percent of the bird's original nesting habitat still remains on the Sunshine Coast.) It was the successful culmination to the court action taken by the Sunshine Coast Conservation Society and WCWC with the financial backing of the West Coast Environment Law's dispute resolution fund, which was heard by the court on January 10, 2002.

Spring 2002 WCWC publishes a four-page full-colour tabloid-sized newspaper titled *Bugaboo Rainforest – CANADA'S INLAND RAINFOREST – NATIONAL PARK PROTECTION IS NEEDED FOR GLOBALLY UNIQUE TREASURE* (Educational Report Vol. 21 No. 2.). Press Run 80,000. It calls for park protection for a 500,000-hectare wilderness link between Canada's Glacier National park and B.C.'s Purcell Wilderness in the southern Columbia Mountain Range.

March 6, 2002 The Haida Nation formally launches its groundbreaking claim to aboriginal title to the Queen Charlotte Islands/Haida Gwaii.

Spring/Summer 2002 WCWC-Manitoba publishes a four-page full-colour tabloid-sized newspaper titled *Manitoba's East Shore Wilderness –Largest Intact Boreal Forest–Wild, Roadless & Threatened–a conservation vision for Lake Winnipeg's East Shore* (Educational Report Vol. 21 No. 3). Press Run 70,000. It calls for the protection of 1,500,000 hectares—the largest single area of intact, roadless boreal wilderness left on Earth—from industrial exploitation including clearcut logging, mining and hydro developments.

April 26, 2002 Ministry of Forests District Manager, Jerry Kennah, Registered Professional Forester, approves logging in spotted owl habitat in Siwash Creek, north of Chilliwack over objections of biologists at Ministry of Water, Land and Air Protection (MWLAP).

April 30, 2002 The B.C. government releases a white paper proposing to change the Forest Practices Code to a "results based code" where the industry would make up their own rules on how to log and would monitor their own conformance to save the companies' money.

May 2002 The B.C. government introduces and passes legislation—Bill 39—that involves taxpayers' money being given to logging companies as compensation for lost cutting rights in 376 provincial parks created since 1995. Environmental groups decry this as a direct payback to companies who had been the biggest donors to the Liberal party the year before (election year).

May 7, 2002 The WCWC three-day-long court hearing of the Forest District Manager's decision to approve logging in the endangered spotted owl old-growth habitat commences.

May 24, 2002 The Greater Vancouver Regional District approves a Watershed Management Plan that puts an end to the threat of logging in the three watersheds that provide drinking water to the Greater Vancouver region. It contains policies to restore areas to their natural states that were disturbed by man in the past and that now negatively impact water quality.

June 20, 2002 Sierra Legal Defence Fund acts for WCWC seeking a judicial review of a decision to allow logging in a Spotted Owl Activity Site. The court grants an interim injunction to stop logging in the spotted owl habitat to provide time for WCWC's lawyers to prepare documents.

June 21, 2002 The Greater Vancouver Water District (GVWD) gives the provincial government 24 months notice, effective June 30, 2004, that it will terminate the Amending Indenture to the district's 999-year lease that allows logging in the drinking watersheds of Greater Vancouver.

Summer 2002 WCWC publishes a four-page three-colour tabloid-sized newspaper titled *Help save B.C. Parks and Wildlife – Liberal Government endangers Beautiful B.C.* (Educational Report Vol. 21 No. 4). Press run 70,000. It outlines the financial cuts made to the Water, Air, Land and Air Protection Ministry that are gutting the parks, closing provincial camp grounds, and even threaten to take away parks like the Southern Chilcotin Mountains Provincial Park.

July 11, 2002 House of Commons passes Canada's first act to protect endangered species in Canada. Many conservationists contend that Canada's Species at Risk Act is extremely weak and because it only covers species on federally controlled land, it will be ineffectual in providing any real protection.

July 13, 2002 The Attorney General of B.C. settles court action without a hearing, consenting to Sierra Legal Defence Fund lawyer's request that the B.C. Forest Service's decision to allow logging in Northern Spotted Owl Activity Areas be rescinded and sent back for further consideration.

July 19, 2002 Order issued by the B.C. Supreme Court rescinding the original approval of the Forest Development Plan that allows logging in the spotted owl habitat.

August 22, 2002 Iisaak, the First Nations' own forestry company in Clayoquot Sound cuts its first tree. WCWC representatives are there to witness.

August 29, 2002 Mr. Justice James Shabbits of the Supreme Court of B.C. dismisses Sierra Legal and the Wilderness Committee's court challenge to allow logging in a critical Northern Spotted Owl habitat area. Although finding that the spotted owl is in grave risk of extinction, he states: *"The [B.C.] Legislature could have enacted legislation that protects the Owl from the risk of extirpation caused it by harvesting of old growth forests. In my opinion, it did not do so with the enactment of s.41(1)(b) of the Forest Practices Code."*

September 4, 2002 Ministry of Forests District Manager, Cindy Stern, RPF, makes new decision regarding logging in a specific spotted owl activity area. She rescinds District Forest Manager Kennah's decision regarding logging in three areas, but approves it in a fourth on the basis that the logging there will *"maintain or enhance"* the owl habitat.

September 2002 The B.C. government re-establishes a Spotted Owl Recovery Team to help determine why populations are declining, and to recommend actions to conserve and recover populations.

September 12, 2002 The B.C. government lifts the 1995 moratorium on establishing new fish farms on the B.C. coast. An 89 percent collapse of the pink salmon runs in the Knight Inlet occurs this same fall. Many believe it is caused by the transference of sea lice from the penned salmon to the wild stock.

September 17, 2002 WCWC, Sierra Legal Defence Fund and Forest Watch release *Logging to Extinction: The Last Stand of the Northern Spotted Owl in Canada*. The report is the result of thousands of hours of research and clearly links the rapid decline of the spotted owl to the destruction of their habitat due to industrial logging. The report also names the top ten destroyers of spotted owl habitat.

November 2002 BC Hydro publishes a "Green Energy" report on tidal current energy prospects for British Columbia, compiled by Triton Consultants of Richmond BC. The report extols the virtues of fuel and pollution-free tidal energy and asserts that BC could be meeting up to 40 percent of its annual generating capacity using present-day technologies.

November 2002 WCWC publishes *The last Voyage of the Black Ship*, a 36-page full-colour epic myth by Haida artist Michael Nicoll Yahgulanaas. It tells the story about logging and the demise of ancient cedars on Haida Gwaii. It is WCWC's first graphic novel publication.

Fall 2002 WCWC co-publishes with the Union of B.C. Indian Chiefs a four-page full-colour tabloid-sized newspaper titled *Wild Fish need Wild Rivers and Oceans – Wild Fisheries Being Destroyed by Government Policy* (Educational Report Vol. 21 No. 5). Press run 50,000. This paper explains how logging has led to a collapse of the wild fisheries on B.C.'s west coast and the rise of salmon fish farming which has furthered that collapse.

December 12, 2002 The federal Species at Risk Act receives Royal Assent. It will come into force in 2003 after regulations are drafted and put in place.

December 17, 2002 Sixty people in a fleet of 14 boats arrive at Ocean Falls to protest the construction of a fish farm hatchery by Omega (owned by financially troubled PANFISH). The protestors are predominantly Heiltsuk and Nuxalk First Nations, but representatives of the Forest Action Network (FAN), Raincoast Conservation Society, commercial fishermen and other local citizens also attend. Once on site, members of FAN, in an act of civil disobedience, successfully deconstruct part of the concrete foundation of the hatchery.

December 17, 2002 The Sierra Legal Defence Fund commences court challenge for WCWC of another Forest Service decision to approve logging in another known spotted owl habitat site.

December 17, 2002 The "streamlining" amendments to the Forest Practices

Code of British Columbia Act and regulations come into effect. According to the B.C. Liberal government, these amendments give licensees "immediate relief from regulatory burden" which licensees can enjoy through the two-year transition period until the government's new Forest and Range Practices Act "performance based code" is fully implemented in April 2005. B.C. environmentalist and conservation groups decry this move that weakens environment standards in logging public forestlands.

2003 The Government of Canada signs an agreement with the government of B.C. to create the Gulf Islands National Park Reserve of Canada and an Inuit Impact and Benefit Agreement with Inuit and the government of Nunavut to create Ukkusiksalik National Park of Canada. This brings the number of national parks that have been created to 41 since Banff, the first national park, was established in 1885.

January 22, 2003 The B.C. government releases its white paper *Working Forest Initiative for British Columbia*. It proposes to legally designate all of BC's forested Crown Lands (45 million hectares) outside existing parks as a guaranteed logging zone, and streamline or facilitate the sell-off (i.e. outright privatization) of Crown lands to private real estate developers.

January 28, 2003 The B.C. government announces the introduction of user parking fees in 28 of the most popular provincial parks in B.C. It also announces that interpretative and educational park programs cut in 2002 will not be reinstated.

January 31, 2003 The B.C. Forest Practices Board issues a report titled *Faster action, incentives needed to protect marbled murrelet habitat*. The report calls for immediate action to save the remaining coastal old-growth forest nesting sites of this endangered species. Board Chair Bill Cafferata says, "Government should designate interim wildlife habitat areas quickly, using the best available information, before the needed habitat is lost."

Winter 2003 WCWC publishes a four-page full-colour demi-tabloid-sized newspaper titled *Beautiful British Columbia – FOR SALE* (Educational Report Vol. 22 No. 1). Press Run 100,000. It alerts B.C. citizens to the proposal of the B.C. government to "protect" 45 million hectares of Crown lands in reserves (Working Forests) where industrial use predominates over other uses. It includes a special report by the Council of Canadians about international trade agreements and the hostile clauses in them that reduce the ability of Canadians to protect their parks and environment.

Spring 2003 WCWC publishes a four-page full-colour tabloid-sized newspaper titled *Fraser Valley Rain Forest ELK CREEK – Chilliwack's Wild Wonder Worthy of Protection* (Educational Report Vol. 22 No. 3). Press run 45,000. This proposed 2,000 hectare Elk Creek protected area has ancient Douglas firs between 7 and 9 feet in diameter with more than two dozen of them towering over 60 metres (22 office tower stories) tall.

March 13, 2003 The Heiltsuk First Nation and Greenpeace, representing several First Nations and environmental groups, petition the implementation body of the Convention for Biological Diversity in Montreal to help end the over-harvesting of western red cedar in British Columbia. They argue that Canada, having signed the Convention in 1992, is obliged to uphold the Convention's provisions to protect biodiversity and cultural traditions.

March 28, 2003 The Minister of Sustainable Resource Management, Stan Hagen, and the Council of the Haida Nation President, Guujaaw, announce that the B.C. government and the Haida Nation have signed a framework agreement to co-manage land use planning on Haida Gwaii/ Queen Charlotte Islands.

Spring 2003 WCWC publishes a four-page full-colour tabloid-sized newspaper titled *British Columbia's Endangered Forests – What government and industry aren't telling you* (Educational Report Vol. 22 No. 2). Press run 110,000. Based on a report by ForestEthics, it is packed with facts pointing to the need to save more of B.C.'s dwindling primary forest ecosystems.

Spring 2003 WCWC publishes a four-page full-colour tabloid-sized newspaper titled *ELK CREEK – Chilliwack's Wild Wonder Worthy of Protection* (Educational Report Vol. 22 No. 3). Press run 45,000 copies. It calls for the protection of the proposed 3,000-hectare Elk Creek area where the tallest giant old-growth Douglas firs in Fraser Valley still stand.

June 2003 WCWC-Manitoba publishes *Manitoba's Species At Risk* an 80-page WCWC report that lists the province's endangered species and the threats to their survival. This report, updated in October 2004, is available online.

Spring/Summer 2003 WCWC Manitoba publishes 50,000 copies of a four-page full-colour tabloid-sized newspaper titled *CARIBOU at the crossroads – Manitoba's threatened caribou herds at risk from proposed developments in the pristine East Shore Wilderness.* (Educational Report Vol. 22 No. 4). It provides compelling reasons why the wilderness on the east shore of Lake Winnipeg should be protected in a series of large interconnected parks.

August 2003 The B.C. government releases its report on the public comments regarding the implementation of "Working Forest" legislation in B.C. Only one percent out of the 2,692 who respond express support for it, 97 percent are opposed.

Summer/Fall 2003 WCWC publishes a four-page full-colour tabloid-sized newspaper titled *Rarest of the Rare – World's Only Inland Rainforest Under Threat – Help save rare ecosystem in southeastern British Columbia* (Educational Report Vol. 22 No. 5). Press run 60,000. It makes a strong case for protecting both the proposed Goat River Protected Area and the proposed Bugaboo Rainforest National Park on the windward side of the Columbia and Rocky Mountains.

October 29, 2003 The B.C. government passes Bill 46 enabling legislation that empowers the B.C. Cabinet to establish various land-use designations, such as the Working Forest Initiative, through Cabinet Orders in Council, which would allow the government to make big land use designations without having to take them through legislative debate.

November 6, 2003 Canada ratifies the UN International Law of the Sea convention, becoming the 144th nation to do so.

2004 The David Suzuki Foundation publishes *Clearcutting Canada's Rainforest – Status Report 2004*. It provides evidence that the commitment by the B.C. government in April 2001 to ensure that forest practices on B.C.'s mid coast are more environmentally responsible is not being fulfilled.

Winter 2004 WCWC publishes a four-page full-colour tabloid-sized newspaper titled *PROPOSED WORKING FOREST DENIES B.C. CITIZENS THEIR PUBLIC LAND* (Educational Report Vol. 23 No. 1). Press Run 50,000. It explains the B.C. government's plan to implement "Working Forest" enabling legislation, a first big step to privatizing B.C.'s publicly owned forest and a move to freeze the protected area land base at its current level of about 12 percent despite massive opposition (97 percent opposition expressed during the government's three-month-long public consultation process).

January 2004 WCWC publishes a 50-page research report titled *An Alternative Recovery Strategy for the Northern Spotted Owl in British Columbia, Canada* by Andrew Miller M.Sc. who resigned earlier this same month from the B.C. government's Spotted Owl Recovery Team in protest over government inaction to protect this critically endangered bird.

January 22, 2004 The B.C. & Yukon Chamber of Mines, the Mining Association of British Columbia and the Council of Tourism Associations in B.C. agree to accommodate each others' interests in regards to land use in B.C., which means the mining sector made a public promise to stay out of B.C.'s parks and protected areas.

January 31, 2004 The B.C. government's new Forest and Range Practices Act comes into force and replaces a weakened Forest Practices Code that is full of specific proscriptions with "performance based" goals with the companies monitoring their own compliance with the Act.

February 5, 2004 An Agreement in Principle is signed between all levels of government and the owners of Burns Bog to purchase over 2,100 hectares of Burns Bog near Boundary Bay (over five times larger that Vancouver's Stanley Park) for $73 million to preserve as a park. This successful deal culminates a 15-year campaign by the Burns Bog Conservation Society to protect the area. This society continues to fight to protect an additional 500 hectares of critical habitat left out of the park acquisition.

February 27, 2004 WCWC, the Sierra Club of Canada, ForestEthics and the David Suzuki Foundation, represented by the Sierra Legal Defence Council, petition Environment Minister David Anderson to invoke the emergency provision of Canada's Species At Risk Act to protect the spotted owl, noting the sharp decline in their population and the failure of the B.C. government to do anything about it.

Spring 2004 WCWC publishes 50,000 copies of a four-page full-colour demi-tabloid newspaper titled *Mining our Parks – can you dig it?* (Educational Report Vol. 23 No. 3). Press run 50,000. It details the threat of mining in our provincial parks, especially in the newly created South Chilcotin Mountains Park and the Tatshenshini Park.

May 2004 The Council of the Haida Nation publishes an 18-page document titled *Haida Land Use Vision* that presents its land use management principles.

May 5, 2004 David Anderson, federal environment minister, in response to the request made by environmental groups on February 27, 2004, declines to intervene using the emergency provisions of the federal Species at Risk Act to help save the spotted owl from going extinct in B.C.

June 3, 2004 The Supreme Court of B.C. orders WCWC to pay costs to Cattermole Timber arising from WCWC's failed attempt to protect endangered spotted owls' habitat from being logged.

June 18, 2004 WCWC launches its "B.C. Parks Lovers' Roving Kiosk Tour." Kicking off in Manning Park, the tour visits 10 parks during the next 13 weeks, educating people about how government cuts, including a 30 percent reduction in park staff, and the introduction of user fees and parking fees, privatization of services and industrial developments are damaging B.C.'s provincial park system and reducing Park visitor numbers.

June 28, 2004 Paul Martin, Liberal, elected Prime Minister of Canada.

June 30, 2004 The B.C. government enacts legislation that re-establishes the Greater Vancouver Water District board's local, autonomous control over the Crown forestlands within 60,000 hectares of Greater Vancouver's three drinking water source watersheds, the Capilano, Seymour and Coquitlam for the purpose of protecting the water resource.

Summer 2004 WCWC publishes a four-page three-colour demi-tab newspaper titled *Last Remnant of B.C.'s vanishing Douglas fir forest... CATHEDRAL GROVE* (Educational Report Vol. 23 No. 6). Press run 20,000 copies. The paper calls on the government to cancel plans to build a parking lot right next to the grove of big Douglas firs and to greatly expand the tiny MacMillan Provincial Park to protect whole bottom valley and all the big old growth left there.

July 2004 The St'at'imc release its preliminary land use plan for the northern portion of the St'at'imc's traditional territory. Lost Valley, along with the Upper Bridge River, Southern Chilcotin Mountains and Melvin Creek Valley are designated "Protection Areas" where industrial activities are not permitted.

July 17, 2004 B.C. Premier Gordon Campbell announces the creation of a special initiative to work towards establishing a Spirit Bear Conservancy that would protect the entire critical, unprotected Green watershed. It is the first time that a B.C. government admits that the spirit bear (white-coated Kermode black bear) is endangered and the first time any B.C. government commits to working towards saving all of the proposed conservancy to protect this unique bear.

July 22, 2004 The B.C. government stuns the environmental community by rolling back the boundaries of the South Chilcotins Mountain Park to accommodate mining interests. In response to heavy lobbying by B.C.'s powerful mining industry, backed by industrial giant Teck Cominco, the government slices off 14,600 hectares from the popular park, which is home to grizzly bears, mountain goats, bighorn sheep and a vibrant tourism industry.

July 29, 2004 The B.C. government announces that it will not implement "Working Forest" legislation through new Orders in Council and that it will continue to manage B.C.'s public forests through existing laws.

September 2004 The B.C. Forest Practices Board issues a special report titled *BC's Mountain Caribou: Last chance for Conservation*. It urges the government to take quick action to implement more effective conservation measures before it is too late.

October 2004 The B.C. government creates a Species at Risk Coordination Office (SARCO) to, it said, address mounting public concerns for the future of three endangered species: spotted owl, mountain caribou and marbled murrelet.

November 8, 2004 An international team of more than 300 scientists releases the results of a four-year study of the arctic climate. Titled *Arctic Climate Impact Assessment Report*, the scientists conclude that the ice in Greenland and the Arctic is melting so rapidly, due to global warming, that much of it could be gone by the end of the 21st century.

November 18, 2004 The Supreme Court of Canada delivers its judgments in the Haida Nation v. British Columbia (Minister of Forests) and the Taku River Tlingit First Nation v. British Columbia. It rules that the province has to enter into honourable, meaningful negotiations with First Nations and balance and accommodate their interests with the interests of society at large. *"The government's duty to consult with Aboriginal peoples and accommodate their interests is grounded in the principle of the honour of the Crown, which must be understood generously."*

November 26, 2004 An estimated 165,000 litres of oil spills from Petro-Canada's Terra Nova platform on the Grand Banks about 350 kilometres from St. John's, Newfoundland. The largest oil spill to arise from Canada's east coast offshore oil industry, it forms a slick occupying a 57-square-kilometre area. Tens of thousands of seabirds congregated in the area are threatened and because of its remote location most of the bird mortality will never be observed or counted.

November 30, 2004 The Valhalla Wilderness Society presents a long-term conservation vision for a landscape 1000 kilometres in length and 200 kilometres in width, spanning nearly 15 million hectares from North of Prince George to the U.S. border, in the Columbia Mountains. It encompasses protection for 55 percent of the area, or 8 million hectares of B.C.'s inland rainforest. The multi-stakeholder plan, prepared with input from top scientists, presents a vision to maintain ecosystems and avoid extinction of the critically endangered mountain caribou.

December 3, 2004 The Manitoba government establishes the Manigotagan River Provincial Park located approximately 150 kilometres northeast of Winnipeg, consisting of a 1500-metre-wide corridor along a 55-kilometre-long stretch of the Manigotagan River.

December 16, 2004 The U.S. government lists the Southern Resident population of orca whales that live in the waters off the coast of B.C. and Washington State as an endangered species under the U.S. Endangered Species Act. The Sierra Legal Defence Fund, acting on behalf of WCWC and other Canadian environmental groups, as well as U.S. environmental groups had been pressing for such a listing for over a year.

December 26, 2004 The largest earthquake to strike on earth in 40 years unleashes a tsunami in the South Asia Sea that kills over 150,000 people.

December 2004 The World Conservation Union, comprising 800 non-government organizations and 10,000 scientists, releases its red list of endangered animals. It contains 15,000 species (an estimated 15 percent of the total number of species that are threatened with extinction).

January 22, 2005 The federal government signs the Labrador Inuit Park Impacts and Benefits Agreement with the Labrador Inuit Association, and a Memorandum of Agreement for a National Park Reserve in the Torngat Mountains with the government of Newfoundland and Labrador. This leads to the formal establishment of Canada's 42nd national park—the Torngat Mountains National Park Reserve of Canada—when the Labrador Inuit Land Claims Agreement is signed.

Winter 2005 WCWC co-publishes with the Seton Lake Band a four-page full-colour tabloid-sized newspaper titled *Save Lost Valley* (Educational Report Vol. 24 No. 1). Press Run 50,000 copies. It contains information about the St'at'imc peoples' efforts to protect the wilderness in their territory. One of these efforts supported by WCWC involves the relocation of a traditional trail in the pristine Lost Valley so that tourists can eventually use it. The trailhead into this 10,000-hectare valley that contains a "pocket rainforest" is located only four kilometres from Seton Portage.

February 2005 WCWC co-publishes with Labour Environmental Alliance Society a four-page tabloid-sized newspaper titled *Turning the Tide – Protecting our health & the marine environment from toxic pollution* – (Educational Report Vol. 24 No. 2). Press Run 85,000 copies. This paper details the connections between the loss of biodiversity, skyrocketing cancer rates and the unprecedented levels of toxic pollutants in the environment.

February 16, 2005 The Kyoto Protocol to reduce heat-trapping emissions that cause global warming to six percent below 1990 level by the year 2012 enters into force.

March 15, 2005 The B.C. government announces that MacMillan Provincial Park, better known as Cathedral Grove, noted for its spectacular stand of old-growth Douglas fir, will nearly double in size from 157 hectares to 280 hectares as a result of a "groundbreaking" agreement between the provincial government, forest industry giant Weyerhaeuser and the Nature Trust of British Columbia. The $5.5-million partnership expands the park on both sides of the highway heading toward Port Alberni.

April 2005 WCWC publishes a four-page, four-colour tabloid-sized newspaper titled *Vote Wild* (Educational Report Vol. 24 No. 3). Press run 125,000. It urges people to consider the environment when choosing who to vote for in the May 16, 2005 election.

May 2005 WCWC co-publishes with Sierra Club and ForestEthics a four-page, four-colour tabloid-sized newspaper titled *CARIBOU NATION* (Educational Report Vol. 24 No. 4). Print run 90,000. This paper looks at the state of the caribou, the only large mammal species to live in all Canadian provinces, and calls for coordinated action by governments to take measures to stop its rapid decline towards extinction and effect its recovery.

June 2005 WCWC publishes an eight-page, four-colour tabloid-sized newspaper titled *Vancouver Island Conservation Vision* (Educational Report Vol. 24 No. 5. Press Run 50,000 copies. It calls for 41 percent of Vancouver Island to be fully protected from logging and mining. It is part of an ongoing campaign initiated in 1993 to establish a viable interconnected network of wilderness areas on Vancouver Island.

July 2005 WCWC publishes a four-page, four-colour tabloid-sized newspaper titled *Squamish Nation Wild Spirit Places* (Educational Report Vol. 24 No. 6). Press run 75,000.

August 3, 2005 WCWC celebrates its 25th birthday – a quarter century of reasonably successful campaigning for wilderness conservation.

Appendix III
Campaigning Insights Summarized

Below are some of the insights from this book. They represent 25 years of lessons learned from many different campaigns and activities including canvassing, trailbuilding, conducting research, educating and mobilizing citizens and producing and distributing voluminous amounts of educational calendars, newspapers, posters, reports and merchandise. The activities all moved forward WCWC's Constitutional Objectives (1) *To educate the public concerning Canada's wilderness heritage and to preserve representative areas for future generations*, (2) *To conduct research concerning wilderness values*, (3) *To obtain and distribute information on areas in Western Canada which have potential for protective status*.

On public education – being your own *free press*

1. The only press that tells your side of the story accurately and fully is your own.
2. The more information you publish about a threatened wilderness area, the less likely it is that that area will perish.
3. Generating your own maps with clearly demarcated boundaries of the proposed protected area is an essential component of a wilderness preservation campaign.
4. Saturating every affected community with accurate information is vital in dispelling myths and softening the opposition.
5. Educational materials should focus on only one campaign issue. It must be understandable and interesting to someone who knows absolutely nothing about the issue.
6. To get through to peoples' hearts pictures speak way louder than words.
7. Images of ugliness created by industrial development contrasted with the magnificent beauty of a wild area are most effective.
8. The secret of marketing is all to do with output volume. Our incoming mail increased in direct proportion to the increase in our mail-outs.
9. Door-to-door canvassing is not about making money. It is a slow but absolutely sure way to develop and maintain a grassroots organization.
10. Door-to-door canvassers are the "front line troops" in the war of ideas between the pro-development fanatics who turn their back on nature and those who believe that we must preserve nature to maintain a healthy planet and have a decent future.

On campaigning generally – how to make it fly

1. Take full advantage as soon as possible of every break you get!
2. If the opposition is focused on responding to your initiatives, you're winning.
3. Nothing is too audacious to try. And nothing is illegal unless a law or court decision makes it so.
4. Campaigns win when the group has a clear vision of the protection it wants, a strong dedicated leader, and never gives up.
5. In our own way, all of us are just winging it. Don't be afraid to improvise and try new tactics.
6. Every new stick added to a campaign fire helps it grow.
7. Make respect for aboriginal rights one of your major principles.
8. Campaigners rely on experts to supply facts, figures and insights.
9. Always double check your facts; even a small error, let alone a big one, can discredit everything you have done.
10. Use the tactic of delay: anything that delays the destruction of a wilderness area increases your chances of saving it.
11. Never undermine the efforts of another wilderness preservation group by selling out or trading off their area to save yours.
12. A big problem is the lack of details regarding the forest industry's destructive impact. It's essential to expose them.
13. Getting people into the wilderness for a transcendent experience empowers people for years, if not for their entire lives.
14. The essential work in all successful social movements has no glory and little, or no, recognition.
15. Avoid being diverted into time wasting activities that distract you from your main goal.
16. A many-pronged, loosely coordinated effort is much harder to thwart. One environmental group's campaign tactic may work better than another's at various times.

General philosophy – motivators in times of despair

1. Everything possible must be done to counter our culture's drift away from nature.
2. Fortunately, most people are naturally emotional about protecting nature and angered by its destruction.
3. Once you personally visit a wild place, it owns a piece of you.
4. It is impossible for those who have never "fallen in love" with a wilderness area to understand the behaviour of those who have.
5. Bushwhacking is the best of extreme sports; every hectare of wild primary forest is utterly unique and astonishingly beautiful beyond words.
6. Constantly hardworking, innovative people with long-term commitment will achieve social change.

Some small warnings – *watch out!*

1. There is no one in charge. No one is looking at the big picture steering us away from the enormous boreal deforestation that is unhinging earth's natural homeostatic mechanisms.
2. Many loggers realize in their hearts that environmentalists are not really their enemy; the big enemy is the multinational companies with excessive executive salaries and shareholder dividends that are not conserving enough for workers, nature and our children's security.
3. No Canadian government has been willing to give citizens a real chance to initiate their own laws democratically (e.g. B.C.'s Recall and Initiative Act is unworkable); so at least for now, we must focus on other approaches.
4. There is a double standard in the Canadian mind-set: U.S. ownership and exploitation of B.C.'s natural resources is OK, but it's not OK for money spent to save B.C.'s wild areas to come from the U.S.
5. There is only one chance in our planet's lifetime to save wilderness. Working only part time at it isn't enough.